W9-BJQ-614

CHAPTER 7

ATCF	after-tax cash flow
BTCF	before-tax cash flow
EVA	Economic Value Added
MACRS	Modified Accelerated Cost Recovery System
NOPAT	Net Operating Profit After Taxes
WACC	Tax adjusted Weighted Average Cost of Capital

CHAPTER 8

A$	actual (current) dollars
f	general inflation rate
R$	real (constant) dollars

CHAPTER 9

EUAC	equivalent uniform annual cost
TC_k	total (marginal) cost for year k

CHAPTER 12

$E(X)$	mean of a random variable
$f(x)$	probability density function of a continuous random variable
$p(x)$	probability mass function of a discrete random variable
$SD(X)$	standard deviation of a random variable
$V(X)$	variance of a random variable

CHAPTER 13

CAPM	Capital Asset Pricing Model
R_F	Risk-Free rate of return
SML	Security Market Line
X_j	binary decision variable in capital allocation problems

ENGINEERING ECONOMY

Fourteenth Edition

ENGINEERING ECONOMY

Fourteenth Edition

William G. Sullivan
Virginia Polytechnic Institute
and State University, Emeritus

Elin M. Wicks
Wicks and Associates

C. Patrick Koelling
Virginia Polytechnic Institute
and State University

PEARSON
Prentice
Hall

Upper Saddle River, New Jersey 07458

Library of Congress Cataloging-in-Publication Data

Sullivan, William G.
  Engineering economy / William G. Sullivan, Elin M. Wicks, C. Patrick Koelling. — 14th ed.
    p. cm.
  Includes bibliographical references and index.
  ISBN-13: 978-0-13-614297-3
  ISBN-10:   0-13-614297-4
  1. Engineering economy   I. Wicks, Elin M.   II. Koelling, C. Patrick
III. Title
  TA177.4.S85 2009
  658.15—dc22   2008010340

Vice President and Editorial Director, ECS: *Marcia J. Horton*
Senior Editor: *Holly Stark*
Associate Editor: *Dee Bernhard*
Editorial Assistant: *Alicia Lucci*
Director of Team-Based Project Management: *Vince O'Brien*
Senior Managing Editor: *Scott Disanno*
Production Liaison: *Jane Bonnell*
Production Editor: *Bharath Parthasarathy, TexTech*
Manufacturing Manager: *Alan Fischer*
Manufacturing Buyer: *Lisa McDowell*
Marketing Manager: *Tim Galligan*

Marketing Assistant: *Mack Patterson*
Art Director: *Kenny Beck*
Cover Designer: *Kristine Carney*
Cover Images: Paper currency: *Paul Paladin/Shutterstock;*
  Hydroelectric dam: *Bryan Busovicki/Shutterstock;*
  Growing market: *Rémi Cauzid/Shutterstock*
Art Editor: *Greg Dulles*
Media Editor: *David Alick*
Media Project Manager: *John M. Cassar*
Composition/Full-Service Project Management:
  TexTech International Pvt. Ltd.

Chapter-opening photos: Chapter 1, Peter Titmuss, Alamy.com; Chapter 2, Antony Nettle, Alamy Images; Chapter 3, Dennis Gilbert, Alamy Images; Chapter 4, Rab Harling, Alamy Images; Chapter 5, David Crausby, Alamy Images; Chapter 6, Getty Images, Inc.—Stockbyte; Chapter 7, Joe Sohm/Visions of America, Getty Images/Digital Vision; Chapter 8, Garry Gay, Alamy Images; Chapter 9, www.photos.com/Jupiter Images; Chapter 10, Andrew Gunners, Getty Images/Digital Vision; Chapter 11, Karen Givens, Shutterstock; Chapter 12, General Electric Company; Chapter 13, George Diebold/Getty Images, Inc.; Chapter 14, Ninette Maumus, Alamy Images

Pearson Education Ltd., London
Pearson Education Singapore, Pte. Ltd.
Pearson Education Canada, Inc.
Pearson Education–Japan
Pearson Education Australia PTY, Limited
Pearson Education North Asia, Ltd., Hong Kong
Pearson Educación de Mexico, S.A. de C.V.
Pearson Education Malaysia, Pte. Ltd.
Pearson Education, Upper Saddle River, New Jersey

10 9 8 7 6 5 4 3

ISBN-13: 978-0-13-614297-3
ISBN-10:   0-13-614297-4

CONTENTS

PART III Additional Topics in Engineering Economy

CHAPTER 10
Evaluating Projects with the Benefit–Cost Ratio Method   431

CHAPTER 11
Breakeven and Sensitivity Analysis   465

PREFACE

We live in a sea of economic decisions.

—Anonymous

About Engineering Economy

A succinct job description for an engineer consists of two words: *problem solver*. Broadly speaking, engineers use knowledge to find new ways of doing things economically. Engineering design solutions do not exist in a vacuum but within the context of a business opportunity. But every problem has multiple solutions, so the issue is, How does one rationally select the design with the most favorable economic result? The answer to this question can also be put forth in two words: *engineering economy*. Engineering economy provides a systematic framework for evaluating the economic aspects of competing design solutions. Just as engineers model the stress on a support column, or the thermodynamic response of a steam turbine, they must also model the economic impact of their recommendations.

Engineering economy—what is it, and why is it important? The initial reaction of many engineering students to these questions is, "Money matters will be handled by someone else. They are not something I need to worry about." In reality, any engineering project must be not only physically realizable but also economically affordable.

Understanding and applying economic principles to engineering have never been more important. Engineering is more than a problem-solving activity focusing on the development of products, systems, and processes to satisfy a need or demand. Beyond function and performance, solutions must also be viable economically. Design decisions affect limited resources such as time, material, labor, capital, natural resources, not only initially (during conceptual design) but also through the remaining phases of the life cycle (e.g., detailed design, manufacture and distribution, service, retirement and disposal). A great solution can die a certain death if it is not profitable.

Fourteenth Edition of *Engineering Economy* Highlights

New or enhanced features of this edition include the following:

- Fifty percent of end-of-chapter problems are new or revised.
- A bank of algorithmically generated test questions is available to adopting instructors.

- Fundamentals of Engineering (FE) exam-style questions are included among the end-of-chapter problem sets.
- Spreadsheet models are integrated throughout.
- An appendix on the basics of accounting is included in Chapter 2.
- Chapter 3 on Cost Estimation appears early in the book.
- An appendix for using Excel in engineering economy is available for reference.
- Numerous comprehensive examples and case studies appear throughout the book.
- Extended learning exercises appear in most chapters.
- Personal finance problems are featured in most chapters.
- Many pointers to relevant Web sites are provided.

Pedagogy of This Book

This book has two primary objectives: (1) to provide students with a sound understanding of the principles, basic concepts, and methodology of engineering economy; and (2) to help students develop proficiency with these methods and with the process for facilitating rational decisions they are likely to encounter in professional practice. Interestingly, an engineering economy course may be a student's only college exposure to the systematic evaluation of alternative investment opportunities. In this regard, *Engineering Economy* is intended to serve as a text for classroom instruction *and* as a basic reference for use by practicing engineers in all specialty areas (e.g., chemical, civil, computer, electrical, industrial, and mechanical engineering). The book is also useful to persons engaged in the management of technical activities.

As a textbook, the fourteenth edition is written principally for the first formal course in engineering economy. A three-credit-hour semester course should be able to cover the majority of topics in this edition, and there is sufficient depth and breadth to enable an instructor to arrange course content to suit individual needs. Representative syllabi for a three-credit and a two-credit semester course in engineering economy are provided in Table P-1. Moreover, because several advanced topics are included, this book can also be used for a second course in engineering economy.

All chapters and appendices have been revised and updated to reflect current trends and issues. Also, numerous exercises that involve open-ended problem statements and iterative problem-solving skills are included throughout the book. A large number of the 750-plus end-of-chapter exercises are new, and many solved examples representing realistic problems that arise in various engineering disciplines are presented. FE exam-style questions have been added to help prepare engineering students for this milestone examination, leading to professional registration. Passing the FE exam is a first step in getting licensed as a professional engineer (PE). Engineering students should seriously consider becoming a PE because it opens many employment opportunities and increases lifetime earning potential.

TABLE P-1 Typical Syllabi for Courses in Engineering Economy

Semester Course (Three Credit Hours)			Semester Course (Two Credit Hours)		
Chapter	Week of the Semester	Topic(s)	Chapter(s)	No. of Class Periods	Topic(s)
1	1	Introduction to Engineering Economy	1	1	Introduction to Engineering Economy
2	2	Cost Concepts and Design Economics	2	4	Cost Concepts, Single Variable Trade-Off Analysis, and Present Economy
3	3	Cost-Estimation Techniques			
4	4–5	The Time Value of Money	4	5	The Time Value of Money
5	6	Evaluating a Single Project	1, 2, 4	1	**Test #1**
6	7	Comparison and Selection among Alternatives	3	3	Developing Cash Flows and Cost-Estimation Techniques
	8	**Midterm Examination**	5	2	Evaluating a Single Project
7	9	Depreciation and Income Taxes	6	4	Comparison and Selection among Alternatives
10	10	Evaluating Projects with the Benefit–Cost Ratio Method	3, 5, 6	1	**Test #2**
8	11	Price Changes and Exchange Rates	11	2	Breakeven and Sensitivity Analysis
11	12	Breakeven and Sensitivity Analysis	7	5	Depreciation and Income Taxes
9	13	Replacement Analysis	14	1	Decision Making Considering Multiattributes
12	14	Probabilistic Risk Analysis			
13–14	15	The Capital Budgeting Process, Decision Making Considering Multiattributes	All the above	1	**Final Examination**
	15	**Final Examination**			

Number of class periods: 45

Number of class periods: 30

It is generally advisable to teach engineering economy at the upper division level. Here, an engineering economy course incorporates the accumulated knowledge students have acquired in other areas of the curriculum and also deals with iterative problem solving, open-ended exercises, creativity in formulating and evaluating feasible solutions to problems, and consideration of realistic constraints (economic, aesthetic, safety, etc.) in problem solving.

Supplements to the Book

The fourteenth edition of *Engineering Economy* is proud to offer adopting instructors **TestGen**, a test generator program with an algorithmic bank of questions. The **TestGen** testbank consists of well-crafted assessment questions that are representative of problems found throughout the textbook. Instructors can regenerate algorithmically generated variables within each problem to offer students a virtually unlimited number of paper or online assessments. Additionally, instructors can view, select, and edit testbank questions or create their own questions. Also available to adopters of this edition is an instructor's Solutions Manual and other classroom resources, including PowerPoint images. Visit www.prenhall.com/sullivan for more information.

Engineering Economy Portfolio

In many engineering economy courses, students are required to design, develop, and maintain an engineering economy portfolio. The purpose of the portfolio is to demonstrate and integrate knowledge of engineering economy beyond the required assignments and tests. This is usually an individual assignment. Professional presentation, clarity, brevity, and creativity are important criteria to be used to evaluate portfolios. Students are asked to keep the audience (i.e., the grader) in mind when constructing their portfolios.

The portfolio should contain a variety of content. To get credit for content, students must display their knowledge. Simply collecting articles in a folder demonstrates very little. To get credit for collected articles, students should read them and write a brief summary of each one. The summary could explain how the article is relevant to engineering economy, it could critique the article, or it could check or extend any economic calculations in the article. The portfolio should include both the summary and the article itself. Annotating the article by writing comments in the margin is also a good idea. Other suggestions for portfolio content follow (note that students are encouraged to be creative):

- Describe and set up or solve an engineering economy problem from your own discipline (e.g., electrical engineering or building construction).
- Choose a project or problem in society or at your university and apply engineering economic analysis to one or more proposed solutions.
- Develop proposed homework or test problems for engineering economy. Include the complete solution. Additionally, state which course objective(s) this problem demonstrates (include text section).

- Reflect upon and write about your progress in the class. You might include a self-evaluation against the course objectives.

- Include a photo or graphic that illustrates some aspects of engineering economy. Include a caption that explains the relevance of the photo or graphic.

- Include completely worked out practice problems. Use a different color pen to show these were checked against the provided answers.

- Rework missed test problems, including an explanation of each mistake.

(The preceding list could reflect the relative value of the suggested items; that is, items at the top of the list are more important than items at the bottom of the list.)

Students should develop an introductory section that explains the purpose and organization of the portfolio. A table of contents and clearly marked sections or headings are highly recommended. Cite the source (i.e., a complete bibliographic entry) of all outside material. Remember, portfolios provide evidence that students know more about engineering economy than what is reflected in the assignments and exams. The focus should be on quality of evidence, not quantity.

Overview of the Book

This book is about making choices among competing engineering alternatives. Most of the cash-flow consequences of the alternatives lie in the future, so our attention is directed toward the future and not the past. In Chapter 2, we examine alternatives when the time value of money is not a complicating factor in the analysis. We then turn our attention in Chapter 3 to how future cash flows are estimated. In Chapter 4 and subsequent chapters, we deal with alternatives where the time value of money is a deciding factor in choosing among competing capital investment opportunities.

Students can appreciate Chapters 2 and 3 and later chapters when they consider alternatives in their personal lives, such as which job to accept upon graduation, which automobile or truck to purchase, whether to buy a home or rent a residence, and many other choices they will face. To be student friendly, we have included many problems throughout this book that deal with personal finance. These problems are timely and relevant to a student's personal and professional success, and these situations incorporate the structured problem-solving process that students will learn from this book.

Chapter 4 concentrates on the concepts of money–time relationships and economic equivalence. Specifically, we consider the time value of money in evaluating the future revenues and costs associated with alternative uses of money. Then, in Chapter 5, the methods commonly used to analyze the economic consequences and profitability of an alternative are demonstrated. These methods, and their proper use in the comparison of alternatives, are primary subjects of Chapter 6, which also includes a discussion of the appropriate time period for an analysis. Thus, Chapters 4, 5, and 6 together develop an essential part of the methodology needed for understanding the remainder of the book and for performing engineering economy studies on a before-tax basis.

In Chapter 7, the additional details required to accomplish engineering economy studies on an after-tax basis are explained. In the private sector, most engineering economy studies are done on an after-tax basis. Therefore, Chapter 7 adds to the basic methodology developed in Chapters 4, 5, and 6.

The effects of inflation (or deflation), price changes, and international exchange rates are the topics of Chapter 8. The concepts for handling price changes and exchange rates in an engineering economy study are discussed both comprehensively and pragmatically from an application viewpoint.

Often, an organization must analyze whether existing assets should be continued in service or replaced with new assets to meet current and future operating needs. In Chapter 9, techniques for addressing this question are developed and presented. Because the replacement of assets requires significant capital, decisions made in this area are important and demand special attention.

Chapter 10 is dedicated to the analysis of public projects with the benefit–cost ratio method of comparison. The development of this widely used method of evaluating alternatives was motivated by the Flood Control Act passed by the U.S. Congress in 1936.

Concern over uncertainty and risk is a reality in engineering practice. In Chapter 11, the impact of potential variation between the estimated economic outcomes of an alternative and the results that may occur is considered. Breakeven and sensitivity techniques for analyzing the consequences of risk and uncertainty in future estimates of revenues and costs are discussed and illustrated.

In Chapter 12, probabilistic techniques for analyzing the consequences of risk and uncertainty in future cash-flow estimates and other factors are explained. Discrete and continuous probability concepts, as well as Monte Carlo simulation techniques, are included in Chapter 12. Real options analysis is also briefly discussed.

Chapter 13 is concerned with the proper identification and analysis of all projects and other needs for capital within an organization. Accordingly, the capital financing and capital allocation process to meet these needs is addressed. This process is crucial to the welfare of an organization, because it affects most operating outcomes, whether in terms of current product quality and service effectiveness or long-term capability to compete in the world market. Finally, Chapter 14 discusses many time-tested methods for including nonmonetary attributes (intangibles) in engineering economy studies.

ACKNOWLEDGMENTS

We would like to extend a heartfelt "thank you" to our colleagues for their many helpful suggestions (and critiques!) for this fourteenth edition of *Engineering Economy*. We owe a special debt of gratitude to Jim Alloway (EMSQ Associates), Dick Bernhard (North Carolina State University), Karen Bursic (University of Pittsburgh), J. Kent Butler (California Polytechnic State University), R. Wai Kiong Chong (University of Kansas), Michael Harnett (Kansas State University), Randy Hudson (Oak Ridge National Laboratory), James Luxhoj (Rutgers University), Behnam Malakooti (Case Western University), Kumar Muthuraman (Purdue University), and Bob Taylor (Virginia Tech).

Without the encouragement, dedication, and creativity of several persons at Prentice Hall, we could not have completed this revision in the exemplary manner expected by our faithful adopters. In particular, we thank Holly Stark, Dee Bernhard, Jane Bonnell, Dave Alick, and Ben Paris for their timely assistance.

Our spouses, Janet, Matt, and Priscilla, deserve our eternal gratitude for their unyielding enthusiasm and support for the successful completion of this work. They are truly the inspiration for all the arduous and time-consuming effort that went into this heavily revised and improved edition of *Engineering Economy*. We all hope you enjoy this edition even more than you have the previous thirteen!

WILLIAM G. SULLIVAN
ELIN M. WICKS
C. PATRICK KOELLING

ENGINEERING ECONOMY

Fourteenth Edition

CHAPTER 1

Introduction to Engineering Economy

The purpose of Chapter 1 is to present the concepts and principles of engineering economy.

Turning Trash into Treasure

Imagine a future in which landfills are obsolete and trash is used as fuel. Although this vision is not new, a southern-based U.S. company has plans to make this a reality. Instead of placing consumer waste in a landfill, trash would be vaporized into steam and a synthetic gas that can be used as a substitute for natural gas. The steam could be sold to neighboring facilities as a power source, and enough synthetic gas could be produced to power 43,000 homes annually.

The company plans to let no by-product go unused. Organic materials would be melted and hardened into materials that could be used for roadway and construction projects. It is projected that the sale of the transformed trash would allow the company to recoup the initial $425 million investment within 20 years. This project deals with not only the environmental concern of waste disposal but also the need for alternative energy sources. Chapter 1 introduces the decision-making process that accompanies "go/no-go" evaluations of this type of new technology.

> [Engineering] is the art of doing that well with one dollar which any bungler can do with two.
>
> —Arthur M. Wellington (1887)

1.1 Introduction

The technological and social environments in which we live continue to change at a rapid rate. In recent decades, advances in science and engineering have made space travel possible, transformed our transportation systems, revolutionized the practice of medicine, and miniaturized electronic circuits so that a computer can be placed on a semiconductor chip. The list of such achievements seems almost endless. In your science and engineering courses, you will learn about some of the physical laws that underlie these accomplishments.

The utilization of scientific and engineering knowledge for our benefit is achieved through the *design* of things we use, such as furnaces for vaporizing trash and structures for supporting magnetic railways. However, these achievements don't occur without a price, monetary or otherwise. Therefore, the purpose of this book is to develop and illustrate the principles and methodology required to answer the basic economic question of any design: Do its benefits exceed its costs?

The Accreditation Board for Engineering and Technology states that engineering "is the profession in which a knowledge of the mathematical and natural sciences gained by study, experience, and practice is applied with judgment to develop ways to utilize, economically, the materials and forces of nature for the benefit of mankind."* In this definition, the economic aspects of engineering are emphasized, as well as the physical aspects. Clearly, it is essential that the economic part of engineering practice be accomplished well. Thus, engineers use knowledge to find new ways of doing things economically.

Engineering economy involves the systematic evaluation of the economic merits of proposed solutions to engineering problems. To be economically acceptable (i.e., affordable), *solutions to engineering problems* must demonstrate a positive balance of long-term benefits over long-term costs, and they must also

- promote the well-being and survival of an organization,
- embody creative and innovative technology and ideas,
- permit identification and scrutiny of their estimated outcomes, and
- translate profitability to the "bottom line" through a valid and acceptable measure of merit.

---

* Accreditation Board of Engineering and Technology, *Criteria for Accrediting Programs in Engineering in the United States* (New York; Baltimore, MD: ABET, 1998).

Engineering economy is the dollars-and-cents side of the decisions that engineers make or recommend as they work to position a firm to be profitable in a highly competitive marketplace. Inherent to these decisions are trade-offs among different types of costs and the performance (response time, safety, weight, reliability, etc.) provided by the proposed design or problem solution. *The mission of engineering economy is to balance these trade-offs in the most economical manner.* For instance, if an engineer at Ford Motor Company invents a new transmission lubricant that increases fuel mileage by 10% and extends the life of the transmission by 30,000 miles, how much can the company afford to spend to implement this invention? Engineering economy can provide an answer.

A few more of the myriad situations in which engineering economy plays a crucial role in the analysis of project alternative come to mind:

1. Choosing the best design for a high-efficiency gas furnace
2. Selecting the most suitable robot for a welding operation on an automotive assembly line
3. Making a recommendation about whether jet airplanes for an overnight delivery service should be purchased or leased
4. Determining the optimal staffing plan for a computer help desk

From these illustrations, it should be obvious that engineering economy includes significant technical considerations. Thus, engineering economy involves technical analysis, with emphasis on the economic aspects, and has the objective of assisting decisions. This is true whether the decision maker is an engineer interactively analyzing alternatives at a computer-aided design workstation or the Chief Executive Officer (CEO) considering a new project. *An engineer who is unprepared to excel at engineering economy is not properly equipped for his or her job.*

1.2 The Principles of Engineering Economy

The development, study, and application of any discipline must begin with a basic foundation. We define the foundation for engineering economy to be a set of principles that provide a comprehensive doctrine for developing the methodology. These principles will be mastered by students as they progress through this book. Once a problem or need has been clearly defined, the foundation of the discipline can be discussed in terms of seven principles.

PRINCIPLE 1	**Develop the Alternatives**

The choice (decision) is among alternatives. The alternatives need to be identified and then defined for subsequent analysis.

A decision situation involves making a choice among two or more alternatives. Developing and defining the alternatives for detailed evaluation is important because of the resulting impact on the quality of the decision. Engineers and managers should place a high priority on this responsibility. Creativity and innovation are essential to the process.

One alternative that may be feasible in a decision situation is making no change to the current operation or set of conditions (i.e., doing nothing). If you judge this option feasible, make sure it is considered in the analysis. However, do not focus on the status quo to the detriment of innovative or necessary change.

PRINCIPLE 2	Focus on the Differences

Only the differences in expected future outcomes among the alternatives are relevant to their comparison and should be considered in the decision.

If all prospective outcomes of the feasible alternatives were exactly the same, there would be no basis or need for comparison. We would be indifferent among the alternatives and could make a decision using a random selection.

Obviously, only the differences in the future outcomes of the alternatives are important. Outcomes that are common to all alternatives can be disregarded in the comparison and decision. For example, if your feasible housing alternatives were two residences with the same purchase (or rental) price, price would be inconsequential to your final choice. Instead, the decision would depend on other factors, such as location and annual operating and maintenance expenses. This example illustrates, in a simple way, Principle 2, which emphasizes the basic purpose of an engineering economic analysis: to recommend a future course of action based on the differences among feasible alternatives.

PRINCIPLE 3	Use a Consistent Viewpoint

The prospective outcomes of the alternatives, economic and other, should be consistently developed from a defined viewpoint (perspective).

The perspective of the decision maker, which is often that of the owners of the firm, would normally be used. However, it is important that the viewpoint for the particular decision be first defined and then used consistently in the description, analysis, and comparison of the alternatives.

As an example, consider a public organization operating for the purpose of developing a river basin, including the generation and wholesale distribution of electricity from dams on the river system. A program is being planned to upgrade and increase the capacity of the power generators at two sites. What perspective should be used in defining the technical alternatives for the program? The "owners of the firm" in this example means the segment of the public that will pay the cost of the program, and their viewpoint should be adopted in this situation.

Now let us look at an example where the viewpoint may not be that of the owners of the firm. Suppose that the company in this example is a private firm and that the problem deals with providing a flexible benefits package for the employees. Also, assume that the feasible alternatives for operating the plan all have the same future costs to the company. The alternatives, however, have differences from the perspective of the employees, and their satisfaction is an important decision criterion. The viewpoint for this analysis should be that of the employees of the company as a group, and the feasible alternatives should be defined from their perspective.

PRINCIPLE 4	Use a Common Unit of Measure

Using a common unit of measurement to enumerate as many of the prospective outcomes as possible will simplify the analysis of the alternatives.

It is desirable to make as many prospective outcomes as possible *commensurable* (directly comparable). For economic consequences, a monetary unit such as dollars is the common measure. You should also try to translate other outcomes (which do not initially appear to be economic) into the monetary unit. This translation, of course, will not be feasible with some of the outcomes, but the additional effort toward this goal will enhance commensurability and make the subsequent analysis of alternatives easier.

What should you do with the outcomes that are not economic (i.e., the expected consequences that cannot be translated (and estimated) using the monetary unit)? First, if possible, quantify the expected future results using an appropriate unit of measurement for each outcome. If this is not feasible for one or more outcomes, describe these consequences explicitly so that the information is useful to the decision maker in the comparison of the alternatives.

PRINCIPLE 5	Consider All Relevant Criteria

Selection of a preferred alternative (decision making) requires the use of a criterion (or several criteria). The decision process should consider both the outcomes enumerated in the monetary unit and those expressed in some other unit of measurement or made explicit in a descriptive manner.

The decision maker will normally select the alternative that will best serve the long-term interests of the owners of the organization. In engineering economic analysis, the primary criterion relates to the long-term financial interests of the owners. This is based on the assumption that available capital will be allocated to provide maximum monetary return to the owners. Often, though, there are other organizational objectives you would like to achieve with your decision, and these should be considered and given weight in the selection of an alternative. These nonmonetary attributes and multiple objectives become the basis for additional criteria in the decision-making process. This is the subject of Chapter 14.

PRINCIPLE 6	Make Risk and Uncertainty Explicit

Risk and uncertainty are inherent in estimating the future outcomes of the alternatives and should be recognized in their analysis and comparison.

The analysis of the alternatives involves projecting or estimating the future consequences associated with each of them. The magnitude and the impact of future outcomes of any course of action are uncertain. Even if the alternative involves no change from current operations, the probability is high that today's estimates of, for example, future cash receipts and expenses will not be what eventually occurs. Thus, dealing with uncertainty is an important aspect of engineering economic analysis and is the subject of Chapters 11 and 12.

PRINCIPLE 7	**Revisit Your Decisions**

Improved decision making results from an adaptive process; to the extent practicable, the initial projected outcomes of the selected alternative should be subsequently compared with actual results achieved.

A good decision-making process can result in a decision that has an undesirable outcome. Other decisions, even though relatively successful, will have results significantly different from the initial estimates of the consequences. Learning from and adapting based on our experience are essential and are indicators of a good organization.

The evaluation of results versus the initial estimate of outcomes for the selected alternative is often considered impracticable or not worth the effort. Too often, no feedback to the decision-making process occurs. Organizational discipline is needed to ensure that implemented decisions are routinely postevaluated and that the results are used to improve future analyses and the quality of decision making. For example, a common mistake made in the comparison of alternatives is the failure to examine adequately the impact of uncertainty in the estimates for selected factors on the decision. Only postevaluations will highlight this type of weakness in the engineering economy studies being done in an organization.

1.3 Engineering Economy and the Design Process

An engineering economy study is accomplished using a structured procedure and mathematical modeling techniques. The economic results are then used in a decision situation that normally includes other engineering knowledge and input.

A sound engineering economic analysis procedure incorporates the basic principles discussed in Section 1.2 and involves several steps. We represent the procedure in terms of the *seven steps* listed in the left-hand column of Table 1-1. There are several feedback loops (not shown) within the procedure. For example, within Step 1, information developed in evaluating the problem will be used as feedback to refine the problem definition. As another example, information from the analysis of alternatives (Step 5) may indicate the need to change one or more of them or to develop additional alternatives.

TABLE 1-1 The General Relationship between the Engineering Economic Analysis Procedure and the Engineering Design Process

Engineering Economic Analysis Procedure	Engineering Design Process (see Figure P1-15 on p. 18)
Step	*Activity*
1. Problem recognition, definition, and evaluation.	1. Problem/need definition.
	2. Problem/need formulation and evaluation.
2. Development of the feasible alternatives.	3. Synthesis of possible solutions (alternatives).
3. Development of the outcomes and cash flows for each alternative.	
4. Selection of a criterion (or criteria).	4. Analysis, optimization, and evaluation.
5. Analysis and comparison of the alternatives.	
6. Selection of the preferred alternative.	5. Specification of preferred alternative.
7. Performance monitoring and post-evaluation of results.	6. Communication.

The seven-step procedure is also used to assist decision making within the engineering design process, shown as the right-hand column in Table 1-1. In this case, activities in the design process contribute information to related steps in the economic analysis procedure. The general relationship between the activities in the design process and the steps of the economic analysis procedure is indicated in Table 1-1.

The engineering design process may be repeated in phases to accomplish a total design effort. For example, in the first phase, a full cycle of the process may be undertaken to select a conceptual or preliminary design alternative. Then, in the second phase, the activities are repeated to develop the preferred detailed design based on the selected preliminary design. The seven-step economic analysis procedure would be repeated as required to assist decision making in each phase of the total design effort. This procedure is discussed next.

1.3.1 Problem Definition

The first step of the engineering economic analysis procedure (problem definition) is particularly important, since it provides the basis for the rest of the analysis. A problem must be well understood and stated in an explicit form before the project team proceeds with the rest of the analysis.

The term *problem* is used here generically. It includes all decision situations for which an engineering economy analysis is required. Recognition of the problem is normally stimulated by internal or external organizational needs or requirements. An operating problem within a company (internal need) or a customer expectation about a product or service (external requirement) are examples.

Once the problem is recognized, its formulation should be viewed from a *systems perspective*. That is, the boundary or extent of the situation needs to be carefully defined, thus establishing the elements of the problem and what constitutes its environment.

Evaluation of the problem includes refinement of needs and requirements, and information from the evaluation phase may change the original formulation of the problem. In fact, redefining the problem until a consensus is reached may be the most important part of the problem-solving process!

1.3.2 Development of Alternatives*

The two primary actions in Step 2 of the procedure are (1) searching for potential alternatives and (2) screening them to select a smaller group of feasible alternatives for detailed analysis. The term *feasible* here means that each alternative selected for further analysis is judged, based on preliminary evaluation, to meet or exceed the requirements established for the situation.

1.3.2.1 Searching for Superior Alternatives In the discussion of Principle 1 (Section 1.2), creativity and resourcefulness were emphasized as being absolutely essential to the development of potential alternatives. The difference between good alternatives and great alternatives depends largely on an individual's or group's *problem-solving efficiency*. Such efficiency can be increased in the following ways:

1. Concentrate on redefining one problem at a time in Step 1.
2. Develop many redefinitions for the problem.
3. Avoid making judgments as new problem definitions are created.
4. Attempt to redefine a problem in terms that are dramatically different from the original Step 1 problem definition.
5. Make sure that the *true problem* is well researched and understood.

In searching for superior alternatives or identifying the true problem, several limitations invariably exist, including (1) lack of time and money, (2) preconceptions of what will and what will not work, and (3) lack of knowledge. Consequently, the engineer or project team will be working with less-than-perfect problem solutions in the practice of engineering.

EXAMPLE 1-1	Defining the Problem and Developing Alternatives

The management team of a small furniture-manufacturing company is under pressure to increase profitability in order to get a much-needed loan from the bank to purchase a more modern pattern-cutting machine. One proposed solution is to sell waste wood chips and shavings to a local charcoal manufacturer instead of using them to fuel space heaters for the company's office and factory areas.

---

* This is sometimes called *option development*. This important step is described in detail in A. B. Van Gundy, *Techniques of Structured Problem Solving*, 2nd ed. (New York: Van Nostrand Reinhold Co., 1988). For additional reading, see E. Lumsdaine and M. Lumsdaine, *Creative Problem Solving—An Introductory Course for Engineering Students* (New York: McGraw-Hill Book Co., 1990), and J. L. Adams, *Conceptual Blockbusting—A Guide to Better Ideas* (Reading, MA: Addison-Wesley Publishing Co., 1986).

(a) Define the company's problem. Next, reformulate the problem in a variety of creative ways.

(b) Develop at least one potential alternative for your reformulated problems in Part (a). (Don't concern yourself with feasibility at this point.)

Solution

(a) The company's problem appears to be that revenues are not sufficiently covering costs. Several reformulations can be posed:

   1. The problem is to increase revenues while reducing costs.

   2. The problem is to maintain revenues while reducing costs.

   3. The problem is an accounting system that provides distorted cost information.

   4. The problem is that the new machine is really not needed (and hence there is no need for a bank loan).

(b) Based only on reformulation 1, an alternative is to sell wood chips and shavings as long as increased revenue exceeds extra expenses that may be required to heat the buildings. Another alternative is to discontinue the manufacture of specialty items and concentrate on standardized, high-volume products. Yet another alternative is to pool purchasing, accounting, engineering, and other white-collar support services with other small firms in the area by contracting with a local company involved in providing these services.

1.3.2.2 Developing Investment Alternatives "It takes money to make money," as the old saying goes. Did you know that in the United States the average firm spends over $250,000 in capital on each of its employees? So, to make money, each firm must invest capital to support its important human resources—but in what else should an individual firm invest? There are usually hundreds of opportunities for a company to make money. Engineers are at the very heart of creating value for a firm by turning innovative and creative ideas into new or reengineered commercial products and services. Most of these ideas require investment of money, and only a few of all feasible ideas can be developed, due to lack of time, knowledge, or resources.

Consequently, most investment alternatives created by good engineering ideas are drawn from a larger population of equally good problem solutions. But how can this larger set of equally good solutions be tapped into? Interestingly, studies have concluded that designers and problem solvers tend to pursue a few ideas that involve "patching and repairing" an old idea.* Truly new ideas are often excluded from consideration! This section outlines two approaches that have found wide acceptance in industry for developing sound investment alternatives by removing

---

* S. Finger and J. R. Dixon, "A Review of Research in Mechanical Engineering Design. Part I: Descriptive, Prescriptive, and Computer-Based Models of Design Processes," in *Research in Engineering Design* (New York: Springer-Verlag, 1990).

some of the barriers to creative thinking: (1) classical brainstorming and (2) the Nominal Group Technique (NGT).

(1) Classical Brainstorming. Classical brainstorming is the most well-known and often-used technique for idea generation. It is based on the fundamental principles of *deferment of judgment* and that *quantity breeds quality*. There are four rules for successful brainstorming:

1. Criticism is ruled out.
2. Freewheeling is welcomed.
3. Quantity is wanted.
4. Combination and improvement are sought.

A. F. Osborn lays out a detailed procedure for successful brainstorming.* A classical brainstorming session has the following basic steps:

1. *Preparation.* The participants are selected, and a preliminary statement of the problem is circulated.
2. *Brainstorming.* A warm-up session with simple unrelated problems is conducted, the relevant problem and the four rules of brainstorming are presented, and ideas are generated and recorded using checklists and other techniques if necessary.
3. *Evaluation.* The ideas are evaluated relative to the problem.

Generally, a brainstorming group should consist of four to seven people, although some suggest larger groups.

(2) Nominal Group Technique. The NGT, developed by Andre P. Delbecq and Andrew H. Van de Ven,[†] involves a structured group meeting designed to incorporate individual ideas and judgments into a group consensus. By correctly applying the NGT, it is possible for groups of people (preferably, 5 to 10) to generate investment alternatives or other ideas for improving the competitiveness of the firm. Indeed, the technique can be used to obtain group thinking (consensus) on a wide range of topics. For example, a question that might be given to the group is, "What are the most important problems or opportunities for improvement of . . .?"

The technique, when properly applied, draws on the creativity of the individual participants, while reducing two undesirable effects of most group meetings: (1) the dominance of one or more participants and (2) the suppression of conflicting ideas. The basic format of an NGT session is as follows:

1. Individual silent generation of ideas
2. Individual round-robin feedback and recording of ideas
3. Group clarification of each idea

---

* A. F. Osborn, *Applied Imagination*, 3rd ed. (New York: Charles Scribner's Sons, 1963). Also refer to P. R. Scholtes, B. L. Joiner, and B. J. Streibel, *The Team Handbook*, 2nd ed. (Madison, WI: Oriel Inc., 1996).
† A. Van de Ven and A. Delbecq, "The Effectiveness of Nominal, Delphi, and Interactive Group Decision Making Processes," *Academy of Management Journal* 17, no. 4 (December 1974): 605–21.

4. Individual voting and ranking to prioritize ideas

5. Discussion of group consensus results

The NGT session begins with an explanation of the procedure and a statement of question(s), preferably written by the facilitator.* The group members are then asked to prepare individual listings of alternatives, such as investment ideas or issues that they feel are crucial for the survival and health of the organization. This is known as the silent-generation phase. After this phase has been completed, the facilitator calls on each participant, in round-robin fashion, to present one idea from his or her list (or further thoughts as the round-robin session is proceeding). Each idea (or opportunity) is then identified in turn and recorded on a flip chart or board by the NGT facilitator, leaving ample space between ideas for comments or clarification. This process continues until all the opportunities have been recorded, clarified, and displayed for all to see. At this point, a voting procedure is used to prioritize the ideas or opportunities. Finally, voting results lead to the development of group consensus on the topic being addressed.

1.3.3 Development of Prospective Outcomes

Step 3 of the engineering economic analysis procedure incorporates Principles 2, 3, and 4 from Section 1.2 and uses the basic *cash-flow approach* employed in engineering economy. A cash flow occurs when money is transferred from one organization or individual to another. Thus, a cash flow represents the economic effects of an alternative in terms of money spent and received.

Consider the concept of an organization having only one "window" to its external environment through which *all* monetary transactions occur—receipts of revenues and payments to suppliers, creditors, and employees. The key to developing the related cash flows for an alternative is estimating what would happen to the revenues and costs, as seen at this window, if the particular alternative were implemented. The *net cash flow* for an alternative is the difference between all cash inflows (receipts or savings) and cash outflows (costs or expenses) during each time period.

In addition to the economic aspects of decision making, *nonmonetary factors (attributes)* often play a significant role in the final recommendation. Examples of objectives other than profit maximization or cost minimization that can be important to an organization include the following:

1. Meeting or exceeding customer expectations

2. Safety to employees and to the public

3. Improving employee satisfaction

4. Maintaining production flexibility to meet changing demands

---

* A good example of the NGT is given in D. S. Sink, "Using the Nominal Group Technique Effectively," *National Productivity Review*, 2 (Spring 1983): 173–84.

5. Meeting or exceeding all environmental requirements
6. Achieving good public relations or being an exemplary member of the community

1.3.4 Selection of a Decision Criterion

The selection of a decision criterion (Step 4 of the analysis procedure) incorporates Principle 5 (consider all relevant criteria). The decision maker will normally select the alternative that will best serve the long-term interests of the owners of the organization. It is also true that the economic decision criterion should reflect a consistent and proper viewpoint (Principle 3) to be maintained throughout an engineering economy study.

1.3.5 Analysis and Comparison of Alternatives

Analysis of the economic aspects of an engineering problem (Step 5) is largely based on cash-flow estimates for the feasible alternatives selected for detailed study. A substantial effort is normally required to obtain reasonably accurate forecasts of cash flows and other factors in view of, for example, inflationary (or deflationary) pressures, exchange rate movements, and regulatory (legal) mandates that often occur. Clearly, the consideration of future uncertainties (Principle 6) is an essential part of an engineering economy study. When cash flow and other required estimates are eventually determined, alternatives can be compared based on their differences as called for by Principle 2. Usually, these differences will be quantified in terms of a monetary unit such as dollars.

1.3.6 Selection of the Preferred Alternative

When the first five steps of the engineering economic analysis procedure have been done properly, the preferred alternative (Step 6) is simply a result of the total effort. Thus, the soundness of the technical-economic modeling and analysis techniques dictates the quality of the results obtained and the recommended course of action. Step 6 is included in Activity 5 of the engineering design process (specification of the preferred alternative) when done as part of a design effort.

1.3.7 Performance Monitoring and Postevaluation of Results

This final step implements Principle 7 and is accomplished during and after the time that the results achieved from the selected alternative are collected. Monitoring project performance during its operational phase improves the achievement of related goals and objectives and reduces the variability in desired results. Step 7 is also the follow-up step to a previous analysis, comparing actual results achieved with the previously estimated outcomes. The aim is to learn how to do better analyses, and the feedback from postimplementation evaluation is important to the continuing improvement of operations in any organization. Unfortunately, like Step 1, this final step is often not done consistently or well in engineering practice; therefore, it needs particular attention to ensure feedback for use in ongoing and subsequent studies.

EXAMPLE 1-2	Application of the Engineering Economic Analysis Procedure

A friend of yours bought a small apartment building for $100,000 in a college town. She spent $10,000 of her own money for the building and obtained a mortgage from a local bank for the remaining $90,000. The *annual* mortgage payment to the bank is $10,500. Your friend also expects that annual maintenance on the building and grounds will be $15,000. There are four apartments (two bedrooms each) in the building that can each be rented for $360 per month.

Refer to the seven-step procedure in Table 1-1 (left-hand side) to answer these questions:

(a) Does your friend have a problem? If so, what is it?

(b) What are her alternatives? (Identify at least three.)

(c) Estimate the economic consequences and other required data for the alternatives in Part (b).

(d) Select a criterion for discriminating among alternatives, and use it to advise your friend on which course of action to pursue.

(e) Attempt to analyze and compare the alternatives in view of at least one criterion in addition to cost.

(f) What should your friend do based on the information you and she have generated?

Solution

(a) A quick set of calculations shows that your friend does indeed have a problem. A lot more money is being spent by your friend each year ($10,500 + $15,000 = $25,500) than is being received (4 × $360 × 12 = $17,280). The problem could be that the monthly rent is too low. She's losing $8,220 per year. (Now, that's a problem!)

(b) Option (1). Raise the rent. (Will the market bear an increase?)
Option (2). Lower maintenance expenses (but not so far as to cause safety problems).
Option (3). Sell the apartment building. (What about a loss?)
Option (4). Abandon the building (bad for your friend's reputation).

(c) Option (1). Raise total monthly rent to $1,440 + $R for the four apartments to cover monthly expenses of $2,125. Note that the minimum increase in rent would be ($2,125 − $1,440)/4 = $171.25 per apartment per month (almost a 50% increase!).
Option (2). Lower monthly expenses to $2,125 − $C so that these expenses are covered by the monthly revenue of $1,440 per month. This would have to be accomplished primarily by lowering the maintenance cost. (There's not much to be done about the annual mortgage cost unless a favorable refinancing opportunity presents itself.) Monthly maintenance expenses would have to be reduced to ($1,440 − $10,500/12) = $565. This represents more than a 50% decrease in maintenance expenses.

Option (3). Try to sell the apartment building for $X, which recovers the original $10,000 investment and (ideally) recovers the $685 per month loss ($8,220 ÷ 12) on the venture during the time it was owned.

Option (4). Walk away from the venture and kiss your investment good-bye. The bank would likely assume possession through foreclosure and may try to collect fees from your friend. This option would also be very bad for your friend's credit rating.

(d) One criterion could be to minimize the expected loss of money. In this case, you might advise your friend to pursue Option (1) or (3).

(e) For example, let's use "credit worthiness" as an additional criterion. Option (4) is immediately ruled out. Exercising Option (3) could also harm your friend's credit rating. Thus, Options (1) and (2) may be her only realistic and acceptable alternatives.

(f) Your friend should probably do a market analysis of comparable housing in the area to see if the rent could be raised (Option 1). Maybe a fresh coat of paint and new carpeting would make the apartments more appealing to prospective renters. If so, the rent can probably be raised while keeping 100% occupancy of the four apartments.

A tip to the wise—as an aside to Example 1-2, your friend would need a good credit report to get her mortgage approved. In this regard, there are three major credit bureaus in the United States: Equifax, Experian, and TransUnion. It's a good idea to regularly review your own credit report for unauthorized activity. You are entitled to a free copy of your report once per year from each bureau. Consider getting a report every four months from www.annualcreditreport.com.

EXAMPLE 1-3 **Stan Moneymaker: Personal Interest Problem**

Stan Moneymaker prides himself on how much money he can save by being frugal. Today, Stan needs 15 gallons of gasoline to top off his automobile's gas tank. If he drives eight miles (round trip) to a gas station on the outskirts of town, Stan can save $0.10 per gallon on the price of gasoline. Suppose gasoline costs $3.00 per gallon and Stan's car gets 25 miles per gallon for in-town driving.

Should Stan make the trip to get less-expensive gasoline? Each mile that Stan drives creates one pound of carbon dioxide. Each pound of CO_2 has a cost impact of $0.02 on the environment. What other factors (cost and otherwise) should Stan consider in his decision making?

Solution

Because each pound of CO_2 has a penalty of $0.02, the savings from Stan's trip = (15 gallons × $0.10) − (8 miles × $0.02) = $1.34. If Stan can drive his car for less than $1.34/8 = $0.1675 per mile, he should make the trip. The cost of gasoline only for the trip is (8 miles ÷ 25 miles per gallon) ($3.00 per gallon) = $0.96, but other costs of driving, such as insurance, maintenance, and depreciation, may also influence Stan's decision. What is the cost of an accident, should Stan have one during his trip to purchase less-expensive gasoline? If Stan makes the trip weekly for a year, should this influence his decision?

1.4 Using Spreadsheets in Engineering Economic Analysis

Spreadsheets are a useful tool for solving engineering economy problems. Most engineering economy problems are amenable to spreadsheet solution for the following reasons:

1. They consist of structured, repetitive calculations that can be expressed as formulas that rely on a few functional relationships.
2. The parameters of the problem are subject to change.
3. The results and the underlying calculations must be documented.
4. Graphical output is often required, as well as control over the format of the graphs.

Spreadsheets allow the analyst to develop an application rapidly, without being inundated by the housekeeping details of programming languages. They relieve the analyst of the drudgery of number crunching but still focus on problem formulation. Computer spreadsheets created in Excel are integrated throughout all chapters in this book. More on spreadsheet modeling can be found in Appendix A.

1.5 Summary

In this chapter, we defined engineering economy and presented the fundamental concepts in terms of seven basic principles (see pages 3–6). Experience has shown that most errors in engineering economic analyses can be traced to some violation of these principles. We will continue to stress these principles in the chapters that follow.

The seven-step engineering economic analysis procedure described in this chapter (see page 7) has direct ties to the engineering design process. Following this systematic approach will assist engineers in designing products and systems and in providing technical services that promote the economic welfare of the company they work for. This same approach will also help you as an individual make sound financial decisions in your personal life.

In summary, engineering economy is a collection of problem-solving tools and techniques that are applied to engineering, business, and environmental issues. Common, yet often complex, problems involving money are easier to understand and solve when you have a good grasp on the engineering economy approach to problem solving and decision making. The problem-solving focus of this text will enable you to master the theoretical and applied principles of engineering economy.

Problems

The number(s) in color at the end of a problem refer to the section(s) in that chapter most closely related to the problem.

1-1. To illustrate the law of large numbers in an economic setting, consider the *One Laptop per Child* project pioneered by a team of engineers from M.I.T. This computer will be sold in remote parts of the world (e.g., rural Cambodia and Nepal) for $50 apiece. If the profit per computer is a mere $0.10 and the estimated market size in 2009 is 100 million students, what is the total profit on this humanitarian effort? (1.3)

1-2. Explain why the subject of engineering economy is important to the practicing engineer. (1.1–1.3)

1-3. Assume that your employer is a manufacturing firm that produces several different electronic consumer products. What are five nonmonetary factors (attributes) that may be important when a significant change is considered in the design of the current bestselling product? (1.2, 1.3)

1-4. Tyler just wrecked his new Nissan, and the accident was his fault. The owner of the other vehicle got two estimates for the repairs: one was for $803 and the other was for $852. Tyler is thinking of keeping the insurance companies out of the incident to keep his driving record "clean." Tyler's deductible on his comprehensive coverage insurance is $500, and he does not want his premium to increase because of the accident. In this regard, Tyler estimates that his semi-annual premium will rise by $60 if he files a claim against his insurance company. In view of the above information, Tyler's initial decision is to write a personal check for $803 payable to the owner of the other vehicle. Did Tyler make the most economical decision? What other options should Tyler have explored? In your answer, be sure to state your assumptions and quantify your thinking. (1.3)

1-5. Henry Ford's Model T was originally designed and built to be run on ethanol. Today, ethanol (190-proof alcohol) can be produced with domestic stills for about $0.75 per gallon. When blended with gasoline costing $3.00 per gallon, a 20% ethanol and 80% gasoline mixture costs $2.55 per gallon. Assume fuel consumption at 25 miles per gallon and engine performance in general are not adversely affected with this 20–80 blend (called E20). (1.3)

a. How much money can be saved for 15,000 miles of driving per year?

b. How much gasoline per year is being conserved if one million people use the E20 fuel?

1-6. Describe the outcomes that should be expected from a feasible alternative. What are the differences between potential alternatives and feasible alternatives? (1.3)

1-7. Define uncertainty. What are some of the basic causes of uncertainty in engineering economy studies? (1.2)

1-8. Studies have concluded that a college degree is a very good investment. Suppose that a college graduate earns about 75% more money per hour than a high-school graduate. If the lifetime earnings of a high-school graduate average $1,200,000, what is the expected value of earning a college degree? (1.3)

1-9. Automobile repair shops typically recommend that their customers change their oil and oil filter every 3,000 miles. Your automobile user's manual suggests changing your oil every 5,000–7,000 miles. If you drive your car 15,000 miles each year and an oil and filter change costs $30, how much money would you save each year if you had this service performed every 5,000 miles? (1.3)

1-10. Describe how it might be feasible in an engineering economic analysis to consider the following different situations in terms of the monetary unit: (1.2)

a. A piece of equipment that is being considered as a replacement for an existing item has greater reliability; that is, the Mean Time Between Failures (MTBF) during operation of the new equipment has been increased 40% in comparison with the present item.

b. A company manufactures wrought iron patio furniture for the home market. Some changes in material and metal treatment, which involve increased manufacturing costs, are being considered to reduce the rusting problem significantly.

c. A large foundry operation has been in the same location in a metropolitan area for the past 35 years. Even though it is in compliance with current air pollution regulations, the continuing residential and commercial development of that area is causing an increasing expectation on the part of local residents for improved environmental control by the foundry. The company considers community relations to be important.

1-11. Explain the relationship between engineering economic analysis and engineering design. How does economic analysis assist decision making in the design process? (1.3)

1-12. During your first month as an employee at Greenfield Industries (a large drill-bit manufacturer), you are asked to evaluate alternatives for producing a newly designed drill bit on a turning machine. Your boss' memorandum to you has practically no information about what the alternatives are and what criteria should be used. The same task was posed to a previous employee who could not finish the analysis, but she has given you the following information: An old turning machine valued at $350,000 exists (in the warehouse) that can be modified for the new drill bit. The in-house technicians have given an estimate of $40,000 to modify this machine, and they assure you that they will have the machine ready before the projected start date (although they have never done any modifications of this type). It is hoped that the old turning machine will be able to meet production requirements at full capacity. An outside company, McDonald Inc., made the machine seven years ago and can easily do the same modifications for $60,000. The cooling system used for this machine is not environmentally safe and would require some disposal costs. McDonald Inc. has offered to build a new turning machine with more environmental safeguards and higher capacity for a price of $450,000. McDonald Inc. has promised this machine before the startup date and is willing to pay any late costs. Your company has $100,000 set aside for the start-up of the new product line of drill bits. For this situation,

a. Define the problem.
b. List key assumptions.
c. List alternatives facing Greenfield Industries.
d. Select a criterion for evaluation of alternatives.
e. Introduce risk into this situation.
f. Discuss how nonmonetary considerations may impact the selection.
g. Describe how a post-audit could be performed.

1-13. *The Almost-Graduating Senior Problem.* Consider this situation faced by a first-semester senior in civil engineering who is exhausted from extensive job interviewing and penniless from excessive partying.

Mary's impulse is to accept immediately a highly attractive job offer to work in her brother's successful manufacturing company. She would then be able to relax for a year or two, save some money, and then return to college to complete her senior year and graduate. Mary is cautious about this impulsive desire, because it may lead to no college degree at all!

a. Develop at least two formulations for Mary's problem.
b. Identify feasible solutions for each problem formulation in Part (a). *Be creative!*

1-14. While studying for the engineering economy final exam, you and two friends find yourselves craving a fresh pizza. You can't spare the time to pick up the pizza and must have it delivered. "Pick-Up-Sticks" offers a 1-1/4-inch-thick (including toppings), 20-inch square pizza with your choice of two toppings for $15 plus 5% sales tax and a $1.50 delivery charge (no sales tax on delivery charge). "Fred's" offers the round, deep-dish Sasquatch, which is 20 inches in diameter. It is 1-3/4 inches thick, includes two toppings, and costs $17.25 plus 5% sales tax and free delivery.

a. What is the problem in this situation? Please state it in an explicit and precise manner.
b. Systematically apply the seven principles of engineering economy (pp. 3–6) to the problem you have defined in Part (a).
c. Assuming that your common unit of measure is dollars (i.e., cost), what is the better value for getting a pizza based on the criterion of *minimizing cost per unit of volume?*
d. What other criteria might be used to select which pizza to purchase?

1-15. Storm doors have been installed on 50% of all homes in Anytown, USA. The remaining 50% of homeowners without storm doors think they may have a problem that a storm door could solve, but they're not sure. Use Activities 1, 2, and 3 in the engineering design process (Table 1-1) to help these homeowners systematically think through the definition of their need (Activity 1), a formal statement of their problem (Activity 2), and the generation of alternatives (Activity 3).

The design process begins in Figure P1-15 with a statement of need and terminates with the specifications for a means of fulfilling that need.

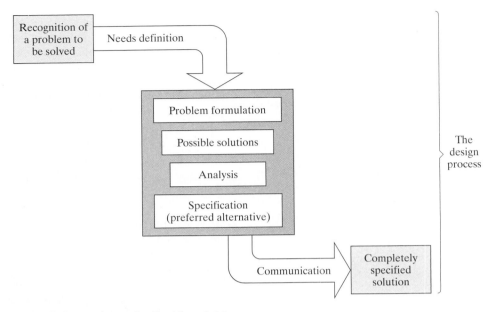

Figure P1-15 Figure for Problem 1-15

1-16. *Extended Learning Exercise.*
Bad news: You have just wrecked your car! You need another car immediately because you have decided that walking, riding a bike, and taking a bus are not acceptable. An automobile wholesaler offers you $2,000 for your wrecked car "as is." Also, your insurance company's claims adjuster estimates that there is $2,000 in damages to your car. Because you have collision insurance with a $1,000 deductibility provision, the insurance company mails you a check for $1,000. The odometer reading on your wrecked car is 58,000 miles.

What should you do? Use the seven-step procedure from Table 1-1 to analyze your situation. Also, identify which principles accompany each step.

1-17. "What you do at work is your boss' business" is a timely warning for all employees to heed. Last year, your company installed a new computer surveillance program in an effort to improve office productivity. As a courtesy, all employees were informed of this change. The license for the software costs $30,000 per year. After a year of use, productivity has risen 10%, which translates into a savings of $30,000. Discuss other factors, in addition to productivity, that could have been used to justify the surveillance software.

1-18. Owing to the rising cost of copper, in 1982 the U.S. Mint changed the composition of pennies from 95% copper (and 5% zinc) to 2.5% copper (and 97.5% zinc) to save money. Your favorite aunt has a collection of 5,000 pennies minted before 1982, and she intends on gifting the collection to you.

a. What is the collection's value based on metal content alone? Copper sells for $3.50 per pound and zinc for $1 per pound. It takes approximately 130 pre-1982 pennies to add up to one pound of total weight.

b. If it cost the U.S. Mint $0.017 to produce a penny in 2007, is it time to eliminate pennies and round off all financial transactions to the nearest 5 cents (nickel)? As a matter of interest, it cost the government almost 10 cents to produce a nickel in 2007.

CHAPTER 2

Cost Concepts and Design Economics

The objective of Chapter 2 is to analyze short-term alternatives when the time value of money is not a factor.

The A380 Superjumbo's Breakeven Point

When Europe's Airbus Company approved the A380 program in 2000, it was estimated that only 250 of the giant, 555-seat aircraft needed to be sold to break even. The program was initially based on expected deliveries of 751 aircraft over its life cycle. Long delays and mounting costs however, have dramatically changed the original breakeven figure. In 2005, this figure was updated to 270 aircraft. According to an article in the *Financial Times* (October 20, 2006, p. 18), Airbus would have to sell 420 aircraft to break even—a 68% increase over the original estimate. To date, only 159 firm orders for the aircraft have been received. The topic of breakeven analysis is an integral part of this chapter.

> The correct solution to any problem depends primarily on a true understanding of what the problem really is.
>
> —Arthur M. Wellington (1887)

2.1 Cost Terminology

There are a variety of costs to be considered in an engineering economic analysis.* These costs differ in their frequency of occurrence, relative magnitude, and degree of impact on the study. In this section, we define a number of cost categories and illustrate how they should be treated in an engineering economic analysis.

2.1.1 Fixed, Variable, and Incremental Costs

Fixed costs are those unaffected by changes in activity level over a feasible range of operations for the capacity or capability available. Typical fixed costs include insurance and taxes on facilities, general management and administrative salaries, license fees, and interest costs on borrowed capital.

Of course, any cost is subject to change, but fixed costs tend to remain constant over a specific range of operating conditions. When larger changes in usage of resources occur, or when plant expansion or shutdown is involved, fixed costs can be affected.

Variable costs are those associated with an operation that vary in total with the quantity of output or other measures of activity level. For example, the costs of material and labor used in a product or service are variable costs, because they vary in total with the number of output units, even though the costs per unit stay the same.

An *incremental cost* (or *incremental revenue*) is the additional cost (or revenue) that results from increasing the output of a system by one (or more) units. Incremental cost is often associated with "go–no go" decisions that involve a limited change in output or activity level. For instance, the incremental cost per mile for driving an automobile may be $0.49, but this cost depends on considerations such as total mileage driven during the year (normal operating range), mileage expected for the next major trip, and the age of the automobile. Also, it is common to read about the "incremental cost of producing a barrel of oil" and "incremental cost to the state for educating a student." As these examples indicate, the incremental cost (or revenue) is often quite difficult to determine in practice.

EXAMPLE 2-1	Fixed and Variable Costs

In connection with surfacing a new highway, a contractor has a choice of two sites on which to set up the asphalt-mixing plant equipment. The contractor estimates that it will cost $1.15 per cubic yard mile (yd^3-mile) to haul the asphalt-paving material from the mixing plant to the job location. Factors relating to the two mixing sites are as follows (production costs at each site are the same):

---

* For the purposes of this book, the words *cost* and *expense* are used interchangeably.

Cost Factor	Site A	Site B
Average hauling distance	6 miles	4.3 miles
Monthly rental of site	$1,000	$5,000
Cost to set up and remove equipment	$15,000	$25,000
Hauling expense	$1.15/yd^3-mile	$1.15/yd^3-mile
Flagperson	Not required	$96/day

The job requires 50,000 cubic yards of mixed-asphalt-paving material. It is estimated that four months (17 weeks of five working days per week) will be required for the job. Compare the two sites in terms of their fixed, variable, and total costs. Assume that the cost of the return trip is negligible. Which is the better site? For the selected site, how many cubic yards of paving material does the contractor have to deliver before starting to make a profit if paid $8.05 per cubic yard delivered to the job location?

Solution

The fixed and variable costs for this job are indicated in the table shown next. Site rental, setup, and removal costs (and the cost of the flagperson at Site B) would be constant for the total job, but the hauling cost would vary in total amount with the distance and thus with the total output quantity of yd^3-miles (x).

Cost	Fixed	Variable	Site A	Site B
Rent	✓		= $4,000	= $20,000
Setup/removal	✓		= 15,000	= 25,000
Flagperson	✓		= 0	5(17)($96) = 8,160
Hauling		✓	6(50,000)($1.15) = 345,000	4.3(50,000)($1.15) = 247,250
			Total: $364,000	$300,410

Site B, which has the larger fixed costs, has the smaller total cost for the job. Note that the extra fixed costs of Site B are being "traded off" for reduced variable costs at this site.

The contractor will begin to make a profit at the point where total revenue equals total cost as a function of the cubic yards of asphalt pavement mix delivered. Based on Site B, we have

$$4.3(\$1.15) = \$4.945 \text{ in variable cost per yd}^3 \text{ delivered}$$

$$\text{Total cost} = \text{total revenue}$$

$$\$53,160 + \$4.945x = \$8.05x$$

$$x = 17,121 \text{ yd}^3 \text{ delivered.}$$

Therefore, by using Site B, the contractor will begin to make a profit on the job after delivering 17,121 cubic yards of material.

2.1.2   Direct, Indirect, and Standard Costs

These frequently encountered cost terms involve most of the cost elements that also fit into the previous overlapping categories of fixed and variable costs and recurring and nonrecurring costs. *Direct costs* are costs that can be reasonably measured and allocated to a specific output or work activity. The labor and material costs directly associated with a product, service, or construction activity are direct costs. For example, the materials needed to make a pair of scissors would be a direct cost.

Indirect costs are costs that are difficult to attribute or allocate to a specific output or work activity. Normally, they are costs allocated through a selected formula (such as proportional to direct labor hours, direct labor dollars, or direct material dollars) to the outputs or work activities. For example, the costs of common tools, general supplies, and equipment maintenance in a plant are treated as indirect costs.

Overhead consists of plant operating costs that are not direct labor or direct material costs. In this book, the terms *indirect costs, overhead,* and *burden* are used interchangeably. Examples of overhead include electricity, general repairs, property taxes, and supervision. Administrative and selling expenses are usually added to direct costs and overhead costs to arrive at a unit selling price for a product or service. (Appendix 2-A provides a more detailed discussion of cost accounting principles.)

Standard costs are planned costs per unit of output that are established in advance of actual production or service delivery. They are developed from anticipated direct labor hours, materials, and overhead categories (with their established costs per unit). Because total overhead costs are associated with a *certain level of production*, this is an important condition that should be remembered when dealing with standard cost data (for example, see Section 2.4.3). Standard costs play an important role in cost control and other management functions. Some typical uses are the following:

1. Estimating future manufacturing costs
2. Measuring operating performance by comparing actual cost per unit with the standard unit cost
3. Preparing bids on products or services requested by customers
4. Establishing the value of work in process and finished inventories

2.1.3   Cash Cost versus Book Cost

A cost that involves payment of cash is called a *cash cost* (and results in a cash flow) to distinguish it from one that does not involve a cash transaction and is reflected in the accounting system as a *noncash cost*. This noncash cost is often referred to as a *book cost*. Cash costs are estimated from the perspective established for the analysis (Principle 3, Section 1.2) and are the future expenses incurred for the alternatives being analyzed. Book costs are costs that do not involve cash payments but rather represent the recovery of past expenditures over a fixed period of time. The most common example of book cost is the *depreciation* charged for the use of assets such as plant and equipment. In engineering economic analysis, only those costs that are cash flows or potential cash flows from the defined perspective for the analysis need to be considered. *Depreciation, for example, is not a cash flow* and is important in an analysis only because it affects income taxes, which are cash flows. We discuss the topics of depreciation and income taxes in Chapter 7.

2.1.4 Sunk Cost

A *sunk cost* is one that has occurred in the past and has no relevance to estimates of future costs and revenues related to an alternative course of action. Thus, a sunk cost is common to all alternatives, is not part of the future (prospective) cash flows, and can be disregarded in an engineering economic analysis. For instance, sunk costs are nonrefundable cash outlays, such as earnest money on a house or money spent on a passport.

The concept of sunk cost is illustrated in the next simple example. Suppose that Joe College finds a motorcycle he likes and pays $40 as a down payment, which will be applied to the $1,300 purchase price, but which must be forfeited if he decides not to take the cycle. Over the weekend, Joe finds another motorcycle he considers equally desirable for a purchase price of $1,230. For the purpose of deciding which cycle to purchase, the $40 is a sunk cost and thus would not enter into the decision, except that it lowers the remaining cost of the first cycle. The decision then is between paying an additional $1,260 ($1,300 − $40) for the first motorcycle versus $1,230 for the second motorcycle.

In summary, sunk costs are irretrievable consequences of past decisions and therefore are irrelevant in the analysis and comparison of alternatives that affect the future. Even though it is sometimes emotionally difficult to do, sunk costs should be ignored, except possibly to the extent that their existence assists you to anticipate better what will happen in the future.

EXAMPLE 2-2	Sunk Costs in Replacement Analysis

A classic example of sunk cost involves the replacement of assets. Suppose that your firm is considering the replacement of a piece of equipment. It originally cost $50,000, is presently shown on the company records with a value of $20,000, and can be sold for an estimated $5,000. For purposes of replacement analysis, the $50,000 is a sunk cost. However, one view is that the sunk cost should be considered as the difference between the value shown in the company records and the present realizable selling price. According to this viewpoint, the sunk cost is $20,000 minus $5,000, or $15,000. Neither the $50,000 nor the $15,000, however, should be considered in an engineering economic analysis, except for the manner in which the $15,000 may affect income taxes, which will be discussed in Chapter 9.

2.1.5 Opportunity Cost

An *opportunity cost* is incurred because of the use of limited resources, such that the opportunity to use those resources to monetary advantage in an alternative use is foregone. Thus, it is the cost of the best rejected (i.e., foregone) opportunity and is often hidden or implied.

Consider a student who could earn $20,000 for working during a year, but chooses instead to go to school for a year and spend $5,000 to do so. The opportunity cost of going to school for that year is $25,000: $5,000 cash outlay and $20,000 for income foregone. (This figure neglects the influence of income taxes and assumes that the student has no earning capability while in school.)

EXAMPLE 2-3	**Opportunity Cost in Replacement Analysis**

The concept of an opportunity cost is often encountered in analyzing the replacement of a piece of equipment or other capital asset. Let us reconsider Example 2-2, in which your firm considered the replacement of an existing piece of equipment that originally cost $50,000, is presently shown on the company records with a value of $20,000, but has a present market value of only $5,000. For purposes of an engineering economic analysis of whether to replace the equipment, the present investment in that equipment should be considered as $5,000, because, by keeping the equipment, the firm is giving up the *opportunity* to obtain $5,000 from its disposal. Thus, the $5,000 immediate selling price is really the investment cost of not replacing the equipment and is based on the opportunity cost concept.

2.1.6 Life-Cycle Cost

In engineering practice, the term *life-cycle cost* is often encountered. This term refers to a summation of all the costs related to a product, structure, system, or service during its life span. The *life cycle* is illustrated in Figure 2-1. The life cycle begins with identification of the economic need or want (the requirement) and ends with retirement and disposal activities. It is a time horizon that must be defined in the context of the specific situation—whether it is a highway bridge, a jet engine for commercial aircraft, or an automated flexible manufacturing cell for a factory. The end of the life cycle may be projected on a functional or an economic basis. For example, the amount of time that a structure or piece of equipment is able to perform economically may be shorter than that permitted by its physical capability. Changes in the design efficiency of a boiler illustrate this situation. The old boiler may be able to produce the steam required, but not economically enough for the intended use.

The life cycle may be divided into two general time periods: the acquisition phase and the operation phase. As shown in Figure 2-1, each of these phases is further subdivided into interrelated but different activity periods.

The acquisition phase begins with an analysis of the economic need or want—the analysis necessary to make explicit the requirement for the product, structure, system, or service. Then, with the requirement explicitly defined, the other activities in the acquisition phase can proceed in a logical sequence. The conceptual design activities translate the defined technical and operational requirements into a preferred preliminary design. Included in these activities are development of the feasible alternatives and engineering economic analyses to assist in selection of the preferred preliminary design. Also, advanced development

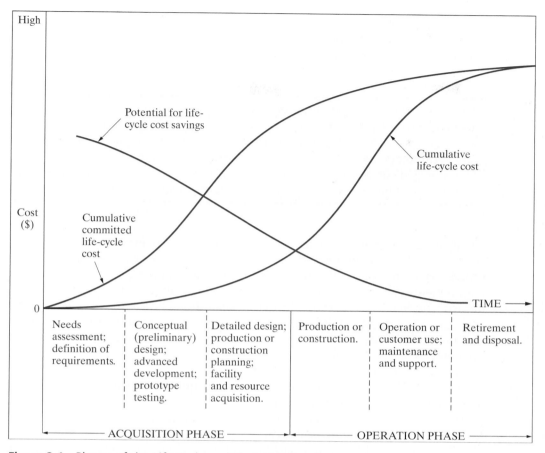

Figure 2-1   Phases of the Life Cycle and Their Relative Cost

and prototype-testing activities to support the preliminary design work occur during this period.

The next group of activities in the acquisition phase involves detailed design and planning for production or construction. This step is followed by the activities necessary to prepare, acquire, and make ready for operation the facilities and other resources needed. *Again, engineering economy studies are an essential part of the design process to analyze and compare alternatives and to assist in determining the final detailed design.*

In the operation phase, the production, delivery, or construction of the end item(s) or service and their operation or customer use occur. This phase ends with retirement from active operation or use and, often, disposal of the physical assets involved. The priorities for engineering economy studies during the operation phase are (1) achieving efficient and effective support to operations, (2) determining whether (and when) replacement of assets should occur, and (3) projecting the timing of retirement and disposal activities.

Figure 2-1 shows relative cost profiles for the life cycle. The greatest potential for achieving life-cycle cost savings is early in the acquisition phase. How much of the life-cycle costs for a product (for example) can be saved is dependent on many factors. However, effective engineering design and economic analysis during this phase are critical in maximizing potential savings.

> The cumulative committed life-cycle cost curve increases rapidly during the acquisition phase. In general, approximately 80% of life-cycle costs are "locked in" at the end of this phase by the decisions made during requirements analysis and preliminary and detailed design. In contrast, as reflected by the cumulative life-cycle cost curve, only about 20% of actual costs occur during the acquisition phase, with about 80% being incurred during the operation phase.

Thus, one purpose of the life-cycle concept is to make explicit the interrelated effects of costs over the total life span for a product. An objective of the design process is to minimize the life-cycle cost—while meeting other performance requirements—by making the right trade-offs between prospective costs during the acquisition phase and those during the operation phase.

The cost elements of the life cycle that need to be considered will vary with the situation. Because of their common use, however, several basic life-cycle cost categories will now be defined.

The *investment cost* is the capital required for most of the activities in the acquisition phase. In simple cases, such as acquiring specific equipment, an investment cost may be incurred as a single expenditure. On a large, complex construction project, however, a series of expenditures over an extended period could be incurred. This cost is also called a *capital investment*.

The term *working capital* refers to the funds required for current assets (i.e., other than fixed assets such as equipment, facilities, etc.) that are needed for the start-up and support of operational activities. For example, products cannot be made or services delivered without having materials available in inventory. Functions such as maintenance cannot be supported without spare parts, tools, trained personnel, and other resources. Also, cash must be available to pay employee salaries and the other expenses of operation. The amount of working capital needed will vary with the project involved, and some or all of the investment in working capital is usually recovered at the end of a project's life.

Operation and maintenance cost (O&M) includes many of the recurring annual expense items associated with the operation phase of the life cycle. The direct and indirect costs of operation associated with the five primary resource areas—people, machines, materials, energy, and information—are a major part of the costs in this category.

Disposal cost includes those nonrecurring costs of shutting down the operation and the retirement and disposal of assets at the end of the life cycle. Normally, costs associated with personnel, materials, transportation, and one-time special activities can be expected. These costs will be offset in some instances by receipts from the sale of assets with remaining market value. A classic example of a disposal cost is that associated with cleaning up a site where a chemical processing plant had been located.

2.2 The General Economic Environment

There are numerous general economic concepts that must be taken into account in engineering studies. In broad terms, economics deals with the interactions between people and wealth, and engineering is concerned with the cost-effective use of scientific knowledge to benefit humankind. This section introduces some of these basic economic concepts and indicates how they may be factors for consideration in engineering studies and managerial decisions.

2.2.1 Consumer and Producer Goods and Services

The goods and services that are produced and utilized may be divided conveniently into two classes. *Consumer goods and services* are those products or services that are directly used by people to satisfy their wants. Food, clothing, homes, cars, television sets, haircuts, opera, and medical services are examples. The providers of consumer goods and services must be aware of, and are subject to, the changing wants of the people to whom their products are sold.

Producer goods and services are used to produce consumer goods and services or other producer goods. Machine tools, factory buildings, buses, and farm machinery are examples. The amount of producer goods needed is determined indirectly by the amount of consumer goods or services that are demanded by people. However, because the relationship is much less direct than for consumer goods and services, the demand for and production of producer goods may greatly precede or lag behind the demand for the consumer goods that they will produce.

2.2.2 Measures of Economic Worth

Goods and services are produced and desired because they have *utility*—the power to satisfy human wants and needs. Thus, they may be used or consumed directly, or they may be used to produce other goods or services. Utility is most commonly measured in terms of *value*, expressed in some medium of exchange as the *price* that must be paid to obtain the particular item.

Much of our business activity, including engineering, focuses on increasing the utility (value) of materials and products by changing their form or location. Thus, iron ore, worth only a few dollars per ton, significantly increases in value by being processed, combined with suitable alloying elements, and converted into razor blades. Similarly, snow, worth almost nothing when high in distant mountains, becomes quite valuable when it is delivered in melted form several hundred miles away to dry southern California.

2.2.3 Necessities, Luxuries, and Price Demand

Goods and services may be divided into two types: *necessities* and *luxuries*. Obviously, these terms are relative, because, for most goods and services, what one person considers a necessity may be considered a luxury by another. For example, a person living in one community may find that an automobile is a necessity to get

Figure 2-2 General Price–Demand Relationship. (Note that price is considered to be the independent variable but is shown as the vertical axis. This convention is commonly used by economists.)

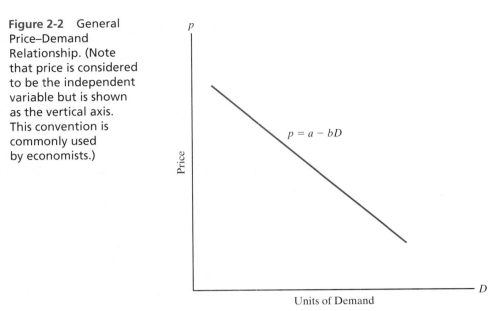

to and from work. If the same person lived and worked in a different city, adequate public transportation might be available, and an automobile would be a luxury. For all goods and services, there is a relationship between the price that must be paid and the quantity that will be demanded or purchased. This general relationship is depicted in Figure 2-2. As the selling price per unit (p) is increased, there will be less demand (D) for the product, and as the selling price is decreased, the demand will increase. The relationship between price and demand can be expressed as the linear function

$$p = a - bD \qquad \text{for } 0 \le D \le \frac{a}{b}, \text{ and } a > 0, b > 0, \tag{2-1}$$

where a is the intercept on the price axis and $-b$ is the slope. Thus, b is the amount by which demand increases for each unit decrease in p. Both a and b are constants. It follows, of course, that

$$D = \frac{a - p}{b} \qquad (b \ne 0). \tag{2-2}$$

2.2.4 Competition

Because economic laws are general statements regarding the interaction of people and wealth, they are affected by the economic environment in which people and wealth exist. Most general economic principles are stated for situations in which *perfect competition* exists.

Perfect competition occurs in a situation in which any given product is supplied by a large number of vendors and there is no restriction on additional suppliers entering the market. Under such conditions, there is assurance of complete freedom on the part of both buyer and seller. Perfect competition may never occur in actual practice, because of a multitude of factors that impose some degree of limitation upon the actions of buyers or sellers, or both. However, with conditions of perfect competition assumed, it is easier to formulate general economic laws.

Monopoly is at the opposite pole from perfect competition. A perfect monopoly exists when a unique product or service is only available from a single supplier and that vendor can prevent the entry of all others into the market. Under such conditions, the buyer is at the complete mercy of the supplier in terms of the availability and price of the product. Perfect monopolies rarely occur in practice, because (1) few products are so unique that substitutes cannot be used satisfactorily and (2) governmental regulations prohibit monopolies if they are unduly restrictive.

2.2.5 The Total Revenue Function

The total revenue, TR, that will result from a business venture during a given period is the product of the selling price per unit, p, and the number of units sold, D. Thus,

$$TR = \text{price} \times \text{demand} = p \cdot D. \tag{2-3}$$

If the relationship between price and demand as given in Equation (2-1) is used,

$$TR = (a - bD)D = aD - bD^2 \qquad \text{for } 0 \le D \le \frac{a}{b} \text{ and } a > 0, b > 0. \tag{2-4}$$

The relationship between total revenue and demand for the condition expressed in Equation (2-4) may be represented by the curve shown in Figure 2-3. From calculus, the demand, $\hat{D}$, that will produce maximum total revenue can be obtained by solving

$$\frac{d\text{TR}}{dD} = a - 2bD = 0. \tag{2-5}$$

Figure 2-3 Total Revenue Function as a Function of Demand

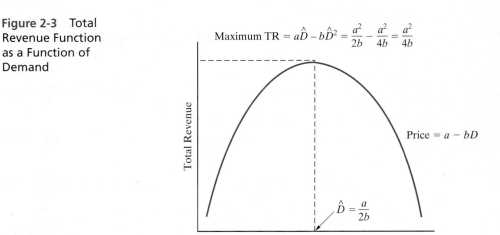

$$\text{Maximum TR} = a\hat{D} - b\hat{D}^2 = \frac{a^2}{2b} - \frac{a^2}{4b} = \frac{a^2}{4b}$$

Price $= a - bD$

$$\hat{D} = \frac{a}{2b}$$

Total Revenue

Demand

Thus,*

$$\hat{D} = \frac{a}{2b}. \tag{2-6}$$

It must be emphasized that, because of cost–volume relationships (discussed in the next section), *most businesses would not obtain maximum profits by maximizing revenue.* Accordingly, the cost–volume relationship must be considered and related to revenue, because cost reductions provide a key motivation for many engineering process improvements.

2.2.6 Cost, Volume, and Breakeven Point Relationships

Fixed costs remain constant over a wide range of activities, but variable costs vary in total with the volume of output (Section 2.1.1). Thus, at any demand D, total cost is

$$C_T = C_F + C_V, \tag{2-7}$$

where C_F and C_V denote fixed and variable costs, respectively. For the linear relationship assumed here,

$$C_V = c_v \cdot D, \tag{2-8}$$

where c_v is the variable cost per unit. In this section, we consider two scenarios for finding breakeven points. In the first scenario, demand is a function of price. The second scenario assumes that price and demand are independent of each other.

Scenario 1 When total revenue, as depicted in Figure 2-3, and total cost, as given by Equations (2-7) and (2-8), are combined, the typical results as a function of demand are depicted in Figure 2-4. At *breakeven point D'_1*, total revenue is equal

Figure 2-4 Combined Cost and Revenue Functions, and Breakeven Points, as Functions of Volume, and Their Effect on Typical Profit (Scenario 1)

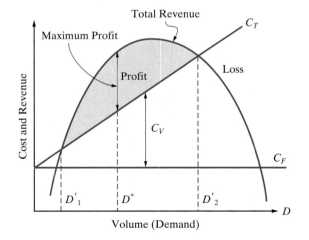

---

* To guarantee that $\hat{D}$ maximizes total revenue, check the second derivative to be sure it is negative:

$$\frac{d^2\text{TR}}{dD^2} = -2b.$$

Also, recall that in cost-minimization problems a positively signed second derivative is necessary to guarantee a minimum-value optimal cost solution.

to total cost, and an increase in demand will result in a profit for the operation. Then at optimal demand, D^*, profit is maximized [Equation (2-10)]. At breakeven point D_2', total revenue and total cost are again equal, but additional volume will result in an operating loss instead of a profit. Obviously, the conditions for which breakeven and maximum profit occur are our primary interest. First, at any volume (demand), D,

$$\text{Profit (loss)} = \text{total revenue} - \text{total costs}$$

$$= (aD - bD^2) - (C_F + c_vD)$$

$$= -bD^2 + (a - c_v)D - C_F \quad \text{for } 0 \le D \le \frac{a}{b} \text{ and } a > 0,\ b > 0. \quad (2\text{-}9)$$

In order for a profit to occur, based on Equation (2-9), and to achieve the typical results depicted in Figure 2-4, two conditions must be met:

1. $(a - c_v) > 0$; that is, the price per unit that will result in no demand has to be greater than the variable cost per unit. (This avoids negative demand.)
2. TR must exceed total cost (C_T) for the period involved.

If these conditions are met, we can find the optimal demand at which maximum profit will occur by taking the first derivative of Equation (2-9) with respect to D and setting it equal to zero:

$$\frac{d(\text{profit})}{dD} = a - c_v - 2bD = 0.$$

The optimal value of D that maximizes profit is

$$D^* = \frac{a - c_v}{2b}. \quad (2\text{-}10)$$

To ensure that we have *maximized* profit (rather than minimized it), the sign of the second derivative must be negative. Checking this, we find that

$$\frac{d^2(\text{profit})}{dD^2} = -2b,$$

which will be negative for $b > 0$ (as earlier specified).

An economic breakeven point for an operation occurs when total revenue equals total cost. Then for total revenue and total cost, as used in the development of Equations (2-9) and (2-10) and at any demand D,

$$\text{Total revenue} = \text{total cost} \qquad \text{(breakeven point)}$$

$$aD - bD^2 = C_F + c_vD$$

$$-bD^2 + (a - c_v)D - C_F = 0. \quad (2\text{-}11)$$

Because Equation (2-11) is a quadratic equation with one unknown (D), we can solve for the breakeven points D'_1 and D'_2 (the roots of the equation):*

$$D' = \frac{-(a - c_v) \pm [(a - c_v)^2 - 4(-b)(-C_F)]^{1/2}}{2(-b)}. \qquad (2\text{-}12)$$

With the conditions for a profit satisfied [Equation (2-9)], the quantity in the brackets of the numerator (the discriminant) in Equation (2-12) will be greater than zero. This will ensure that D'_1 and D'_2 have real positive, unequal values.

EXAMPLE 2-4   **Optimal Demand When Demand Is a Function of Price**

A company produces an electronic timing switch that is used in consumer and commercial products. The fixed cost (C_F) is $73,000 per month, and the variable cost (c_v) is $83 per unit. The selling price per unit is $p = \$180 - 0.02(D)$, based on Equation (2-1). For this situation,

(a) determine the optimal volume for this product and confirm that a profit occurs (instead of a loss) at this demand.

(b) find the volumes at which breakeven occurs; that is, what is the range of profitable demand? Solve by hand and by spreadsheet.

👆 **Solution by Hand**

(a) $D^* = \dfrac{a - c_v}{2b} = \dfrac{\$180 - \$83}{2(0.02)} = 2{,}425$ units per month [from Equation (2-10)].

Is $(a - c_v) > 0$?

($180 - $83) = $97,   which is greater than 0.

And is (total revenue $-$ total cost) > 0 for $D^* = 2{,}425$ units per month?

[$180(2,425) - 0.02(2,425)^2] - [\$73{,}000 + \$83(2{,}425)] = \$44{,}612$

A demand of $D^* = 2{,}425$ units per month results in a maximum profit of $44,612 per month. Notice that the second derivative is negative (-0.04).

(b) Total revenue = total cost   (breakeven point)

$$-bD^2 + (a - c_v)D - C_F = 0 \qquad \text{[from Equation (2-11)]}$$

$$-0.02D^2 + (\$180 - \$83)D - \$73{,}000 = 0$$

$$-0.02D^2 + 97D - 73{,}000 = 0$$

---

* Given the quadratic equation $ax^2 + bx + c = 0$, the roots are given by $x = \frac{-b \pm \sqrt{b^2 - 4ac}}{2a}$.

And, from Equation (2-12),

$$D' = \frac{-97 \pm [(97)^2 - 4(-0.02)(-73{,}000)]^{0.5}}{2(-0.02)}$$

$$D'_1 = \frac{-97 + 59.74}{-0.04} = 932 \text{ units per month}$$

$$D'_2 = \frac{-97 - 59.74}{-0.04} = 3{,}918 \text{ units per month.}$$

Thus, the range of profitable demand is 932–3,918 units per month.

Spreadsheet Solution

Figure 2-5(a) displays the spreadsheet solution for this problem. This spreadsheet calculates profit for a range of demand values (shown in column A). For a specific value of demand, price per unit is calculated in column B by using Equation (2-1) and Total Revenue is simply demand × price. Total Expense is computed by using Equations (2-7) and (2-8). Finally, Profit (column E) is then Total Revenue − Total Expense.

A quick inspection of the Profit column gives us an idea of the optimal demand value as well as the breakeven points. Note that profit steadily increases as demand increases to 2,500 units per month and then begins to drop off. This tells us that the optimal demand value lies in the range of 2,250 to 2,750 units per month. A more specific value can be obtained by changing the Demand Start point value in cell E1 and the Demand Increment value in cell E2. For example, if the value of cell E1 is set to 2,250 and the increment in cell E2 is set to 10, the optimal demand value is shown to be between 2,420 and 2,430 units per month.

The breakeven points lie within the ranges 750–1,000 units per month and 3,750–4,000 units per month, as indicated by the change in sign of profit. Again, by changing the values in cells E1 and E2, we can obtain more exact values of the breakeven points.

Figure 2-5(b) is a graphical display of the Total Revenue, Total Expense, and Profit functions for the range of demand values given in column A of Figure 2-5(a). This graph enables us to see how profit changes as demand increases. The optimal demand value (maximum point of the profit curve) appears to be around 2,500 units per month.

Figure 2-5(b) is also a graphical representation of the breakeven points. By graphing the total revenue and total cost curves separately, we can easily identify the breakeven points (the intersection of these two functions). From the graph, the range of profitable demand is approximately 1,000 to 4,000 units per month. Notice also that, at these demand values, the profit curve crosses the x-axis ($0).

$$= \$B\$3 - \$B\$4 * A7$$

$$= \$B\$1 + \$B\$2 * A7$$

$$= B7 * A7$$

$$= C7 - D7$$

$$= E1$$

	A	B	C	D	E
1	Fixed cost/ mo. =	$ 73,000		Demand Start point (D) =	0
2	Variable cost/unit =	$ 83		Demand Increment =	250
3	a =	$ 180			
4	b =	$ 0.02			
5					
6	Monthly Demand	Price per Unit	Total Revenue	Total Expense	Profit
7	0	$ 180	$ -	$ 73,000	$ (73,000)
8	250	$ 175	$ 43,750	$ 93,750	$ (50,000)
9	500	$ 170	$ 85,000	$ 114,500	$ (29,500)
10	750	$ 165	$ 123,750	$ 135,250	$ (11,500)
11	1000	$ 160	$ 160,000	$ 156,000	$ 4,000
12	1250	$ 155	$ 193,750	$ 176,750	$ 17,000
13	1500	$ 150	$ 225,000	$ 197,500	$ 27,500
14	1750	$ 145	$ 253,750	$ 218,250	$ 35,500
15	2000	$ 140	$ 280,000	$ 239,000	$ 41,000
16	2250	$ 135	$ 303,750	$ 259,750	$ 44,000
17	2500	$ 130	$ 325,000	$ 280,500	$ 44,500
18	2750	$ 125	$ 343,750	$ 301,250	$ 42,500
19	3000	$ 120	$ 360,000	$ 322,000	$ 38,000
20	3250	$ 115	$ 373,750	$ 342,750	$ 31,000
21	3500	$ 110	$ 385,000	$ 363,500	$ 21,500
22	3750	$ 105	$ 393,750	$ 384,250	$ 9,500
23	4000	$ 100	$ 400,000	$ 405,000	$ (5,000)
24	4250	$ 95	$ 403,750	$ 425,750	$ (22,000)
25	4500	$ 90	$ 405,000	$ 446,500	$ (41,500)
26	4750	$ 85	$ 403,750	$ 467,250	$ (63,500)
27	5000	$ 80	$ 400,000	$ 488,000	$ (88,000)
28	5250	$ 75	$ 393,750	$ 508,750	$ (115,000)
29	5500	$ 70	$ 385,000	$ 529,500	$ (144,500)

$$= A7 + \$E\$2$$

(a) Table of profit values for a range of demand values

Figure 2-5 Spreadsheet Solution, Example 2-4

Comment

As seen in the hand solution to this problem, Equations (2-10) and (2-12) can be used directly to solve for the optimal demand value and breakeven points.

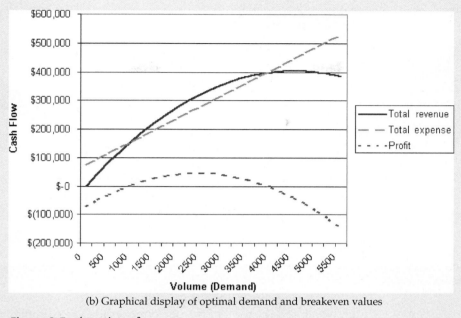

(b) Graphical display of optimal demand and breakeven values

Figure 2-5 (*continued*)

The power of the spreadsheet in this example is the ease with which graphical displays can be generated to support your analysis. Remember, a picture really can be worth a thousand words. Spreadsheets also facilitate sensitivity analysis (to be discussed more fully in Chapter 11). For example, what is the impact on the optimal demand value and breakeven points if variable costs are reduced by 10% per unit? (The new optimal demand value is increased to 2,632 units per month, and the range of profitable demand is widened to 822 to 4,443 units per month.)

Scenario 2 When the price per unit (p) for a product or service can be represented more simply as being independent of demand [versus being a linear function of demand, as assumed in Equation (2-1)] and is greater than the variable cost per unit (c_v), a single breakeven point results. Then, under the assumption that demand is immediately met, total revenue (TR) $= p \cdot D$. If the linear relationship for costs in Equations (2-7) and (2-8) is also used in the model, the typical situation is depicted in Figure 2-6. This scenario is typified by the Airbus example presented at the beginning of the chapter.

Figure 2-6 Typical Breakeven Chart with Price (p) a Constant (Scenario 2)

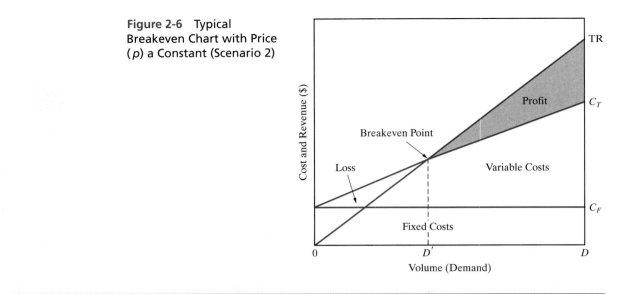

| EXAMPLE 2-5 | Breakeven Point When Price Is Independent of Demand |

An engineering consulting firm measures its output in a standard service hour unit, which is a function of the personnel grade levels in the professional staff. The variable cost (c_v) is $62 per standard service hour. The charge-out rate [i.e., selling price (p)] is $85.56 per hour. The maximum output of the firm is 160,000 hours per year, and its fixed cost (C_F) is $2,024,000 per year. For this firm,

(a) what is the breakeven point in standard service hours and in percentage of total capacity?

(b) what is the percentage reduction in the breakeven point (sensitivity) if fixed costs are reduced 10%; if variable cost per hour is reduced 10%; and if the selling price per unit is increased by 10%?

Solution

(a)

$$\text{Total revenue} = \text{total cost} \qquad (\text{breakeven point})$$

$$pD' = C_F + c_v D'$$

$$D' = \frac{C_F}{(p - c_v)}, \qquad (2\text{-}13)$$

and

$$D' = \frac{\$2{,}024{,}000}{(\$85.56 - \$62)} = 85{,}908 \text{ hours per year}$$

$$D' = \frac{85{,}908}{160{,}000} = 0.537,$$

or 53.7% of capacity.

(b) A 10% reduction in C_F gives

$$D' = \frac{0.9(\$2,024,000)}{(\$85.56 - \$62)} = 77,318 \text{ hours per year}$$

and

$$\frac{85,908 - 77,318}{85,908} = 0.10,$$

or a 10% reduction in D'.
   A 10% reduction in c_v gives

$$D' = \frac{\$2,024,000}{[\$85.56 - 0.9(\$62)]} = 68,011 \text{ hours per year}$$

and

$$\frac{85,908 - 68,011}{85,908} = 0.208,$$

or a 20.8% reduction in D'.
   A 10% increase in p gives

$$D' = \frac{\$2,024,000}{[1.1(\$85.56) - \$62]} = 63,021 \text{ hours per year}$$

and

$$\frac{85,908 - 63,021}{85,908} = 0.266,$$

or a 26.6% reduction in D'.

Thus, the breakeven point is more sensitive to a reduction in variable cost per hour than to the same percentage reduction in the fixed cost. Furthermore, notice that the breakeven point in this example is highly sensitive to the selling price per unit, p.

Market competition often creates pressure to lower the breakeven point of an operation; the lower the breakeven point, the less likely that a loss will occur during market fluctuations. Also, if the selling price remains constant (or increases), a larger profit will be achieved at any level of operation above the reduced breakeven point.

2.3   Cost-Driven Design Optimization

As discussed in Section 2.1.6, engineers must maintain a *life-cycle* (i.e., "cradle to grave") viewpoint as they design products, processes, and services. Such a complete perspective ensures that engineers consider initial investment costs, operation and maintenance expenses and other annual expenses in later years, and environmental and social consequences over the life of their designs. In fact, a movement called *Design for the Environment* (DFE), or "green engineering," has prevention of waste, improved materials selection, and reuse and recycling of resources among its goals. Designing for energy conservation, for example, is a subset of green engineering. Another example is the design of an automobile bumper that can be easily recycled. As you can see, *engineering design is an economically driven art*.

Examples of cost minimization through effective design are plentiful in the practice of engineering. Consider the design of a heat exchanger in which tube material and configuration affect cost and dissipation of heat. The problems in this section designated as "cost-driven design optimization" are simple design models intended to illustrate the importance of cost in the design process. These problems show the procedure for determining an optimal design, using cost concepts. We will consider discrete and continuous optimization problems that involve a single design variable, X. This variable is also called a *primary cost driver*, and knowledge of its behavior may allow a designer to account for a large portion of total cost behavior.

For cost-driven design optimization problems, the two main tasks are as follows:

1. Determine the optimal value for a certain alternative's design variable. For example, what velocity of an aircraft minimizes the total annual costs of owning and operating the aircraft?

2. Select the best alternative, each with its own unique value for the design variable. For example, what insulation thickness is best for a home in Virginia: R11, R19, R30, or R38?

In general, the cost models developed in these problems consist of three types of costs:

1. fixed cost(s)
2. cost(s) that vary *directly* with the design variable
3. cost(s) that vary *indirectly* with the design variable

A simplified format of a cost model with one design variable is

$$\text{Cost} = aX + \frac{b}{X} + k, \tag{2-14}$$

where a is a parameter that represents the directly varying cost(s),
        b is a parameter that represents the indirectly varying cost(s),
        k is a parameter that represents the fixed cost(s), and
        X represents the design variable in question (e.g., weight or velocity).

In a particular problem, the parameters a, b, and k may actually represent the sum of a group of costs in that category, and the design variable may be raised to some power for either directly or indirectly varying costs.*

The following steps outline a general approach for optimizing a design with respect to cost:

1. Identify the design variable that is the primary cost driver (e.g., pipe diameter or insulation thickness).

2. Write an expression for the cost model in terms of the design variable.

3. Set the first derivative of the cost model with respect to the continuous design variable equal to zero. For discrete design variables, compute the value of the cost model for each discrete value over a selected range of potential values.

4. Solve the equation found in Step 3 for the optimum value of the continuous design variable.[†] For discrete design variables, the optimum value has the minimum cost value found in Step 3. This method is analogous to taking the first derivative for a continuous design variable and setting it equal to zero to determine an optimal value.

5. For continuous design variables, use the second derivative of the cost model with respect to the design variable to determine whether the optimum value found in Step 4 corresponds to a global maximum or minimum.

EXAMPLE 2-6	**How Fast Should the Airplane Fly?**

The cost of operating a jet-powered commercial (passenger-carrying) airplane varies as the three-halves (3/2) power of its velocity; specifically, $C_O = knv^{3/2}$, where n is the trip length in miles, k is a constant of proportionality, and v is velocity in miles per hour. It is known that at 400 miles per hour the *average* cost of operation is $300 per mile. The company that owns the aircraft wants to minimize the cost of operation, but that cost must be balanced against the cost of the passengers' time (C_C), which has been set at $300,000 per hour.

(a) At what velocity should the trip be planned to minimize the total cost, which is the sum of the cost of operating the airplane and the cost of passengers' time?

(b) How do you know that your answer for the problem in Part (a) minimizes the total cost?

---

* A more general model is the following: Cost $= k + ax + b_1 x^{e_1} + b_2 x^{e_2} + \cdots$, where $e_1 = -1$ reflects costs that vary inversely with X, $e_2 = 2$ indicates costs that vary as the square of X, and so forth.
† If multiple optima (stationary points) are found in Step 4, finding the global optimum value of the design variable will require a little more effort. One approach is to systematically use each root in the second derivative equation and assign each point as a maximum or a minimum based on the sign of the second derivative. A second approach would be to use each root in the objective function and see which point best satisfies the cost function.

Solution

(a) The equation for total cost (C_T) is

$$C_T = C_O + C_C = knv^{3/2} + (\$300,000 \text{ per hour}) \left(\frac{n}{v}\right),$$

where n/v has time (hours) as its unit.
Now we solve for the value of k:

$$\frac{C_O}{n} = kv^{3/2}$$

$$\frac{\$300}{\text{mile}} = k \left(400 \frac{\text{miles}}{\text{hour}}\right)^{3/2}$$

$$k = \frac{\$300/\text{mile}}{\left(400 \dfrac{\text{miles}}{\text{hour}}\right)^{3/2}}$$

$$k = \frac{\$300/\text{mile}}{8000 \left(\dfrac{\text{miles}^{3/2}}{\text{hour}^{3/2}}\right)}$$

$$k = \$0.0375 \frac{\text{hours}^{3/2}}{\text{miles}^{5/2}}.$$

Thus,

$$C_T = \left(\$0.0375 \frac{\text{hours}^{3/2}}{\text{miles}^{5/2}}\right)(n \text{ miles}) \left(v \frac{\text{miles}}{\text{hour}}\right)^{3/2} + \left(\frac{\$300,000}{\text{hour}}\right) \left(\frac{n \text{ miles}}{v \dfrac{\text{miles}}{\text{hour}}}\right)$$

$$C_T = \$0.0375 nv^{3/2} + \$300,000 \left(\frac{n}{v}\right).$$

Next, the first derivative is taken:

$$\frac{dC_T}{dv} = \frac{3}{2}(\$0.0375)nv^{1/2} - \frac{\$300,000n}{v^2} = 0.$$

So,

$$0.05625v^{1/2} - \frac{300,000}{v^2} = 0$$

$$0.05625v^{5/2} - 300,000 = 0$$

$$v^{5/2} = \frac{300,000}{0.05625} = 5,333,333$$

$$v^* = (5,333,333)^{0.4} = 490.68 \text{ mph}.$$

(b) Finally, we check the second derivative to confirm a minimum cost solution:

$$\frac{d^2C_T}{dv^2} = \frac{0.028125}{v^{1/2}} + \frac{600,000}{v^3} \quad \text{for } v > 0, \text{ and therefore, } \frac{d^2C_T}{dv^2} > 0.$$

The company concludes that $v = 490.68$ mph minimizes the total cost of this particular airplane's flight.

| EXAMPLE 2-7 | **Energy Savings through Increased Insulation** |

This example deals with a discrete optimization problem of determining the most economical amount of attic insulation for a large single-story home in Virginia. In general, the heat lost through the roof of a single-story home is

$$\begin{array}{c}\text{Heat loss} \\ \text{in Btu} \\ \text{per hour}\end{array} = \left(\begin{array}{c}\Delta\,\text{Temperature} \\ \text{in }°\text{F}\end{array}\right)\left(\begin{array}{c}\text{Area} \\ \text{in} \\ \text{ft}^2\end{array}\right)\left(\begin{array}{c}\text{Conductance in} \\ \text{Btu/hour} \\ \overline{\text{ft}^2 - °\text{F}}\end{array}\right),$$

or

$$Q = (T_{\text{in}} - T_{\text{out}}) \cdot A \cdot U.$$

In southwest Virginia, the number of heating days per year is approximately 230, and the annual heating degree-days equals $230\,(65°\text{F}-46°\text{F}) = 4{,}370$ degree-days per year. Here $65°\text{F}$ is assumed to be the average inside temperature and $46°\text{F}$ is the average outside temperature each day.

Consider a 2,400-ft^2 single-story house in Blacksburg. The typical annual space-heating load for this size of a house is 100×10^6 Btu. That is, with no insulation in the attic, we lose about 100×10^6 Btu per year.* Common sense dictates that the "no insulation" alternative is not attractive and is to be avoided.

With insulation in the attic, the amount of heat lost each year will be reduced. The value of energy savings that results from adding insulation and reducing heat loss is dependent on what type of residential heating furnace is installed. For this example, we assume that an electrical resistance furnace is installed by the builder, and its efficiency is near 100%.

Now we're in a position to answer the following question: What amount of insulation is most economical? An additional piece of data we need involves the cost of electricity, which is $0.074 per kWh. This can be converted to dollars per 10^6 Btu as follows (1 kWh = 3,413 Btu):

$$\frac{\text{kWh}}{3{,}413\,\text{Btu}} = 293\,\text{kWh per million Btu}$$

---

* 100×10^6 Btu/yr $\cong \left(\dfrac{4{,}370\,°\text{F-days per year}}{1.00\,\text{efficiency}}\right)(2{,}400\,\text{ft}^2)(24\,\text{hours/day})\left(\dfrac{0.397\,\text{Btu/hr}}{\text{ft}^2-°\text{F}}\right)$, where 0.397 is the U-factor with no insulation.

$$\frac{293 \text{ kWh}}{10^6 \text{ Btu}} \left(\frac{\$0.074}{\text{kWh}} \right) \cong \$21.75/10^6 \text{ Btu.}$$

The cost of several insulation alternatives and associated space-heating loads for this house are given in the following table:

	Amount of Insulation			
	R11	R19	R30	R38
Investment cost ($)	600	900	1,300	1,600
Annual heating load (Btu/year)	74×10^6	69.8×10^6	67.2×10^6	66.2×10^6

In view of these data, which amount of attic insulation is most economical? The life of the insulation is estimated to be 25 years.

Solution

Set up a table to examine total life-cycle costs:

		R11	R19	R30	R38
A.	Investment cost	$600	$900	$1,300	$1,600
B.	Cost of heat loss per year	$1,609.50	$1,518.15	$1,461.60	$1,439.85
C.	Cost of heat loss over 25 years	$40,237.50	$37,953.75	$36,540	$35,996.25
D.	Total life cycle costs (A + C)	$40,837.50	$38,853.75	$37,840	$37,596.25

Answer: To minimize total life-cycle costs, select R38 insulation.

Caution

This conclusion may change when we consider the time value of money (i.e., an interest rate greater than zero) in Chapter 4. In such a case, it will not necessarily be true that adding more and more insulation is the optimal course of action.

2.4   Present Economy Studies

When alternatives for accomplishing a specific task are being compared over *one year or less* and the influence of time on money can be ignored, engineering economic analyses are referred to as *present economy studies.* Several situations involving present economy studies are illustrated in this section. The rules, or criteria, shown next will be used to select the preferred alternative when defect-free output (yield) is variable *or* constant among the alternatives being considered.

RULE 1:	When revenues and other economic benefits are present and vary among alternatives, choose the alternative that *maximizes* overall profitability based on the number of defect-free units of a product or service produced.

RULE 2:	When revenues and other economic benefits are *not* present *or* are constant among all alternatives, consider only the costs and select the alternative that *minimizes* total cost per defect-free unit of product or service output.

2.4.1   Total Cost in Material Selection

In many cases, economic selection among materials cannot be based solely on the costs of materials. Frequently, a change in materials will affect the design and processing costs, and shipping costs may also be altered.

EXAMPLE 2-8	Choosing the Most Economic Material for a Part

A good example of this situation is illustrated by a part for which annual demand is 100,000 units. The part is produced on a high-speed turret lathe, using 1112 screw-machine steel costing $0.30 per pound. A study was conducted to determine whether it might be cheaper to use brass screw stock, costing $1.40 per pound. Because the weight of steel required per piece was 0.0353 pounds and that of brass was 0.0384 pounds, the material cost per piece was $0.0106 for steel and $0.0538 for brass. However, when the manufacturing engineering department was consulted, it was found that, although 57.1 defect-free parts per hour were being produced by using steel, the output would be 102.9 defect-free parts per hour if brass were used. Which material should be used for this part?

Solution

The machine attendant was paid $15.00 per hour, and the variable (i.e., traceable) overhead costs for the turret lathe were estimated to be $10.00 per hour. Thus, the total-cost comparison for the two materials was as follows:

	1112 Steel	Brass
Material	$0.30 \times 0.0353 = $0.0106	$1.40 \times 0.0384 = $0.0538
Labor	$15.00/57.1 = 0.2627	$15.00/102.9 = 0.1458
Variable overhead	$10.00/57.1 = 0.1751	$10.00/102.9 = 0.0972
Total cost per piece	$0.4484	$0.2968

Saving per piece by use of brass = $0.4484 − $0.2968 = $0.1516

Because 100,000 parts are made each year, revenues are constant across the alternatives. Rule 2 would select brass, and its use will produce a savings of $151.60 per thousand (a total of $15,160 for the year). It is also clear that costs other than the cost of material were important in the study.

Care should be taken in making economic selections between materials to ensure that any differences in shipping costs, yields, or resulting scrap are taken into account. Commonly, alternative materials do not come in the same stock sizes, such as sheet sizes and bar lengths. This may considerably affect the yield obtained from a given weight of material. Similarly, the resulting scrap may differ for various materials.

In addition to deciding what material a product should be made of, there are often alternative methods or machines that can be used to produce the product, which, in turn, can impact processing costs. Processing times may vary with the machine selected, as may the product yield. As illustrated in Example 2-9, these considerations can have important economic implications.

EXAMPLE 2-9   **Choosing the Most Economical Machine for Production**

Two currently owned machines are being considered for the production of a part. The capital investment associated with the machines is about the same and can be ignored for purposes of this example. The important differences between the machines are their production capacities (production rate × available production hours) and their reject rates (percentage of parts produced that cannot be sold). Consider the following table:

	Machine A	Machine B
Production rate	100 parts/hour	130 parts/hour
Hours available for production	7 hours/day	6 hours/day
Percent parts rejected	3%	10%

The material cost is $6.00 per part, and all defect-free parts produced can be sold for $12 each. (Rejected parts have negligible scrap value.) For either machine, the operator cost is $15.00 per hour and the variable overhead rate for traceable costs is $5.00 per hour.

(a) Assume that the daily demand for this part is large enough that all defect-free parts can be sold. Which machine should be selected?

(b) What would the percent of parts rejected have to be for Machine B to be as profitable as Machine A?

Solution

(a) Rule 1 applies in this situation because total daily revenues (selling price per part times the number of parts sold per day) and total daily costs will vary depending on the machine chosen. Therefore, we should select the machine that will maximize the profit per day:

$$\text{Profit per day} = \text{Revenue per day} - \text{Cost per day}$$

$$= (\text{Production rate})(\text{Production hours})(\$12/\text{part})$$

$$\times [1 - (\%\text{rejected}/100)]$$

$$- (\text{Production rate})(\text{Production hours})(\$6/\text{part})$$

$$- (\text{Production hours})(\$15/\text{hour} + \$5/\text{hour}).$$

$$\text{Machine } A: \text{Profit per day} = \left(\frac{100 \text{ parts}}{\text{hour}}\right)\left(\frac{7 \text{ hours}}{\text{day}}\right)\left(\frac{\$12}{\text{part}}\right)(1 - 0.03)$$

$$- \left(\frac{100 \text{ parts}}{\text{hour}}\right)\left(\frac{7 \text{ hours}}{\text{day}}\right)\left(\frac{\$6}{\text{part}}\right)$$

$$- \left(\frac{7 \text{ hours}}{\text{day}}\right)\left(\frac{\$15}{\text{hour}} + \frac{\$5}{\text{hour}}\right)$$

$$= \$3,808 \text{ per day}.$$

$$\text{Machine } B: \text{Profit per day} = \left(\frac{130 \text{ parts}}{\text{hour}}\right)\left(\frac{6 \text{ hours}}{\text{day}}\right)\left(\frac{\$12}{\text{part}}\right)(1 - 0.10)$$

$$- \left(\frac{130 \text{ parts}}{\text{hour}}\right)\left(\frac{6 \text{ hours}}{\text{day}}\right)\left(\frac{\$6}{\text{part}}\right)$$

$$- \left(\frac{6 \text{ hours}}{\text{day}}\right)\left(\frac{\$15}{\text{hour}} + \frac{\$5}{\text{hour}}\right)$$

$$= \$3,624 \text{ per day}.$$

Therefore, *select Machine A* to maximize profit per day.

(b) To find the breakeven percent of parts rejected, X, for Machine B, set the profit per day of Machine A equal to the profit per day of Machine B, and solve for X:

$$\$3,808/\text{day} = \left(\frac{130 \text{ parts}}{\text{hour}}\right)\left(\frac{6 \text{ hours}}{\text{day}}\right)\left(\frac{\$12}{\text{part}}\right)(1 - X) - \left(\frac{130 \text{ parts}}{\text{hour}}\right)$$

$$\times \left(\frac{6 \text{ hours}}{\text{day}}\right)\left(\frac{\$6}{\text{part}}\right) - \left(\frac{6 \text{ hours}}{\text{day}}\right)\left(\frac{\$15}{\text{hour}} + \frac{\$5}{\text{hour}}\right).$$

Thus, $X = 0.08$, so the percent of parts rejected for Machine B can be no higher than 8% for it to be as profitable as Machine A.

2.4.2   Alternative Machine Speeds

Machines can frequently be operated at various speeds, resulting in different rates of product output. However, this usually results in different frequencies of machine downtime to permit servicing or maintaining the machine, such as resharpening or adjusting tooling. Such situations lead to present economy studies to determine the preferred operating speed. We *first* assume that there is an unlimited amount of work to be done in Example 2-10. Secondly, Example 2-11 illustrates how to deal with a fixed (limited) amount of work.

EXAMPLE 2-10	Best Operating Speed for an Unlimited Amount of Work

A simple example of alternative machine speeds involves the planing of lumber. Lumber put through the planer increases in value by $0.10 per board foot. When the planer is operated at a cutting speed of 5,000 feet per minute, the blades have to be sharpened after 2 hours of operation, and the lumber can be planed at the rate of *1,000 board-feet per hour*. When the machine is operated at 6,000 feet per minute, the blades have to be sharpened after $1\frac{1}{2}$ hours of operation, and the rate of planing is *1,200 board-feet per hour*. Each time the blades are changed, the machine has to be shut down for 15 minutes. The blades, unsharpened, cost $50 per set and can be sharpened 10 times before having to be discarded. Sharpening costs $10 per occurrence. The crew that operates the planer changes and resets the blades. At what speed should the planer be operated?

Solution

Because the labor cost for the crew would be the same for either speed of operation and because there was no discernible difference in wear upon the planer, these factors did not have to be included in the study.

In problems of this type, the operating time plus the delay time due to the necessity for tool changes constitute a cycle time that determines the output from the machine. The time required for a complete cycle determines the number of cycles that can be accomplished in a period of available time (e.g., one day), and a certain portion of each complete cycle is productive. The actual productive time will be the product of the productive time per cycle and the number of cycles per day.

	Value per day ($)
At 5,000 feet per minute	
Cycle time = 2 hours + 0.25 hour = 2.25 hours	
Cycles per day = 8 ÷ 2.25 = 3.555	
Value added by planing = 3.555 × 2 × 1,000 × $0.10 =	711.00
Cost of resharpening blades = 3.555 × $10 = $35.55	
Cost of blades = 3.555 × $50/10 = 17.78	
Total cost cash flow	−53.33
Net increase in value (profit) per day	657.67

At 6,000 feet per minute
   Cycle time = 1.5 hours + 0.25 hour = 1.75 hours
   Cycles per day = 8 ÷ 1.75 = 4.57
   Value added by planing = 4.57 × 1.5 × 1,200 × $0.10 =   822.60*
   Cost of resharpening blades = 4.57 × $10 = $45.70
   Cost of blades = 4.57 × $50/10 = 22.85
      Total cost cash flow   −68.55
   Net increase in value (profit) per day   754.05

* The units work out as follows: (cycles/day)(hours/cycle)(board feet/hour) (dollar value/board-foot) = dollars/day.

Thus, in Example 2-10 it is more economical according to Rule 1 to operate at 6,000 feet per minute, in spite of the more frequent sharpening of blades that is required.

EXAMPLE 2-11   **Fixed Amount of Work: Now Which Speed Is Best?**

Example 2-10 assumed that every board-foot of lumber that is planed can be sold. If there is limited demand for the lumber, a correct choice of machining speeds can be made with Rule 2 by *minimizing total cost* per unit of output. Suppose now we want to know the better machining speed when only one job requiring 6,000 board-feet of planing is considered. Solve by using a spreadsheet.

Spreadsheet Solution

For a fixed planing requirement of 6,000 board-feet, the value added by planing is 6,000 ($0.10) = $600 for either cutting speed. Hence, we want to minimize total cost per board-foot planed.

The total cost per board-foot planed is a combination of the blade cost and resharpening cost. These costs are most easily stated on a per cycle basis (blade cost/cycle = $50/10 cycles and resharpening cost/cycle = $10/cycle). Now the total cost for a fixed job length can be determined by the number of cycles required.

Figure 2-7 presents a spreadsheet model for this problem. The cell formulas were developed by using the cycle time solution approach of Example 2-10. The production rate per hour (cells E1 and F1) is converted to a production rate per cycle (cells B10 and C10). This value is used to determine the number of cycles required to complete the fixed length job (cells B11 and C11).

For a 6,000-board-foot job, select the slower cutting speed (5,000 feet per minute) to minimize cost. During the 0.92 hour of time savings for the 6,000-feet-per-minute cutting speed, we assume that the operator is idle.

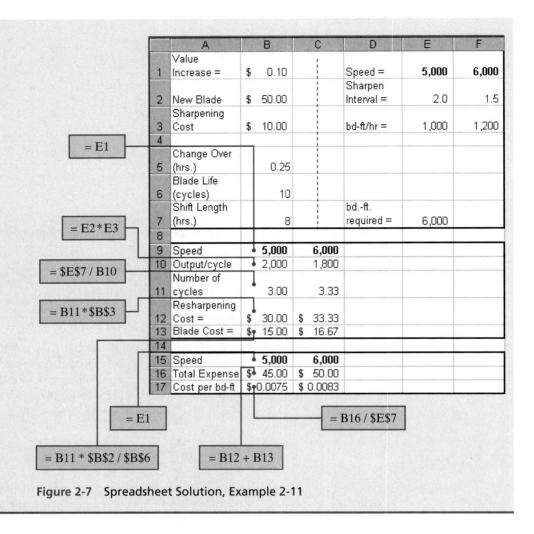

	A	B	C	D	E	F
1	Value Increase =	$ 0.10		Speed =	**5,000**	**6,000**
2	New Blade	$ 50.00		Sharpen Interval =	2.0	1.5
3	Sharpening Cost	$ 10.00		bd-ft/hr =	1,000	1,200
4						
5	Change Over (hrs.)	0.25				
6	Blade Life (cycles)	10				
7	Shift Length (hrs.)	8		bd.-ft. required =	6,000	
8						
9	Speed	**5,000**	**6,000**			
10	Output/cycle	2,000	1,800			
11	Number of cycles	3.00	3.33			
12	Resharpening Cost =	$ 30.00	$ 33.33			
13	Blade Cost =	$ 15.00	$ 16.67			
14						
15	Speed	**5,000**	**6,000**			
16	Total Expense	$ 45.00	$ 50.00			
17	Cost per bd-ft	$ 0.0075	$ 0.0083			

= E1

= E2 * E3

= $E$7 / B10

= B11 * $B$3

= E1

= B16 / $E$7

= B11 * $B$2 / $B$6

= B12 + B13

Figure 2-7   Spreadsheet Solution, Example 2-11

2.4.3   Making versus Purchasing (Outsourcing) Studies*

In the short run, say, one year or less, a company may consider producing an item in-house even though the item can be purchased (outsourced) from a supplier at a price lower than the company's standard production costs. (See Section 2.1.2.) This could occur if (1) direct, indirect, and overhead costs are incurred regardless of whether the item is purchased from an outside supplier and (2) the *incremental* cost of producing an item in the short run is less than the supplier's price. Therefore, the relevant short-run costs of make versus purchase decisions are the *incremental costs* incurred and the *opportunity costs* of the resources involved.

---

* Much interest has been shown in outsourcing decisions. For example, see P. Chalos, "Costing, Control, and Strategic Analysis in Outsourcing Decisions," *Journal of Cost Management*, 8, no. 4 (Winter 1995): 31–37.

Opportunity costs may become significant when in-house manufacture of an item causes other production opportunities to be forgone (often because of insufficient capacity). But in the long run, capital investments in additional manufacturing plant and capacity are often feasible alternatives to outsourcing. (Much of this book is concerned with evaluating the economic worthiness of proposed capital investments.) Because engineering economy often deals with *changes* to existing operations, standard costs may not be too useful in make-versus-purchase studies. In fact, if they are used, standard costs can lead to uneconomical decisions. Example 2-12 illustrates the correct procedure to follow in performing make-versus-purchase studies based on incremental costs.

EXAMPLE 2-12   **To Produce or Not to Produce?—That Is the Question**

A manufacturing plant consists of three departments: A, B, and C. Department A occupies 100 square meters in one corner of the plant. Product X is one of several products being produced in Department A. The daily production of Product X is 576 pieces. The cost accounting records show the following average daily production costs for Product X:

Direct labor	(1 operator working 4 hours per day at $22.50/hr, including fringe benefits, plus a part-time foreman at $30 per day)	$120.00
Direct material		86.40
Overhead	(at $0.82 per square meter of floor area)	82.00
	Total cost per day =	$288.40

The department foreman has recently learned about an outside company that sells Product X at $0.35 per piece. Accordingly, the foreman figured a cost per day of $0.35(576) = 201.60, resulting in a daily savings of $288.40 - $201.60 = 86.80. Therefore, a proposal was submitted to the plant manager for shutting down the production line of Product X and buying it from the outside company.

However, after examining each component separately, the plant manager decided not to accept the foreman's proposal based on the unit cost of Product X:

1. *Direct labor:* Because the foreman was supervising the manufacture of other products in Department A in addition to Product X, the only possible savings in labor would occur if the operator working 4 hours per day on Product X were not reassigned after this line is shut down. That is, a maximum savings of $90.00 per day would result.

2. *Materials:* The maximum savings on direct material will be $86.40. However, this figure could be lower if some of the material for Product X is obtained from scrap of another product.

3. ***Overhead:*** Because other products are made in Department *A*, no reduction in total floor space requirements will probably occur. Therefore, no reduction in overhead costs will result from discontinuing Product *X*. It has been estimated that there will be daily savings in the variable overhead costs traceable to Product *X* of about $3.00 due to a reduction in power costs and in insurance premiums.

Solution

If the manufacture of Product *X* is discontinued, the firm will save at most $90.00 in direct labor, $86.40 in direct materials, and $3.00 in variable overhead costs, which totals $179.40 per day. This estimate of actual cost savings per day is less than the potential savings indicated by the cost accounting records ($288.40 per day), and it would not exceed the $201.60 to be paid to the outside company if Product *X* is purchased. For this reason, the plant manager used Rule 2 and rejected the proposal of the foreman and continued the manufacture of Product *X*.

In conclusion, Example 2-12 shows how an erroneous decision might be made by using the unit cost of Product *X* from the cost accounting records without detailed analysis. The fixed cost portion of Product *X*'s unit cost, which is present even if the manufacture of Product *X* is discontinued, was not properly accounted for in the original analysis by the foreman.

2.4.4   Trade-Offs in Energy Efficiency Studies

Energy efficiency affects the annual expense of operating an electrical device such as a pump or motor. Typically, a more energy-efficient device requires a higher capital investment than does a less energy-efficient device, but the extra capital investment usually produces annual savings in electrical power expenses relative to a second pump or motor that is less energy efficient. This important trade-off between capital investment and annual electric power consumption will be considered in several chapters of this book. Hence, the purpose of Section 2.4.4 is to explain how the annual expense of operating an electrical device is calculated and traded off against capital investment cost.

If an electric pump, for example, can deliver a given horsepower (hp) or kiloWatt (kW) rating to an industrial application, the *input* energy requirement is determined by dividing the given output by the energy efficiency of the device. The input requirement in hp or kW is then multiplied by the annual hours that the device operates and the unit cost of electric power. You can see that the higher the efficiency of the pump, the lower the annual cost of operating the device is relative to another less-efficient pump.

| EXAMPLE 2-13 | **Investing In Electrical Efficiency** |

Two pumps capable of delivering 100 hp to an agricultural application are being evaluated in a present economy study. The selected pump will only be utilized for one year, and it will have no market value at the end of the year. Pertinent data are summarized as follows:

	ABC Pump	XYZ Pump
Purchase price	$2,900	$6,200
Annual maintenance	$170	$510
Efficiency	80%	90%

If electric power costs $0.10 per kWh and the pump will be operated 4,000 hours per year, which pump should be chosen? Recall that 1 hp = 0.746 kW.

Solution

The annual expense of electric power for the ABC pump is

$$(100 \text{ hp}/0.80)(0.746 \text{ kW/hp})(\$0.10/\text{kWh})(4{,}000 \text{ hours/yr}) = \$37{,}300.$$

For the XYZ Pump, the annual expense of electric power is

$$(100 \text{ hp}/0.90)(0.746 \text{ kW/hp})(\$0.10/\text{kWh})(4{,}000 \text{ hours/yr}) = \$33{,}156.$$

Thus, the total annual cost of owning and operating the ABC pump is $40,370, while the total cost of owning and operating the XYZ pump for one year is $39,866. Consequently, the more energy-efficient XYZ pump should be selected to minimize total annual cost. Notice the difference in annual energy expense ($4,144) that results from a 90% efficient pump relative to an 80% efficient pump. This cost reduction more than balances the extra $3,300 in capital investment and $340 in annual maintenance required for the XYZ pump.

2.5 CASE STUDY—The Economics of Daytime Running Lights

The use of Daytime Running Lights (DRLs) has increased in popularity with car designers throughout the world. In some countries, motorists are required to drive with their headlights *on* at all times. U.S. car manufacturers now offer models equipped with daytime running lights. Most people would agree that driving with the headlights on at night is cost effective with respect to extra *fuel* consumption and safety considerations (not to mention required by law!). Cost effective means that benefits outweigh (exceed) the costs. However, some consumers have questioned whether it is cost effective to drive with your headlights on during the day.

In an attempt to provide an answer to this question, let us consider the following suggested data:

75% of driving takes place during the daytime.

2% of fuel consumption is due to accessories (radio, headlights, etc.).

Cost of fuel = $3.00 per gallon.

Average distance traveled per year = 15,000 miles.

Average cost of an accident = $2,800.

Purchase price of headlights = $25.00 per set (2 headlights).

Average time car is in operation per year = 350 hours.

Average life of a headlight = 200 operating hours.

Average fuel consumption = 1 gallon per 30 miles.

Let's analyze the cost-effectiveness of driving with headlights on during the day by considering the following set of questions:

- What are the extra costs associated with driving with headlights on during the day?
- What are the benefits associated with driving with headlights on during the day?
- What additional assumptions (if any) are needed to complete the analysis?
- Is it cost effective to drive with headlights on during the day?

Solution

After some reflection on the above questions, you could reasonably contend that the extra costs of driving with headlights on during the day include increased fuel consumption and more frequent headlight replacement. Headlights increase visibility to other drivers on the road. Another possible benefit is the reduced chance of an accident.

Additional assumptions needed to consider during our analysis of the situation include:

1. the percentage of fuel consumption due to headlights alone
2. how many accidents can be avoided per unit time.

Selecting the dollar as our common unit of measure, we can compute the extra cost associated with daytime use of headlights and compare it to the expected benefit (also measured in dollars). As previously determined, the extra costs include increased fuel consumption and more frequent headlight replacement. Let's develop an estimate of the annual fuel cost:

Annual fuel cost = (15,000 mi/yr)(1 gal/30 mi)($3.00/gal) = $1,500/yr.

Assume (worst case) that 2% of fuel consumption is due to normal (night-time) use of headlights.

Fuel cost due to normal use of headlights = ($1,500/yr)(0.02) = $30/yr.

Fuel cost due to continuous use of headlights = (4)($30/yr) = $120/yr.

$$\text{Headlight cost for normal use} = (0.25)\left(\frac{350 \text{ hours/yr}}{200 \text{ hours/set}}\right)\left(\frac{\$25}{\text{set}}\right) = \$10.94/\text{yr}.$$

$$\text{Headlight cost for continuous use} = \left(\frac{350 \text{ hours/yr}}{200 \text{ hours/set}} \right) \left(\frac{\$25}{\text{set}} \right) = \$43.75/\text{yr}.$$

$$\text{Total cost associated with daytime use} = (\$120 - \$30) + (\$43.75 - \$10.94)$$

$$= \$122.81/\text{yr}.$$

If driving with your headlights on during the day results in at least one accident being avoided during the next ($2,800)/($122.81) = 22.8 years, then the continuous use of your headlights is cost effective. Although in the short term, you may be able to contend that the use of DRLs lead to increased fuel and replacement bulb costs, the benefits of increased personal safety and mitigation of possible accident costs in the long-run more than offset the apparent short-term cost savings.

2.6 Summary

In this chapter, we have discussed cost terminology and concepts important in engineering economy. It is important that the meaning and use of various cost terms and concepts be understood in order to communicate effectively with other engineering and management personnel. A listing of important abbreviations and notation, by chapter, is provided in Appendix B.

Several general economic concepts were discussed and illustrated. First, the ideas of consumer and producer goods and services, measures of economic growth, competition, and necessities and luxuries were covered. Then, some relationships among costs, price, and volume (demand) were discussed. Included were the concepts of optimal volume and breakeven points. Important economic principles of design optimization were also illustrated in this chapter.

The use of present-economy studies in engineering decision making can provide satisfactory results and save considerable analysis effort. When an adequate engineering economic analysis can be accomplished by considering the various monetary consequences that occur in a short time period (usually one year or less), a present-economy study should be used.

Problems

The number(s) in color at the end of a problem refer to the section(s) in that chapter most closely related to the problem.

2-1. A company in the process industry produces a chemical compound that is sold to manufacturers for use in the production of certain plastic products. The plant that produces the compound employs approximately 300 people. Develop a list of six different cost elements that would be *fixed* and a similar list of six cost elements that would be *variable*. (2.1)

2-2. Classify each of the following cost items as mostly fixed or variable: (2.1)

Raw materials	Administrative salaries
Direct labor	Payroll taxes
Depreciation	Insurance (building and
Supplies	equipment)
Utilities	Clerical salaries
Property taxes	Sales commissions
Interest on borrowed	Rent
money	

2-3. A group of enterprising engineering students has developed a process for extracting combustible methane gas from cow manure (don't worry, the exhaust is odorless). With a specially adapted internal combustion engine, the students claim that an automobile can be propelled 15 miles per day from the "cow gas" produced by a single cow. Their experimental car can travel 60 miles per day for an estimated cost of $5 (this is the allocated cost of the methane process equipment—the cow manure is essentially free). (2.1)

a. How many cows would it take to fuel 1,000,000 miles of annual driving by a fleet of cars? What is the annual cost?

b. How does your answer to Part (a) compare to a gasoline-fueled car averaging 30 miles per gallon when the cost of gasoline is $3.00 per gallon?

2-4. A municipal solid-waste site for a city must be located at Site *A* or Site *B*. After sorting, some of the solid refuse will be transported to an electric power plant where it will be used as fuel. Data for the hauling of refuse from each site to the power plant are shown in Table P2-4.

TABLE P2-4	Table for Problem 2-4	
	Site *A*	Site *B*
Average hauling distance	4 miles	3 miles
Annual rental fee for solid-waste site	$5,000	$100,000
Hauling cost	$1.50/yd^3-mile	$1.50/yd^3-mile

If the power plant will pay $8.00 per cubic yard of sorted solid waste delivered to the plant, where should the solid-waste site be located? Use the city's viewpoint and assume that 200,000 cubic yards of refuse will be hauled to the plant for one year only. One site must be selected. (2.1)

2-5. Stan Moneymaker presently owns a 10-year-old automobile with low mileage (78,000 miles). The NADA "blue book" value of the car is $2,500. Unfortunately, the car's transmission just failed, and Stan decided to spend $1,500 to have it repaired. Now, six months later, Stan has decided to sell the car, and he reasons that his asking price should be $2,500 + $1,500 = $4,000.

Comment on the wisdom of Stan's logic. If he receives an offer for $3,000, should he accept it? Explain your reasoning. (2.1)

2-6. You have been invited by friends to fly to Germany for Octoberfest next year. For international travel, you apply for a passport that costs $97 and is valid for 10 years. After you receive your passport, your travel companions decide to cancel the trip because of "insufficient funds." You decide to also cancel your travel plans because traveling alone is no fun. Is your passport expense a sunk cost or an opportunity cost? Explain your answer. (2.1)

2-7. Suppose your company has just discovered $100,000 worth (this is the original manufacturing cost) of obsolete inventory in an old warehouse. Your boss asks you to evaluate two options: (1) remachine the obsolete parts at a cost of $30,000 and then hopefully resell them for $60,000, or (2) scrap them for $15,000 cash (which is certain) through a secondhand market. What recommendation would you make to your boss? Explain your reasoning. (2.1)

2-8. A friend of yours has been thinking about quitting her regular day job and going into business for herself. She currently makes $60,000 per year as an employee of the Ajax Company, and she anticipates no raise for at least another year. She believes she can make $200,000 as an independent consultant in six-sigma "black belt" training for large corporations. Her start-up expenses are expected to be $120,000 over the next year. If she decides to keep her current job, what is the expected opportunity cost of this decision? Attempt to balance the pros and cons of the option that your friend is turning away from. (2.1)

2-9. Suppose your wealthy aunt has given you a gift of $25,000. You have come up with three options for spending (or investing) the money. First, you'd like (but do not need) a new car to brighten up your home and social life. Second, you can invest the money in a high-tech firm's common stock. It is expected to increase in value by 20% per year, but this option is fairly risky. Third, you can put the money into a three-year certificate of deposit with a local bank and earn 6% per year. There is little risk in the third option. (2.1)

a. If you decide to purchase the new car, what is the opportunity cost of this choice? Explain your reasoning.

b. If you invest in the high-tech common stock, what is the opportunity cost of this choice? Explain your reasoning.

2-10. In your own words, describe the life-cycle cost concept. Why is the potential for achieving life-cycle cost savings greatest in the acquisition phase of the life cycle? (2.1)

2-11. A large, profitable commercial airline company flies 737-type aircraft, each with a maximum seating capacity of 132 passengers. Company literature states that the economic breakeven point with these aircraft is 62 passengers. (2.2)

a. Draw a conceptual graph to show total revenue and total costs that this company is experiencing.

b. Identify three types of fixed costs that the airline should carefully examine to lower its breakeven point. Explain your reasoning.

c. Identify three types of variable costs that can possibly be reduced to lower the breakeven point. Why did you select these cost items?

2-12. A company produces circuit boards used to update outdated computer equipment. The fixed cost is $42,000 per month, and the variable cost is $53 per circuit board. The selling price per unit is $p = \$150 - 0.02D$. Maximum output of the plant is 4,000 units per month. (2.2)

a. Determine optimum demand for this product.

b. What is the maximum profit per month?

c. At what volumes does breakeven occur?

d. What is the company's range of profitable demand?

2-13. A local defense contractor is considering the production of fireworks as a way to reduce dependence on the military. The variable cost per unit is $40. The fixed cost that can be allocated to the production of fireworks is negligible. The price charged per unit will be determined by the equation $p = \$180 - (5)D$, where D represents demand in units sold per week. (2.2)

a. What is the optimum number of units the defense contractor should produce in order to maximize profit per week?

b. What is the profit if the optimum number of units are produced?

2-14. A large wood products company is negotiating a contract to sell plywood overseas. The fixed cost that can be allocated to the production of plywood is $900,000 per month. The variable cost per thousand board feet is $131.50. The price charged will be determined by $p = \$600 - (0.05)D$ per 1,000 board feet. (2.2)

a. For this situation determine the optimal monthly sales volume for this product and calculate the profit (or loss) at the optimal volume.

b. What is domain of profitable demand during a month?

2-15. A company produces and sells a consumer product and is able to control the demand for the product by varying the selling price. The approximate relationship between price and demand is

$$p = \$38 + \frac{2,700}{D} - \frac{5,000}{D^2}, \text{ for } D > 1,$$

where p is the price per unit in dollars and D is the demand per month. The company is seeking to maximize its profit. The fixed cost is $1,000 per month and the variable cost (c_v) is $40 per unit. (2.2)

a. What is the number of units that should be produced and sold each month to maximize profit?

b. Show that your answer to Part (a) maximizes profit.

2-16. An electric power plant uses solid waste for fuel in the production of electricity. The cost Y in dollars per hour to produce electricity is $Y = 12 + 0.3X + 0.27X^2$, where X is in megawatts. Revenue in dollars per hour from the sale of electricity is $15X - 0.2X^2$. Find the value of X that gives maximum profit. (2.2)

2-17. The annual fixed costs for a plant are $100,000, and the variable costs are $140,000 at 70% utilization of available capacity, with net sales of $280,000. What is the breakeven point in units of production if the selling price per unit is $40? (2.2)

2-18. A plant has a capacity of 4,100 hydraulic pumps per month. The fixed cost is $504,000 per month. The variable cost is $166 per pump, and the sales price is $328 per pump. (Assume that sales equal output volume.) What is the breakeven point in number of pumps per month? What percentage reduction will occur in the breakeven point if fixed costs are reduced by 18% and unit variable costs by 6%? (2.2)

2-19. A cell phone company has a fixed cost of $1,000,000 per month and a variable cost of $20 per month per subscriber. The company charges $29.95 per month to its cell phone customers. (2.2)

a. What is the breakeven point for this company?

b. The company currently has 95,000 subscribers and proposes to raise its monthly fees to $39.95 to cover add-on features such as text messaging, song downloads, game playing, and video watching. What is the new breakeven point if the variable cost increases to $25 per customer per month?

c. If 20,000 subscribers will drop their service because of the monthly fee increase in Part (b), will the company still be profitable?

2-20. A plant operation has fixed costs of $2,000,000 per year, and its output capacity is 100,000 electrical appliances per year. The variable cost is $40 per unit, and the product sells for $90 per unit.

a. Construct the economic breakeven chart.

b. Compare annual profit when the plant is operating at 90% of capacity with the plant operation at 100% capacity. Assume that the first 90% of capacity output is sold at $90 per unit and that the remaining 10% of production is sold at $70 per unit. (2.2)

2-21. A regional airline company estimated four years ago that each pound of aircraft weight adds $30 per year to its fuel expense. Now the cost of jet fuel has doubled from what it was four years ago. A recent engineering graduate employed by the company has made a recommendation to reduce fuel consumption of an aircraft by installing leather seats as part of a "cabin refurbishment program." The total reduction in weight would be approximately 600 pounds per aircraft. If seats are replaced annually (a worst-case situation), how much can this airline afford to spend on the cabin refurbishments? What nonmonetary advantages might be associated with the refurbishments? Would you support the engineer's recommendation? (2.1)

2-22. Jerry Smith's residential air conditioning (AC) system has not been able to keep his house cool enough in 90°F weather. He called his local AC maintenance person, who discovered a leak in the evaporator. The cost to recharge the AC unit is $40 for gas and $45 for labor, but the leak will continue and perhaps grow worse. The AC person cautioned that this service would have to be repeated each year unless the evaporator is replaced. A new evaporator would run about $800–$850.

Jerry reasons that fixing the leak in the evaporator on an annual basis is the way to go. "After all, it will take ten years of leak repairs to equal the evaporator's replacement cost." Comment on Jerry's logic. What would you do? (2.1)

2-23. Ethanol blended with gasoline can be used to power a "flex-fueled" car. One particular blend that is gaining in popularity is E85, which is 85% ethanol and 15% gasoline. E85 is 80% cleaner burning than gasoline alone, and it reduces our dependency on foreign oil. But a flex-fueled car costs $1,000 more than a conventional gasoline-fueled car. Additionally, E85 fuel gets 10% less miles per gallon than a conventional automobile.

Consider a 100% gasoline-fueled car that averages 30 miles per gallon. The E85-fueled car will average about 27 miles per gallon. If either car will be driven 81,000 miles before being traded in, how much will the E85 fuel have to cost (per gallon) to make the flex-fueled car as economically attractive as a conventional gasoline-fueled car? Gasoline costs $2.89 per gallon. Work this problem without considering the time value of money. (2.1)

2-24. The fixed cost for a steam line per meter of pipe is $450X + $50 per year. The cost for loss of heat from the pipe per meter is $4.8/X^{1/2}$ per year. Here X represents the thickness of insulation in meters, and X is a continuous design variable. (2.3)

a. What is the optimum thickness of the insulation?

b. How do you know that your answer in Part (a) minimizes total cost per year?

c. What is the basic trade-off being made in this problem?

2-25. A farmer estimates that if he harvests his soybean crop now, he will obtain 1,000 bushels, which he can sell at $3.00 per bushel. However, he estimates that this crop will increase by an additional 1,200 bushels of soybeans for each week he delays harvesting, but the price will drop at a rate of 50 cents per bushel per week; in addition, it is likely that he will experience spoilage of approximately 200 bushels per week for each week he delays harvesting. When should he harvest his crop to obtain the largest net cash return, and how much will be received for his crop at that time? (2.3)

2-26. The cost of operating a large ship (C_O) varies as the square of its velocity (v); specifically, $C_O = knv^2$, where n is the trip length in miles and k is a constant of proportionality. It is known that at 12 miles/hour the *average* cost of operation is $100 per mile. The owner of the ship wants to minimize the cost of operation, but it must be balanced against the cost of the perishable cargo (C_c), which the customer has set at $1,500 per hour. At what velocity should the trip be planned to minimize the total cost (C_T), which is the sum of the cost of operating the ship and the cost of perishable cargo? (2.3)

2-27. Refer to Example 2-7 on pages 41–42. Which alternative (insulation thickness) would be most economical if the cost of insulation triples? Show all your work. (2.3)

2-28. Suppose you are going on a long trip to your grandmother's home in Seattle, 3,000 miles away. You have decided to drive your old Ford out there, which gets approximately 18 miles per gallon when cruising at 70 mph. Because grandma is an excellent cook and you can stay and eat at her place as long as you want (for free), you want to get to Seattle as economically as possible. However, you are also worried about your fuel consumption rate at high speeds. You also have cost of food, snacks, and lodging to balance against the cost of fuel.

What is the optimum average speed you should use so as to minimize your total trip cost, C_T? (2.3)

$$C_T = C_G + C_{FSS},$$

where $C_G = n \times p_g \times f$ $(C_G = \text{cost of gas})$,
$C_{FSS} = n \times p_{fss} \times v^{-1}$
$(C_{FSS} = \text{cost of food, snacks, and lodging})$,

n: trip length (miles),
p_g: gas price, $3.00/gallon,
p_{fss}: average hourly spending money, $5/hour,
(motel, breakfast, snacks, etc., $120 per 24 hours!),
v: average Ford velocity (mph),

$$f = k \times v,$$

where k is a constant of proportionality and f is the fuel consumption rate in gallons per mile.

2-29. One component of a system's life-cycle cost is the cost of system failure. Failure costs can be reduced by designing a more reliable system. A simplified expression for system life-cycle cost, C, can be written as a function of the system's failure rate:

$$C = \frac{C_I}{\lambda} + C_R \cdot \lambda \cdot t.$$

Here C_I = investment cost ($ per hour per failure),
C_R = system repair cost,
λ = system failure rate
(failures/operating hour),
t = operating hours.

a. Assume that C_I, C_R, and t are constants. Derive an expression for λ, say λ^*, that optimizes C. (2.3)

b. Does the equation in Part (a) correspond to a maximum or minimum value of C? Show all work to support your answer.

c. What trade-off is being made in this problem?

2-30. Stan Moneymaker has been shopping for a new car. He is interested in a certain 4-cylinder sedan that averages 28 miles per gallon. But the salesperson tried to persuade Stan that the 6-cylinder model of the same automobile only costs $2,500 more and is really a "more sporty and responsive" vehicle. Stan is impressed with the zip of the 6-cylinder car and reasons that $2,500 is not too much to pay for the extra power.

How much extra is Stan really paying if the 6-cylinder car averages 22 miles per gallon? Assume that Stan will drive either automobile 100,000 miles, gasoline will average $3.00 per gallon, and maintenance is roughly the same for both cars. State other assumptions you think are appropriate. (2.4)

2-31. A producer of synthetic motor oil for automobiles and light trucks has made the following statement: "One quart of Dynolube added to your next oil change will increase fuel mileage by one percent. This one-time additive will improve your fuel mileage over 50,000 miles of driving." (2.4)

a. Assume the company's claim is correct. How much money will be saved by adding one quart of Dynolube if gasoline costs $3.00 per gallon and your car averages 20 miles per gallon without the Dynolube?

b. If a quart of Dynolube sells for $19.95, would you use this product in your automobile?

2-32. An automobile dealership offers to fill the four tires of your new car with 100% nitrogen for a cost of $20. The dealership claims that nitrogen-filled tires run cooler than those filled with compressed air, and they advertise that nitrogen extends tire mileage (life) by 25%. If new tires cost $50 each and are guaranteed to get 50,000 miles (filled with air) before they require replacement, is the dealership's offer a good deal? (2.4)

2-33. In the design of an automobile radiator, an engineer has a choice of using either a brass–copper alloy casting or a plastic molding. Either material provides the same service. However, the brass–copper alloy casting weighs 25 pounds, compared with 20 pounds for the plastic molding. Every pound of extra weight in the automobile has been assigned a penalty of $6 to account for increased fuel consumption during the life cycle of the car. The brass–copper alloy casting costs

$3.35 per pound, whereas the plastic molding costs $7.40 per pound. Machining costs per casting are $6.00 for the brass–copper alloy. Which material should the engineer select, and what is the difference in unit costs? (2.4)

2-34. Rework Example 2-9 for the case where the capacity of each machine is further reduced by 25% because of machine failures, materials shortages, and operator errors. In this situation, 30,000 units of good (nondefective) product must be manufactured during the next three months. Assume one shift per day and five work days per week. (2.4)

a. Can the order be delivered on time?

b. If only one machine (A or B) can be used in Part (a), which one should it be?

2-35. Two alternative designs are under consideration for a tapered fastening pin. The fastening pins are sold for $0.70 each. Either design will serve equally well and will involve the same material and manufacturing cost except for the lathe and drill operations.
Design A will require 16 hours of lathe time and 4.5 hours of drill time per 1,000 units. Design B will require 7 hours of lathe time and 12 hours of drill time per 1,000 units. The variable operating cost of the lathe, including labor, is $18.60 per hour. The variable operating cost of the drill, including labor, is $16.90 per hour. Finally, there is a sunk cost of $5,000 for Design A and $9,000 for Design B due to obsolete tooling. (2.4)

a. Which design should be adopted if 125,000 units are sold each year?

b. What is the annual saving over the other design?

2-36. A bicycle component manufacturer produces hubs for bike wheels. Two processes are possible for manufacturing, and the parameters of each process are as follows:

	Process 1	Process 2
Production rate	35 parts/hour	15 parts/hour
Daily production time	4 hours/day	7 hours/day
Percent of parts rejected based on visual inspection	20%	9%

Assume that the daily demand for hubs allows all defect-free hubs to be sold. Additionally, tested or rejected hubs cannot be sold.
Find the process that maximizes profit per day if each part is made from $4 worth of material and can be

sold for $30. Both processes are fully automated, and variable overhead cost is charged at the rate of $40 per hour. (2.4)

2-37. The speed of your automobile has a huge effect on fuel consumption. Traveling at 65 miles per hour (mph) instead of 55 mph can consume almost 20% more fuel. As a general rule, for every mile per hour over 55 you lose 2% in fuel economy. For example, if your automobile gets 30 miles per gallon at 55 mph, the fuel consumption is 21 miles per gallon at 70 mph.
If you take a 400-mile trip and your average speed is 80 mph rather than the posted speed limit of 70 mph, what is the extra cost of fuel if gasoline costs $3.00 per gallon? Your car gets 30 miles per gallon (mpg) at 60 mph. (2.4)

2-38. The following results were obtained after analyzing the operational effectiveness of a production machine at two different speeds:

Speed	Output (pieces per hour)	Time between tool grinds (hours)
A	400	15
B	540	10

A set of unsharpened tools costs $1,000 and can be ground 20 times. The cost of each grinding is $25. The time required to change and reset the tools is 1.5 hours, and such changes are made by a tool-setter who is paid $18/hour. The production machine operator is paid $15/hour, including the time that the machine is down for tool sharpening. Variable overhead on the machine is charged at the rate of $25/hour, including tool-changing time. A fixed-size production run will be made (independent of machine speed). (2.4)

a. At what speed should the machine be operated to minimize the total cost per piece? State your assumptions.

b. What is the basic trade-off in this problem?

2-39. An automatic machine can be operated at three speeds, with the following results:

Speed	Output (pieces per hour)	Time between tool grinds (hours)
A	400	15
B	480	12
C	540	10

A set of unsharpened tools costs $500 and can be ground 20 times. The cost of each grinding is $25. The time required to change and reset the tools is 1.5 hours, and such changes are made by a tool-setter who is paid $8.00 per hour. Variable overhead on the machine is charged at the rate of $3.75 per hour, including tool-changing time. At which speed should the machine be operated to minimize total cost per piece? The basic trade-off in this problem is between the rate of output and tool usage. (2.4)

2-40. A company is analyzing a make-versus-purchase situation for a component used in several products, and the engineering department has developed these data:

Option *A*: Purchase 10,000 items per year at a fixed price of $8.50 per item. The cost of placing the order is negligible according to the present cost accounting procedure.

Option *B*: Manufacture 10,000 items per year, using available capacity in the factory. Cost estimates are direct materials = $5.00 per item and direct labor = $1.50 per item. Manufacturing overhead is allocated at 200% of direct labor (= $3.00 per item).

Based on these data, should the item be purchased or manufactured? (2.4)

2-41. A national car rental agency asks, "Do you want to bring back the economy-class car full of gas or with an empty tank? If we fill up the tank for you, we'll charge you $2.70 per gallon, which is 30 cents less than the local price for gasoline." Which choice should you make? State your assumptions. (2.4)

2-42. One method for developing a mine containing an estimated 100,000 tons of ore will result in the recovery of 62% of the available ore deposit and will cost $23 per ton of material removed. A second method of development will recover only 50% of the ore deposit, but it will cost only $15 per ton of material removed. Subsequent processing of the removed ore recovers 300 pounds of metal from each ton of processed ore and costs $40 per ton of ore processed. The recovered metal can be sold for $0.80 per pound. Which method for developing the mine should be used if your objective is to maximize total profit from the mine? (2.4)

2-43. Ocean water contains 0.9 ounce of gold per ton. Method *A* costs $220 per ton of water processed and will recover 85% of the metal. Method *B* costs $160 per ton of water processed and will recover 65% of the

metal. The two methods require the same investment and are capable of producing the same amount of gold each day. If the extracted gold can be sold for $350 per ounce, which method of extraction should be used? Assume that the supply of ocean water is unlimited. Work this problem on the basis of profit per ounce of gold extracted. (2.4)

2-44. Which of the following statements are true and which are false? (all sections)

a. Working capital is a variable cost.

b. The greatest potential for cost savings occurs in the operation phase of the life cycle.

c. If the capacity of an operation is significantly changed (e.g., a manufacturing plant), the fixed costs will also change.

d. A noncash cost is a cash flow.

e. Goods and services have utility because they have the power to satisfy human wants and needs.

f. The demand for necessities is more inelastic than the demand for luxuries.

g. Indirect costs can normally be allocated to a specific output or work activity.

h. Present economy studies are often done when the time value of money is not a significant factor in the situation.

i. Overhead costs normally include all costs that are not direct costs.

j. Optimal volume (demand) occurs when total costs equal total revenues.

k. Standard costs per unit of output are established in advance of actual production or service delivery.

l. A related sunk cost will normally affect the prospective cash flows associated with a situation.

m. The life cycle needs to be defined within the context of the specific situation.

n. The greatest commitment of costs occurs in the acquisition phase of the life cycle.

o. High breakeven points in capital intensive industries are desirable.

p. The fixed return on borrowed capital (i.e., interest) is more risky than profits paid to equity investors (i.e., stockholders) in a firm.

q. There is no D^* for this Scenario 1 situation: $p = 40 - 0.2D$ and $C_T = \$100 + \$50D$.

r. Most decisions are based on differences that are perceived to exist among alternatives.

s. A nonrefundable cash outlay (e.g., money spent on a passport) is an example of an opportunity cost.

2-45. A hot water leak in one of the faucets of your apartment can be very wasteful. A continuous leak of one quart per hour (a "slow" leak) at 155°F causes a loss of 1.75 million Btu per year. Suppose your water is heated with electricity. (2.3)

a. How many pounds of coal delivered to your electric utility does this leak equate to if one pound of coal contains 12,000 Btu and the boiler combustion process and water distribution system have an overall efficiency of 30%?

b. If a pound of coal produces 1.83 pounds of CO_2 during the combustion process, how much extra carbon dioxide does the leaky faucet produce in a year?

2-46. *Extended Learning Exercise* The student chapter of the American Society of Mechanical Engineers is planning a six-day trip to the national conference in Albany, NY. For transportation, the group will rent a car from either the State Tech Motor Pool or a local car dealer. The Motor Pool charges $0.36 per mile, has no daily fee, and the motor pool pays for the gas. The car dealer charges $30 per day and $0.20 per mile, but the group must pay for the gas. The car's fuel rating is 20 miles per gallon, and the price of gas is estimated to be $2.00 per gallon. (2.2)

a. At what point, in miles, is the cost of both options equal?

b. The car dealer has offered a special student discount and will give the students 100 free miles per day. What is the new breakeven point?

c. Suppose now that the Motor Pool reduces its all-inclusive rate to $0.34 per mile and that the car dealer increases its rate to $30 per day and $0.28 per mile. In this case, the car dealer wants to encourage student business, so he offers 900 free miles for the entire six-day trip. He claims that if more than 750 miles are driven, students will come out ahead with one of his rental cars. If the students anticipate driving 2,000 miles (total), from whom should they rent a car? Is the car dealer's claim entirely correct?

2-47. *Web Exercise* Home heating accounts for approximately one-third of energy consumption in a typical U.S. household. Despite soaring prices of oil, coal, and natural gas, one can make his/her winter heating bill noninflationary by installing an ultraconvenient corn burning stove that costs in the neighborhood of $2,400. That's right—a small radiant-heating stove that burns corn and adds practically nothing to global warming or air pollution can be obtained through www.magnumfireplace.com. Its estimated annual savings per household in fuel is $300 in a regular U.S. farming community.

Conduct research on this means of home heating by accessing the above Web site. Do the annual savings you determine in your locale for a 2,000-square foot ranch-style house more than offset the cost of installing and maintaining a corn-burning stove? What other factors besides dollars might influence your decision to use corn for your home heating requirements? Be specific with your suggestions. (2.4)

Spreadsheet Exercises

2-48. Refer to Example 2-4. If your focus was on reducing expenses, would it be better to reduce the fixed cost (B1) or variable cost (B2) component? What is the effect of a $+/-$ 10% change in both of these factors? (2.2)

2-49. Refer to Example 2-10. A competitor of your current blade supplier would like you to switch to their new model, which costs $5 less than your current supplier. After a long lunch, it comes out that this new blade has a life that is two cycles less than the current blade. Is the new blade a good deal? Remember that our current focus is on an *unlimited* amount of work. (2.4)

2-50. Refer to Example 2-11. Confirm that for *any* limited job size (in required board-feet), the cost of operating the planer at 5,000 ft/min is less than operating the planer at 6,000 ft/min. Why? (2.4)

2-51. Refer to Example 2-7. If the average inside temperature of this house in Virginia is increased from 65°F to 72°F, what is the most economical insulation amount? Assume that 100,000,000 Btu are lost with no insulation when the thermostat is set at 65°F. The cost of electricity is now $0.086 per kWh. In addition, the cost of insulation has increased by 50%. Develop a spreadsheet to solve this problem. (2.3)

Case Study Exercises

2-52. What are the key factors in this analysis, and how would your decision change if the assumed value of these factors changes? For example, what impact does rising fuel costs have on this analysis? Or, what if studies have shown that drivers can expect to avoid at least one accident every 10 years due to daytime use of headlights? (2.5)

2-53. Visit your local car dealer (either in person or online) to determine the cost of the daytime running lights option. How many accidents (per unit time) would have to be avoided for this option to be cost effective? (2.5)

FE Practice Problems

A company has determined that the price and the monthly demand of one of its products are related by the equation

$$D = \sqrt{(400 - p)},$$

where p is the price per unit in dollars and D is the monthly demand. The associated fixed costs are $1,125/month, and the variable costs are $100/unit. Use this information to answer Problems 2-54 and 2-55. Select the closest answer. (2.2)

2-54. What is the optimal number of units that should be produced and sold each month?

(a) 10 units   (b) 15 units   (c) 20 units
(d) 25 units

2-55. Which of the following values of D represents the breakeven point?

(a) 10 units   (b) 15 units   (c) 20 units
(d) 25 units

A manufacturing company leases a building for $100,000 per year for its manufacturing facilities. In addition, the machinery in this building is being paid for in installments of $20,000 per year. Each unit of the product produced costs $15 in labor and $10 in materials. The product can be sold for $40. Use this information to answer Problems 2-56 through 2-58. Select the closest answer. (2.2)

2-56. How many units per year must be sold for the company to breakeven?

(a) 4,800   (b) 3,000   (c) 8,000
(d) 6,667   (e) 4,000

2-57. If 10,000 units per year are sold, what is the annual profit?

(a) $280,000   (b) $50,000   (c) $150,000
(d) −$50,000   (e) $30,000

2-58. If the selling price is lowered to $35 per unit, how many units must be sold each year for the company to earn a profit of $60,000 per year?

(a) 12,000   (b) 10,000   (c) 16,000
(d) 18,000   (e) 5,143

2-59. A recent engineering graduate was given the job of determining the best production rate for a new type of casting in a foundry. After experimenting with many combinations of hourly production rates and total production cost per hour, he summarized his findings in Table I. (See Table P2-59.) The engineer then talked to the firm's marketing specialist, who provided estimates of selling price per casting as a function of production output (see Table II on page 62). There are 8,760 hours in a year. What production rate would you recommend to maximize total profits per year? (2.3)

(a) 100   (b) 200   (c) 300
(d) 400   (e) 500

2-60. A manufacturer makes 7,900,000 memory chips per year. Each chip takes 0.4 minutes of direct labor at the rate of $8 per hour. The overhead costs are estimated at $11 per direct labor hour. A new process will reduce the unit production time by 0.01 minutes. If the overhead cost will be reduced by $5.50 for each hour by which total direct hours are reduced, what is the maximum amount you will pay for the new process? Assume that the new process must pay for itself by the end of the first year. (2.4)

(a) $25,017   (b) $1,066,500   (c) $10,533
(d) $17,775   (e) $711,000

TABLE P2-59	Table for Problem 2-59					
Table I	Total cost per hour	$1,000	$2,600	$3,200	$3,900	$4,700
	Castings produced per hour	100	200	300	400	500
Table II	Selling price per casting	$20.00	$17.00	$16.00	$15.00	$14.50
	Castings produced per hour	100	200	300	400	500

Appendix 2-A   Accounting Fundamentals

Accounting is often referred to as the language of business. Engineers should make serious efforts to learn about a firm's accounting practice so that they can better communicate with top management. This section contains an extremely brief and simplified exposition of the elements of financial accounting in recording and summarizing transactions affecting the finances of the enterprise. These fundamentals apply to any entity (such as an individual or a corporation) called here a *firm*.

2-A.1   The Accounting Equation

All accounting is based on the *fundamental accounting equation*, which is

$$\text{Assets} = \text{liabilities} + \text{owners' equity,} \qquad \text{(2-A-1)}$$

where *assets* are those things of monetary value that the firm possesses, *liabilities* are those things of monetary value that the firm owes, and *owners' equity* is the worth of what the firm owes to its stockholders (also referred to as *equities, net worth,* etc.). For example, typical accounts in each term of Equation (2-A-1) are as follows:

Asset Accounts = Liability Accounts + Owner's Equity Accounts		
Cash	Short-term debt	Capital stock
Receivables	Payables	Retained earnings (income
Inventories	Long-term debt	retained in the firm)
Equipment		
Buildings		
Land		

The fundamental accounting equation defines the format of the *balance sheet*, which is one of the two most common accounting statements and which shows the financial position of the firm at any given point in time.

Another important, and rather obvious, accounting relationship is

$$\text{Revenues} - \text{expenses} = \text{profit (or loss).} \qquad \text{(2-A-2)}$$

This relationship defines the format of the *income statement* (also commonly known as a *profit-and-loss statement*), which summarizes the revenue and expense results of operations *over a period of time.* Equation (2-A-1) can be expanded to take into account profit as defined in Equation (2-A-2):

$$\text{Assets} = \text{liabilities} + (\text{beginning owners' equity} + \text{revenue} - \text{expenses}). \qquad \text{(2-A-3)}$$

Profit is the increase in money value (not to be confused with cash) that results from a firm's operations and is available for distribution to stockholders. It therefore represents the return on owners' invested capital.

A useful analogy is that a balance sheet is like a snapshot of the firm at an instant in time, whereas an income statement is a summarized moving picture of the firm over an interval of time. It is also useful to note that revenue serves to increase owners' interests in a firm, but an expense serves to decrease the owners' equity amount for a firm.

To illustrate the workings of accounts in reflecting the decisions and actions of a firm, suppose that an individual decides to undertake an investment opportunity and the following sequence of events occurs over a period of one year:

1. Organize XYZ firm and invest $3,000 cash as capital.
2. Purchase equipment for a total cost of $2,000 by paying cash.
3. Borrow $1,500 through a note to the bank.
4. Manufacture year's supply of inventory through the following:
    (a) Pay $1,200 cash for labor.
    (b) Incur $400 accounts payable for material.
    (c) Recognize the partial loss in value (depreciation) of the equipment amounting to $500.
5. Sell on credit all goods produced for year, 1,000 units at $3 each. Recognize that the accounting cost of these goods is $2,100, resulting in an increase in equity (through profits) of $900.
6. Collect $2,200 of accounts receivable.
7. Pay $300 of accounts payable and $1,000 of bank note.

A simplified version of the accounting entries recording the same information in a format that reflects the effects on the fundamental accounting equation (with a "+" denoting an increase and a "−" denoting a decrease) is shown in Figure 2-A-1. A summary of results is shown in Figure 2-A-2.

It should be noted that the profit for a period serves to increase the value of the owners' equity in the firm by that amount. Also, it is significant that the net cash flow from operation of $700(= $2,200 − $1,200 − $300) is not the same as profit. This amount was recognized in transaction 4(c), in which capital consumption (depreciation) for equipment of $500 was declared. Depreciation serves to convert part of an asset into an expense, which is then reflected in a firm's profits, as seen in Equation (2-A-2). Thus, the profit was $900, or $200 more than the net cash flow. For our purposes, revenue is recognized when it is earned, and expenses are recognized when they are incurred.

One important and potentially misleading indicator of after-the-fact financial performance that can be obtained from Figure 2-A-2 is "annual rate of return." If the invested capital is taken to be the owners' (equity) investment, the annual rate of return at the end of this particular year is $900/$3,900 = 23%.

Financial statements are usually most meaningful if figures are shown for two or more years (or other reporting periods such as quarters or months) or for two or more individuals or firms. Such comparative figures can be used to reflect trends or financial indications that are useful in enabling investors and management to determine the effectiveness of investments *after* they have been made.

2-A.2 Cost Accounting

Cost accounting, or management accounting, is a phase of accounting that is of particular importance in engineering economic analysis because it is concerned principally with

			Transaction					
Account	1	2	3	4	5	6	7	Balances at End of Year
Assets {								
Cash	+$3,000	−$2,000	+$1,500	−$1,200		+$2,200	−$1,300	+$2,200
Accounts receivable					+$3,000	−2,200		+800
Inventory		+2,000		+2,100	−2,100			0
Equipment				−500				+1,500
equals								$4,500
Liabilities {								
Accounts payable				+400			−300	+100
Bank note			+1,500				−1,000	+500
plus								
Owners' equity { Equity	+3,000				+900			+3,900
								$4,500

Figure 2-A-1 Accounting Effects of Transactions: XYZ Firm

XYZ Firm Balance Sheet
as of December 31, 2006

Assets		Liabilities and Owners' Equity	
Cash	$2,200	Bank note	$500
Accounts receivable	800	Accounts payable	100
Equipment	1,500	Equity	3,900
Total	$4,500	Total	$4,500

XYZ Firm Income Statement
for Year Ending December 31, 2006

			Cash Flow
Operating revenues (Sales)		$3,000	$2,200
Operating costs (Inventory depleted)			
Labor	$1,200		−1,200
Material	400		−300
Depreciation	500		0
		$2,100	
Net income (Profits)		$900	$700

Figure 2-A-2 Balance Sheet and Income Statement Resulting from Transactions Shown in Figure 2-A-1

decision making and control in a firm. Consequently, cost accounting is the source of much of the cost data needed in making engineering economy studies. Modern cost accounting may satisfy any or all of the following objectives:

1. Determination of the actual cost of products or services
2. Provision of a rational basis for pricing goods or services
3. Provision of a means for allocating and controlling expenditures
4. Provision of information on which operating decisions may be based and by means of which operating decisions may be evaluated

Although the basic objectives of cost accounting are simple, the exact determination of costs usually is not. As a result, some of the procedures used are arbitrary devices that make it possible to obtain reasonably accurate answers for most cases but that may contain a considerable percentage of error in other cases, particularly with respect to the actual cash flow involved.

2-A.3 The Elements of Cost

One of the first problems in cost accounting is that of determining the elements of cost that arise in the production of a product or the rendering of a service. A study of how these costs occur gives an indication of the accounting procedure that must be established to give satisfactory cost information. Also, an understanding of the procedure that is used to account for these costs makes it possible to use them more intelligently.

From an engineering and managerial viewpoint in manufacturing enterprises, it is common to consider the general elements of cost to be *direct materials, direct labor,* and *overhead.* Such terms as *burden* and *indirect costs* are often used synonymously with *overhead,* and overhead costs are often divided into several subcategories.

Ordinarily, the materials that can be conveniently and economically charged directly to the cost of the product are called *direct materials.* Several guiding principles are used when we decide whether a material is classified as a direct material. In general, direct materials should be readily measurable, be of the same quantity in identical products, and be used in economically significant amounts. Those materials that do not meet these criteria are classified as *indirect materials* and are a part of the charges for overhead. For example, the exact amount of glue and sandpaper used in making a chair would be difficult to determine. Still more difficult would be the measurement of the exact amount of coal that was used to produce the steam that generated the electricity that was used to heat the glue. Some reasonable line must be drawn beyond which no attempt is made to measure directly the material that is used for each unit of production.

labor costs also are ordinarily divided into *direct* and *indirect* categories. Direct labor costs are those that can be conveniently and easily charged to the product or service in question. Other labor costs, such as for supervisors, material handlers, and design engineers, are charged as indirect labor and are thus included as part of overhead costs. It is often imperative to know what is included in direct labor and direct material cost data before attempting to use them in engineering economy studies.

In addition to indirect materials and indirect labor, there are numerous other cost items that must be incurred in the production of products or the rendering of services. Property taxes must be paid; accounting and personnel departments must be maintained; buildings and equipment must be purchased and maintained; supervision must be provided. It is essential that these necessary *overhead* costs be attached to each unit produced in proper proportion to the benefits received. Proper allocation of these overhead costs is not easy, and some factual, yet reasonably simple, method of allocation must be used.

As might be expected, where solutions attempt to meet conflicting requirements such as exist in overhead-cost allocation, the resulting procedures are empirical approximations that are accurate in some cases and less accurate in others.

There are many methods of allocating overhead costs among the products or services produced. The most commonly used methods involve allocation in proportion to direct labor cost, direct labor hours, direct materials cost, sum of direct labor and direct materials cost, or machine hours. In these methods, it is necessary to estimate what the total overhead costs will be if standard costs are being determined. Accordingly, *total overhead costs are customarily associated with a certain level of production,* which is an important condition that should always be remembered when dealing with unit-cost data. These costs can be correct only for the conditions for which they were determined.

To illustrate one method of allocation of overhead costs, consider the method that assumes that overhead is incurred in direct proportion to the cost of direct labor used. With this method, the overhead rate (overhead per dollar of direct labor) and the overhead cost per unit would respectively be

$$\text{Overhead rate} = \frac{\text{total overhead in dollars for period}}{\text{direct labor in dollars for period}}$$

$$\text{Overhead cost/unit} = \text{overhead rate} \times \text{direct labor cost/unit.} \qquad \text{(2-A-4)}$$

Suppose that for a future period (say, a quarter), the total overhead cost is expected to be $100,000 and the total direct labor cost is expected to be $50,000. From this, the overhead rate = \$100,000/\$50,000 = \$2 per dollar of direct labor cost. Suppose further that for a given unit of production (or job) the direct labor cost is expected to be \$60. From Equation (2-A-4), the overhead cost for the unit of production would be \$60 × 2 = \$120.

This method obviously is simple and easy to apply. In many cases, it gives quite satisfactory results. However, in many other instances, it gives only very approximate results because some items of overhead, such as depreciation and taxes, have very little relationship to labor costs. Quite different total costs may be obtained for the same product when different procedures are used for the allocation of overhead costs. The magnitude of the difference will depend on the extent to which each method produces or fails to produce results that realistically capture the facts.

2-A.4 Cost Accounting Example

This relatively simple example involves a job-order system in which costs are assigned to work by job number. Schematically, this process is illustrated in the following diagram:

Costs are assigned to jobs in the following manner:

1. Raw materials attach to jobs via material requisitions.
2. Direct labor attaches to jobs via direct labor tickets.
3. Overhead cannot be attached to jobs directly but must have an allocation procedure that relates it to one of the resource factors, such as direct labor, which is already accumulated by the job.

Consider how an order for 100 tennis rackets accumulates costs at the Bowling Sporting Goods Company:

Job #161	100 tennis rackets
Labor rate	\$7 per hour
Leather	50 yards at \$2 per yard
Gut	300 yards at \$0.50 per yard
Graphite	180 pounds at \$3 per pound
Labor hours for the job	200 hours
Total annual factory overhead costs	\$600,000
Total annual direct labor hours	200,000 hours

The three major costs are now attached to the job. Direct labor and material expenses are straightforward:

Job #161		
Direct labor	$200 \times \$7$	= $1,400
Direct material	leather: $50 \times \$2$ =	100
	gut: $300 \times \$0.5$ =	150
	graphite: $180 \times \$3$ =	540
Prime costs (direct labor + direct materials)		$2,190

Notice that this cost is not the total cost. We must somehow find a way to attach (allocate) factory costs that cannot be directly identified to the job but are nevertheless involved in producing the 100 rackets. Costs such as the power to run the graphite molding machine, the depreciation on this machine, the depreciation of the factory building, and the supervisor's salary constitute overhead for this company. These overhead costs are part of the cost structure of the 100 rackets but cannot be directly traced to the job. For instance, do we really know how much machine obsolescence is attributable to the 100 rackets? Probably not. Therefore, we must allocate these overhead costs to the 100 rackets by using the overhead rate determined as follows:

$$\text{Overhead rate} = \frac{\$600,000}{200,000} = \$3 \text{ per direct labor hour.}$$

This means that $600 ($3 × 200) of the total annual overhead cost of $600,000 would be allocated to Job #161. Thus, the total cost of Job #161 would be as follows:

Direct labor	$1,400
Direct materials	790
Factory overhead	600
	$2,790

The cost of manufacturing each racket is thus $27.90. If selling expenses and administrative expenses are allocated as 40% of the cost of goods sold, the total expense of a tennis racket becomes 1.4($27.90) = $39.06.

Appendix 2-A Problems

2-A-1.* Jill Smith opens an apartment-locater business near a college campus. She is the sole owner of the proprietorship, which she names Campus Apartment Locators. During the first month of operations, July 2007, she engages in the following transactions:

a. Smith invests $35,000 of personal funds to start the business.

b. She purchases on account office supplies costing $350.

c. Smith pays cash of $30,000 to acquire a lot next to the campus. She intends to use the land as a future building site for her business office.

d. Smith locates apartments for clients and receives cash of $1,900.

---

* Adapted from C. T. Horngren, W. T. Harrison Jr., and L. S. Bamber, *Accounting*, 6th ed. (Upper Saddle River, NJ: Prentice Hall, 2005), 23 and 32. Reprinted by permission of the publisher.

TABLE P2-A-2 Data for Problem 2-A-2

	Assets			=	Liabilities	+	Owner's Equity
	Accounts				Accounts		Daniel Peavy,
Cash +	Receivable +	Supplies +	Land	=	Payable	+	Capital
Bal. 1,720	3,240		24,100		5,400		23,660

e. She pays $100 on the account payable she created in transaction (b).

f. She pays $2,000 of personal funds for a vacation.

g. She pays cash expenses for office rent, $400, and utilities, $100.

h. The business sells office supplies to another business for its cost of $150.

i. Smith withdraws cash of $1,200 for personal use.

Required

(a) Analyze the preceding transactions in terms of their effects on the accounting equation of Campus Apartment Locators. Use Figure 2-A-1 as a guide.

(b) Prepare the income statement and balance sheet of the business after recording the transactions. Use Figure 2-A-2 as a guide.

2-A-2. * Daniel Peavy owns and operates an architectural firm called Peavy Design. Table P2-A-2 summarizes the financial position of his business on April 30, 2007.

During May 2007, the following events occurred:

a. Peavy received $12,000 as a gift and deposited the cash in the business bank account.

b. He paid off the beginning balance of accounts payable.

c. He performed services for a client and received cash of $1,100.

d. He collected cash from a customer on account, $750.

e. Peavy purchased supplies on account, $720.

f. He consulted on the interior design of a major office building and billed the client for services rendered, $5,000.

g. He invested personal cash of $1,700 in the business.

h. He recorded the following business expenses for the month:

1. Paid office rent, $1,200.
2. Paid advertising, $660.

i. Peavy sold supplies to another interior designer for $80 cash, which was the cost of the supplies.

j. He withdrew cash of $4,000 for personal use.

Required

(a) Analyze the effects of the preceding transactions on the accounting equation of Peavy Design. Adapt the format of Figure 2-A-1.

(b) Prepare the income statement of Peavy Design for the month ended May 31, 2007. List expenses in decreasing order by amount.

(c) Prepare the balance sheet of Peavy Design at May 31, 2007.

2-A-3. * Lubbock Engineering Consultants is a firm of professional civil engineers. It mostly does surveying jobs for the heavy construction industry throughout Texas. The firm obtains its jobs by giving fixed-price quotations, so profitability depends on the ability to predict the time required for the various subtasks on the job. (This situation is similar to that in the auditing profession, where times are budgeted for such audit steps as reconciling cash and confirming accounts receivable.)

A client may be served by various professional staff, who hold positions in the hierarchy from partners to managers to senior engineers to assistants. In addition, there are secretaries and other employees.

Lubbock Engineering has the following budget for 2008:

Compensation of professional staff	$3,600,000
Other costs	1,449,000
Total budgeted costs	$5,049,000

* Adapted from C. T. Horngren, G. L. Sundem, and W. O. Stratton, *Introduction to Management Accounting*, 11th ed. (Upper Saddle River, NJ: Prentice Hall, 1999), 528–529. Reprinted by permission of the publisher.

Each professional staff member must submit a weekly time report, which is used for charging hours to a client job-order record. The time report has seven columns, one for each day of the week. Its rows are as follows:

- Chargeable hours
  Client 156
  Client 183
  Etc.

- Nonchargeable hours
  Attending seminar on new equipment
  Unassigned time
  Etc.

In turn, these time reports are used for charging hours and costs to the client job-order records. The managing partner regards these job records as absolutely essential for measuring the profitability of various jobs and for providing an "experience base for improving predictions on future jobs."

a. The firm applies overhead to jobs at a budgeted percentage of the professional compensation charged directly to the job ("direct labor"). For all categories of professional personnel, chargeable hours average 85% of available hours. Nonchargeable hours are regarded as additional overhead. What is the overhead rate as a percentage of "direct labor," the chargeable professional compensation cost?

b. A senior engineer works 48 weeks per year, 40 hours per week. His compensation is $60,000. He has worked on two jobs during the past week, devoting 10 hours to Job 156 and 30 hours to Job 183. How much cost should be charged to Job 156 because of his work there?

CHAPTER 3

Cost-Estimation Techniques

The objective of Chapter 3 is to present an assortment of methods for estimating important factors in an engineering economy study.

The High Cost of Office Rental Space

In 2008, your multinational firm needs to acquire a total of 50,000 square feet of additional office space in London, Hong Kong, and New York City. You have been tasked with developing a cost estimate for 20,000 square feet of space in London, 10,000 square feet of space in Hong Kong, and 20,000 square feet of space in New York City. While researching this topic, you find an article in *USA Today* (April 10, 2007, page B-1) that provides the following annual cost estimates per square foot of rental space in the three cities: London, $233.57; Hong Kong, $201.29; and New York City, $175.00. How much should your firm expect to pay per year for the 50,000 square feet of office space? In Chapter 3, you will learn about several techniques that are highly useful in estimating the costs of engineering projects.

Decisions, both great and small, depend in part on estimates. "Looking down the barrel" need not be an embarrassment for the engineer if newer techniques, professional staffing, and a greater awareness are assigned to the engineering cost estimating function.

—Phillip F. Ostwald (1992)

3.1 Introduction

In Chapter 1, we described the engineering economic analysis procedure in terms of seven steps. In this chapter, we address Step 3, development of the outcomes and cash flows for each alternative. Because engineering economy studies deal with outcomes that extend into the future, estimating the future cash flows for feasible alternatives is a critical step in the analysis procedure. Often, the most difficult, expensive, and time-consuming part of an engineering economy study is the estimation of costs, revenues, useful lives, residual values, and other data pertaining to the alternatives being analyzed. A decision based on the analysis is economically sound only to the extent that these cost and revenue estimates are representative of what subsequently will occur. In this chapter, we introduce the role of *cost estimating* in engineering practice. Definitions and examples of important cost concepts were provided in Chapter 2.

Whenever an engineering economic analysis is performed for a major capital investment, the cost-estimating effort for that analysis should be an integral part of a comprehensive planning and design process requiring the active participation of not only engineering designers but also personnel from marketing, manufacturing, finance, and top management. Results of cost estimating are used for a variety of purposes, including the following:

1. Providing information used in setting a selling price for quoting, bidding, or evaluating contracts

2. Determining whether a proposed product can be made and distributed at a profit (for simplicity, price = cost + profit)

3. Evaluating how much capital can be justified for process changes or other improvements

4. Establishing benchmarks for productivity improvement programs

There are two fundamental approaches to cost estimating: the "top-down" approach and the "bottom-up" approach. The top-down approach basically uses historical data from similar engineering projects to estimate the costs, revenues, and other data for the current project by modifying these data for changes in inflation or deflation, activity level, weight, energy consumption, size, and other factors. This approach is best used early in the estimating process when alternatives are still being developed and refined.

The bottom-up approach is a more detailed method of cost estimating. This method breaks down a project into small, manageable units and estimates their economic consequences. These smaller unit costs are added together with other types of costs to obtain an overall cost estimate. This approach usually works best when the detail concerning the desired output (a product or a service) has been defined and clarified.

EXAMPLE 3-1	**Estimating the Cost of a College Degree**

A simple example of cost estimating is to forecast the expense of getting a Bachelor of Science (B.S.) from the university you are attending. In our solution, we outline the two basic approaches just discussed for estimating these costs.

Solution

A top-down approach would take the published cost of a four-year degree at the same (or a similar) university and adjust it for inflation and extraordinary items that an incoming student might encounter, such as fraternity/sorority membership, scholarships, and tutoring. For example, suppose that the published cost of attending your university is $15,750 for the current year. This figure is anticipated to increase at the rate of 6% per year and includes full-time tuition and fees, university housing, and a weekly meal plan. Not included are the costs of books, supplies, and other personal expenses. For our initial estimate, these "other" expenses are assumed to remain constant at $5,000 per year.

The total estimated cost for four years can now be computed. We simply need to adjust the published cost for inflation each year and add in the cost of "other" expenses.

Year	Tuition, Fees, Room and Board	"Other" Expenses	Total Estimated Cost for Year
1	$15,750 × 1.06 = $16,695	$5,000	$21,695
2	16,695 × 1.06 = 17,697	5,000	22,697
3	17,697 × 1.06 = 18,759	5,000	23,759
4	18,759 × 1.06 = 19,885	5,000	24,885
		Grand Total	$93,036

In contrast with the top-down approach, a bottom-up approach to the same cost estimate would be to first break down anticipated expenses into the typical categories shown in Figure 3-1 for each of the four years at the university. Tuition and fees can be estimated fairly accurately in each year, as can books and supplies. For example, suppose that the average cost of a college textbook is $100. You can estimate your annual textbook cost by simply multiplying the average cost per book by the number of courses you plan to take. Assume that you plan on taking five courses each semester during the first year. Your estimated textbook costs would be

$$\left(\frac{5 \text{ courses}}{\text{semester}}\right)(2 \text{ semesters})\left(\frac{1 \text{ book}}{\text{course}}\right)\left(\frac{\$100}{\text{book}}\right) = \$1,000.$$

The other two categories, living expenses and transportation, are probably more dependent on your lifestyle. For example, whether you own and operate an automobile and live in a "high-end" apartment off-campus can dramatically affect the estimated expenses during your college years.

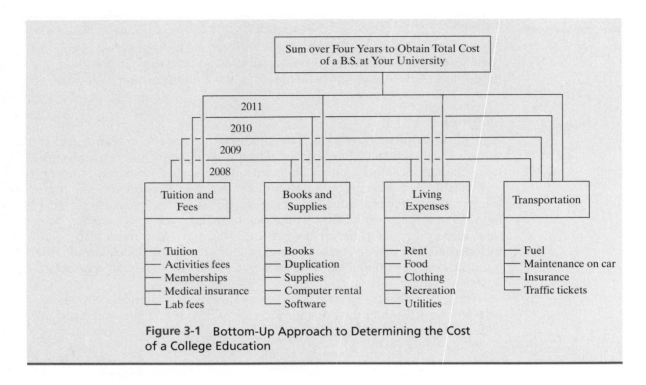

Figure 3-1 Bottom-Up Approach to Determining the Cost of a College Education

3.2 An Integrated Approach

An integrated approach to developing the net cash flows for feasible project alternatives is shown in Figure 3-2. This integrated approach includes three basic components:

1. *Work breakdown structure (WBS)* This is a technique for explicitly defining, at successive levels of detail, the work elements of a project and their interrelationships (sometimes called a *work element structure*).

2. *Cost and revenue structure (classification)* Delineation of the cost and revenue categories and elements is made for estimates of cash flows at each level of the WBS.

3. *Estimating techniques (models)* Selected mathematical models are used to estimate the future costs and revenues during the analysis period.

These three basic components, together with integrating procedural steps, provide an organized approach for developing the cash flows for the alternatives.

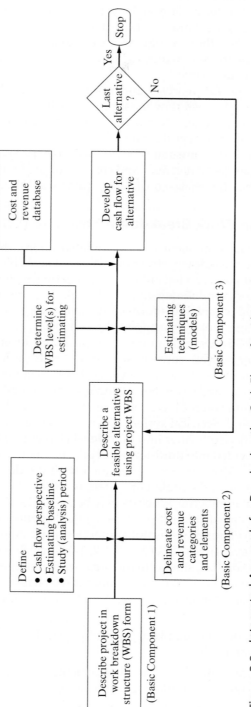

Figure 3-2 Integrated Approach for Developing the Cash Flows for Alternatives

As shown in Figure 3-2, the integrated approach begins with a description of the project in terms of a WBS. WBS is used to describe the project and each alternative's unique characteristics in terms of design, labor, material requirements, and so on. Then these variations in design, resource requirements, and other characteristics are reflected in the estimated future costs and revenues (net cash flow) for that alternative.

To estimate future costs and revenues for an alternative, the perspective (viewpoint) of the cash flow must be established and an estimating baseline and analysis period defined. Normally, cash flows are developed from the owner's viewpoint. The net cash flow for an alternative represents what is estimated to happen to future revenues and costs from the perspective being used. Therefore, the estimated changes in revenues and costs associated with an alternative have to be relative to a baseline that is consistently used for all the alternatives being compared.

3.2.1 The Work Breakdown Structure (WBS)

The first basic component in an integrated approach to developing cash flows is the work breakdown structure (WBS). The WBS is a basic tool in project management and is a vital aid in an engineering economy study. The WBS serves as a framework for defining all project work elements and their interrelationships, collecting and organizing information, developing relevant cost and revenue data, and integrating project management activities.

Figure 3-3 shows a diagram of a typical four-level WBS. It is developed from the top (project level) down in successive levels of detail. The project is divided into its major work elements (Level 2). These major elements are then divided to develop Level 3, and so on. For example, an automobile (first level of the WBS) can be divided into second-level components (or work elements) such as the chassis, drive train, and electrical system. Then each second-level component of the WBS can be subdivided further into third-level elements. The drive train, for example, can be subdivided into third-level components such as the engine, differential, and transmission. This process is continued until the desired detail in the definition and description of the project or system is achieved.

Different numbering schemes may be used. The objectives of numbering are to indicate the interrelationships of the work elements in the hierarchy. The scheme illustrated in Figure 3-3 is an alphanumeric format. Another scheme often used is all numeric—Level 1: 1-0; Level 2: 1-1, 1-2, 1-3; Level 3: 1-1-1, 1-1-2, 1-2-1, 1-2-2, 1-3-1, 1-3-2; and so on (i.e., similar to the organization of this book). Usually, the level is equal (except for Level 1) to the number of characters indicating the work element.

Other characteristics of a project WBS are as follows:

1. Both functional (e.g., planning) and physical (e.g., foundation) work elements are included in it:

   (a) Typical functional work elements are logistical support, project management, marketing, engineering, and systems integration.

   (b) Physical work elements are the parts that make up a structure, product, piece of equipment, or similar item; they require labor, materials, and other resources to produce or construct.

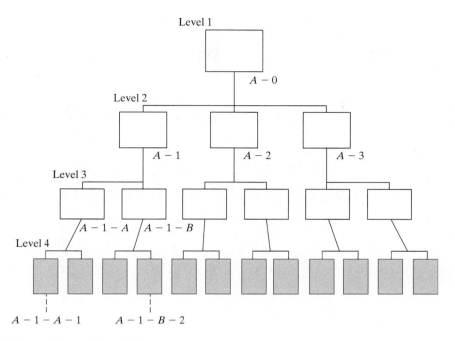

Figure 3-3 The WBS Diagram

2. The content and resource requirements for a work element are the sum of the activities and resources of related subelements below it.

3. A project WBS usually includes recurring (e.g., maintenance) and nonrecurring (e.g., initial construction) work elements.

EXAMPLE 3-2	**A WBS for a Construction Project**

You have been appointed by your company to manage a project involving construction of a small commercial building with two floors of 15,000 gross square feet each. The ground floor is planned for small retail shops, and the second floor is planned for offices. Develop the first three levels of a representative WBS adequate for all project efforts from the time the decision was made to proceed with the design and construction of the building until initial occupancy is completed.

Solution

There would be variations in the WBSs developed by different individuals for a commercial building. A representative three-level WBS is shown in Figure 3-4. Level 1 is the total project. At Level 2, the project is divided into seven major physical work elements and three major functional work elements. Then each of these major elements is divided into subelements as required (Level 3). The numbering scheme used in this example is all numeric.

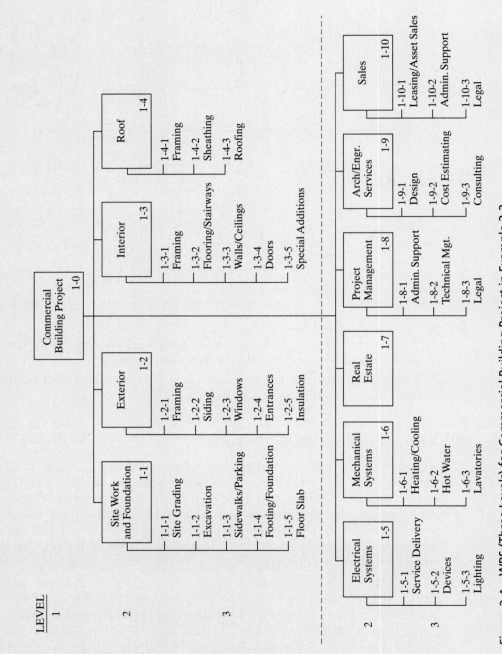

LEVEL

1   Commercial Building Project   1-0

2

Site Work and Foundation   1-1
- 1-1-1   Site Grading
- 1-1-2   Excavation
- 1-1-3   Sidewalks/Parking
- 1-1-4   Footing/Foundation
- 1-1-5   Floor Slab

Exterior   1-2
- 1-2-1   Framing
- 1-2-2   Siding
- 1-2-3   Windows
- 1-2-4   Entrances
- 1-2-5   Insulation

Interior   1-3
- 1-3-1   Framing
- 1-3-2   Flooring/Stairways
- 1-3-3   Walls/Ceilings
- 1-3-4   Doors
- 1-3-5   Special Additions

Roof   1-4
- 1-4-1   Framing
- 1-4-2   Sheathing
- 1-4-3   Roofing

3

2

Electrical Systems   1-5
- 1-5-1   Service Delivery
- 1-5-2   Devices
- 1-5-3   Lighting

Mechanical Systems   1-6
- 1-6-1   Heating/Cooling
- 1-6-2   Hot Water
- 1-6-3   Lavatories

Real Estate   1-7

Project Management   1-8
- 1-8-1   Admin. Support
- 1-8-2   Technical Mgt.
- 1-8-3   Legal

Arch/Engr. Services   1-9
- 1-9-1   Design
- 1-9-2   Cost Estimating
- 1-9-3   Consulting

Sales   1-10
- 1-10-1   Leasing/Asset Sales
- 1-10-2   Admin. Support
- 1-10-3   Legal

3

Figure 3-4   WBS (Three Levels) for Commercial Building Project in Example 3-2

3.2.2 The Cost and Revenue Structure

The second basic component of the integrated approach for developing cash flows (Figure 3-2) is the cost and revenue structure. This structure is used to identify and categorize the costs and revenues that need to be included in the analysis. Detailed data are developed and organized within this structure for use with the estimating techniques of Section 3.3 to prepare the cash-flow estimates.

The life-cycle concept and the WBS are important aids in developing the cost and revenue structure for a project. The life cycle defines a maximum time period and establishes a range of cost and revenue elements that need to be considered in developing cash flows. The WBS focuses the analyst's effort on the specific functional and physical work elements of a project and on its related costs and revenue.

> Perhaps the most serious source of errors in developing cash flows is overlooking important categories of costs and revenues. The cost and revenue structure, prepared in tabular or checklist form, is a good means of preventing such oversights. Technical familiarity with the project is essential in ensuring the completeness of the structure, as are using the life-cycle concept and the WBS in its preparation.

The following is a brief listing of some categories of costs and revenues that are typically needed in an engineering economy study:

1. Capital investment (fixed and working)
2. Labor costs
3. Material costs
4. Maintenance costs
5. Property taxes and insurance
6. Overhead costs
7. Disposal costs
8. Revenues
9. Quality (and scrap) costs
10. Market (or salvage) values

3.2.3 Estimating Techniques (Models)

The third basic component of the integrated approach (Figure 3-2) involves estimating techniques (models). These techniques, together with the detailed cost and revenue data, are used to develop individual cash-flow estimates and the overall net cash-flow for each alternative.

> The purpose of estimating is to develop cash-flow projections—*not to produce exact data* about the future, which is virtually impossible. Neither a preliminary estimate nor a final estimate is expected to be exact; rather, it should adequately suit the need at a reasonable cost and is often presented as a range of numbers.

Cost and revenue estimates can be classified according to detail, accuracy, and their intended use as follows:

1. *Order-of-magnitude estimates:* used in the planning and initial evaluation stage of a project.
2. *Semidetailed, or budget, estimates:* used in the preliminary or conceptual design stage of a project.
3. *Definitive (detailed) estimates:* used in the detailed engineering/construction stage of a project.

Order-of-magnitude estimates are used in selecting the feasible alternatives for the study. They typically provide accuracy in the range of ±30 to 50% and are developed through semiformal means such as conferences, questionnaires, and generalized equations applied at Level 1 or 2 of the WBS.

Budget (semidetailed) estimates are compiled to support the preliminary design effort and decision making during this project period. Their accuracy usually lies in the range of ±15%. These estimates differ in the fineness of cost and revenue breakdowns and the amount of effort spent on the estimate. Estimating equations applied at Levels 2 and 3 of the WBS are normally used.

Detailed estimates are used as the basis for bids and to make detailed design decisions. Their accuracy is ±5%. They are made from specifications, drawings, site surveys, vendor quotations, and in-house historical records and are usually done at Level 3 and successive levels in the WBS.

Thus, it is apparent that a cost or revenue estimate can vary from a "back of the envelope" calculation by an expert to a very detailed and accurate prediction of the future prepared by a project team. The level of detail and accuracy of estimates should depend on the following:

1. Time and effort available as justified by the importance of the study
2. Difficulty of estimating the items in question
3. Methods or techniques employed
4. Qualifications of the estimator(s)
5. Sensitivity of study results to particular factor estimates

As estimates become more detailed, accuracy typically improves, but the cost of making the estimate increases dramatically. This general relationship is shown in Figure 3-5 and illustrates the idea that cost and revenue estimates should be prepared with full recognition of how accurate a particular study requires them to be.

Figure 3-5
Accuracy of Cost and Revenue Estimates versus the Cost of Making Them

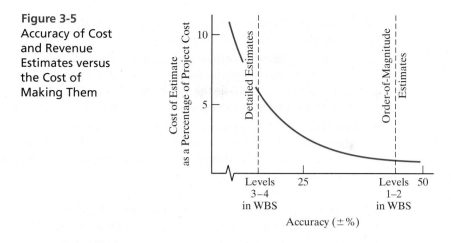

3.2.3.1 Sources of Estimating Data The information sources useful in cost and revenue estimating are too numerous to list completely. The following four major sources of information are listed roughly in order of importance:

1. *Accounting records.* Accounting records are a prime source of information for economic analyses; however, they are often not suitable for direct, unadjusted use.

   A brief discussion of the accounting process and information was given in Appendix 2-A. In its most basic sense, accounting consists of a series of procedures for keeping a detailed record of monetary transactions between established categories of assets. Accounting records are a good source of historical data but have some limitations when used in making prospective estimates for engineering economic analyses. Moreover, accounting records rarely contain direct statements of incremental costs or opportunity costs, both of which are essential in most engineering economic analyses. Incremental costs are decision specific for one-of-a-kind investments and are unlikely to be available from the accounting system.

2. *Other sources within the firm.* The typical firm has a number of people and records that may be excellent sources of estimating information. Examples of functions within firms that keep records useful to economic analyses are engineering, sales, production, quality, purchasing, and personnel.

3. *Sources outside the firm.* There are numerous sources outside the firm that can provide helpful information. The main problem is in determining those that are most beneficial for particular needs. The following is a listing of some commonly used outside sources:

   (a) *Published information.* Technical directories, buyer indexes, U.S. government publications, reference books, and trade journals offer a wealth of information. For instance, *Standard and Poor's Industry Surveys* gives monthly information regarding key industries. *The Statistical Abstract of the United States* is a remarkably comprehensive source of cost indexes and data. The Bureau of Labor Statistics publishes many periodicals that are good sources of labor costs, such as the *Monthly Labor Review, Employment and Earnings,*

Current Wage Developments, Handbook of Labor Statistics, and the *Chartbook on Wages, Prices and Productivity.*

(b) *Personal contacts* are excellent potential sources. Vendors, salespeople, professional acquaintances, customers, banks, government agencies, chambers of commerce, and even competitors are often willing to furnish needed information on the basis of a serious and tactful request.

4. ***Research and development (R&D).*** If the information is not published and cannot be obtained by consulting someone, the only alternative may be to undertake R&D to generate it. Classic examples are developing a pilot plant and undertaking a test market program.

The Internet can also be a source of cost-estimating data, though you should assure yourself that the information is from a reputable source. The following Web sites may be useful to you both professionally and personally.

www.enr.com	*Engineering News-Record*	Construction and labor costs
www.kbb.com	Kelley Blue Book	Automobile pricing
www.factsonfuel.com	American Petroleum Institute	Fuel costs

3.2.3.2   How Estimates Are Accomplished Estimates can be prepared in a number of ways, such as the following examples:

1. A *conference* of various people who are thought to have good information or bases for estimating the quantity in question. A special version of this is the *Delphi method*, which involves cycles of questioning and feedback in which the opinions of individual participants are kept anonymous.
2. *Comparison* with similar situations or designs about which there is more information and from which estimates for the alternatives under consideration can be extrapolated. This is sometimes called *estimating by analogy.* The comparison method may be used to approximate the cost of a design or product that is new. This is done by taking the cost of a more complex design for a similar item as an upper bound and the cost of a less complex item of similar design as a lower bound. The resulting approximation may not be very accurate, but the comparison method does have the virtue of setting bounds that might be useful for decision making.
3. *Using quantitative techniques,* which do not always have standardized names. Some selected techniques, with the names used being generally suggestive of the approaches, are discussed in the next section.

3.3   Selected Estimating Techniques (Models)

The estimating models discussed in this section are applicable for order-of-magnitude estimates and for many semidetailed or budget estimates. They are useful in the initial selection of feasible alternatives for further analysis and in the conceptual or preliminary design phase of a project. Sometimes, these models can be used in the detailed design phase of a project.

3.3.1 Indexes

Costs and prices* vary with time for a number of reasons, including (1) technological advances, (2) availability of labor and materials, and (3) inflation. An *index* is a dimensionless number that indicates how a cost or a price has changed with time (typically escalated) with respect to a base year. Indexes provide a convenient means for developing present and future cost and price estimates from historical data. An estimate of the cost or selling price of an item in year n can be obtained by multiplying the cost or price of the item at an earlier point in time (year k) by the ratio of the index value in year n ($\bar{I}_n$) to the index value in year k ($\bar{I}_k$);† that is,

$$C_n = C_k \left(\frac{\bar{I}_n}{\bar{I}_k} \right), \tag{3-1}$$

where k = reference year (e.g., 2000) for which cost or price of item is known;
n = year for which cost or price is to be estimated $(n > k)$;
C_n = estimated cost or price of item in year n;
C_k = cost or price of item in reference year k.

Equation (3-1) is sometimes referred to as the *ratio technique* of updating costs and prices. Use of this technique allows the cost or potential selling price of an item to be taken from historical data with a specified base year and updated with an index. This concept can be applied at the lower levels of a WBS to estimate the cost of equipment, materials, and labor, as well as at the top level of a WBS to estimate the total project cost of a new facility, bridge, and so on.

EXAMPLE 3-3	Indexing the Cost of a New Boiler

A certain index for the cost of purchasing and installing utility boilers is keyed to 1988, where its baseline value was arbitrarily set at 100. Company XYZ installed a 50,000-lb/hour boiler for $525,000 in 2000 when the index had a value of 468. This same company must install another boiler of the same size in 2007. The index in 2007 is 542. What is the approximate cost of the new boiler?

Solution

In this example, n is 2007 and k is 2000. From Equation (3-1), an approximate cost of the boiler in 2007 is

$$C_{2007} = \$525,000(542/468) = \$608,013.$$

Indexes can be created for a single item or for multiple items. For a single item, the index value is simply the ratio of the cost of the item in the current year to the cost of the same item in the reference year, multiplied by the reference year factor

---

* The terms *cost* and *price* are often used together. The cost of a product or service is the total of the resources, direct and indirect, required to produce it. The price is the value of the good or service in the marketplace. In general, price is equal to cost plus a profit.
† In this section only, k is used to denote the reference year.

(typically, 100). A composite index is created by averaging the ratios of selected item costs in a particular year to the cost of the same items in a reference year. The developer of an index can assign different weights to the items in the index according to their contribution to total cost. For example, a general weighted index is given by

$$\bar{I}_n = \frac{W_1(C_{n1}/C_{k1}) + W_2(C_{n2}/C_{k2}) + \cdots + W_M(C_{nM}/C_{kM})}{W_1 + W_2 + \cdots + W_M} \times \bar{I}_k, \qquad (3\text{-}2)$$

where M = total number of items in the index $(1 \leq m \leq M)$;
 C_{nm} = unit cost (or price) of the mth item in year n;
 C_{km} = unit cost (or price) of the mth item in year k;
 W_m = weight assigned to the mth item;
 $\bar{I}_k$ = composite index value in year k.

The weights $W_1, W_2, \ldots, W_M$ can sum to any positive number, but typically sum to 1.00 or 100. Almost any combination of labor, material, products, services, and so on can be used for a composite cost or price index.

| EXAMPLE 3-4 | Weighted Index for Gasoline Cost |

Based on the following data, develop a weighted index for the price of a gallon of gasoline in 2006, when 1992 is the reference year having an index value of 99.2. The weight placed on *regular unleaded* gasoline is three times that of either premium or unleaded plus, because roughly three times as much regular unleaded is sold compared with premium or unleaded plus.

	Price (Cents/Gal) in Year		
	1992	1996	2006
Premium	114	138	240
Unleaded plus	103	127	230
Regular unleaded	93	117	221

Solution

In this example, k is 1992 and n is 2006. From Equation (3-2), the value of $\bar{I}_{2006}$ is

$$\frac{(1)(240/114) + (1)(230/103) + (3)(221/93)}{1 + 1 + 3} \times 99.2 = 227.5.$$

Now, if the index in 2008, for example, is estimated to be 253, it is a simple matter to determine the corresponding 2008 prices of gasoline from $\bar{I}_{2006} = 227.5$:

$$\text{Premium: } 240 \text{ cents/gal} \left(\frac{253}{227.5} \right) = 267 \text{ cents/gal,}$$

$$\text{Unleaded plus: 230 cents/gal} \left(\frac{253}{227.5} \right) = 256 \text{ cents/gal,}$$

$$\text{Regular unleaded: 221 cents/gal} \left(\frac{253}{227.5} \right) = 246 \text{ cents/gal.}$$

Many indexes are periodically published, including the *Engineering News Record* Construction Index (www.enr.com), which incorporates labor and material costs and the Marshall and Stevens cost index.

3.3.2 Unit Technique

The *unit technique* involves using a per unit factor that can be estimated effectively. Examples are as follows:

1. Capital cost of plant per kilowatt of capacity
2. Revenue per mile
3. Capital cost per installed telephone
4. Revenue per customer served
5. Temperature loss per 1,000 feet of steam pipe
6. Operating cost per mile
7. Construction cost per square foot

Such factors, when multiplied by the appropriate unit, give a total estimate of cost, savings, or revenue.

As a simple example, suppose that we need a preliminary estimate of the cost of a particular house. Using a unit factor of, say, $95 per square foot and knowing that the house is approximately 2,000 square feet, we estimate its cost to be $95 × 2,000 = $190,000.

While the unit technique is very useful for preliminary estimating purposes, such average values can be misleading. In general, more detailed methods will result in greater estimation accuracy.

3.3.3 Factor Technique

The *factor technique* is an extension of the unit method in which we sum the product of several quantities or components and add these to any components estimated directly. That is,

$$C = \sum_d C_d + \sum_m f_m U_m, \tag{3-3}$$

where C = cost being estimated;

C_d = cost of the selected component d that is estimated directly;

f_m = cost per unit of component m;

U_m = number of units of component m.

As a simple example, suppose that we need a slightly refined estimate of the cost of a house consisting of 2,000 square feet, two porches, and a garage. Using a unit factor of $85 per square foot, $10,000 per porch, and $8,000 per garage for the two directly estimated components, we can calculate the total estimate as

$$(\$10,000 \times 2) + \$8,000 + (\$85 \times 2,000) = \$198,000.$$

We can also answer the question posed at the beginning of the chapter (page 71) using the factor technique. The rental expense per year will be approximately $20,000 \times \$233.57$ in London plus $10,000 \times \$201.29$ in Hong Kong plus $20,000 \times \$175.00$ in New York City. The total expense will be $10,184,300 per year for the additional office space.

The factor technique is particularly useful when the complexity of the estimating situation does not require a WBS, but several different parts are involved. Example 3-5 and the product cost-estimating example to be presented in Section 3.5.1 further illustrate this technique.

| EXAMPLE 3-5 | **Analysis of Building Space Cost Estimates** |

The detailed design of the commercial building described in Example 3-2 affects the utilization of the gross square feet (and, thus, the net rentable space) available on each floor. Also, the size and location of the parking lot and the prime road frontage available along the property may offer some additional revenue sources. As project manager, analyze the potential revenue impacts of the following considerations.

The first floor of the building has 15,000 gross square feet of retail space, and the second floor has the same amount planned for office use. Based on discussions with the sales staff, develop the following additional information:

(a) The retail space should be designed for two different uses—60% for a restaurant operation (utilization = 79%, yearly rent = $23/sq.ft.) and 40% for a retail clothing store (utilization = 83%, yearly rent = $18/sq.ft.).

(b) There is a high probability that all the office space on the second floor will be leased to one client (utilization = 89%, yearly rent = $14/sq.ft.).

(c) An estimated 20 parking spaces can be rented on a long-term basis to two existing businesses that adjoin the property. Also, one spot along the road frontage can be leased to a sign company, for erection of a billboard, without impairing the primary use of the property.

Solution

Based on this information, you estimate annual project revenue ($\hat{R}$) as

$$\hat{R} = W(r_1)(12) + \Upsilon(r_2)(12) + \sum_{j=1}^{3} S_j(u_j)(d_j),$$

where W = number of parking spaces;
   Υ = number of billboards;
   r_1 = rate per month per parking space = \$22;
   r_2 = rate per month per billboard = \$65;
   j = index of type of building space use;
   S_j = space (gross square feet) being used for purpose j;
   u_j = space j utilization factor (% net rentable);
   d_j = rate per (rentable) square foot per year of building space used for purpose j.

Then,

$$\hat{R} = [20(\$22)(12) + 1(\$65)(12)] + [9,000(0.79)(\$23)$$
$$+ 6,000(0.83)(\$18) + 15,000(0.89)(\$14)]$$
$$\hat{R} = \$6,060 + \$440,070 = \$446,130.$$

A breakdown of the annual estimated project revenue shows that

1.4% is from miscellaneous revenue sources;
98.6% is from leased building space.

From a detailed design perspective, changes in annual project revenue due to changes in building space utilization factors can be easily calculated. For example, an average 1% improvement in the ratio of rentable space to gross square feet would change the annual revenue ($\Delta\hat{R}$) as follows:

$$\Delta\hat{R} = \sum_{j=1}^{3} S_j(u_j + 0.01)(d_j) - (\$446,130 - \$6,060)$$
$$= \$445,320 - \$440,070$$
$$= \$5,250 \text{ per year.}$$

3.4 Parametric Cost Estimating

Parametric cost estimating is the use of historical cost data and statistical techniques to predict future costs. Statistical techniques are used to develop cost estimating relationships (CERs) that tie the cost or price of an item (e.g., a product, good,

TABLE 3-1	Examples of Cost Drivers Used in Parametric Cost Estimates
Product	**Cost Driver (Independent Variable)**
Construction	Floor space, roof surface area, wall surface area
Trucks	Empty weight, gross weight, horsepower
Passenger car	Curb weight, wheel base, passenger space, horsepower
Turbine engine	Maximum thrust, cruise thrust, specific fuel consumption
Reciprocating engine	Piston displacement, compression ratio, horsepower
Aircraft	Empty weight, speed, wing area
Electrical power plants	KiloWatts
Motors	Horsepower
Software	Number of lines of code

service, or activity) to one or more independent variables (i.e., cost drivers). Recall from Chapter 2 that cost drivers are design variables that account for a large portion of total cost behavior. Table 3-1 lists a variety of items and associated cost drivers. The unit technique described in the previous section is a simple example of parametric cost estimating.

Parametric models are used in the early design stages to get an idea of how much the product (or project) will cost, on the basis of a few physical attributes (such as weight, volume, and power). The output of the parametric models (an estimated cost) is used to gauge the impact of design decisions on the total cost.

Various statistical and other mathematical techniques are used to develop the CERs. For example, simple linear regression and multiple linear regression models, which are standard statistical methods for estimating the value of a dependent variable (the unknown quantity) as a function of one or more independent variables, are often used to develop estimating relationships. This section describes two commonly used estimating relationships, the power-sizing technique and the learning curve, followed by an overview of the procedure used to develop CERs.

3.4.1 Power-Sizing Technique

The *power-sizing technique*, which is sometimes referred to as an *exponential model*, is frequently used for developing capital investment estimates for industrial plants and equipment. This CER recognizes that cost varies as some power of the change in capacity or size. That is,

$$\frac{C_A}{C_B} = \left(\frac{S_A}{S_B}\right)^X,$$

$$C_A = C_B \left(\frac{S_A}{S_B}\right)^X, \tag{3-4}$$

where C_A = cost for plant A $\Big\}$ (both in $ as of the point in
C_B = cost for plant B time for which the estimate is
desired);

S_A = size of plant A $\Big\}$ (both in same physical units);
S_B = size of plant B

$X = cost\text{-}capacity\ factor$ to reflect economies of scale.*

The value of the cost-capacity factor will depend on the type of plant or equipment being estimated. For example, $X = 0.68$ for nuclear generating plants and 0.79 for fossil-fuel generating plants. Note that $X < 1$ indicates decreasing economies of scale (each additional unit of capacity costs less than the previous unit), $X > 1$[†] indicates increasing economies of scale (each additional unit of capacity costs more than the previous unit), and $X = 1$ indicates a linear cost relationship with size.

| EXAMPLE 3-6 | **Power-Sizing Model for Cost Estimating** |

Suppose that an aircraft manufacturer desires to make a preliminary estimate of the cost of building a 600-MW fossil-fuel plant for the assembly of its new long-distance aircraft. It is known that a 200-MW plant cost $100 million 20 years ago when the approximate cost index was 400, and that cost index is now 1,200. The cost-capacity factor for a fossil-fuel power plant is 0.79.

Solution

Before using the power-sizing model to estimate the cost of the 600-MW plant (C_A), we must first use the cost index information to update the known cost of the 200-MW plant 20 years ago to a current cost. Using Equation (3-1), we find that the current cost of a 200-MW plant is

$$C_B = \$100\ \text{million}\left(\frac{1,200}{400}\right) = \$300\ \text{million}.$$

So, using Equation (3-4), we obtain the following estimate for the 600-MW plant:

$$C_A = \$300\ \text{million}\left(\frac{600\text{-MW}}{200\text{-MW}}\right)^{0.79}$$

$$C_A = \$300\ \text{million} \times 2.38 = \$714\ \text{million}.$$

Note that Equation (3-4) can be used to estimate the cost of a larger plant (as in Example 3-6) or the cost of a smaller plant. For example, suppose we need to estimate the cost of building a 100-MW plant. Using Equation (3-4) and the data

---

* This may be calculated or estimated from experience by using statistical techniques. For typical factors, see W. R. Park, *Cost Engineering Analysis* (New York: John Wiley & Sons, 1973), 137.

† Precious gems are all example to increasing economies of scale. For example, a one-carat diamond typically costs more than four quarter-carat diamonds.

for the 200-MW plant in Example 3-6, we find that the current cost of a 100-MW plant is

$$C_A = \$300 \text{ million} \left(\frac{100 \text{ MW}}{200 \text{ MW}} \right)^{0.79}$$

$$C_A = \$300 \text{ million} \times 0.58 = \$174 \text{ million}.$$

3.4.2 Learning and Improvement

A *learning curve* is a mathematical model that explains the phenomenon of increased worker efficiency and improved organizational performance with repetitive production of a good or service. The learning curve is sometimes called an *experience curve* or a *manufacturing progress function*; fundamentally, it is an estimating relationship. The learning (improvement) curve effect was first observed in the aircraft and aerospace industries with respect to labor hours per unit. However, it applies in many different situations. For example, the learning curve effect can be used in estimating the professional hours expended by an engineering staff to accomplish successive detailed designs within a family of products, as well as in estimating the labor hours required to assemble automobiles.

The basic concept of learning curves is that some input resources (e.g., energy costs, labor hours, material costs, engineering hours) decrease, on a per-output-unit basis, as the number of units produced increases. Most learning curves are based on the assumption that a constant percentage reduction occurs in, say, labor hours, as the number of units produced is *doubled*. For example, if 100 labor hours are required to produce the first output unit and a 90% learning curve is assumed, then $100(0.9) = 90$ labor hours would be required to produce the second unit. Similarly, $100(0.9)^2 = 81$ labor hours would be needed to produce the fourth unit, $100(0.9)^3 = 72.9$ hours to produce the eighth unit, and so on. Therefore, a 90% learning curve results in a 10% reduction in labor hours each time the production quantity is doubled.

Equation (3-5) can be used to compute resource requirements assuming a constant percentage reduction in input resources each time the output quantity is doubled.

$$Z_u = K(u^n), \tag{3-5}$$

where $u =$ the output unit number;

$\quad Z_u =$ the number of input resource units needed to produce output unit u;

$\quad K =$ the number of input resource units needed to produce the first output unit;

$\quad s =$ the learning curve slope parameter expressed as a decimal ($s = 0.9$ for a 90% learning curve);

$\quad n = \dfrac{\log s}{\log 2} =$ the learning curve exponent.

EXAMPLE 3-7   **Learning Curve for a Formula Car Design Team**

The Mechanical Engineering department has a student team that is designing a formula car for national competition. The time required for the team to assemble the first car is 100 hours. Their improvement (or learning rate) is 0.8, which means that as output is doubled, their time to assemble a car is reduced by 20%. Use this information to determine

(a) the time it will take the team to assemble the 10th car.

(b) the *total time* required to assemble the first 10 cars.

(c) the estimated *cumulative average* assembly time for the first 10 cars.

Solve by hand and by spreadsheet.

 Solution by Hand

(a) From Equation (3-5), and assuming a proportional decrease in assembly time for output units between doubled quantities, we have

$$Z_{10} = 100(10)^{\log 0.8/\log 2}$$

$$= 100(10)^{-0.322}$$

$$= \frac{100}{2.099} = 47.6 \text{ hours.}$$

(b) The total time to produce x units, T_x, is given by

$$T_x = \sum_{u=1}^{x} Z_u = \sum_{u=1}^{x} K(u^n) = K \sum_{u=1}^{x} u^n. \qquad (3\text{-}6)$$

Using Equation (3-6), we see that

$$T_{10} = 100 \sum_{u=1}^{10} u^{-0.322} = 100[1^{-0.322} + 2^{-0.322} + \cdots + 10^{-0.322}] = 631 \text{ hours.}$$

(c) The cumulative average time for x units, C_x, is given by

$$C_x = T_x/x. \qquad (3\text{-}7)$$

Using Equation (3-7), we get

$$C_{10} = T_{10}/10 = 631/10 = 63.1 \text{ hours.}$$

Spreadsheet Solution

Figure 3-6 shows the spreadsheet solution for this example problem. For each unit number, the unit time to complete the assembly, the cumulative total time, and the cumulative average time are computed with the use of Equations (3-5),

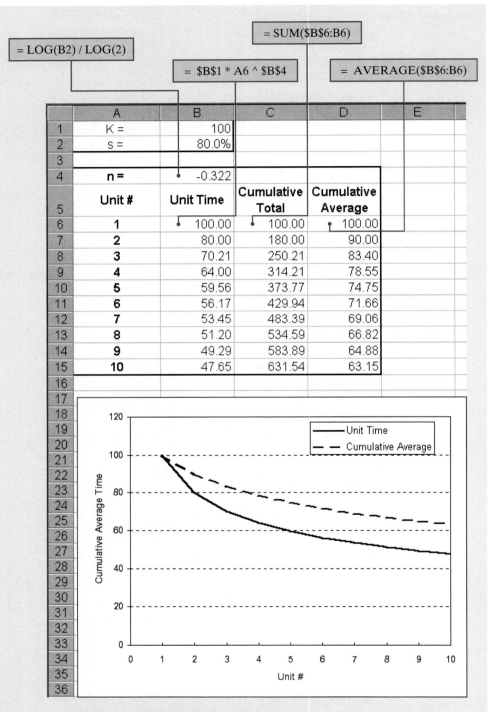

Figure 3-6 Spreadsheet Solution, Example 3-7

(3-6), and (3-7), respectively. Note that these formulas are entered once in row 6 of the spreadsheet and are then simply copied into rows 7 through 15.

A plot of unit time and cumulative average time is easily created with the spreadsheet software. This spreadsheet model can be used to examine the impact of a different learning slope parameter (e.g., $s = 90\%$) on predicted car assembly times by changing the value of cell B2.

3.4.3 Developing a Cost Estimating Relationship (CER)

A CER is a mathematical model that describes the cost of an engineering project as a function of one or more design variables. CERs are useful tools because they allow the estimator to develop a cost estimate quickly and easily. Furthermore, estimates can be made early in the design process before detailed information is available. As a result, engineers can use CERs to make early design decisions that are cost effective in addition to meeting technical requirements.

There are four basic steps in developing a CER:

1. Problem definition
2. Data collection and normalization
3. CER equation development
4. Model validation and documentation

3.4.3.1 Problem Definition The first step in any engineering analysis is to define the problem to be addressed. A well-defined problem is much easier to solve. For the purposes of cost estimating, developing a WBS is an excellent way of describing the elements of the problem. A review of the completed WBS can also help identify potential cost drivers for the development of CERs.

3.4.3.2 Data Collection and Normalization The collection and normalization of data is the most critical step in the development of a CER. We're all familiar with the adage "garbage in—garbage out." Without reliable data, the cost estimates obtained by using the CER would be meaningless. The WBS is also helpful in the data collection phase. The WBS helps to organize the data and ensure that no elements are overlooked.

Data can be obtained from both internal and external sources. Costs of similar projects in the past are one source of data. Published cost information is another source of data. Once collected, data must be normalized to account for differences due to inflation, geographical location, labor rates, and so on. For example, cost indexes or the price inflation techniques to be described in Chapter 8 can be used to normalize costs that occurred at different times. Consistent definition of the data is another important part of the normalization process.

3.4.3.3 CER Equation Development The next step in the development of a CER is to formulate an equation that accurately captures the relationship between

TABLE 3-2 Typical Equation Forms	
Type of Relationship	Generalized Equation
Linear	$\text{Cost} = b_0 + b_1 x_1 + b_2 x_2 + b_3 x_3 + \cdots$
Power	$\text{Cost} = b_0 + b_1 x_1^{b_{11}} x_2^{b_{12}} + \cdots$
Logarithmic	$\text{Cost} = b_0 + b_1 \log(x_1) + b_2 \log(x_2) + b_3 \log(x_3) + \cdots$
Exponential	$\text{Cost} = b_0 + b_1 \exp^{b_{11} x_1} + b_2 \exp^{b_{22} x_2} + \cdots$

the selected cost driver(s) and project cost. Table 3-2 lists four general equation types commonly used in CER development. In these equations, b_0, b_1, b_2, and b_3 are constants, while x_1, x_2, and x_3 represent design variables.

A simple, yet very effective, way to determine an appropriate equation form for the CER is to plot the data. If a plot of the data on regular graph paper appears to follow a straight line, then a linear relationship is suggested. If a curve is suggested, then try plotting the data on semilog or log-log paper. If a straight line results on semilog paper, then the relationship is logarithmic or exponential. If a straight line results on log-log paper, then the relationship is a power curve.

Once we have determined the basic equation form for the CER, the next step is to determine the values of the coefficients in the CER equation. The most common technique used to solve for the coefficient values is the method of least squares. Basically, this method seeks to determine a straight line through the data that minimizes the total deviation of the actual data from the predicted values. (The line itself represents the CER.) This method is relatively easy to apply manually and is also available in many commercial software packages. (Most spreadsheet packages are capable of performing a least-squares fit of data.) The primary requirement for the use of the least-squares method is a linear relationship between the independent variable (the cost driver) and the dependent variable (project cost).*

All of the equation forms presented in Table 3-2 can easily be transformed into a linear form. The following equations can be used to calculate the values of the coefficients b_0 and b_1 in the simple linear equation $y = b_0 + b_1 x$:

$$b_1 = \frac{n \sum_{i=1}^{n} x_i y_i - \left(\sum_{i=1}^{n} x_i\right)\left(\sum_{i=1}^{n} y_i\right)}{n \sum_{i=1}^{n} x_i^2 - \left(\sum_{i=1}^{n} x_i\right)^2}, \tag{3-8}$$

$$b_0 = \frac{\sum_{i=1}^{n} y_i - b_1 \sum_{i=1}^{n} x_i}{n}. \tag{3-9}$$

Note that the variable n in the foregoing equations is equal to the number of data sets used to estimate the values of b_0 and b_1.

---

* In addition, the observations should be independent. The difference between predicted and actual values is assumed to be normally distributed with an expected value of zero. Furthermore, the variance of the dependent variable is assumed to be equal for each value of the independent variable.

| EXAMPLE 3-8 | Cost Estimating Relationship (CER) for a Spacecraft |

In the early stages of design, it is believed that the cost of a Martian rover spacecraft is related to its weight. Cost and weight data for six spacecraft have been collected and normalized and are shown in the next table. A plot of the data suggests a linear relationship. Use a spreadsheet model to determine the values of the coefficients for the CER.

Spacecraft i	Weight (lb) x_i	Cost ($ million) y_i
1	400	278
2	530	414
3	750	557
4	900	689
5	1,130	740
6	1,200	851

Spreadsheet Solution

Figure 3-7 displays the spreadsheet model for determining the coefficients of the CER. This example illustrates the basic regression features of Excel. No formulas

(a) Regression Dialog Box

Figure 3-7 Spreadsheet Solution, Example 3-8

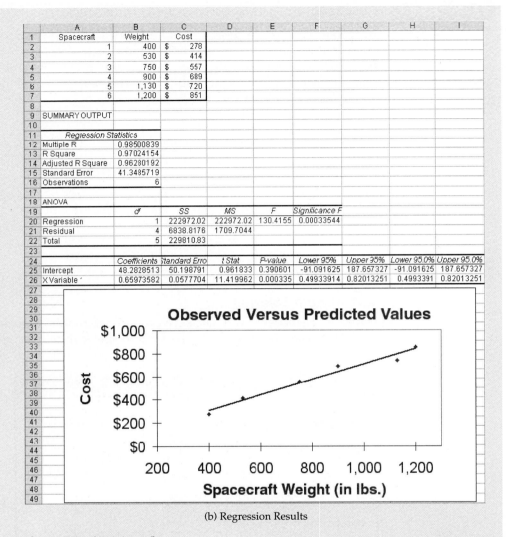

	A	B	C	D	E	F	G	H	I
1	Spacecraft	Weight	Cost						
2	1	400	$ 278						
3	2	530	$ 414						
4	3	750	$ 557						
5	4	900	$ 689						
6	5	1,130	$ 720						
7	6	1,200	$ 851						
8									
9	SUMMARY OUTPUT								
10									
11	*Regression Statistics*								
12	Multiple R	0.98500839							
13	R Square	0.97024154							
14	Adjusted R Square	0.96280192							
15	Standard Error	41.3485719							
16	Observations	6							
17									
18	ANOVA								
19		*df*	*SS*	*MS*	*F*	*Significance F*			
20	Regression	1	222972.02	222972.02	130.4155	0.00033544			
21	Residual	4	6838.8176	1709.7044					
22	Total	5	229810.83						
23									
24		*Coefficients*	*Standard Erro*	*t Stat*	*P-value*	*Lower 95%*	*Upper 95%*	*Lower 95.0%*	*Upper 95.0%*
25	Intercept	48.2828513	50.198791	0.961833	0.390601	-91.091625	187.657327	-91.091625	187.657327
26	X Variable '	0.65973582	0.0577704	11.419962	0.000335	0.49933914	0.82013251	0.4993391	0.82013251

Observed Versus Predicted Values

(b) Regression Results

Figure 3-7 *(continued)*

are entered, only the cost and weight data for the spacecraft. The challenge in spreadsheet regression lies in making sure that the underlying regression assumptions are satisfied and in interpreting the output properly.

The *Tools | Data Analysis | Regression* menu command brings up the Regression dialog box shown in Figure 3-7(a) and shows the values used for this model. The results of the analysis are generated by Excel and are displayed beginning in cell A9 of Figure 3-7(b). For the purposes of this example, the coefficients b_0 and b_1 of the CER are found in cells B25 and B26, respectively.

The resulting CER relating spacecraft cost (in millions of dollars) to spacecraft weight is

$$\text{Cost} = 48.28 + 0.6597x,$$

where x represents the weight of the spacecraft in pounds, and $400 \leq x \leq 1{,}200$.

3.4.3.4 Model Validation and Documentation

Once the CER equation has been developed, we need to determine how well the CER can predict cost (i.e., model validation) and document the development and appropriate use of the CER. Validation can be accomplished with statistical "goodness of fit" measures such as standard error and the correlation coefficient. Analysts must use goodness of fit measures to infer how well the CER predicts cost as a function of the selected cost driver(s). Documenting the development of the CER is important for future use of the CER. It is important to include in the documentation the data used to develop the CER and the procedures used for normalizing the data.

The standard error (SE) measures the average amount by which the actual cost values and the predicted cost values vary. The SE is calculated by

$$SE = \sqrt{\frac{\sum_{i=1}^{n}(y_i - \text{Cost}_i)^2}{n-2}}, \tag{3-10}$$

where Cost_i is the cost predicted by using the CER with the independent variable values for data set i and y_i is the actual cost. A small value of SE is preferred.

The correlation coefficient (R) measures the closeness of the actual data points to the regression line $(y = b_0 + b_1 x)$. It is simply the ratio of explained deviation to total deviation.

$$R = \frac{\sum_{i=1}^{n}(x_i - \bar{x})(y_i - \bar{y})}{\sqrt{\left[\sum_{i=1}^{n}(x_i - \bar{x})^2\right]\left[\sum_{i=1}^{n}(y_i - \bar{y})^2\right]}}, \tag{3-11}$$

where $\bar{x} = \frac{1}{n}\sum_{i=1}^{n} x_i$ and $\bar{y} = \frac{1}{n}\sum_{i=1}^{n} y_i$. The sign $(+/-)$ of R will be the same as the sign of the slope (b_1) of the regression line. Values of R close to one (or minus one) are desirable in that they indicate a strong linear relationship between the dependent and independent variables.

In cases where it is not clear which is the "best" cost driver to select or which equation form is best, you can use the goodness of fit measures to make a selection.

In general, all other things being equal, the CER with better goodness of fit measures should be selected.

EXAMPLE 3-9	**Regression Statistics for the Spacecraft CER**

Determine the SE and the correlation coefficient for the CER developed in Example 3-8.

Solution

From the spreadsheet for Example 3-8 (Figure 3-7), we find that the SE is 41.35 (cell B15) and that the correlation coefficient is 0.985 (cell B12). The value of the correlation coefficient is close to one, indicating a strong positive linear relationship between the cost of the spacecraft and the spacecraft's weight.

In summary, CERs are useful for a number of reasons. First, given the required input data, they are quick and easy to use. Second, a CER usually requires very little detailed information, making it possible to use the CER early in the design process. Finally, a CER is an excellent predictor of cost if correctly developed from good historical data.

3.5   Cost Estimation in the Design Process

Today's companies are faced with the challenge of providing quality goods and services at competitive prices. The price of their product is based on the overall cost of making the item plus a built-in profit. To ensure that products can be sold at competitive prices, cost must be a major factor in the design of the product.

In this section, we will discuss both a *bottom-up* approach and a *top-down* approach to determining product costs and selling price. Used together with the concepts of target costing, design-to-cost, and value engineering, these techniques can assist engineers in the design of cost-effective systems and competitively priced products.

3.5.1   The Elements of Product Cost and "Bottom-Up" Estimating

As discussed in Chapter 2, product costs are classified as direct or indirect. Direct costs are easily assignable to a specific product, while indirect costs are not easily allocated to a certain product. For instance, direct labor would be the wages of a machine operator; indirect labor would be supervision.

Manufacturing costs have a distinct relationship to production volume in that they may be fixed, variable, or step-variable. Generally, administrative costs are fixed, regardless of volume; material costs vary directly with volume; and equipment cost is a step function of production level.

The primary costs within the manufacturing expense category include engineering and design, development costs, tooling, manufacturing labor, materials, supervision, quality control, reliability and testing, packaging, plant overhead, general and administrative, distribution and marketing, financing, taxes, and insurance. Where do we start?

Major types of costs that must be estimated are as follows:

1. Tooling costs, which consist of repair and maintenance plus the cost of any new equipment.

2. Manufacturing labor costs, which are determined from standard data, historical records, or the accounting department. Learning curves are often used for estimating direct labor.

3. Materials costs, which can be obtained from historical records, vendor quotations, and the bill of materials. Scrap allowances must be included.

4. Supervision, which is a fixed cost based on the salaries of supervisory personnel.

5. Factory overhead, which includes utilities, maintenance, and repairs. As discussed in Chapter 2 and Appendix 2-A, there are various methods used to allocate overhead, such as in proportion to direct labor hours or machine hours.

6. General and administrative costs, which are sometimes included with the factory overhead.

A "bottom-up" procedure for estimating total product cost is commonly used by companies to help them make decisions about what to produce and how to price their products. The term "bottom-up" is used because the procedure requires estimates of cost elements at the lower levels of the cost structure, which are then added together to obtain the total cost of the product. The following simple example shows the general bottom-up procedure for making a per-unit product cost estimate and illustrates the use of a typical spreadsheet form of the cost structure for preparing the estimate.

The spreadsheet in Figure 3-8 shows the determination of the cost of a throttle assembly. Column A shows typical cost elements that contribute to total product cost. The list of cost elements can easily be modified to meet a company's needs. This spreadsheet allows for per-unit estimates (column B), factor estimates (column C), and direct estimates (column D). The shaded rows are selected subtotals.

Typically, direct labor costs are estimated via the unit technique. The manufacturing process plan is used to estimate the total number of direct labor hours required per output unit. This quantity is then multiplied by the composite labor rate to obtain the total direct labor cost. In this example, 36.48 direct labor hours are required to produce 50 throttle assemblies and the composite labor rate is $10.54 per hour, which yields a total direct labor cost of $384.50.

Indirect costs, such as quality control and planning labor, are often allocated to individual products by using factor estimates. Estimates are obtained by expressing the cost as a percentage of another cost. In this example, planning labor and quality control are expressed as 12% and 11% of direct labor cost (row A), respectively. This gives a total labor cost of $472.93. Factory overhead and general and administrative expenses are estimated as percentages of the total labor cost (row D).

Entries for cost elements for which direct estimates are available are placed in column D. The total production materials cost for the 50 throttle assemblies is

Column A	Column B		Column C		Column D	Column E
	Unit Estimate		Factor Estimate		Direct	Row
	Unit	Cost/Unit	Factor	of Row	Estimate	Total
A: Factory Labor	36.48	$ 10.54				$ 384.50
B: Planning Labor			12%	A		46.14
C: Quality Control			11%	A		42.29
D: TOTAL LABOR						472.93
E: Factory Overhead			105%	D		496.58
F: General & Admin. Expense			15%	D		70.94
G: Production Material					$ 167.17	167.17
H: Outside Manufacture					28.00	28.00
I: SUBTOTAL						1,235.62
J: Packing Costs			5%	I		61.78
K: TOTAL DIRECT CHARGE						1,297.41
L: Other Direct Charge			1%	K		12.97
M: Facility Rental						--
N: TOTAL MANUFACTURING COST						1,310.38
O: Quantity (lot size)						50
P: MANUFACTURING COST / UNIT						26.21
Q: Profit/Fee			10%	P		2.62
R: UNIT SELLING PRICE						$ 28.83

Figure 3-8 Manufacturing Cost-Estimating Spreadsheet

$167.17. A direct estimate of $28.00 applies to the outside manufacture of required components. The subtotal of cost elements at this point is $1,235.62.

Packing costs are estimated as 5% of all previous costs (row I), giving a total direct charge of $1,297.41. The cost of other miscellaneous direct charges are figured in as 1% of the current subtotal (row K). This results in a total manufacturing cost of $1,310.38 for the entire lot of 50 throttle assemblies. The manufacturing cost per assembly is $1,310.38/50 = $26.21.

The price of a product is based on the overall cost of making the item plus a built-in profit. The bottom of the spreadsheet in Figure 3-8 shows the computation of unit selling price based on this strategy. In this example, the desired profit (often called the profit margin) is 10% of the unit manufacturing cost, which corresponds to a profit of $2.62 per throttle assembly. The total selling price of a throttle assembly is then $26.21 + $2.62 = $28.83.

3.5.2 Target Costing and Design to Cost: A "Top-Down" Approach

Traditionally, American firms determine an initial estimate of a new product's selling price by using the bottom-up approach described in the previous section. That is, the estimated selling price is obtained by accumulating relevant fixed and variable costs and then adding a profit margin, which is a percentage of total manufacturing costs. This process is often termed *design to price*. The estimated selling price is then used by the marketing department to determine whether the new product can be sold.

In contrast, Japanese firms apply the concept of *target costing*, which is a top-down approach. The focus of target costing is "what *should* the product cost" instead of "what *does* the product cost." The objective of target costing is to design costs out of products before those products enter the manufacturing process. In this top-down approach, cost is viewed as an input to the design process, not as an outcome of it.

Target costing is initiated by conducting market surveys to determine the selling price of the best competitor's product. A target cost is obtained by deducting the desired profit from the best competitor's selling price:

$$\text{Target cost} = \text{Competitor's price} - \text{Desired profit}. \qquad (3\text{-}12)$$

Desired profit is often expressed as a percentage of the total manufacturing cost, called the profit margin. For a specific profit margin (e.g., 10%), the target cost can be computed from the following equation:

$$\text{Target cost} = \frac{\text{Competitor's price}}{(1 + \text{Profit margin})}. \qquad (3\text{-}13)$$

This target cost is obtained prior to the design of the product and is used as a goal for engineering design, procurement, and production.

EXAMPLE 3-10 | **Target Costing at the Throttle Assembly Plant**

Recall the throttle assemblies discussed in the previous example. Suppose that a market survey has shown that the best competitor's selling price is $27.50 per assembly. If a profit margin of 10% (based on total manufacturing cost) is desired, determine a target cost for the throttle assembly.

Solution

Since the desired profit has been expressed as a percentage of total manufacturing costs, we can use Equation (3-13) to determine the target cost:

$$\text{Target cost} = \frac{\$27.50}{(1 + 0.10)} = \$25.00.$$

Note that the total manufacturing cost calculated in Figure 3-8 was $26.21 per assembly. Since this cost exceeds the target cost, a redesign of the product itself or of the manufacturing process is needed to achieve a competitive selling price.

Figure 3-9 illustrates the relationship between target costing and the design process. It is important to note that target costing takes place before the design process begins. Engineers use the target cost as a cost performance requirement. Now the final product must satisfy both technical and cost performance requirements. The practice of considering cost performance as important as technical performance during the design process is termed *design to cost*. The design to cost procedure begins with the target cost as the cost goal for the entire product. This target cost is then broken down into cost goals for major subsystems, components, and subassemblies. These cost goals would cover targets for direct material costs and direct labor.

Once cost goals have been established, the preliminary engineering design process is initiated. Conventional tools, such as WBS and cost estimating, are

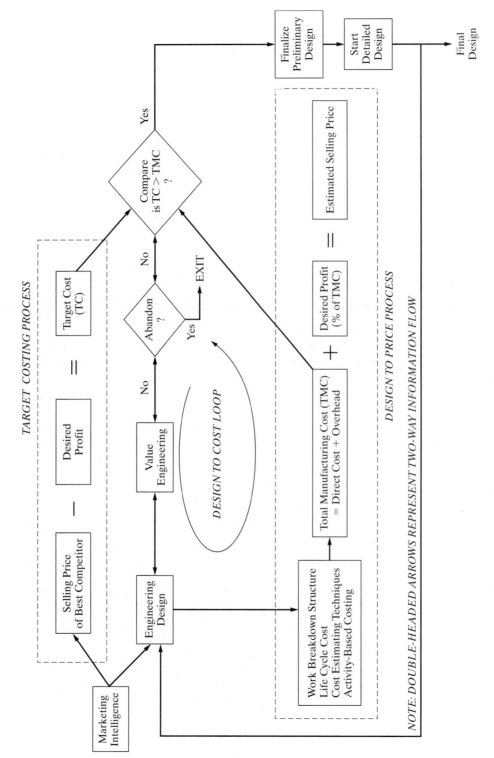

Figure 3-9 The Concept of Target Costing and Its Relationship to Design

used to prepare a bottom-up total manufacturing cost projection (discussed in the previous section). The total manufacturing cost represents an *initial* appraisal of what it would cost the firm to design and manufacture the product being considered. The total manufacturing cost is then compared with the top-down target cost. If the total manufacturing cost is more than the target cost, then the design must be fed back into the *value engineering process* (to be discussed in Section 3.5.3) to challenge the functionality of the design and attempt to reduce the cost of the design. This iterative process is a key feature of the design to cost procedure. If the total manufacturing cost can be made less than the target cost, the design process continues into detailed design, culminating in the final design to be produced. If the total manufacturing cost cannot be reduced to the target cost, the firm should seriously consider abandoning the product.

Figure 3-10 illustrates the use of the manufacturing cost-estimating spreadsheet to compute both a target cost and the cost reductions necessary to achieve the target cost. As computed in Example 3-10, the target cost for the throttle assembly is $25.00. Since the initial total manufacturing cost (determined to be $26.21 in Figure 3-8) is greater than the target cost, we must work backward from the total manufacturing cost, changing values of a chosen (single) cost element to the level required to reduce the cost to the desired target. This method of establishing new cost targets for individual elements can be accomplished by trial and error (manually manipulating the spreadsheet values) or by making use of the "solver" feature of the software package (if one is available). Figure 3-10 shows one possible result of this process. If the process of assembling the throttles could be made more efficient, such that

Column A		Column B		Column C		Column D	Column E
		Unit Estimate		Factor Estimate		Direct	Row
MANUFACTURING COST ELEMENTS		Units	Cost/Unit	Factor	of Row	Estimate	Total
A: Factory labor		34.48	$ 10.54				$ 363.42
B: Planning labor				12%	A		43.61
C: Quality control				11%	A		39.98
D: TOTAL LABOR							447.01
E: Factory overhead				105%	D		469.36
F: General & admin. expense				15%	D		67.05
G: Production material						$ 167.17	167.17
H: Outside manufacture						28.00	28.00
I: SUBTOTAL							1,178.58
J: Packing costs				5%	I		58.93
K: TOTAL DIRECT CHARGE							1,237.51
L: Other direct charge				1%	K		12.38
M: Facility rental						-	-
N: TOTAL MANUFACTURING COST							1,249.89
O: Quantity (lot size)							50
P: MANUFACTURING COST / UNIT							25.00
Competitor's selling price		$ 27.50					
Desired return on sales		10%					
Target cost		$ 25.00					

Figure 3-10   Manufacturing Cost Estimating and Target Costing

total direct labor requirements are reduced to 34.48 hours (instead of 36.48), the target cost could be realized. Now the challenge is to find a way, either through product or process redesign, to reduce the direct labor requirements.

EXAMPLE 3-11	Cost Goal for Production Material at the Throttle Assembly Plant

Given the current estimated total manufacturing cost of $26.21, as shown in Figure 3-8, determine a cost goal for production material that would allow us to achieve a target cost of $25.00.

Solution

With the spreadsheet in Figure 3-8 as a starting point, one approach to determining a cost goal for production material would be to iteratively change the value entered in Row G and Column D until the desired total manufacturing cost of $25.00 is obtained. The following table shows a series of costs for production material and the resulting total manufacturing cost per assembly:

Production Material Cost per 50 Assemblies ($)	Total Manufacturing Cost per Assembly ($)
167.17	26.21
150.00	25.84
140.00	25.63
130.00	25.42
120.00	25.21
110.00	25.00

As seen in the table, a production material cost of $110 per batch of 50 assemblies would result in a total manufacturing cost of $25.00, the target cost. Now it is left to design engineers to determine if a different, less costly material could be used or if process improvements could be made to reduce material scrap. Other possibilities would be to negotiate a new purchase price with the supplier of the material or seek a new supplier.

The chapter case study on the demanufacturing of computers illustrates the versatility of target costing concepts.

3.5.3 Value Engineering

This section introduces the topic of value engineering (VE). The objective of VE is very similar to that of design to cost. The VE objective is to provide the required product functions at a minimum cost. VE necessitates a detailed examination of a product's functions, and the cost of each, in addition to a thorough review of product specifications. VE is performed by a team of specialists from a variety of disciplines (design, manufacturing, marketing, etc.), and the team focuses on determining the most cost-effective way to provide high value at an acceptable cost to the customer. VE is most appropriately applied early in the life cycle where there is more potential for cost savings. It is applied repeatedly during the design phase as new information becomes available. Note that, in Figure 3-9, the VE function appears within the design to cost loop and is a critical part of obtaining a total manufacturing cost that is less than the target cost.

The key to successful VE is to ask critical questions and seek creative answers. Table 3-3 lists some sample questions that should be included in a VE study. It is important to question everything and not take anything for granted. Cost-reduction opportunities are sometimes so simple that they get overlooked. Creative solutions can be obtained by using classical brainstorming or the Nominal Group Technique (discussed in Chapter 1). Alternatives that seem promising should be analyzed to determine if cost reduction is possible without compromising functionality. For example, many airlines today are looking for creative ways to reduce fuel costs. Some of these ways include using winglets, or upturned tabs on wingtips, to reduce air resistance when flying and to cut fuel costs. It has been estimated that winglets reduce fuel consumption by 5%. Airlines are also reducing their fuel reserves aboard aircraft, reducing aircraft weight by removing ovens on short flights on which hot meals are no longer served or by placing manuals on laptops for pilots, and reducing air speed. One airline estimates that it saved $10.6 million a year without compromising safety by reducing fuel reserves on its transoceanic flights.

The examples that follow illustrate how VE has been used to obtain enhanced functionality and improved value. The redesign of the remote control console (for a TV/DVD) is a classic example of VE. In Figure 3-11, the product on the left looks more like a calculator than a remote control console. VE was used to identify many unnecessary functions that customers did not want and would not pay for. By eliminating these unwanted functions, we were able to cut significant production costs. The redesigned remote control on the right-hand side projects a highly simplified, user-friendly appearance. This design, therefore, enhances use value and esteem value even before considering the effect of production cost savings on the product's selling price.

TABLE 3-3   Value Engineering Checklist

Are all the functions provided required by the customer?	Are all the machined surfaces necessary?
Can less-expensive material be used?	Would product redesign eliminate a quality problem?
Can the number of different materials used be reduced?	Is the current level of packaging necessary?
Can the design be simplified to reduce the number of parts?	Can a part designed for another product be used?

Figure 3-11 Remote Control Consoles

3.6 CASE STUDY—Demanufacturing of Computers

Let's consider a case that deals with a timely important issue concerning environmental economics. What are companies to do with all of the old computers that typically accumulate at their facilities?

As one possible solution to this problem, a number of states and companies have jointly established "resource recovery" facilities to process the handling of old electronic equipment, such as used computers. Demanufacturing of computers involves disassembly, refurbishing, and either donating or reselling the units. However, there are some residuals, or remaining components, that cannot be reused, are harmful to the environment, and contribute to the cost of demanufacturing.

Let's consider the case of one resource recovery center in the Northeast that currently handles approximately 2,000 computers per year. Of these computers, approximately 50% are refurbished and donated, 40% are remanufactured and resold, and 10% are "residuals." What to do with the residuals is a challenging problem.

The facility currently has a contract with an outside source to take the monitors and CPUs. This leads to storage requirements in their facility, and there are issues concerning the determination of the optimal truckload capacity. The labor costs for the outside contractor are $10.00 per unit, and transportation costs are $1.70 per unit, for a total of $11.70 per unit, based on a full capacity truck load.

The recovery facility is seeking an alternative solution to the handling of residuals. One alternative under consideration is to establish a method for demanufacturing of the computers that will consider the CPUs only, since there are environmental issues associated with computer monitors.

The recovery facility recently completed a work measurement study to develop a method for demanufacture of the CPUs. The method is as follows:

1. Remove fasteners or screws.
2. Remove metal casing.

3. Remove power and data cables from floppy drive.
4. Unfasten floppy drive.
5. Unfasten supporting bar.
6. Remove support and floppy drive.
7. Remove power and data cables from power supply.
8. Remove motherboard.
9. Unfasten and remove power supply.
10. Unfasten drive housing.
11. Remove rubber feet.

From the predetermined time study, it is determined that the demanufacturing rate is 7.35 minutes/unit. The facility uses an internal labor rate of $12.00/hour, and it is estimated that training costs for the demanufacturing method will be 10% of total labor costs. Setup costs and transportation costs are estimated at 150% and 20% of total labor costs, respectively. The facility's goal is to demanufacture 200 units during regular operations. Determine the estimated cost per unit for the proposed demanufacturing method. What percentage reduction is this over the per unit cost of using the outside contractor?

Solution

As one possible solution, the industrial engineer performs the following calculations. For 200 units, the labor quantity is estimated as

$$7.35 \text{ minutes/unit} \times 200 \text{ units} = 1,470 \text{ minutes} = 24.5 \text{ hours.}$$

The engineer develops the cost template provided below:

	Demanufacturing Cost Elements	Units	Cost/Unit	Factor	of Row	Row Total
		Unit Elements		Factor Estimates		
A:	Factory labor	24.5 hrs	$12.00/hr			$294.00
B:	Quality costs—Training			10%	A	$29.40
C:	Total Labor					$323.40
D:	Factory overhead—Setup costs			150%	C	$485.10
E:	Transportation cost			20%	C	$64.68
F:	Total Direct Charge					$550.08
G:	Facility rental					—
H:	Total Demanufacturing Cost					$873.48
I:	Quantity—Lot size					200
J:	Demanufacturing cost per unit					$4.37
	Outside cost per unit—Target cost	$ 11.70				

The per unit cost of the proposed internal demanufacturing method is $4.37, while the per unit cost of using the outside contractor (i.e., the target cost) is $11.70. Should the proposed demanufacturing method be adopted, the estimated per unit cost savings is $7.33 for a 62.6% reduction over the per unit cost for the outside contractor.

This case illustrates that developing creative solutions for internal demanufacturing methods not only results in cost savings of approximately 63% from current practices for the case company, but proper demanufacturing methods will minimize the number of computer residuals entering the waste stream. What makes economic sense is also good for the environment!   ————■

3.7   Summary

Developing the cash flow for each alternative in a study is a pivotal, and usually the most difficult, step in the engineering economic analysis procedure. An integrated approach for developing cash flows includes three major components: (1) a WBS definition of the project, (2) a cost and revenue structure that identifies all the cost and revenue elements involved in the study, and (3) estimating techniques (models).

Estimating techniques (models) are used to develop the cash flows for the alternatives as they are defined by the WBS. Thus, the estimating techniques form a bridge between the WBS and detailed cost and revenue data and the estimated cash flows for the alternatives.

The results of the cost-estimating process are a set of cash flows for a proposed engineering project. Since these cash flows will occur at different points in time over the life cycle of the project, Chapter 4 will demonstrate how to account for the time value of money in our analysis. Then, in Chapter 5, we will learn procedures for determining the profitability, or economic feasibility, of the proposed project.

Problems

The number in parentheses that follows each problem refers to the section from which the problem is taken.

3-1. Develop an estimate for the cost of washing and drying a 12-pound load of laundry. Remember to consider all the costs. Your time is worth nothing unless you have an opportunity to use it for making (or saving) money on other activities. (3.2)

3-2. You are planning to build a new home with approximately 2,000–2,500 gross square feet of living space on one floor. In addition, you are planning an attached two-car garage (with storage space) of approximately 450 gross square feet. Develop a cost and revenue structure for designing and constructing, operating (occupying) for 10 years, and then selling the home at the end of the 10th year. (3.2)

3-3. Estimate the cost of an oil change (5 quarts of oil) and a new oil filter for your automobile at a local service station. It takes a technician 20 minutes to do this job. Compare your estimate with the actual cost of an oil change at the service station. What percent markup is being made by the service station? (3.3)

3-4. A large electric utility company spews 62 million tons of greenhouse gases (mostly carbon dioxide) into the environment each year. This company has committed to spending $1.2 billion in capital over the next five years to reduce its annual emissions by 5%. More will be spent after five years to reduce greenhouse gases further. (3.3)

a. What is the implicit cost of a ton of greenhouse gas?

b. If the United States produces 3 billion tons of greenhouse gases per year, how much capital must be spent to reduce total emissions by 3% over the next five years based on your answer in Part (a)?

3-5. The proposed small office building in Example 3-2 has 24,000 net square feet of area heated by a natural gas furnace. The owner of the building wants to know the approximate cost of heating the structure from October through March (six months) because she will lease heated space to the building's occupants. During the heating season, the building will require roughly 60,000 Btu per square foot of heated area. Natural gas has 1,000 Btu per cubic foot, and natural gas costs $9.00 per thousand cubic foot. What will the owner pay to heat her building? (3.3)

3-6. An electric power distributor charges residential customers $0.10 per kilowatt-hour (kWh). The company advertises that "green power" is available in 150 kWh blocks for an additional $4 per month. (Green power is generated from solar, wind power, and methane sources.) (3.3)

a. If a certain customer uses an average of 400 kWh per month and commits to one monthly 150 kWh block of green power, what is her annual power bill?

b. What is the average cost per kWh with green power during the year?

c. Why does green power cost more than conventional power?

3-7. Suppose that your brother-in-law has decided to start a company that produces synthetic lawns for senior homeowners. He anticipates starting production in 18 months. In estimating future cash flows of the company, which of the following items would be relatively easy versus relatively difficult to obtain? Also, suggest how each might be estimated with reasonable accuracy. (3.2)

a. Cost of land for a 10,000-square-foot building.

b. Cost of the building (cinder block construction).

c. Initial working capital.

d. Total capital investment cost.

e. First year's labor and material costs.

f. First year's sales revenues.

3-8. Transportation equipment that was purchased in 2004 for $200,000 must be replaced at the end of 2009. What is the estimated cost of the replacement, based on the following equipment cost index? (3.3)

Year	Index	Year	Index
2004	223	2007	257
2005	238	2008	279
2006	247	2009	293

3-9. Prepare a composite (weighted) index for housing construction costs in 2008, using the following data: (3.3)

Type of Housing	Percent	Reference year ($I = 100$)	2008
Single Units	70	41 ⎫	62 ⎫
Duplex Units	5	38 ⎬ $/ft^2	57 ⎬ $/ft^2
Multiple	25	33 ⎭	53 ⎭

3-10. A microbrewery was built in 2004 at a total cost of $650,000. Additional information is given in the accompanying table (all 2000 indices = 100). (3.3)

Cost Element	Average Percentage of Total Brewery Cost	Index (2004)	Index (2008)
Labor	30	160	200
Materials	20	145	175
Equipment	50	135	162

a. Calculate a weighted index for microbrewery construction in 2008.

b. Prepare a budget estimate for a microbrewery in 2008.

3-11. The purchase price of a natural gas-fired commercial boiler (capacity X) was $181,000 eight years ago. Another boiler of the same basic design, except with capacity $1.42X$, is currently being considered for purchase. If it is purchased, some optional features presently costing $28,000 would be added for your application. If the cost index was 162 for this type of equipment when the capacity X boiler was purchased and is 221 now, and the applicable cost capacity factor is 0.8, what is your estimate of the purchase price for the new boiler? (3.3, 3.4)

3-12. The manufacturer of commercial jets has a cost index equal to 94.9 per aircraft in 2005. The anticipated cost index for the airplane in 2010 is 106.8. The average compound rate of growth should hold steady for the next 15 years. If an aircraft costs $10.2 million to build in 2006, what is its expected cost in 2008? State your assumptions. (3.3)

3-13. Six years ago, an 80-kW diesel electric set cost $160,000. The cost index for this class of equipment six years ago was 187 and is now 194. The cost-capacity factor is 0.6. (3.4)

a. The plant engineering staff is considering a 120-kW unit of the same general design to power a small isolated plant. Assume we want to add a precompressor, which (when isolated and estimated separately) currently costs $18,000. Determine the total cost of the 120-kW unit.

b. Estimate the cost of a 40-kW unit of the same general design. Include the cost of the $18,000 precompressor.

3-14. In a new 55-and-over gated community, light poles and underground cables need to be installed. It is estimated that 25 miles of underground cabling will be needed and that each mile of cabling costs $15,000 (includes labor). A light pole is to be installed every 50 yards, on average, and the cost of the pole and installation is $2,500. What is the estimated cost for the entire job? (3.4)

3-15. In a building construction project, 7,500 feet of insulated ductwork is required. The ductwork is made from 14-gauge steel costing $8.50 per pound. The 24-inch-diameter duct weighs 15 pounds per foot. Insulation for the ductwork costs $10 per foot. Engineering design will cost $16,000, and labor to install the ductwork will amount to $180,000. What is the total cost of the installed ductwork for this project? (3.3)

3-16. Four hundred pounds of copper go into a 2,000-square-foot, newly constructed house. Today's price of copper is $3.50 per pound. If the cost of copper is expected to increase 4.5% per year into the foreseeable future, what is the cost of copper going to be in a new 2,400 square foot house 10 years from now. Assume the cost capacity factor for increases of copper in houses equals 1.0. (3.4)

3-17. The structural engineering design section within the engineering department of a regional electrical utility corporation has developed several standard designs for a group of similar transmission line towers. The detailed design for each tower is based on one of the

standard designs. A transmission line project involving 50 towers has been approved. The estimated number of engineering hours needed to accomplish the first detailed tower design is 126. Assuming a 95% learning curve,

a. What is your estimate of the number of engineering hours needed to design the eighth tower and to design the last tower in the project?

b. What is your estimate of the cumulative average hours required for the first five designs? (3.4)

3-18. The overhead costs for a company are presently $X per month. The management team of the company, in cooperation with the employees, is ready to implement a comprehensive improvement program to reduce these costs. If you (a) consider an observation of actual overhead costs for one month analogous to an output unit, (b) estimate the overhead costs for the first month of program implementation to be 1.15X due to extra front-end effort, and (c) consider a 90% improvement curve applicable to the situation, what is your estimate of the percentage reduction in present overhead costs per month after 30 months of program implementation? (3.4)

3-19. In a learning curve application, 846.2 work-hours are required for the third production unit and 783.0 work-hours are required for the fifth production unit. Determine the value of n (and therefore s) in Equation (3-5). (3.4)

3-20. You have been asked to estimate the cost of 100 prefabricated structures to be sold to a local school district. Each structure provides 1,000 square feet of floor space, with 8-feet ceilings. In 1999, you produced 70 similar structures consisting of the same materials and having the same ceiling height, but each provided only 800 square feet of floor space. The material cost for each structure was $25,000 in 1999, and the cost capacity factor is 0.65. The cost index values for 1999 and 2006 are 200 and 289, respectively. The estimated manufacturing cost for the first 1,000-square-foot structure is $12,000. Assume a learning curve of 88% and use the cost of the 50th structure as your standard time for estimating manufacturing cost. Estimate the total material cost and the total manufacturing cost for the 100 prefabricated structures. (3.3, 3.4)

3-21. The cost of building a supermarket is related to the total area of the building. Data for the last 10 supermarkets built for Regork, Inc., are shown in the accompanying table.

Building	Area (ft²)	Cost ($)
1	14,500	800,000
2	15,000	825,000
3	17,000	875,000
4	18,500	972,000
5	20,400	1,074,000
6	21,000	1,250,000
7	25,000	1,307,000
8	26,750	1,534,000
9	28,000	1,475,500
10	30,000	1,525,000

a. Develop a CER for the construction of supermarkets. Use the CER to estimate the cost of Regork's next store, which has a planned area of 23,000 square feet. (3.4)

b. Compute the standard error and correlation coefficient for the CER developed in Part (a). (3.4)

3-22. In the packaging department of a large aircraft parts distributor, a fairly reliable estimate of packaging and processing costs can be determined by knowing the weight of an order. Thus, the weight is a cost driver that accounts for a sizable fraction of the packaging and processing costs at this company. Data for the past 10 orders are given as follows: (3.4)

Packaging and Processing Costs ($), y	Weight (Pounds), x
97	230
109	280
88	210
86	190
123	320
114	300
112	280
102	260
107	270
86	190

a. Estimate the b_0 and b_1 coefficients, and determine the linear regression equation to fit these data.

b. What is the correlation coefficient (R)?

c. If an order weighs 250 lb, how much should it cost to package and process it?

3-23. Using the costing worksheet provided in this chapter (Figure 3-8), estimate the unit cost and selling price of manufacturing metal wire cutters in lots of 100 when these data have been obtained: (3.5)

Factory (direct) labor:	4.2 hours at $11.15/hour
Factory overhead:	150% of factory labor
Outside manufacture:	$74.87
Production material:	$26.20
Packing costs:	7% of factory labor
Desired profit:	12% of total manufacturing cost

3-24. You have been asked to *estimate the per unit selling price* of a new line of clothing. Pertinent data are as follows:

Direct labor rate:	$15.00 per hour
Production material:	$375 per 100 items
Factory overhead:	125% of direct labor
Packing costs:	75% of direct labor
Desired profit:	20% of total manufacturing cost

Past experience has shown that an 80% learning curve applies to the labor required for producing these items. The time to complete the first item has been estimated to be 1.76 hours. Use the estimated time to complete the 50th item as your standard time for the purpose of estimating the unit selling price. (3.4, 3.5)

3-25. An electronics manufacturing company is planning to introduce a new product in the market. The best competitor sells a similar product at $420/unit. Other pertinent data are as follows:

Direct labor cost:	$15.00/hour
Factory overhead:	120% of direct labor
Production materials:	$300/unit
Packaging costs:	20% of direct labor

It has been found that an 85% learning curve applies to the labor required. The time to complete the first unit has been estimated to be 5.26 hours. The company decides to use the time required to complete the 20th unit as a standard for cost estimation purposes. The profit margin is based on the total manufacturing costs.

a. Using the information given, determine the maximum profit margin that the company can have so as to remain competitive. (3.4, 3.5)

TABLE P3-29	Table for Problem 3-29			
Equipment	Reference Size	Unit Reference Cost	Cost-Capacity Factors	New Design Size
Two boilers	6 MW	$300,000	0.80	10 MW
Two generators	6 MW	400,000	0.60	9 MW
Tank	80,000 gal	106,000	0.66	91,500 gal

b. If the company desires a profit margin of 15%, can the target cost be achieved? If not, suggest two ways in which the target cost can be achieved.

3-26. Given the following information, how many units must be sold to achieve a profit of $25,000? [Note that the units sold must account for total production costs (direct and overhead) plus desired profit.] (3.4, 3.5)

Direct labor hours:	0.2 hour/unit
Direct labor costs:	$21.00/hour
Direct materials cost:	$4.00/unit
Overhead costs:	120% of direct labor
Packaging and shipping:	$1.20/unit
Selling price:	$20.00/unit

3-27. Today a proposed community college is estimated to cost $34.6 million, which is $127 per square foot of space multiplied by 272,310 square feet. If construction costs are expected to increase by 19% per year because of high demand for construction labor and materials, how much would a 320,000-square-foot community college cost five years from now? The cost capacity factor is 1.0. (3.3)

3-28. Your FICO score is a commonly used measure of credit risk (see www.myfico.com). A score of 850 is the best (highest) score possible. Thirty-five percent of the FICO score is based on payment history for credit cards, car loans, home mortgages, and so on. Suppose your current FICO score is 720 and you have just missed a credit card payment due date and have incurred a late-payment fee! If your FICO score will drop 10% in the "payment history" category because of the late payment on your credit card, what is your new FICO score? (3.3)

3-29. A small plant has been constructed and the costs are known. A new plant is to be estimated with the use of the exponential (power sizing) costing model. Major equipment, costs, and factors are as shown in Table P3-29. (Note MW = 10^6 Watts.)

If ancillary equipment will cost an additional $200,000, find the cost for the proposed plant. (3.4)

3-30. If 1,702.5 labor hours are required for the first production unit and 873.0 labor hours are required for the fifth production unit, determine the learning curve parameter s. (3.4)

Spreadsheet Exercises

3-31. Refer to Example 3-7. Construct a graph to show how the time to complete the 10th car changes as the learning curve slope parameter is varied from 75% to 95%. (3.4)

3-32. The Betterbilt Construction Company designs and builds residential mobile homes. The company is ready to construct, in sequence, 16 new homes of 2,400 square feet each. The successful bid for the construction materials in the first home is $64,800, or $27 per square foot. The purchasing manager believes that several actions can be taken to reduce material costs by 8% each time the number of homes constructed doubles. Based on this information,

a. What is the estimated cumulative average material cost per square foot for the first five homes?

b. What is the estimated material cost per square foot for the last (16th) home? (3.4)

3-33. Refer to Example 3-8. While cleaning out an old file, someone uncovers the first spacecraft manufactured by your company—30 years ago! It weighed 100 pounds and cost $600. Extend the spreadsheet to include this data point. How does adding this observation affect R and the standard error? How about the regression coefficients? Should this new data point be included in the model used for predicting future costs? (3.4)

3-34. Refer to the throttle assembly problem described in Section 3.5.1. For a target cost of $25.00, determine a production material cost goal for the throttle assembly. (3.5)

Case Study Exercises

3-35. What other cost factors might you include in such an economic analysis? (3.6)

3-36. What cost factor is the per unit demanufacturing cost most sensitive to and why? (3.6)

3-37. What is the projected impact on the per unit demanufacturing cost of a 50% increase in training costs coupled with a 90% increase in transportation costs? What is the revised cost reduction percentage? (3.6)

FE Practice Problems

3-38. Find the average time per unit required to produce the first 30 units, if the slope parameter of the learning rate is 92% and the first unit takes 460 hours.

(a) $-3.30693E-11$ (b) 305.5404   (c) 245
(d) 347.3211

3-39. A student is considering the purchase of two alternative cars. Car A initially costs $1,500 more than Car B, but uses 0.05 gallons per mile, versus 0.07 gallons per mile for Car B. Both cars will last for 10 years, and B's market value is $800 less than A's. Fuel costs $3.00 per gallon. If all else is equal, at how many miles driven per year does Car A become preferable to Car B?

(a) 1,167   (b) 1,723   (c) 1,892
(d) 2,243

3-40. An automatic process controller will eliminate the current manual control operation. Annual cost of the current method is $4,000. If the controller has a service life of 13 years and an expected market value of 11% of the first cost, what is the maximum economical price for the controller? Ignore interest.

(a) $28,869   (b) $58,426   (c) $26,358
(d) $25,694   (e) $53,344

3-41. A foreman supervises A, B, and eight other employees. The foreman states that he spends twice as much time supervising A and half as much time supervising B, compared with the average time spent supervising his other subordinates. All employees have the same production rate. On the basis of equal cost per unit production, what monthly salary is justified for B if the foreman gets $3,800 per month and A gets $3,000 per month?

(a) $3,543   (b) $3,800   (c) $3,000
(d) $2,457   (e) $3,400

3-42. A car rental agency is considering a modification in its oil change procedure. Currently, it uses a Type X filter, which costs $5 and must be changed every 7,000 miles along with the oil (5 quarts). Between each oil change, one quart of oil must be added after each 1,000 miles. The proposed filter (Type Y) has to be replaced every 5,000 miles (along with 5 quarts of oil) but does not require any additional oil between filter changes. If the oil costs $1.08 per quart, what is the maximum acceptable price for the Type Y filter?

(a) $12.56   (b) $7.43   (c) $11.48
(d) $6.66

CHAPTER 4

The Time Value of Money

The primary focus of Chapter 4 is to explain time value of money calculations and to illustrate economic equivalence.

Avoiding a Catastrophe

In 2007, the Wolf Creek dam in Kentucky was leaking a lot of water through its limestone foundation and was dangerously weakened. The U.S. Corp of Engineers has reduced the water level by 40 feet to relieve pressure on the aging concrete and earthen dam. If the dam were to fail suddenly, thousands of people would be in danger of losing their lives, and property damage along the Cumberland River basin would be extensive. The Corp will need to budget (accumulate) $309 million over five years to make repairs by pumping grout (concrete) into the cavities under and around the earthen part of the dam. In this chapter, you will learn to determine how much money the Corp of Engineers has to budget each month to pay for these repairs as a lump sum in 2012.

If you would know the value of money, go and try to borrow some.

—Benjamin Franklin

4.1 Introduction

The term *capital* refers to wealth in the form of money or property that can be used to produce more wealth. The majority of engineering economy studies involve commitment of capital for extended periods of time, so the effect of time must be considered. In this regard, it is recognized that a dollar today is worth more than a dollar one or more years from now because of the interest (or profit) it can earn. Therefore, money has a *time value*. It has been said that often the riskiest thing a person can do with money is nothing! Money has value, and if money remains uninvested (like in a large bottle), value is lost. Money changes in value not only because of interest rates (Chapter 4)—inflation (or deflation) and currency exchange rates also cause money to change in value. The latter two topics will be discussed in Chapter 8.

4.1.1 Why Consider Return to Capital?

There are fundamental reasons why return to capital in the form of interest and profit is an essential ingredient of engineering economy studies. First, interest and profit pay the providers of capital for forgoing its use during the time the capital is being used. The fact that the supplier can realize a return on capital acts as an *incentive* to accumulate capital by savings, thus postponing immediate consumption in favor of creating wealth in the future. Second, interest and profit are payments for the *risk* the investor takes in permitting another person, or an organization, to use his or her capital.

In typical situations, investors must decide whether the expected return on their capital is sufficient to justify buying into a proposed project or venture. If capital is invested in a project, investors would expect, as a minimum, to receive a return at least equal to the amount they have sacrificed by not using it in some other available opportunity of comparable risk. This interest or profit available from an alternative investment is the *opportunity cost* of using capital in the proposed undertaking. Thus, whether borrowed capital or equity capital is involved, there is a cost for the capital employed in the sense that the project and venture must provide a sufficient return to be financially attractive to suppliers of money or property.

In summary, whenever capital is required in engineering and other business projects and ventures, it is essential that proper consideration be given to its cost (i.e., time value). The remainder of this chapter deals with time value of money principles, which are vitally important to the proper evaluation of engineering projects that form the foundation of a firm's competitiveness, and hence to its very survival.

4.1.2   The Origins of Interest

Like taxes, interest has existed from earliest recorded human history. Records reveal its existence in Babylon in 2000 B.C. In the earliest instances, interest was paid in money for the use of grain or other commodities that were borrowed; it was also paid in the form of grain or other goods. Many existing interest practices stem from early customs in the borrowing and repayment of grain and other crops.

History also reveals that the idea of interest became so well established that a firm of international bankers existed in 575 B.C., with home offices in Babylon. The firm's income was derived from the high interest rates it charged for the use of its money for financing international trade.

Throughout early recorded history, typical annual rates of interest on loans of money were in the neighborhood of 6 to 25%, although legally sanctioned rates as high as 40% were permitted in some instances. The charging of exorbitant interest rates on loans was termed *usury*, and prohibition of usury is found in the Bible. (See *Exodus* 22: 21–27.)

During the Middle Ages, interest taking on loans of money was generally outlawed on scriptural grounds. In 1536, the Protestant theory of usury was established by John Calvin, and it refuted the notion that interest was unlawful. Consequently, interest taking again became viewed as an essential and legal part of doing business. Eventually, published interest tables became available to the public.

4.2   Simple Interest

When the total interest earned or charged is linearly proportional to the initial amount of the loan (principal), the interest rate, and the number of interest periods for which the principal is committed, the interest and interest rate are said to be *simple*. Simple interest is not used frequently in modern commercial practice.

When simple interest is applicable, the total interest, I, earned or paid may be computed using the formula

$$I = (P)(N)(i), \tag{4-1}$$

where P = principal amount lent or borrowed;
N = number of interest periods (e.g., years);
i = interest rate per interest period.

The total amount repaid at the end of N interest periods is $P + I$. Thus, if \$1,000 were loaned for three years at a simple interest rate of 10% per year, the interest earned would be

$$I = \$1,000 \times 3 \times 0.10 = \$300.$$

The total amount owed at the end of three years would be \$1,000 + \$300 = \$1,300. Notice that the cumulative amount of interest owed is a linear function of time until the principal (and interest) is repaid (usually not until the end of period N).

The Importance of Interest in Your Daily Life

In 2005, total debt (credit cards, auto loans, home mortgages, etc.) amounted to *more* than 100% of total disposable income for the average U.S. household. If your total disposable income is $50,000, how much interest can you expect to pay when the average interest rate on your debt is 12% per year?

Answer: You can expect to pay $50,000 (0.12) = $6,000 per year!

4.3   Compound Interest

Whenever the interest charge for any interest *period* (a year, for example) is based on the remaining principal amount plus any accumulated interest charges up to the *beginning* of that period, the interest is said to be *compound*. The effect of compounding of interest can be seen in the following table for $1,000 loaned for three periods at an interest rate of 10% compounded each period:

Period	(1) Amount Owed at Beginning of Period	(2) = (1) × 10% Interest Amount for Period	(3) = (1) + (2) Amount Owed at End of Period
1	$1,000	$100	$1,100
2	$1,100	$110	$1,210
3	$1,210	$121	$1,331

As you can see, a total of $1,331 would be due for repayment at the end of the third period. If the length of a period is one year, the $1,331 at the end of three periods (years) can be compared with the $1,300 given earlier for the same problem with simple interest. A graphical comparison of simple interest and compound interest is given in Figure 4-1. The difference is due to the effect of *compounding,* which is essentially the calculation of interest on previously earned interest. This difference would be much greater for larger amounts of money, higher interest rates, or greater numbers of interest periods. Thus, simple interest does consider the time value of money but does not involve compounding of interest. Compound interest is much more common in practice than simple interest and is used throughout the remainder of this book.

4.4   The Concept of Equivalence

Alternatives should be compared when they produce similar results, serve the same purpose, or accomplish the same function. This is not always possible in some types of economy studies (as we shall see later), but now our attention is directed at answering the question: How can alternatives for providing the same service or accomplishing the same function be compared when interest is involved over extended periods of time? Thus, we should consider the comparison of alternative

Figure 4-1
Illustration of Simple versus Compound Interest

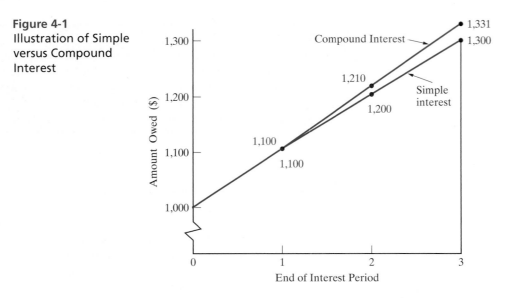

options, or proposals, by reducing them to an *equivalent basis* that is dependent on (1) the interest rate, (2) the amounts of money involved, and (3) the timing of the monetary receipts or expenses.

To better understand the mechanics of interest and to explain the concept of equivalence, suppose you have a $17,000 balance on your credit card. "This has got to stop!" you say to yourself. So you decide to repay the $17,000 debt in four months. An unpaid credit card balance at the beginning of a month will be charged interest at the rate of 1% by your credit card company. For this situation, we have selected three plans to repay the $17,000 principal plus interest owed.* These three plans are illustrated in Table 4-1, and we will demonstrate that they are equivalent (i.e., the same) when the interest rate is 1% per month on the unpaid balance of principal.

Plan 1 indicates that none of the principal is repaid until the end of the fourth month. The monthly payment of interest is $170, and all of the principal is also repaid at the end of month four. Because interest does not accumulate in Plan 1, compounding of interest is not present in this situation. In Table 4-1, there are 68,000 dollar-months of borrowing ($17,000 × 4 months) and $680 total interest. Therefore, the monthly interest rate is ($680 ÷ 68,000 dollar-months) × 100% = 1%.

Plan 2 stipulates that we repay $4,357.10 per month. Later we will show how this number is determined (Section 4.9). For our purposes here, you should observe that interest is being compounded and that the $17,000 principal is completely repaid over the four months. From Table 4-1, you can see that the monthly interest rate is ($427.10 ÷ 42,709.5 dollar-months of borrowing) × 100% = 1%. There are fewer dollar-months of borrowing in Plan 2 (as compared with Plan 1) because principal is being repaid every month and the total amount of interest paid ($427.10) is less.

———————————

* These repayment plans are for demonstration purposes *only*. It is very unlikely that a credit card company would agree to either Plan 1 or Plan 3 without additional charges and/or damaging your credit history.

TABLE 4-1	**Three Plans for Repayment of $17,000 in Four Months with Interest at 1% per Month**				
(1) Month	(2) Amount Owed at Beginning of Month	(3) = 1% × (2) Interest Accrued for Month	(4) = (2) + (3) Total Money Owed at End of Month	(5) Principal Payment	(6) = (3) + (5) Total End-of-Month Payment (Cash Flow)
Plan 1: Pay interest due at end of each month and principal at end of fourth month.					
1	$17,000	$170	$17,170	$0	$170
2	17,000	170	17,170	0	170
3	17,000	170	17,170	0	170
4	17,000	170	17,170	17,000	17,170
	68,000 $-mo.	$680 (total interest)			
Plan 2: Pay off the debt in four equal end-of-month installments (principal and interest).					
1	$17,000	$170	$17,170	$4,187.10	$4,357.10
2	12,812.90	128.13	12,941.03	4,228.97	4,357.10
3	8,583.93	85.84	8,669.77	4,271.26	4,357.10
4	4,312.67	43.13	4,355.80	4,313.97	4,357.10
	42,709.5 $-mo.	$427.10 (total interest)	Difference = $1.30 due to roundoff		
Plan 3*: Pay principal and interest in one payment at end of fourth month.					
1	$17,000	$170	$17,170	$0	$0
2	17,170	171.70	17,341.70	0	0
3	17,341.70	173.42	17,515.12	0	0
4	17,515.12	175.15	17,690.27	17,000	17,690.27
	69,026.8 $-mo.	$690.27 (total interest)			

* Here, column 6 ≠ column 3 + column 5.

Finally, Plan 3 shows that no interest and no principal are repaid in the first three months. Then at the end of month four, a single lump-sum amount of $17,690.27 is repaid. This includes the original principal and the accumulated (compounded) interest of $690.27. The dollar-months of borrowing are very large for Plan 3 (69,026.8) because none of the principal and accumulated interest is repaid until the end of the fourth month. Again, the ratio of total interest paid to dollar-months is 0.01.

This brings us back to the concept of economic equivalence. If the interest rate remains at 1% per month, you should be indifferent as to which plan you use to repay the $17,000 to your credit card company. This assumes that you are charged 1% of the outstanding principal balance (which includes any unpaid interest) each month for the next four months. If interest rates in the economy go up and increase your credit card rate to say, $1\frac{1}{4}$% per month, the plans are no longer equivalent. What varies among the three plans is the rate at which principal is repaid and how interest is repaid.

4.5   Notation and Cash-Flow Diagrams and Tables

The following notation is utilized in formulas for compound interest calculations:

i = effective interest rate per interest period;

N = number of compounding (interest) periods;

P = present sum of money; the *equivalent* value of one or more cash flows at a reference point in time called the present;

F = future sum of money; the *equivalent* value of one or more cash flows at a reference point in time called the future;

A = end-of-period cash flows (or *equivalent* end-of-period values) in a uniform series continuing for a specified number of periods, starting at the end of the first period and continuing through the last period.

The use of cash-flow (time) diagrams or tables is strongly recommended for situations in which the analyst needs to clarify or visualize what is involved when flows of money occur at various times. In addition, viewpoint (remember Principle 3?) is an essential feature of cash-flow diagrams.

The difference between total cash inflows (receipts) and cash outflows (expenditures) for a specified period of time (e.g., one year) is the net cash-flow for the period. As discussed in Chapters 2 and 3, cash-flows are important in engineering economy because they form the basis for evaluating alternatives. Indeed, the usefulness of a cash-flow diagram for economic analysis problems is analogous to that of the free-body diagram for mechanics problems.

Figure 4-2 shows a cash-flow diagram for Plan 3 of Table 4-1, and Figure 4-3 depicts the net cash flows of Plan 2. These two figures also illustrate the definition of the preceding symbols and their placement on a cash-flow diagram. Notice that all cash flows have been placed at the end of the month to correspond with the convention used in Table 4-1. In addition, a viewpoint has been specified.

The cash-flow diagram employs several conventions:

1. The horizontal line is a *time scale*, with progression of time moving from left to right. The period (e.g., year, quarter, month) labels can be applied to intervals of time rather than to points on the time scale. Note, for example, that the end of Period 2 is coincident with the beginning of Period 3. When the end-of-period cash-flow convention is used, period numbers are placed at the end of each time interval, as illustrated in Figures 4-2 and 4-3.

Figure 4-2   Cash-Flow Diagram for Plan 3 of Table 4-1 (Credit Card Company's Viewpoint)

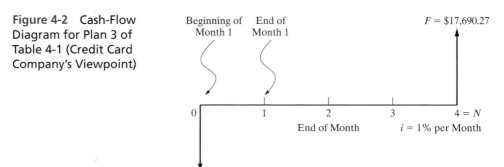

Figure 4-3   Cash-Flow
Diagram for Plan 2 of Table
4-1 (Lender's Viewpoint)

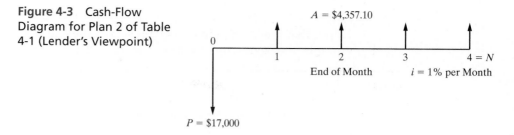

2. The arrows signify cash flows and are placed at the end of the period. If a distinction needs to be made, downward arrows represent expenses (negative cash flows or cash outflows) and upward arrows represent receipts (positive cash flows or cash inflows).

3. The cash-flow diagram is dependent on the point of view. For example, the situations shown in Figures 4-2 and 4-3 were based on cash flow as seen by the lender (the credit card company). If the directions of all arrows had been reversed, the problem would have been diagrammed from the borrower's viewpoint.

EXAMPLE 4-1   Cash-Flow Diagramming

Before evaluating the economic merits of a proposed investment, the XYZ Corporation insists that its engineers develop a cash-flow diagram of the proposal. An investment of $10,000 can be made that will produce uniform annual revenue of $5,310 for five years and then have a market (recovery) value of $2,000 at the end of year (EOY) five. Annual expenses will be $3,000 at the end of each year for operating and maintaining the project. Draw a cash-flow diagram for the five-year life of the project. Use the corporation's viewpoint.

Solution

As shown in the figure below, the initial investment of $10,000 and annual expenses of $3,000 are cash outflows, while annual revenues and the market value are cash inflows.

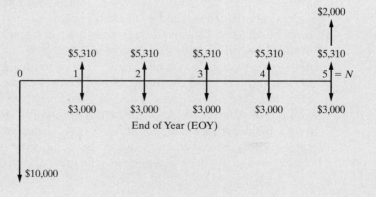

Notice that the beginning of a given year is the end of the preceding year. For example, the beginning of year two is the end of year one.

Example 4-2 presents a situation in which cash flows are represented in tabular form to facilitate the analysis of plans and designs.

EXAMPLE 4-2	Developing a Net Cash-Flow Table

In a company's renovation of a small office building, two feasible alternatives for upgrading the heating, ventilation, and air conditioning (HVAC) system have been identified. Either Alternative *A or* Alternative *B* must be implemented. The costs are as follows:

Alternative *A*   *Rebuild (overhaul) the existing HVAC system*
- Equipment, labor, and materials to rebuild $18,000
- Annual cost of electricity   32,000
- Annual maintenance expenses   2,400

Alternative *B*   *Install a new HVAC system that utilizes existing ductwork*
- Equipment, labor, and materials to install $60,000
- Annual cost of electricity   9,000
- Annual maintenance expenses   16,000
- Replacement of a major component four years hence . .   9,400

At the end of eight years, the estimated market value for Alternative *A* is $2,000 and for Alternative *B* it is $8,000. Assume that both alternatives will provide comparable service (comfort) over an eight-year period, and assume that the major component replaced in Alternative *B* will have no market value at EOY eight. (1) Use a cash-flow table and end-of-year convention to tabulate the net cash flows for both alternatives. (2) Determine the annual net cash-flow difference between the alternatives ($B - A$).

Solution

The cash-flow table (company's viewpoint) for this example was developed by using a spreadsheet and is shown in Figure 4-4. On the basis of these results, several points can be made: (1) Doing nothing is not an option—either A or B must be selected; (2) even though positive and negative cash flows are included in the table, on balance we are investigating two "cost-only" alternatives; (3) a decision between the two alternatives can be made just as easily on the *difference* in cash flows (i.e., on the avoidable difference) as it can on the stand-alone net cash flows for Alternatives *A* and *B*; (4) Alternative *B* has cash flows identical to those of Alternative *A*, *except for* the differences shown in the table, so if the avoidable difference can "pay its own way," Alternative *B* is the recommended choice; (5) cash-flow changes caused by inflation or other suspected influences could have easily been inserted into the table and included in the analysis; and

	= − 25000 − 9400		= C3 − B3	= SUM(D$3:D3)

	A	B	C	D	E
1		Alternative A	Alternative B	Difference	Cumulative
2	End of Year	Net Cash Flow	Net Cash Flow	(B-A)	Difference
3	0 (now)	$ (18,000)	$ (60,000)	$ (42,000)	$ (42,000)
4	1	$ (34,400)	$ (25,000)	$ 9,400	$ (32,600)
5	2	$ (34,400)	$ (25,000)	$ 9,400	$ (23,200)
6	3	$ (34,400)	$ (25,000)	$ 9,400	$ (13,800)
7	4	$ (34,400)	$ (34,400)	$ -	$ (13,800)
8	5	$ (34,400)	$ (25,000)	$ 9,400	$ (4,400)
9	6	$ (34,400)	$ (25,000)	$ 9,400	$ 5,000
10	7	$ (34,400)	$ (25,000)	$ 9,400	$ 14,400
11	8	$ (32,400)	$ (17,000)	$ 15,400	$ 29,800
12	Total	$ (291,200)	$ (261,400)		

= − 34400 + 2000		= − 25000 + 8000

= SUM(B3:B11)

Figure 4-4   Cash-Flow Table, Example 4-2

(6) it takes six *years* for the extra $42,000 investment in Alternative B to generate sufficient cumulative savings in annual expenses to justify the higher investment. (This ignores the time value of money.) So, which alternative is better? We'll be able to answer this question later when we consider the time value of money in order to recommend choices between alternatives.

Comment

Cash-flow tables are invaluable when using a spreadsheet to model engineering economy problems.

It should be apparent that a cash-flow table clarifies the timing of cash flows, the assumptions that are being made, and the data that are available. A cash-flow table is often useful when the complexity of a situation makes it difficult to show all cash-flow amounts on a diagram.

The remainder of Chapter 4 deals with the development and illustration of equivalence (time value of money) principles for assessing the economic attractiveness of investments, such as those proposed in Examples 4-1 and 4-2.

Viewpoint: In most examples presented in this chapter, the company's (investor's) viewpoint will be taken.

4.6   Relating Present and Future Equivalent Values of Single Cash Flows

Figure 4-5 shows a cash-flow diagram involving a present single sum, P, and a future single sum, F, separated by N periods with interest at $i\%$ per period. Throughout this chapter, a *dashed arrow*, such as that shown in Figure 4-5, indicates the quantity to be determined.

4.6.1   Finding *F* when Given *P*

If an amount of P dollars is invested at a point in time and $i\%$ is the interest (profit or growth) rate per period, the amount will grow to a future amount of $P + Pi = P(1 + i)$ by the end of one period; by the end of two periods, the amount will grow to $P(1 + i)(1 + i) = P(1 + i)^2$; by the end of three periods, the amount will grow to $P(1 + i)^2(1 + i) = P(1 + i)^3$; and by the end of N periods the amount will grow to

$$F = P(1 + i)^N. \tag{4-2}$$

EXAMPLE 4-3	Future Equivalent of a Present Sum

Suppose that you borrow $8,000 now, promising to repay the loan principal plus accumulated interest in four years at $i = 10\%$ per year. How much would you repay at the end of four years?

Solution

Year	Amount Owed at Start of Year	Interest Owed for Each Year	Amount Owed at End of Year	Total End-of-Year Payment
1	P = $ 8,000	iP = $ 800	$P(1 + i)$ = $ 8,800	0
2	$P(1 + i)$ = $ 8,800	$iP(1 + i)$ = $ 880	$P(1 + i)^2$ = $ 9,680	0
3	$P(1 + i)^2$ = $ 9,680	$iP(1 + i)^2$ = $ 968	$P(1 + i)^3$ = $10,648	0
4	$P(1 + i)^3$ = $10,648	$iP(1 + i)^3$ = $1,065	$P(1 + i)^4$ = $11,713	F = $11,713

In general, we see that $F = P(1 + i)^N$, and the total amount to be repaid is $11,713.

The quantity $(1 + i)^N$ in Equation (4-2) is commonly called the *single payment compound amount factor*. Numerical values for this factor are given in the second column from the left in the tables of Appendix C for a wide range of values of i and N. In this book, we shall use the functional symbol $(F/P, i\%, N)$ for $(1 + i)^N$. Hence, Equation (4-2) can be expressed as

$$F = P(F/P, i\%, N), \tag{4-3}$$

where the factor in parentheses is read "find F given P at $i\%$ interest per period for N interest periods." Note that the sequence of F and P in F/P is the same as in

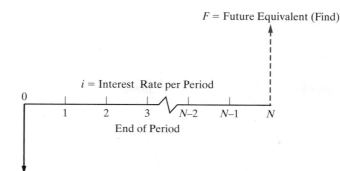

Figure 4-5 General Cash-Flow Diagram Relating Present Equivalent and Future Equivalent of Single Payments

the initial part of Equation (4-3), where the unknown quantity, F, is placed on the left-hand side of the equation. This sequencing of letters is true of all functional symbols used in this book and makes them easy to remember.

Let's look at Example 4-3 again. Using Equation (4-3) and Appendix C, we have

$$F = \$8{,}000(F/P, 10\%, 4)$$
$$= \$8{,}000(1.4641)$$
$$= \$11{,}713.$$

This, of course, is the same result obtained in Example 4-3 since $(F/P,\ 10\%,\ 4) = (1 + 0.10)^4 = 1.4641$.

Another example of finding F when given P, together with a cash-flow diagram and solution, appears in Table 4-2. Note in Table 4-2 that, for each of the six common discrete compound interest circumstances covered, two problem statements are given—(1) *in borrowing–lending terminology* and (2) *in equivalence terminology*—but they both represent the same cash-flow situation. Indeed, there are generally many ways in which a given cash-flow situation can be expressed.

In general, a good way to interpret a relationship such as Equation (4-3) is that the calculated amount, F, at the point in time at which it occurs, *is equivalent to* (i.e., can be traded for) the known value, P, at the point in time at which it occurs, for the given interest or profit rate, i.

4.6.2   Finding *P* when Given *F*

From Equation (4-2), $F = P(1 + i)^N$. Solving this for P gives the relationship

$$P = F\left(\frac{1}{1+i}\right)^N = F(1 + i)^{-N}. \tag{4-4}$$

The quantity $(1 + i)^{-N}$ is called the *single payment present worth factor*. Numerical values for this factor are given in the third column of the tables in Appendix C for

TABLE 4-2 Discrete Cash-Flow Examples Illustrating Equivalence

Example Problems (All Using an Interest Rate of $i = 10\%$ Per Year—See Table C-13 of Appendix C)

To Find:	Given:	(a) In Borrowing– Lending Terminology:	(b) In Equivalence Terminology:	Cash-Flow Diagram[a]	Solution
For single cash flows:					
F	P	A firm borrows $1,000 for eight years. How much must it repay in a lump sum at the end of the eighth year?	What is the future equivalent at the end of eight years of $1,000 at the beginning of those eight years?	$P = \$1,000$ $N = 8$ 0 $F = ?$	$F = P(F/P, 10\%, 8)$ $= \$1,000(2.1436)$ $= \$2,143.60$
P	F	A firm wishes to have $2,143.60 eight years from now. What amount should be deposited now to provide for it?	What is the present equivalent of $2,143.60 received eight years from now?	$F = \$2,143.60$ 0 $N = 8$ $P = ?$	$P = F(P/F, 10\%, 8)$ $= \$2,143.60(0.4665)$ $= \$1,000.00$
For uniform series:					
F	A	If eight annual deposits of $187.45 each are placed in an account, how much money has accumulated immediately after the last deposit?	What amount at the end of the eighth year is equivalent to eight EOY payments of $187.45 each?	$F = ?$ $1\,2\,3\,4\,5\,6\,7\,8$ $A = \$187.45$	$F = A(F/A, 10\%, 8)$ $= \$187.45(11.4359)$ $= \$2,143.60$

TABLE 4-2 (Continued)

Find	Given		Diagram	
P	A	How much should be deposited in a fund now to provide for eight EOY withdrawals of $187.45 each?	What is the present equivalent of eight EOY payments of $187.45 each?	$P = A(P/A, 10\%, 8)$ $= \$187.45(5.3349)$ $= \$1,000.00$
A	F	What uniform annual amount should be deposited each year in order to accumulate $2,143.60 at the time of the eighth annual deposit?	What uniform payment at the end of eight successive years is equivalent to $2,143.60 at the end of the eighth year?	$A = F(A/F, 10\%, 8)$ $= \$2,143.60(0.0874)$ $= \$187.45$
A	P	What is the size of eight equal annual payments to repay a loan of $1,000? The first payment is due one year after receiving the loan.	What uniform payment at the end of eight successive years is equivalent to $1,000 at the beginning of the first year?	$A = P(A/P, 10\%, 8)$ $= \$1,000(0.18745)$ $= \$187.45$

[a] The cash-flow diagram represents the example as stated in borrowing-lending terminology.

a wide range of values of i and N. We shall use the functional symbol $(P/F, i\%, N)$ for this factor. Hence,

$$P = F(P/F, i\%, N). \tag{4-5}$$

EXAMPLE 4-4   Present Equivalent of a Future Amount of Money

An investor (owner) has an option to purchase a tract of land that will be worth $10,000 in six years. If the value of the land increases at 8% each year, how much should the investor be willing to pay now for this property?

Solution

The purchase price can be determined from Equation (4-5) and Table C-11 in Appendix C as follows:

$$P = \$10,000(P/F, 8\%, 6)$$
$$P = \$10,000(0.6302)$$
$$= \$6,302.$$

Another example of this type of problem, together with a cash-flow diagram and solution, is given in Table 4-2.

Based on Equations (4-2) and (4-4), the following three simple rules apply when performing arithmetic calculations with cash flows:

Rule A. Cash flows cannot be added or subtracted unless they occur at the same point in time.

Rule B. To move a cash flow forward in time by one time unit, multiply the magnitude of the cash flow by $(1 + i)$, where i is the interest rate that reflects the time value of money.

Rule C. To move a cash flow backward in time by one time unit, divide the magnitude of the cash flow by $(1 + i)$.

4.6.3   Finding the Interest Rate Given *P*, *F*, and *N*

There are situations in which we know two sums of money (P and F) and how much time separates them (N), but we don't know the interest rate (i) that makes them equivalent. For example, if we want to turn $500 into $1,000 over a period of 10 years, at what interest rate would we have to invest it? We can easily solve Equation (4-2) to obtain an expression for i.

$$i = \sqrt[N]{F/P} - 1 \tag{4-6}$$

So, for our simple example, $i = \sqrt[10]{\$1,000/\$500} - 1 = 0.0718$ or 7.18% per year.

Inflation is another example of when it may be necessary to solve for an interest rate. Suppose you are interested in determining the annual rate of increase in the price of gasoline. Given the average prices in different years, you can use the relationship between P and F to solve for the interest rate.

| EXAMPLE 4-5 | The Inflating Price of Gasoline |

The average price of gasoline in 2005 was $2.31 per gallon. In 1993, the average price was $1.07.* What was the average annual rate of increase in the price of gasoline over this 12-year period?

Solution

With respect to the year 1993, the year 2005 is in the future. Thus, $P = \$1.07$, $F = \$2.31$, and $N = 12$. Using Equation (4-6), we find $i = \sqrt[12]{2.31/1.07} - 1 = 0.0662$ or 6.62% per year.

---

* This data was obtained from the Energy Information Administration of the Department of Energy. Historical prices of gasoline and other energy sources can be found at www.eia.doe.gov.

4.6.4 Finding *N* when Given *P*, *F*, and *i*

Sometimes we are interested in finding the amount of time needed for a present sum to grow into a future sum at a specified interest rate. For example, how long would it take for $500 invested today at 15% interest per year to be worth $1,000? We can use the equivalence relationship given in Equation (4-2) to obtain an expression for N.

$$F = P(1 + i)^N$$

$$(1 + i)^N = (F/P)$$

Using logarithms,

$$N \log(1 + i) = \log(F/P)$$

and

$$N = \frac{\log(F/P)}{\log(1 + i)}. \tag{4-7}$$

For our simple example, $N = \log(\$1{,}000/\$500)/\log(1.15) = 4.96 \cong 5$ years.

| EXAMPLE 4-6 | When Will Gasoline Cost $5.00 per Gallon? |

In Example 4-5, the average price of gasoline was given as $2.31 in 2005. We computed the average annual rate of increase in the price of gasoline to be 6.62%. If we assume that the price of gasoline will continue to inflate at this rate, how long will it be before we are paying $5.00 per gallon?

Solution

We have $P = \$2.31$, $F = \$5.00$, and $i = 6.62\%$ per year. Using Equation (4-7), we find

$$N = \frac{\log(\$5.00/\$2.31)}{\log(1 + 0.0662)} = \frac{\log(2.1645)}{\log(1.0662)} = 12.05 \text{ years.}$$

So, if gasoline prices continue to increase at the same rate, we can expect to be paying $5.00 per gallon in 2017.

4.7 Relating a Uniform Series (Annuity) to Its Present and Future Equivalent Values

Figure 4-6 shows a general cash-flow diagram involving a series of uniform (equal) receipts, each of amount A, occurring at the end of each period for N periods with interest at $i\%$ per period. Such a uniform series is often called an *annuity*. It should be noted that the formulas and tables to be presented are derived such that A occurs at the end of each period, and thus,

1. P (present equivalent value) occurs one interest period before the first A (uniform amount),
2. F (future equivalent value) occurs at the same time as the last A, and N periods after P, and
3. A (annual equivalent value) occurs at the end of periods 1 through N, inclusive.

The timing relationship for P, A, and F can be observed in Figure 4-6. Four formulas relating A to F and P will be developed.

4.7.1 Finding *F* when Given *A*

If a cash flow in the amount of A dollars occurs at the end of each period for N periods and $i\%$ is the interest (profit or growth) rate per period, the future equivalent value, F, at the end of the Nth period is obtained by summing the future equivalents

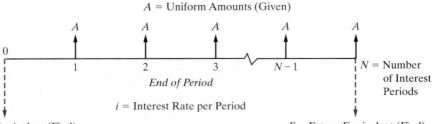

Figure 4-6 General Cash-Flow Diagram Relating Uniform Series (Ordinary Annuity) to Its Present Equivalent and Future Equivalent Values

of each of the cash flows. Thus,

$$F = A(F/P, i\%, N-1) + A(F/P, i\%, N-2) + A(F/P, i\%, N-3) + \cdots$$

$$+ A(F/P, i\%, 1) + A(F/P, i\%, 0)$$

$$= A[(1+i)^{N-1} + (1+i)^{N-2} + (1+i)^{N-3} + \cdots + (1+i)^1 + (1+i)^0].$$

The bracketed terms comprise a geometric sequence having a common ratio of $(1+i)^{-1}$. Recall that the sum of the first N terms of a geometric sequence is

$$S_N = \frac{a_1 - b a_N}{1 - b} \quad (b \neq 1),$$

where a_1 is the first term in the sequence, a_N is the last term, and b is the common ratio. If we let $b = (1+i)^{-1}$, $a_1 = (1+i)^{N-1}$, and $a_N = (1+i)^0$, then

$$F = A \left[\frac{(1+i)^{N-1} - \dfrac{1}{(1+i)}}{1 - \dfrac{1}{(1+i)}} \right],$$

which reduces to

$$F = A \left[\frac{(1+i)^N - 1}{i} \right]. \tag{4-8}$$

The quantity $\{[(1+i)^N - 1]/i\}$ is called the *uniform series compound amount factor*. It is the starting point for developing the remaining three uniform series interest factors.

Numerical values for the uniform series compound amount factor are given in the fourth column of the tables in Appendix C for a wide range of values of i and N. We shall use the functional symbol $(F/A, i\%, N)$ for this factor. Hence, Equation (4-8) can be expressed as

$$F = A(F/A, i\%, N). \tag{4-9}$$

Examples of this type of "wealth accumulation" problem based on the $(F/A, i\%, N)$ factor are provided here and in Table 4-2.

EXAMPLE 4-7	**Future Value of a College Degree**

A recent government study reported that a college degree is worth an extra $23,000 per year in income (A) compared to what a high-school graduate makes. If the interest rate (i) is 6% per year and you work for 40 years (N), what is the future compound amount (F) of this extra income?

Solution

The viewpoint we will use to solve this problem is that of "lending" the $23,000 of extra annual income to a savings account (or some other investment vehicle). The future equivalent is the amount that can be withdrawn after the 40th deposit is made.

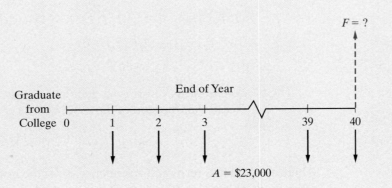

Notice that the future equivalent occurs at the *same time* as the last deposit of $23,000.

$$F = \$23,000(F/A, 6\%, 40)$$

$$= \$23,000(154.762)$$

$$= \$3,559,526$$

The bottom line is "Get your college degree!"

| EXAMPLE 4-8 | **Become a Millionaire by Saving $1.00 a Day!** |

To illustrate further the amazing effects of compound interest, we consider the credibility of this statement: "If you are 20 years of age and save $1.00 each day for the rest of your life, you can become a millionaire." Let's assume that you live to age 80 and that the annual interest rate is 10% ($i = 10\%$). Under these specific conditions, we compute the future compound amount (F) to be

$$F = \$365/\text{year} \,(F/A, 10\%, 60 \text{ years})$$

$$= \$365 \,(3{,}034.81)$$

$$= \$1{,}107{,}706.$$

Thus, the statement is true for the assumptions given! The moral is to *start saving early* and let the "magic" of compounding work on your behalf!

A few words to the wise: Saving money early and preserving resources through frugality (avoiding waste) are extremely important ingredients of *wealth creation* in general. Often, being frugal means postponing the satisfaction of immediate material wants for the creation of a better tomorrow. In this regard, be very *cautious* about spending tomorrow's cash today by undisciplined borrowing (e.g., with credit cards). The (F/A, $i\%$, N) factor also demonstrates how *fast* your debt can accumulate!

4.7.2 Finding P when Given A

From Equation (4-2), $F = P(1 + i)^N$. Substituting for F in Equation (4-8) we determine that

$$P(1 + i)^N = A \left[\frac{(1 + i)^N - 1}{i} \right].$$

Dividing both sides by $(1 + i)^N$, we get

$$P = A \left[\frac{(1 + i)^N - 1}{i(1 + i)^N} \right]. \qquad (4\text{-}10)$$

Thus, Equation (4-10) is the relation for finding the present equivalent value (as of the beginning of the first period) of a uniform series of end-of-period cash flows of amount A for N periods. The quantity in brackets is called the *uniform series present worth factor*. Numerical values for this factor are given in the fifth column of the tables in Appendix C for a wide range of values of i and N. We shall use the functional symbol $(P/A, i\%, N)$ for this factor. Hence,

$$P = A(P/A, i\%, N). \qquad (4\text{-}11)$$

EXAMPLE 4-9 **Present Equivalent of an Annuity (Uniform Series)**

If a certain machine undergoes a major overhaul now, its output can be increased by 20%—which translates into additional cash flow of $20,000 at the end of each year for five years. If $i = 15\%$ per year, how much can we afford to invest to overhaul this machine?

Solution

In the cash-flow diagram below, notice that the present equivalent, P, occurs one time period (year) *before* the first cash flow of $20,000.

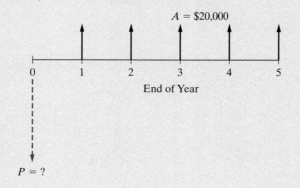

The increase in cash flow is $20,000 per year, and it continues for five years at 15% annual interest. The upper limit on what we can afford to spend now is

$$P = \$20,000(P/A, 15\%, 5)$$

$$= \$20,000(3.3522)$$

$$= \$67,044.$$

EXAMPLE 4-10   **How Much Is a Lifetime Oil Change Offer Worth?**

"Make your best deal with us on a new automobile and we'll change your oil for free for as long as you own the car!" If you purchase a car from this dealership, you expect to have four free oil changes per year during the five years you keep the car. Each oil change would normally cost you $30. If you save your money in a mutual fund earning 2% per quarter, how much are the oil changes worth to you at the time you buy the car?

Solution

In this example, we need to find the present equivalent of the cost of future oil changes. The cash-flow diagram is shown below. Notice that P occurs one time period (a quarter of a year, in this example) before the first oil change cash flow (A).

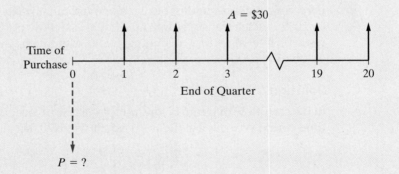

The interest rate is 2% per quarter, and a total of (4 oil changes/year × 5 years) = 20 oil changes (cash flows) are anticipated.

$$P = \$30(P/A, 2\%, 20)$$

$$= \$30(16.3514)$$

$$= \$490.54$$

Now you are in a position to determine how great of a deal you are being offered. If the best price of another dealership is more than $490.54 cheaper than what you are being offered at this dealership, maybe this deal isn't so great.

4.7.3   Finding *A* when Given *F*

Taking Equation (4-8) and solving for *A*, we find that

$$A = F\left[\frac{i}{(1+i)^N - 1}\right]. \tag{4-12}$$

Thus, Equation (4-12) is the relation for finding the amount, *A*, of a uniform series of cash flows occurring at the end of *N* interest periods that would be equivalent to (have the same value as) its future value occurring at the end of the last period. The quantity in brackets is called the *sinking fund factor*. Numerical values for this factor are given in the sixth column of the tables in Appendix C for a wide range of values of *i* and *N*. We shall use the functional symbol $(A/F, i\%, N)$ for this factor. Hence,

$$A = F(A/F, i\%, N). \tag{4-13}$$

The Wolf Creek dam situation described at the beginning of this chapter is an example of a sinking fund. Using Equation (4-13) we can determine the monthly amount that has to be placed into a sinking fund by the Corp of Engineers. The amount to be accumulated after 60 monthly deposits is $309 million, and the opportunity cost of capital is 0.5% per month. The cash-flow diagram for this situation is shown below.

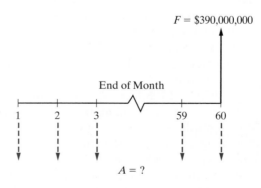

Note that the accumulated amount, *F*, occurs at the same time as the last payment into the fund. The monthly sinking fund amount (see Table C-2) is

$$A = \$309 \text{ million } (A/F, 0.5\% \text{ per month, } 60 \text{ months})$$

$$= \$309 \text{ million } (0.0143)$$

$$= \$4.4187 \text{ million per month.}$$

Notice that ($4.4187 million per month × 60 months) = $265.122 million is less than $309 million because interest is being earned on deposits made to the sinking fund during the 60-month construction period.

Another example of this type of problem, together with a cash-flow diagram and solution, is given in Table 4-2.

4.7.4   Finding *A* when Given *P*

Taking Equation (4-10) and solving for *A*, we find that

$$A = P \left[\frac{i(1+i)^N}{(1+i)^N - 1} \right]. \tag{4-14}$$

Thus, Equation (4-14) is the relation for finding the amount, *A*, of a uniform series of cash flows occurring at the end of each of *N* interest periods that would be equivalent to, or could be traded for, the present equivalent *P*, occurring at the beginning of the first period. The quantity in brackets is called the *capital recovery factor*.* Numerical values for this factor are given in the seventh column of the tables in Appendix C for a wide range of values of *i* and *N*. We shall use the functional symbol $(A/P, i\%, N)$ for this factor. Hence,

$$A = P(A/P, i\%, N). \tag{4-15}$$

An example that uses the equivalence between a present lump-sum loan amount and a series of equal uniform monthly payments starting at the end of month one and continuing through month four was provided in Table 4-1 as Plan 2. Equation (4-15) yields the equivalent value of *A* that repays the $17,000 loan plus 1% interest per month over four months:

$$A = \$17{,}000(A/P, 1\%, 4) = \$17{,}000(0.2563) = \$4{,}357.10$$

The entries in columns three and five of Plan 2 in Table 4-1 can now be better understood. Interest owed at the end of month one equals $17,000(0.01), and therefore the principal repaid out of the total end-of-month payment of $4,357.10 is the difference, $4,187.10. At the beginning of month two, the amount of principal owed is $17,000 − $4,187.10 = $12,812.90. Interest owed at the end of month two is $12,812.90(0.01) = $128.13, and the principal repaid at that time is $4,357.10 − $128.13 = $4,228.97. The remaining entries in Plan 2 are obtained by performing these calculations for months three and four.

A graphical summary of Plan 2 is given in Figure 4-7. Here it can be seen that 1% interest is being paid on the beginning-of-month amount owed and that month-end payments of $4,357.10, consisting of interest and principal, bring the amount owed to $0 at the end of the fourth month. (The exact value of *A* is $4,356.78 and produces an exact value of $0 at the end of four months.) It is important to note that all the uniform series interest factors in Table 4-2 involve the same concept as the one illustrated in Figure 4-7.

---

* The capital recovery factor is more conveniently expressed as $i/[1 - (1+i)^{-N}]$ for computation with a hand-held calculator.

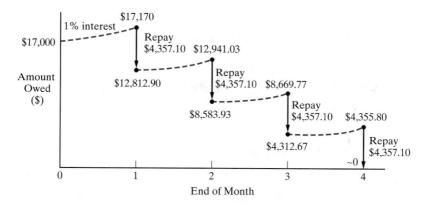

Figure 4-7 Relationship of Cash Flows for Plan 2 of Table 4-1 to Repayment of the $17,000 Loan Principal

---

EXAMPLE 4-11	**Computing Your Monthly Car Payment**

You borrow $15,000 from your credit union to purchase a used car. The interest rate on your loan is 0.25% per month* and you will make a total of 36 monthly payments. What is your monthly payment?

Solution

The cash-flow diagram shown below is drawn from the viewpoint of the bank. Notice that the present amount of $15,000 occurs one month (interest period) *before* the first cash flow of the uniform repayment series.

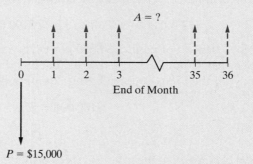

The amount of the car payment is easily calculated using Equation (4-15).

$$A = \$15,000(A/P, 1/4\%, 36)$$

$$= \$15,000(0.0291)$$

$$= \$436.50 \text{ per month}$$

---

* A good credit score (rating) can help you secure lower interest rates on loans. The Web site www. annualcreditreport.com allows you to check your credit score once per year at no cost.

Another example of a problem where we desire to compute an equivalent value for A, from a given value of P and a known interest rate and number of compounding periods, is given in Table 4-2.

For an annual interest rate of 10%, the reader should now be convinced from Table 4-2 that $1,000 at the beginning of year one is equivalent to $187.45 at the EOYs one through eight, which is then equivalent to $2,143.60 at EOY eight.

4.7.5 Finding the Number of Cash Flows in an Annuity Given *A*, *P*, and *i*

Sometimes we may have information about a present amount of money (P), the magnitude of an annuity (A), and the interest rate (i). The unknown factor in this case is the number of cash flows in the annuity (N). Although there is not a closed-form equation for finding N, we can use the relationship between P and A to determine N.

EXAMPLE 4-12 | **Prepaying a Loan—Finding *N***

Your company has a $100,000 loan for a new security system it just bought. The annual payment is $8,880 and the interest rate is 8% per year for 30 years. Your company decides that it can afford to pay $10,000 per year. After how many payments (years) will the loan be paid off?

Solution

The original loan payment was found using Equation (4-15).

$$A = \$100{,}000 \ (A/P, 8\%, 30) = \$100{,}000 \ (0.0888) = \$8{,}800 \text{ per year}$$

Now, instead of paying $8,880 per year, your company is going to pay $10,000 per year. Common sense tells us that less than 30 payments will be necessary to pay off the $100,000 loan. Using Equation (4-11), we find

$$\$100{,}000 = \$10{,}000 \ (P/A, 8\%, N)$$

$$(P/A, 8\%, N) = 10.$$

We can now use the interest tables provided in Appendix C to find N. Looking down the Present Worth Factor column (P/A) of Table C-11, we see that

$$(P/A, 8\%, 20) = 9.8181$$

and

$$(P/A, 8\%, 21) = 10.0168.$$

So, if $10,000 is paid per year, the loan will be paid off after 21 years instead of 30. The exact amount of the 21st payment will be slightly less than $10,000 (but we'll save that solution for another example).

Spreadsheet Solution

There is a financial function in Excel that would allow us to solve for the unknown number of periods. NPER(*rate, pmt, pv*) will compute the number of payments of magnitude *pmt* required to pay off a present amount (*pv*) at a fixed interest rate (*rate*).

$$N = \text{NPER}(0.08, -10000, 100000) = 20.91$$

Note that from your company's viewpoint, it received $100,000 (a cash inflow) at time 0 and is making $10,000 payments (cash outflows). Hence the annuity is expressed as a negative number in NPER( ) and the present amount as a positive number. If we were to reverse the signs—which would represent the lender's viewpoint—the same result would be obtained, namely NPER(0.08, 10000, −100000) = 20.91.

Comment

Prepaying loans can save thousands of dollars in interest. For example, look at the total interest paid under these two repayment plans.

Original payment plan ($8,880 per year for 30 years):

$$\text{Total interest paid} = \$8{,}880 \times 30 - \$100{,}000 = \$166{,}400$$

New payment plan ($10,000 per year for 21 years):

$$\text{Total interest paid} = \$10{,}000 \times 21 - \$100{,}000 = \$110{,}000$$

Prepaying the loan in this way would save over $56,000 in interest!

4.7.6   Finding the Interest Rate, *i*, Given *A*, *F*, and *N*

Now let's look at the situation in which you know the amount (*A*) and duration (*N*) of a uniform payment series. You also know the desired future value of the series (*F*). What you don't know is the interest rate that makes them equivalent. As was the case for an unknown *N*, there is no single equation to determine *i*. However, we can use the known relationships between *i*, *A*, *F*, and *N* and the method of linear interpolation to approximate the interest rate.

EXAMPLE 4-13   Finding the Interest Rate to Meet an Investment Goal

After years of being a poor, debt-encumbered college student, you decide that you want to pay for your dream car in cash. Not having enough money now, you decide to specifically put money away each year in a "dream car" fund. The car you want to buy will cost $60,000 in eight years. You are going to put aside $6,000 each year (for eight years) to save for this. At what interest rate must you invest your money to achieve your goal of having enough to purchase the car after eight years?

Solution

We can use Equation (4-9) to show our desired equivalence relationship.

$$\$60,000 = \$6,000\,(F/A,\ i\%,\ 8)$$

$$(F/A,\ i\%,\ 8) = 10$$

Now we can use the interest tables in Appendix C to help track down the unknown value of i. What we are looking for are two interest rates, one with an $(F/A,\ i\%,\ 8)$ value greater than 10 and one with an $(F/A,\ i\%,\ 8)$ less than 10. Thumbing through Appendix C, we find

$$(F/A,\ 6\%,\ 8) = 9.8975 \quad \text{and} \quad (F/A,\ 7\%,\ 8) = 10.2598,$$

which tells us that the interest rate we are looking for is between 6% and 7% per year. Even though the function $(F/A, i\%, N)$ is nonlinear, we can use linear interpolation to approximate the value of i.

The dashed curve in Figure 4-8 is what we are linearly approximating. The answer, i', can be determined by using the similar triangles dashed in Figure 4-8.

$$\frac{\text{line dA}}{\text{line ed}} = \frac{\text{line BA}}{\text{line CB}}$$

$$\frac{i' - 0.06}{10 - 9.8975} = \frac{0.07 - 0.06}{10.2598 - 9.8975}$$

$$i' = 0.0628 \quad \text{or} \quad 6.28\% \text{ per year}$$

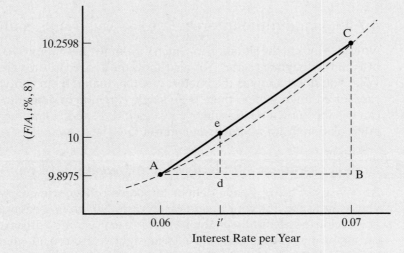

Figure 4-8   Using Linear Interpolation to Approximate i in Example 4-13

So if you can find an investment account that will earn at least 6.28% interest per year, you'll have the $60,000 you need to buy your dream car in eight years.

Spreadsheet Solution

Excel has another financial function that allows you to solve for an unknown interest rate. RATE(*nper, pmt, pv, fv*) will return the fixed interest rate that equates an annuity of magnitude *pmt* that lasts for *nper* periods to either its present value (*pv*) or its future value (*fv*).

$$i' = \text{RATE}(8, -6000, 0, 60000) = 0.0629 \text{ or } 6.29\% \text{ per year}$$

Note that a 0 was entered for *pv* since we were working with a known future value in this example.

4.8 Summary of Interest Formulas and Relationships for Discrete Compounding

Table 4-3 provides a summary of the six most common discrete compound interest factors, utilizing notation of the preceding sections. The formulas are for *discrete compounding*, which means that the interest is compounded at the end of each finite-length period, such as a month or a year. Furthermore, the formulas also assume discrete (i.e., lump sum) cash flows spaced at the end of equal time intervals on a cash-flow diagram. Discrete compound interest factors are given in Appendix C, where the assumption is made that i remains constant during the N compounding periods.

There are also several useful relationships between the compound interest factors. These relationships are summarized in the following equations.

$$(P/F, i\%, N) = \frac{1}{(F/P, i\%, N)}; \tag{4-16}$$

$$(A/P, i\%, N) = \frac{1}{(P/A, i\%, N)}; \tag{4-17}$$

$$(A/F, i\%, N) = \frac{1}{(F/A, i\%, N)}; \tag{4-18}$$

$$(F/A, i\%, N) = (P/A, i\%, N)(F/P, i\%, N); \tag{4-19}$$

$$(P/A, i\%, N) = \sum_{k=1}^{N} (P/F, i\%, k); \tag{4-20}$$

TABLE 4-3   Discrete Compounding-Interest Factors and Symbols[a]

To Find:	Given:	Factor by which to Multiply "Given"[a]	Factor Name	Factor Functional Symbol[b]
For single cash flows:				
F	P	$(1+i)^N$	Single payment compound amount	$(F/P, i\%, N)$
P	F	$\dfrac{1}{(1+i)^N}$	Single payment present worth	$(P/F, i\%, N)$
For uniform series (annuities):				
F	A	$\dfrac{(1+i)^N-1}{i}$	Uniform series compound amount	$(F/A, i\%, N)$
P	A	$\dfrac{(1+i)^N-1}{i(1+i)^N}$	Uniform series present worth	$(P/A, i\%, N)$
A	F	$\dfrac{i}{(1+i)^N-1}$	Sinking fund	$(A/F, i\%, N)$
A	P	$\dfrac{i(1+i)^N}{(1+i)^N-1}$	Capital recovery	$(A/P, i\%, N)$

[a] i equals effective interest rate per interest period; N, number of interest periods; A, uniform series amount (occurs at the end of each interest period); F, future equivalent; P, present equivalent.
[b] The functional symbol system is used throughout this book.

$$(F/A, i\%, N) = \sum_{k=1}^{N}(F/P, i\%, N-k); \qquad (4\text{-}21)$$

$$(A/F, i\%, N) = (A/P, i\%, N) - i. \qquad (4\text{-}22)$$

4.9   Deferred Annuities (Uniform Series)

All annuities (uniform series) discussed to this point involve the first cash flow being made at the end of the first period, and they are called *ordinary annuities*. If the cash flow does not begin until some later date, the annuity is known as a *deferred annuity*. If the annuity is deferred for J periods ($J < N$), the situation is as portrayed in Figure 4-9, in which the entire framed ordinary annuity has been moved away from "time present," or "time zero," by J periods. Remember that, in an annuity deferred for J periods, the first payment is made at the end of period $(J + 1)$, assuming that all periods involved are equal in length.

The present equivalent at the end of period J of an annuity with cash flows of amount A is, from Equation (4-9), $A(P/A, i\%, N - J)$. The present equivalent of the single amount $A(P/A, i\%, N - J)$ as of time zero will then be

$$P_0 = A(P/A, i\%, N - J)(P/F, i\%, J).$$

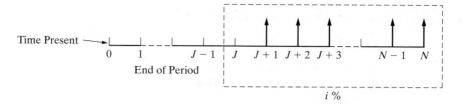

Figure 4-9 General Cash-Flow Representation of a Deferred Annuity (Uniform Series)

| EXAMPLE 4-14 | **Present Equivalent of a Deferred Annuity** |

To illustrate the preceding discussion, suppose that a father, on the day his son is born, wishes to determine what lump amount would have to be paid into an account bearing interest of 12% per year to provide withdrawals of $2,000 on each of the son's 18th, 19th, 20th, and 21st birthdays.

Solution

The problem is represented in the following cash-flow diagram. One should first recognize that an ordinary annuity of four withdrawals of $2,000 each is involved and that the present equivalent of this annuity occurs at the 17th birthday when a $(P/A, i\%, N - J)$ factor is utilized. In this problem, $N = 21$ and $J = 17$. It is often helpful to use a *subscript* with P or F to denote the respective point in time. Hence,

$$P_{17} = A(P/A, 12\%, 4) = \$2,000(3.0373) = \$6,074.60.$$

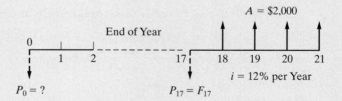

Note the dashed arrow in the cash-flow diagram denoting P_{17}. Now that P_{17} is known, the next step is to calculate P_0. With respect to P_0, P_{17} is a future equivalent, and hence it could also be denoted F_{17}. Money at a given point in time, such as the end of period 17, is the same regardless of whether it is called a present equivalent or a future equivalent. Hence,

$$P_0 = F_{17}(P/F, 12\%, 17) = \$6,074.60(0.1456) = \$884.46,$$

which is the amount that the father would have to deposit on the day his son is born.

EXAMPLE 4-15	Deferred Future Value of an Annuity

When you take your first job, you decide to start saving right away for your retirement. You put $5,000 per year into the company's 401(k) plan, which averages 8% interest per year. Five years later, you move to another job and start a new 401(k) plan. You never get around to merging the funds in the two plans. If the first plan continued to earn interest at the rate of 8% per year for 35 years after you stopped making contributions, how much is the account worth?

Solution

The following cash-flow diagram clarifies the timing of the cash flows for the original 401(k) investment plan.

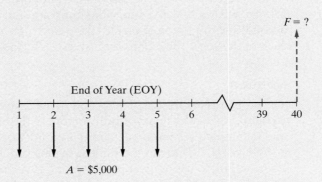

The easiest way to approach this is to first find the future equivalent of the annuity as of time 5.

$$F_5 = \$5{,}000\,(F/A,\,8\%,\,5) = \$5{,}000\,(5.8666) = \$29{,}333.$$

To determine F_{40}, F_5 can now be denoted P_5, and

$$F_{40} = P_5\,(F/P,\,8\%,\,35) = \$29{,}333\,(14.7853) = \$433{,}697.$$

4.10 Equivalence Calculations Involving Multiple Interest Formulas

You should now be comfortable with equivalence problems that involve discrete compounding of interest and discrete cash flows. All compounding of interest takes place once per time period (e.g., a year), and to this point cash flows also occur once per time period. This section provides examples involving two or more equivalence calculations to solve for an unknown quantity. The end-of-year cash-flow convention is used. Again, the interest rate is constant over the N time periods.

EXAMPLE 4-16	Calculating Equivalent *P*, *F*, and *A* Values

Figure 4-10 depicts an example problem with a series of year-end cash flows extending over eight years. The amounts are $100 for the first year, $200 for the second year, $500 for the third year, and $400 for each year from the fourth through the eighth. These could represent something like the expected maintenance expenditures for a certain piece of equipment or payments into a fund. Note that the payments are shown at the end of each year, which is a standard assumption (convention) for this book and for economic analyses in general, unless we have information to the contrary. It is desired to find

(a) the present equivalent expenditure, P_0;

(b) the future equivalent expenditure, F_8;

(c) the annual equivalent expenditure, A;

of these cash flows if the annual interest rate is 20%. Solve by hand and by using a spreadsheet.

Solution by Hand

(a) To find the equivalent P_0, we need to sum the equivalent values of all payments as of the beginning of the first year (time zero). The required movements of money through time are shown graphically in Figure 4-10(a).

$$
\begin{aligned}
P_0 = F_1(P/F, 20\%, 1) \quad &= \$100(0.8333) \quad &= \$83.33 \\
+ F_2(P/F, 20\%, 2) \quad &+ \$200(0.6944) \quad &+ 138.88 \\
+ F_3(P/F, 20\%, 3) \quad &+ \$500(0.5787) \quad &+ 289.35 \\
+ A(P/A, 20\%, 5) \times (P/F, 20\%, 3) \quad &+ \$400(2.9900) \times (0.5787) \quad &\underline{+ 692.26} \\
& & \$1{,}203.82.
\end{aligned}
$$

(b) To find the equivalent F_8, we can sum the equivalent values of all payments as of the end of the eighth year (time eight). Figure 4-10(b) indicates these movements of money through time. However, since the equivalent P_0 is already known to be $1,203.82, we can directly calculate

$$F_8 = P_0(F/P, 20\%, 8) = \$1{,}203.82(4.2998) = \$5{,}176.19.$$

(c) The equivalent A of the irregular cash flows can be calculated directly from either P_0 or F_8 as

$$A = P_0(A/P, 20\%, 8) = \$1{,}203.82(0.2606) = \$313.73$$

or

$$A = F_8(A/F, 20\%, 8) = \$5{,}176.19(0.0606) = \$313.73.$$

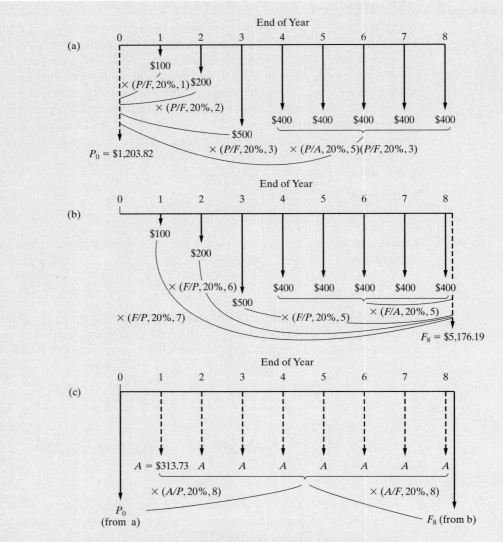

Figure 4-10   Example 4-16 for Calculating the Equivalent *P*, *F*, and *A* Values

The computation of *A* from P_0 and F_8 is shown in Figure 4-10(c). Thus, we find that the irregular series of payments shown in Figure 4-10 is equivalent to $1,203.82 at time zero, $5,176.19 at time eight, or a uniform series of $313.73 at the end of each of the eight years.

Spreadsheet Solution

Figure 4-11 displays a spreadsheet solution for this example. The present equivalent (P_0) of the tabulated cash flows is easily computed by using the NPV

Figure 4-11
Spreadsheet
Solution,
Example 4-16

	A	B	C	D	E
1	i / yr	20%		EOY	Cash Flow
2	Years in series	8		0	$ -
3				1	$ 100
4				2	$ 200
5				3	$ 500
6				4	$ 400
7				5	$ 400
8				6	$ 400
9				7	$ 400
10				8	$ 400
11					
12	$P_0 =$	$ 1,204			
13	$F_8 =$	$ 5,176			
14	$A =$	$ 314			

= NPV(B1, E3:E10)

= PMT(B1, B2, −B12)

= B12 * (1 + B1) ^ B2

function with the stated interest rate (20% in cell B1). The future equivalent (F_8) is determined from the present equivalent by using the $(F/P, i\%, N)$ relationship. The annual equivalent is also determined from the present equivalent by applying the PMT function. The slight differences in results when compared to the hand solution are due to rounding of the interest factor values in the hand solution.

EXAMPLE 4-17   **How Much Is that Last Payment? (Example 4-12 Revisited)**

In Example 4-12, we looked at paying off a loan early by increasing the annual payment. The $100,000 loan was to be repaid in 30 annual installments of $8,880 at an interest rate of 8% per year. As part of the example, we determined that the loan could be paid in full after 21 years if the annual payment was increased to $10,000.

As with most real-life loans, the final payment will be something different (usually less) than the annuity amount. This is due to the effect of rounding in the interest calculations—you can't pay in fractions of a cent! For this example, determine the amount of the 21st (and final) payment on the $100,000 loan when 20 payments of $10,000 have already been made. The interest rate remains at 8% per year.

Solution

The cash-flow diagram for this example is shown below. It is drawn from the lender's viewpoint.

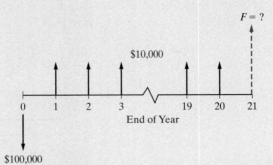

We need to determine the value of F that will make the present equivalent of all loan payments equal to the amount borrowed. We can do this by discounting all of the payments to time 0 (including the final payment, F) and setting their value equal to $100,000.

$$\$10,000 \, (P/A, 8\%, 20) + F \, (P/F, 8\%, 21) = \$100,000$$

$$\$10,000 \, (9.8181) + F \, (0.1987) = \$100,000$$

$$F = \$9,154.50$$

Thus, a payment of $9,154.50 is needed at the end of year 21 to pay off the loan.

EXAMPLE 4-18   **Equivalence Calculations Involving Unknown Quantities**

Transform the cash flows on the left-hand side of Figure 4-12 to their equivalent cash flows on the right-hand side. That is, take the left-hand quantities as given and determine the unknown value of Q in terms of H in Figure 4-12. The interest rate is 10% per year. (Notice that "$\Longleftrightarrow$" means "equivalent to.")

Solution

If all cash flows on the left are discounted to year zero, we have $P_0 = 2H(P/A, 10\%, 4) + H(P/A, 10\%, 3)(P/F, 10\%, 5) = 7.8839H$. When cash flows on the right are also discounted to year zero, we can solve for Q in terms of H. [Notice that Q at EOY two is positive, Q at EOY seven is negative, and the two Q values must be equal in amount.] So,

$$7.8839H = Q(P/F, 10\%, 2) - Q(P/F, 10\%, 7),$$

or

$$Q = 25.172H.$$

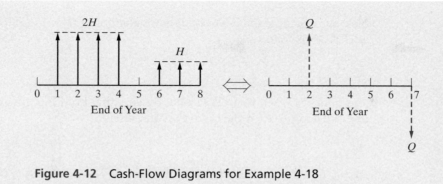

Figure 4-12   Cash-Flow Diagrams for Example 4-18

EXAMPLE 4-19   **Determining an Unknown Annuity Amount**

Two receipts of $1,000 each are desired at the EOYs 10 and 11. To make these receipts possible, four EOY annuity amounts will be deposited in a bank at EOYs 2, 3, 4, and 5. The bank's interest rate (i) is 12% per year.

(a) Draw a cash-flow diagram for this situation.

(b) Determine the value of A that establishes equivalence in your cash-flow diagram.

(c) Determine the lump-sum value at the end of year 11 of the completed cash-flow diagram based on your answers to Parts (a) and (b).

Solution

(a) Cash-flow diagrams can make a seemingly complex problem much clearer. The cash-flow diagram for this example is shown below.

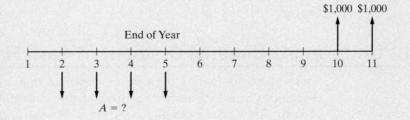

(b) Because the unknown annuity, A, begins at EOY two, it makes sense to establish our reference year for the equivalence calculations at EOY one (remember that the first annuity amount follows its P-equivalent amount by one year). So the P-equivalent at EOY 1 of the four A amounts is

$$P_1 = A(P/A, 12\%, 4).$$

Next we calculate the EOY one P-equivalent of \$1,000 at EOY 10 and \$1,000 at EOY 11 as follows:

$$P'_1 = \$1,000\,(P/A,\,12\%,\,2)\,(P/F,\,12\%,\,8).$$

The $(P/F, 12\%, 8)$ factor is needed to discount the equivalent value of the A amounts at EOY nine to EOY one. By equating both P-equivalents at EOY one, we can solve for the unknown amount, A.

$$P_1 = P'_1$$

$$A\,(P/A,\,12\%,\,4) = \$1,000\,(P/A,\,12\%,\,2)\,(P/F,\,12\%,\,8),$$

or

$$3.0373\,A = \$682.63$$

and

$$A = \$224.75.$$

Therefore, we conclude that deposits of \$224.75 at EOYs two, three, four, and five are equivalent to \$1,000 at EOYs 10 and 11 if the interest rate is 12% per year.

(c) Now we need to calculate the F-equivalent at time 11 of the $-\$224.75$ annuity in years 2 through 5 and the \$1,000 annuity in years 10 and 11.

$$-\$224.75\,(F/A,\,12\%,\,4)\,(F/P,\,12\%,\,6) + \$1,000\,(F/A,\,12\%,\,2)$$

$$= -\$224.75\,(4.7793)\,(1.9738) + \$1,000\,(2.1200)$$

$$= -\$0.15$$

This value should be zero, but round-off error in the interest factors causes a small difference of \$0.15.

4.11   Uniform (Arithmetic) Gradient of Cash Flows

Some problems involve receipts or expenses that are projected to increase or decrease by a uniform *amount* each period, thus constituting an arithmetic sequence of cash flows. For example, because of leasing a certain type of equipment, maintenance and repair savings relative to purchasing the equipment may increase by a roughly constant amount each period. This situation can be modeled as a *uniform gradient* of cash flows.

Figure 4-13 is a cash-flow diagram of a sequence of end-of-period cash flows increasing by a constant amount, G, in each period. The G is known as the *uniform gradient amount*. Note that the timing of cash flows on which the derived formulas and tabled values are based is as follows:

End of Period	Cash Flows
1	$(0)G$
2	$(1)G$
3	$(2)G$
.	.
.	.
.	.
$N-1$	$(N-2)G$
N	$(N-1)G$

Notice that the first uniform gradient cash flow, G, occurs at the end of period two.

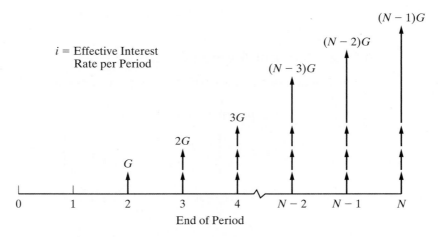

Figure 4-13 Cash-Flow Diagram for a Uniform Gradient Increasing by G Dollars per Period

4.11.1   Finding *P* when Given *G*

The present equivalent, P, of the arithmetic sequence of cash flows shown in Figure 4-13 is

$$P = \frac{G(1)}{(1+i)^2} + \frac{G(2)}{(1+i)^3} + \frac{G(3)}{(1+i)^4} + \cdots + \frac{G(N-2)}{(1+i)^{N-1}} + \frac{G(N-1)}{(1+i)^N}.$$

If we add in the dummy term $G(0)/(1+i)^1$ to represent the "missing" cash flow at time one, we can rewrite the above equation as:

$$P = G \sum_{n=1}^{N} \frac{(n-1)}{(1+i)^n}.$$

Recognizing the above equation as the summation of a geometric sequence, we can make the appropriate substitutions as we did in the development of Equation (4-6). After some algebraic manipulation, we have

$$P = G \left\{ \frac{1}{i} \left[\frac{(1+i)^N - 1}{i(1+i)^N} - \frac{N}{(1+i)^N} \right] \right\}. \tag{4-23}$$

The term in braces in Equation (4-23) is called the *gradient to present equivalent conversion factor*. It can also be expressed as $(1/i)[(P/A, i\%, N) - N(P/F, i\%, N)]$. Numerical values for this factor are given in column 8 of Appendix C for a wide assortment of i and N values. We shall use the functional symbol $(P/G, i\%, N)$ for this factor. Hence,

$$P = G\,(P/G,\ i\%,\ N). \tag{4-24}$$

4.11.2 Finding *A* when Given *G*

From Equation (4-23), it is easy to develop an expression for A as follows:

$$
\begin{aligned}
A &= P(A/P, i\%, N) \\
&= G \left\{ \frac{1}{i} \left[\frac{(1+i)^N - 1}{i(1+i)^N} - \frac{N}{(1+i)^N} \right] \right\} (A/P, i\%, N) \\
&= \frac{G}{i} \left[(P/A, i\%, N) - \frac{N}{(1+i)^N} \right] (A/P, i\%, N) \\
&= \frac{G}{i} \left[1 - \frac{Ni(1+i)^N}{(1+i)^N \left[(1+i)^N - 1 \right]} \right] \\
&= \frac{G}{i} - G \left[\frac{N}{(1+i)^N - 1} \right] \\
&= G \left[\frac{1}{i} - \frac{N}{(1+i)^N - 1} \right].
\end{aligned}
\tag{4-25}
$$

The term in brackets in Equation (4-25) is called the *gradient to uniform series conversion factor*. Numerical values for this factor are given on the right-hand side of Appendix C for a range of i and N values. We shall use the functional symbol $(A/G, i\%, N)$ for this factor. Thus,

$$A = G\,(A/G,\ i\%,\ N) \tag{4-26}$$

4.11.3 Finding *F* when Given *G*

We can develop an equation for the future equivalent, F, of an arithmetic series using Equation (4-23):

$$F = P(F/P, i\%, N)$$

$$= G \left\{ \frac{1}{i} \left[\frac{(1+i)^N - 1}{i(1+i)^N} - \frac{N}{(1+i)^N} \right] \right\} (1+i)^N$$

$$= G \left\{ \frac{1}{i} \left[\frac{(1+i)^N - 1}{i} - N \right] \right\}$$

$$= \frac{G}{i} (F/A, i\%, N) - \frac{NG}{i}. \tag{4-27}$$

It is usually more practical to deal with present and annual equivalents of arithmetic series.

4.11.4   Computations Using *G*

Be sure to notice that the direct use of gradient conversion factors applies when there is no cash flow at the end of period one, as in Example 4-20. There may be an *A* amount at the end of period one, but it is treated separately, as illustrated in Examples 4-21 and 4-22. A major advantage of using gradient conversion factors (i.e., computational time savings) is realized when *N* becomes large.

| EXAMPLE 4-20 | Using the Gradient Conversion Factors to Find *P* and *A* |

As an example of the straightforward use of the gradient conversion factors, suppose that certain EOY cash flows are expected to be $1,000 for the *second* year, $2,000 for the third year, and $3,000 for the fourth year and that, if interest is 15% per year, it is desired to find

(a) present equivalent value at the beginning of the first year,

(b) uniform annual equivalent value at the end of each of the four years.

Solution

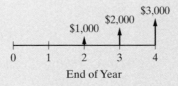

Observe that this schedule of cash flows fits the model of the arithmetic gradient formulas with $G = \$1,000$ and $N = 4$. Note that there is no cash flow at the end of the first period.

(a) The present equivalent can be calculated as

$$P_0 = G(P/G, 15\%, 4) = \$1,000(3.79) = \$3,790.$$

(b) The annual equivalent can be calculated from Equation (4-26) as

$$A = G(A/G, 15\%, 4) = \$1{,}000(1.3263) = \$1{,}326.30.$$

Of course, once P_0 is known, the value of A can be calculated as

$$A = P_0(A/P, 15\%, 4) = \$3{,}790(0.3503) = \$1{,}326.30.$$

EXAMPLE 4-21   **Present Equivalent of an Increasing Arithmetic Gradient Series**

As a further example of the use of arithmetic gradient formulas, suppose that we have cash flows as follows:

End of year	Cash flows ($)
1	5,000
2	6,000
3	7,000
4	8,000

Also, assume that we wish to calculate their present equivalent at $i = 15\%$ per year, using gradient conversion factors.

Solution

The schedule of cash flows is depicted in the left-hand diagram of Figure 4-14. The right two diagrams of Figure 4-14 show how the original schedule can be broken into two separate sets of cash flows, an annuity series of $5,000 plus an arithmetic gradient of $1,000 that fits the general gradient model for which factors are tabled. The summed present equivalents of these two separate sets of cash flows equal the present equivalent of the original problem. Thus, using the symbols shown in Figure 4-14, we have

$$P_{0T} = P_{0A} + P_{0G}$$

$$= A(P/A, 15\%, 4) + G(P/G, 15\%, 4)$$

$$= \$5{,}000(2.8550) + \$1{,}000(3.79) = \$14{,}275 + 3{,}790 = \$18{,}065.$$

The annual equivalent of the original cash flows could be calculated with the aid of Equation (4-26) as follows:

$$A_T = A + A_G$$

$$= \$5{,}000 + \$1{,}000(A/G, 15\%, 4) = \$6{,}326.30.$$

A_T is equivalent to P_{0T} because $\$6{,}326.30(P/A, 15\%, 4) = \$18{,}061$, which is the same value obtained previously (subject to round-off error).

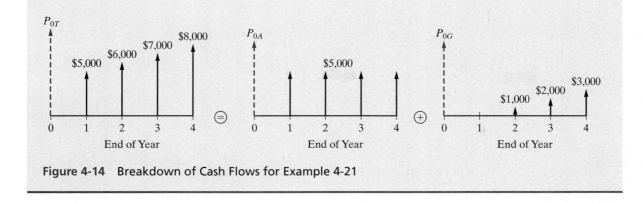

Figure 4-14   Breakdown of Cash Flows for Example 4-21

| EXAMPLE 4-22 | **Present Equivalent of a Decreasing Arithmetic Gradient Series** |

For another example of the use of arithmetic gradient formulas, suppose that we have cash flows that are timed in exact reverse of the situation depicted in Example 4-21. The left-hand diagram of Figure 4-15 shows the following sequence of cash flows:

End of year	Cash flows ($)
1	8,000
2	7,000
3	6,000
4	5,000

Calculate the present equivalent at $i = 15\%$ per year, using arithmetic gradient interest factors.

Solution

The right two diagrams of Figure 4-15 show how the uniform gradient can be broken into two separate cash-flow diagrams. In this example, we are *subtracting* an arithmetic gradient of $1,000 from an annuity series of $8,000.
     So,

$$P_{0T} = P_{0A} - P_{0G}$$

$$= A(P/A, 15\%, 4) - G(P/G, 15\%, 4)$$

$$= \$8,000(2.8550) - \$1,000(3.79)$$

$$= \$22,840 - \$3,790 = \$19,050.$$

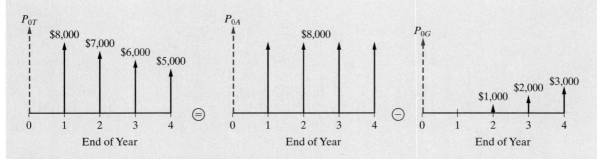

Figure 4-15 Breakdown of Cash Flows for Example 4-22

Again, the annual equivalent of the original decreasing series of cash flows can be calculated by the same rationale:

$$A_T = A - A_G$$

$$= \$8{,}000 - \$1{,}000(A/G, 15\%, 4)$$

$$= \$6{,}673.70.$$

Note from Examples 4-21 and 4-22 that the present equivalent of $18,065 for an increasing arithmetic gradient series of payments is different from the present equivalent of $19,050 for an arithmetic gradient of payments of identical amounts, but with reversed timing (decreasing series of payments). This difference would be even greater for higher interest rates and gradient amounts and exemplifies the marked effect of the timing of cash flows on equivalent values.

4.12 Geometric Sequences of Cash Flows

Some economic equivalence problems involve projected cash-flow patterns that are changing at an average *rate*, $\bar{f}$, each period. A fixed amount of a commodity that inflates in price at a constant rate each year is a typical situation that can be modeled with a geometric sequence of cash flows. The resultant EOY cash-flow pattern is referred to as a *geometric gradient series* and has the general appearance shown in Figure 4-16. Notice that *the initial cash flow in this series, A_1, occurs at the end of period* one and that $A_k = (A_{k-1})(1 + \bar{f})$, $2 \leq k \leq N$. The Nth term in this geometric sequence is $A_N = A_1(1 + \bar{f})^{N-1}$, and the common ratio throughout the sequence is $(A_k - A_{k-1})/A_{k-1} = \bar{f}$. Be sure to notice that $\bar{f}$ can be positive *or* negative.

Each term in Figure 4-16 could be discounted, or compounded, at interest rate i per period to obtain a value of P or F, respectively. However, this becomes quite tedious for large N, so it is convenient to have a single equation instead.

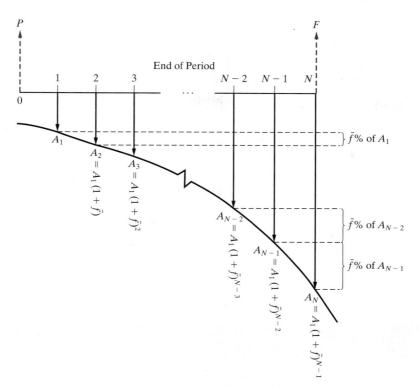

Figure 4-16 Cash-Flow Diagram for a Geometric Sequence of Payments Increasing at a Constant Rate of $\bar{f}$ per Period

The present equivalent of the geometric gradient series shown in Figure 4-16 is

$$P = A_1(P/F, i\%, 1) + A_2(P/F, i\%, 2) + A_3(P/F, i\%, 3)$$

$$+ \cdots + A_N(P/F, i\%, N)$$

$$= A_1(1 + i)^{-1} + A_2(1 + i)^{-2} + A_3(1 + i)^{-3} + \cdots + A_N(1 + i)^{-N}$$

$$= A_1(1 + i)^{-1} + A_1(1 + \bar{f})(1 + i)^{-2} + A_1(1 + \bar{f})^2(1 + i)^{-3}$$

$$+ \cdots + A_1(1 + \bar{f})^{N-1}(1 + i)^{-N}$$

$$= A_1(1 + i)^{-1}[1 + x + x^2 + \cdots + x^{N-1}], \tag{4-28}$$

where $x = (1 + \bar{f})/(1 + i)$. The expression in brackets in Equation (4-28) reduces to $(1 + x^N)/(1 - x)$ when $x \neq 1$ or $\bar{f} \neq i$. If $\bar{f} = i$, then $x = 1$ and the expression in brackets reduces to N, the number of terms in the summation. Hence,

$$P = \begin{cases} A_1(1 + i)^{-1}(1 - x^N)/(1 - x) & \bar{f} \neq i \\ A_1 N(1 + i)^{-1} & \bar{f} = i, \end{cases}$$

which reduces to

$$P = \begin{cases} \dfrac{A_1[1 - (1+i)^{-N}(1+\bar{f})^N]}{i - \bar{f}} & \bar{f} \neq i \\ A_1 N(1+i)^{-1} & \bar{f} = i, \end{cases} \qquad (4\text{-}29)$$

or

$$P = \begin{cases} \dfrac{A_1[1 - (P/F, i\%, N)(F/P, \bar{f}\%, N)]^*}{i - \bar{f}} & \bar{f} \neq i \\ A_1 N(P/F, i\%, 1) & \bar{f} = i. \end{cases} \qquad (4\text{-}30)$$

Once we know the present equivalent of a geometric gradient series, we can easily compute the equivalent uniform series or future amount using the basic interest factors $(A/P, i\%, N)$ and $(F/P, i\%, N)$.

Additional discussion of geometric sequences of cash flows is provided in Chapter 8, which deals with price changes and exchange rates.

| EXAMPLE 4-23 | Equivalence Calculations for an Increasing Geometric Gradient Series |

Consider the following EOY geometric sequence of cash flows and determine the P, A, and F equivalent values. The rate of increase is 20% per year after the first year, and the interest rate is 25% per year.

Solution

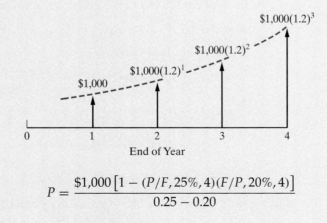

$$P = \frac{\$1,000\,[1 - (P/F, 25\%, 4)(F/P, 20\%, 4)]}{0.25 - 0.20}$$

---

* Equation (4-30) for $\bar{f} \neq i$ is mathematically equivalent to the following:

$$P = \frac{A_1}{(1+\bar{f})}\left(P/A, \frac{1+i}{1+\bar{f}} - 1, N\right).$$

$$P = \frac{\$1,000}{0.05}\left[1 - (0.4096)(2.0736)\right]$$

$$= \$20,000(0.15065)$$

$$= \$3,013;$$

$$A = \$3,013(A/P, 25\%, 4) = \$1,275.70;$$

$$F = \$3,013(F/P, 25\%, 4) = \$7,355.94.$$

EXAMPLE 4-24 | **Equivalence Calculations for a Decreasing Geometric Gradient Series**

Suppose that the geometric gradient in Example 4-23 begins with $1,000 at EOY one and *decreases* by 20% per year after the first year. Determine P, A, and F under this condition.

Solution

The value of $\bar{f}$ is −20% in this case. The desired quantities are as follows:

$$P = \frac{\$1,000[1 - (P/F, 25\%, 4)(F/P, -20\%, 4)]}{0.25 - (-0.20)}$$

$$= \frac{\$1,000}{0.45}\left[1 - (0.4096)(1 - 0.20)^4\right]$$

$$= \$2,222.22(0.83222)$$

$$= \$1,849.38;$$

$$A = \$1,849.38(A/P, 25\%, 4) = \$783.03;$$

$$F = \$1,849.38(F/P, 25\%, 4) = \$4,515.08.$$

EXAMPLE 4-25 | **Using a Spreadsheet to Model Geometric Gradient Series**

Create a spreadsheet to solve

(a) Example 4-23

(b) Example 4-24.

Solution

Figure 4-17 displays the spreadsheet solution for this problem. A formula is used to compute the actual cash flow for each year on the basis of EOY one cash flow (A_1) and the yearly rate of change ($\bar{f}$). Once we have the set of EOY cash flows, P can be computed by using the NPV function. The values of A and F are easily

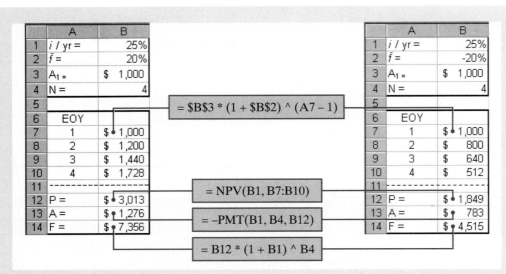

(a) Solution to Example 4-23   (b) Solution to Example 4-24

Figure 4-17   Spreadsheet Solution, Example 4-25

derived from the value of P. Note that the structure of the spreadsheets for Parts (a) and (b) are the same—the only difference is the value of $\bar{f}$ in cell B2.

EXAMPLE 4-26	A Retirement Savings Plan

On your 23rd birthday you decide to invest $4,500 (10% of your annual salary) in a mutual fund earning 7% per year. You will continue to make annual deposits equal to 10% of your annual salary until you retire at age 62 (40 years after you started your job). You expect your salary to increase by an average of 4% each year during this time. How much money will you have accumulated in your mutual fund when you retire?

Solution

Since the amount of your deposit is 10% of your salary, each year the amount you deposit will increase by 4% as your salary increases. Thus, your deposits constitute a geometric gradient series with $f = 4\%$ per year. We can use Equation (4-30) to determine the present equivalent amount of the deposits.

$$P = \frac{\$4,500[1 - (P/F, 7\%, 40)(F/P, 4\%, 40)]}{0.07 - 0.04}$$

$$= \frac{\$4,500[1 - (0.0668)(4.8010)]}{0.03}$$

$$= \$101,894.$$

Now the future worth at age 62 can be determined.

$$F = \$101,894(F/P, 7\%, 40)$$
$$= \$101,894(14.9745)$$
$$= \$1,525,812.$$

This savings plan will make you a millionaire when you retire. *Moral:* Start saving early!

4.13   Interest Rates that Vary with Time

Student loans under the U.S. government's popular Stafford program let students borrow money up to a certain amount each year (based on their year in school and financial need). Stafford loans are the most common type of educational loan, and they have a floating interest rate that readjusts every year (but cannot exceed 8.25% per year). When the interest rate on a loan can vary with time, it is necessary to take this into account when determining the future equivalent value of the loan. Example 4-27 demonstrates how this situation is treated.

EXAMPLE 4-27   Compounding with Changing Interest Rates

Ashea Smith is a 22-year-old senior who used the Stafford loan program to borrow $4,000 four years ago when the interest rate was 4.06% per year. $5,000 was borrowed three years ago at 3.42%. Two years ago she borrowed $6,000 at 5.23%, and last year $7,000 was borrowed at 6.03% per year. Now she would like to consolidate her debt into a single 20-year loan with a 5% fixed annual interest rate. If Ashea makes annual payments (starting in one year) to repay her total debt, what is the amount of each payment?

Solution

The following cash-flow diagram clarifies the timing of Ashea's loans and the applicable interest rates. The diagram is drawn using Ashea's viewpoint.

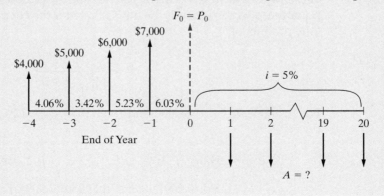

Before we can find the annual repayment amount, we need to find the current (time 0) equivalent value of the four loans. This problem can be solved by compounding the amount owed at the beginning of each year by the interest rate that applies to each individual year and repeating this process over the four years to obtain the total current equivalent value.

$$F_{-3} = \$4,000(F/P, 4.06\%, 1) + \$5,000 = \$4,000(1.0406)$$

$$+ \$5,000 = \$9,162.40$$

$$F_{-2} = \$9,162.40(F/P, 3.42\%, 1) + \$6,000 = \$15,475.75$$

$$F_{-1} = \$15,475.75(F/P, 5.23\%, 1) + \$7,000 = \$23,285.13$$

$$F_0 = \$23,285.13(F/P, 6.03\%, 1) = \$24,689.22$$

Notice that it was a simple matter to substitute $(F/P, i\%, N) = (1 + i)^N$ for the noninteger values of i.

Now that we have the current equivalent value of the amount Ashea borrowed ($F_0 = P_0$), we can easily compute her annual repayment amount over 20 years when the interest rate is fixed at 5% per year.

$$A = \$24,689.22(A/P, 5\%, 20) = \$24,689.22(0.0802) = \$1,980.08 \text{ per year}$$

Comment

The total principal borrowed was $\$4,000 + \$5,000 + \$6,000 + \$7,000 = \$22,000$. Notice that a total of $\$17,601.60$ ($20 \times \$1,980.08 - \$22,000$) in interest is repaid over the entire 20-year loan period. This interest amount is close to the amount of principal originally borrowed. *Moral:* Borrow as little as possible and repay as quickly as possible to reduce interest expense! See www.finaid.com.

To obtain the present equivalent of a series of future cash flows subject to varying interest rates, a procedure similar to the preceding one would be utilized with a sequence of $(P/F, i_k\%, k)$ factors. In general, the present equivalent value of a cash flow occurring at the end of period N can be computed using Equation (4-31), where i_k is the interest rate for the kth period (the symbol $\prod$ means "the product of"):

$$P = \frac{F_N}{\prod_{k=1}^{N}(1 + i_k)}. \tag{4-31}$$

For instance, if $F_4 = \$1,000$ and $i_1 = 10\%$, $i_2 = 12\%$, $i_3 = 13\%$, and $i_4 = 10\%$, then

$$P = \$1,000[(P/F, 10\%, 1)(P/F, 12\%, 1)(P/F, 13\%, 1)(P/F, 10\%, 1)]$$

$$= \$1,000[(0.9091)(0.8929)(0.8850)(0.9091)] = \$653.$$

4.14   Nominal and Effective Interest Rates

Very often the interest period, or time between successive compounding, is less than one year (e.g., daily, weekly, monthly, or quarterly). It has become customary to quote interest rates on an annual basis, followed by the compounding period if different from one year in length. For example, if the interest rate is 6% per interest period and the interest period is six months, it is customary to speak of this rate as "12% compounded semiannually." Here the annual rate of interest is known as the *nominal rate*, 12% in this case. A nominal interest rate is represented by r. But the actual (or effective) annual rate on the principal is not 12%, but something greater, because compounding occurs twice during the year.

Consequently, the frequency at which a nominal interest rate is compounded each year can have a pronounced effect on the dollar amount of total interest earned. For instance, consider a principal amount of $1,000 to be invested for three years at a nominal rate of 12% compounded semiannually. The interest earned during the first six months would be $1,000 × (0.12/2) = $60.

Total principal and interest at the beginning of the second six-month period is

$$P + Pi = \$1,000 + \$60 = \$1,060.$$

The interest earned during the second six months would be

$$\$1,060 \times (0.12/2) = \$63.60.$$

Then total interest earned during the year is

$$\$60.00 + \$63.60 = \$123.60.$$

Finally, the *effective* annual interest rate for the entire year is

$$\frac{\$123.60}{\$1,000} \times 100 = 12.36\%.$$

If this process is repeated for years two and three, the *accumulated* (compounded) *amount of interest* can be plotted as in Figure 4-18. Suppose that the same $1,000 had been invested at 12% compounded *monthly*, which is 1% per month.

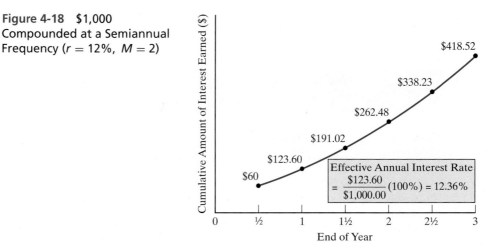

Figure 4-18 $1,000 Compounded at a Semiannual Frequency ($r = 12\%$, $M = 2$)

The accumulated interest over three years that results from monthly compounding is shown in Figure 4-19.

The actual or exact rate of interest earned on the principal during one year is known as the *effective rate*. It should be noted that effective interest rates are always expressed on an annual basis, unless specifically stated otherwise. In this text, the effective interest rate per year is customarily designated by i and the nominal interest rate per year by r. In engineering economy studies in which compounding is annual, $i = r$. The relationship between effective interest, i, and nominal interest, r, is

$$i = (1 + r/M)^M - 1, \qquad (4\text{-}32)$$

where M is the number of compounding periods per year. It is now clear from Equation (4-32) why $i > r$ when $M > 1$.

The effective rate of interest is useful for describing the compounding effect of interest earned on interest during one year. Table 4-4 shows effective rates for various nominal rates and compounding periods.

Interestingly, the federal truth-in-lending law now requires a statement regarding the annual percentage rate (APR) being charged in contracts involving

Figure 4-19
$1,000 Compounded at a Monthly Frequency ($r = 12\%$, $M = 12$)

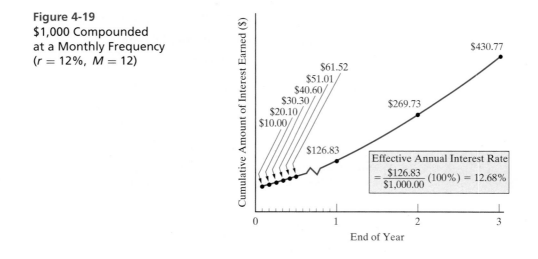

TABLE 4-4   Effective Interest Rates for Various Nominal Rates and Compounding Frequencies

Compounding Frequency	Number of Compounding Periods per Year, M	Effective Rate (%) for Nominal Rate of					
		6%	8%	10%	12%	15%	24%
Annually	1	6.00	8.00	10.00	12.00	15.00	24.00
Semiannually	2	6.09	8.16	10.25	12.36	15.56	25.44
Quarterly	4	6.14	8.24	10.38	12.55	15.87	26.25
Bimonthly	6	6.15	8.27	10.43	12.62	15.97	26.53
Monthly	12	6.17	8.30	10.47	12.68	16.08	26.82
Daily	365	6.18	8.33	10.52	12.75	16.18	27.11

borrowed money. The APR is a nominal interest rate and *does not* account for compounding that may occur, or be appropriate, during a year. Before this legislation was passed by Congress in 1969, creditors had no obligation to explain how interest charges were determined or what the true cost of money on a loan was. As a result, borrowers were generally unable to compute their APR and compare different financing plans.

EXAMPLE 4-28	Effective Annual Interest Rate

A credit card company charges an interest rate of 1.375% per month on the unpaid balance of all accounts. The annual interest rate, they claim, is 12(1.375%) = 16.5%. What is the effective rate of interest per year being charged by the company?

Solution

Equation (4-32) is used to compute the effective rate of interest in this example:

$$i = \left(1 + \frac{0.165}{12}\right)^{12} - 1$$

$$= 0.1781, \text{ or } 17.81\%/\text{year.}$$

Note that $r = 12(1.375\%) = 16.5\%$, which is the APR. In general, it is true that $r = M(r/M)$, where r/M is the interest rate per period.

Numerous Web sites are available to assist you with personal finance decisions. Take a look at www.dinkytown.net and www.bankrate.com.

4.15   Compounding More Often than Once per Year

4.15.1   Single Amounts

If a nominal interest rate is quoted and the number of compounding periods per year and number of years are known, any problem involving future, annual, or present equivalent values can be calculated by straightforward use of Equations (4-3) and (4-32), respectively.

EXAMPLE 4-29	Future Equivalent when Interest Is Compounded Quarterly

Suppose that a $100 lump-sum amount is invested for 10 years at a nominal interest rate of 6% compounded quarterly. How much is it worth at the end of the 10th year?

Solution

There are four compounding periods per year, or a total of $4 \times 10 = 40$ interest periods. The interest rate per interest period is $6\%/4 = 1.5\%$. When the values are used in Equation (4-3), one finds that

$$F = P(F/P, 1.5\%, 40) = \$100.00(1.015)^{40} = \$100.00(1.814) = \$181.40.$$

Alternatively, the effective interest rate from Equation (4-32) is 6.14%. Therefore,

$$F = P(F/P, 6.14\%, 10) = \$100.00(1.0614)^{10} = \$181.40.$$

4.15.2   Uniform Series and Gradient Series

When there is more than one compounded interest period per year, the formulas and tables for uniform series and gradient series can be used *as long as* there is a cash flow at the end of each interest period, as shown in Figures 4-6 and 4-13 for a uniform annual series and a uniform gradient series, respectively.

EXAMPLE 4-30   **Computing a Monthly Auto Payment**

Stan Moneymaker has a bank loan for $10,000 to pay for his new truck. This loan is to be repaid in equal *end-of-month* installments for five years with a nominal interest rate of 12% compounded monthly. What is the amount of each payment?

Solution

The number of installment payments is $5 \times 12 = 60$, and the interest rate per month is $12\%/12 = 1\%$. When these values are used in Equation (4-15), one finds that

$$A = P(A/P, 1\%, 60) = \$10,000(0.0222) = \$222.$$

Notice that there is a cash flow at the end of each month (interest period), including month 60, in this example.

EXAMPLE 4-31   **Uniform Gradient Series and Semiannual Compounding**

Certain operating savings are expected to be 0 at the end of the first six months, to be $1,000 at the end of the second six months, and to increase by $1,000 at the end of each six-month period thereafter, for a total of four years. It is desired to find the equivalent uniform amount, A, at the end of each of the eight six-month periods if the nominal interest rate is 20% compounded semiannually.

Solution

A cash-flow diagram is given below, and the solution is

$$A = G(A/G, 10\%, 8) = \$1,000(3.0045) = \$3,004.50.$$

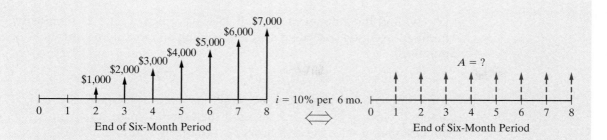

The symbol "$\Longleftrightarrow$" in between the cash-flow diagrams indicates that the left-hand cash-flow diagram is *equivalent to* the right-hand cash-flow diagram when the correct value of A has been determined. In Example 4-31, the interest rate per six-month period is 10%, and cash flows occur every six months.

EXAMPLE 4-32   **Finding the Interest Rate on a Loan**

A loan of $15,000 requires monthly payments of $477 over a 36-month period of time. These payments include both principal and interest.

(a) What is the nominal interest rate (APR) for this loan?

(b) What is the effective interest rate per year?

(c) Determine the amount of unpaid loan principal after 20 months.

Solution

(a) We can set up an equivalence relationship to solve for the unknown interest rate since we know that $P = \$15,000$, $A = \$477$, and $N = 36$ months.

$$\$477 = \$15,000(A/P, i_{mo}, 36)$$

$$(A/P, i_{mo}, 36) = 0.0318$$

We can now look through Appendix C to find values of i that have an $(A/P, i, 36)$ value close to 0.0318. From Table C-3 ($i = \frac{3}{4}\%$), we find $(A/P, \frac{3}{4}\%, 36) = 0.0318$. Therefore,

$$i_{mo} = 0.75\% \text{ per month}$$

and

$$r = 12 \times 0.75\% = 9\% \text{ per year, compounded monthly.}$$

(b) Using Equation (4-32),

$$i_{eff} = \left(1 + \frac{0.09}{12}\right)^{12} - 1 = 0.0938 \text{ or } 9.38\% \text{ per year.}$$

(c) We can find the amount of the unpaid loan principal after 20 months by finding the equivalent value of the remaining 16 monthly payments as of month 20.

$$P_{20} = \$477(P/A, \tfrac{3}{4}\%, 16) = \$477(15.0243) = \$7{,}166.59$$

After 20 payments have been made, almost half of the original principal amount remains. Notice that we used the monthly interest rate of $\tfrac{3}{4}\%$ in our calculation since the cash flows are occurring monthly.

4.16   Interest Formulas for Continuous Compounding and Discrete Cash Flows

In most business transactions and economy studies, interest is compounded at the end of discrete periods, and, as has been discussed previously, cash flows are assumed to occur in discrete amounts at the end of such periods. *This practice will be used throughout the remaining chapters of this book.* However, it is evident that in most enterprises, cash is flowing in and out in an almost continuous stream. Because cash, whenever it's available, can usually be used profitably, this situation creates opportunities for very frequent compounding of the interest earned. So that this condition can be dealt with (modeled) when continuously compounded interest rates are available, the concept of continuous compounding is sometimes used in economy studies. Actually, the effect of this procedure, compared with that of discrete compounding, is rather small in most cases.

Continuous compounding assumes that cash flows occur at discrete intervals (e.g., once per year), but that compounding is continuous throughout the interval. For example, with a nominal rate of interest per year of r, if the interest is compounded M times per year, one unit of principal will amount to $[1 + (r/M)]^M$ at the end of one year. Letting $M/r = p$, we find that the foregoing expression becomes

$$\left[1 + \frac{1}{p}\right]^{rp} = \left[\left(1 + \frac{1}{p}\right)^{p}\right]^{r}. \tag{4-33}$$

Because

$$\lim_{p \to \infty} \left(1 + \frac{1}{p}\right)^{p} = e^{1} = 2.71828\ldots,$$

Equation (4-33) can be written as e^r. Consequently, the *continuously compounded compound amount factor (single cash flow)* at $\underline{r}\%$ nominal interest for N years is e^{rN}. Using our functional notation, we express this as

$$(F/P, \underline{r}\%, N) = e^{rN}. \tag{4-34}$$

Note that the symbol $\underline{r}$ is directly comparable to that used for discrete compounding and discrete cash flows ($i\%$), except that $\underline{r}\%$ is used to denote the nominal rate *and* the use of continuous compounding.

Since e^{rN} for continuous compounding corresponds to $(1 + i)^N$ for discrete compounding, e^r is equal to $(1 + i)$. Hence, we may correctly conclude that

$$i = e^r - 1. \tag{4-35}$$

By using this relationship, the corresponding values of (P/F), (F/A), and (P/A) for continuous compounding may be obtained from Equations (4-4), (4-8), and (4-10), respectively, by substituting $e^r - 1$ for i in these equations. Thus, for continuous compounding and discrete cash flows,

$$(P/F, \underline{r}\%, N) = \frac{1}{e^{rN}} = e^{-rN}; \tag{4-36}$$

$$(F/A, \underline{r}\%, N) = \frac{e^{rN} - 1}{e^r - 1}; \tag{4-37}$$

$$(P/A, \underline{r}\%, N) = \frac{1 - e^{-rN}}{e^r - 1} = \frac{e^{rN} - 1}{e^{rN}(e^r - 1)}. \tag{4-38}$$

Values for $(A/P, \underline{r}\%, N)$ and $(A/F, \underline{r}\%, N)$ may be derived through their inverse relationships to $(P/A, \underline{r}\%, N)$ and $(F/A, \underline{r}\%, N)$, respectively. Numerous continuous compounding, discrete cash-flow interest factors, and their uses are summarized in Table 4-5.

Because continuous compounding is used infrequently in this text, detailed values for $(A/F, \underline{r}\%, N)$ and $(A/P, \underline{r}\%, N)$ are not given in Appendix D. However, the tables in Appendix D do provide values of $(F/P, \underline{r}\%, N)$, $(P/F, \underline{r}\%, N)$, $(F/A, \underline{r}\%, N)$, and $(P/A, \underline{r}\%, N)$ for a limited number of interest rates.

TABLE 4-5   Continuous Compounding and Discrete Cash Flows: Interest Factors and Symbols[a]

To Find:	Given:	Factor by which to Multiply "Given"	Factor Name	Factor Functional Symbol
For single cash flows:				
F	P	e^{rN}	Continuous compounding compound amount (single cash flow)	$(F/P, \underline{r}\%, N)$
P	F	e^{-rN}	Continuous compounding present equivalent (single cash flow)	$(P/F, \underline{r}\%, N)$
For uniform series (annuities):				
F	A	$\frac{e^{rN} - 1}{e^r - 1}$	Continuous compounding compound amount (uniform series)	$(F/A, \underline{r}\%, N)$
P	A	$\frac{e^{rN} - 1}{e^{rN}(e^r - 1)}$	Continuous compounding present equivalent (uniform series)	$(P/A, \underline{r}\%, N)$
A	F	$\frac{e^r - 1}{e^{rN} - 1}$	Continuous compounding sinking fund	$(A/F, \underline{r}\%, N)$
A	P	$\frac{e^{rN}(e^r - 1)}{e^{rN} - 1}$	Continuous compounding capital recovery	$(A/P, \underline{r}\%, N)$

[a] $\underline{r}$, nominal annual interest rate, compounded continuously; N, number of periods (years); A, annual equivalent amount (occurs at the end of each year); F, future equivalent; P, present equivalent.

Note that tables of interest and annuity factors for continuous compounding are tabulated in terms of nominal annual rates of interest.

EXAMPLE 4-33	Continuous Compounding and Single Amounts

You have $10,000 to invest for two years. Your bank offers 5% interest, compounded continuously for funds in a money market account. Assuming no additional deposits or withdrawals, how much money will be in that account at the end of two years?

Solution

$$F = \$10,000 \ (F/P, \underline{r} = 5\%, 2) = \$10,000 \ e^{(0.05)(2)} = \$10,000 \ (1.1052) = \$11,052$$

Comment

If the interest rate was 5% compounded annually, the account would have been worth

$$F = \$10,000 \ (F/P, 5\%, 2) = \$10,000 \ (1.1025) = \$11,025.$$

EXAMPLE 4-34	Continuous Compounding and Annual Payments

Suppose that one has a present loan of $1,000 and desires to determine what equivalent uniform EOY payments, A, could be obtained from it for 10 years if the nominal interest rate is 20% compounded continuously ($M = \infty$).

Solution

Here we utilize the formulation

$$A = P(A/P, \underline{r}\%, N).$$

Since the (A/P) factor is not tabled for continuous compounding, we substitute its inverse (P/A), which is tabled in Appendix D. Thus,

$$A = P \times \frac{1}{(P/A, \underline{20}\%, 10)} = \$1,000 \times \frac{1}{3.9054} = \$256.$$

Note that the answer to the same problem, with discrete annual compounding ($M = 1$), is

$$A = P(A/P, 20\%, 10)$$

$$= \$1,000(0.2385) = \$239.$$

EXAMPLE 4-35	**Continuous Compounding and Semiannual Payments**

An individual needs $12,000 immediately as a down payment on a new home. Suppose that he can borrow this money from his insurance company. He must repay the loan in equal payments every six months over the next eight years. The nominal interest rate being charged is 7% compounded continuously. What is the amount of each payment?

Solution

The nominal interest rate per six months is 3.5%. Thus, A each six months is $12,000(A/P, \underline{r} = 3.5\%, 16)$. By substituting terms in Equation (4-38) and then using its inverse, we determine the value of A per six months to be $997:

$$A = \$12,000 \left[\frac{1}{(P/A, \underline{r} = 3.5\%, 16)} \right] = \frac{\$12,000}{12.038} = \$997.$$

4.17 CASE STUDY—Understanding Economic "Equivalence"

Enrico Suarez just graduated with a B.S. in engineering and landed a new job with a starting annual salary of $48,000. There are a number of things that he would like to do with his newfound "wealth." For starters, he needs to begin repaying his student loans (totaling $20,000) and he'd like to reduce some outstanding balances on credit cards (totaling $5,000). Enrico also needs to purchase a car to get to work and would like to put money aside to purchase a condo in the future. Last, but not least, he wants to put some money aside for his eventual retirement.

Our recent graduate needs to do some financial planning for which he has selected a 10-year time frame. At the end of 10 years, he'd like to have paid off his current student loan and credit card debt, as well as have accumulated $40,000 for a down payment on a condo. If possible, Enrico would like to put aside 10% of his take home salary for retirement. He has gathered the following information to assist him in his planning.*

- Student loans are typically repaid in equal monthly installments over a period of 10 years. The interest rate on Enrico's loan is 8% compounded monthly.

- Credit cards vary greatly in the interest rate charged. Typical APR rates are close to 17%, and monthly minimum payments are usually computed using a 10-year repayment period. The interest rate on Enrico's credit card is 18% compounded monthly.

- Car loans are usually repaid over three, four, or five years. The APR on a car loan can be as low as 2.9% (if the timing is right) or as high as 12%. As a first-time car buyer, Enrico can secure a $15,000 car loan at 9% compounded monthly to be repaid over 60 months.

---

* The stated problem data are current as of 2007.

- A 30-year, fixed rate mortgage is currently going for 5.75–6.0% per year. If Enrico can save enough to make a 20% down payment on the purchase of his condo, he can avoid private mortgage insurance that can cost as much as $60 per month.

- Investment opportunities can provide variable returns. "Safe" investments can guarantee 7% per year, while "risky" investments could return 30% or more per year.

- Enrico's parents and older siblings have reminded him that his monthly take home pay will be reduced by income taxes and benefit deductions. He should not count on being able to spend more than 80% of his gross salary.

As Enrico's friend (and the one who took Engineering Economy instead of Appreciating the Art of Television Commercials), you have been asked to review his financial plans. How reasonable are his goals?

Solution

Since all repayments are done on a monthly basis, it makes sense to adopt the month as Enrico's unit of time. There are five categories for his cash flows: debt repayment, transportation costs, housing costs, other living expenses, and savings. The following paragraphs summarize his estimates of monthly expenses in each of these categories.

Debt Repayment

Enrico's student loan debt is $20,000 and is to be repaid over the next 120 months (10 years) at a nominal interest rate of 8% compounded monthly ($i_{month} = 8/12\% = 2/3\%$). His monthly loan payment is then

$$A_{Student\ Loan} = \$20,000(A/P, 2/3\%, 120) = \$20,000(0.01213) = \$242.60 \text{ per month.}$$

Enrico's credit card debt is $5,000 and is to be completely repaid over the next 120 months at a nominal interest rate of 18% compounded monthly ($i_{month} = 1.5\%$). His monthly credit card payment, assuming no additional usage, is then

$$A_{Credit\ Card} = \$5,000(A/P, 1.5\%, 120) = \$5,000(0.01802) = \$90.10 \text{ per month.}$$

Enrico's monthly debt repayment totals $242.60 + $90.10 = $332.70.

Transportation Costs

The certified pre-owned vehicle Enrico would like to buy costs $15,000. The best rate he can find as a first-time car buyer with no assets or credit history is 9% compounded monthly with a 60-month repayment period. Based on these figures, his monthly car payment will be

$$A_{Car} = \$15,000(A/P, 0.75\%, 60) = \$15,000(0.02076) = \$311.40 \text{ per month.}$$

Even though the car will be completely repaid after five years, Enrico knows that he will eventually need to replace the car. To accumulate funds for the replacement of the car, he wants to continue to set aside this amount each month for the second five years.

Insurance for this vehicle will cost $1,200 per year, and Enrico has budgeted $100 per month to cover fuel and maintenance. Thus, his monthly transportation costs total $311.40 + $1,200/12 + $100 = $511.40.

Housing Costs

A nice two-bedroom apartment near Enrico's place of work has a monthly rent of $800. The rental office provided him with a monthly utility cost estimate of $150 to cover water and electricity. Based on this information, $800 + $150 = $950 per month has been budgeted for his housing costs.

Other Living Expenses

This expense category has troubled Enrico the most. While all the previous categories are pretty straightforward, he knows that his day-to-day spending will have the most variability and that it will be tempting to overspend. Nonetheless, he has developed the following estimates of monthly living expenses not already covered in the other categories:

Food	$200
Phone	$70
Entertainment	$100
Miscellaneous	
(clothing, household items)	$150
Subtotal	$520

Savings

Enrico wants to accumulate $40,000 over the next 10 years to be used as a down payment on a condo. He feels that if he chooses relatively "safe" investments (CDs and bonds), he should be able to earn 6% compounded monthly on his savings. He must then set aside

$$A_{Condo} = \$40,000(A/F, 0.5\%, 120) = \$40,000(0.00610) = \$244.00 \text{ per month}$$

to reach his goal.

Enrico's gross monthly starting salary is $48,000/12 = $4,000. Based on the information gathered from his family, he estimates his monthly net (take home) pay to be $4,000(0.80) = $3,200. His monthly retirement savings will then be $3,200(0.10) = $320. Thus, the total amount to be set aside each month for the future is $244 + $320 = $564.

Monthly Financial Plan

Based on the preceding calculations, the following table summarizes Enrico's monthly financial plan.

	Net Income	Expense
Salary	$3,200	
Debt Repayment		$332.70
Transportation Costs		511.40
Housing Costs		950.00
Living Expenses		520.00
Savings		564.00
Total	$3,200	$2,878.10

Enrico is aware that he has not explicitly accounted for price increases over the next 10 years; however, neither has he increased his annual salary. He is hopeful that, if he works hard, his annual salary increases will at least cover the increase in the cost of living.

You congratulate your friend on his efforts. You feel that he has covered the major areas of expenses, as well as savings for the future. You note that he has $3,200 − $2,878.10 = $321.90 of "extra" money each month to cover unanticipated costs. With this much leeway, you feel that he should have no problem meeting his financial goals.

──────────■

4.18 Summary

Chapter 4 presented the fundamental time value of money relationships that are used throughout the remainder of this book. Considerable emphasis has been placed on the concept of equivalence—if two cash flows (or series of cash flows) are equivalent for a stated interest rate, you are willing to trade one for the other. Formulas relating present and future amounts to their equivalent uniform, arithmetic, and geometric series have been presented. You should feel comfortable with the material in this chapter before embarking on your journey through subsequent chapters. Important abbreviations and notation are listed in Appendix B.

In the next chapter, we will see how to apply time value of money relationships to the estimated cash flows of a project (the topic of Chapter 3) to arrive at a measure of its economic worth.

Problems

The number(s) in color at the end of a problem refer to the section(s) in that chapter most closely related to the problem.

4-1. What lump-sum amount of interest will be paid on a $10,000 loan that was made on August 1, 2008, and repaid on November 1, 2012, with ordinary simple interest at 10% per year? (4.2)

4-2. You borrow $200 from a family member and agree to pay it back in six months. Because you are part of the family, you are only being charged simple interest at the rate of 1% per month. How much will you owe after six months? How much of this is interest? (4.2)

4-3. What is the future equivalent of $1,000 invested at 8% simple interest per year for $2\frac{1}{2}$ years? (4.2)

4-4. Compare the interest earned by P dollars at $i\%$ per year simple interest with that earned by the same amount P for five years at $i\%$ compounded annually. (4.2)

4-5. How much interest is *payable each year* on a loan of $2,000 if the interest rate is 10% per year when half of the loan principal will be repaid as a lump sum at the end of four years and the *other half* will be repaid in one lump-sum amount at the end of eight years? How much interest will be paid over the eight-year period? (4.4)

4-6. Suppose that, in Plan 1 of Table 4-1, $8,500 of the original unpaid balance is to be repaid at the end of months two and four only. How much total interest would have been paid by the end of month four? (4.4)

4-7. Consider the following loan information and repayment table:

Loan principal = $10,000

Interest rate = 8% per year

Duration of loan = 3 years

Annual payment = $3,880

EOY k	Interest Paid	Principal Repayment
1	$800	?
2	$553.60	$3,326.40
3	?	?

a. Determine the value of each "?" in the table.

b. What is the amount of principal owed at the beginning of year three?

4-8. Jim loans Juanita $10,000 with interest compounded at a rate of 6% per year. How much money will Juanita owe Jim if she repays the entire loan at the end of five years? (4.6)

4-9. Your spendthrift cousin wants to buy a fancy watch for $425. Instead, you suggest that she buy an inexpensive watch for $25 and save the difference of $400. Your cousin eventually agrees with your idea and invests $400 for 40 years in an account earning 9% interest per year. How much will she accumulate in this account after 40 years have passed? (4.6)

4-10. A lump-sum loan of $5,000 is needed by Chandra to pay for college expenses. She has obtained small consumer loans with 12% interest per year in the past to help pay for college. But her father has advised

Chandra to apply for a PLUS student loan charging only 8.5% interest per year. If the loan will be repaid in full in five years, what is the difference in total interest accumulated by these two types of student loans? (4.6)

4-11. You invest $25,000 in a stock-based mutual fund. This fund should earn (on average) 10% per year over a long period of time. How much should your investment be worth in 25 years? (4.6)

4-12. What is the present equivalent of $18,000 to be received in 15 years when the interest rate is 7% per year? (4.6)

4-13. You just inherited $10,000. While you plan to squander some of it away, how much should you deposit in an account earning 5% interest per year if you'd like to have $10,000 in the account in 10 years? (4.6)

4-14. The price of oil in 2005 was $67 per barrel. "This price is still lower than the price of oil in 1981" says a government publication. If inflation has averaged 3.2% per year from 1981 to 2005, what was the price per barrel of oil in 1981? (4.6)

4-15. Ashlea purchased 100 shares of Microsoft at a price of $25 per share. She hopes to double her investment. How long will she have to wait if the stock price increases at a rate of 10% per year? (4.6)

4-16. Use the rule of 72 to determine how long it takes to accumulate $10,000 in a savings account when $P = $5,000 and $i = 10\%$ per year. (4.6)

Rule of 72: The time (years) required to double the value of a lump-sum investment that is allowed to compound is approximately

72 ÷ annual interest rate (as a %).

4-17. How long does it take (to the nearest whole year) for $1,000 to quadruple in value when the interest rate is 15% per year? (4.6)

4-18. Suppose that you have $10,000 cash today and can invest it at an interest rate of 10% compounded each year. How many years will it take you to become a millionaire? (4.6)

4-19. An enterprising student invests $1,000 at an annual interest rate that will grow the original investment to $2,000 in four years. In four more years, the amount will grow to $4,000, and this pattern of doubling every four years repeats over a total time span of 36 years. How much money will the student *gain* in 36 years? What is the magical annual interest rate the student is earning? (4.6)

4-20. In 1885, first-class postage for a one-ounce letter cost $0.02. The same postage in 2007 costs $0.41. What compounded annual increase in the cost of first-class postage has been experienced over this period of time? (4.6)

4-21. A good stock-based mutual fund should earn at least 10% per year over a long period of time. Consider the case of Barney and Lynn, who were overheard gloating (for all to hear) about how well they had done with their mutual fund investment. "We turned a $25,000 investment of money in 1982 into $100,000 in 2007." (4.6)

a. What return (interest rate) did they really earn on their investment? Should they have been bragging about how investment-savvy they were?

b. Instead, if $1,000 had been invested each year for 25 years to accumulate $100,000, what return did Barney and Lynn earn?

4-22. In 1972, the maximum earnings of a worker subject to social security tax was $9,000. The maximum earnings subject to social security tax in 2008 is $102,000. What compounded annual increase has been experienced over this 36-year period of time? How does it compare with a 3% annual increase in the consumer price index (CPI) over this same time interval? (4.6)

4-23. At a certain state-supported university, annual tuition and fees have risen dramatically in recent years as shown in the table below. (4.6)

Year	Tuition and Fees	Consumer Price Index
1982–1983	$827	96.5
1987–1988	$1,404	113.6
1993–1994	$2,018	144.5
2003–2004	$4,450	184.0
2005–2006	$5,290	198.1 (est.)

Source: www.bls.gov

a. If all tuition and fees are paid at the beginning of each academic year, what is the compound annual rate of increase from 1982 to 2005?

b. What is the annual rate of increase from 1993 to 2005?

c. How do the increases in Parts (a) and (b) compare with the CPI for the same period of time?

4-24. The cost of 1,000 cubic feet of natural gas has increased from $6 in 2000 to $15 in 2006. What compounded annual increase in cost is this? How does the increase in the cost of natural gas compare to a 3% annual rate of inflation during the same period of time? (4.6)

4-25. Suppose you make 15 equal annual deposits of $1,000 each into a bank account paying 5% interest per year. The first deposit will be made one year from today. How much money can be withdrawn from this bank account immediately after the 15th deposit? (4.7)

4-26. Albert Einstein once noted that "compounding of interest is one of humanity's greatest inventions." To illustrate the mind-boggling effects of compounding, suppose $100 is invested at the end of each year for 25 years at $i = 50\%$ per year. In this case, the accumulated sum is $5,050,000! Now it is your turn. What amount is accumulated after 25 years if the interest rate is 30% per year? (4.7)

4-27. A credit card company wants your business. If you accept their offer and use their card, they will deposit 1% of your monetary transactions into a savings account that will earn a guaranteed 5% per year. If your annual transactions total an average of $20,000, how much will you have in this savings plan after 15 years? (4.7)

4-28. One of life's great lessons is to start early and save all the money you can! If you save two dollars today and two dollars each and every day thereafter until you are 60 years old, how much money will you accumulate (say, $730 per year for 35 years) if the annual interest rate is 7%? (4.7)

4-29. Liam O'Kelly is 20 years old and is thinking about buying a term life insurance policy with his wife as the beneficiary. The quoted annual premium for Liam is $8.48 per thousand dollars of insurance coverage. Because Liam wants a $100,000 policy (which is 2.5 times his annual salary), the annual premium would be $848, with the first payment due immediately (i.e., at age 21).

A friend of Liam's suggests that the $848 annual premium should be deposited in a good mutual fund rather than in the insurance policy. "If the mutual fund earns 10% per year, you can become a millionaire by the time you retire at age 65," the friend advises. (4.7)

a. Is the friend's statement really true?

b. Discuss the trade-off that Liam is making if he decides to invest his money in a mutual fund.

4-30. A basic lesson to be learned in life is to minimize annual maintenance fees (and sales commissions) on portfolios of securities (stocks and bonds). To illustrate this lesson, consider an actively managed stock market fund that charges 1% per year as its management fee. A less actively managed fund (e.g., an index fund) charges 0.5% per year. For a one-time investment of $100,000, how much will this decrease in annual management fee (i.e., 0.5%) save during a 20-year period? Let $i = 8\%$ per year and express your answer as a percentage of the original $100,000. (4.7)

4-31. Twelve payments of $10,000 each are to be repaid monthly at the end of each month. The monthly interest rate is 2%. (4.7)

a. What is the present equivalent (i.e., P_0) of these payments?

b. Repeat Part (a) when the payments are made at the beginning of the month. Note that the present equivalent will be at the same time as the first monthly payment.

c. Explain why the present equivalent amounts in Parts (a) and (b) are different.

4-32. Frank has been making annual payments of $2,500 to repay a college loan. He wishes to pay off the loan immediately after having made an annual payment, and he has four payments on the loan remaining. With an annual compound interest rate of 5%, how much should Frank pay? (4.7)

4-33. Automobiles of the future will most likely be manufactured largely with carbon fibers made from recycled plastics, wood pulp, and cellulose. Replacing half the ferrous metals in current automobiles could reduce a vehicle's weight by 60% and fuel consumption by 30%. One impediment to using carbon fibers in cars is *cost*. If the justification for the extra sticker price of carbon-fiber cars is solely based on fuel savings, how much extra sticker price can be justified over a six-year life span if the carbon-fiber car would average 39 miles per gallon of gasoline compared to a conventional car averaging 30 miles per gallon? Assume that gasoline costs $3.00 per gallon, the interest rate is 20% per year, and 117,000 miles are driven uniformly over six years. (4.7)

4-34. It is estimated that a certain piece of equipment can save $22,000 per year in labor and materials costs. The equipment has an expected life of five years

and no market value. If the company must earn a 15% annual return on such investments, how much could be justified now for the purchase of this piece of equipment? Draw a cash-flow diagram from the company's viewpoint. (4.7)

4-35. Gerard just won $2,000,000 in his state lottery. He is informed the payout will be $100,000 per year for 20 years with the first payment occurring one year from now, or he can accept a lump-sum payment of $1,000,000 right away. Gerard is disappointed that he will be receiving less than an immediate lump-sum payout of $2,000,000. If his opportunity cost of capital (i) is 8% per year, how much less than $2,000,000 will Gerard be receiving if he chooses to receive $100,000 per year for 20 years? (4.7)

4-36. Suppose that installation of low-loss thermal windows in your area is expected to save $350 a year on your home heating bill for the next 18 years. If you can earn 8% per year on other investments, how much could you afford to spend now for these windows? (4.7)

4-37. A large retailer is going to run an experiment at one of its stores to see how cost effective it will be to imbed merchandise with small radio frequency identification (RFID) chips. Shoppers will select items from the store and walk out without stopping to pay at a checkout lane. The RFID chips will record what items are taken and will automatically deduct their price from shoppers' bank accounts. The retailer expects to save the cost of staffing 12 checkout lanes, which amounts to $100,000 per year. How much can the retailer afford to spend on its RFID investment if the system has a life of 10 years and no residual value? The retailer's interest rate is 15% per year. (4.7)

4-38. A future amount of $150,000 is to be accumulated through annual payments, A, over 20 years. The last payment of A occurs simultaneously with the future amount at EOY 20. If the interest rate is 9% per year, what is the value of A? (4.7)

4-39. A 40-year-old person wants to accumulate $500,000 by age 65. How much will she need to save each month, starting one month from now, if the interest rate is 0.5% per month? (4.7)

4-40. Qwest Airlines has implemented a program to recycle all plastic drink cups used on their aircraft. Their goal is to generate $5 million by the end of the recycle program's five-year life. Each recycled cup can be sold for $0.005 (1/2 cent). (4.7)

a. How many cups must be recycled annually to meet this goal? Assume uniform annual plastic cup usage and a 0% interest rate.

b. Repeat Part (a) when the annual interest rate is 15%.

c. Why is the answer to Part (b) less than the answer to Part (a)?

4-41. Show that the following relationship is true: $(A/P, i\%, N) = i/[1 - (P/F, i\%, N)]$. (4.7)

4-42. A large electronic retailer is considering the purchase of software that will minimize shipping expenses in its supply chain network. This software, including installation and training, would be a $10-million investment for the retailer. If the firm's effective interest rate is 15% per year and the life of the software is four years, what annual savings in shipping expenses must there be to justify the purchase of the software? (4.7)

4-43. A large automobile manufacturer has developed a continuous variable transmission (CVT) that provides smooth shifting and enhances fuel efficiency by two miles per gallon of gasoline. The extra cost of a CVT is $800 on the sticker price of a new car. For a particular model averaging 28 miles per gallon with the CVT, what is the cost of gasoline (dollars per gallon) that makes this option affordable when the buyer's interest rate is 10% per year? The car will be driven 100,000 miles uniformly over an eight-year period. (4.7)

4-44. The Dell Corporation borrowed $10,000,000 at 7% interest per year, which must be repaid in equal EOY amounts (including both interest and principal) over the next six years. How much must Dell repay at the end of each year? How much of the total amount repaid is interest? (4.7)

4-45. The cost to equip a large cruise ship with security cameras is $500,000. If the interest rate is 15% per year and the cameras have a life of six years, what is the equivalent annual cost (A) of the security cameras? (4.7)

4-46. A piece of airport baggage handling equipment can be purchased for $80,000 cash or for $84,000 to be financed over 60 months at 0% interest. This special offer is good for only the next two days. The salesperson states that at least $20,000 can be saved by the 0% offer compared to their "traditional" financing plan at 0.75% per month over 60 months. Is this claim really true? Show all your work. (4.7)

4-47. The Golden Gate Bridge in San Francisco was financed with construction bonds sold for $35 million in 1931. These were 40-year bonds, and the $35 million principal plus almost $39 million in interest were repaid in total in 1971. If interest was repaid as a lump sum, what interest rate was paid on the construction bonds? (4.6)

4-48. Rework Problem 4-47 assuming the construction bonds had been retired as an annuity (i.e., equal uniform annual payments had been made to repay the $35 million). (4.7)

4-49. Consider the accompanying cash-flow diagram. (See Figure P4-49.) (4.7)

a. If $P = \$1,000$, $A = \$200$, and $i\% = 12\%$ per year, then $N = ?$

b. If $P = \$1,000$, $A = \$200$, and $N = 10$ years, then $i = ?$

c. If $A = \$200$, $i\% = 12\%$ per year, and $N = 5$ years, then $P = ?$

d. If $P = \$1,000$, $i\% = 12\%$ per year, and $N = 5$ years, then $A = ?$

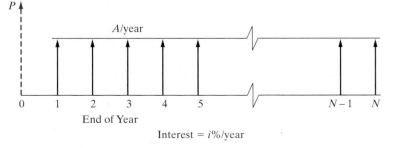

Figure P4-49 Figure for Problem 4-49

4-50. Suppose that your rich uncle has $1,000,000 that he wishes to distribute to his heirs at the rate of $100,000 per year. If the $1,000,000 is deposited in a bank account that earns 6% interest per year, how many years will it take to completely deplete the account? How long will it take if the account earns 8% interest per year instead of 6%? (4.7)

4-51. Today the average undergraduate student is responsible for paying off a $3,500 balance on his/her credit card. Suppose the monthly interest rate is 1.75% (21% APR). How many months will it take to repay the $3,500 balance, assuming monthly payments of $100 are made and no additional expenses are charged to the credit card? [What is the APR on your personal credit card(s)?] (4.7)

4-52. You have just inherited $100,000 as a lump-sum amount from your distant aunt. After depositing the money in a practically risk-free certificate of deposit (CD) earning 5% per year, you plan to withdraw $10,000 per year for your living expenses. How many years will your $100,000 last in view of these withdrawals? (*Hint:* It is longer than 10 years!) (4.7)

4-53. Jason makes six EOY deposits of $2,000 each in a savings account paying 5% compounded annually. If the accumulated account balance is withdrawn four years after the last deposit, how much money will be withdrawn? (4.7)

4-54. Kris borrows money in her senior year to buy a new car. The car dealership allows her to defer payments for 12 months, and Kris makes 36 end-of-month payments thereafter. If the original note (loan) is for $24,000 and interest is 1/2% per month on the unpaid balance, how much will Kris' payments be? (4.9)

4-55. Leon and Heidi decided to invest $3,000 annually for only the first eight years of their marriage. The first payment was made at age 25. If the annual interest rate is 10%, how much accumulated interest and principal will they have at age 65? (4.9)

4-56. How much money should be deposited each year for 12 years if you wish to withdraw $309 each year for five years, beginning at the end of the 14th year? Let $i = 8\%$ per year. (4.9)

4-57. What lump sum of money must be deposited into a bank account at the present time so that $500 per month can be withdrawn for five years, with the first withdrawal scheduled for six years from today? The interest rate is 3/4% per month. (*Hint:* Monthly withdrawals begin at the end of the month 72.) (4.9)

4-58. You can buy a machine for $100,000 that will produce a net income, after operating expenses, of $10,000 per year. If you plan to keep the machine for four years, what must the market (resale) value be at the end of four years to justify the investment? You must make a 15% annual return on your investment. (4.7)

4-59. Major overhaul expenses of $5,000 each are anticipated for a large piece of earthmoving equipment. The expenses will occur at EOY four and will continue every three years thereafter up to and including year 13. The interest rate is 12% per year. (4.10)

a. Draw a cash-flow diagram.

b. What is the present equivalent of the overhaul expenses at time 0?

c. What is the annual equivalent expense during only years 5–13?

4-60. Maintenance costs for a small bridge with an expected 50-year life are estimated to be $1,000 each year for the first 5 years, followed by a $10,000 expenditure in the year 15 and a $10,000 expenditure in the year 30. If $i = 10\%$ per year, what is the equivalent uniform annual cost over the entire 50-year period? (4.10)

4-61. It is estimated that you will pay about $80,000 into the social security system (FICA) over your 40-year work span. For simplicity, assume this is an annuity of $2,000 per year, starting with your 26th birthday and continuing through your 65th birthday. (4.10)

a. What is the future equivalent worth of your social security savings when you retire at age 65 if the government's interest rate is 6% per year?

b. What annual withdrawal can you make if you expect to live 20 years in retirement? Let $i = 6\%$ per year.

4-62. Solve for the value of Z in the accompanying figure, so that the top cash-flow diagram is equivalent to the bottom one. Let $i = 8\%$ per year. (4.10)

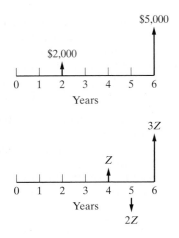

4-63. The Turners have 10 years to save a lump-sum amount for their child's college education. Today a four-year college education costs $75,000, and this is expected to increase by 10% per year into the foreseeable future. (4.10)

a. If the Turners can earn 6% per year on a conservative investment in a highly rated tax-free municipal bond, how much money must they save each year for the next 10 years to afford to send their child to college?

b. If a certain college will "freeze" the cost of education in 10 years for a lump-sum of current value $150,000, is this a good deal?

4-64. A loan of $10,000 is to be repaid over a period of eight years. During the first four years, exactly half of the *loan principal* is to be repaid (along with accumulated compound interest) by a uniform series of payments of A_1 dollars per year. The other half of the loan principal is to be repaid over four years, with accumulated interest, by a uniform series of payments of A_2 dollars per year. If $i = 9\%$ per year, what are A_1 and A_2? (4.10)

4-65. An auto dealership is running a promotional deal whereby they will replace your tires free of charge for the life of the vehicle when you purchase your car from them. You expect the original tires to last for 30,000 miles, and then they will need replacement every 30,000 miles thereafter. Your driving mileage averages 15,000 miles per year. A set of new tires costs $400. If you trade in the car at 150,000 miles with new tires then, what is the lump-sum present value of this deal if your personal interest rate is 12% per year? (4.10)

4-66. Transform the cash flows on the left-hand side of the accompanying diagram (see Figure P4-66) to their equivalent amount, F, shown on the right-hand side. The interest rate is 8% per year. (4.10)

4-67. In recent years, the United States has gone from being a "positive savings" nation to a "negative savings" nation (i.e., we spend more money than we earn). Suppose a typical American household spends $10,000 more than it makes and it does this for eight consecutive years. If this debt will be financed at an interest rate of 15% per year, what annual repayment will be required to repay the debt over a 10-year period (repayments will start at EOY 9)? (4.10)

4-68. Determine the value of W on the right-hand side of the accompanying diagram (see Figure P4-68) that makes the two cash-flow diagrams equivalent when $i = 12\%$ per year. (4.10)

4-69. A friend of yours just bought a new sports car with a $5,000 down payment, and her $30,000 car loan is financed at an interest rate of 0.75% per month for 48 months. After 2 years, the "Blue Book" value of her vehicle in the used-car marketplace is $15,000. (4.10)

a. How much does your friend still owe on the car loan immediately after she makes her 24th payment?

b. Compare your answer to Part (a) to $15,000. This situation is called being "upside down." What can she do about it? Discuss your idea(s) with your instructor.

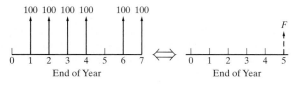

Figure P4-66   Figure for Problem 4-66

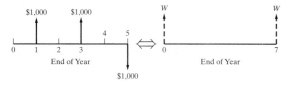

Figure P4-68   Figure for Problem 4-68

4-70. A certain fluidized-bed combustion vessel has an investment cost of $100,000, a life of 10 years, and negligible market (resale) value. Annual costs of materials, maintenance, and electric power for the vessel are expected to total $10,000. A major relining of the combustion vessel will occur during the fifth year at a cost of $30,000. If the interest rate is 15% per year, what is the lump-sum equivalent cost of this project at the present time? (4.10)

4-71. It costs $30,000 to retrofit the gasoline pumps at a certain filling station so the pumps can dispense E85 fuel (85% ethanol and 15% gasoline). If the station makes a profit of $0.08 per gallon from selling E85 and sells an average of 20,000 gallons of E85 per month, how many months will it take for the owner to recoup her $30,000 investment in the retrofitted pumps? The interest rate is 1% per month. (4.10)

4-72. An expenditure of $20,000 is made to modify a material-handling system in a small job shop. This modification will result in first-year savings of $2,000, a second-year savings of $4,000, and a savings of $5,000 per year thereafter. How many years must the system last if an 18% return on investment is required? The system is tailor made for this job shop and has no market (salvage) value at any time. (4.10)

4-73. Determine the value of P_0, as a function of H, for these two investment alternatives to be equivalent at an interest rate of $i = 15\%$ per year: (4.10)

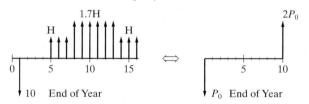

4-74. Find the uniform annual amount that is equivalent to a uniform gradient series in which the first year's payment is $500, the second year's payment is $600, the third year's payment is $700, and so on, and there is a total of 20 payments. The annual interest rate is 8%. (4.11)

4-75. Suppose that annual income from a rental property is expected to start at $1,300 per year and decrease at a uniform amount of $50 each year after the first year for the 15-year expected life of the property. The investment cost is $8,000, and i is 9% per year. Is this a good investment? Assume that the investment occurs at time zero (now) and that the annual income is first received at EOY one. (4.11)

4-76. For a repayment schedule that starts at EOY four at $Z and proceeds for years 4 through 10 at $2Z, $3Z, \ldots$, what is the value of Z if the principal of this loan is $10,000 and the interest rate is 7% per year? Use a uniform gradient amount (G) in your solution. (4.11)

4-77. Refer to the accompanying cash-flow diagram (see Figure P4-77), and solve for the unknown quantity in Parts (a) through (d) that makes the equivalent value of cash outflows equal to the equivalent value of the cash inflow, F. (4.11)

a. If $F = \$10,000$, $G = \$600$, and $N = 6$, then $i = ?$
b. If $F = \$10,000$, $G = \$600$, and $i = 5\%$ per period, then $N = ?$
c. If $G = \$1,000$, $N = 12$, and $i = 10\%$ per period, then $F = ?$
d. If $F = \$8,000$, $N = 6$, and $i = 10\%$ per period, then $G = ?$

4-78. You owe your best friend $2,000. Because you are short on cash, you offer to repay the loan over 12 months

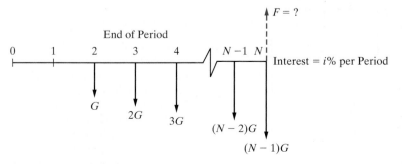

Figure P4-77 Figure for Problem 4-77

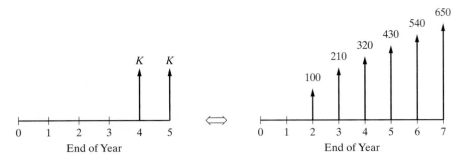

Figure P4-79 Figure for Problem 4-79

under the following condition. The first payment will be $100 at the end of month one. The second payment will be $100 + G at the end of month two. At the end of month three, you'll repay $100 + 2G. This pattern of increasing G amounts will continue for all remaining months. (4.11)

a. What is the value of G if the interest rate is 0.5% per month?

b. What is the equivalent uniform monthly payment?

c. Repeat Part (a) when the first payment is $150 (i.e., determine G).

4-79. In the accompanying diagram, Figure P4-79, what is the value of K on the left-hand cash-flow diagram that is equivalent to the right-hand cash-flow diagram? Let $i = 12\%$ per year. (4.11)

4-80. Calculate the future equivalent at the end of 2005, at 8% per year, of the following series of cash flows in Figure P4-80: [Use a uniform gradient amount (G) in your solution.] (4.11)

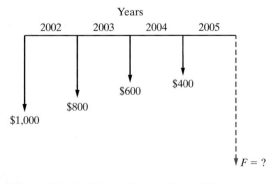

Figure P4-80 Figure for Problem 4-80

4-81. Suppose that the parents of a young child decide to make annual deposits into a savings account, with the first deposit being made on the child's fifth birthday and the last deposit being made on the 15th birthday. Then, starting on the child's 18th birthday, the withdrawals as shown will be made. If the effective annual interest rate is 8% during this period of time, what are the annual deposits in years 5 through 15? Use a uniform gradient amount (G) in your solution. (See Figure P4-81.) (4.11)

4-82. The heat loss through the exterior walls of a certain poultry processing plant is estimated to cost the owner $3,000 next year. A salesman from Superfiber Insulation, Inc., has told you, the plant engineer, that he can reduce the heat loss by 80% with the installation of $18,000 worth of Superfiber now. If the cost of heat loss rises by $200 per year (uniform gradient) after the next year and the owner plans to keep the present building for 15 more years, what would you recommend if the interest rate is 10% per year? (4.11)

4-83. What value of N comes closest to making the left-hand cash-flow diagram of the accompanying figure, Figure P4-83, equivalent to the one on the right? Let $i = 15\%$ per year. Use a uniform gradient amount (G) in your solution. (4.11)

4-84. A retail outlet is being designed in a strip mall in Nebraska. For this outlet, the installed fiberglass insulation to protect against heat loss in the winter and heat gain in the summer will cost an estimated $100,000. The annual savings in energy due to the insulation will be $18,000 at EOY one in the 10-year life of the outlet, and these savings will increase by 12% each year thereafter. If the annual interest rate is 15%, is the cost of the proposed amount of insulation justified? (4.12)

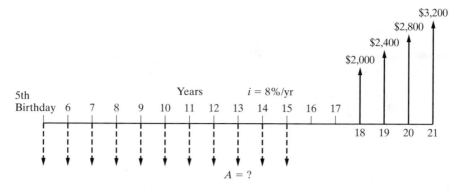

Figure P4-81 Figure for Problem 4-81

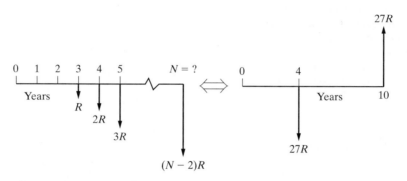

Figure P4-83 Figure for Problem 4-83

4-85. You are the manager of a large crude-oil refinery. As part of the refining process, a certain heat exchanger (operated at high temperatures and with abrasive material flowing through it) must be replaced every year. The replacement and downtime cost in the first year is $175,000. This cost is expected to increase due to inflation at a rate of 8% per year for five years, at which time this particular heat exchanger will no longer be needed. If the company's cost of capital is 18% per year, how much could you afford to spend for a higher quality heat exchanger so that these annual replacement and downtime costs could be eliminated? (4.12)

4-86. A geometric gradient has a positive cash flow of $1,000 at EOY zero (now), and it increases 5% per year for the following five years. Another geometric gradient has a positive value of $2,000 at EOY one, and it decreases 6% per year for years two through five. If the annual interest rate is 10%, which geometric gradient would you prefer? (4.12)

4-87. A geometric gradient that increases at $\bar{f} = 6\%$ per year for 15 years is shown in the accompanying diagram. The annual interest rate is 12%. What is the present equivalent value of this gradient? (4.12)

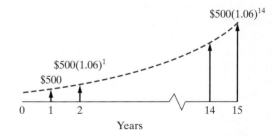

4-88. A small company heats its building and spends $8,000 per year on natural gas for this purpose. Cost increases of natural gas are expected to be 10% per year starting one year from now (i.e., the first cash flow is $8,800 at EOY one). Their maintenance on the gas furnace is $345 per year, and this expense is expected to increase by 15% per year starting one year from now. If the planning horizon is 15 years, what is the total annual equivalent expense for operating and maintaining the furnace? The interest rate is 18% per year. (4.12)

4-89. In a geometric sequence of annual cash flows starting at EOY *zero*, the value of A_0 is $1,304.35 (which *is* a cash flow). The value of the last term in the series, A_{10}, is $5,276.82. What is the equivalent value of A for years 1 through 10? Let $i = 20\%$ per year. (4.12)

4-90. Amy Parker, a 22-year-old and newly hired marine biologist, is quick to admit that she does not plan to keep close tabs on how her 401(k) retirement plan will grow with time. This sort of thing does not really interest her. Amy's contribution, plus that of her employer, amounts to $2,200 per year starting at age 23. Amy expects this amount to increase by 3% each year until she retires at the age of 62 (there will be 40 EOY payments). What is the compounded future value of Amy's 401(k) plan if it earns 7% per year? (4.12)

4-91. An electronic device is available that will reduce this year's labor costs by $10,000. The equipment is expected to last for eight years. If labor costs increase at an average rate of 7% per year and the interest rate is 12% per year. (4.12)

a. What is the maximum amount that we could justify spending for the device?

b. What is the uniform annual equivalent value (A) of labor costs over the eight-year period?

4-92. An amount, P, must be invested now to allow withdrawals of $1,000 per year for the next 15 years and to permit $300 to be withdrawn starting at the end of year 6 and continuing over the remainder of the 15-year period as the $300 increases by 6% per year thereafter. That is, the withdrawal at EOY seven will be $318, $337.08 at EOY eight, and so forth for the remaining years. The interest rate is 12% per year. *Hint:* Draw a cash-flow diagram. (4.12)

4-93. Consider an EOY geometric gradient, which lasts for eight years, whose initial value at EOY one is $5,000 and $\bar{f} = 6\%$ per year thereafter. Find the equivalent uniform gradient amount over the same period if the initial value of the cash flows at the end of year one is $4,000. Complete the following questions to determine the value of the gradient amount, G. The interest rate is 8% per year. (4.12)

a. What is P_0 for the geometric gradient series?

b. What is P_0' of the uniform (arithmetic) gradient of cash flows?

c. What is the value of G?

4-94. An individual makes five annual deposits of $2,000 in a savings account that pays interest at a rate of 4% per year. One year after making the last deposit, the interest rate changes to 6% per year. Five years after the last deposit, the accumulated money is withdrawn from the account. How much is withdrawn? (4.13)

4-95. A person has made an arrangement to borrow $1,000 now and another $1,000 two years hence. The entire obligation is to be repaid at the end of four years. If the projected interest rates in years one, two, three, and four are 10%, 12%, 12%, and 14%, respectively, how much will be repaid as a lump-sum amount at the end of four years? (4.13)

4-96. Suppose that you have a money market certificate earning an annual rate of interest, which varies over time as follows:

Year k	1	2	3	4	5
i_k	14%	12%	10%	10%	12%

If you invest $10,000 in this certificate at the beginning of year one and do not add or withdraw any money for five years, what is the value of the certificate at the end of the fifth year? (4.13)

4-97. Determine the present equivalent value of the cash-flow diagram of Figure P4-97 when the annual interest rate, i_k, varies as indicated. (4.13)

4-98. Mary's credit card situation is out of control because she cannot afford to make her monthly payments. She has three credit cards with the following loan balances and APRs: Card 1, $4,500, 21%; Card 2, $5,700, 24%; and Card 3, $3,200, 18%. Interest compounds monthly on all loan balances. A credit card loan consolidation company has captured Mary's attention by stating they can save Mary 25% per month on her credit card payments. This company charges 16.5% APR. Is the company's claim correct? (4.14)

4-99. Compute the effective annual interest rate in each of these situations: (4.14)

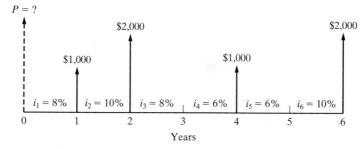

Figure P4-97 Figure for Problem 4-97

a. 10% nominal interest, compounded semiannually.

b. 10% nominal interest compounded quarterly.

c. 10% nominal interest compounded weekly.

4-100. An APR of 5.875% produces an effective annual interest rate of 6.04%. What is the compounding frequency (M) in this situation? (4.14)

4-101. What extra semiannual expenditure for five years would be justified for the maintenance of a machine in order to avoid an overhaul costing $3,000 at the end of five years? Assume nominal interest at 8%, compounded semiannually. (4.15)

4-102. A mortgage banking company has been evaluating the merits of a 50-year mortgage (in addition to their popular 30-year mortgage). The basic idea is to reduce the monthly payment and make home ownership more affordable. The APR of either mortgage is 6% and the compounding is monthly. (4.15)

a. For a mortgage loan of $300,000, what is the difference in monthly payment for the 30-year mortgage and the 50-year mortgage?

b. What is the difference in total interest paid between the two mortgages?

4-103. Determine the current amount of money that must be invested at 12% nominal interest, compounded monthly, to provide an annuity of $10,000 (per year) for 6 years, starting 12 years from now. The interest rate remains constant over this entire period of time. (4.15)

4-104. To pay off $50,000,000 worth of new construction bonds when they come due in 20 years, a water municipality must deposit money into a sinking fund. Payments to the fund will be made quarterly, starting three months from now. If the interest rate for the sinking fund is 8% compounded quarterly, how much will each deposit be? (4.15)

4-105. Suppose that you have just borrowed $7,500 at 12% nominal interest compounded quarterly. What is the total lump-sum, compounded amount to be paid by you at the end of a 10-year loan period? (4.15)

4-106. A light-duty pickup truck has a manufacturer's suggested retail price (MSRP) of $14,000 on its window. After haggling with the salesperson for several days, the prospective buyer is offered the following deal: "You pay a $1,238 down payment now and $249 each month thereafter for 39 months and the truck will be yours." The APR at this dealership is 2.4% compounded monthly. How good a deal is this relative to the MSRP? (4.15)

4-107. You borrow $10,000 from a bank for three years at an annual interest rate, or annual percentage rate (APR), of 12%. Monthly payments will be made until all the principal and interest have been repaid. (4.14, 4.15)

a. What is your monthly payment?

b. If you must pay two points up front, meaning that you get only $9,800 from the bank, what is your true APR on the loan?

4-108. Many people get ready for retirement by depositing money into a monthly or annual savings plan. (4.15)

a. If $200 per month is deposited in a bank account paying a 6% APR compounded monthly, how much will be accumulated in the account after 30 years?

b. If inflation is expected to average 2% per year into the foreseeable future, what is today's equivalent spending power for your answer to Part (a)?

4-109. The longer a loan schedule lasts, the more interest you will pay. To illustrate, a $20,000-car-loan at 9% APR (compounded monthly) for three years (36 monthly payments) will incur total interest of

$2,896. But a $20,000 car loan at 9% APR over six years (72 payments) will have a total interest cost of $5,920. (4.15)

a. Verify that the difference in total interest of $3,024 is correct.

b. Why would you be willing to pay this extra interest?

4-110. On January 1, 2005, a person's savings account was worth $200,000. Every month thereafter, this person makes a cash contribution of $676 to the account. If the fund is expected to be worth $400,000 on January 1, 2010, what annual rate of interest is being earned on this fund? (4.15)

4-111. Suppose you owe $1,100 on your credit card. The annual percentage rate (APR) is 18%, compounded monthly. The credit card company says your minimum monthly payment is $19.80. (4.15)

a. If you make only this minimum payment, how long will it take for you to repay the $1,100 balance (assuming no more charges are made)?

b. If you make the minimum payment plus $10 extra each month (for a total of $29.80), how long will it take to repay the $1,100 balance?

c. Compare the total interest paid in Part (a) with the total interest paid in Part (b).

4-112. An effective annual interest rate of 35% has been determined with continuous compounding. What is the nominal interest rate that was compounded continuously to get this number? (4.16)

4-113. A nominal interest rate of 11.333%, continuously compounded, yields an effective annual interest rate of how much? (4.16)

4-114. A bank offers a nominal interest rate (APR) of 6%, continuously compounded. What is the effective interest rate? (4.16)

4-115. If a nominal interest rate of 8% is compounded continuously, determine the unknown quantity in each of the following situations: (4.16)

a. What uniform EOY amount for 10 years is equivalent to $8,000 at EOY 10?

b. What is the present equivalent value of $1,000 per year for 12 years?

c. What is the future equivalent at the end of the sixth year of $243 payments made every six months during the six years? The first payment occurs six months from the present and the last occurs at the end of the sixth year.

d. Find the equivalent lump-sum amount at EOY nine when $P_0 = $1,000$.

4-116. Find the value of the unknown quantity Z in the following diagram, such that the equivalent cash outflow equals the equivalent cash inflows when $\underline{r} = 20\%$ compounded continuously: (4.16)

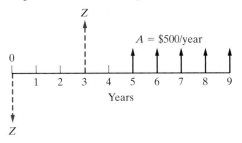

4-117. Juan deposits $5,000 into a savings account that pays 7.2% per year, continuously compounded. What is the effective annual interest rate? Determine the value of his account at the end of two years. (4.16)

4-118. A man deposited $10,000 in a savings account when his son was born. The nominal interest rate was 8% per year, compounded continuously. On the son's 18th birthday, the accumulated sum is withdrawn from the account. How much will this accumulated amount be? (4.16)

4-119. A person needs $18,000 immediately as a down payment on a new home. Suppose that she can borrow this money from her company credit union. She will be required to repay the loan in equal payments *made every six months* over the next 12 years. The annual interest rate being charged is 10% compounded continuously. What is the amount of each payment? (4.16)

4-120. What is the present equivalent of a uniform series of annual payments of $3,500 each for five years if the interest rate, compounded continuously, is 10%? (4.16)

4-121. The amount of $7,000 is invested in a Certificate of Deposit (CD) and will be worth $16,000 in nine years. What is the continuously compounded nominal (annual) interest rate for this CD? (4.16)

4-122. Many persons prepare for retirement by making monthly contributions to a savings program. Suppose that $2,000 is set aside each year and invested in a savings account that pays 10% interest per year, compounded continuously. (4.16)

a. Determine the accumulated savings in this account at the end of 30 years.

b. In Part (a), suppose that an annuity will be withdrawn from savings that have been accumulated at the EOY 30. The annuity will extend from the EOY 31 to the EOY 40. What is the value of this annuity if the interest rate and compounding frequency in Part (a) do not change?

4-123. Indicate whether each of the following statements is true (T) or false (F). (all sections)

a. T F Interest is money paid for the use of equity capital.

b. T F $(A/F, i\%, N) = (A/P, i\%, N) + i$.

c. T F Simple interest ignores the time value of money principle.

d. T F Cash-flow diagrams are analogous to free-body diagrams for mechanics problems.

e. T F $1,791 10 years from now is equivalent to $900 now if the interest rate equals 8% per year.

f. T F It is always true that $i > r$ when $M \geq 2$.

g. T F Suppose that a lump sum of $1,000 is invested at $\underline{r} = 10\%$ for eight years. The future equivalent is greater for daily compounding than it is for continuous compounding.

h. T F For a fixed amount, F dollars, that is received at EOY N, the "A equivalent" increases as the interest rate increases.

i. T F For a specified value of F at EOY N, P at time zero will be larger for $\underline{r} = 10\%$ per year than it will be for $r = 10\%$ per year, compounded monthly.

4-124. Mark each statement true (T) or false (F), and fill in the blanks in Part (e). (all sections)

a. T F The nominal interest rate will always be less than the effective interest rate when $r = 10\%$ and $M = \infty$.

b. T F A certain loan involves monthly repayments of $185 over a 24-month period. If $\underline{r} = 10\%$ per year, more than half of the principal is still owed on this loan after the 10th monthly payment is made.

c. T F $1,791 in 10 years is equivalent to $900 now if the nominal interest rate is 8% compounded semiannually.

d. T F The $(P/A, i\%, N)$ factor equals $N \cdot (P/F, i\%, 1)$.

e. Fill in the missing interest factor:

    i. $(P/A, i\%, N)(\text{————}) = (F/A, i\%, N)$.

    ii. $(A/G, i\%, N)(P/A, i\%, N) = (\text{————})$.

4-125. A mutual fund investment is expected to earn 11% per year for the next 25 years. If inflation will average 3% per year during this 25-year period of time, what is the compounded value (in today's dollars) of this savings vehicle when $10,000 is invested now? (4.6)

4-126. Javier just bought a condominium in Collegetown, USA. His $100,000 mortgage is 6% compounded monthly, and Javier will make monthly payments on his loan for 30 years. In addition, property taxes and title insurance amount to $400 per month. (4.15)

a. What is the total mortgage-related amount of Javier's monthly condo payment?

b. Develop an estimate of Javier's total monthly expenses (maintenance, utilities, and so on) for his condominium.

c. If Javier qualifies for a 15-year mortgage having an APR of 5.8% compounded monthly, what will his monthly mortgage payment be (there will be 180 payments)?

4-127. Analyze the truth of this statement, assuming you are 20 years old: "For every five years that you wait to start accumulating money for your retirement, it takes twice as much savings per year to catch up." Base your analysis on the target of having $1,000,000 when you retire at age 60. Be sure to state your assumptions. (4.7)

4-128. A mortgage company advertises that their 6% APR is an effective annual rate of 6.58% with monthly payments and compounding. Let's assume this is made possible by paying points on a mortgage (one point is 1% of the loan amount). How many points are being paid up front on a $100,000 mortgage loan over 15 years to arrive at an effective interest rate of 6.58%? (4.15)

Spreadsheet Exercises

4-129. Create a spreadsheet that duplicates Table 4-1. Make it flexible enough that you can investigate the impact of different interest rates and principal loan amounts without changing the structure of the spreadsheet. (4.7)

4-130. A $15,000 investment is to be made with anticipated annual returns as shown in the spreadsheet in Figure P4-130. If the investor's time value of money is 10% per year, what should be entered in cells B11, B12, and B13 to obtain present, annual, and future equivalent values for the investment? (4.10)

4-131. Refer to Example 4-23. Suppose the cash-flow sequence continues for 10 years (instead of four). Determine the new values of P, A, and F. (4.12)

4-132. Higher interest rates won't cost you as much to drive a car as do higher gasoline prices. This is because automobile loans have payment schedules that are only modestly impacted by Federal Reserve Board interest-rate increases. Create a spreadsheet to compare the difference in monthly payments for a $25,000 loan having 60 monthly payments for a select number of interest rates. Use 3% as the base APR and go as high

as an APR of 12%. *Challenge:* Make your spreadsheet flexible enough to be able to look at the impact of the different interest rates for different loan amounts and different repayment periods. (4.15)

	A	B
1	EOY	Cash flow
2	0	−$15,000
3	1	$2,000
4	2	$2,500
5	3	$3,000
6	4	$3,500
7	5	$4,000
8	6	$4,000
9	7	$4,000
10	8	$4,000
11	P	
12	A	
13	F	

Figure P4-130   Spreadsheet for Problem 4-130

Case Study Exercises

4-133. After Enrico's car is paid off, he plans to continue setting aside the amount of his car payment to accumulate funds for the car's replacement. If he invests this amount at a rate of 3% compounded monthly, how much will he have saved by the end of the initial 10-year period? (4.17)

4-134. Enrico has planned to have $40,000 at the end of 10 years to place a down payment on a condo. Property taxes and insurance can be as much as 30%

of the monthly principal and interest payment (i.e., for a principal and interest payment of $1,000, taxes and insurance would be an additional $300). What is the maximum purchase price he can afford if he'd like to keep his housing costs at $950 per month? (4.17)

4-135. If Enrico is more daring with his retirement investment savings and feels he can average 10% per year, how much will he have accumulated for retirement at the end of the 10-year period? (4.17)

FE Practice Problems

4-136. If you borrow $3,000 at 6% simple interest per year for seven years, how much will you have to repay at the end of seven years? (4.2)

(a) $3,000   (b) $4,511   (c) $1,260
(d) $1,511   (e) $4,260

4-137. When you were born, your grandfather established a trust fund for you in the Cayman

Islands. The account has been earning interest at the rate of 10% per year. If this account will be worth $100,000 on your 25th birthday, how much did your grandfather deposit on the day you were born? (4.6)

(a) $4,000   (b) $9,230   (c) $10,000
(d) $10,150   (e) $10,740

4-138. Every year you deposit $2,000 into an account that earns 2% interest per year. What will be the balance of your account immediately after the 30th deposit? (4.7)

(a) $44,793   (b) $60,000   (c) $77,385
(d) $81,136   (e) $82,759

4-139. Your monthly mortgage payment (principal plus interest) is $1,500. If you have a 30-year loan with a fixed interest rate of 0.5% per month, how much did you borrow from the bank to purchase your house? Select the closest answer. (4.7)

(a) $154,000   (b) $180,000   (c) $250,000
(d) $300,000   (e) $540,000

4-140. Consider the following sequence of year-end cash flows:

EOY	1	2	3	4	5
Cash Flow	$8,000	$15,000	$22,000	$29,000	$36,000

What is the uniform annual equivalent if the interest rate is 12% per year? (4.11)

(a) $20,422   (b) $17,511   (c) $23,204
(d) $22,000   (e) $12,422

4-141. A cash flow at time zero (now) of $9,982 is equivalent to another cash flow that is an EOY annuity of $2,500 over five years. Each of these two cash-flow series is equivalent to a third series, which is a uniform gradient series. What is the value of G for this third series over the same five-year time interval? (4.11)

(a) $994   (b) $1,150   (c) $1,250
(d) $1,354   (e) Not enough information given

4-142. What is the monthly payment on a loan of $15,000 for five years at a nominal rate of interest of 9% compounded monthly? (4.15)

(a) $214.   (b) $250.   (c) $312.
(d) $324.   (e) $381.

4-143. Sixty monthly deposits are made into an account paying 6% nominal interest compounded monthly. If the objective of these deposits is to accumulate $100,000 by the end of the fifth year, what is the amount of each deposit? (4.15)

(a) $1,930.   (b) $1,478.   (c) $1,667.
(d) $1,430.   (e) $1,695.

4-144. What is the principal remaining after 20 monthly payments have been made on a $20,000 five-year loan? The annual interest rate is 12% nominal compounded monthly. (4.15)

(a) $10,224.   (b) $13,333.   (c) $14,579.
(d) $16,073.   (e) $17,094.

4-145. If you borrow $5,000 to buy a car at 12% compounded monthly, to be repaid over the next four years, what is your monthly payment? (4.15)

(a) $131   (b) $137   (c) $1,646
(d) $81   (e) $104

4-146. The effective annual interest rate is given to be 19.2%. What is the nominal interest rate per year (r) if continuous compounding is being used? Choose the closest answer below. (4.16)

(a) 19.83%   (b) 18.55%   (c) 17.56%
(d) 16.90%

4-147. A bank advertises mortgages at 12% compounded continuously. What is the effective annual interest? (4.16)

(a) 12.36%   (b) 12.55%   (c) 12.75%
(d) 12.68%   (e) 12.00%

4-148. If you invest $7,000 at 12% compounded continuously, how much would it be worth in three years? (4.16)

(a) $9,449   (b) $4,883   (c) $10,033
(d) $9,834   (e) $2,520

CHAPTER 5

Evaluating a Single Project

The objective of Chapter 5 is to discuss and critique contemporary methods for determining project profitability.

Solar-Powered Air-Conditioning*

For $25,000 a consumer can purchase a solar salt water–evaporative system for home cooling and heating. The technology stores solar energy in a slurry of salt and water. It was developed in Sweden and is capable of trimming $130 per month from typical air-conditioning and heating energy bills. Additionally, the system can reduce carbon dioxide output by 13 tons a year compared with traditional electrical cooling/heating of a standard home. The big question is, will home (and building) owners be willing to make the trade-off between reduced electricity consumption and reduced carbon dioxide emissions versus the extra capital investment of about $10,000 for a standard home? The firm that manufactures and markets the solar-based system intends to lobby vigorously in favor of more sustainable life styles that will dovetail with the adoption of their system. This change in life style probably lies at the heart of the firm's marketing challenge—the company's CEO has stated that "changing behavior is more difficult than changing technology." In this chapter, you will learn how to evaluate the economic advisability of this solar-powered system.

---

* Source: Adam Smith, *Time*, January 8, 2007, page A10.

5.1 Introduction

All engineering economy studies of capital projects should consider the return that a given project will or should produce. A basic question this book addresses is whether a proposed capital investment and its associated expenditures can be recovered by revenue (or savings) over time *in addition to* a return on the capital that is sufficiently attractive in view of the risks involved and the potential alternative uses. The interest and money–time relationships discussed in Chapter 4 emerge as essential ingredients in answering this question, and they are applied to many different types of problems in this chapter.

Because patterns of capital investment, revenue (or savings) cash flows, and expense cash flows can be quite different in various projects, there is no single method for performing engineering economic analyses that is ideal for all cases. Consequently, several methods are commonly used. A project focus will be taken as we introduce ways of gauging profitability.

In this chapter, we concentrate on the correct use of five methods for evaluating the economic profitability of a single proposed problem solution (i.e., *alternative*).* Later, in Chapter 6, multiple alternatives are evaluated. The five methods described in Chapter 5 are Present Worth (PW), Future Worth (FW), Annual Worth (AW), Internal Rate of Return (IRR), and External Rate of Return (ERR). The first three methods convert cash flows resulting from a proposed problem solution into their equivalent worth at some point (or points) in time by using an interest rate known as the *Minimum Attractive Rate of Return (MARR)*. The concept of a MARR, as well as the determination of its value, is discussed in the next section. The IRR and ERR methods compute annual rates of profit, or returns, resulting from an investment and are then compared to the MARR.

The payback period is also discussed briefly in this chapter. The payback period is a measure of the *speed* with which an investment is recovered by the cash inflows it produces. This measure, in its most common form, ignores time value of money principles. For this reason, the payback method is often used to supplement information produced by the five primary methods featured in this chapter.

Unless otherwise specified, the end-of-period cash flow convention and discrete compounding of interest are used throughout this and subsequent chapters. A planning horizon, or study (analysis) period, of N compounding periods (usually years) is used to evaluate prospective investments throughout the remainder of the book.

---

* The analysis of engineering projects using the benefit–cost ratio method is discussed in Chapter 10.

5.2 Determining the Minimum Attractive Rate of Return (MARR)

The Minimum Attractive Rate of Return (MARR) is usually a policy issue resolved by the top management of an organization in view of numerous considerations. Among these considerations are the following:

1. The amount of money available for investment, and the source and cost of these funds (i.e., equity funds or borrowed funds)

2. The number of good projects available for investment and their purpose (i.e., whether they sustain present operations and are *essential,* or whether they expand on present operations and are *elective*)

3. The amount of perceived risk associated with investment opportunities available to the firm and the estimated cost of administering projects over short planning horizons versus long planning horizons.

4. The type of organization involved (i.e., government, public utility, or private industry).

In theory, the MARR, which is sometimes called the *hurdle rate,* should be chosen to maximize the economic well-being of an organization, subject to the types of considerations just listed. How an individual firm accomplishes this in practice is far from clear-cut and is frequently the subject of discussion. One popular approach to establishing a MARR involves the *opportunity cost* viewpoint described in Chapter 2, and it results from the phenomenon of *capital rationing.* This situation may arise when the amount of available capital is insufficient to sponsor all worthy investment opportunities. The subject of capital rationing is covered in Chapter 13.

A simple example of capital rationing is given in Figure 5-1, where the cumulative investment requirements of seven acceptable projects are plotted against the prospective annual rate of profit of each. Figure 5-1 shows a limit of $600 million on available capital. In view of this limitation, the last funded project would be *E*, with a prospective rate of profit of 19% per year, and the best *rejected* project is *F*. In this case, the MARR by the opportunity cost principle would be 16% per year. By *not* being able to invest in project *F*, the firm would presumably be forfeiting the chance to realize a 16% annual return. As the amount of investment capital and opportunities available change over time, the firm's MARR will also change.

Superimposed on Figure 5-1 is the approximate cost of obtaining the $600 million, illustrating that project *E* is acceptable only as long as its annual rate of profit exceeds the cost of raising the last $100 million. As shown in Figure 5-1, the cost of capital will tend to increase gradually as larger sums of money are acquired through increased borrowing (debt) or new issuances of common stock (equity). Determining the MARR is discussed further in Chapter 13.

5.3 The Present Worth Method

The PW method is based on the concept of equivalent worth of all cash flows relative to some base or beginning point in time called the present. That is, all cash inflows and outflows are discounted to the present point in time at an interest rate

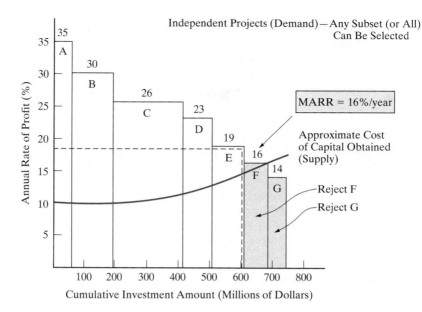

Figure 5-1 Determination of the MARR Based on the Opportunity Cost Viewpoint. (A popular measure of annual rate of profit is "Internal Rate of Return," discussed later in this chapter.)

that is generally the MARR. A positive PW for an investment project is a dollar amount of profit over the minimum amount required by investors. It is assumed that cash generated by the alternative is available for other uses that earn interest at a rate equal to the MARR.

To find the PW as a function of $i\%$ (per interest period) of a series of cash inflows and outflows, it is necessary to discount future amounts to the present by using the interest rate over the appropriate study period (years, for example) in the following manner:

$$\text{PW}(i\%) = F_0(1+i)^0 + F_1(1+i)^{-1} + F_2(1+i)^{-2}$$
$$+ \cdots + F_k(1+i)^{-k} + \cdots + F_N(1+i)^{-N}$$
$$= \sum_{k=0}^{N} F_k(1+i)^{-k}. \tag{5-1}$$

Here, i = effective interest rate, or MARR, per compounding period;
k = index for each compounding period ($0 \le k \le N$);
F_k = future cash flow at the end of period k;
N = number of compounding periods in the planning horizon (i.e., study period).

The relationship given in Equation (5-1) is based on the assumption of a *constant interest rate* throughout the life of a particular project. If the interest rate is assumed to change, the PW must be computed in two or more steps, as was illustrated in Chapter 4.

To apply the PW method of determining a project's economic worthiness, we simply compute the present equivalent of all cash flows using the MARR as the interest rate. If the present worth is greater than or equal to zero, the project is acceptable.

PW Decision Rule: If PW (i = MARR) ≥ 0, the project is economically justified.

It is important to observe that the higher the interest rate and the farther into the future a cash flow occurs, the lower its PW is. This is shown graphically in Figure 5-2. The PW of $1,000 10 years from now is $613.90 when i = 5% per year. However, if i = 10%, that same $1,000 is only worth $385.50 now.

Figure 5-2   PW of $1,000 Received at the End of Year k at an Interest Rate of i% per Year

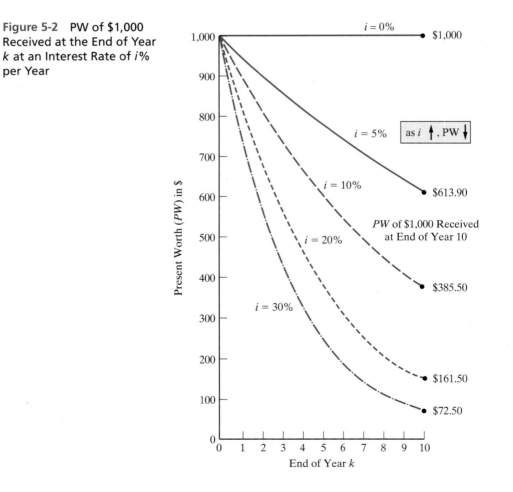

EXAMPLE 5-1	Evaluation of New Equipment Purchase Using PW

A piece of new equipment has been proposed by engineers to increase the productivity of a certain manual welding operation. The investment cost is $25,000, and the equipment will have a market value of $5,000 at the end of a study period of five years. Increased productivity attributable to the equipment will amount to $8,000 per year after extra operating costs have been subtracted from the revenue generated by the additional production. A cash-flow diagram for this investment opportunity is given below. If the firm's MARR is 20% per year, is this proposal a sound one? Use the PW method.

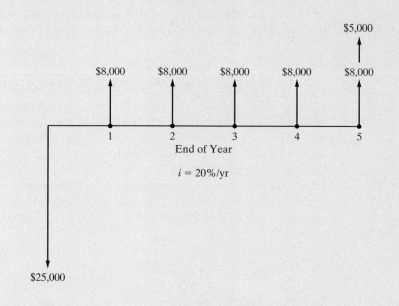

Solution

$$PW = PW \text{ of cash inflows} - PW \text{ of cash outflows,}$$

or

$$PW(20\%) = \$8,000(P/A, 20\%, 5) + \$5,000(P/F, 20\%, 5) - \$25,000$$

$$= \$934.29.$$

Because PW(20%) $\geq$ 0, this equipment is economically justified.

The MARR in Example 5-1 (and in other examples throughout this chapter) is to be interpreted as an effective interest rate (i). Here, $i = 20\%$ per year. Cash flows are discrete, end-of-year (EOY) amounts. If *continuous compounding* had been specified for a nominal interest rate (r) of 20% per year, the PW would have been calculated by using the interest factors presented in Appendix D:

$$PW(\underline{r} = 20\%) = -\$25,000 + \$8,000(P/A, \underline{r} = 20\%, 5)$$

$$+\$5,000(P/F, \underline{r} = 20\%, 5)$$

$$= -\$25,000 + \$8,000(2.8551) + \$5,000(0.3679)$$

$$= -\$319.60.$$

Consequently, with continuous compounding, the equipment would not be economically justifiable. The reason is that the higher effective annual interest rate ($e^{0.20} - 1 = 0.2214$) reduces the PW of future positive cash flows but does not affect the PW of the capital invested at the beginning of year one.

EXAMPLE 5-2	Present Worth of a Space-Heating System

A retrofitted space-heating system is being considered for a small office building. The system can be purchased and installed for $110,000, and it will save an estimated 300,000 kilowatt-hours (kWh) of electric power each year over a six-year period. A kilowatt-hour of electricity costs $0.10, and the company uses a MARR of 15% per year in its economic evaluations of refurbished systems. The market value of the system will be $8,000 at the end of six years, and additional annual operating and maintenance expenses are negligible. Use the PW method to determine whether this system should be installed.

Solution

To find the PW of the proposed heating system, we need to find the present equivalent of all associated cash flows. The estimated annual savings in electrical power is worth 300,000 kWh × $0.10/kWh = $30,000 per year. At a MARR of 15%, we get

$$PW(15\%) = -\$110,000 + \$30,000 \ (P/A, 15\%, 6) + \$8,000 \ (P/F, 15\%, 6)$$

$$= -\$110,000 + \$30,000(3.7845) + \$8,000(0.4323)$$

$$= \$6,993.40.$$

Since PW(15%) ≥ 0, we conclude that the retrofitted space-heating system should be installed.

Now that we know how to apply the PW method, we can use it to evaluate the economic advisability of the solar-powered cooling and heating system described in the chapter opener. Should a homeowner's incremental investment of $10,000 be traded off for energy savings of $130 per month? Let's assume a MARR of

0.5% per month and a 20-year useful life of the solar-powered system. In this case, the PW of the system is

$$PW = -\$10,000 + \$130(P/A, 0.5\%, \text{ per month, } 240 \text{ months})$$
$$= -\$10,000 + \$130(139.5808)$$
$$= \$8,145.50.$$

The positive-valued PW signals a favorable investment. Additionally, 13 tons/ year × 20 years = 260 tons of carbon dioxide will be avoided. Can you rework this problem when the MARR is 1% per month? Is the system still a judicious choice?

5.3.1   Assumptions of the PW Method

There are several noteworthy assumptions that we make when using PW to model the wealth-creating promise of a capital investment opportunity. First, it is assumed we know the future with certainty (we don't live in a certain world!). For example, we presume to know with certainty future interest rates and other factors. Second, it is assumed we can borrow and lend money at the same interest rate (i.e., capital markets are perfect). Regrettably, the real world has neither certainty nor perfect (*frictionless*) capital markets.

The PW (and FW and AW, to follow) model is built on these seemingly restrictive assumptions, but it is cost-beneficial in the sense that the cost of using the PW model is less than the benefits of improved decisions resulting from PW analysis. More sophisticated models exist, but they usually do not reverse decisions made with the PW model. Therefore, our goal is to cost-beneficially recommend capital investments that maximize the wealth of a firm to its owners (i.e., stockholders). A positive-valued PW (and FW and AW) means that accepting a project will increase the worth, or value, of the firm.

5.3.2   Bond Value

A bond provides an excellent example of commercial value as being the PW of the future net cash flows that are expected to be received through ownership of an interest-bearing certificate. Thus, the value of a bond, at any time, is the PW of future cash receipts. For a bond, let

Z = face, or par, value;
C = redemption or disposal price (usually equal to Z);
r = bond rate (nominal interest) per interest period;
N = number of periods before redemption;
i = bond *yield* rate per period;
V_N = value (price) of the bond N interest periods prior to
      redemption—this is a PW measure of merit.

The owner of a bond is paid two types of payments by the borrower. The first consists of the series of periodic interest payments he or she will receive until the bond is retired. There will be N such payments, each amounting to rZ. These constitute an annuity of N payments. In addition, when the bond is retired or sold, the bondholder will receive a single payment equal in amount to C. The PW of the bond is the sum of PWs of these two types of payments at the bond's yield rate ($i\%$):

$$V_N = C(P/F, i\%, N) + rZ(P/A, i\%, N). \tag{5-2}$$

The most common situations faced by you as a potential investor in bonds are (1) for a desired yield rate, how much should you be willing to pay for the bond and (2) for a stated purchase price, what will your yield be? Examples 5-3 and 5-4 demonstrate how to solve these types of problems.

EXAMPLE 5-3	**Stan Moneymaker Wants to Buy a Bond**

Stan Moneymaker has the opportunity to purchase a certain U.S. Treasury bond that matures in eight years and has a face value of $10,000. This means that Stan will receive $10,000 cash when the bond's maturity date is reached. The bond stipulates a fixed nominal interest rate of 8% per year, but interest payments are made to the bondholder every three months; therefore, each payment amounts to 2% of the face value.

Stan would like to earn 10% nominal interest (compounded quarterly) per year on his investment, because interest rates in the economy have risen since the bond was issued. How much should Stan be willing to pay for the bond?

Solution

To establish the value of this bond, in view of the stated conditions, the PW of future cash flows during the next eight years (the study period) must be evaluated. Interest payments are quarterly. Because Stan Moneymaker desires to obtain 10% *nominal interest per year* on the investment, the PW is computed at $i = 10\%/4 = 2.5\%$ per quarter for the remaining $8(4) = 32$ quarters of the bond's life:

$$V_N = \$10,000(P/F, 2.5\%, 32) + \$10,000(0.02)(P/A, 2.5\%, 32)$$
$$= \$4,537.71 + \$4,369.84 = \$8,907.55.$$

Thus, Stan should pay no more than $8,907.55 when 10% nominal interest per year is desired.

EXAMPLE 5-4	**Current Price and Annual Yield of Bond Calculations**

A bond with a face value of $5,000 pays interest of 8% per year. This bond will be redeemed at par value at the end of its 20-year life, and the first interest payment is due one year from now.

(a) How much should be paid now for this bond in order to receive a yield of 10% per year on the investment?

(b) If this bond is purchased now for $4,600, what annual yield would the buyer receive?

Solution

(a) By using Equation (5-2), the value of V_N can be determined:

$$V_N = \$5,000(P/F, 10\%, 20) + \$5,000(0.08)(P/A, 10\%, 20)$$
$$= \$743.00 + \$3,405.44 = \$4,148.44.$$

(b) Here, we are given $V_N = \$4,600$, and we must find the value of $i\%$ in Equation (5-2):

$$\$4,600 = \$5,000(P/F, i'\%, 20) + \$5,000(0.08)(P/A, i'\%, 20).$$

To solve for $i'\%$, we can resort to an iterative trial-and-error procedure (e.g., try 8.5%, 9.0%), to determine that $i'\% = 8.9\%$ per year.

5.3.3 The Capitalized-Worth Method

One special variation of the PW method discussed in Section 5.3 involves determining the PW of all revenues or expenses over an infinite length of time. This is known as the *Capitalized-Worth* (CW) method. If only expenses are considered, results obtained by this method are sometimes referred to as *capitalized cost*. As will be demonstrated in Chapter 6, the CW method is a convenient basis for comparing mutually exclusive alternatives when the period of needed service is indefinitely long.

The CW of a perpetual series of end-of-period uniform payments A, with interest at $i\%$ per period, is $A(P/A, i\%, \infty)$. From the interest formulas, it can be seen that $(P/A, i\%, N) \to 1/i$ as N becomes very large. Thus, CW $= A/i$ for such a series, as can also be seen from the relation

$$\mathrm{CW}(i\%) = \mathrm{PW}_{N\to\infty} = A(P/A, i\%, \infty) = A\left[\lim_{N\to\infty} \frac{(1+i)^N - 1}{i(1+i)^N}\right] = A\left(\frac{1}{i}\right).$$

Hence, the CW of a project with interest at $i\%$ per year is the annual equivalent of the project over its useful life divided by i (as a decimal).

The AW of a series of payments of amount $\$X$ at the end of each kth period with interest at $i\%$ per period is $\$X(A/F, i\%, k)$. The CW of such a series can thus be calculated as $\$X(A/F, i\%, k)/i$.

EXAMPLE 5-5 | **Determining the Capitalized Worth of an Endowment**

Suppose that a firm wishes to endow an advanced biological processes laboratory at a university. The endowment principal will earn interest that averages 8% per year, which will be sufficient to cover all expenditures incurred in the establishment and maintenance of the laboratory for an indefinitely long period of time (forever). Cash requirements of the laboratory are estimated to be $100,000 now (to establish it), $30,000 per year indefinitely, and $20,000 at the end of every fourth year (forever) for equipment replacement.

(a) For this type of problem, what analysis period (N) is, practically speaking, defined to be "forever"?

(b) What amount of endowment principal is required to establish the laboratory and then earn enough interest to support the remaining cash requirements of this laboratory forever?

Solution

(a) A practical approximation of "forever" (infinity) is dependent on the interest rate. By examining the $(P/A, i\%, N)$ factor as N increases, we observe that this factor approaches a value of $1/i$. For $i = 8\%$ ($1/i = 12.5000$), note that the $(P/A, 8\%, N)$ factor equals 12.4943 when $N = 100$. Therefore, $N = 100$ is essentially forever (∞) when $i = 8\%$. As the interest rate gets larger, the approximation of forever drops dramatically. For instance, when $i = 20\%$ ($1/i = 5.0000$), forever can be approximated by using about 40 years; the $(P/A, 20\%, N)$ factor equals 4.9966 when $N = 40$.

(b) The CW of the cash requirements is synonymous with the endowment principal initially needed to establish and then support the laboratory forever. By using the relationship that CW $= A/i =$ (equivalent annual cost)$/i$, we can compute the amount of the endowment as

$$\text{CW}(8\%) = \frac{-\$100,000(A/P, 8\%, \infty) - \$30,000 - \$20,000(A/F, 8\%, 4)}{0.08}$$

$$= \frac{-\$8,000 - \$30,000 - \$4,438}{0.08}$$

$$= -\$530,475,$$

where the factor value for $(A/P, 8\%, \infty)$ is given in Table C-11 (Appendix C) as equal to 0.0800.

Another way of considering the amount of endowment principal needed in this example is to have enough money to establish the facility ($100,000) and then have enough principal left in the fund to earn a return that will meet the annual maintenance costs ($30,000) and the periodic replacement of equipment needs ($20,000 at the end of each fourth year). Using this logic, we have

$$\text{CW}(8\%) = -\$100,000 - \left[\frac{\$30,000 + \$20,000(A/F, 8\%, 4)}{0.08}\right]$$

$$= -\$100,000 - \left[\frac{(\$30,000 + \$4,438)}{0.08}\right]$$

$$= -\$530,475,$$

which, of course, is the same CW amount as our previous calculation.

5.4   The Future Worth Method

Because a primary objective of all time value of money methods is to maximize the future wealth of the owners of a firm, the economic information provided by the FW method is very useful in capital investment decision situations. The FW is based on the equivalent worth of all cash inflows and outflows at the end of the planning horizon (study period) at an interest rate that is generally the MARR. Also, the FW of a project is equivalent to its PW; that is, FW = PW$(F/P, i\%, N)$. If FW ≥ 0 for a project, it would be economically justified.

FW Decision Rule: If FW $(i = \text{MARR}) \geq 0$, the project is economically justified.

Equation (5-3) summarizes the general calculations necessary to determine a project's FW:

$$\text{FW}(i\%) = F_0(1+i)^N + F_1(1+i)^{N-1} + \cdots + F_N(1+i)^0$$

$$= \sum_{k=0}^{N} F_k(1+i)^{N-k}. \tag{5-3}$$

EXAMPLE 5-6   **The Relationship between FW and PW**

Evaluate the FW of the potential improvement project described in Example 5-1. Show the relationship between FW and PW for this example.

Solution

$$\text{FW}(20\%) = -\$25,000(F/P, 20\%, 5)$$
$$+ \$8,000(F/A, 20\%, 5) + \$5,000$$
$$= \$2,324.80.$$

Again, the project is shown to be a good investment (FW ≥ 0). The PW is a multiple of the equivalent FW value:

$$\text{PW}(20\%) = \$2,324.80(P/F, 20\%, 5) = \$934.29.$$

To this point, the PW and FW methods have used a known and constant MARR over the study period. Each method produces a measure of merit expressed in dollars and is equivalent to the other. The difference in economic information provided is relative to the point in time used (i.e., the present for the PW versus the future, or end of the study period, for the FW).

EXAMPLE 5-7   **Sensitivity Analysis Using FW (Example 5-2 Revisited)**

In Example 5-2, the $110,000 retrofitted space-heating system was projected to save $30,000 per year in electrical power and be worth $8,000 at the end of the six-year study period. Use the FW method to determine whether the project is still economically justified if the system has zero market value after six years. The MARR is 15% per year.

Solution

In this example, we need to find the future equivalent of the $110,000 investment and the $30,000 annual savings at an interest rate of 15% per year.

$$FW(15\%) = -\$110,000\,(F/P, 15\%, 6) + \$30,000\,(F/A, 15\%, 6)$$

$$= -\$110,000\,(2.3131) + \$30,000\,(8.7537)$$

$$= \$8,170.$$

The heating system is still a profitable project (FW $\geq$ 0) even if it has no market value at the end of the study period.

5.5 The Annual Worth Method

The AW of a project is an equal annual series of dollar amounts, for a stated study period, that is *equivalent* to the cash inflows and outflows at an interest rate that is generally the MARR. Hence, the AW of a project is annual equivalent revenues or savings ($\underline{R}$) minus annual equivalent expenses ($\underline{E}$), less its annual equivalent capital recovery (CR) amount, which is defined in Equation (5-5). An annual equivalent value of $\underline{R}$, $\underline{E}$, and CR is computed for the study period, N, which is usually in years. In equation form, the AW, which is a function of $i\%$, is

$$AW(i\%) = \underline{R} - \underline{E} - CR(i\%). \qquad (5\text{-}4)$$

Also, we need to notice that the AW of a project is equivalent to its PW and FW. That is, AW = PW$(A/P, i\%, N)$, and AW = FW$(A/F, i\%, N)$. Hence, it can be easily computed for a project from these other equivalent values.

As long as the AW evaluated at the MARR is greater than or equal to zero, the project is economically attractive; otherwise, it is not. An AW of zero means that an annual return exactly equal to the MARR has been earned. Many decision makers prefer the AW method because it is relatively easy to interpret when they are accustomed to working with annual income statements and cash-flow summaries.

AW Decision Rule: If AW $(i = MARR) \geq 0$, the project is economically justified.

The CR amount for a project is the equivalent uniform annual *cost* of the capital invested. It is an annual amount that covers the following two items:

1. Loss in value of the asset
2. Interest on invested capital (i.e., at the MARR)

As an example, consider a device that will cost $10,000, last five years, and have a salvage (market) value of $2,000. Thus, the loss in value of this asset over five years is $8,000. Additionally, the MARR is 10% per year.

It can be shown that, no matter which method of calculating an asset's loss in value over time is used, the equivalent annual CR amount is the same. For example, if a uniform loss in value is assumed ($8,000/5 years = $1,600 per year), the equivalent annual CR amount is calculated to be $2,310, as shown in Table 5-1.

There are several convenient formulas by which the CR amount (cost) may be calculated to obtain the result in Table 5-1. Probably the easiest formula to understand involves finding the annual equivalent of the initial capital investment and then subtracting the annual equivalent of the salvage value.* Thus,

$$CR(i\%) = I(A/P, i\%, N) - S(A/F, i\%, N), \qquad (5\text{-}5)$$

where I = initial investment for the project,[†]
S = salvage (market) value at the end of the study period,
N = project study period.

Thus, by substituting the CR(i%) expression of Equation (5-5) into the AW expression, Equation (5-4) becomes

$$AW(i\%) = \underline{R} - \underline{E} - I(A/P, i\%, N) + S(A/F, i\%, N).$$

			Interest on	CR	
	Value of		Beginning-	Amount	
	Investment	Uniform	of-Year	for	
	at Beginning	Loss in	Investment	Year	PW of CR Amount
Year	of Year[a]	Value	at $i = 10\%$		at $i = 10\%$
1	$10,000	$1,600	$1,000	$2,600	$2,600(P/F, 10\%, 1) = \$2,364$
2	8,400	1,600	840	2,440	$2,440(P/F, 10\%, 2) = \$2,016$
3	6,800	1,600	680	2,280	$2,280(P/F, 10\%, 3) = \$1,713$
4	5,200	1,600	520	2,120	$2,120(P/F, 10\%, 4) = \$1,448$
5	3,600	1,600	360	1,960	$1,960(P/F, 10\%, 5) = \$1,217$

TABLE 5-1 Calculation of Equivalent Annual CR Amount

$8,758

$CR(10\%) = \$8,758(A/P, 10\%, 5) = \$2,310.$

[a] This is also referred to later as the *beginning-of-year unrecovered investment.*

---

* The following two equations are alternative ways of calculating the CR amount:

$$CR(i\%) = (I - S)(A/F, i\%, N) + I(i\%);$$

$$CR(i\%) = (I - S)(A/P, i\%, N) + S(i\%).$$

It is left as a student exercise to show that the above equations are equivalent to Equation (5-5).
[†] In some cases, the investment will be spread over several periods. In such situations, I is the PW of all investment amounts.

When Equation (5-5) is applied to the example in Table 5-1, the CR cost is

$$CR(10\%) = \$10,000(A/P, 10\%, 5) - \$2,000(A/F, 10\%, 5)$$

$$= \$10,000(0.2638) - \$2,000(0.1638) = \$2,310.$$

EXAMPLE 5-8	**Using AW to Evaluate the Purchase of New Equipment (Example 5-1 Revisited)**

By using the AW method and Equation (5-4), determine whether the equipment described in Example 5-1 should be recommended.

Solution

The AW Method applied to Example 5-1 yields the following:

$$AW(20\%) = \overbrace{\$8,000}^{R-E} - \overbrace{[\$25,000(A/P, 20\%, 5) - \$5,000(A/F, 20\%, 5)]}^{\text{CR amount [Equation (5-5)]}}$$

$$= \$8,000 - \$8,359.50 + \$671.90$$

$$= \$312.40.$$

Because its AW(20%) is positive, the equipment more than pays for itself over a period of five years, while earning a 20% return per year on the unrecovered investment. In fact, the annual equivalent "surplus" is $312.40, which means that the equipment provided more than a 20% return on beginning-of-year unrecovered investment. This piece of equipment should be recommended as an attractive investment opportunity. Also, we can confirm that the AW(20%) is equivalent to PW(20%) = $934.29 in Example 5-1 and FW(20%) = $2,324.80 in Example 5-6. That is,

$$AW(20\%) = \$934.29(A/P, 20\%, 5) = \$312.40, \text{ and also}$$

$$AW(20\%) = \$2,324.80(A/F, 20\%, 5) = \$312.40.$$

When revenues are *absent* in Equation (5-4), we designate this metric as EUAC($i\%$) and call it "equivalent uniform annual cost." A low-valued EUAC($i\%$) is preferred to a high-valued EUAC($i\%$).

EXAMPLE 5-9	**Equivalent Uniform Annual Cost of a Corporate Jet**

A corporate jet costs $1,350,000 and will incur $200,000 per year in fixed costs (maintenance, licenses, insurance, and hangar rental) and $277 per hour in variable costs (fuel, pilot expense, etc.). The jet will be operated for 1,200 hours per year for five years and then sold for $650,000. The MARR is 15% per year.

(a) Determine the capital recovery cost of the jet.
(b) What is the EUAC of the jet?

Solution

(a) CR = $1,350,000 $(A/P, 15\%, 5)$ − $650,000 $(A/F, 15\%, 5)$ = $306,310
(b) The total annual expense for the jet is the sum of the fixed costs and the variable costs.

$$E = \$200,000 + (1,200 \text{ hours})(\$277/\text{hour}) = \$532,400$$

$$\text{EUAC}(15\%) = \$532,400 + \$306,310 = \$838,710$$

EXAMPLE 5-10	**Determination of Annual Savings by Using the AW Method (Example 5-2 Revisited)**

Consider the retrofitted space-heating system described in Example 5-2. Given the investment of $110,000 and market value of $8,000 at the end of the six-year study period, what is the minimum annual electrical power savings (in kWh) required to make this project economically acceptable? The MARR = 15% per year and electricity costs $0.10 per kWh.

Solution

To make this project acceptable, the annual power savings must be at least as great as the annual CR amount. Using Equation (5-5),

$$\text{CR} = \$110,000(A/P, 15\%, 6) - \$8,000 (A/F, 15\%, 6) = \$28,148.40.$$

This value ($28,148.40) is the minimum annual dollar savings needed to justify the space-heating system. This equates to

$$\frac{\$28,148.40}{\$0.10/\text{kWh}} = 281,480 \text{ kWh per year.}$$

If the space-heating system can save 281,480 kWh per year, it is economically justified (exactly 15% is earned on the beginning-of-year unrecovered investment). Any savings greater than 281,480 kWh per year (such as the original estimate of 300,000 kWh per year) will serve to make this project even more attractive.

EXAMPLE 5-11	Avoid Getting Fleeced on an Auto Lease

Automobile leases are built around three factors: negotiated sales price, residual value, and interest rate. The residual value is what the dealership expects the car's value will be when the vehicle is returned at the end of the lease period. The monthly cost of the lease is the capital recovery amount determined by using these three factors.

(a) Determine the monthly lease payment for a car that has an agreed-upon sales price of $34,995, an APR of 9% compounded monthly, and an estimated residual value of $20,000 at the end of a 36-month lease.* An up-front payment of $3,000 is due when the lease agreement (contract) is signed.

(b) If the estimated residual value is raised to $25,000 by the dealership to get your business, how much will the monthly payment be?

Solution

(a) The effective sales price is $31,995 ($34,995 less the $3,000 due at signing). The monthly interest rate is 9%/12 = 0.75% per month. So the capital recovery amount is:

$$CR = \$31,995(A/P, 0.75\%, 36) - \$20,000(A/F, 0.75\%, 36)$$
$$= \$1,017.44 - \$486$$
$$= \$531.44 \text{ per month.}$$

(b) The capital recovery amount is now $1,017.44 − $25,000 $(A/F, 0.75\%, 36) =$ $409.94 per month. But the customer might experience an actual residual value of less than $25,000 and have to pay the difference in cash when the car is returned after 36 months. This is the "trap" that many experience when they lease a car, so be careful not to drive the car excessively or to damage it in any way.

---

* There are many excellent Web sites offering free calculators for car loans, home mortgages, life insurance, and credit cards. In this regard, check out www.choosetosave.org.

5.6   The Internal Rate of Return Method

The IRR method is the most widely used rate-of-return method for performing engineering economic analyses. It is sometimes called by several other names, such as the *investor's method*, the *discounted cash flow method*, and the *profitability index*.

This method solves for the interest rate that equates the equivalent worth of an alternative's cash inflows (receipts or savings) to the equivalent worth of cash outflows (expenditures, including investment costs). Equivalent worth may be computed using any of the three methods discussed earlier. The resultant interest rate is termed the *Internal Rate of Return (IRR)*. The *IRR* is sometimes referred to as the breakeven interest rate.

For a single alternative, from the lender's viewpoint, the IRR is not positive unless (1) both receipts and expenses are present in the cash-flow pattern, and (2) the sum of receipts exceeds the sum of all cash outflows. Be sure to check both of these conditions in order to avoid the unnecessary work involved in finding that the IRR is *negative.* (Visual inspection of the total net cash flow will determine whether the IRR is zero or less.)

Using a PW formulation, we see that the IRR is the $i'\%^*$ at which

$$\sum_{k=0}^{N} R_k(P/F, i'\%, k) = \sum_{k=0}^{N} E_k(P/F, i'\%, k), \tag{5-6}$$

where R_k = net revenues or savings for the kth year;
  E_k = net expenditures, including any investment costs for the kth year;
  N = project life (or study period).

Once i' has been calculated, it is compared with the MARR to assess whether the alternative in question is acceptable. If $i' \geq$ MARR, the alternative is acceptable; otherwise, it is not.

IRR Decision Rule: If IRR $\geq$ MARR, the project is economically justified.

A popular variation of Equation (5-6) for computing the IRR for an alternative is to determine the i' at which its *net* PW is zero. In equation form, the IRR is the value of i' at which

$$PW = \sum_{k=0}^{N} R_k(P/F, i'\%, k) - \sum_{k=0}^{N} E_k(P/F, i'\%, k) = 0. \tag{5-7}$$

For an alternative with a single investment cost at the present time followed by a series of positive cash inflows over N, a graph of PW versus the interest rate typically has the general convex form shown in Figure 5-3. The point at which PW = 0 in Figure 5-3 defines $i'\%$, which is the project's IRR. The value of $i'\%$ can also be determined as the interest rate at which FW = 0 or AW = 0.

Another way to interpret the IRR is through an *investment-balance diagram.* Figure 5-4 shows how much of the original investment in an alternative is still to be recovered as a function of time. The downward arrows in Figure 5-4 represent annual returns, $(R_k - E_k)$ for $1 \leq k \leq N$, against the unrecovered

Figure 5-3 Plot of PW versus Interest Rate

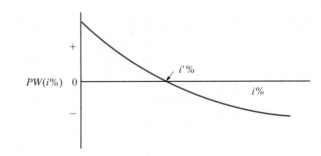

---

* i' is often used in place of i to mean the interest rate that is to be determined.

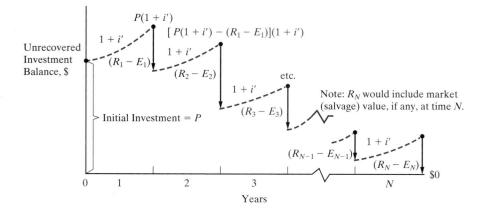

Figure 5-4 Investment-Balance Diagram Showing IRR

investment, and the dashed lines indicate the opportunity cost of interest, or profit, on the beginning-of-year investment balance. *The IRR is the value of i' in Figure 5-4 that causes the unrecovered investment balance to exactly equal zero at the end of the study period (year N)* and thus represents the *internal* earning rate of a project. It is important to notice that i'% is calculated on the *beginning-of-year unrecovered* investment through the life of a project rather than on the total initial investment.

The method of solving Equations (5-6) and (5-7) normally involves trial-and-error calculations until the i'% is converged upon or can be interpolated. Example 5-12 presents a typical solution. We also demonstrate how a spreadsheet application significantly assists in the computation of the IRR.

EXAMPLE 5-12	Economic Desirability of a Project Using the IRR Method

AMT, Inc., is considering the purchase of a digital camera for the maintenance of design specifications by feeding digital pictures directly into an engineering workstation where computer-aided design files can be superimposed over the digital pictures. Differences between the two images can be noted, and corrections, as appropriate, can then be made by design engineers. The capital investment requirement is $345,000 and the estimated market value of the system after a six-year study period is $115,000. Annual revenues attributable to the new camera system will be $120,000, whereas additional annual expenses will be $22,000. You have been asked by management to determine the IRR of this project and to make a recommendation. The corporation's MARR is 20% per year. Solve first by using linear interpolation and then by using a spreadsheet.

Solution by Linear Interpolation

In this example, we can easily see that the sum of positive cash flows ($835,000) exceeds the sum of negative cash flows ($455,000). Thus, it is likely that a positive-valued IRR can be determined. By writing an equation for the PW of the project's total net cash flow and setting it equal to zero, we can compute the IRR:

$$PW = 0 = -\$345{,}000 + (\$120{,}000 - \$22{,}000)(P/A, i'\%, 6)$$
$$+ \$115{,}000(P/F, i'\%, 6)$$
$$i'\% = ?$$

To use linear interpolation, we first need to try a few values for i'. A good starting point is to use the MARR.

At $i' = 20\%$: $PW = -\$345{,}000 + \$98{,}000(3.3255) + \$115{,}000(0.3349)$
$$= +\$19{,}413$$

Since the PW is positive at 20%, we know that $i' > 20\%$.

At $i' = 25\%$: $PW = -\$345{,}000 + \$98{,}000(2.9514) + \$115{,}000(0.2621)$
$$= -\$25{,}621$$

Now that we have both a positive and a negative PW, the answer is bracketed ($20\% \leq i'\% \leq 25\%$). The dashed curve in Figure 5-5 is what we are linearly approximating. The answer, $i'\%$, can be determined by using the similar triangles represented by dashed lines in Figure 5-5.

$$\frac{\text{line } BA}{\text{line } BC} = \frac{\text{line } dA}{\text{line } de}.$$

Here, BA is the line segment $B - A = 25\% - 20\%$. Thus,

$$\frac{25\% - 20\%}{\$19{,}413 - (-\$25{,}621)} = \frac{i'\% - 20\%}{\$19{,}413 - \$0}$$
$$i' \approx 22.16\%.$$

Figure 5-5 Use of Linear Interpolation to Find the Approximation of IRR for Example 5-12

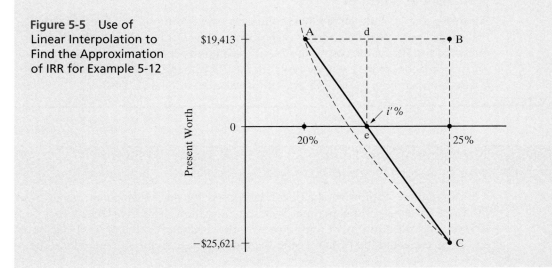

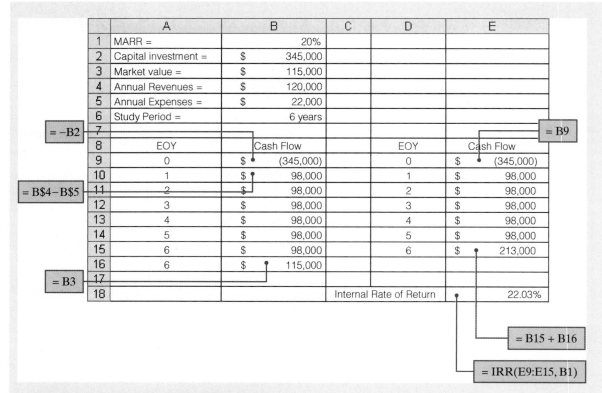

Figure 5-6 Spreadsheet Solution, Example 5-12

Because the IRR of the project (22.16%) is greater than the MARR, the project is acceptable.

Spreadsheet Solution

Figure 5-6 displays the spreadsheet solution for this example. The Excel function IRR(*range, guess*) requires the net cash flows for the study period and an initial guess for the IRR value (using the MARR is a good idea). Unlike the trial-and-error approach required when solving for the IRR by hand, the Excel IRR function is direct and simple. In cell E18, the IRR is calculated to be 22.03%.

| EXAMPLE 5-13 | **Evaluation of New Equipment Purchase, Using the Internal Rate of Return Method (Example 5-1 Revisited)** |

A piece of new equipment has been proposed by engineers to increase the productivity of a certain manual welding operation. The investment cost is $25,000, and the equipment will have a market (salvage) value of $5,000 at the end of its expected life of five years. Increased productivity attributable to the equipment will amount to $8,000 per year after extra operating costs have been subtracted from the value of the additional production. Use a spreadsheet to evaluate the IRR of the proposed equipment. Is the investment a good one? Recall that the MARR is 20% per year.

Spreadsheet Solution

The spreadsheet solution for this problem is shown in Figure 5-7. In column E of Figure 5-7(a), the individual EOY cash flows for year five (net annual savings and market value) are combined into a single entry for the direct computation of the IRR via the IRR function. The IRR for the proposed piece of equipment is 21.58%, which is greater than the MARR of 20%. Thus, we conclude that the new equipment is economically feasible. We reached the same conclusion by using the equivalent worth method in previous examples, summarized as follows:

$$\text{Example 5-1} \quad \text{PW}(20\%) = \$934.29;$$

$$\text{Example 5-6} \quad \text{FW}(20\%) = \$2,324.80; \text{and}$$

$$\text{Example 5-8} \quad \text{AW}(20\%) = \$312.40.$$

Figure 5-7(b) is a graph of the PW of the proposed equipment as a function of the interest rate, i. For the given problem data, the graph shows the IRR to be approximately 22%. This graph can be used to get a feel for how the IRR

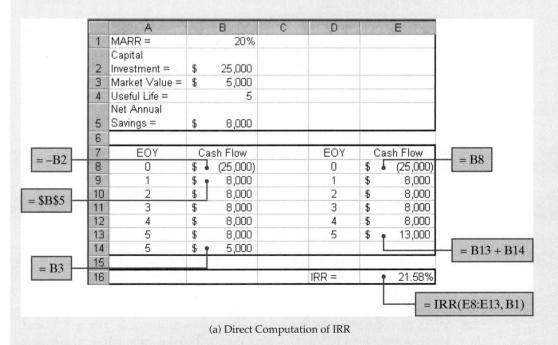

(a) Direct Computation of IRR

Figure 5-7 Spreadsheet Solution, Example 5-13

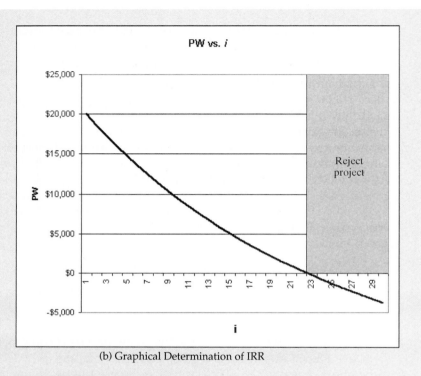

(b) Graphical Determination of IRR

Figure 5-7 *(Continued)*

fluctuates with changes in the original cash-flow estimates. For example, if the net annual savings estimate [cell B5 of Figure 5-7(a)] is revised to be $7,500, the graph would update and show the IRR to be 19%, and the project would no longer be economically viable.

A final point needs to be made for Example 5-13. The investment-balance diagram is provided in Figure 5-8, and the reader should notice that $i' = 21.577\%$ is a rate of return calculated on the beginning-of-year unrecovered investment. The IRR is *not* an average return each year based on the total investment of $25,000.

5.6.1 Installment Financing

A rather common application of the IRR method is in so-called *installment financing* types of problems. These problems are associated with financing arrangements for purchasing merchandise "on time." The total interest, or finance, charge is often paid by the borrower on the basis of what amount is owed at the beginning of the loan *instead* of on the unpaid loan balance, as illustrated in Figure 5-8. Usually, the average unpaid loan balance is about one-half of the initial amount borrowed. Clearly, a finance charge based solely on the *entire* amount of money borrowed involves payment of interest on money not actually borrowed for the full term.

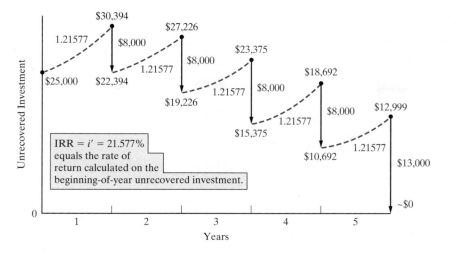

Figure 5-8 Investment-Balance Diagram for Example 5-13

This practice leads to an actual interest rate that often greatly exceeds the stated interest rate. To determine the true interest rate being charged in such cases, the IRR method is frequently employed. Examples 5-14, 5-15, and 5-16 are representative installment-financing problems.

EXAMPLE 5-14 | **Do You Know What Your Effective Interest Rate Is?**

In 1915, Albert Epstein allegedly borrowed $7,000 from a large New York bank on the condition that he would repay 7% of the loan every three months, until a total of 50 payments had been made. At the time of the 50th payment, the $7,000 loan would be completely repaid. Albert computed his annual interest rate to be [0.07($7,000) × 4]/$7,000 = 0.28 (28%).

(a) What true *effective* annual interest rate did Albert pay?
(b) What, if anything, was wrong with his calculation?

Solution

(a) The true interest rate per quarter is found by equating the equivalent value of the amount borrowed to the equivalent value of the amounts repaid. Equating the AW amounts per quarter, we find

$$\$7,000(A/P, i'\%/\text{quarter}, 50 \text{ quarters}) = 0.07(\$7,000) \text{ per quarter},$$

$$(A/P, i'\%, 50) = 0.07.$$

Linearly interpolating to find $i'\%$/quarter by using similar triangles is the next step:

$$(A/P, 6\%, 50) = 0.0634,$$

$$(A/P, 7\%, 50) = 0.0725.$$

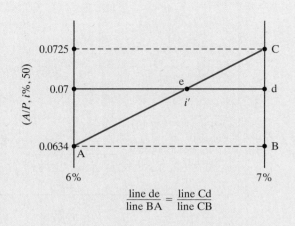

$$\frac{\text{line } de}{\text{line } BA} = \frac{\text{line } Cd}{\text{line } CB}$$

$$\frac{7\% - i'\%}{7\% - 6\%} = \frac{0.0725 - 0.07}{0.0725 - 0.0634},$$

$$i'\% = 7\% - 1\% \left(\frac{0.0025}{0.0091} \right),$$

or $i'\% \simeq 6.73\%$ per quarter.

Now we can compute the effective $i'\%$ per year that Albert was paying:

$$i'\% = [(1.0673)^4 - 1]100\%$$

$$\simeq 30\% \text{ per year.}$$

(b) Even though Albert's answer of 28% is close to the true value of 30%, his calculation is insensitive to how long his payments were made. For instance, he would get 28% for an answer when 20, 50, or 70 quarterly payments of \$490 were made! For 20 quarterly payments, the true effective interest rate is 14.5% per year, and for 70 quarterly payments, it is 31% per year. As more payments are made, the true annual effective interest rate being charged by the bank will increase, but Albert's method would not reveal by how much.

Another Solution

When the initial loan amount, the payment amount, and the number of payments are known, Excel has a useful financial function, RATE (*nper, pmt, pv*), that will return the interest rate per period. For this example,

$$\text{RATE}(50, -490, 7000) = 6.73\%.$$

This is the same quarterly interest rate we obtained via linear interpolation in Part (a).

EXAMPLE 5-15	Be Careful with "Fly-by-Night" Financing!

The Fly-by-Night finance company advertises a "bargain 6% plan" for financing the purchase of automobiles. To the amount of the loan being financed, 6% is added for each year money is owed. This total is then divided by the number of months over which the payments are to be made, and the result is the amount of the monthly payments. For example, a woman purchases a $10,000 automobile under this plan and makes an initial cash payment of $2,500. She wishes to pay the $7,500 balance in 24 monthly payments:

Purchase price	= $10,000
−Initial payment	= 2,500
= Balance due, (P_0)	= 7,500
+6% finance charge = 0.06×2 years $\times$ \$7,500	= 900
= Total to be paid	= 8,400
$\therefore$ Monthly payments $(A) = \$8,400/24$	= $350

What effective annual rate of interest does she actually pay?

Solution

Because there are to be 24 payments of $350 each, made at the end of each month, these constitute an annuity (A) at some unknown rate of interest, $i'\%$, that should be computed only upon the unpaid balance instead of on the entire $7,500 borrowed. A cash-flow diagram of this situation is shown below.

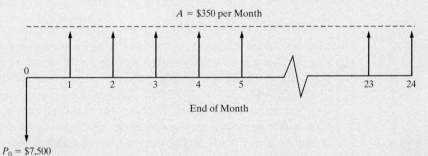

In this example, the amount owed on the automobile (i.e., the initial unpaid balance) is $7,500, so the following equivalence expression is employed to compute the unknown monthly interest rate:

$$P_0 = A(P/A, i'\%, N),$$

$$\$7,500 = \$350/\text{month}(P/A, i'\%, 24 \text{ months}),$$

$$(P/A, i'\%, 24) = \frac{\$7,500}{\$350} = 21.43.$$

Examining the interest tables for P/A factors at $N = 24$ that come closest to 21.43, we find that $(P/A, 3/4\%, 24) = 21.8891$ and $(P/A, 1\%, 24) = 21.2434$.

Using the linear interpolation procedure, the IRR is computed as 0.93% per month*, since payments are monthly. The nominal rate paid on the borrowed money is 0.93%(12) = 11.16% compounded monthly. This corresponds to an effective annual interest rate of $[(1 + 0.0093)^{12} - 1] \times 100\% \cong 12\%$. What appeared at first to be a real bargain turns out to involve effective annual interest at twice the stated rate. The reason is that, on the average, only $3,750 is borrowed over the two-year period, but interest on $7,500 over 24 months was charged by the finance company.

---

* Using Excel, RATE(24, −350, 7500) = 0.93%.

EXAMPLE 5-16   **Effective Interest Rate for Purchase of New Aircraft Equipment**

A small airline executive charter company needs to borrow $160,000 to purchase a prototype synthetic vision system for one of its business jets. The SVS is intended to improve the pilots' situational awareness when visibility is impaired. The local (and only) banker makes this statement: "We can loan you $160,000 at a very favorable rate of 12% per year for a five-year loan. However, to secure this loan, you must agree to establish a checking account (with no interest) in which the *minimum* average balance is $32,000. In addition, your interest payments are due at the end of each year, and the principal will be repaid in a lump-sum amount at the end of year five." What is the true effective annual interest rate being charged?

Solution

The cash-flow diagram from the banker's viewpoint appears below. When solving for an unknown interest rate, it is good practice to draw a *cash-flow diagram* prior to writing an equivalence relationship. Notice that $P_0 = \$160,000 - \$32,000 = \$128,000$. Because the bank is requiring the company to open an account worth $32,000, the bank is only $128,000 out of pocket. This same principal applies to F_5 in that the company only needs to repay $128,000 since the $32,000 on deposit can be used to repay the original principal.

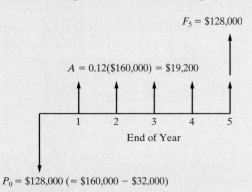

The interest rate (IRR) that establishes equivalence between positive and negative cash flows can now easily be computed:*

$$P_0 = F_5(P/F, i'\%, 5) + A(P/A, i'\%, 5),$$

$$\$128{,}000 = \$128{,}000(P/F, i'\%, 5) + \$19{,}200(P/A, i'\%, 5).$$

If we try $i' = 15\%$, we discover that $\$128{,}000 = \$128{,}000$. Therefore, the true effective interest rate is 15% per year.

---

* Several Web sites provide excellent tutorials for equivalent worth (e.g., PW) and rate-of-return calculations. For example, see www.investopedia.com and www.datadynamica.com/IRR.asp.

5.6.2 Difficulties Associated with the IRR Method

The PW, AW, and FW methods assume that net receipts less expenses (positive recovered funds) each period are reinvested at the MARR during the study period, N. However, the IRR method is not limited by this common assumption and measures the internal earning rate of an investment.*

Two difficulties with the IRR method are its computational difficulty and the occurrence of multiple IRRs in some types of problems. A procedure for dealing with seldom-experienced multiple rates of return is discussed and demonstrated in Appendix 5-A. Generally speaking, multiple rates are not meaningful for decision-making purposes, and another method of evaluation (e.g., PW) should be utilized.

Another possible drawback to the IRR method is that it must be carefully applied and interpreted in the analysis of two or more alternatives when only one of them is to be selected (i.e., mutually exclusive alternatives). This is discussed further in Chapter 6. The key advantage of the method is its widespread acceptance by industry, where various types of rates of return and ratios are routinely used in making project selections. The difference between a project's IRR and the required return (i.e., MARR) is viewed by management as a measure of investment safety. A large difference signals a wider margin of safety (or less relative risk).

5.7 The External Rate of Return Method†

The reinvestment assumption of the IRR method may not be valid in an engineering economy study. For instance, if a firm's MARR is 20% per year and the IRR for a project is 42.4%, it may not be possible for the firm to reinvest net cash proceeds from the project at much more than 20%. This situation, coupled with the computational

---

* See H. Bierman and S. Smidt, *The Capital Budgeting Decision: Economic Analysis of Investment Projects*, 8th ed. (Upper Saddle River, NJ: Prentice Hall, 1993). The term *internal* rate of return means that the value of this measure depends only on the cash flows from an investment and not on any assumptions about reinvestment rates: "One does not need to know the reinvestment rates to compute the IRR. However, one may need to know the reinvestment rates to compare alternatives" (p. 60).

† This method is also known as the "modified internal rate of return" (MIRR) method. For example, see C. S. Park and G. P. Sharp-Bette, *Advanced Engineering Economy*. (New York: John Wiley & Sons, 1990), 223–226.

demands and possible multiple interest rates associated with the IRR method, has given rise to other rate of return methods that can remedy some of these weaknesses.

One such method is the ERR method. It directly takes into account the interest rate ($\in$) external to a project at which net cash flows generated (or required) by the project over its life can be reinvested (or borrowed). If this external reinvestment rate, which is usually the firm's MARR, happens to equal the project's IRR, then the ERR method produces results identical to those of the IRR method.

In general, *three* steps are used in the calculating procedure. First, all net cash *outflows* are discounted to time zero (the present) at $\in$% per compounding period. Second, all net cash *inflows* are compounded to period N at $\in$%. Third, the ERR, which is the interest rate that establishes equivalence between the two quantities, is determined. The *absolute value* of the present equivalent worth of the net cash outflows at $\in$% (first step) is used in this last step. In equation form, the ERR is the i'% at which

$$\sum_{k=0}^{N} E_k(P/F, \in \%, k)(F/P, i'\%, N) = \sum_{k=0}^{N} R_k(F/P, \in \%, N - k), \qquad (5\text{-}8)$$

where R_k = excess of receipts over expenses in period k;

$\quad\quad\;\; E_k$ = excess of expenditures over receipts in period k;

$\quad\quad\;\; N$ = project life or number of periods for the study;

$\quad\quad\;\; \in$ = external reinvestment rate per period.

Graphically, we have the following (the numbers relate to the three steps):

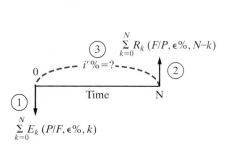

A project is acceptable when i'% of the ERR method is greater than or equal to the firm's MARR.

ERR Decision Rule: If ERR $\geq$ MARR, the project is economically justified.

The ERR method has two basic advantages over the IRR method:

1. It can usually be solved for directly, without needing to resort to trial and error.
2. It is not subject to the possibility of multiple rates of return. (*Note:* The multiple-rate-of-return problem with the IRR method is discussed in Appendix 5-A.)

| EXAMPLE 5-17 | **Calculation of the ERR** |

Referring to Example 5-13, suppose that $\epsilon = \text{MARR} = 20\%$ per year. What is the project's ERR, and is the project acceptable?

Solution

By utilizing Equation (5-8), we have the following relationship to solve for i':

$$\$25,000(F/P, i'\%, 5) = \$8,000(F/A, 20\%, 5) + \$5,000,$$

$$(F/P, i'\%, 5) = \frac{\$64,532.80}{\$25,000} = 2.5813 = (1 + i')^5,$$

$$i' = 20.88\%.$$

Because $i' > \text{MARR}$, the project is justified, but just barely.

| EXAMPLE 5-18 | **Determining the Acceptability of a Project, Using ERR** |

When $\epsilon = 15\%$ and MARR $= 20\%$ per year, determine whether the project (whose net cash-flow diagram appears next) is acceptable. Notice in this example that the use of an $\epsilon\%$ different from the MARR is illustrated. This might occur if, for some reason, part or all of the funds related to a project are "handled" outside the firm's normal capital structure.

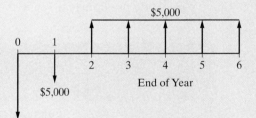

Solution

$$E_0 = \$10,000 \ (k = 0),$$

$$E_1 = \$5,000 \ (k = 1),$$

$$R_k = \$5,000 \text{ for } k = 2, 3, \ldots, 6,$$

$$[\$10,000 + \$5,000(P/F, 15\%, 1)](F/P, i'\%, 6) = \$5,000(F/A, 15\%, 5);$$

$$i'\% = 14.2\%.$$

The $i'\%$ is less than the MARR $= 20\%$; therefore, this project would be unacceptable according to the ERR method.

5.8   The Payback (Payout) Period Method

All methods presented thus far reflect the *profitability* of a proposed alternative for a study period of N. The payback method, which is often called the *simple payout method*, mainly indicates a project's *liquidity* rather than its profitability. Historically, the payback method has been used as a measure of a project's riskiness, since liquidity deals with how fast an investment can be recovered. A low-valued payback period is considered desirable. Quite simply, the payback method calculates the number of years required for cash inflows to just equal cash outflows. Hence, the simple payback period is the *smallest* value of $\theta (\theta \leq N)$ for which this relationship is satisfied under our normal EOY cash-flow convention. For a project where all capital investment occurs at time 0, we have

$$\sum_{k=1}^{\theta}(R_k - E_k) - I \geq 0. \tag{5-9}$$

The *simple* payback period, θ, ignores the time value of money and all cash flows that occur after θ. If this method is applied to the investment project in Example 5-13, the number of years required for the undiscounted sum of cash inflows to exceed the initial investment is four years. This calculation is shown in Column 3 of Table 5-2. Only when $\theta = N$ (the last time period in the planning horizon) is the market (salvage) value included in the determination of a payback period. As can be seen from Equation (5-9), the payback period does not indicate anything about project desirability except the speed with which the investment will be recovered. The payback period can produce misleading results, and it is recommended as supplemental information only in conjunction with one or more of the five methods previously discussed.

TABLE 5-2   Calculation of the Simple Payback Period (θ) and the Discounted Payback Period (θ') at MARR $= 20\%$ for Example 5-13[a]

Column 1 End of Year k	Column 2 Net Cash Flow	Column 3 Cumulative PW at $i = 0\%$/yr through Year k	Column 4 PW of Cash Flow at $i = 20\%$/yr	Column 5 Cumulative PW at $i = 20\%$/yr. through Year k
0	−$25,000	−$25,000	−$25,000	−$25,000
1	8,000	−17,000	6,667	−18,333
2	8,000	−9,000	5,556	−12,777
3	8,000	−1,000	4,630	−8,147
4	8,000	+7,000	3,858	−4,289
5	13,000		5,223	+934

$\theta = 4$ years because the cumulative balance turns positive at EOY 4.

$\theta' = 5$ years because the cumulative discounted balance turns positive at EOY 5.

[a] Notice that $\theta' \geq \theta$ for MARR $\geq 0\%$.

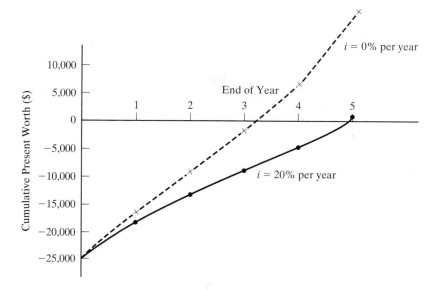

Figure 5-9   Graph of Cumulative PW for Example 5-13

Sometimes, the *discounted* payback period, $\theta'(\theta' \leq N)$, is calculated so that the time value of money is considered. In this case,

$$\sum_{k=1}^{\theta'}(R_k - E_k)(P/F,i\%,k) - I \geq 0, \qquad (5\text{-}10)$$

where $i\%$ is the MARR, I is the capital investment usually made at the present time ($k = 0$), and θ' is the smallest value that satisfies Equation (5-10). Table 5-2 (Columns 4 and 5) also illustrates the determination of θ' for Example 5-13. Notice that θ' is the first year in which the cumulative discounted cash inflows exceed the $25,000 capital investment. Payback periods of three years or less are often desired in U.S. industry, so the project in Example 5-13 could be *rejected*, even though it is profitable. [IRR = 21.58%, PW(20%) = $934.29]. The simple and discounted payback periods are shown graphically in Figure 5-9.

This variation (θ') of the simple payback period produces the *breakeven life* of a project, in view of the time value of money. However, neither payback period calculation includes cash flows occurring after θ (or θ'). This means that θ (or θ') may not take into consideration the entire useful life of physical assets. Thus, these methods will be misleading if one alternative has a longer (less desirable) payback period than another but produces a higher rate of return (or PW) on the invested capital.

Using the payback period to make investment decisions should generally be avoided except as a secondary measure of how quickly invested capital will be recovered, which is an indicator of project risk. The simple payback and discounted payback period methods tell us how long it takes cash inflows from a project to accumulate to equal (or exceed) the project's cash outflows. The longer it takes to recover invested monies, the greater is the perceived riskiness of a project.

5.9 CASE STUDY—A Proposed Capital Investment to Improve Process Yield

Many engineering projects aim at improving facility utilization and process yields. This case study illustrates an engineering economy analysis related to the redesign of a major component in the manufacture of semiconductors.

Semiconductor manufacturing involves taking a flat disc of silicon, called a wafer, and depositing many layers of material on top of it. Each layer has a pattern on it that, upon completion, defines the electrical circuits of the finished microprocessor. Each 8-inch wafer has up to 100 microprocessors on it. However, the typical average yield of the production line is 75% good microprocessors per wafer.

At one local company, the process engineers responsible for the chemical-vapor-deposition (CVD) tool (i.e., process equipment) that deposits *one* of the many layers have an idea for improving overall yield. They propose to improve this tool's vacuum with a redesign of one of its major components. The engineers believe the project will result in a 2% increase in the average production yield of nondefective microprocessors per wafer.

This company has only one CVD tool, and it can process 10 wafers per hour. The process engineers have determined that the CVD tool has an average utilization rate (i.e., "time running") of 80%. A wafer costs $5,000 to manufacture, and a good microprocessor can be sold for $100. These semiconductor fabrication plants ("fabs") operate 168 hours per week, and all good microprocessors produced can be sold.

The capital investment required for the project is $250,000, and maintenance and support expenses are expected to be $25,000 per month. The lifetime of the modified tool will be five years, and the company uses a 12% MARR per year (compounded monthly) as its "hurdle rate."

Before implementing the proposed engineering solution, top management has posed the following questions to you (hired as an independent consultant) to evaluate the merits of the proposal:

(a) Based on the PW method, should the project be approved?

(b) If the achievable improvement in production yield has been overestimated by the process engineers, at what percent yield improvement would the project breakeven?

Solution

You start your economic evaluation of the engineering proposal by first calculating the production rate of wafers. The average number of wafers per week is

$$(10 \text{ wafers/hour}) \times (168 \text{ hours/week}) \times (0.80) = 1,344.$$

Since the cost per wafer is $5,000 and good microprocessors can be sold for $100 each, you determine that profit is earned on each microprocessor produced and sold after the 50th microprocessor on each wafer. Thus, the 2% increase in production yield is all profit (i.e., from 75 good microprocessors per wafer on the average to 76.5). The corresponding additional profit per wafer is $150. The added profit per month, assuming a month is (52 weeks/year ÷ 12 months per year) = 4.333 weeks, is

$$(1,344 \text{ wafers/week})(4.333 \text{ weeks/month})(\$150/\text{wafer}) = \$873,533.$$

Therefore, the PW of the project is

$$\text{PW}(1\%) = -\$250,000 - \$25,000(P/A, 1\% \text{ per month, 60 months})$$
$$+ \$873,533(P/A, 1\%, 60)$$
$$= \$37,898,813.$$

You advise company management that, based on PW, the project *should* be undertaken.

It is known that at the breakeven point, profit equals zero. That is, the PW of the project is equal to zero, or PW of costs = PW of revenues. In other words,

$$\$1,373,875 = (1,344 \text{ wafers/week}) \times (4.333 \text{ weeks/month}) \times (\$X/\text{wafer})$$
$$\times (P/A, 1\%, 60),$$

where $X = \$100$ times the number of extra microprocessors per wafer. Then,

$$\frac{\$1,373,875}{(1,344)(4.333)(44.955)} = X, \text{ or } X \cong \$5.25 \text{ per wafer.}$$

Thus, ($5.25/$100) = 0.0525 *extra* microprocessors per wafer (total of 75.0525) equates PW of costs to PW of revenues. This corresponds to a breakeven increase in yield of

$$\frac{1.5 \text{ die per wafer}}{0.0525 \text{ die per wafer}} = \frac{2.0\% \text{ increase}}{\text{breakeven increase}},$$

or breakeven increase in yield = 0.07%.

You advise management that an increase of only 0.07% in process yield would enable the project to breakeven. Thus, although management may believe that the process engineers have overestimated projected process yield improvements in the past, there is quite a bit of "economic safety margin" provided by the engineers in their current projection of process yield improvement as long as the other assumptions concerning the average utilization rate of the CVD tool, the wafer production rate, and the plant operating hours are valid. ∎

5.10 Summary

Throughout this chapter, we have examined five basic methods for evaluating the financial *profitability* of a single project: PW, AW, FW, IRR, and ERR. These methods lead to the use of simple decision rules for economic evaluation of projects as

TABLE 5-3	Summary of Decision Rules	
Economic Measure of Merit	Notes	Decision Rule
AW, PW, FW	All are functions of the MARR	If PW (or AW, FW) ≥ 0, accept the project; otherwise, reject it
IRR, ERR	Solve for unknown interest rate, i'	If $i' \geq$ MARR, accept the project; otherwise, reject it

presented in Table 5-3. Two supplemental methods for assessing a project's *liquidity* were also presented: the simple payback period and the discounted payback period. Computational procedures, assumptions, and acceptance criteria for all methods were discussed and illustrated with examples. Appendix B provides a listing of new abbreviations and notations that have been introduced in this chapter.

Problems

Unless stated otherwise, discrete compounding of interest and end-of-period cash flows should be assumed in all problem exercises for the remainder of the book. All MARRs are "per year." The number in parentheses at the end of a problem refers to the chapter section(s) most closely related to the problem.

5-1. "The higher the MARR, the higher the price that a company should be willing to pay for equipment that reduces annual operating expenses." Do you agree with this statement? Explain your answer. (5.2)

5-2. You are faced with making a decision on a large capital investment proposal. The capital investment amount is $640,000. Estimated annual revenue at the end of each year in the eight-year study period is $180,000. The estimated annual year-end expenses are $42,000 starting in year one. These expenses begin *decreasing* by $4,000 per year at EOY four and continue decreasing through EOY eight. Assuming a $20,000 market value at EOY eight and a MARR = 12% per year, answer the following questions: (5.3)

a. What is the PW of this proposal?

b. What is your conclusion about the acceptability of this proposal?

5-3. Josh Ritchey has just been hired as a cost engineer by a large airlines company. Josh's first idea is to quit giving complimentary cocktails, wine, and beer to the international flying public. He calculates this will save 5,000,000 drinks per year, and each drink costs $0.50, for a total of $2.5 million per year. Instead of complimentary drinks, Josh estimates that the airlines company can sell 2,000,000 drinks at $5.00 per drink. The net savings

would amount to $12.5 million per year! Josh's boss really likes the idea and agrees to give Josh a lump-sum bonus now equaling 0.1% of the present equivalent worth of three years of net savings. If the company's MARR is 20% per year, what is Josh's bonus? (5.3)

5-4. Evaluate machine XYZ on the basis of the PW method when the MARR is 12% per year. Pertinent cost data are as follows: (5.3)

	Machine *XYZ*
Investment cost	$13,000
Useful life	15 years
Market value	$3,000
Annual operating expenses	$100
Overhaul cost—end of 5th year	$200
Overhaul cost—end of 10th year	$550

5-5. To join an upscale country club, an individual must first purchase a membership bond for $20,000. In addition, monthly membership dues are $250. Suppose an individual wants to put aside a lump sum of money now to pay for her basic country club membership expenses (including the $20,000 bond) over the next 30 years. She can earn an APR of 6%, compounded monthly, on her investments. What amount must this person now commit to the membership? (5.3)

5-6. A large induced-draft fan is needed for an upgraded industrial process. The motor to drive this fan is rated at 100 horsepower, and the motor will operate at full load for 8,760 hours per year. The motor's efficiency is 90%. Because the motor is fairly large, a demand charge of $90 per kilowatt per year will be incurred in addition to an energy charge of $0.08 per kilowatt-hour. If the installed cost of the motor is $3,500, what is the present worth of the motor over a 10-year period when the MARR is 15% per year? (5.3)

5-7. A new municipal refuse-collection truck can be purchased for $84,000. Its expected useful life is six years, at which time its market value will be zero. Annual receipts less expenses will be approximately $18,000 per year over the six-year study period. Use the PW method and a MARR of 18% to determine whether this is a good investment. (5.3)

5-8. The winner of a state lottery will receive $5,000 per week for the rest of her life. If the winner's interest rate is 6.5% per year compounded weekly, what is the present worth of this jackpot? (5.3)

5-9. A new six-speed automatic transmission for automobiles offers an estimated 4% improvement in fuel economy compared to traditional four-speed transmissions in front-wheel drive cars. If a four-speed transmission car averages 30 miles per gallon and gasoline costs $3.00 per gallon, how much *extra* can a motorist pay for a fuel-efficient six-speed transmission? Assume that the car will be driven for 120,000 miles over its lifetime of 10 years. The motorist can earn 6% per year on investments. (5.3)

5-10. What is the maximum price you will pay for a bond with a face value of $1,000 and a coupon rate of 14%, paid annually, if you want a yield to maturity of 10%? Assume that the bond will mature in 10 years and the first payment will be received in one year. (5.3)

5-11. Last month Jim purchased $10,000 of U.S. Treasury bonds (their face value was $10,000). These bonds have a 30-year maturity period, and they pay 1.5% interest every three months (i.e., the APR is 6%, and Jim receives a check for $150 every three months). But interest rates for similar securities have since risen to a 7% APR because of interest rate increases by the Federal Reserve Board. In view of the interest-rate increase to 7%, what is the current value of Jim's bonds? (5.3)

5-12. A company has issued 10-year bonds, with a face value of $1,000,000, in $1,000 units. Interest at 8% is paid quarterly. If an investor desires to earn 12% nominal interest (compounded quarterly) on $10,000

worth of these bonds, what would the purchase price have to be? (5.3)

5-13. A company sold a $1,000,000 issue of bonds with a 15-year life, paying 4% interest per year. The bonds were sold at par value. If the company paid a selling fee of $50,000 and has an annual expense of $70,256 for mailing and record keeping, what is the true rate of interest that the company is paying for the borrowed money? (5.3)

5-14. A U.S. government bond matures in 10 years. Its quoted price is now 97.8, which means the buyer will pay $97.80 per $100 of the bond's face value. The bond pays 5% interest on its face value each year. If $10,000 (the face value) worth of these bonds is purchased now, what is the yield to the investor who holds the bonds for 10 years? (5.3)

5-15. A small company bought a BMI bond at its face value on January 1, 1995. This bond pays interest of 7.25% every six months (14.5% per year). The face value of the bond is $100,000, and it matures on December 31, 2006. On January 1, 2005, this bond was sold for $110,000. What interest rate (per six months) was earned by the company on the BMI bond? (5.3)

5-16.

a. What is the CW, when $i = 10\%$ per year, of $1,500 per year, starting in year one and continuing forever; and $10,000 in year five, repeating every four years thereafter, and continuing forever? (5.3)

b. When $i = 10\%$ per year in this type of problem, what value of N, practically speaking, defines "forever"? (5.3)

5-17. A city is spending $20 million on a new sewage system. The expected life of the system is 40 years, and it will have no market value at the end of its life. Operating and maintenance expenses for the system are projected to average $0.6 million per year. If the city's MARR is 8% per year, what is the capitalized worth of the system? (5.3)

5-18. A foundation was endowed with $10,000,000 in July 2000. In July 2004, $3,000,000 was expended for facilities, and it was decided to provide $250,000 at the end of each year forever to cover operating expenses. The first operating expense was in July 2005, and the first replacement expense is in July 2009. If all money earns interest at 5% after the time of endowment, what amount would be available for capital replacements at the end of every fifth year forever? (*Hint:* Draw a cash-flow diagram first.) (5.3)

5-19. Vidhi is investing in some rental property in Collegeville and is investigating her income from the investment. She knows the rental revenue will increase each year, but so will the maintenance expenses. She has been able to generate the data that follow regarding this investment opportunity. Assume that all cash flows occur at the end of each year and that the purchase and sale of this property are not relevant to the study. If Vidhi's MARR = 6% per year, what is the FW of Vidhi's projected net income? (5.4)

Year	Revenue	Year	Expenses
1	$6,000	1	$3,100
2	6,200	2	3,300
3	6,300	3	3,500
4	6,400	4	3,700
5	6,500	5	3,900
6	6,600	6	6,100
7	6,700	7	4,300
8	6,800	8	4,500
9	6,900	9	4,700
10	7,000	10	4,900

5-20. After graduation, you have been offered an engineering job with a large company that has offices in Tennessee and Ohio. The salary is $50,000 per year at either location. Tennessee's tax burden (state and local taxes) is 8.6% of income, while Ohio's is 12%. If you accept the position in Tennessee and stay with this company for 10 years, what is the FW of the tax savings? Your personal MARR is 12% per year. (5.4)

5-21. Determine the FW of the following engineering project when the MARR is 15% per year. Is the project acceptable? (5.4)

	Proposal A
Investment cost	$10,000
Expected life	5 years
Market (salvage) value[a]	−$1,000
Annual receipts	$8,000
Annual expenses	$4,000

[a] A negative market value means that there is a net cost to dispose of an asset.

5-22. Your uncle has almost convinced you to invest in his peach farm. It would require a $10,000 initial investment on your part. He promises you revenue (before expenses) of $1,800 per year the first year, and increasing by $100 per year thereafter. Your share of the estimated annual expenses is $500. You are planning to invest for six years. Your uncle has promised to buy out your share of the business at that time for $12,000. You have decided to set a personal MARR of 15% per year. Use the FW method to determine the profitability of this investment project. Include a cash-flow diagram. (5.4)

5-23. Fill in Table P5-23 on page 227 when $P = \$10,000$, $S = \$2,000$ (at the end of four years), and $i = 15\%$ per year. Complete the accompanying table and show that the equivalent uniform CR amount equals $3,102.12. (5.5)

5-24. Given that the purchase price of a machine is $1,000 and its market value at EOY four is $300, complete Table P5-24 on page 227 [values (a) through (f)], using an opportunity cost of 5% per year. Compute the equivalent uniform CR amount, using information from the completed table. (5.5)

5-25. A drug store is looking into the possibility of installing a 24/7-automated prescription refill system to increase its projected revenues by $20,000 per year over the next five years. Annual expenses to maintain the system are expected to be $5,000. The system will cost $50,000 and will have no market value at the end of the five-year study period. The store's MARR is 20% per year. Use the AW method to evaluate this investment. (5.5)

5-26. An office supply company has purchased a light-duty delivery truck for $15,000. It is anticipated that the purchase of the truck will increase the company's revenue by $10,000 annually, while the associated operating expenses are expected to be $3,000 per year. The truck's market value is expected to decrease by $2,500 each year it is in service. If the company plans to keep the truck for only two years, what is the annual worth of this investment? The MARR = 18% per year. (5.5)

5-27. A company is considering constructing a plant to manufacture a proposed new product. The land costs $300,000, the building costs $600,000, the equipment costs $250,000, and $100,000 additional working capital is required. It is expected that the product will result in sales of $750,000 per year for 10 years, at which

TABLE P5-23 Table for Problem 5-23

Year	Investment Beginning of Year	Opportunity Cost of Interest ($i = 15\%$)	Loss in Value of Asset During Year	Capital Recovery Amount for Year
1	$10,000	☐	$3,000	☐
2	☐	☐	$2,000	☐
3	☐	☐	$2,000	☐
4	☐	☐	☐	☐

TABLE P5-24 Table for Problem 5-24

Year	Investment at Beginning of Year	Opportunity Cost (5% per Year)	Loss in Value of Asset During Year	Capital Recovery Amount for Year
1	$1,000	$50	$(a)	$250
2	(b)	(c)	200	240
3	600	30	200	230
4	(d)	20	(e)	(f)

time the land can be sold for $400,000, the building for $350,000, and the equipment for $50,000. All of the working capital would be recovered at the EOY 10. The annual expenses for labor, materials, and all other items are estimated to total $475,000. If the company requires a MARR of 15% per year on projects of comparable risk, determine if it should invest in the new product line. Use the AW method. (5.5)

5-28. A 50-kilowatt gas turbine has an investment cost of $40,000. It costs another $14,000 for shipping, insurance, site preparation, fuel lines, and fuel storage tanks. The operation and maintenance expense for this turbine is $450 per year. Additionally, the hourly fuel expense for running the turbine is $7.50 per hour, and the turbine is expected to operate 3,000 hours each year. The cost of dismantling and disposing of the turbine at the end of its 8-year life is $8,000. (5.5)

a. If the MARR is 15% per year, what is the annual equivalent life-cycle cost of the gas turbine?

b. What percent of annual life-cycle cost is related to fuel?

5-29. Anderson County has 35 older-model school buses that will be salvaged for $5,000 each. These buses cost $144,000 per year for fuel and maintenance. Now

the county will purchase 35 new school buses for $40,000 each. They will travel an average of 2,000 miles per day for a total of 360,000 miles per year. These new buses will save $10,000 per year in fuel (compared with the older buses) for the entire group of 35 buses. If the new buses will be driven for 15 years and the county's MARR is 6% per year, what is the equivalent uniform annual cost of the new buses if they have negligible market value after 15 years?

5-30. An apartment complex wishes to establish a fund at the end of 2004 that, by the end of the year 2021, will grow to an amount large enough to place new roofs on its 39 apartment units. Each new roof is estimated to cost $2,500 in 2019, at which time 13 apartments will be reroofed. In 2020, another 13 apartments will be reroofed, but the unit cost will be $2,625. The last 13 apartments will be reroofed in 2021 at a unit cost of $2,750.

The annual effective interest rate that can be earned on this fund is 4%. How much money *each year* must be put aside (saved), starting at the end of 2005, to pay for all 39 new roofs? State any assumptions you make. (5.5)

5-31. An environmentally friendly green home (99% air tight) costs about 8% more to construct than a conventional home. Most green homes can save 15%

per year on energy expenses to heat and cool the dwelling. For a $250,000 conventional home, how much would have to be saved in energy expenses per year when the life of the home is 30 years and the interest rate is 10% per year? Assume the additional cost of a green home has no value at the end of 30 years. (5.5)

5-32. Your company is considering the introduction of a new product line. The initial investment required for this project is $500,000, and annual maintenance costs are anticipated to be $35,000. Annual operating costs will be directly proportional to the level of production at $7.50 per unit, and each unit of product can be sold for $50. If the MARR is 15% and the project has a life of five years, what is the minimum annual production level for which this project is economically viable? (5.5)

5-33. Stan Moneymaker has been informed of a major automobile manufacturer's plan to conserve on gasoline consumption through improved engine design. The idea is called "engine displacement," and it works by switching from 8-cylinder operation to 4-cylinder operation at approximately 40 miles per hour. Engine displacement allows enough power to accelerate from a standstill and to climb hills while also permitting the automobile to cruise at speeds over 40 miles per hour with little loss in driving performance.

The trade literature studied by Stan makes the claim that the engine displacement option will cost the customer an extra $1,200 on the automobile's sticker price. This option is expected to save 4 miles per gallon (an average of in-town and highway driving). A regular 8-cylinder engine in the car that Stan is interested in buying gets an average of 20 miles per gallon of gasoline. If Stan drives approximately 1,200 miles per month, how many months of ownership will be required to make this $1,200 investment pay for itself? Stan's opportunity cost of capital (*i*) is 0.5% per month, and gasoline costs $2.75 per gallon. (5.3)

5-34. The world's largest carpet maker has just completed a feasibility study of what to do with the 16,000 tons of overruns, rejects, and remnants it produces every year. The company's CEO launched the feasibility study by asking, why pay someone to dig coal out of the ground and then pay someone else to put our waste into a landfill? Why not just burn our own waste? The company is proposing to build a $10-million power plant to burn its waste as fuel, thereby saving $2.8 million a year in coal purchases. Company engineers have determined that the waste-burning plant will be environmentally sound, and after its four-year study

period the plant can be sold to a local electric utility for $5 million. (5.6)

a. What is the IRR of this proposed power plant?

b. If the firm's MARR is 15% per year, should this project be undertaken?

5-35. A small business owner estimates that the heat loss through the exterior walls of her warehouse will cost approximately $4,000 next year. A local company is offering insulation that can reduce the heat loss by 80%, with an installation cost of $17,000 now. She intends to keep the warehouse for 10 years. If the cost of the heat loss increases by $300 per year, after next year, what is the IRR? (5.6)

5-36. An assembly operation at a software company now requires $100,000 per year in labor costs. A robot can be purchased and installed to automate this operation. The robot will cost $200,000 and will have no market value at the end of the 10-year study period. Maintenance and operation expenses of the robot are estimated to be $64,000 per year. Invested capital must earn at least 12% per year. Use the IRR method to determine if the robot is a justifiable investment. (5.6)

5-37. You recently purchased a Sirius satellite receiver. The monthly subscription rate is $12.95. You can either pay on a monthly basis or prepay an entire year and receive the 13th month free. Assuming you will maintain your subscription for at least 13 months, what is the effective annual rate of return on your investment if you choose to prepay for the entire year? All payments are due at the beginning of the month. (5.6)

5-38. A loan of $3,000 for a new, high-end laptop computer is to be repaid in 15 end-of-month payments (starting one month from now). The monthly payments are determined as follows.

Loan principal	$3,000
Interest for 15 months at	
1.5% per month	675
Loan application fee	150
Total	$3,825

Monthly payment = $3,825/15 = $255

What nominal and effective interest rates per year are actually being paid? Hint: Draw a cash-flow diagram from the perspective of the lender. (5.6)

5-39. The "Arctic Express" mutual fund company invests its monies in Russia. They claim that money invested with them will quadruple in value over the next 10 years. Suppose you decide to invest $200 per month for the next 10 years in this fund. If their claim is true, what effective annual rate of return (IRR) will you have earned in this fund? (5.6)

5-40. A zero-coupon certificate involves payment of a fixed sum of money now with a future lump-sum withdrawal of an accumulated amount. Earned interest is not paid out periodically but instead compounds to become the major component of the accumulated amount paid when the zero-coupon certificate matures. Consider a certain zero-coupon certificate that was issued on March 25, 1993 and matures on January 30, 2010. A person who purchases a certificate for $13,500 will receive a check for $54,000 when the certificate matures. What is the annual interest rate (yield) that will be earned on this certificate? Assume monthly compounding. (5.3)

5-41. In 2000, Ramon Rodriquez purchased 1,000 shares of common stock of a company in the Internet search engine business. He paid $20 per share for the stock. Ramon believed the business was going to grow nicely in the years ahead, so he bought an additional 500 shares of this company's stock in 2005 for $32 per share. In 2009, Ramon decides to take his profit, and he sold all of his stock in the company for $40 per share. How profitable was Ramon's investment? Use the IRR method to calculate a compound annual rate of interest in this situation. (5.6)

5-42. Sergeant Jess Frugal has the problem of running out of money near the end of each month (he gets paid once a month). Near his army base there is a payday lender company, called Predatory Lenders, Inc., that will give Jess a cash advance of $350 if he will repay the loan a month later with a post-dated check for $375. Almost as soon as Frugal's check for $375 clears the bank, he unfortunately must again borrow $350 to make ends meet. Jess's wife has gotten a bit concerned that her husband might be paying an exorbitant interest rate to this payday lender. Assuming Jess has repeated this borrowing and repayment scheme for 12 months in a row, what effective annual interest rate is he really paying? Is Jess's wife correct in her worry? Hint: Draw a cash-flow diagram from the viewpoint of the lender. (5.6)

5-43. At a military base in Texas, Corporal Stan Moneymaker has been offered a wonderful savings plan. These are the salesperson's words: "During your 48-month tour of duty, you will invest $200 per month

for the first 45 months. We will make the 46th, 47th, and 48th payments of $200 each for you. When you leave the service, we will pay you $10,000 cash."

Is this a good deal for Corporal Moneymaker? Use the IRR method in developing your answer. What assumptions are being made by Corporal Moneymaker if he enters into this contract? (5.6)

5-44. To purchase a used automobile, you borrow $8,000 from Loan Shark Enterprises. They tell you the interest rate being charged is 1% per month for 35 months. They also charge you $200 for a credit investigation, so you leave with $7,800 in your pocket. The monthly payment they calculated for you is

$$\frac{\$8,000(0.01)(35) + \$8,000}{35} = \$308.57/\text{month}.$$

If you agree to these terms and sign their contract, what is the actual APR (annual percentage rate) that you are paying? (5.6)

5-45. Your boss has just presented you with the summary in the accompanying table of projected costs and annual receipts for a new product line. He asks you to calculate the IRR for this investment opportunity. What would you present to your boss, and how would you explain the results of your analysis? (It is widely known that the boss likes to see graphs of PW versus interest rate for this type of problem.) The company's MARR is 10% per year. (5.6)

End of Year	Net Cash Flow
0	−$450,000
1	−42,500
2	+92,800
3	+386,000
4	+614,600
5	−$202,200

5-46. A small company purchased now for $23,000 will lose $1,200 each year the first four years. An additional $8,000 invested in the company during the fourth year will result in a profit of $5,500 each year from the fifth year through the fifteenth year. At the end of 15 years, the company can be sold for $33,000.

a. Determine the IRR. (5.6)

b. Calculate the FW if MARR = 12%. (5.4)

c. Calculate the ERR when ϵ = 12%. (5.7)

5-47. A company has the opportunity to take over a redevelopment project in an industrial area of a city. No immediate investment is required, but it must raze the existing buildings over a four-year period and, at the end of the fourth year, invest $2,400,000 for new construction. It will collect all revenues and pay all costs for a period of 10 years, at which time the entire project, and properties thereon, will revert to the city. The net cash flows are estimated to be as follows:

Year End	Net Cash Flow
1	$500,000
2	300,000
3	100,000
4	−2,400,000
5	150,000
6	200,000
7	250,000
8	300,000
9	350,000
10	400,000

Tabulate the PW versus the interest rate and determine whether multiple IRRs exist. If so, use the ERR method when $\epsilon = 8\%$ per year to determine a rate of return. (5.7)

5-48. The prospective exploration for oil in the outer continental shelf by a small, independent drilling company has produced a rather curious pattern of cash flows, as follows:

End of Year	Net Cash Flow
0	−$520,000
1–10	+200,000
10	−1,500,000

The $1,500,000 expense at EOY 10 will be incurred by the company in dismantling the drilling rig.

a. Over the 10-year period, plot PW versus the interest rate (i) in an attempt to discover whether multiple rates of return exist. (5.6)

b. Based on the projected net cash flows and results in Part (a), what would you recommend regarding the pursuit of this project? Customarily, the company expects to earn at least 20% per year on invested capital before taxes. Use the ERR method ($\epsilon = 20\%$). (5.7)

5-49. An hospital plans to invest $14,000 in a portable medical fluoroscope. The estimated cash outflows and inflows are shown below. What is the simple payback period for this proposed investment? (5.8)

EOY k	Expected Cash Flows
0	−$10,000
1	−$4,000
2–7	$3,000
8–12	$6,000
12[a]	$4,000

[a] Recovery of working capital

5-50. An integrated, combined cycle power plant produces 285 MW of electricity by gasifying coal. The capital investment for the plant is $570 million, spread evenly over two years. The operating life of the plant is expected to be 20 years. Additionally, the plant will operate at full capacity 75% of the time (downtime is 25% of any given year). (5.8)

a. If this plant will make a profit of three cents per kilowatt-hour of electricity sold to the power grid, what is the simple payback period of the plant? Is it a low-risk venture?

b. What is the IRR for the plant? Is it profitable?

5-51. A computer call center is going to replace all of its incandescent lamps with more energy-efficient fluorescent lighting fixtures. The total energy savings are estimated to be $1,875 per year, and the cost of purchasing and installing the fluorescent fixtures is $4,900. The study period is five years, and terminal market values for the fixtures are negligible. (5.8)

a. What is the IRR of this investment?

b. What is the simple payback period of the investment?

c. Is there a conflict in the answers to Parts (a) and (b)? List your assumptions.

d. The simple payback "rate of return" is $1/\theta$. How close does this metric come to matching your answer in Part (a)?

5-52. A hospital is considering an energy conservation project with a life of six years. This project will save 15,000 kilowatt-hours of electric energy per year. The cost of a kWh of electricity is $0.09. Maintenance for the project is expected to cost $300 per year. The hospital's MARR is 15% per year.

a. If the simple payback period must be two years (or less), how much money can be invested in this project?

b. What is the annual worth of the project when its required investment is the answer to Part (a)?

5-53. Advanced Modular Technology (AMT) typically exhibits net annual revenues that increase over a fairly long period. In the long run, an AMT project may be profitable as measured by IRR, but its simple payback period may be unacceptable. Evaluate this AMT project using the IRR method when the company MARR is 15% per year and its maximum allowable payback period is three years. What is your recommendation? (5.6, 5.8)

Capital investment at time 0	$100,000
Net revenues in year k	$20,000+
	$10,000 \cdot (k-1)$
Market (salvage) value	$10,000
Life	5 years

5-54. The Injection Molding Company (IMC) has a policy that all capital investments must have a four-year or less discounted payback period in order to be considered for funding. The MARR at IMC is 8% per year. Is the following project able to meet this benchmark for funding? (5.8)

End of Year	Cash Flow
0	−$300,000
1	−$50,000
2	$50,000
3	$150,000
4	$250,000
5	$350,000
6–10	$100,000

5-55. The International Parcel Service has installed a new radio frequency identification system to help reduce the number of packages that are incorrectly delivered. The capital investment in the system is $65,000, and the projected annual savings are tabled below. The system's market value at the EOY five is negligible, and the MARR is 18% per year.

End of Year	Savings
1	$25,000
2	30,000
3	30,000
4	40,000
5	46,000

a. What is the FW of this investment?

b. What is the IRR of the system?

c. What is the discounted payback period for this investment?

5-56. The Fischer-Tropsch (F-T) process was developed in Germany in 1923 to convert synthesis gas (i.e., a mixture of hydrogen and carbon monoxide) into liquid with some gaseous hydrocarbons. Interestingly, the F-T process was used in World War II to make gasoline and other fuels. (5.8)

a. If the U.S. military can save one billion gallons per year of foreign oil by blending its jet fuel with F-T products, how much can the government afford to invest (through tax breaks and subsidies) in the F-T industry in the United States? The government's MARR is 7% per year, the study period is 40 years, and one gallon of jet fuel costs $2.50.

b. If Congress appropriates $25 billion to support the F-T process industry, what is the simple payback period?

5-57. A small start-up biotech firm anticipates it will have cash outflows of $200,000 per year at the end of the next three years. Then the firm expects a positive cash flow of $50,000 at the EOY four and positive cash flows of $250,000 at the EOYs five through nine. Based on these estimates, would you invest money in this company if your MARR is 15% per year? (all sections).

5-58. A remotely situated fuel cell has an installed cost of $2,000 and will reduce existing surveillance expenses by $350 per year for eight years. The border security agency's MARR is 10% per year. (5.6)

a. What is the minimum salvage (market) value after eight years that makes the fuel cell worth purchasing?

b. What is the fuel cell's IRR if the salvage value is negligible?

5-59. In your own words, explain to your grandmother why the values of various government bonds go up when interest rates in the U.S. economy drop. (5.3)

5-60.

a. Calculate the IRR for each of the three cash-flow diagrams that follow. Use EOY zero for (i) and EOY four for (ii) and (iii) as the reference points in time. What can you conclude about "reference year shift" and "proportionality" issues of the IRR method?

b. Calculate the PW at MARR = 10% per year at EOY zero for (i) and (ii) and EOY four for (ii) and (iii). How do the IRR and PW methods compare?

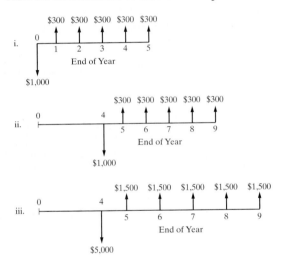

5-61. A group of private investors borrowed $30 million to build 300 new luxury apartments near a large university. The money was borrowed at 6% annual interest, and the loan is to be repaid in equal annual amounts (principal and interest) over a 40-year period. Annual operating, maintenance, and insurance expenses are estimated to be $4,000 per apartment, and these expenses are incurred independently of the occupancy rate for the apartments. The rental fee for each apartment will be $12,000 per year, and the worst-case occupancy rate is projected to be 80%. (5.5)

a. How much profit (or loss) will the investors make each year with 80% occupancy?

b. Repeat Part (a) when the occupancy rate is 95%.

5-62. Javier borrows $50,000 from a local bank at an APR of 9%, compounded monthly. His monthly payments amount to $50,000 (A/P, 0.75%, 60) = $1,040 for a 60-month loan. If Javier makes an extra payment on the first month of each year, his repayment duration for the loan will be reduced by how many months? (5.5)

5-63. *Extended Learning Exercise* A company is producing a high-volume item that sells for $0.75 per unit. The variable production cost is $0.30 per unit. The company is able to produce and sell 10,000,000 items per year when operating at full capacity.

The critical attribute for this product is weight. The target value for weight is 1,000 grams, and the specification limits are set at ±50 grams. The filling machine used to dispense the product is capable of weights following a normal distribution with an average (μ) of 1,000 grams and a standard deviation (σ) of 40 grams. Because of the large standard deviation (with respect to the specification limits), 21.12% of all units produced are not within the specification limits. (They either weigh less than 950 grams or more than 1,050 grams.) This means that 2,112,000 out of 10,000,000 units produced are nonconforming and cannot be sold without being reworked.

Assume that nonconforming units can be reworked to specification at an additional fixed cost of $0.10 per unit. Reworked units can be sold for $0.75 per unit. It has been estimated that the demand for this product will remain at 10,000,000 units per year for the next five years.

To improve the quality of this product, the company is considering the purchase of a new filling machine. The new machine will be capable of dispensing the product with weights following a normal distribution with μ = 1,000 grams and σ = 20 grams. As a result, the percent of nonconforming units will be reduced to 1.24% of production. The new machine will cost $710,000 and will last for at least five years. At the end of five years, this machine can be sold for $100,000.

a. If the company's MARR is 15% per year, is the purchase of the new machine to improve quality (reduce variability) economically attractive? Use the AW method to make your recommendation.

b. Compute the IRR, simple payback period, and discounted payback period of the proposed investment.

c. What other factors, in addition to reduced total rework costs, may influence the company's decision about quality improvement?

Spreadsheet Exercises

5-64. Jane Roe's plan is to accumulate $250,000 in her personal savings account by the time she retires at 60. The longer she stalls on getting started, the tougher it will be to meet her objective. Jane is now 25 years old. Create a spreadsheet to complete the following table to show how much Jane will have to save each year (assuming she waits N years to start saving) to have $250,000 in her account when she is 60. Comment on the pattern you observe in the table. (5.4)

	Interest Rate per Year		
	4%	8%	12%
N = 5			
N = 10			
N = 15			
N = 20			

5-65. An investor has a principal amount of $P. If he desires a payout (return) of 0.1P each year, how many years will it take to deplete an account that earns 8% per year?

$$0.1P = P(A/P, 8\%, N), \text{ so } N \cong 21 \text{ years.}$$

A payout duration table can be constructed for select payout percentages and compound interest rates. Complete the following table. (Note: table entries are years.) Summarize your conclusions about the pattern observed in the table. (5.5)

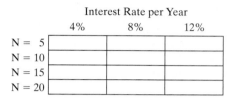

		Interest Rate Per Year			
		4%	6%	8%	10%
Payout per Year	10%			20.9	
(% of principal)	20%				
	30%				

5-66. Refer to Example 5-13. Create a single spreadsheet that calculates PW, FW, AW, IRR, and ERR for the proposed investment. Assume that $\epsilon = $ MARR $= 20\%$ per year. Does your recommendation change if the MARR decreases to 18%? Increases to 22%? (5.6, 5.7)

5-67. A certain medical device will result in an estimated $15,000 reduction in hospital labor expenses during its first year of operation. Labor expenses (and thus savings) are projected to increase at a rate of 7% per year after the first year. Additional operating expenses for the device (maintenance, electric power, etc.) are $3,500 annually, and they increase by $250 per year thereafter (i.e., $3,750 in year two and so on). It is anticipated that the device will last for 10 years and will have no market value at that time. If the MARR is 10% per year, how much can the hospital afford to pay for this device? Use an Excel spreadsheet in your solution. (5.3)

5-68. Refer to Problem 5-61. Develop a spreadsheet to investigate the sensitivity of annual profit (loss) to changes in the occupancy rate and the annual rental fee. (5.5)

Case Study Exercises

5-69. Suppose that the industrial engineers are able to increase the average utilization rate of the CVD tool from 80% to 90%. What is the projected impact of this 10% increase on the PW of the project? (5.9)

5-70. Suppose that the mechanical engineer in the plant has retrofitted the CVD tool so that it now produces 15 wafers per hour. What is the new breakeven point? (5.9)

5-71. Suppose that the average utilization of the CVD tool increased to 90%; however, the operating hours of the fabrication plant were decreased from 168 hours to 150 hours. What are the corresponding impacts on the PW and breakeven point, respectively? (5.9)

FE Practice Problems

5-72. Doris Wade purchased a condominium for $50,000 in 1987. Her down payment was $20,000. She financed the remaining amount as a $30,000, 30-year mortgage at 7%, compounded monthly. Her monthly payments are $200. It is now 2007 (20 years later) and Doris has sold the condominium for $100,000, immediately after making her 240th payment on the unit. Her effective annual internal rate of return on this investment is closest to which answer below? (5.6)

(a) 3.6%   (b) 8.5%   (c) 5.3%   (d) 1.5%

5-73. Elin purchased a used car for $10,000. She wrote a check for $2,000 as a down payment for the car and financed the $8,000 balance. The annual percentage rate (APR) is 9% compounded monthly, and the loan is to be repaid in equal monthly installments over the next four years. Which of the following is most near to Elin's monthly car payment? (5.5)

(a) $167   (b) $172   (c) $188
(d) $200   (e) $218

5-74. A specialized automatic machine costs $300,000 and is expected to save $111,837.50 per year while in operation. Using a 12% interest rate, what is the discounted payback period? (5.8)

(a) 4   (b) 5   (c) 6
(d) 7   (e) 8

5-75. With interest at 8% compounded annually, how much money is required today to provide a perpetual income of $14,316 per year? (5.3)

(a) $178,950   (b) $96,061   (c) $175,134
(d) $171,887

5-76. What is the IRR in the following cash flow? (5.6)

Year End	0	1	2	3	4
Cash Flow ($)	−3,345	1,100	1,100	1,100	1,100

(a) 12.95%   (b) 11.95%   (c) 9.05%
(d) 10.05%   (e) 11.05%

5-77. A bond has a face value of $1,000, is redeemable in eight years, and pays interest of $100 at the end of each of the eight years. If the bond can be purchased for $981, what is the rate of return if the bond is held until maturity? (5.3)

(a) 10.65%   (b) 12.65%   (c) 10.35%
(d) 11.65%

5-78. If you invest $5,123 in a venture, you will receive $1,110 per year for the next 20 years. Assuming 10% interest, what is the discounted payback period for your investment? (5.8)

(a) 7   (b) 8   (c) 5
(d) 9   (e) 6

5-79. What is the equivalent AW of a two-year contract that pays $5,000 at the beginning of the first month and increases by $500 for each month thereafter? MARR = 12% compounded monthly. (5.5)

(a) $10,616   (b) $131,982   (c) $5,511
(d) $5,235   (e) $134,649

5-80. A new machine was bought for $9,000 with a life of six years and no salvage value. Its annual operating costs were as follows:

$$\$7{,}000,\ \$7{,}350,\ \$7{,}717.50, \ldots, \$8{,}933.97.$$

If the MARR = 12%, what was the annual equivalent cost of the machine? (5.5)

(a) $7,809   (b) $41,106   (c) $9,998
(d) $2,190   (e) $9,895

5-81. You want to deposit enough money in a bank account for your son's education. You estimate that he will need $8,000 per year for four years, starting on his 18th birthday. Today is his first birthday. If you earn 12% interest, how much lump sum should you deposit in the bank today to provide for his education? (5.3)

(a) $24,298   (b) $3,538   (c) $32,000
(d) $3,963

5-82. An automobile dealership offers a car with $0 down payment, $0 first month's payment, and $0 due at signing. The monthly payment, starting at the end of month two, is $386 and there are a total of 38 payments. If the APR is 12% compounded monthly, the negotiated sales prices is closest to which answer below? (5.3)

(a) $14,668   (b) $12,035   (c) $12,415
(d) $13,175

Appendix 5-A   The Multiple Rate of Return Problem with the IRR Method

Whenever the IRR method is used and the cash flows reverse sign (from net cash outflow to net cash inflow or the opposite) more than once over the study period, we should be aware of the rather remote possibility that either no interest rate or multiple interest rates may exist. According to Descartes' rule of signs, the *maximum* number of possible IRRs in the $(-1, \infty)$ interval for any given project is equal to the number of cash flow sign reversals during the study period. Nordstrom's criterion says if there is only one sign change in the cumulative cash flows over time, a unique interest rate exists. Two or more sign changes in the cumulative cash flow indicates the possibility of multiple interest rates. Descartes' rule of signs and Nordstrom's criterion guarantee a unique internal rate of return if there is only one sign change in the cash-flow sequence and in the cumulative cash flows over time, respectively. *The simplest way to check for multiple IRRs is to plot equivalent worth (e.g., PW) against the interest rate.* If the resulting plot crosses the interest rate axis more than once, multiple IRRs are present and another equivalence method is recommended for determining project acceptability.

As an example, consider the following project for which the IRR is desired.

EXAMPLE 5-A-1

Plot the PW versus interest rate for the following cash flows. Are there multiple IRRs? If so, what do they mean? Notice there are two sign changes in net cash flows (positive, negative, positive), so at most two IRRs exist for this situation. Nordstrom's criterion also suggests a maximum of two interest rates because there are two sign changes in cumulative cash flow.

Year, k	Net Cash Flow	$i\%$	PW($i\%$)	Year, k	Cumulative Cash Flow
0	$500	0	$250	0	$500
1	−1,000	10	105	1	−500
2	0	20	32	2	−500
3	250	30	~0	3	−250
4	250	40	−11	4	0
5	250	62	~0	5	250
		80	24		

Thus, the PW of the net cash flows equals zero at interest rates of about 30% and 62%, so multiple IRRs do exist. Whenever there are multiple IRRs, which is seldom, it is likely that none are correct.

In this situation, the ERR method (see Section 5.7) could be used to decide whether the project is worthwhile. Or, we usually have the option of using an equivalent worth method. In Example 5-A-1, if the external reinvestment rate (ϵ) is 10% per year, we see that the ERR is 12.4%.

$$\$1,000(P/F, 10\%, 1)(F/P, i'\%, 5) = \$500(F/P, 10\%, 5) + \$250(F/A, 10\%, 3)$$

$$(P/F, 10\%, 1)(F/P, i', 5) = 1.632$$

$$i' = 0.124 \ (12.4\%).$$

In addition, PW(10%) = $105, so both the ERR and PW methods indicate that this project is acceptable when the MARR is 10% per year.

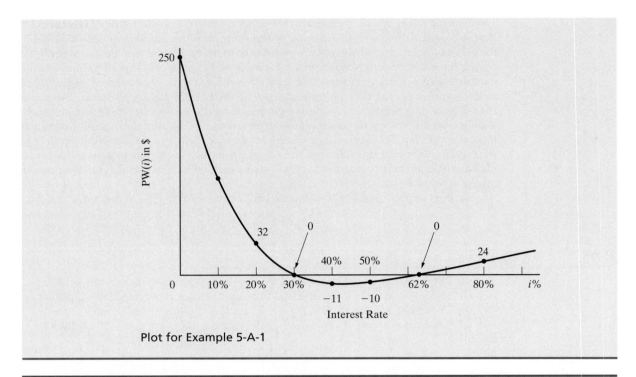

Plot for Example 5-A-1

EXAMPLE 5-A-2

Use the ERR method to analyze the cash-flow pattern shown in the accompanying table. The IRR is indeterminant (none exists), so the IRR is not a workable procedure. The external reinvestment rate (ϵ) is 12% per year, and the MARR equals 15%.

Year	Cash Flows
0	$5,000
1	−7,000
2	2,000
3	2,000

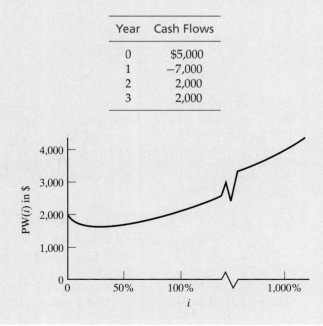

Solution

The ERR method provides this result:

$$\$7,000(P/F, 12\%, 1)(F/P, i'\%, 3) = \$5,000(F/P, 12\%, 3)$$
$$+\$2,000(F/P, 12\%, 1) + \$2,000$$
$$(F/P, i', 3) = 1.802$$
$$i' = 21.7\%.$$

Thus, the ERR is greater than the MARR. Hence, the project having this cash-flow pattern would be acceptable. The PW at 15% is equal to $1,740.36, which confirms the acceptability of the project.

Appendix 5-A Problems

5-A-1. Use the ERR method with $\epsilon = 8\%$ per year to solve for a unique rate of return for the following cash-flow diagram. How many IRRs (the maximum) are suggested by Descartes' rule of signs?

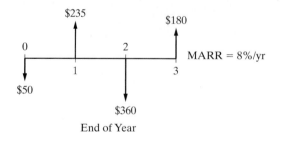

End of Year

5-A-2. Apply the ERR method with $\epsilon = 12\%$ per year to the following series of cash flows. Is there a single, unique IRR for these cash flows? What is the maximum number of IRRs suggested by Nordstrom's criterion?

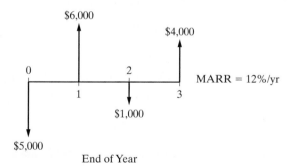

End of Year

5-A-3. Are there multiple IRRs for the following cash-flow sequence? How many are possible according to Descartes' rule of signs? If $\epsilon = 10\%$ per year, what is the ERR for the cash flows of this project? Let MARR = 10% per year.

EOY	0	1	2	3	4	5	6	7	8	9	10
Cash Flow ($)	120	90	60	30	−1,810	600	500	400	300	200	100

CHAPTER 6

Comparison and Selection among Alternatives

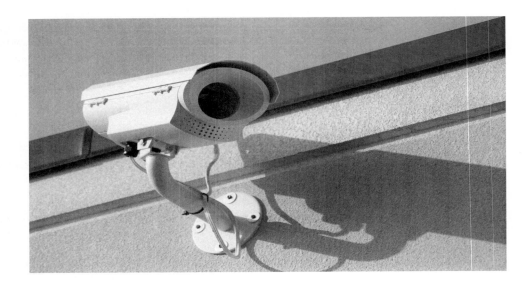

Our aim in Chapter 6 is to compare capital investment alternatives when the time value of money is a key influence.

Alternatives for Border Surveillance

The U.S. Border patrol is considering two remotely operated, solar-powered surveillance systems for a popular border-crossing point. One system must be chosen, and the study period is indefinitely long. Because revenues do not exist, only negative cash flows and market values will be considered. The MARR is 15% per year. Based on the following cash flows, which system should be selected?

	System RX-1	System ABY
Capital investment	$24,000	$45,000
Market value at end of useful life	$1,000	$4,000
Annual expenses	$2,500	$1,500
Useful life (years)	5	10

In this chapter, you will learn how to evaluate alternatives with long study periods such as these. We will return to this problem in Example 6-10.

> The best alternative may be the one you haven't yet uncovered.
>
> —Anonymous

6.1 Introduction

Most engineering projects can be accomplished by more than one feasible design alternative. When the selection of one of these alternatives excludes the choice of any of the others, the alternatives are called *mutually exclusive*. Typically, the alternatives being considered require the investment of different amounts of capital, and their annual revenues and costs may vary. Sometimes the alternatives may have different useful lives. Because different levels of investment normally produce varying economic outcomes, we must perform an analysis to determine which one of the mutually exclusive alternatives is preferred and, consequently, how much capital should be invested.

A seven-step procedure for accomplishing engineering economy studies was discussed in Chapter 1. In this chapter, we address Step 5 (analysis and comparison of the feasible alternatives) and Step 6 (selection of the preferred alternative) of this procedure, and we compare mutually exclusive alternatives on the basis of economic considerations alone.

Five of the basic methods discussed in Chapter 5 for analyzing cash flows are used in the analyses in this chapter [present worth (PW), annual worth (AW), future worth (FW), internal rate of return (IRR) and external rate of return (ERR)]. These methods provide a *basis for economic comparison* of alternatives for an engineering project. When correctly applied, these methods result in the correct selection of a preferred alternative from a set of mutually exclusive alternatives. The comparison of mutually exclusive alternatives by means of the benefit–cost ratio method is discussed in Chapter 10.

6.2 Basic Concepts for Comparing Alternatives

Principle 1 (Chapter 1) emphasized that a choice (decision) is among alternatives. Such choices must incorporate the fundamental purpose of capital investment; namely, to obtain at least the minimum attractive rate of return (MARR) for each dollar invested. In practice, there are usually a limited number of feasible alternatives to consider for an engineering project. The problem of deciding which mutually exclusive alternative should be selected is made easier if we adopt this rule based on Principle 2 (focus on the differences) in Chapter 1: *The alternative that requires the minimum investment of capital and produces satisfactory functional results will be chosen unless the incremental capital associated with an alternative having a larger investment can be justified with respect to its incremental benefits.*

Under this rule, we consider the acceptable alternative that requires the least investment of capital to be the *base alternative*. The investment of additional capital over that required by the base alternative usually results in increased capacity, increased quality, increased revenues, decreased operating expenses, or increased life. Therefore, before additional money is invested, it must be shown that each avoidable increment of capital can pay its own way relative to other available investment opportunities.

In summary, *if* the extra benefits obtained by investing additional capital are better than those that could be obtained from investment of the same capital elsewhere in the company at the MARR, the investment should be made.

If this is not the case, we obviously would not invest more than the minimum amount of capital required, and we may even do nothing at all. Stated simply, our rule will keep as much capital as possible invested at a rate of return equal to or greater than the MARR.

6.2.1   Investment and Cost Alternatives

This basic policy for the comparison of mutually exclusive alternatives can be demonstrated with two examples. The *first example* involves an investment project situation. Alternatives *A* and *B* are two mutually exclusive *investment alternatives* with estimated net cash flows,* as shown. *Investment alternatives are those with initial (or front-end) capital investment(s) that produce positive cash flows from increased revenue, savings through reduced costs, or both.* The useful life of each alternative in this example is four years.

	Alternative		
	A	*B*	$\Delta(B - A)$
Capital investment	−$60,000	−$73,000	−$13,000
Annual revenues less expenses	22,000	26,225	4,225

The cash-flow diagrams for Alternatives *A* and *B*, and for the year-by-year differences between them (i.e., *B* minus *A*), are shown in Figure 6-1. These diagrams typify those for investment project alternatives. In this first example, at MARR = 10% per year, the PW values are

$$\text{PW}(10\%)_A = -\$60,000 + \$22,000(P/A, 10\%, 4) = \$9,738,$$

$$\text{PW}(10\%)_B = -\$73,000 + \$26,225(P/A, 10\%, 4) = \$10,131.$$

Since the PW_A is greater than zero at $i = \text{MARR}$, Alternative *A* is the base alternative and would be selected *unless* the additional (incremental) capital associated with Alternative *B* ($13,000) is justified. In this case, Alternative *B* is preferred to *A* because it has a greater PW value. Hence, *the extra benefits obtained*

---

* In this book, the terms *net cash flow* and *cash flow* will be used interchangeably when referring to periodic cash inflows and cash outflows for an alternative.

Figure 6-1 Cash-Flow Diagrams for Alternatives *A* and *B* and Their Difference

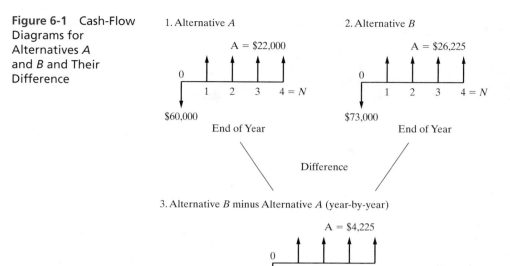

by investing the additional $13,000 of capital in B (diagram 3, Figure 6-1) have a PW of $10,131 − $9,738 = $393. That is,

$$PW(10\%)_{Diff} = -\$13,000 + \$4,225(P/A, 10\%, 4) = \$393,$$

and the additional capital invested in *B* is justified.

The *second example* involves a cost project situation. Alternatives *C* and *D* are two mutually exclusive *cost alternatives* with estimated net cash flows, as shown, over a three-year life. *Cost alternatives are those with all negative cash flows, except for a possible positive cash flow element from disposal of assets at the end of the project's useful life.* This situation occurs when the organization must take some action, and the decision involves the most economical way of doing it (e.g., the addition of environmental control capability to meet new regulatory requirements). It also occurs when the expected revenues are the same for each alternative.

| | Alternative | | |
End of Year	C	D	Δ(D − C)
0	−$380,000	−$415,000	−$35,000
1	−38,100	−27,400	10,700
2	−39,100	−27,400	11,700
3	−40,100	−27,400	12,700
3[a]	0	26,000	26,000

[a] Market value.

The cash-flow diagrams for Alternatives *C* and *D*, and for the year-by-year differences between them (i.e., *D* minus *C*), are shown in Figure 6-2. These diagrams typify those for cost project alternatives. In this "must take action"

Figure 6-2   Cash-Flow
Diagrams for
Alternatives C
and D and Their
Difference

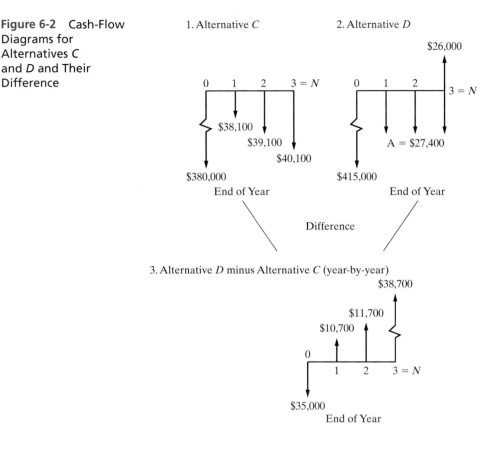

situation, Alternative C, which has the lesser capital investment, is automatically the base alternative and would be selected *unless* the additional (incremental) capital associated with Alternative D ($35,000) is justified. With the greater capital investment, Alternative D in this illustration has smaller annual expenses. Otherwise, it would not be a feasible alternative. (It would not be logical to invest more capital in an alternative without obtaining additional revenues or savings.) Note, in diagram 3, Figure 6-2, that the difference between two feasible cost alternatives is an investment alternative.

   In this second example, at MARR = 10% per year, the PW(10%)$_C$ = −$477,077 and the PW(10%)$_D$ = −$463,607. Alternative D is preferred to C because it has the less negative PW (minimizes costs). Hence, *the lower annual expenses obtained by investing the additional $35,000 of capital in Alternative D* have a PW of −$463,607 − (−$477,077) = $13,470. That is, the PW(10%)$_{Diff}$ = $13,470, and the additional capital invested in Alternative D is justified.

6.2.2   Ensuring a Comparable Basis

Each feasible, mutually exclusive alternative selected for detailed analysis meets the functional requirements established for the engineering project (Section 1.3.2). Differences among the alternatives, however, may occur in many forms. Ensuring a comparable basis for their analysis requires that any economic impacts of these differences be included in the estimated cash flows for the alternatives (as well as comparing them over the same analysis period—see Section 6.3). Otherwise, the wrong design alternative may be selected for implementing the project. The following are examples of the types of differences that may occur:

1. Operational performance factors such as output capacity, speed, thrust, heat dissipation rate, reliability, fuel efficiency, setup time, and so on
2. Quality factors such as the number of defect-free (nondefective) units produced per period or the percentage of defective units (reject rate)
3. Useful life, capital investment required, revenue changes, various annual expenses or cost savings, and so on

This list of examples could be expanded. The specific differences, however, for each engineering project and its design alternatives must be identified. Then, the cash-flow estimates for the alternatives must include the economic impact of these differences. *This is a fundamental premise for comparing alternatives in Chapter 6 and in the chapters that follow.*

Two rules were given in Section 2.4 for facilitating the correct analysis and comparison of mutually exclusive alternatives when the time value of money is not a factor (present economy studies). For convenience, these rules are repeated here and *extended to account for the time value of money*:

Rule 1: When revenues and other economic benefits are present and vary among the alternatives, choose the alternative that *maximizes* overall profitability. That is, select the alternative that has the greatest positive equivalent worth at $i =$ MARR and satisfies all project requirements.

Rule 2: When revenues and other economic benefits are *not* present *or* are constant among the alternatives, consider only the costs and select the alternative that *minimizes* total cost. That is, select the alternative that has the least negative equivalent worth at $i =$ MARR and satisfies all project requirements.

6.3 The Study (Analysis) Period

The study (analysis) period, sometimes called the *planning horizon*, is the selected time period over which mutually exclusive alternatives are compared. The determination of the study period for a decision situation may be influenced by several factors—for example, the service period required, the useful life* of the shorter-lived alternative, the useful life of the longer-lived alternative and company policy. *The key point is that the selected study period must be appropriate for the decision situation under investigation.*

---

* The useful life of an asset is the period during which it is kept in productive use in a trade or business.

The useful lives of alternatives being compared, relative to the selected study period, can involve two situations:

1. Useful lives are the same for all alternatives and equal to the study period.
2. Useful lives are unequal among the alternatives, and at least one does not match the study period.

> Unequal lives among alternatives somewhat complicate their analysis and comparison. To conduct engineering economy analyses in such cases, we adopt the rule of comparing mutually exclusive alternatives over the same period of time.

The repeatability assumption and the coterminated assumption are the two types of assumptions used for these comparisons.

The *repeatability assumption* involves two main conditions:

1. The study period over which the alternatives are being compared is either indefinitely long or equal to a common multiple of the lives of the alternatives.
2. The economic consequences that are estimated to happen in an alternative's initial useful life span will also happen in all succeeding life spans (replacements).

Actual situations in engineering practice seldom meet both conditions. This has tended to limit the use of the repeatability assumption, except in those situations where the difference between the AW of the first life cycle and the AW over more than one life cycle of the assets involved is quite small.*

The *coterminated assumption* uses a finite and identical study period for all alternatives. This planning horizon, combined with appropriate adjustments to the estimated cash flows, puts the alternatives on a common and comparable basis. For example, if the situation involves providing a service, the same time period requirement applies to each alternative in the comparison. To force a match of cash-flow durations to the cotermination time, adjustments (based on additional assumptions) are made to cash-flow estimates of project alternatives having useful lives different from the study period. For example, if an alternative has a useful life shorter than the study period, the estimated annual cost of contracting for the activities involved might be assumed and used during the remaining years. Similarly, if the useful life of an alternative is longer than the study period, a reestimated market value is normally used as a terminal cash flow at the end of a project's coterminated life.

---

* T. G. Eschenbach and A. E. Smith, "Violating the Identical Repetition Assumption of EAC," *Proceedings, International Industrial Engineering Conference* (May 1990), The Institute of Industrial Engineers, Norcross, GA, pp. 99–104.

6.4   Useful Lives Are Equal to the Study Period

When the useful life of an alternative is equal to the selected study period, adjustments to the cash flows are not required. In this section, we discuss the comparison of mutually exclusive alternatives, using equivalent-worth methods and rate-of-return methods when the useful lives of all alternatives are equal to the study period.

6.4.1   Equivalent-Worth Methods

In Chapter 5, we learned that the equivalent-worth methods convert all relevant cash flows into equivalent present, annual, or future amounts. When these methods are used, consistency of alternative selection results from this equivalency relationship. Also, the economic ranking of mutually exclusive alternatives will be the same when using the three methods. Consider the general case of two alternatives A and B. If $\text{PW}(i\%)_A < \text{PW}(i\%)_B$, then the FW and AW analyses will result in the same preference for Alternative B.

The most straightforward technique for comparing mutually exclusive alternatives when all useful lives are equal to the study period is to determine the equivalent worth of each alternative based on total investment at $i = \text{MARR}$. Then, for investment alternatives, the one with the greatest positive equivalent worth is selected. And, in the case of cost alternatives, the one with the least negative equivalent worth is selected.

EXAMPLE 6-1   **Analyzing Investment Alternatives by Using Equivalent Worth**

Best Flight, Inc., is considering three mutually exclusive alternatives for implementing an automated passenger check-in counter at its hub airport. Each alternative meets the same service requirements, but differences in capital investment amounts and benefits (cost savings) exist among them. The study period is 10 years, and the useful lives of all three alternatives are also 10 years. Market values of all alternatives are assumed to be zero at the end of their useful lives. If the airline's MARR is 10% per year, which alternative should be selected in view of the cash-flow diagrams shown on page 246?

Solution by the PW Method

$$\text{PW}(10\%)_A = -\$390,000 + \$69,000(P/A, 10\%, 10) = \$33,977,$$

$$\text{PW}(10\%)_B = -\$920,000 + \$167,000(P/A, 10\%, 10) = \$106,148,$$

$$\text{PW}(10\%)_C = -\$660,000 + \$133,500(P/A, 10\%, 10) = \$160,304.$$

Based on the PW method, Alternative C would be selected because it has the largest PW value ($\$160,304$). The order of preference is $C \succ B \succ A$, where $C \succ B$ means C is preferred to B.

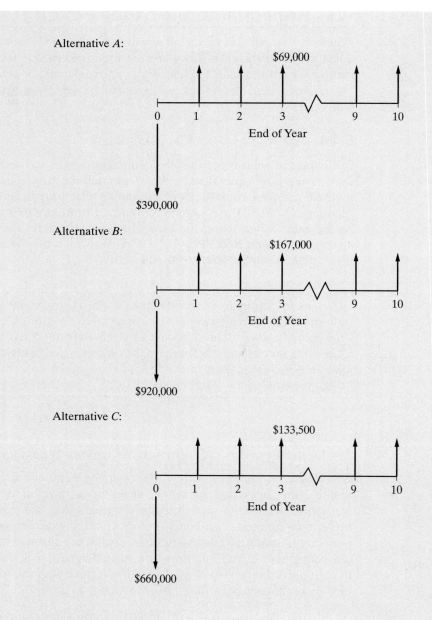

Alternative *A*:

Alternative *B*:

Alternative *C*:

Solution by the AW Method

$$AW(10\%)_A = -\$390{,}000(A/P, 10\%, 10) + \$69{,}000 = \$5{,}547,$$

$$AW(10\%)_B = -\$920{,}000(A/P, 10\%, 10) + \$167{,}000 = \$17{,}316,$$

$$AW(10\%)_C = -\$660{,}000(A/P, 10\%, 10) + \$133{,}500 = \$26{,}118.$$

Alternative *C* is again chosen because it has the largest AW value ($26,118).

Solution by the FW Method

$$\text{FW}(10\%)_A = -\$390,000(F/P, 10\%, 10) + \$69,000(F/A, 10\%, 10) = \$88,138,$$

$$\text{FW}(10\%)_B = -\$920,000(F/P, 10\%, 10) + \$167,000(F/A, 10\%, 10) = \$275,342,$$

$$\text{FW}(10\%)_C = -\$660,000(F/P, 10\%, 10) + \$133,500(F/A, 10\%, 10) = \$415,801.$$

Based on the FW method, the choice is again Alternative C because it has the largest FW value ($415,801). For all three methods (PW, AW, and FW) in this example, notice that $C > B > A$ because of the equivalency relationship among the methods. Also, notice that Rule 1 (Section 6.2.2) applies in this example, since the economic benefits (cost savings) vary among the alternatives.

Example 6-2 and Example 6-3 illustrate the impact that estimated differences in the capability of alternatives to produce defect-free products have on the economic analysis. In Example 6-2, each of the plastic-molding presses produces the same total amount of output units, all of which are defect-free. Then, in Example 6-3, each press still produces the same total amount of output units, but the percentage of defective units (reject rate) varies among the presses.

| EXAMPLE 6-2 | **Analyzing Cost-Only Alternatives, Using Equivalent Worth** |

A company is planning to install a new automated plastic-molding press. Four different presses are available. The initial capital investments and annual expenses for these four mutually exclusive alternatives are as follows:

	Press			
	P1	P2	P3	P4
Capital investment	$24,000	$30,400	$49,600	$52,000
Useful life (years)	5	5	5	5
Annual expenses				
Power	2,720	2,720	4,800	5,040
Labor	26,400	24,000	16,800	14,800
Maintenance	1,600	1,800	2,600	2,000
Property taxes and insurance	480	608	992	1,040
Total annual expenses	$31,200	$29,128	$25,192	$22,880

Assume that each press has the same output capacity (120,000 units per year) and has no market value at the end of its useful life; the selected analysis period is five years; and any additional capital invested is expected to earn at least 10% per year. Which press should be chosen if 120,000 nondefective units per year are produced by each press and all units can be sold? The selling price is $0.375 per unit. Solve by hand and by spreadsheet.

Solution by Hand

Since the same number of nondefective units per year will be produced and sold using each press, revenue can be disregarded (Principle 2, Chapter 1). The end-of-year cash-flow diagrams of the four presses are:

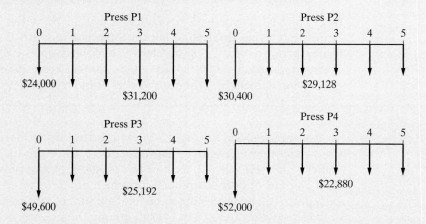

The preferred alternative will minimize the equivalent worth of total costs over the five-year analysis period (Rule 2, page 243). That is, the four alternatives can be compared as cost alternatives. The PW, AW, and FW calculations for Alternative P1 are

$$PW(10\%)_{P1} = -\$24,000 - \$31,200(P/A, 10\%, 5) = -\$142,273,$$

$$AW(10\%)_{P1} = -\$24,000(A/P, 10\%, 5) - \$31,200 = -\$37,531,$$

$$FW(10\%)_{P1} = -\$24,000(F/P, 10\%, 5) - \$31,200(F/A, 10\%, 5)$$

$$= -\$229,131.$$

The PW, AW, and FW values for Alternatives P2, P3, and P4 are determined with similar calculations and shown for all four presses in Table 6-1. Alternative P4 minimizes all three equivalent-worth values of total costs and is the preferred alternative. The preference ranking (P4 > P2 > P1 > P3) resulting from the analysis is the same for all three methods.

		Press (Equivalent-Worth Values)		
TABLE 6-1	Comparison of Four Molding Presses, Using the PW, AW, and FW Methods to Minimize Total Costs			
Method	P1	P2	P3	P4
Present worth	−$142,273	−$140,818	−$145,098	−$138,734
Annual worth	−37,531	−37,148	−38,276	−36,598
Future worth	−229,131	−226,788	−233,689	−223,431

Spreadsheet Solution

Figure 6-3 presents a spreadsheet solution for identifying the press that minimizes total equivalent costs. The top section of the spreadsheet displays the problem data. These data are then tabulated as total end-of-year (EOY) cash flows in the middle section of the spreadsheet. Finally, the PW, AW, and FW amounts are computed and the results displayed at the bottom of the spreadsheet. Note that the AW and FW cell formulas make use of the PW result in row 22. The results are the same (except for rounding) as those computed by hand and displayed in Table 6-1.

NOTE: This spreadsheet model allows us to easily do what the solution by hand does not—evaluate how our recommendation will change if data values change. For example, if MARR changes from 10% to 15%, then Press 2 becomes the lowest cost alternative.

	A	B	C	D	E
1	MARR =	10%		Annual Output Capacity	120,000
2	Useful Life =	5		Selling price = $	0.375
3					
4	**Expenses**	**P1**	**P2**	**P3**	**P4**
5	Capital Investment =	$ 24,000	$ 30,400	$ 49,600	$ 52,000
6	Power =	$ 2,720	$ 2,720	$ 4,800	$ 5,040
7	Labor =	$ 26,400	$ 24,000	$ 16,800	$ 14,800
8	Maintenance =	$ 1,600	$ 1,800	$ 2,600	$ 2,000
9	Tax & Insurance =	$ 480	$ 608	$ 992	$ 1,040
10					
11					
12					
13					
14	**EOY**	**P1**	**P2**	**P3**	**P4**
15	0	$ (24,000)	$ (30,400)	$ (49,600)	$ (52,000)
16	1	$ (31,200)	$ (29,128)	$ (25,192)	$ (22,880)
17	2	$ (31,200)	$ (29,128)	$ (25,192)	$ (22,880)
18	3	$ (31,200)	$ (29,128)	$ (25,192)	$ (22,880)
19	4	$ (31,200)	$ (29,128)	$ (25,192)	$ (22,880)
20	5	$ (31,200)	$ (29,128)	$ (25,192)	$ (22,880)
21					
22	PW =	$ (142,273)	$ (140,818)	$ (145,098)	$ (138,733)
23	AW =	$ (37,531)	$ (37,147)	$ (38,276)	$ (36,597)
24	FW =	$ (229,131)	$ (226,789)	$ (233,681)	$ (223,431)

$= -B5$

$= -SUM(B6:B9)$

$= B\$16$

$= NPV(\$B\$1, B16:B20) + B15$

$= B22*(1 + \$B\$1)^\$B\$2$

$= PMT(\$B\$1, \$B\$2, -B22)$

Figure 6-3   Spreadsheet Solution, Example 6-2

| EXAMPLE 6-3 | **Analyzing Alternatives with Different Reject Rates** |

Consider the four plastic molding presses of Example 6-2. Suppose that each press is still capable of producing 120,000 total units per year, but the estimated reject rate is different for each alternative. This means that the expected revenue will differ among the alternatives since only nondefective units can be sold. The data for the four presses are summarized below. The life of each press (and the study period) is five years.

	Press			
	P1	P2	P3	P4
Capital investment	$24,000	$30,400	$49,600	$52,000
Total annual expenses	$31,200	$29,128	$25,192	$22,880
Reject rate	8.4%	0.3%	2.6%	5.6%

If all nondefective units can be sold for $0.375 per unit, which press should be chosen? Solve by hand and by spreadsheet.

Solution by Hand

In this example, each of the four alternative presses produces 120,000 units per year, but they have different estimated reject rates. Therefore, the number of nondefective output units produced and sold per year, as well as the annual revenues received by the company, vary among the alternatives. But the annual expenses are assumed to be unaffected by the reject rates. In this situation, the preferred alternative will maximize overall profitability (Rule 1, Section 6.2.2). That is, the four presses need to be compared as investment alternatives. The PW, AW, and FW calculations for Alternative P4 are given below:

$$PW(10\%)_{P4} = -\$52,000 + [(1-0.056)(120,000)(\$0.375)-\$22,880](P/A, 10\%, 5)$$

$$= \$22,300,$$

$$AW(10\%)_{P4} = -\$52,000(A/P, 10\%, 5) + [(1 - 0.056)(120,000)(\$0.375) - \$22,880]$$

$$= \$5,882,$$

$$FW(10\%)_{P4} = -\$52,000(F/P, 10\%, 5)$$

$$+ [(1 - 0.056)(120,000)(\$0.375) - \$22,800](F/A, 10\%, 5)$$

$$= \$35,914.$$

The PW, AW, and FW values for Alternatives P1, P2, and P3 are determined with similar calculations and shown for all four alternatives in Table 6-2. Alternative P2 maximizes all three equivalent-worth measures of overall profitability and is preferred [versus P4 in Example 6-2]. The preference ranking (P2 > P4 > P3 > P1) is the same for the three methods but is different from the ranking in Example 6-2. The different preferred alternative and preference ranking are the result of the varying capability among the presses to produce nondefective output units.

TABLE 6-2 **Comparison of Four Molding Presses, Using the PW, AW, and FW Methods to Maximize Overall Profitability**

	Press (Equivalent-Worth Values)			
Method	P1	P2	P3	P4
Present worth	$13,984	$29,256	$21,053	$22,300
Annual worth	3,689	7,718	5,554	5,882
Future worth	22,521	47,117	33,906	35,914

Spreadsheet Solution

Figure 6-4 displays the spreadsheet solution for identifying the preferred press when the impact of different annual revenues among the alternatives is included. The data section of the spreadsheet includes the reject rate for each press, which is used to compute expected annual revenues in row 12. The revenue values are then combined with annual expenses to arrive at the EOY net cash flows for each alternative. The resulting equivalent-worth amounts are the same (except for rounding) as those computed by hand and shown in Table 6-2.

$= (1 - B11) *\$E\$1 *\$E\$2$

$= -SUM(B6:B9) + B\$12$

	A	B	C	D	E
1	MARR =	10%		Annual Output Capacity	120,000
2	Useful Life =	5		Selling price = $	0.375
3					
4	**Expenses**	P1	P2	P3	P4
5	Capital Investment	$ 24,000	$ 30,400	$ 49,600	$ 52,000
6	Power	$ 2,720	$ 2,720	$ 4,800	$ 5,040
7	Labor	$ 26,400	$ 24,000	$ 16,800	$ 14,800
8	Maintenance	$ 1,600	$ 1,800	$ 2,600	$ 2,000
9	Tax & Insurance	$ 480	$ 608	$ 992	$ 1,040
10					
11	Reject Rate	8.4%	0.3%	2.6%	5.6%
12	Revenue	$ 41,220	$ 44,865	$ 43,830	$ 42,480
13					
14	**EOY**	P1	P2	P3	P4
15	0	$ (24,000)	$ (30,400)	$ (49,600)	$ (52,000)
16	1	$ 10,020	$ 15,737	$ 18,638	$ 19,600
17	2	$ 10,020	$ 15,737	$ 18,638	$ 19,600
18	3	$ 10,020	$ 15,737	$ 18,638	$ 19,600
19	4	$ 10,020	$ 15,737	$ 18,638	$ 19,600
20	5	$ 10,020	$ 15,737	$ 18,638	$ 19,600
21					
22	PW =	$ 13,984	$ 29,256	$ 21,053	$ 22,299
23	AW =	$ 3,689	$ 7,718	$ 5,554	$ 5,883
24	FW =	$ 22,521	$ 47,116	$ 33,906	$ 35,913

Figure 6-4 Spreadsheet Solution, Example 6-3

6.4.2 Rate-of-Return Methods

Annual return on investment is a popular metric of profitability in the United States. When using rate-of-return methods to evaluate mutually exclusive alternatives, the best alternative produces satisfactory functional results and requires the minimum investment of capital. This is true unless a larger investment can be justified in terms of its incremental benefits and costs. Accordingly, these three guidelines are applicable to rate-of-return methods:

1. Each increment of capital must justify itself by producing a sufficient rate of return (greater than or equal to MARR) on that increment.
2. Compare a higher investment alternative against a lower investment alternative only when the latter is acceptable. The difference between the two alternatives is usually an *investment alternative* and permits the better one to be determined.
3. Select the alternative that requires the largest investment of capital, as long as the incremental investment is justified by benefits that earn at least the MARR. This maximizes equivalent worth on total investment at $i = $ MARR.

Do not compare the IRRs of mutually exclusive alternatives (or IRRs of the differences between mutually exclusive alternatives) against those of other alternatives. Compare an IRR only against MARR (IRR $\geq$ MARR) in determining the acceptability of an alternative.

These guidelines can be implemented using the *incremental investment analysis technique* with rate-of-return methods.* First, however, we will discuss the *inconsistent ranking problem* that can occur with incorrect use of rate-of-return methods in the comparison of alternatives.

6.4.2.1 The Inconsistent Ranking Problem In Section 6.2, we discussed a small investment project involving two alternatives, A and B. The cash flow for each alternative is restated here, as well as the cash flow (incremental) difference.

	Alternative		Difference
	A	*B*	$\Delta(B - A)$
Capital investment	$60,000	$73,000	$13,000
Annual revenues less expenses	22,000	26,225	4,225

The useful life of each alternative (and the study period) is four years. Also, assume that MARR = 10% per year. First, check to see if the sum of positive cash flows

---

* The IRR method is the most celebrated time value-of-money–based profitability metric in the United States. The incremental analysis technique must be learned so that the IRR method can be correctly applied in the comparison of mutually exclusive alternatives.

exceeds the sum of negative cash flows. This is the case here, so the IRR and PW(10%) of each alternative are calculated and shown as follows:

Alternative	IRR	PW(10%)
A	17.3%	$9,738
B	16.3	10,131

If, at this point, a choice were made based on maximizing the IRR of the total cash flow, Alternative A would be selected. But, based on maximizing the PW of the total investment at i = MARR, Alternative B is preferred. Obviously, here we have an inconsistent ranking of the two mutually exclusive investment alternatives.

Now that we know Alternative A is acceptable (IRR > MARR; PW at MARR > 0), we will analyze the incremental cash flow between the two alternatives, which we shall refer to as $\Delta(B - A)$. The IRR of this increment, IRR_Δ, is 11.4%. This is greater than the MARR of 10%, and the incremental investment of $13,000 is justified. This outcome is confirmed by the PW of the increment, PW_Δ (10%), which is equal to $393. Thus, when the IRR of the incremental cash flow is used, the rankings of A and B are consistent with that based on the PW on total investment.

The fundamental role that the incremental net cash flow, $\Delta(B - A)$, plays in the comparison of two alternatives (where B has the greater capital investment) is based on the following relationship:

Cash flow of B = Cash flow of A + Cash flow of the difference.

Clearly, the cash flow of B is made up of two parts. The first part is equal to the cash flow of Alternative A, and the second part is the incremental cash flow between A and B, $\Delta(B - A)$. Obviously, if the equivalent worth of the difference is greater than or equal to zero at i = MARR, then Alternative B is preferred. Otherwise, given that Alternative A is justified (an acceptable *base alternative*), Alternative A is preferred. It is always true that if $PW_\Delta \geq 0$, then $IRR_\Delta \geq$ MARR.

Figure 6-5 illustrates how ranking errors can occur when a selection among mutually exclusive alternatives is based wrongly on maximization of IRR on the total cash flow. When MARR lies to the left of IRR_Δ (11.4% in this case), an incorrect choice will be made by selecting an alternative that maximizes IRR. This is because the IRR method assumes reinvestment of cash flows at the calculated rate of return (17.3% and 16.3%, respectively, for Alternatives A and B in this case), whereas the PW method assumes reinvestment at MARR (10%).

Figure 6-5 shows our previous results with $PW_B > PW_A$ at MARR = 10%, even though $IRR_A > IRR_B$. Also, the figure shows how to avoid this ranking inconsistency by examining the IRR of the increment, IRR_Δ, which correctly leads to the selection of Alternative B, the same as with the PW method.

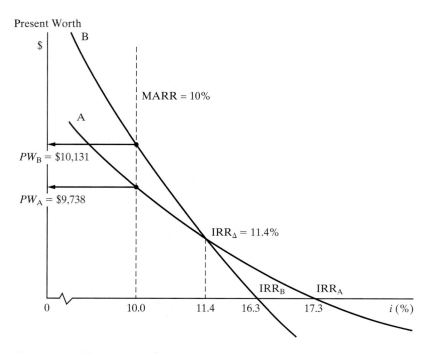

Figure 6-5   Illustration of the Ranking Error in Studies Using the IRR Method

6.4.2.2   The Incremental Investment Analysis Procedure

> We recommend the incremental investment analysis procedure to avoid incorrect ranking of mutually exclusive alternatives when using rate-of-return methods. We will use this procedure in the remainder of the book.

The incremental analysis procedure for the comparison of mutually exclusive alternatives is summarized in three basic steps (and illustrated in Figure 6-6):

1. Arrange (rank-order) the feasible alternatives based on increasing capital investment.*
2. Establish a base alternative.
   (a) Cost alternatives—the first alternative (least capital investment) is the base.
   (b) Investment alternatives—if the first alternative is acceptable (IRR ≥ MARR; PW, FW, or AW at MARR ≥ 0), select it as the base. If the first alternative is not acceptable, choose the next alternative in order of increasing

---

* This ranking rule assumes a logical set of mutually exclusive alternatives. That is to say, for investment or cost alternatives, increased initial investment results in additional economic benefits, whether from added revenues, reduced costs, or a combination of both. Also, this rule assumes that for any nonconventional investment cash flow, the PW, AW, FW, or ERR analysis method would be used instead of IRR. Simply stated, a nonconventional investment cash flow involves multiple sign changes or positive cash flow at time zero, or both. For a more detailed discussion of ranking rules, see C. S. Park and G. P. Sharp-Bette, *Advanced Engineering Economy* (New York: John Wiley & Sons, 1990).

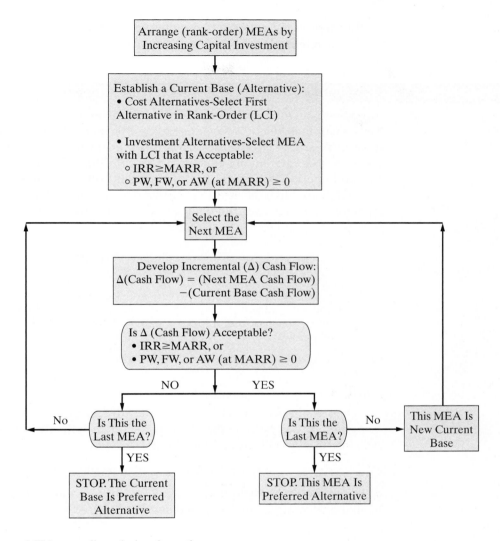

MEA: mutually exclusive alternative
LCI: least capital investment

Figure 6-6 Incremental Investment Analysis Procedure

capital investment and check the profitability criterion (PW, etc.) values. Continue until an acceptable alternative is obtained. If none is obtained, the do-nothing alternative is selected.

3. Use iteration to evaluate differences (incremental cash flows) between alternatives until all alternatives have been considered.

(a) If the incremental cash flow between the next (higher capital investment) alternative and the current selected alternative is acceptable, choose the next alternative as the current best alternative. Otherwise, retain the last acceptable alternative as the current best.

(b) Repeat and select as the preferred alternative the last one for which the incremental cash flow was acceptable.

EXAMPLE 6-4   **Incremental Analysis: Investment Alternatives**

Suppose that we are analyzing the following six mutually exclusive alternatives for a small investment project, using the IRR method. The useful life of each alternative is 10 years, and the MARR is 10% per year. Also, net annual revenues less expenses vary among all alternatives, and Rule 1, Section 6.2.2, applies. If the study period is 10 years, and the market (salvage) values are zero, which alternative should be chosen? Notice that the alternatives have been *rank-ordered* from low capital investment (LCI) to high capital investment.

	Alternative					
	A	B	C	D	E	F
Capital investment	$900	$1,500	$2,500	$4,000	$5,000	$7,000
Annual revenues less expenses	150 LCI	276	400	925	1,125	1,425

Solution

For each of the feasible alternatives, the IRR on the total cash flow can be computed by determining the interest rate at which the PW, FW, or AW equals zero (use of AW is illustrated for Alternative A):*

$$0 = -\$900(A/P, i'_A\%, 10) + \$150; \quad i'\% = ?$$

By trial and error, we determine that $i'_A\% = 10.6\%$. In the same manner, the IRRs of all the alternatives are computed and summarized:

	A	B	C	D	E	F
IRR on total cash flow	10.6%	13.0%	9.6%	19.1%	18.3%	15.6%

At this point, *only Alternative C is unacceptable* and can be eliminated from the comparison because its IRR is less than MARR of 10% per year. Also, A is the base alternative from which to begin the incremental investment analysis procedure, because it is the mutually exclusive alternative with the lowest capital investment whose IRR (10.6%) is equal to or greater than MARR (10%).

---

* The three steps of the incremental analysis procedure previously discussed (and illustrated in Figure 6-6) do not require the calculation of the IRR value for each alternative. In this example, the IRR of each alternative is used for illustrating common errors made with the IRR method.

TABLE 6-3 **Comparison of Five Acceptable Investment Alternatives Using the IRR Method (Example 6-4)**

Increment Considered	A	Δ(B − A)	Δ(D − B)	Δ(E − D)	Δ(F − E)
Δ Capital investment	$900	$600	$2,500	$1,000	$2,000
Δ Annual revenues less expenses	$150	$126	$649	$200	$300
IRR$_\Delta$	10.6%	16.4%	22.6%	15.1%	8.1%
Is increment justified?	Yes	Yes	Yes	Yes	No

This pre-analysis of the feasibility of each alternative is not required by the incremental analysis procedure. It is useful, however, when analyzing a larger set of mutually exclusive alternatives. You can immediately eliminate nonfeasible (nonprofitable) alternatives, as well as easily identify the base alternative.

As discussed in Section 6.4.2.1, it is not necessarily correct to select the alternative that maximizes the IRR on total cash flow. That is to say, Alternative D may not be the best choice, since *maximization of IRR does not guarantee maximization of equivalent worth on total investment at the MARR.* Therefore, to make the correct choice, we must examine each increment of capital investment to see if it will pay its own way. Table 6-3 provides the analysis of the five remaining alternatives, and the IRRs on incremental cash flows are again computed by setting AW$_\Delta$ $(i') = 0$ for cash-flow differences between alternatives.

From Table 6-3, it is apparent that Alternative E will be chosen (not D) because it requires the largest investment for which the last increment of capital investment is justified. That is, we desire to invest additional increments of the $7,000 presumably available for this project as long as each avoidable increment of investment can earn 10% per year or better.

It was assumed in Example 6-4 (and in all other examples involving mutually exclusive alternatives, unless noted to the contrary) that available capital for a project *not* committed to one of the feasible alternatives is invested in some other project where it will earn an annual return equal to the MARR. Therefore, in this case, the $2,000 left over by selecting Alternative E instead of F is assumed to earn 10% per year elsewhere, which is more than we could obtain by investing it in F.

In sum, three errors commonly made in this type of analysis are to choose the mutually exclusive alternative (1) with the highest overall IRR on total cash flow, (2) with the highest IRR on an incremental capital investment, or (3) with the largest capital investment that has an IRR greater than or equal to the MARR. None of these criteria are generally correct. For instance, in Example 6-4, we might erroneously choose Alternative D rather than E because the IRR for the increment from B to D is 22.6% and that from D to E is only 15.1% (error 2). A more obvious error, as previously discussed, is the temptation to maximize the IRR on total cash flow and select Alternative D (error 1). The third error would be committed by selecting Alternative F for the reason that it has the largest total investment with an IRR greater than the MARR (15.6% > 10%).

EXAMPLE 6-5	**Incremental Analysis: Cost-Only Alternatives**

The estimated capital investment and the annual expenses (based on 1,500 hours of operation per year) for four alternative designs of a diesel-powered air compressor are shown, as well as the estimated market value for each design at the end of the common five-year useful life. The perspective (Principle 3, Chapter 1) of these cost estimates is that of the typical user (construction company, plant facilities department, government highway department, and so on). The study period is five years, and the MARR is 20% per year. One of the designs must be selected for the compressor, and each design provides the same level of service. On the basis of this information,

(a) determine the preferred design alternative, using the IRR method
(b) show that the PW method (i = MARR), using the incremental analysis procedure, results in the same decision.

Solve by hand and by spreadsheet.

	Design Alternative			
	D1	D2	D3	D4
Capital investment	$100,000	$140,600	$148,200	$122,000
Annual expenses	29,000	16,900	14,800	22,100
Useful life (years)	5	5	5	5
Market value	10,000	14,000	25,600	14,000

Observe that this example is a *cost-type situation with four mutually exclusive cost alternatives*. The following solution demonstrates the use of the incremental analysis procedure to compare cost alternatives and applies Rule 2 in Section 6.2.2.

Solution by Hand

The first step is to arrange (rank-order) the four mutually exclusive cost alternatives on the basis of their increasing capital investment costs. Therefore, the order of the alternatives for incremental analysis is D1, D4, D2, and D3.

Since these are cost alternatives, the one with the least capital investment, D1, is the base alternative. Therefore, the base alternative will be preferred unless additional increments of capital investment can produce cost savings (benefits) that lead to a return equal to or greater than the MARR.

The first incremental cash flow to be analyzed is that between designs D1 and D4, Δ(D4 − D1). The results of this analysis, and of subsequent differences between the cost alternatives, are summarized in Table 6-4, and the incremental investment analysis for the IRR method is illustrated in Figure 6-7. These results show the following:

TABLE 6-4	Comparison of Four Cost (Design) Alternatives Using the IRR and PW Methods with Incremental Analysis (Example 6-5)		
Increment Considered	Δ(D4 − D1)	Δ(D2 − D4)	Δ(D3 − D4)
Δ Capital investment	$22,000	$18,600	$26,200
Δ Annual expense (savings)	6,900	5,200	7,300
Δ Market value	4,000	0	11,600
Useful life (years)	5	5	5
IRR$_\Delta$	20.5%	12.3%	20.4%
Is increment justified?	Yes	No	Yes
PW$_\Delta$(20%)	$243	−$3,049	$293
Is increment justified?	Yes	No	Yes

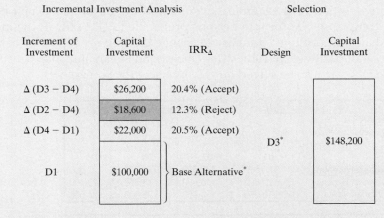

* Since these are cost alternatives, the IRR of D3 cannot be determined.

Figure 6-7 Representation of Capital Investment Increments and IRR on Increments Considered in Selecting Design 3 (D3) in Example 6-5

1. The incremental cash flows between the cost alternatives are, in fact, investment alternatives.

2. The first increment, Δ(D4 − D1), is justified (IRR$_\Delta$ = 20.5% is greater than MARR = 20%, and PW$_\Delta$(20%) = $243 > 0); the increment Δ(D2 − D4) is not justified; and the last increment, Δ(D3 − D4)—not Δ(D3 − D2), because Design D2 has already been shown to be unacceptable—is justified, resulting in the selection of Design D3 for the air compressor. It is the highest investment for which each increment of investment capital is justified from the user's perspective.

3. The same capital investment decision results from the IRR method and the PW method, using the incremental analysis procedure, because *when the equivalent worth of an investment at i = MARR is greater than zero, its IRR is greater than MARR* (from the definition of the IRR; Chapter 5).

Spreadsheet Solution

Figure 6-8 shows the complete spreadsheet solution for this example. The first set of incremental EOY cash flows has two EOY five entries: one for the difference in annual expense savings and one for the difference in market value. These two values are combined in the second set of incremental EOY cash flows for direct computation of the incremental IRR and PW amounts. Note that the IRR function can handle EOY zero through EOY five cash flows as given, while the PW computation needs to add the EOY zero cash flow outside of the NPV function.

As was previously discovered via manual computation, the first increment $\Delta(D4 - D1)$ is justified; the increment $\Delta(D2 - D4)$ is not justified; and the last increment $\Delta(D3 - D4)$ is justified. Thus, D3 is selected as the preferred compressor.

As previously mentioned, spreadsheets make it easy to answer *what if* types of questions. For example, how much would the annual expense have to be for compressor D4 for it to become preferable to D3? (HINT: the incremental PW has to be positive.) It is easy to change the value in cell E6 by hand to bracket a small positive PW value. This corresponds to an annual expense of $22,002 for D4. It can also be solved quickly by using the Solver tool.

	A	B	C	D	E
1	MARR =	20%			
2	Useful Life =	5			
3					
4		**D1**	**D2**	**D3**	**D4**
5	Capital Investment	$ 100,000	$ 140,600	$ 148,200	$ 122,000
6	Annual Expense	$ 29,000	$ 16,900	$ 14,800	$ 22,100
7	Market Value	$ 10,000	$ 14,000	$ 25,600	$ 14,000
8					
9	**EOY**	**Δ(D4-D1)**	**Δ(D2-D4)**	**Δ(D3-D4)**	
10	0	$ (22,000)	$ (18,600)	$ (26,200)	
11	1	$ 6,900	$ 5,200	$ 7,300	
12	2	$ 6,900	$ 5,200	$ 7,300	
13	3	$ 6,900	$ 5,200	$ 7,300	
14	4	$ 6,900	$ 5,200	$ 7,300	
15	5	$ 6,900	$ 5,200	$ 7,300	
16	5	$ 4,000	$ -	$ 11,600	
17	-----	-----	-----	-----	
18	**EOY**	**Δ(D4-D1)**	**Δ(D2-D4)**	**Δ(D3-D4)**	
19	0	$ (22,000)	$ (18,600)	$ (26,200)	
20	1	$ 6,900	$ 5,200	$ 7,300	
21	2	$ 6,900	$ 5,200	$ 7,300	
22	3	$ 6,900	$ 5,200	$ 7,300	
23	4	$ 6,900	$ 5,200	$ 7,300	
24	5	$ 10,900	$ 5,200	$ 18,900	
25					
26	ΔIRR	20.5%	12.3%	20.4%	
27	ΔPW(20%)	$ 243	$ (3,049)	$ 293	

Cell annotations (callouts):
- $= E5 - C5$
- $= B5 - E5$
- $= \$B\$6 - \$E\$6$
- $= E7 - B7$
- $= B10$
- $= B\$11$
- $= B15 + B16$
- $= \$E\$6 - \$C\$6$
- $= E5 - D5$
- $= \$E\$6 - \$D\$6$
- $= D7 - E7$
- $= C7 - E7$
- $= IRR(B19:B24, \$B\$1)$
- $= NPV(\$B\$1, B20:B24) + B19$

Figure 6-8   Spreadsheet Solution, Example 6-5

Now we turn our attention to the ERR method which was explained in Chapter 5. Also, in Appendix 5-A, the ERR method was illustrated as a substitute for the IRR method when analyzing a nonconventional investment type of cash flow. In Example 6-6, the ERR method is applied using the incremental investment analysis procedure to compare the mutually exclusive alternatives for an engineering improvement project.

EXAMPLE 6-6	Incremental Analysis Using ERR

In an automotive parts plant, an engineering team is analyzing an improvement project to increase the productivity of a flexible manufacturing center. The estimated net cash flows for the three feasible alternatives being compared are shown in Table 6-5. The analysis period is six years, and MARR for capital investments at the plant is 20% per year. Using the ERR method, which alternative should be selected? (ϵ = MARR.)

Solution

The procedure for using the ERR method to compare mutually exclusive alternatives is the same as for the IRR method. The only difference is in the calculation methodology.

Table 6-5 provides a tabulation of the calculation and acceptability of each increment of capital investment considered. Since these three feasible alternatives are a mutually exclusive set of investment alternatives, the base alternative is the one with the least capital investment cost that is economically justified. For Alternative A, the PW of the negative cash-flow amounts (at $i = \epsilon$ %)

TABLE 6-5   Comparison of Three Mutually Exclusive Alternatives Using the ERR Method (Example 6-6)

End of Period	Alternative Cash Flows			Incremental Analysis of Alternatives		
	A	B	C	A^a	$\Delta(B - A)$	$\Delta(C - A)$
0	−$640,000	−$680,000	−$755,000	−$640,000	−$40,000	−$115,000
1	262,000	−40,000	205,000	262,000	−302,000	−57,000
2	290,000	392,000	406,000	290,000	102,000	116,000
3	302,000	380,000	400,000	302,000	78,000	98,000
4	310,000	380,000	390,000	310,000	70,000	80,000
5	310,000	380,000	390,000	310,000	70,000	80,000
6	260,000	380,000	324,000	260,000	120,000	64,000
	Incremental analysis:					
	Δ PW of negative cash-flow amounts			640,000	291,657	162,498
	Δ FW of positive cash-flow amounts			2,853,535	651,091	685,082
			ERR	28.3%	14.3%	27.1%
	Is increment justified?			Yes	No	Yes

^{a} The net cash flow for Alternative A, which is the incremental cash flow between making no change ($0) and implementing Alternative A.

is just the $640,000 investment cost. Therefore, the ERR for Alternative A is the following:

$$\$640{,}000(F/P, i'\%, 6) = \$262{,}000(F/P, 20\%, 5) + \cdots + \$260{,}000$$
$$= \$2{,}853{,}535$$
$$(F/P, i'\%, 6) = (1 + i')^6 = \$2{,}853{,}535/\$640{,}000 = 4.4586$$
$$(1 + i') = (4.4586)^{1/6} = 1.2829$$
$$i' = 0.2829, \text{ or ERR} = 28.3\%.$$

Using a MARR $= 20\%$ per year, this capital investment is justified, and Alternative A is an acceptable base alternative. By using similar calculations, the increment $\Delta(B - A)$, earning 14.3%, is not justified and the increment $\Delta(C - A)$, earning 27.1%, is justified. Therefore, Alternative C is the preferred alternative for the improvement project. Note in this example that revenues varied among the alternatives and that Rule 1, Section 6.2.2, was applied.

By this point in the chapter, three key observations are clear concerning the comparison of mutually exclusive alternatives: (1) equivalent-worth methods are computationally less cumbersome to use, (2) both the equivalent-worth and rate-of-return methods, if used properly, will consistently recommend the best alternative, but (3) rate-of-return methods may not produce correct choices if the analyst or the manager insists on maximizing the rate of return on the total cash flow. That is, incremental investment analysis must be used with rate-of-return methods to ensure that the best alternative is selected.

The next example illustrates the comparison of two mutually exclusive alternatives to increase the production capacity of a critical system in a plant by improving its operational availability.

EXAMPLE 6-7   Increasing Production Capacity by Improving System Availability

The reliability engineer for an electronic manufacturing plant is trying to decrease the downtime of critical production systems. It is desired to improve the operational availability of these systems so that the potential capacity of the plant will be increased. One critical system under review is used to manufacture an electronic control unit used in major home appliances. A reliability improvement team has developed two mutually exclusive alternatives for improving the operational availability of this system. The alternatives have differences in real-time monitoring technologies (predictive maintenance), preplanned preventive maintenance actions, computer information system support, and personnel training. Also, there are differences in annual maintenance expenses and the amount of increase in system availability. Relative to the current operation of the system, the following estimates have been developed:

| | Alternative | |
Factor	A1	A2
Capital investment	$260,000	$505,000
Annual maintenance expenses:		
Increase	$9,400	0
Decrease (savings)	0	$6,200
Increased system availability	4%	6.5%

Assume that the MARR $= 18\%$ per year, the analysis period is five years, any additional units produced can be immediately sold, the current average availability of the system (80.3%) results in 7,400 units being produced and sold per month, each 1% increase in the average availability of the system results in a 0.7% increase in plant capacity for production of the product, and each additional unit sold increases revenues by $48.20.

(a) Select the preferred alternative by using the IRR method.

(b) Which rule (Section 6.2.2, page 243) was used in the selection? Why?

Solution

(a) Rate-of-return methods require the use of the incremental investment analysis procedure. The rank-order of the alternatives for incremental analysis, based on capital investment, is do nothing, A1, A2. Since these are investment alternatives, the next step is to check whether A1 is an acceptable base alternative:

$$PW(18\%)_{A1} = -\$260,000 - \$9,400(P/A, 18\%, 5)$$

$$+ 4(0.007)(7,400)(12)(\$48.20)(P/A, 18\%, 5)$$

$$= \$85,382.$$

Because the PW(MARR $= 18\%)_{A1} > 0$, we know that the $IRR_{A1} >$ MARR, and the alternative (A1) is an acceptable base alternative. Next, we need to find the IRR of the incremental cash flow between Alternatives A1 and A2:

$$0 = [-\$505,000 - (-\$260,000)] + [\$6,200 - (-\$9,400)](P/A, i'\%, 5)$$

$$+ (6.5 - 4.0)(0.007)(7,400)(12)(\$48.20)(P/A, i'\%, 5).$$

By linear interpolation (Section 5.6), we find that $i'\% = 24.7\%$ per year, which is greater than MARR of 18% per year. Therefore, the additional capital investment in A2 over A1 is economically justified, and Alternative A2 would be selected.

(b) Rule 1, Section 6.2.2, applies in this example because the economic benefits vary among the alternatives, and we want to maximize overall profitability.

6.5   Useful Lives Are Unequal among the Alternatives

When the useful lives of mutually exclusive alternatives are unequal, the *repeatability assumption* may be used in their comparison if the study period can be infinite in length or a common multiple of the useful lives. This assumes that the economic estimates for an alternative's initial useful life cycle will be repeated in all subsequent replacement cycles. As we discussed in Section 6.3, this condition is more robust for practical application than it may appear. Another viewpoint is to consider the repeatability assumption as a modeling convenience for the purpose of making a current decision. *When this assumption is applicable to a decision situation, it simplifies comparison of the mutually exclusive alternatives.*

If the repeatability assumption is not applicable to a decision situation, then an appropriate study period needs to be selected (*coterminated assumption*). This is the approach most frequently used in engineering practice because product life cycles are becoming shorter. Often, one or more of the useful lives will be shorter or longer than the selected study period. When this is the case, cash-flow adjustments based on additional assumptions need to be used *so that all the alternatives are compared over the same study period*. The following guidelines apply to this situation:

1. Useful life < Study period
   (a) Cost alternatives: Because each cost alternative has to provide the same level of service over the study period, contracting for the service or leasing the needed equipment for the remaining years may be appropriate. Another potential course of action is to repeat part of the useful life of the original alternative and then use an estimated market value to truncate it at the end of the study period.
   (b) Investment alternatives: The first assumption is that all cash flows will be reinvested in other opportunities available to the firm at the MARR to the end of the study period. A second assumption involves replacing the initial investment with another asset having possibly different cash flows over the remaining life. A convenient solution method is to calculate the FW of each mutually exclusive alternative at the end of the study period. The PW can also be used for investment alternatives, since the FW at the end of the study period, say N, of each alternative is its PW times a common constant $(F/P, i\%, N)$, where $i\% =$ MARR.

2. Useful life > Study period: The most common technique is to truncate the alternative at the end of the study period, using an estimated market value. This assumes that the disposable assets will be sold at the end of the study period at that value.

The underlying principle, as discussed in Section 6.3, is to compare the mutually exclusive alternatives being considered in a decision situation over the same study (analysis) period.

In this section, we explain how to evaluate mutually exclusive alternatives having unequal useful lives. First we consider equivalent-worth methods for making comparisons of alternatives. Then we turn our attention to the use of the rate-of-return method for performing the analysis.

6.5.1   Equivalent-Worth Methods

When the useful lives of alternatives are not the same, the *repeatability assumption* is appropriate if the study period is infinite (very long in length) or a common multiple of the useful lives. Under this assumption, the cash flows for an alternative's initial life cycle will be repeated (i.e., they are identical) in all subsequent replacement cycles. Because this assumption is applicable in many decision situations, it is extremely useful and greatly simplifies the comparison of mutually exclusive alternatives. With repeatability, we will simply compute the AW of each alternative over its own *useful life* and recommend the one having the most economical value (i.e., the alternative with the highest positive AW for investment alternatives and the alternative with the least negative AW for cost alternatives).

Example 6-8 demonstrates the computational advantage of using the AW method (instead of PW or FW) when the repeatability assumption is applicable. Example 6-9 illustrates the use of the coterminated assumption for the same set of alternatives when the selected study period is not a common multiple of the useful lives.

EXAMPLE 6-8	Useful Lives ≠ Study Period: The Repeatability Assumption

The following data have been estimated for two mutually exclusive investment alternatives, *A* and *B*, associated with a small engineering project for which revenues as well as expenses are involved. They have useful lives of four and six years, respectively. If MARR = 10% per year, show which alternative is more desirable by using equivalent-worth methods (computed by hand and by spreadsheet). Use the repeatability assumption.

	A	B
Capital investment	$3,500	$5,000
Annual cash flow	1,255	1,480
Useful life (years)	4	6
Market value at end of useful life	0	0

Solution

The least common multiple of the useful lives of Alternatives *A* and *B* is 12 years. Using the repeatability assumption and a 12-year study period, the first like (identical) replacement of Alternative *A* would occur at EOY four, and the second would be at EOY eight. For Alternative *B*, one like replacement would occur at EOY six. This is illustrated in Part 1 of Figure 6-9.

Solution by the PW Method

The PW (or FW) solution must be based on the total study period (12 years). The PW of the initial useful life cycle will be different than the PW of

Figure 6-9 Illustration of Repeatability Assumption (Example 6-8) and Coterminated Assumption (Example 6-9)

Part 1: Repeatability Assumption, Example 6-8, Least Common Multiple of Useful Lives Is 12 years.

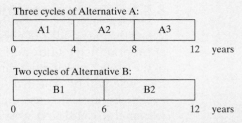

Three cycles of Alternative A:

A1	A2	A3
0	4	8	12 years

Two cycles of Alternative B:

B1	B2
0	6	12 years

Part 2: Coterminated Assumption, Example 6-9, Six-Year Analysis Period.

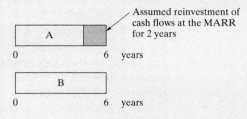

Assumed reinvestment of cash flows at the MARR for 2 years

A

0   6   years

B

0   6   years

subsequent replacement cycles:

$$PW(10\%)_A = -\$3{,}500 - \$3{,}500[(P/F, 10\%, 4) + (P/F, 10\%, 8)]$$
$$+ (\$1{,}255)(P/A, 10\%, 12)$$
$$= \$1{,}028,$$

$$PW(10\%)_B = -\$5{,}000 - \$5{,}000(P/F, 10\%, 6)$$
$$+ (\$1{,}480)(P/A, 10\%, 12)$$
$$= \$2{,}262.$$

Based on the PW method, we would select Alternative *B* because it has the larger value ($2,262).

Solution by the AW Method

The like replacement of assets assumes that the economic estimates for the initial useful life cycle will be repeated in each subsequent replacement cycle. Consequently, the AW will have the same value for each cycle and for the study period (12 years). This is demonstrated in the next AW solution by calculating (1) the AW of each alternative over the 12-year analysis period based on the previous PW values and (2) determining the AW of each alternative over one useful life cycle. Based on the previously calculated PW values, the AW values are

$$AW(10\%)_A = \$1{,}028(A/P, 10\%, 12) = \$151,$$
$$AW(10\%)_B = \$2{,}262(A/P, 10\%, 12) = \$332.$$

Next, the AW of each alternative is calculated over one useful life cycle:

$$AW(10\%)_A = -\$3,500(A/P, 10\%, 4) + (\$1,255) = \$151,$$

$$AW(10\%)_B = -\$5,000(A/P, 10\%, 6) + (\$1,480) = \$332.$$

This confirms that both calculations for each alternative result in the same AW value, and we would again select Alternative B because it has the larger value ($332).

Spreadsheet Solution

Figure 6-10 shows the spreadsheet solution for this example. EOY cash flows are computed for each alternative over the entire 12-year study period. For Alternative A, the annual cash flow of $1,255 is combined with the necessary reinvestment cost ($3,500) at the end of each useful life cycle (at EOY 4 and EOY 8). A similar statement can be made for Alternative B at the end of its first life cycle (EOY 6 = $1,480 − $5,000). As was the case in the previous solution by hand, Alternative B is selected because it has the largest PW (and therefore AW) value. (How much would the annual cash flow for Alternative A have to be for it to be as desirable as Alternative B? Answer: $1,436.)

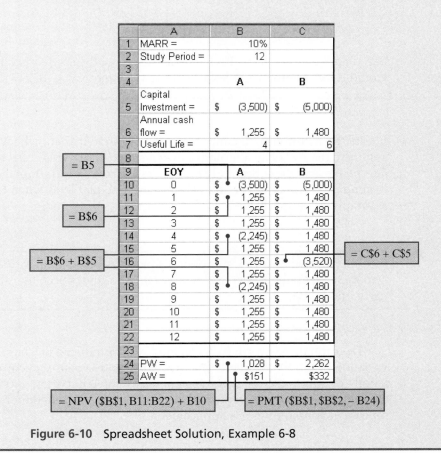

Figure 6-10 Spreadsheet Solution, Example 6-8

EXAMPLE 6-9	**Useful Lives ≠ Study Period: The Coterminated Assumption**

Suppose that Example 6-8 is modified such that an analysis period of six years is used (coterminated assumption) instead of 12 years, which was based on repeatability and the least common multiple of the useful lives. Perhaps the responsible manager did not agree with the repeatability assumption and wanted a six-year analysis period because it is the planning horizon used in the company for small investment projects.

Solution

An assumption used for an investment alternative (when useful life is less than the study period) is that all cash flows will be reinvested by the firm at the MARR until the end of the study period. This assumption applies to Alternative A, which has a four-year useful life (two years less than the study period), and it is illustrated in Part 2 of Figure 6-9. We use the FW method to analyze this situation:

$$\text{FW}(10\%)_A = [-\$3{,}500(F/P, 10\%, 4) + (\$1{,}255)(F/A, 10\%, 4)](F/P, 10\%, 2)$$

$$= \$847,$$

$$\text{FW}(10\%)_B = -\$5{,}000(F/P, 10\%, 6) + (\$1{,}480)(F/A, 10\%, 6)$$

$$= \$2{,}561.$$

Based on the FW of each alternative at the end of the six-year study period, we would select Alternative B because it has the larger value (\$2,561).

In the solution to Example 6-8, it was shown that, when repeatability is assumed, the AW of an alternative over a single life cycle is equal to the AW of the alternative over the entire study period. As a result, we can adopt the following rule to simplify the analysis of alternatives with unequal lives when the repeatability assumption is applicable:

When the repeatability assumption is applied, simply compare the AW amounts of each alternative over its own useful life and select the alternative that maximizes AW.

The capitalized-worth (CW) method was introduced in Chapter 5 as a special variation of the PW method when revenues and expenses occur over an infinite length of time. CW is a convenient basis for comparing mutually exclusive alternatives when the period of needed service is indefinitely long and the repeatability assumption is applicable.

| EXAMPLE 6-10 | Comparing Alternatives Using CW |

We now revisit the problem posed at the beginning of the chapter involving two border-surveillance systems. Because an indefinitely long study period is specified, we use the CW method to compare the two surveillance systems. First we compute the AW of each system over its useful life, and then we determine the capitalized worth (refer to Section 5.3.3) over a very long study period.

$$RX - 1: \quad AW(15\%) = -\$24,000(A/P, 15\%, 5) + \$1,000(A/F, 15\%, 5) - \$2,500$$

$$CW(15\%) = AW(15\%)/0.15 = -\$63,406$$

$$ABY: \quad AW(15\%) = -\$45,000(A/P, 15\%, 10) + \$4,000(A/F, 15\%, 10) - \$1,500$$

$$CW(15\%) = AW(15\%)/0.15 = -\$68,475$$

We recommend the RX-1 system because it has the lesser negative (higher) CW.

Example 6-11 demonstrates how to deal with situations in which multiple machines are required to satisfy a fixed annual demand for a product or service. Such problems can be solved by using Rule 2 and the repeatability assumption.

| EXAMPLE 6-11 | AW and Repeatability: Perfect Together! |

Three products will be manufactured in a new facility at the Apex Manufacturing Company. They each require an identical manufacturing operation, but different production times, on a broaching machine. Two alternative types of broaching machines (M1 and M2) are being considered for purchase. One machine type must be selected.

For the same level of annual demand for the three products, *annual production requirements* (machine hours) and annual operating expenses (per machine) are listed next. Which machine should be selected if the MARR is 20% per year? Solve by hand and by spreadsheet. Show all work to support your recommendation. (Use Rule 2 on page 243 to make your recommendation.)

Product	Machine M1	Machine M2
ABC	1,500 hr	900 hr
MNQ	1,750 hr	1,000 hr
STV	2,600 hr	2,300 hr
	5,850 hr	4,200 hr
Capital investment	$15,000 per machine	$22,000 per machine
Expected life	five years	eight years
Annual expenses	$4,000 per machine	$6,000 per machine

Assumptions: The facility will operate 2,000 hours per year. Machine availability is 90% for Machine M1 and 80% for Machine M2. The yield of Machine M1 is 95%, and the yield of Machine M2 is 90%. Annual operating expenses are based on an assumed operation of 2,000 hours per year, and workers are paid during any idle time of Machine M1 or Machine M2. Market values of both machines are negligible.

Solution by Hand

The company will need 5,850 hours/[2,000 hours (0.90)(0.95)] = 3.42 (four machines of type M1) or 4,200 hours/[2,000 hours (0.80)(0.90)] = 2.92 (three machines of type M2). The maximum operation time of 2,000 hours per year

	A	B	C	D	E	F	G
1	MARR =	20%			Facility Operation Hours/ year =		2,000
2							
3		**M1**	**M2**		**Product**	**M1**	**M2**
4	Capital Investment (ea.)	$ 15,000	$ 22,000		ABC (hrs.)	1,500	900
5	Useful Life	5	8		MNQ (hrs.)	1,750	1,000
6	Annual Expenses (ea.)	$ 4,000	$ 6,000		STV (hrs.)	2,600	2,300
7	Availability	90%	80%				
8	Yield	95%	90%		**Total Hrs.**	5,850	4,200
9	= $G$1 * B7				= –B4 * B15	= SUM(F4:F6)	
10							
11		**M1**	**M2**		**EOY**	**M1**	**M2**
12	Annual Available Hours	1,800	1,600		0	$ (60,000)	$ (66,000)
13	Annual Net Yield Hours	1,710	1,440		1	$ (16,000)	$ (18,000)
14	Theoretical Requirement	3.42	2.92		2	$ (16,000)	$ (18,000)
15	Actual Requirement	4	3		3	$ (16,000)	$ (18,000)
16					4	$ (16,000)	$ (18,000)
17	= B12 * B8				5	$ (16,000)	$ (18,000)
18					6		$ (18,000)
19	= F8/B13				7		$ (18,000)
20					8		$ (18,000)
21							
22					AW =	($36,063)	($35,200)

= CEILING(B14, 1)

= –B$6 * B$15

= –PMT ($B$1, B5, NPV ($B$1, F13:F17) + F12)

= –PMT ($B$1, C5, NPV ($B$1, G13:G20) + G12)

Figure 6-11   Spreadsheet Solution, Example 6-11

in the denominator must be multiplied by the availability of each machine and the yield of each machine, as indicated.

The annual cost of ownership, assuming a MARR = 20% per year, is $15,000(4)(A/P, 20%, 5) = $20,064 for Machine M1 and $22,000(3)(A/P, 20%, 8) = $17,200 for Machine M2.

There is an excess capacity when four Machine M1s and three Machine M2s are used to provide the machine-hours (5,850 and 4,200, respectively) just given. If we assume that the operator is paid for idle time he or she may experience on M1 or M2, the annual expense for the operation of four M1s is 4 machines × $4,000 per machine = $16,000. For three M2s, the annual expense is 3 machines × $6,000 per machine = $18,000.

The total equivalent annual cost for four Machine M1s is $20,064 + $16,000 = $36,064. Similarly, the total equivalent annual expense for three Machine M2s is $17,200 + $18,000 = $35,200. By a slim margin, Machine M2 is the preferred choice to minimize equivalent annual costs with the repeatability assumption.

Spreadsheet Solution

Figure 6-11 on page 270 shows the complete spreadsheet solution for this example. Note the use of the CEILING function in cell B15 to convert the noninteger theoretical number of machines required into an actual requirement. This actual requirement is used to compute the total EOY cash flows for each alternative by multiplying the per-machine costs by the number of machines required.

Since we are assuming repeatability, the EOY cash flows are shown only for the initial useful life span for each alternative. These cash flows are then used to compute the AW of each alternative. As seen previously, purchasing three broaching machines of type M2 is preferred to purchasing four machines of type M1.

EXAMPLE 6-12   **Modeling Estimated Expenses as Arithmetic Gradients**

You are a member of an engineering project team that is designing a new processing facility. Your present design task involves the portion of the catalytic system that requires pumping a hydrocarbon slurry that is corrosive and contains abrasive particles. For final analysis and comparison, you have selected two fully lined slurry pump units, of equal output capacity, from different manufacturers. Each unit has a large-diameter impeller required and an integrated electric motor with solid-state controls. Both units will provide the same level of service (support) to the catalytic system but have different useful lives and costs.

	Pump Model	
	SP240	HEPS9
Capital investment	$33,200	$47,600
Annual expenses:		
Electrical energy	$2,165	$1,720
Maintenance	$1,100 in year 1, and increasing $500/yr thereafter	$500 in year 4, and increasing $100/yr thereafter
Useful life (years)	5	9
Market value (end of useful life)	0	5,000

The new processing facility is needed by your firm at least as far into the future as the strategic plan forecasts operating requirements. The MARR is 20% per year. Based on this information, which slurry pump should you select?

Solution

Notice that the estimates for maintenance expenses involve an arithmetic gradient series (Chapter 4). A cash-flow diagram is very useful in this situation to help keep track of the various cash-flow series. The cash-flow diagrams for pump models SP240 and HEPS9 are shown in Figure 6-12.

The repeatability assumption is a logical choice for this analysis, and a study period of either infinite or 45 years (least common multiple of the useful lives) in length can be used. With repeatability, the AW over the initial *useful life* of each alternative is the same as its AW over the length of either study period:

$$\text{AW}(20\%)_{\text{SP240}} = -\$33,200(A/P, 20\%, 5) - \$2,165$$

$$-[\$1,100 + \$500(A/G, 20\%, 5)]$$

$$= -\$15,187,$$

$$\text{AW}(20\%)_{\text{HEPS9}} = -\$47,600(A/P, 20\%, 9) + \$5,000(A/F, 20\%, 9)$$

$$-\$1,720 - [\$500(P/A, 20\%, 6)$$

$$+ \$100(P/G, 20\%, 6)] \times (P/F, 20\%, 3) \times (A/P, 20\%, 9)$$

$$= -\$13,622.$$

Based on Rule 2 (Section 6.2.2), you should select pump model HEPS9, since the AW over its useful life (nine years) has the smaller negative value (−$13,622).

As additional information, the following two points support in choosing the repeatability assumption in Example 6-12:

1. The repeatability assumption is commensurate with the long planning horizon for the new processing facility and with the design and operating requirements of the catalytic system.

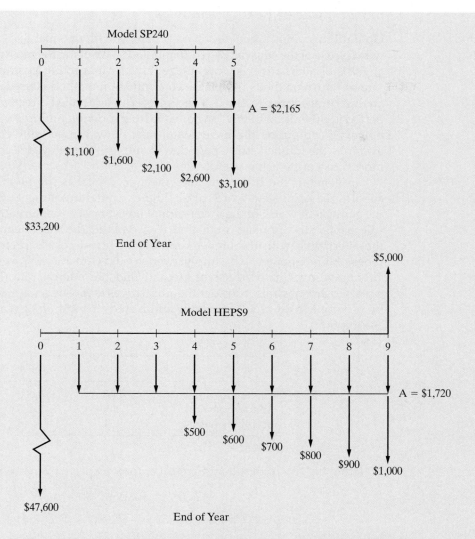

Figure 6-12   Cash-Flow Diagrams for the Pump Models Being Compared in Example 6-12

2. If the initial estimated costs change for future pump-replacement cycles, a logical assumption is that the ratio of the AW values for the two alternatives will remain approximately the same. Competition between the two manufacturers should cause this to happen. Hence, the pump selected (model HEPS9) should continue to be the preferred alternative.

If the existing model is redesigned or new models of slurry pumps become available, however, another study to analyze and compare all feasible alternatives is required before a replacement of the selected pump occurs.

6.5.2   Rate-of-Return Analysis

Up until this point, we have solved problems with unequal lives (see Section 6.5) with the use of the equivalent-worth methods (AW being the most convenient). The analysis of alternatives having unequal lives can also be accomplished by using rate-of-return methods. When the cotermination method is used, the incremental analysis procedure described in Section 6.4 can be applied directly. When the study period is either indefinitely long or equal to a common multiple of the useful lives, however, computing the incremental cash flows can be quite cumbersome. For example, the implicit study period in Example 6-12 is 45 years! In this instance, a more direct approach is useful.

In general, the IRR of an increment of capital is the interest rate, i^*, that equates the equivalent worth of the higher capital investment cost alternative to the equivalent worth of the lower capital investment cost alternative. The decision rule using this approach is that, if $i^* \geq$ MARR, the increment is justified and the alternative with the higher capital investment cost is preferred. So, when repeatability applies, we simply need to develop the AW equation for each alternative over its own useful life and find the interest rate that makes them equal. To demonstrate, consider the alternatives that were analyzed in Example 6-9 by using a MARR of 10% per year and a study period of 12 years (repeatability assumption).

	A	B
Capital investment	$3,500	$5,000
Annual cash flow	1,255	1,480
Useful life (years)	4	6

Equating the AW of the alternatives over their respective lives, we get

$$AW_A(i^*\%) = AW_B(i^*\%)$$

$$-\$3,500(A/P, i^*\%, 4) + \$1,255 = -\$5,000(A/P, i^*\%, 6) + \$1,480.$$

By trial and error, the IRR of the extra capital needed to repeatedly invest in Alternative B (instead of Alternative A) over the study period is $i^* = 26\%$. Since this value is greater than the MARR, the increment is justified and Alternative B is preferred. This is the same decision arrived at in Example 6-8. The interested student is encouraged to verify that, if the spreadsheet previously displayed in Figure 6-10 were to be expanded to include a column of incremental cash flows over the entire 12-year study period, the IRR of these cash flows, using the IRR function, is indeed 26%.

6.5.3   The Imputed Market Value Technique

Obtaining a current estimate from the marketplace for a piece of equipment or another type of asset is the preferred procedure in engineering practice when a market value at time $T <$ (useful life) is required. This approach, however, may not be feasible in some cases. For example, a type of asset may have low turnover in the marketplace, and information for recent transactions is not available. Hence, it is sometimes necessary to estimate the market value for an asset without current and representative historical data.

The *imputed market value* technique, which is sometimes called the *implied market value*, can be used for this purpose as well as for comparison with marketplace values when current data are available. The estimating procedure used in the technique is based on logical assumptions about the value of the remaining useful life for an asset. If an imputed market value is needed for a piece of equipment, say, at the end of year $T <$ (useful life), the estimate is calculated on the basis of the sum of two parts, as follows:

$$MV_T = [\text{PW at EOY } T \text{ of remaining capital recovery (CR) amounts}]$$

$$+[\text{PW at EOY } T \text{ of original market value at end of useful life}],$$

where PW is computed at $i = $ MARR.

The next example uses information from Example 6-12 to illustrate the technique.

EXAMPLE 6-13   **Estimating a New Market Value when Useful Life > Study Period**

Use the imputed market value technique to develop an estimated market value for pump model HEPS9 (Example 6-12) at EOY five. The MARR remains 20% per year.

Solution

The original information from Example 6-12 will be used in the solution: capital investment $= \$47,600$, useful life $=$ nine years, and market value $= \$5,000$ at the end of useful life.

First, compute the PW at EOY five of the remaining CR amounts [Equation (5-5)]:

$$PW(20\%)_{CR} = [\$47,600(A/P, 20\%, 9) - \$5,000(A/F, 20\%, 9)] \times (P/A, 20\%, 4)$$

$$= \$29,949.$$

Next, compute the PW at EOY five of the original MV at the end of useful life (nine years):

$$PW(20\%)_{MV} = \$5,000(P/F, 20\%, 4) = \$2,412.$$

Then, the estimated market value at EOY five $(T = 5)$ is as follows:

$$MV_5 = PW_{CR} + PW_{MV}$$

$$= \$29,949 + \$2,412 = \$32,361.$$

In summary, utilizing the repeatability assumption for unequal lives among alternatives reduces to the simple rule of "comparing alternatives over their useful lives using the AW method, at $i = $ MARR." This simplification, however, may not apply when a study period, selected to be shorter or longer than the common multiple of lives (coterminated assumption), is more appropriate for the decision situation. When utilizing the coterminated assumption, cash flows of alternatives need to be adjusted to terminate at the end of the study period. Adjusting these cash flows usually requires estimating the market value of assets at the end of the study period or extending service to the end of the study period through leasing or some other assumption.

6.6   Personal Finances

Sound financial planning is all about making wise choices for your particular circumstances (e.g., your amount of personal savings, your job security, your attitude toward risk). Thus far in Chapter 6, we have focused on facilitating good decision making from the perspective of a corporation. Now we apply these same principles (remember them from Chapter 1?) to several problems you are likely to face soon in your personal decision making.

Two of the largest investments you'll ever make involve houses and automobiles. This section presents examples of acquiring these assets, usually with borrowed money. Another concern is the extensive use of credit cards (maybe we think that nothing is expensive on a credit card). It turns out that people who use credit cards (almost all of us!) tend to spend more money than others who pay cash or write checks. An enlightening exercise to see how addicted you are to credit cards is to go cold turkey for two months.

A fundamental lesson underlying this section is to save now rather than spending on luxury purchases. By choosing to save now, we are making an attempt to minimize the risk of making poor decisions later on.

EXAMPLE 6-14   **Automobile Financing Options**

You have decided to purchase a new automobile with a hybrid-fueled engine and a six-speed transmission. After the trade-in of your present car, the purchase price of the new automobile is $30,000. This balance can be financed by an auto dealer at 2.9% APR with payments over 48 months. Alternatively, you can get a $2,000 discount on the purchase price if you finance the loan balance at an APR of 8.9% over 48 months. Should you accept the 2.9% financing plan or accept the dealer's offer of a $2,000 rebate with 8.9% financing? Both APRs are compounded monthly.

Solution

In this example, we assume that your objective is to minimize your monthly car payment.

2.9% financing monthly payment:

$30,000 $(A/P, 2.9\%/12, 48 \text{ months}) = \$30,000(0.0221) = \$663.00$ per month

8.9% financing monthly payment:

$$\$28,000 \; (A/P, \; 8.9\%/12, 48 \text{ months}) = \$28,000(0.0248) = \$694.90 \text{ per month}$$

Therefore, to minimize your monthly payment, you should select the 2.9% financing option.

What If Questions?

When shopping for an automobile you'll find that there are many financing options like the ones in this example available to you. You may find it useful to ask yourself questions such as "how high would the rebate have to be for me to prefer the rebate option," or "how low would the APR have to be for me to select the rebate option?" The answers to these questions can be found through simple equivalence calculations.

(a) How much would the rebate have to be?

Let X = rebate amount. Using the monthly payment of $663 from the 2.9% financing option, we can solve for the rebate amount that would yield the same monthly payment.

$$(\$30,000 - X)(A/P, \; 8.9\%/12 \text{ months}, \; 48 \text{ months}) = \$663$$

$$(\$30,000 - X)(0.0248) = \$663$$

$$X = \$3,266$$

(b) How low would the interest rate have to be?

Now we want to find the interest rate that equates borrowing $28,000 to 48 monthly payments of $663. This question is easily solved using a spreadsheet package.

$$\text{RATE } (48, -663, \; 28000) = 0.535\% \text{ per month}$$

$$\text{APR} = 0.535\% \times 12 = 6.42\%$$

EXAMPLE 6-15	Mortgage Financing Options

A general rule of thumb is that your monthly mortgage payment should not exceed 28% of your household's gross monthly income. Consider the situation of Jerry and Tracy, who just committed to a $300,000 mortgage on their dream home. They have reduced their financing choices to a 30-year conventional mortgage at 6% APR, or a 30-year interest-only mortgage at 6% APR.

(a) Which mortgage, if either, do they qualify for if their combined gross annual income is $70,000?

(b) What is the disadvantage in an interest-only mortgage compared to the conventional mortgage?

Solution

(a) Using the general rule of thumb, Jerry and Tracy can afford a monthly mortgage payment of $(0.28)(\$70,000/12) = \$1,633$. The monthly payment for the conventional mortgage is

$$\$300,000 \, (A/P, 0.5\%, 360) = \$1,800.$$

For the interest-only mortgage, the monthly payment is

$$(0.005)(\$300,000) = \$1,500.$$

Thus, the conventional mortgage payment is larger than what the guideline suggests is affordable. This type of loan is marginal because it stretches their budget too much. They easily qualify for the interest-only mortgage because the $1,500 payment is less than $1,633.

(b) If home prices fall in the next several years, Jerry and Tracy may have "negative equity" in their home because no principal has been repaid in their monthly interest-only payments. They will not have any buffer to fall back on should they have to sell their house for less than they purchased it for.

　　It is important to note that interest-only loans don't remain interest-only for the entire loan period. The length of time that interest-only payments may be made is defined in the mortgage contract and can be as short as 5 years or as long as 15 years. After the interest-only period is over, the monthly payment adjusts to include principal and interest. It is calculated to repay the entire loan by the end of the loan period. Suppose Jerry and Tracy's interest-only period was five years. After this time, the monthly payment would become

$$\$300,000(A/P, 0.5\%, 300) = \$1,932.90.$$

Before Jerry and Tracy accept this type of loan, they should be confident that they will be able to afford the $1,932.90 monthly payment in five years.

EXAMPLE 6-16	**Comparison of Two Savings Plans**

Suppose you start a savings plan in which you save $500 each year for 15 years. You make your first payment at age 22 and then leave the accumulated sum in the savings plan (and make no more annual payments) until you reach age 65, at which time you withdraw the total accumulated amount. The average annual interest rate you'll earn on this savings plan is 10%.

　　A friend of yours (exactly your age) from Minnesota State University waits 10 years to start her savings plan. (That is, she is 32 years old.) She decides to save $2,000 each year in an account earning interest at the rate of 10% per year. She will make these annual payments until she is 65 years old, at which time she will withdraw the total accumulated amount.

How old will you be when your friend's *accumulated* savings amount (including interest) exceeds yours? State any assumptions you think are necessary.

Solution

Creating cash-flow diagrams for Example 6-16 is an important first step in solving for the unknown number of years, N, until the future equivalent values of both savings plans are equal. The two diagrams are shown below. The future equivalent (F) of your plan is $500(F/A, 10\%, 15)(F/P, 10\%, N - 36)$ and that of your friend is $F' = \$2,000(F/A, 10\%, N - 31)$. It is clear that N, the age at which $F = F'$, is greater than 32. Assuming that the interest rate remains constant at 10% per year, the value of N can be determined by trial and error:

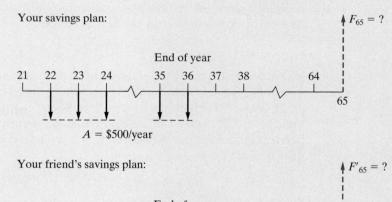

Your savings plan:

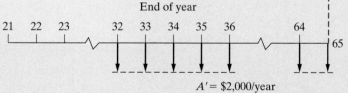

Your friend's savings plan:

N	Your Plan's F	Friend's F'
36	$15,886	$12,210
38	$19,222	$18,974
39	$21,145	$22,872
40	$23,259	$27,159

By the time you reach age 39, your friend's accumulated savings will exceed yours. (If you had deposited $1,000 instead of $500, you would be over 76 years when your friend's plan surpassed yours. Moral: Start saving early!)

EXAMPLE 6-17	Credit Card Offers

Randy just cancelled his credit card with a large bank. A week later, a representative of the bank called Randy with an offer of a "better" credit card that will advance Randy $2,000 when he accepts it. Randy could not refuse the offer and several days later receives a check for $2,000 from the bank. With this money, Randy decides to buy a new computer.

At the next billing cycle (a month later), the $2,000 advance appears as a charge against Randy's account, and the APR is stated to be 21% (compounded monthly). At this point in time, Randy elects to pay the minimum monthly payment of $40 and cuts up the credit card so that he cannot make any additional purchases.

(a) Over what period of time does this payment extend in order to repay the $2,000 principal?

(b) If Randy decides to repay all remaining principal after having made 15 monthly payments, how much will he repay?

Solution

(a) You may be surprised to know that the majority of credit card companies determine the minimum monthly payment based on a repayment period of 10 years. The monthly interest rate being charged for Randy's card is 21%/12 = 1.75%. We can solve the following equivalence relationship to determine the number of $40 monthly payments required to pay off a loan principal of $2,000.

$$\$2,000 = \$40(P/A, 1.75\%, N)$$

This is easily solved using the NPER (rate, payment, principal) function in Excel.

$$NPER(1.75\%, -40, 2000) = 119.86$$

Sure enough, it will take Randy 120 months to pay off this debt.

(b) After having made 15 payments, Randy has 105 payments remaining. To find the single sum payoff for this loan, we simply have to determine the present worth of the remaining payments.

$$\text{Payoff} = \$40(P/A, 1.75\%, 105) = \$1,915.96$$

This payoff amount assumes no penalty for early repayment of the loan (which is typically the case when it comes to credit cards). Notice how very little principal ($84.06) was repaid in the early part of the loan.

6.7 CASE STUDY—Ned and Larry's Ice Cream Company

Ned and Larry's Ice Cream Company produces specialty ice cream and frozen yogurt in pint-sized containers. The latest annual performance report praised the firm for its progressive policies but noted that environmental issues like packaging disposal were a concern. In an effort to reduce the effects of consumer disposal of product packaging, the report stated that Ned and Larry's should consider the following proposals:

Proposal A—Package all ice cream and frozen yogurt in quart containers.

Proposal B—Package all ice cream and frozen yogurt in half-gallon containers.

By packaging the product in containers larger than the current pints, the plastic-coated bleached sulfate board containers will hold more ounces of product per square inch of surface area. The net result is less discarded packaging per ounce of product consumed. Additional advantages to using larger containers include lower packaging costs per ounce and less handling labor per ounce.

Changing to a larger container requires redesign of the packaging and modifications to the filling production line. The existing material-handling equipment can handle the pints and quarts, but additional equipment will be required to handle half-gallons. Any new equipment purchased for proposals A and B has an expected useful life of six years. The total capital investment for each proposal is shown in the accompanying table. The table summarizes the details of these proposals, as well as the current production of pints.

	Current (Pints)	(A) Quarts	(B) Half-Gallons
Capital investment	$0	$1,200,000	$1,900,000
Packaging cost per gallon	$0.256	$0.225	$0.210
Handling labor cost per gallon	$0.128	$0.120	$0.119
Postconsumer landfill contribution from discarded packaging (yd^3/yr)	6,500	5,200	4,050

Because Ned and Larry's promotes partnering with suppliers, customers, and the community, they wish to include a portion of the cost to society when evaluating these alternatives. They will consider 50% of the postconsumer landfill cost as part of the costs for each alternative. They have estimated landfill costs to average $20 per cubic yard nationwide.

Ned and Larry's uses a MARR of 15% per year and IRR analyses to evaluate capital investments. A study period of six years will be used, at which time the equipment purchased for proposals A and B will have negligible market value. Production will remain constant at 10,625,000 gallons per year. Determine whether Ned and Larry's should package ice cream and frozen yogurt in pints, quarts, or half-gallons.

Solution

Assuming that Ned and Larry's is able to sell all ice cream and frozen yogurt produced, we can focus on the *differences* in costs associated with the three packaging alternatives. Since our recommendation is to be supported by IRR analysis, we must use the *incremental analysis* procedure.

The first step in the incremental analysis procedure is to rank-order the alternatives by increasing capital investment:

$$\text{Current(pints)} \rightarrow \text{A(quarts)} \rightarrow \text{B(half-gallons)}.$$

Now we can compute the incremental difference between producing in quarts (A) instead of pints (current) and determine whether the incremental capital investment is justified.

	Δ (A − current)
Δ Capital investment	−$1,200,000
Δ Packaging cost savings	[$0.031/gal](10,625,000 gal/yr) = $329,375/yr
Δ Handling cost savings	[$0.008/gal](10,625,000 gal/yr) = $85,000/yr
Δ Landfill cost savings	[1,300 yd^3/yr](0.5)($20/yd^3) = $13,000/yr

To determine the IRR of the incremental investment, we can find the interest rate at which the PW of the incremental cash flows is zero.

$$PW(i'_\Delta\%) = 0 = -\$1,200,000 + (\$329,375 + \$85,000 + \$13,000)(P/A, i'_\Delta\%, 6).$$

By trial and error, $i'_\Delta\% = 27.2\% >$ MARR. Thus, the investment required to produce the quart-size containers for ice cream is economically justified. Our decision is currently to produce quart-size containers unless the extra investment required to produce in half-gallon-size containers (B) earns at least the MARR based on projected savings.

	Δ (B − A)
Δ Capital investment	−$700,000
Δ Packaging cost savings	[$0.015/gal](10,625,000 gal/yr) = $159,375/yr
Δ Handling cost savings	[$0.001/gal](10,625,000 gal/yr) = $10,625/yr
Δ Landfill cost savings	[1,150 yd^3/yr](0.5)($20/yd^3) = $11,500/yr

$$PW(i'_\Delta\%) = 0 = -\$700,000 + (\$159,375 + \$10,625 + \$11,500)(P/A, i'_\Delta\%, 6)$$

By trial and error, $i'_\Delta\% = 14.3\% <$ MARR. Therefore, the extra investment required to produce in half-gallon-size containers (B) is not justified by the quantified extra savings.

Based on the preceding incremental IRR analysis, the final recommendation for Ned and Larry's Ice Cream Company is to produce ice cream and frozen yogurt in quart containers (i.e., Proposal A). The foregoing incremental analysis procedure indicates that this recommendation will yield an $i'_\Delta\% = 27.2\%$, which is greater than the MARR of 15%. So, by making the recommended change to packaging all ice cream and frozen yogurt in quart containers, not only will Ned and Larry's be delivering great tasting products that are economically attractive to the company, but the company will be environmentally conscious as well! A win all around! What flavor would you like?

6.8   Postevaluation of Results

After comparing alternatives and identifying the preferred course of action (Step 6), we return to the important Step 7 of the economic analysis procedure in Table 1-1. Postaudit reviews, which are appraisals of how a project is progressing against critical milestones, should be conducted during and after a project, recommended by the procedures in Chapter 6, has been approved and funded. This periodic feedback will inform management about the following:

1. Are the planned objectives and milestones being (or have been) attained by the project?
2. Is corrective action required to bring the project in line with the expectations?
3. What can be learned from the selected (and implemented) project that will improve estimating for future projects?

Learning from past decisions (both good and bad!) is critical to improving future decisions.

Periodic postaudit reviews serve to reduce, and often correct for, possible bias in favor of what individual corporate units, divisions, and subsidiaries would like to accomplish to serve their own self-interests. The tendency to estimate optimistically future cash flows and other conditions for pet projects seems to be human nature, but a fair and above-board audit during and after each major project should keep this in check. As a result, estimating project outcomes will be taken more seriously when projects are tracked and monitored over their life cycles. However, a balance must be struck between overly aggressive postaudits and the counter behavior of causing estimators to become overly conservative in their project appraisals.

6.9   Summary

Chapter 6 built on the previous chapters, in which the principles and applications of money—time relationships were developed. Specifically, Chapter 6 has (1) introduced several difficulties associated with selecting the best alternative from a mutually exclusive set of feasible candidates when using time value-of-money concepts and (2) demonstrated the application of the profitability analysis methods discussed in Chapter 5 to select the preferred alternative. Moreover, alternatives with unequal lives and cost-only versus different revenues and costs were considered in deciding how to maximize the productivity of invested capital based on the MARR.

If a rate-of-return method is being used to analyze mutually exclusive alternatives, each avoidable increment of additional capital must earn at least the MARR to ensure that the best alternative is chosen. Therefore, alternatives are rank-ordered from the least capital investment to the greatest capital investment. Examples were provided to illustrate correct computational procedures for avoiding the ranking inconsistency that sometimes occurs when equivalent-worth and rate-of-return methods are applied to the same set of mutually exclusive alternatives. We also considered projects with perpetual lives, applying the capitalized-worth (CW) method of economic evaluation.

Problems

The number in parentheses () that follows each problem refers to the section from which the problem is taken.

6-1. Four mutually exclusive alternatives are being evaluated, and their costs and revenues are itemized in Table P6-1. (6.4)

a. If the MARR is 15% per year and the analysis period is 12 years, use the PW method to determine which alternatives are economically acceptable and which one should be selected.

b. If the total capital investment budget available is $200,000, which alternative should be selected?

c. Which rule (Section 6.2.2) applies? Why?

6-2. The Consolidated Oil Company must install antipollution equipment in a new refinery to meet federal clean-air standards. Four design alternatives are being considered, which will have capital investment and annual operating expenses as shown in Table P6-2. Assuming a useful life of 10 years for each design, no market value, a desired MARR of 12% per

year, and an analysis period of 10 years, determine which design should be selected on the basis of the PW method. Confirm your selection by using the FW and AW methods. Which rule (Section 6.2.2) applies? Why? (6.4)

6-3. At the Motorola plant in Mesa, Arizona, it is desired to determine whether one-inch-thick insulation or two-inch-thick insulation should be used to reduce heat loss from a long section of steam pipe. The heat loss from this pipe without any insulation would cost $2.00 per linear foot per year. The one-inch insulation will eliminate 88% of the heat loss and will cost $0.60 per foot. Two-inch insulation will eliminate 92% of the heat loss and will cost $1.10 per foot. The steam pipe is 1,000 feet in length and will last for 10 years. MARR = 6% per year. Which insulation thickness should be recommended? (6.4)

6-4. Three mutually exclusive design alternatives are being considered. The estimated sales and cost data for each alternative are given on page 285. The MARR is 20% per year.

TABLE P6-1 Table for Problem 6-1				
	Mutually Exclusive Alternative			
	I	II	III	IV
Capital investment	$100,000	$152,000	$184,000	$220,000
Annual revenues less expenses	15,200	31,900	35,900	41,500
Market value (end of useful life)	10,000	0	15,000	20,000
Useful life (years)	12	12	12	12

TABLE P6-2 Table for Problem 6-2				
	Alternative Design			
	D1	D2	D3	D4
Capital investment	$600,000	$760,000	$1,240,000	$1,600,000
Annual expenses:				
Power	68,000	68,000	120,000	126,000
Labor	40,000	45,000	65,000	50,000
Maintenance	660,000	600,000	420,000	370,000
Taxes and insurance	12,000	15,000	25,000	28,000

	A	B	C
Investment cost	$30,000	$60,000	$40,000
Estimated units to be sold/year	15,000	20,000	18,000
Unit selling price, $/unit	$3.10	$4.40	$3.70
Variable costs, $/unit	$1.00	$1.40	$0.90
Annual expenses (fixed costs)	$15,000	$30,000	$25,000
Market value	$10,000	$10,000	$10,000
Useful life	10 years	10 years	10 years

Annual revenues are based on the number of units sold and the selling price. Annual expenses are based on fixed and variable costs. Determine which selection is preferable based on AW. State your assumptions. (6.4)

6-5. Which mutually exclusive design (if any) should be chosen from the designs listed here? A 10-year study period is to be used, and the MARR is 10% per year. All market values are negligible.

	Design A	Design B	Design C
Capital investment	$170,000	$330,000	$300,000
Annual receipts	114,000	147,000	130,000
Annual expenses	70,000	79,000	64,000

Use the FW method. Confirm your recommendation, using the PW and AW methods. (6.4)

6-6. You have been asked to evaluate the economic implications of various methods for cooling condenser effluents from a 200-MW steam-electric plant. In this regard, cooling ponds and once-through cooling systems have been eliminated from consideration because of their adverse ecological effects. It has been decided to use cooling towers to dissipate waste heat to the atmosphere. There are two basic types of cooling towers: wet and dry. Furthermore, heat may be removed from condenser water by (1) forcing (mechanically) air through the tower or (2) allowing heat transfer to occur by making use of natural draft. Consequently, there are four basic cooling tower designs that could be considered. Assuming that the cost of capital to the utility company is 12% per year, your job is to recommend the best alternative (i.e., the least expensive during the service life) in view of the data in Table P6-6. Further, assume that each alternative is capable of

satisfactorily removing waste heat from the condensers of a 200-MW power plant. What *non-economic* factors can you identify that might also play a role in the decision-making process? (6.4)

6-7. Three mutually exclusive design alternatives are being considered. The estimated cash flows for each alternative are given next. The MARR is 20% per year. At the conclusion of the useful life, the investment will be sold.

	A	B	C
Investment cost	$28,000	$55,000	$40,000
Annual expenses	$15,000	$13,000	$22,000
Annual revenues	$23,000	$28,000	$32,000
Market value	$6,000	$8,000	$10,000
Useful life	10 years	10 years	10 years
IRR	26.4%	24.7%	22.4%

A decision-maker can select one of these alternatives or decide to select none of them. Make a recommendation using the PW method. (6.4)

6-8. An old, heavily used warehouse currently has an incandescent lighting system. The lights run essentially 24 hours per day, 365 days per year, and draw 10 kW (input) of power. Consideration is being given to replacing these lights with fluorescent lights to save on electricity. It is estimated that the *same* level of lighting can be achieved with 4.5 kW (input) of fluorescent lights. Replacement of the lights will cost about $11,000. Bulb replacement and other maintenance are not expected to be significantly different. Electricity for the lights currently costs $0.045/kWh. The warehouse is scheduled for demolition in five years to make way for a more modern facility. The company has a MARR of 15% per year. Should the company replace the incandescent lights with fluorescent lights? Use the PW method and state any assumptions you make. (6.4)

6-9. Consider the following mutually exclusive alternatives:

	Alternative A	Alternative B
Capital investment	$780,000	$1,840,000
Net annual receipts	$138,060	$311,000

Both alternatives have a useful life of 10 years and no market value at that time. The MARR is 10% per year. Use the FW method to identify the most profitable course of action. (6.4)

TABLE P6-6   Alternative Types of Cooling Towers for a 200-Megawatt Fossil-Fired Power Plant Operating at Full Capacity[a] in Problem 6-6

	Alternative			
	Wet Tower Mech. Draft	Wet Tower Natural Draft	Dry Tower Mech. Draft	Dry Tower Natural Draft
Initial cost	$3 million	$8.7 million	$5.1 million	$9.0 million
Power for I.D. fans	40 200-hp induced-draft fans	None	20 200-hp I.D. fans	None
Power for pumps	20 150-hp pumps	20 150-hp pumps	40 100-hp pumps	40 100-hp pumps
Mechanical maintenance/year	$0.15 million	$0.10 million	$0.17 million	$0.12 million
Service life	30 years	30 years	30 years	30 years
Market value	0	0	0	0

[a] 100 hp = 74.6 kW; cost of power to plant is 2.2 cents per kWh or kilowatt-hour; induced-draft fans and pumps operate around the clock for 365 days/year (continuously). Assume that electric motors for pumps and fans are 90% efficient.

6-10. A steam generator is needed in the design of a new power plant. This generator can be fired by three different fuels, A, B, or C, with the following cost implications.

	A	B	C
Installed investment cost	$30,000	$55,000	$180,000
Annual fuel expense	X + $7,500	X	X − $1,500
Residual value	None	None	None

If the study period is 20 years and MARR is 8% per year, which type of fuel is most economical? (6.4)

6-11. Refer to Example 6-7. Assume the company's marketing department estimates that 91,000 units of the electronic control unit is the maximum that can be sold in any year. Based on this assumption, should the project still be implemented? If yes, which alternative (A1, A2) should be selected? Why? (*Note:* Use the AW method of analysis in your solution.) (6.4)

6-12. A new highway is to be constructed. Design *A* calls for a *concrete* pavement costing $90 per foot with a 20-year life; two paved ditches costing $3 per foot each; and three box culverts every mile, each costing $9,000 and having a 20-year life. Annual maintenance will cost $1,800 per mile; the culverts must be cleaned every five years at a cost of $450 each per mile.

Design *B* calls for a *bituminous* pavement costing $45 per foot with a 10-year life; two sodded ditches costing $1.50 per foot each; and three pipe culverts every mile, each costing $2,250 and having a 10-year life. The replacement culverts will cost $2,400 each. Annual maintenance will cost $2,700 per mile; the culverts must be cleaned yearly at a cost of $225 each per mile; and the annual ditch maintenance will cost $1.50 per foot per ditch.

Compare the two designs on the basis of equivalent worth per mile for a 20-year period. Find the most economical design on the basis of AW and PW if the MARR is 6% per year. (6.3, 6.4)

6-13. The alternatives for an engineering project to recover most of the energy presently being lost in the primary cooling stage of a chemical processing system have been reduced to three designs. The estimated capital investment amounts and annual expense *savings* are as follows:

	Design		
EOY	ER1	ER2	ER3
0	−$98,600	−$115,000	−$81,200
1	25,800	29,000	19,750
2			
3	$\bar{f}=6\%$[a]	G=$150[b]	c
4			
5			
6	34,526	29,750	19,750

[a] After year one, the annual savings are estimated to increase at the rate of 6% per year.

[b] After year one, the annual savings are estimated to increase $150 per year.

[c] Uniform sequence of annual savings.

Assume that the MARR is 12% per year, the study period is six years, and the market value is zero for all three designs. Apply an equivalent-worth analysis method to determine the preferred alternative. (6.4)

6-14. The following cash-flow estimates have been developed for two small, mutually exclusive investment alternatives:

End of Year	Alternative 1	Alternative 2
0	−$2,500	−$4,000
1	750	1,200
2	750	1,200
3	750	1,200
4	750	1,200
5	2,750	3,200

The MARR = 12% per year. (6.4)

a. What is the AW of Alternative 1?
b. What is the AW of Alternative 2?
c. What is the AW of the incremental net cash flow?
d. Given your answers for Parts (a) through (c), which alternative should be selected?

6-15. Consider the three mutually exclusive projects that follow. The firm's MARR is 10% per year. (6.4)

EOY	Project 1	Project 2	Project 3
0	−$10,000	−$8,500	−$11,000
1–3	$5,125	$4,450	$5,400

a. Calculate each project's PW.
b. Determine the IRR of each project.
c. Which project would you recommend?
d. Why might one project have the highest PW while a different project has the largest IRR?

6-16. A special-purpose 30-horsepower electric motor has an efficiency of 90%. Its purchase and installation price is $2,200. A second 30-horsepower high-efficiency motor can be purchased for $3,200, and its efficiency is 93%. Either motor will be operated 4,000 hours per year at full load, and electricity costs $0.10 per kilowatt-hour (kWh). MARR = 15% per year, and neither motor will have a market value at the end of the eight-year study period. (6.4)

a. Which motor should be chosen?
b. For an incremental investment of $1,000 in the more efficient motor, what is the PW of the energy savings over the eight-year period?

6-17. Refer to the situation in Problem 6-16. Most motors are operated at a fraction of their rated capacity (i.e., full load) in industrial applications. Suppose the average usage (load factor) of the motor in Problem 6-16 is expected to be 60%. Which motor should be recommended under this condition? (6.4)

6-18. A stem cell research project requires expensive specialized laboratory equipment. For this purpose, three pieces of equipment and their associated cash flows (listed below) are under consideration. One piece of equipment must be selected, and the laboratory's MARR is 15% per year. (6.4)

| EOY | Equipment | | |
	A	B	C
0	−$136,500	−$84,000	−$126,000
1 – 4	−$12,500	−$31,500	−$15,500
5	−$15,000	−$31,500	−$15,500

a. Use the PW method to rank-order the economic attractiveness of the three projects.
b. Determine the interest rate at which the laboratory would be indifferent between Equipment B and Equipment C.

6-19. Consider the following two mutually exclusive alternatives for reclaiming a deteriorating inner-city neighborhood (one of them must be chosen). Notice that the IRR for both alternatives is 27.19%. (6.4)

| EOY | Alternative | |
	X	Y
0	−$100,000	−$100,000
1	$50,000	0
2	$51,000	0
3	$60,000	$205,760
IRR	27.19%	27.19%

TABLE P6-20	Data for Problem 6-20					
	End of Year					
Alternative	0	1	2	3	IRR	PW(0%)
M	−$52,000	$24,000	$24,000	$24,000	18.2%	$20,000
N	−$51,000	0	$25,000	$50,000	?	$24,000
Δ(N−M)	$1,000	−$24,000	$1,000	$26,000	8.7%	$4,000

a. If MARR is 15% per year, which alternative is better?

b. What is the IRR on the incremental cash flow [i.e., $\Delta(Y - X)$]?

c. If MARR is 27.5% per year, which alternative is better?

d. What is the simple payback period for each alternative?

e. Which alternative would you recommend?

6-20. Consider the two mutually exclusive alternatives in Table P6-20. (6.4)

a. Compute the IRR for Alternative N.

b. Assume MARR to be 8% per year, and capital is available for either project. Use the IRR method to determine which project should be recommended.

c. Now let MARR be equal to 15% per year. Which project would be recommended with the IRR method?

6-21. A movie theatre is considering the purchase of a new digital projection system (each showing is as perfect as when it was first filmed). The new ticket price per person would be $7.50, which is $1.00 higher than for the traditional cellulose film projection system. The new projection system would cost $50,000. (6.4)

a. If the theatre expects to sell 20,000 tickets per year, how many years (as an integer) will it take for the theatre to recover the $50,000 investment of the new system (i.e., what is the simple payback period)?

b. What is the discounted payback period (as an integer) if MARR is 20% per year?

c. If the digital system has a life of five years and a salvage value of $5,000 at that time, what is the IRR of the new system?

6-22. Four mutually exclusive projects are being considered for a new two-mile jogging track. The life of the track is expected to be 80 years, and the sponsoring agency's MARR is 12% per year. Annual benefits to the public have been estimated by an advisory committee and are shown below. Use the IRR method (incrementally) to select the best jogging track. (6.4)

	Alternative			
	A	B	C	D
Initial cost	$62,000	$52,000	$150,000	$55,000
Annual benefits	$10,000	$8,000	$20,000	$9,000
Rate of return on investment	16.1%	15.4%	13.3%	16.4%

6-23. An industrial coal-fired boiler for process steam is equipped with a 10-year-old electrostatic precipitator (ESP). Changes in coal quality have caused stack emissions to be in noncompliance with federal standards for particulates. Two mutually exclusive alternatives have been proposed to rectify this problem (doing nothing is not an option).

	New Baghouse	New ESP
Capital investment	$1,140,000	$992,500
Annual operating expenses	115,500	73,200

The life of both alternatives is 10 years, and the MARR is 15% per year. Use the IRR method (incrementally) to make a recommendation regarding which alternative to select. Can you list some nonmonetary factors that would favor the new baghouse? (6.4)

6-24. Two mutually exclusive harvesting combines are being considered for purchase by a wheat farmer. He must buy a new combine now. Information relevant to his comparison of the alternatives is summarized below.

	Deere	FMC
Capital investment	$240,000	$200,000
Market value at end of service life	$80,000	$20,000
Annual fuel and maintenance expenses	$5,000	$10,000
Service life	12 years	12 years

Use the IRR method to determine the better machine to purchase. The farmer's MARR is 12% per year. (6.4)

6-25. There are three mutually exclusive alternatives that are candidates for implementation by the Yellow Freight Company's sorting operations center, and doing nothing is not an option. All alternatives have a life of 10 years, and they have negligible market (salvage) value after 10 years. Use the IRR method (incrementally) to make your recommendation. The firm's MARR is 10% per year. (6.4)

Alternative	Capital Investment	Annual Expenses
A	$740,000	$361,940
B	$1,840,000	$183,810
C	$540,000	$420,000

6-26. In the Rawhide Company (a leather products distributor), decisions regarding approval of proposals for capital investment are based upon a stipulated MARR of 18% per year. The five packaging devices listed in Table P6-26 were compared, assuming a 10-year life and zero market value for each at that time. Which one (if any) should be selected? Make any additional calculations that you think are needed to make a comparison, using the ERR method. Let $\epsilon = 18\%$. (6.4)

6-27. Refer to Problem 6-2. Solve this problem using the ERR method. Let $\epsilon = 12\%$ per year. (6.4)

6-28. Consider the following EOY cash flows for two mutually exclusive alternatives (one must be chosen):

	Machine X	Machine Y
Capital investment	$6,000	$14,000
Annual expenses	$2,500	$2,400
Useful life	12 years	18 years
Market value at end of useful life	$0	$2,800

The MARR is 5% per year. (6.5)

a. Determine which alternative should be selected if the repeatability assumption applies.

b. Determine which alternative should be selected if the analysis period is 18 years, the repeatability assumption does *not* apply, and a machine can be leased for $8,000 per year after the useful life of either machine is over.

6-29. Two electric motors are being considered to drive a centrifugal pump. One of the motors must be selected. Each motor is capable of delivering 60 horsepower (output) to the pumping operation. It is expected that the motors will be in use 800 hours per year. The following data are available: (6.5)

	Motor A	Motor B
Capital investment	$1,200	$1,000
Electrical efficiency	0.90	0.80
Annual maintenance	$160	$100
Useful life	3 years	5 years

a. If electricity costs $0.07 per kilowatt-hour, which motor should be selected if the MARR is 8% per year? Recall that 1 hp = 0.746 kW. Assume repeatability.

TABLE P6-26   Table for Problem 6-26

	Packaging Equipment				
	A	B	C	D	E
Capital investment	$38,000	$50,000	$55,000	$60,000	$70,000
Annual revenues less expenses	11,000	14,100	16,300	16,800	19,200
External rate of return (ERR)	21.1%	20.8%	21.4%	20.7%	20.5%

b. What is the basic trade-off being made in this problem?

6-30. The National Park Service is considering two plans for rejuvenating the forest and landscape of a large tract of public land. The study period is indefinitely long, and the Park Service's MARR is 10% per year. You have been asked to compare the two plans using the CW method.

The first plan (*Skyline*) calls for an initial investment of $500,000, with expenses of $20,000 per year for the first 20 years and $30,000 per year thereafter. Skyline also requires an expenditure of $200,000 twenty years after the initial investment, and this will repeat every 20 years thereafter. The second plan (*Prairie View*) has an initial investment of $700,000 followed by a single (one-time) investment of $300,000 thirty years later. Prairie View will incur annual expenses of $10,000 forever.

Based on the CW measure, which plan would you recommend? (6.5)

6-31. A new manufacturing facility will produce two products, each of which requires a drilling operation during processing. Two alternative types of drilling machines (D1 and D2) are being considered for purchase. One of these machines must be selected. For the same annual demand, the *annual* production requirements (machine hours) and the annual operating expenses (per machine) are listed in Table P6-31. *Which machine should be selected if the MARR is 15% per year?* Show all your work to support your recommendation. (6.5)

Assumptions: The facility will operate 2,000 hours per year. Machine availability is 80% for Machine D1 and 75% for Machine D2. The yield of D1 is 90%, and the yield of D2 is 80%. Annual operating expenses are based on an assumed operation of 2,000 hours per year, and workers are paid during any idle time of Machine D1 or Machine D2. State any other assumptions needed to solve the problem.

6-32. As the supervisor of a facilities engineering department, you consider mobile cranes to be critical equipment. The purchase of a new medium-sized, truck-mounted crane is being evaluated. The economic estimates for the two best alternatives are shown in the table below. You have selected the longest useful life (nine years) for the study period and would lease a crane for the final three years under Alternative *A*. On the basis of previous experience, the estimated annual leasing cost at that time will be $66,000 per year (plus the annual expenses of $28,800 per year). The MARR is 15% per year. Show that the same selection is made with

(a) the PW method.

(b) the IRR method.

(c) the ERR method.

(d) Would leasing crane A for nine years, assuming the same costs per year as for three years, be preferred over your present selection? (ϵ = MARR = 15%). (6.4, 6.5)

	Alternatives	
	A	B
Capital investment	$272,000	$346,000
Annual expenses[a]	28,800	19,300
Useful life (years)	6	9
Market value (at end of life)	$25,000	$40,000

[a] Excludes the cost of an operator, which is the same for both alternatives.

6-33. Estimates for a proposed small public facility are as follows: Plan *A* has a first cost of $50,000, a life of 25 years, a $5,000 market value, and annual maintenance expenses of $1,200. Plan *B* has a first cost of $90,000, a life of 50 years, no market value, and annual maintenance

TABLE P6-31	Table for Problem 6-31	
Product	Machine D1	Machine D2
R-43	1,200 hours	800 hours
T-22	2,250 hours	1,550 hours
	3,450 hours	2,350 hours
Capital investment	$16,000/machine	$24,000/machine
Useful life	six years	eight years
Annual expenses	$5,000/machine	$7,500/machine
Market value	$3,000/machine	$4,000/machine

expenses of $6,000 for the first 15 years and $1,000 per year for years 16 through 50. Assuming interest at 10% per year, compare the two plans, using the CW method. (6.4)

6-34. Potable water is in short supply in many countries. To address this need, two mutually exclusive water purification systems are being considered for implementation in China. Doing nothing is not an option. Refer to the data below and state your key assumptions in working this problem. (6.5)

	System 1	System 2
Capital investment	$100,000	$150,000
Annual revenues	50,000	70,000
Annual expenses	22,000	40,000
MV at end of useful life	20,000	0
Useful life	5 years	10 years
IRR	16.5%	15.1%

a. Use the PW method to determine which system should be selected when MARR = 8% per year.

b. Which system should be selected when MARR = 15% per year?

6-35. Your plant must add another boiler to its steam generating system. Bids have been obtained from two boiler manufacturers as follows:

	Boiler A	Boiler B
Capital investment	$50,000	$120,000
Useful life, N	20 years	40 years
Market value at EOY N	$10,000	$20,000
Annual operating costs	$9,000	$3,000, increasing $300 per year after the first year

If the MARR is 10% per year, which boiler would you recommend? Use the repeatability assumption. (6.5)

6-36. Three mutually exclusive alternatives are being considered for the production equipment at a tissue paper factory. The estimated cash flows for each alternative are given here. (All costs are in thousands.)

	A	B	C
Capital investment	$2,000	$4,200	$7,000
Annual revenues	3,200	6,000	8,000
Annual costs	2,100	4,000	5,100
Market value at end of useful life	100	420	600
Useful life (years)	5	10	10

Which equipment alternative, if any, should be selected? The firm's MARR is 20% per year. Please state your assumptions. (6.5)

6-37. In the design of a special-use structure, two mutually exclusive alternatives are under consideration. These design alternatives are as follows:

	D1	D2
Capital investment	$50,000	$120,000
Annual expenses	$9,000	$5,000
Useful life (years)	20	50
Market value (at end of useful life)	$10,000	$20,000

If perpetual service from the structure is assumed, which design alternative do you recommend? The MARR is 10% per year. (6.4)

6-38. Two insulation thickness alternatives have been proposed for a process steam line subject to severe weather conditions. One alternative must be selected. Estimated savings in heat loss and installation cost are given below.

Thickness	Installed Cost	Annual Savings	Life
2 cm.	$20,000	$5,000	4 years
5 cm.	$40,000	$7,500	6 years

Which thickness would you recommend for a MARR = 15% per year and negligible market (salvage) values? The study period is 12 years. (6.5)

6-39. Refer to Problem 6-38. Which thickness would you recommend if the study period was four years?

Use the imputed market value technique to estimate the market value of the 5 cm alternative after four years. (6.5)

6-40.

a. Compare the probable part cost from Machine *A* and Machine *B* below, assuming that each will make the part to the same specification. Which machine yields the lowest part cost? Assume that the MARR $= 10\%$ per year. (6.5)

b. If the cost of labor can be cut in half by using part-time employees, which machine should be recommended? (6.5)

	Machine A	Machine B
Initial capital investment	$35,000	$150,000
Life	10 years	8 years
Market value	$3,500	$15,000
Parts required per year	10,000	10,000
Labor cost per hour	$16	$20
Time to make one part	20 minutes	10 minutes
Maintenance cost per year	$1,000	$3,000

6-41. Two mutually exclusive alternatives are being considered for the environmental protection equipment at a petroleum refinery. One of these alternatives must be selected. The estimated cash flows for each alternative are as follows: (6.5)

	Alternative A	Alternative B
Capital investment	$20,000	$38,000
Annual expenses	5,500	4,000
Market value at end of useful life	1,000	4,200
Useful life	5 years	10 years

a. Which environmental protection equipment alternative should be selected? The firm's MARR is 20% per year. Assume the equipment will be needed indefinitely.

b. Assume the study period is shortened to five years. The market value of Alternative B after five years is estimated to be $15,000. Which alternative would you recommend?

6-42. A "greenway" walking trail has been proposed by the city of Richmond. Two mutually exclusive alternative locations for the 2-meter-wide trail have been proposed: one is on flat terrain and is 14 kilometers in length, and the other is in hilly terrain and is 12 kilometers in length.

Planning and site preparation cost is 20% of the asphalt-paving cost, which is $3.00 per square meter. Annual maintenance for the flat terrain trail is 5% of the paving cost, and annual maintenance for the hilly trail is 8% of the paving cost.

If the city's MARR is 6% per year and perpetual life of the trail is assumed, which trail should be recommended to minimize the capitalized cost of this project? (6.5)

6-43. Your rich uncle has offered you two living trust payouts for the rest of your life. The positive cash flows of each trust are shown below. If your personal MARR is 10% per year, which trust should you select? (6.5)

End of Year	Trust A	Trust B
0	$0	$0
1–10	$1,000	$1,200
11–∞	$1,500	$1,300

Note: At 10% per year interest, infinity occurs when N= 70 years.

6-44. Three models of baseball bats will be manufactured in a new plant in Pulaski. Each bat requires some manufacturing time at either Lathe 1 or Lathe 2, according to the following table. Your task is to help decide which type of lathe to install. Show and explain all work to support your recommendation. (6.5)

Machining Hours for the Production of Baseball Bats		
Product	Lathe 1 (L1)	Lathe 2 (L2)
Wood bat	1,600 hr	950 hr
Aluminum bat	1,800 hr	1,100 hr
Kevlar bat	2,750 hr	2,350 hr
Total machining hours	6,150 hr	4,400 hr

The plant will operate 3,000 hours per year. Machine availability is 85% for Lathe 1 and 90% for Lathe 2. Scrap rates for the two lathes are 5% versus 10% for L1 and L2, respectively. Cash flows and expected lives for the two lathes are as follows:

Cash Flows and Expected Lives for L1 and L2		
	Lathe 1 (L1)	Lathe 2 (L2)
Capital investment	$18,000 per lathe	$25,000 per lathe
Expected life	7 years	11 years
Annual expenses	$5,000 per lathe	$9,500 per lathe

Annual operating expenses are based on an assumed operation of 3,000 hours per year, and workers are paid during any idle time of L1 and L2. Upper management has decided that MARR = 18% per year.

a. How many type L1 lathes will be required to meet the machine-hour requirement?

b. What is the CR cost of the required type L2 lathes?

c. What is the annual operating expense of the type L2 lathes?

d. Which type of lathe should be selected on the basis of lowest *total* equivalent annual cost?

6-45. A piece of production equipment is to be replaced immediately because it no longer meets quality requirements for the end product. The two best alternatives are a used piece of equipment (E1) and a new automated model (E2). The economic estimates for each are shown in the accompanying table.

	Alternative	
	E1	E2
Capital investment	$14,000	$65,000
Annual expenses	$14,000	$9,000
Useful life (years)	5	20
Market value (at end of useful life)	$8,000	$13,000

The MARR is 15% per year.

a. Which alternative is preferred, based on the repeatability assumption? (6.5)

b. Show, for the coterminated assumption with a five-year study period and an imputed market value for Alternative *B*, that the AW of *B* remains the same as it was in Part (a). [And obviously, the selection is the same as in Part (a)]. Explain why that occurs in this problem. (6.5)

6-46. A high-pressure pump at a methane gas (biofuel) plant in Memphis costs $30,000 for installation and has an estimated life of 12 years. By the addition of a specialized piece of auxiliary equipment, an annual savings of $400 in operating expense for the pump can be realized, and the estimated life of the pump can be doubled to 24 years. The salvage (market) value of either alternative is negligible at any time. If MARR is 6% per year, what present expenditure for the auxiliary equipment can plan managers justify spending? (6.5)

6-47. Use the CW method to determine which mutually exclusive bridge design (*L* or *H*) to recommend, based on the data provided in the accompanying table. The MARR is 15% per year. (6.4)

	Bridge Design *L*	Bridge Design *H*
Capital investment	$274,000	$326,000
Annual expenses	$10,000	$8,000
Periodic upgrade cost	$50,000 (every sixth year)	$42,000 (every seventh year)
Market value	0	0
Useful life (years)	83	92

6-48. You are involved in an equipment selection task. The operating requirement to be met by the selected piece of equipment is for the next six years *only*. The economic decision criterion presently being used by your company is 15% per year. Your choice has been reduced to two alternatives, E1 or E2. (See Table P6-48 on page 294).

Which alternative ought to be selected? Use the PW method of analysis. Which rule, Section 6.2.2, did you use? (6.2, 6.5)

6-49. Two proposals have been offered for streamlining the business operations of a customer call center. Proposal A has an investment cost of $30,000, an expected life of five years, property taxes of $450 per year, and no market value. Annual expenses are estimated to be $6,000.

Proposal B has an investment cost of $38,000, an expected life of four years, property taxes of $600 per year, and no market value. Its annual operating expenses are expected to be $4,000.

Using a MARR = 10% per year, which proposal should be recommended? Use the AW method and state your assumption(s). (6.5)

TABLE P6-48 Table for Problem 6-48		
Factor	E1	E2
Capital investment	$210,000	$264,000
Useful life (years)	6	10
Annual expenses	$31,000 in the first year and increasing $2,000 per year thereafter	$19,000 in the first year and increasing 5.7% per year thereafter
Market value (end of useful life)	$21,000	$38,000

6-50. Consider these mutually exclusive alternatives. MARR = 8% per year, so all the alternatives are acceptable. (6.5)

	Alternative		
	A	B	C
Capital investment (thousands)	$250	$375	$500
Uniform annual savings (thousands)	$40.69	$44.05	$131.90
Useful life (years)	10	20	5
Computed IRR (over useful life)	10%	10%	10%

a. At the end of their useful lives, alternatives A and C will be replaced with identical replacements (the repeatability assumption) so that a 20-year service requirement (study period) is met. Which alternative should be chosen and why?

b. Now suppose that at the end of their useful lives, alternatives A and C will be replaced with replacement alternatives having an 8% internal rate of return. Which alternative should be chosen and why? (Note: This assumption allows MEAs to be directly compared with the PW method over their individual useful lives—which violates the repeatability assumption implicit to Chapter 6.)

6-51. There is a continuing requirement for standby electrical power at a public utility service facility. Equipment alternative S1 involves an initial cost of $72,000, a nine-year useful life, annual expenses of $2,200 the first year and increasing $300 per year thereafter, and a net market value of $8,400 at the end of the useful life. Alternative S2 has an initial cost of $90,000, a 12-year useful life, annual expenses of $2,100 the first year and increasing at the rate of 5% per year thereafter, and a net market value of $13,000 at the end of the useful life. The current interest rate is

10% per year. Which alternative is preferred, using the CW method of analysis? (6.4)

6-52. A firm is considering three mutually exclusive alternatives as part of an upgrade to an existing transportation network.

	I	II	III
Installed cost	$40,000	$30,000	$20,000
Net annual revenue	$6,400	$5,650	$5,250
Salvage value	0	0	0
Useful life	20 years	20 years	10 years
Calculated IRR	15.0%	18.2%	22.9%

At EOY 10, alternative III would be replaced with another alternative III having the same installed cost and net annual revenues.

If MARR is 10% per year, which alternative (if any) should be chosen? Use the incremental IRR procedure. (6.5)

6-53. Three mutually exclusive investment alternatives are being considered. The estimated cash flows for each alternative are given in the accompanying chart. The study period is 30 years, and the firm's MARR is 20% per year.

	Alternative 1	Alternative 2	Alternative 3
Capital investment	$30,000	$60,000	$40,000
Annual costs	16,000	30,000	25,000
Annual revenues	28,000	53,500	38,000
Market value at end of useful life	10,000	10,000	10,000
Useful life	5 years	5 years	6 years
IRR	33.0%	29.9%	26.0%

Use the IRR method to make a recommendation. State all assumptions to support your analysis. (6.5)

6-54. Use the ERR method incrementally to decide whether project A or B should be recommended. These are two mutually exclusive cost alternatives, and one of them must be selected. MARR = ϵ = 10% per year. Assume repeatability is appropriate for this comparison. (6.5)

	Alternative	
	A	B
Capital investment	$100,000	$5,000
Annual operating expense	$5,000	$17,500
Useful life	20 years	10 years
Market (salvage) value	None	None

6-55. Three mutually exclusive equipment alternatives are being considered. The estimated cash flows for each alternative are as follows:

	Caterpillar	Deere	Komatsu
Capital investment	$22,000	$26,200	$17,000
Annual costs	7,000	7,500	5,800
Annual revenues	14,000	15,000	12,000
Market value at end of useful life	4,000	5,000	3,500
Useful life	4 years	10 years	5 years
IRR	15.6%	26.3%	27.2%

The firm's MARR is 15% per year. Which of the three alternatives, if any, should be adopted, on the basis of an IRR analysis? Be sure to state your assumptions. (6.5)

6-56. You have been requested to offer a recommendation of one of the mutually exclusive industrial sanitation control systems that follow. If the MARR is 15% per year, which system would you select? Use the IRR method. (6.5)

	Gravity-fed	Vacuum-led
Capital investment	$24,500	$37,900
Annual receipts less expenses	8,000	8,000
Life in years	5	10
IRR	18.9%	16.5%

6-57. A single-stage centrifugal blower is to be selected for an engineering design application. Suppliers have been consulted, and the choice has been narrowed down to two new models, both made by the same company and both having the same rated capacity and pressure. Both are driven at 3,600 rpm by identical 90-hp electric motors (output).

One blower has a guaranteed efficiency of 72% at full load and is offered installation for $42,000. The other is more expensive because of aerodynamic refinement, which gives it a guaranteed efficiency of 81% at full load.

Except for these differences in efficiency and installed price, the units are equally desirable in other operating characteristics such as durability, maintenance, ease of operation, and quietness. In both cases, plots of efficiency versus amount of air handled are flat in the vicinity of full rated load. The application is such that, whenever the blower is running, it will be at full load.

Assume that both blowers have negligible market values at the end of the useful life, and the firm's MARR is 20% per year. Develop a formula for calculating how much the user could afford to pay for the more efficient unit. (*Hint:* You need to specify important parameters and use them in your formula, and remember 1 hp = 0.746 kW.) (6.4)

6-58. Winfield County is attempting to develop a site for a convention center. Under the previous county government, a site was acquired and preliminary grading and land development were done, at a total cost of $284,000. The present government is faced with the problem of what to do. The development could be continued, in which case it is anticipated that $320,000 more will be spent during the course of one year. (Treat this as if it were paid at the end of the year.) At the end of that year, the site could be rented, and the county believes that it would produce an income stream of $24,000 per year (payable at the beginning of each year) for the foreseeable future. The county thinks that this estimate is very close to the actual amounts that would be produced. Alternatively, the land could be sold as is, now. Because of the extensive grading work, the sale price would be only $20,000. The county's MARR for decision-making is 7% per year.

One of the county commissioners has argued for abandonment, saying, "The total cost of the development will be $584,000, and at 7% this means we should be bringing in nearly $41,000 per year to justify the investment. We're not even close. Let's cut our losses and at least recover $20,000 out of this mess."

The county has asked you to advise them. Explain what you recommend and justify your answer. (6.4)

6-59. Today you have $80,000 to invest. Two investment alternatives are available to you. One would require you to invest your $80,000 now; the other would require the $80,000 two years from now. In either case, the alternatives will end five years from now. The cash flows for each alternative are as follows:

Year	Alternative 1	Alternative 2
0	−$80,000	
1	20,000	
2	20,000	−$80,000
3	20,000	33,000
4	24,000	33,000
5	26,000	33,000

Using a MARR of 10% per year, what should you do with the $80,000? (6.5)

6-60. Amy makes her first $10,000 payment into a conservative savings plan, earning 8% per year into her savings account when she is 28 years old. After the 10th payment, Amy makes no additional payments nor does she withdraw any money from the account.

One of Amy's friends, Frank, starts a savings plan that is identical to Amy's when he is 38 years old (Frank's account also earns 8% per year). He continues making $10,000 annual payments until he is 60 years old. If both Amy and Frank retire when they are 60 years old, who will have more money in their savings plan at that time? (6.6)

6-61. Automobile repair shops typically remind customers to change their oil and oil filter every 3,000 miles. Let's call this strategy 1. Usually you can save money by following another legitimate guideline—your automobile user's manual. It suggests changing your oil every 5,000–7,000 miles. We'll call this Strategy 2.

To conserve a nonrenewable and valuable resource, you decide to follow Strategy 2 and change your oil every 5,000 miles (you are conservative). You drive an average of 15,000 miles per year, and you expect each oil change to cost $30. Usually you drive your car 90,000 miles before trading it in on a new one. (6.6)

a. Develop an EOY cash-flow diagram for each strategy.

b. If your personal MARR is 10% per year, how much will you save with Strategy 2 over Strategy 1, expressed as a lump sum at the present time?

6-62. On your student loans, if possible, try to make interest-only payments while you are still in school. If interest is not repaid, it folds into principal after graduation and can cost you hundreds (or thousands) of extra dollars in finance charges.

For example, Sara borrowed $5,000 at the beginning of her freshman year and another $5,000 at the beginning of her junior year. The interest rate (APR) is 9% per year, compounded monthly, so Sara's interest accumulates at $3/4$% per month. Sara will repay what she owes as an ordinary annuity over 60 months, starting one month after she graduates in the summer term of her fourth full year of college. (6.6)

a. How much money does Sara owe upon graduation if she pays off monthly interest during school?

b. How much money does Sara owe if she pays no interest at all during her school years?

c. After graduation, what is the amount of the monthly loan repayment in Parts (a) and (b)?

d. How much interest does Sara repay without interest payments during school and with interest payments while in college?

6-63. A 30-year fixed rate–mortgage is available now at 7% APR. Monthly payments (360 of them) will be made on a $200,000 loan. The realtor says that if you wait several months to obtain a mortgage, the APR could be as high as 8%, "and the monthly payments will jump by 14%." Is the realtor's claim true? If not, what percent increase in monthly payment will result from this 1% increase in APR? (6.6)

6-64. A certain U.S. government savings bond can be purchased for $7,500. This bond will be worth $10,000 when it matures in five years. As an alternative, a 60-month certificate of deposit (CD) can be purchased for $7,500 from a local bank, and the CD yields 6.25% per year. Which is the better investment if your personal MARR is 5% per year? (6.6)

6-65. Your university advertises that next year's "discounted" tuition of $8,000 can be paid in full by July 1. The other option is to pay the first semester's tuition of $4,200 by July 1, with the remaining $4,200 due on January 1. Assuming you and/or your parents have $8,000 available for the full year's tuition, which tuition plan is more economical if you (or your parents) can earn 6% annually on a six-month CD? (6.6)

6-66. Bob and Sally have just bought a new house and they are considering the installation of 10 compact fluorescent bulb (CFB) fixtures instead of the 10 conventional incandescent lighting fixtures (which cost a total of $500) typically installed by the builder. According to the home builder, CFBs use 70% less electricity than incandescent bulbs, and they last 10 times longer before the bulb needs replacement. Bob calculates that 1,000 watts of incandescent lighting will be required in the house for 3,000 hours of usage per year. The cost of electricity is $0.10 per kWh. Also, each CFB will save 150 pounds of CO_2 per year. Each pound of CO_2 has a penalty of $0.02. If incandescent bulbs cost $0.75 each and last for one year, use a 10-year study period to determine the maximum cost of CFB fixtures and bulbs that can be justified in this house. MARR = 8% per year. (6.4)

6-67. *Extended Learning Exercise* The management of your company is considering adding a SO_2 scrubber unit to your present plant to remove SO_2 from stack gases, and you have conceived four designs to accomplish this task. Management does not believe that cleaning the gas just to reduce air pollution is worthwhile unless the government forces this to occur.

Management, however, would make the investment if more than a 10% annual rate of return on the capital investment can be earned (before taxes) by SO_2 recovery savings. Assume that your company can always invest money and earn a 10% return (MARR = 10%) and they are not pollution-control oriented. Thus, the recommended system, if any, will have to "pay its own way."

Following are the results of your calculations for the four designs. Which, if any, would you recommend to your management, and why? The life of each design is 10 years.

Design No.	Total Installed Cost for the Recovery Unit	Added Annual Operating Costs Due to Unit	Value of SO_2 Saved Per Year	Market Value
1	$100,000	$21,000	$41,000	$0
2	160,000	33,000	60,000	20,000
3	200,000	41,000	69,000	40,000
4	260,000	53,000	98,500	10,000

Spreadsheet Exercises

6-68. Refer to Example 6-3. Re-evaluate the recommended alternative if (a) the MARR = 15% per year; (b) the selling price is $0.50 per good unit; and (c) rejected units can be sold as scrap for $0.10 per unit. Evaluate each change individually. (d) What is the recommended alternative if all three of these changes occur simultaneously? (6.4)

6-69. Wilbur is a college student who desires to establish a long-term Roth IRA account with $4,000 that his grandmother gifted to him. He intends to invest the money in a mutual fund that earns an expected 5% per year on his account (this is a very conservative estimate). The sales agent says "There is a 5% up-front commission (payable now) on a Class A account, which every year thereafter then charges

a 0.61% management fee. A Class B account has no up-front commission, but its management fee is 2.35% in year one, 0.34% in year two, and 1.37% per year thereafter." Set up a spreadsheet to determine how many years are required before the worth of the Class A account overtakes (is preferred to) the Class B account. (6.6)

6-70. Create a spreadsheet to solve Example 6-12. What would the capital investment amount for pump SP240 have to be such that the firm would be indifferent as to which pump model is selected? (6.5)

6-71. Using the data from Problem 6-7, create a spreadsheet to identify the best design using an IRR analysis. (6.4)

Case Study Exercises

6-72. What other factors might you include in such an analysis? Should projected revenues be included in the analysis? (6.7)

6-73. Would the recommendation change if landfill costs double to $40 per cubic yard? (6.7)

6-74. Go to your local grocery store to obtain the price of ice cream packaged in different-sized containers. Use this information (along with additional assumptions you feel necessary) to make a recommendation. (6.7)

FE Practice Problems

6-75. Bill just won the sweepstakes in Tennessee! He has the option of either receiving a check right now for $125,000 or receiving a check for $50,000 each year for three years. Bill would receive the first check for $50,000 immediately. At what interest rate would Bill have to invest his winnings for him to be indifferent as to how he receives his winnings? Choose the closest answer. (6.5)

(a) 15.5%   (b) 20.0%   (c) 21.6%
(d) 23.3%

6-76. The Tree Top Airline (TTA) is a small feeder-freight line started with very limited capital to serve the independent petroleum operators in the arid Southwest. All of its planes are identical, although they are painted different colors. TTA has been contracting its overhaul work to Alamo Airmotive for $40,000 per plane per year. TTA estimates that, by building a $500,000 maintenance facility with a life of 15 years and a residual (market) value of $100,000 at the end of its life, they could handle their own overhaul at a cost of only $30,000 per plane per year. What is the minimum number of planes they must operate to make it economically feasible to build this facility? The MARR is 10% per year. (6.4)

(a) 7   (b) 4   (c) 5
(d) 3   (e) 8

6-77. Complete the following analysis of investment alternatives and select the preferred alternative. The study period is three years and the MARR = 15% per year. (6.4)

	Alternative A	Alternative B	Alternative C
Capital investment	$11,000	$16,000	$13,000
Annual revenues	4,000	6,000	5,540
Annual costs	250	500	400
Market value at EOY 3	5,000	6,150	2,800
PW(15%)	850	???	577

(a) Do Nothing   (b) Alternative A
(c) Alternative B   (d) Alternative C

6-78. Complete the following analysis of cost alternatives and select the preferred alternative. The study period is 10 years and the MARR = 12% per year. "Do Nothing" is not an option. (6.4)

	A	B	C	D
Capital investment	$15,000	$16,000	$13,000	$18,000
Annual costs	250	300	500	100
Market value at EOY 10	1,000	1,300	1,750	2,000
FW(12%)	−$49,975	−$53,658	???	−$55,660

(a) Alternative A   (b) Alternative B
(c) Alternative C   (d) Alternative D

6-79. The Ford Motor Company is considering three mutually exclusive electronic stability control systems for protection against rollover of its automobiles. The investment period is four years (equal lives), and the MARR is 12% per year. Data for fixturing costs of the systems are given below.

Alternative	IRR	Capital Investment	Annual Receipts Less Expenses	Salvage Value
A	19.2%	$12,000	$4,000	$3,000
B	18%	$15,800	$5,200	$3,500
C	23%	$8,000	$3,000	$1,500

Which alternative should the company select? (6.4)

(a) Alternative A   (b) Alternative B
(c) Alternative C   (d) Do nothing

6-80. For the following table, assume a MARR of 10% per year and a useful life for each alternative of six years that equals the study period. The rank-order of alternatives from least capital investment to greatest capital investment is Do Nothing → A → C → B.

Complete the IRR analysis by selecting the preferred alternative. (6.4)

	Do Nothing → A	A → C	C → B
Δ Capital investment	−$15,000	−$2,000	−$3,000
Δ Annual revenues	4,000	900	460
Δ Annual costs	−1,000	−150	100
Δ Market value	6,000	−2,220	3,350
Δ IRR	12.7%	10.9%	???

(a) Do nothing (b) Alternative A
(c) Alternative B (d) Alternative C

6-81. For the following table, assume a MARR of 15% per year and a useful life for each alternative of eight years which equals the study period. The rank-order of alternatives from least capital investment to greatest capital investment is Z → Y → W → X. Complete the incremental analysis by selecting the preferred alternative. "Do nothing" is not an option. (6.4)

	Z → Y	Y → W	W → X
Δ Capital investment	−$250	−$400	−$550
Δ Annual cost savings	70	90	15
Δ Market value	100	50	200
Δ PW(15%)	97	20	???

(a) Alternative W (b) Alternative X
(c) Alternative Y (d) Alternative Z

The following mutually exclusive investment alternatives have been presented to you. (See Table P6-82.) The life of all alternatives is 10 years. Use this information to solve problems 6-82 through 6-85. (6.4)

6-82. After the base alternative has been identified, the first comparison to be made in an incremental analysis should be which of the following?
(a) C → B (b) A → B (c) D → E
(d) C → D (e) D → C

6-83. Using a MARR of 15%, the PW of the investment in A when compared *incrementally* to C is most nearly:
(a) −$69,000 (b) −$21,000 (c) $20,000
(d) $61,000 (e) $53,000

6-84. The IRR for Alternative A is most nearly:
(a) 30% (b) 15% (c) 36%
(d) 10% (e) 20%

6-85. Using a MARR of 15%, the preferred Alternative is:
(a) Do nothing (b) Alt. A (c) Alt. B
(d) Alt. C (e) Alt. D (f) Alt. E

6-86. Consider the mutually exclusive alternatives given in the table below. MARR is 10% per year.

	Alternative		
	X	Y	Z
Capital investment (thousands)	$500,000	$250,000	$400,000
Uniform annual savings (thousands)	$131,900	$40,690	$44,050
Useful life (years)	5	10	20

Assuming repeatability, which alternative should the company select? (6.5)
(a) Alternative X (b) Alternative Y
(c) Alternative Z (d) Do nothing

TABLE P6-82	Data for Problems 6-82 through 6-85				
	A	B	C	D	E
Capital investment	$60,000	$90,000	$40,000	$30,000	$70,000
Annual expenses	30,000	40,000	25,000	15,000	35,000
Annual revenues	50,000	52,000	38,000	28,000	45,000
Market value at EOY 10	10,000	15,000	10,000	10,000	15,000
IRR	???	7.4%	30.8%	42.5%	9.2%

CHAPTER 7

Depreciation and Income Taxes

The objective of Chapter 7 is to explain how depreciation affects income taxes and how income taxes affect economic decision making.

The After-Tax Cost of Producing Electricity

T he PennCo Electric Utility Company is going to construct an 800 megawatt (MW) coal-fired plant that will have a 30-year operating life. The $1.12-billion construction cost will be depreciated with the straight-line method over 30 years to a terminal book value of zero. Its market value will also be negligible after 30 years. This plant's efficiency will be 35%, its capacity factor is estimated to be 80%, and annual operating and maintenance expenses are expected to be $0.02 per kilowatt hour (kWh). In addition, the cost of coal is estimated to average $3.50 per million Btu, and the tax on carbon dioxide emissions will be $15 per metric ton. Burning coal emits 90 metric tons of CO_2 per billion Btu produced. If the effective income tax rate for PennCo is 40%, what is the after-tax cost of electricity per year for this plant? The firm's after-tax minimum attractive rate of return (MARR) is 10% per year. After studying this chapter, you will be able to include the impact of income taxes on engineering projects and answer this question (see Example 7-14).

. . . Nothing in this world is certain but death and taxes.

—Benjamin Franklin (1789)

7.1 Introduction

Taxes have been collected since the dawn of civilization. Interestingly, in the United States, a federal income tax did not exist until March 13, 1913, when Congress enacted the Sixteenth Amendment to the Constitution.* Most organizations consider the effect of income taxes on the financial results of a proposed engineering project because income taxes usually represent a significant cash outflow that cannot be ignored in decision making. Based on the previous chapters, we can now describe how income tax liabilities (or credits) and after-tax cash flows are determined in engineering practice. In this chapter, an *After-Tax Cash Flow (ATCF) procedure* will be used in capital investment analysis because it avoids innumerable problems associated with measuring corporate net income. This theoretically sound procedure provides a quick and relatively easy way to define a project's profitability.†

Because the amount of material included in the Internal Revenue Code (and in state and municipal laws, where they exist) is very extensive, only selected parts of the subject can be discussed within the scope of this textbook. Our focus in this chapter is *federal corporate income taxes* and their general effect on the financial results of proposed engineering projects. The material presented is for educational purposes. In practice, you should seek expert counsel when analyzing a specific project.

Due to the effect of *depreciation* on the ATCFs of a project, this topic is discussed first. The selected material on depreciation will then be used in the remainder of the chapter for accomplishing after-tax analyses of engineering projects.

7.2 Depreciation Concepts and Terminology

Depreciation is the decrease in value of physical properties with the passage of time and use. More specifically, depreciation is an *accounting concept* that establishes an annual deduction against before-tax income such that the effect of time and use on an asset's value can be reflected in a firm's financial statements.

---

* During the Civil War, a federal income tax rate of 3% was initially imposed in 1862 by the Commissioner of Internal Revenue to help pay for war expenditures. The federal rate was later raised to 10%, but eventually eliminated in 1872.
† ATCF is used throughout this book as a measure of the after-tax profitability of an investment project. Other measures, however, are employed for this purpose by management accountants (net income after taxes) and finance professionals (free cash flow). For additional information, refer to Horngren, Sundem and Stratton (*Management Accounting*) and Bodie and Merton (*Finance*) in Appendix F.

Depreciation is a noncash cost that is intended to "match" the yearly fraction of value used by an asset in the production of income over the asset's life. The actual amount of depreciation can never be established until the asset is retired from service. Because depreciation is a *noncash cost* that affects income taxes, we must consider it properly when making after-tax engineering economy studies.

Depreciable property is property for which depreciation is allowed under federal, state, or municipal income tax laws and regulations. To determine whether depreciation deductions can be taken, the classification of various types of property must be understood. In general, property is depreciable if it meets the following basic requirements:

1. It must be used in business or held to produce income.
2. It must have a determinable useful life (defined in Section 7.2.2), and the life must be longer than one year.
3. It must be something that wears out, decays, gets used up, becomes obsolete, or loses value from natural causes.
4. It is not inventory, stock in trade, or investment property.

Depreciable property is classified as either *tangible* or *intangible*. Tangible property can be seen or touched, and it includes two main types called *personal property* and *real property*. Personal property includes assets such as machinery, vehicles, equipment, furniture, and similar items. In contrast, real property is land and generally anything that is erected on, growing on, or attached to land. Land itself, however, is not depreciable, because it does not have a determinable life.

Intangible property is personal property such as a copyright, patent, or franchise. We will not discuss the depreciation of intangible assets in this chapter because engineering projects rarely include this class of property.

A company can begin to depreciate property it owns when the property is *placed in service* for use in the business and for the production of income. Property is considered to be placed in service when it is ready and available for a specific use, even if it is not actually used yet. Depreciation stops when the cost of placing an asset in service has been recovered or when the asset is sold, whichever occurs first.

7.2.1 Depreciation Methods and Related Time Periods

The depreciation methods permitted under the Internal Revenue Code have changed with time. In general, the following summary indicates the *primary methods used for property placed in service during three distinct time periods:*

Before 1981 Several methods could be elected for depreciating property placed in service before 1981. The primary methods used were Straight-Line (SL), Declining-Balance (DB), and Sum of the Years Digits (SYD). We will refer to these methods, collectively, as the *classical*, or *historical, methods* of depreciation.

After 1980 and before 1987 For federal income taxes, tangible property placed in service during this period must be depreciated by using the Accelerated Cost Recovery System (ACRS). This system was implemented by the Economic Recovery Tax Act of 1981 (ERTA).

After 1986 The Tax Reform Act of 1986 (TRA 86) was one of the most extensive income tax reforms in the history of the United States. This act modified the previous ACRS implemented under ERTA and requires the use of the Modified Accelerated Cost Recovery System (MACRS) for the depreciation of tangible property placed in service after 1986.

A description of the selected (historical) methods of depreciation is included in the chapter for several important reasons. They apply directly to property placed in service prior to 1981, as well as for property, such as intangible property (which requires the SL method), that does not qualify for ACRS or MACRS. Also, these methods are often specified by the tax laws and regulations of state and municipal governments in the United States and are used for depreciation purposes in other countries. In addition, as we shall see in Section 7.4, the DB and the SL methods are used in determining the annual recovery rates under MACRS.

We do not discuss the application of ACRS in the chapter, but readily available Internal Revenue Service (IRS) publications describe its use.* Selected parts of the MACRS, however, are described and illustrated because this system applies to depreciable property in present and future engineering projects.

7.2.2   Additional Definitions

Because this chapter uses many terms that are not generally included in the vocabulary of engineering education and practice, an abbreviated set of definitions is presented here. The following list is intended to supplement the previous definitions provided in this section:

Adjusted (cost) basis The original cost basis of the asset, adjusted by allowable increases or decreases, is used to compute depreciation deductions. For example, the cost of any improvement to a capital asset with a useful life *greater* than one year increases the original cost basis, and a casualty or theft loss decreases it. If the basis is altered, the depreciation deduction may need to be adjusted.

Basis or cost basis The initial cost of acquiring an asset (purchase price plus any sales taxes), including transportation expenses and other normal costs of making the asset serviceable for its intended use; this amount is also called the *unadjusted cost basis*.

Book value (BV) The worth of a depreciable property as shown on the accounting records of a company. It is the original cost basis of the property, including any adjustments, less all allowable depreciation deductions. It thus represents the amount of capital that remains invested in the property and must be recovered in the future through the accounting process. The BV of a property may not be an accurate measure of its market value. In general, the BV of a property at the end of year k is

$$(\text{Book value})_k = \text{adjusted cost basis} - \sum_{j=1}^{k} (\text{depreciation deduction})_j. \quad (7\text{-}1)$$

---

* Useful references on material in this chapter, available from the Internal Revenue Service in an annually updated version, are Publication 534 (*Depreciation*), Publication 334 (*Tax Guide for Small Business*), Publication 542 (*Tax Information on Corporations*), and Publication 544 (*Sales and Other Dispositions of Assets*).

Market value (MV) The amount that will be paid by a willing buyer to a willing seller for a property, where each has equal advantage and is under no compulsion to buy or sell. The MV approximates the present value of what will be received through ownership of the property, including the time value of money (or profit).

Recovery period The number of years over which the basis of a property is recovered through the accounting process. For the classical methods of depreciation, this period is normally the useful life. Under MACRS, this period is the *property class* for the General Depreciation System (GDS), and it is the *class life* for the Alternative Depreciation System (ADS) (see Section 7.4).

Recovery rate A percentage (expressed in decimal form) for each year of the MACRS recovery period that is utilized to compute an annual depreciation deduction.

Salvage value (SV) The estimated value of a property at the end of its useful life.* It is the expected selling price of a property when the asset can no longer be used productively by its owner. The term *net salvage value* is used when the owner will incur expenses in disposing of the property, and these cash outflows must be deducted from the cash inflows to obtain a final net SV. Under MACRS, the SV of a depreciable property is defined to be zero.

Useful life The expected (estimated) period that a property will be used in a trade or business to produce income. It is not how long the property will last but how long the owner expects to productively use it.

7.3   The Classical (Historical) Depreciation Methods

This section describes and illustrates the SL and DB methods of calculating depreciation deductions. As mentioned in Section 7.2, these historical methods continue to apply, directly and indirectly, to the depreciation of property. Also included is a discussion of the units of production method.

7.3.1   Straight-Line (SL) Method

SL depreciation is the simplest depreciation method. It assumes that a constant amount is depreciated each year over the depreciable (useful) life of the asset. If we define

$$N = \text{depreciable life of the asset in years;}$$
$$B = \text{cost basis, including allowable adjustments;}$$
$$d_k = \text{annual depreciation deduction in year } k (1 \leq k \leq N);$$
$$\text{BV}_k = \text{book value at end of year } k;$$
$$\text{SV}_N = \text{estimated salvage value at end of year } N; \text{ and}$$
$$d_k^* = \text{cumulative depreciation through year } k,$$

---

* We often use the term market value (MV) in place of salvage value (SV).

$$\text{then} \quad d_k = (B - SV_N)/N, \tag{7-2}$$

$$d_k^* = k \cdot d_k \text{ for } 1 \le k \le N, \tag{7-3}$$

$$BV_k = B - d_k^*. \tag{7-4}$$

Note that, for this method, you must have an estimate of the final SV, which will also be the final BV at the end of year N. In some cases, the estimated SV_N may not equal an asset's actual terminal MV.

EXAMPLE 7-1   **SL Depreciation**

A laser surgical tool has a cost basis of $200,000 and a five-year depreciable life. The estimated SV of the laser is $20,000 at the end of five years. Determine the annual depreciation amounts using the SL method. Tabulate the annual depreciation amounts and the book value of the laser at the end of each year.

Solution

The depreciation amount, cumulative depreciation, and BV for each year are obtained by applying Equations (7-2), (7-3), and (7-4). Sample calculations for year three are as follows:

$$d_3 = \frac{\$200,000 - \$20,000}{5} = \$36,000$$

$$d_3^* = 3 \left(\frac{\$200,000 - \$20,000}{5} \right) = \$108,000$$

$$BV_3 = \$200,000 - \$108,000 = \$92,000$$

The depreciation and BV amounts for each year are shown in the following table.

EOY, k	d_k	BV_k
0	—	$200,000
1	$36,000	$164,000
2	$36,000	$128,000
3	$36,000	$92,000
4	$36,000	$56,000
5	$36,000	$20,000

Note that the BV at the end of the depreciable life is equal to the SV used to calculate the yearly depreciation amount.

7.3.2 Declining-Balance (DB) Method

In the DB method, sometimes called the *constant-percentage method* or the *Matheson formula*, it is assumed that the annual cost of depreciation is a fixed percentage of the BV at the *beginning* of the year. The ratio of the depreciation in any one year to the BV at the beginning of the year is constant throughout the life of the asset and is designated by R ($0 \leq R \leq 1$). In this method, $R = 2/N$ when a 200% DB is being used (i.e., twice the SL rate of $1/N$), and N equals the depreciable (useful) life of an asset. If the 150% DB method is specified, then $R = 1.5/N$. The following relationships hold true for the DB method:

$$d_1 = B(R), \tag{7-5}$$

$$d_k = B(1 - R)^{k-1}(R), \tag{7-6}$$

$$d_k^* = B[1 - (1 - R)^k], \tag{7-7}$$

$$BV_k = B(1 - R)^k. \tag{7-8}$$

Notice that Equations (7-5) through (7-8) do not contain a term for SV_N.

EXAMPLE 7-2 **DB Depreciation**

A new electric saw for cutting small pieces of lumber in a furniture manufacturing plant has a cost basis of $4,000 and a 10-year depreciable life. The estimated SV of the saw is zero at the end of 10 years. Use the DB method to calculate the annual depreciation amounts when

(a) $R = 2/N$ (200% DB method)
(b) $R = 1.5/N$ (150% DB method).

Tabulate the annual depreciation amount and BV for each year.

Solution

Annual depreciation, cumulative depreciation, and BV are determined by using Equations (7-6), (7-7), and (7-8), respectively. Sample calculations for year six are as follows:

(a)

$$R = 2/10 = 0.2,$$

$$d_6 = \$4,000(1 - 0.2)^5(0.2) = \$262.14,$$

$$d_6^* = \$4,000[1 - (1 - 0.2)^6] = \$2,951.42,$$

$$BV_6 = \$4,000(1 - 0.2)^6 = \$1,048.58.$$

(b)

$$R = 1.5/10 = 0.15,$$

$$d_6 = \$4,000(1 - 0.15)^5(0.15) = \$266.22,$$

$$d_6^* = \$4{,}000[1 - (1 - 0.15)^6] = \$2{,}491.40,$$

$$BV_6 = \$4{,}000(1 - 0.15)^6 = \$1{,}508.60.$$

The depreciation and BV amounts for each year, when $R = 2/N = 0.2$, are shown in the following table:

200% DB Method Only		
EOY, k	d_k	BV_k
0	—	$4,000
1	$800	3,200
2	640	2,560
3	512	2,048
4	409.60	1,638.40
5	327.68	1,310.72
6	262.14	1,048.58
7	209.72	838.86
8	167.77	671.09
9	134.22	536.87
10	107.37	429.50

7.3.3   DB with Switchover to SL

Because the DB method never reaches a BV of zero, it is permissible to switch from this method to the SL method so that an asset's BV_N will be zero (or some other determined amount, such as SV_N). Also, this method is used in calculating the MACRS recovery rates (shown later in Table 7-3).

Table 7-1 illustrates a switchover from double DB depreciation to SL depreciation for Example 7-2. The switchover occurs in the year in which a larger depreciation amount is obtained from the SL method. From Table 7-1, it is apparent that $d_6 = \$262.14$. The BV at the end of year six (BV_6) is $1,048.58. Additionally, observe that BV_{10} is $4,000 − $3,570.50 = $429.50 without switchover to the SL method in Table 7-1. With switchover, BV_{10} equals zero. It is clear that this asset's d_k, d_k^*, and BV_k in years 7 through 10 are established from the SL method, which permits the full cost basis to be depreciated over the 10-year recovery period.

7.3.4   Units-of-Production Method

All the depreciation methods discussed to this point are based on elapsed time (years) on the theory that the decrease in value of property is mainly a function of time. When the decrease in value is mostly a function of use, depreciation may be based on a method not expressed in terms of years. The units-of-production method is normally used in this case.

TABLE 7-1   The 200% DB Method with Switchover to the SL Method (Example 7-2)

		Depreciation Method		
	(1)	(2)	(3)	(4)
Year, k	Beginning-of-Year BV^a	200% DB Method^b	SL Method^c	Depreciation Amount Selected^d
1	$4,000.00	$800.00	>$400.00	$800.00
2	3,200.00	640.00	>355.56	640.00
3	2,560.00	512.00	>320.00	512.00
4	2,048.00	409.60	>292.57	409.60
5	1,638.40	327.68	>273.07	327.68
6	1,310.72	262.14	=262.14	262.14 (switch)
7	1,048.58	209.72	<262.14	262.14
8	786.44	167.77	<262.14	262.14
9	524.30	134.22	<262.14	262.14
10	262.16	107.37	<262.14	262.14
		$3,570.50		$4,000.00

^{a} Column 1 for year k less column 4 for year k equals the entry in column 1 for year $k+1$.
^{b} 200%(= 2/10) of column 1.
^{c} Column 1 minus estimated SV_N divided by the remaining years from the beginning of the year through the 10th year.
^{d} Select the larger amount in column 2 or column 3.

This method results in the cost basis (minus final SV) being allocated equally over the estimated number of units produced during the useful life of the asset. The depreciation rate is calculated as

$$\frac{\text{Depreciation}}{\text{per unit of production}} = \frac{B - SV_N}{(\text{Estimated lifetime production units})}. \quad (7\text{-}9)$$

EXAMPLE 7-3   Depreciation Based on Activity

A piece of equipment used in a business has a basis of $50,000 and is expected to have a $10,000 SV when replaced after 30,000 hours of use. Find its depreciation rate per hour of use, and find its BV after 10,000 hours of operation.

Solution

$$\text{Depreciation per unit of production} = \frac{\$50,000 - \$10,000}{30,000 \text{ hours}} = \$1.33 \text{ per hour.}$$

After 10,000 hours, BV $= \$50,000 - \dfrac{\$1.33}{\text{hour}}(10,000 \text{ hours})$, or BV $= \$36,700.$

7.4 The Modified Accelerated Cost Recovery System

As we discussed in Section 7.2.1, the MACRS was created by TRA 86 and is now the principal method for computing depreciation deductions for property in engineering projects. MACRS applies to most tangible depreciable property placed in service after December 31, 1986. Examples of assets that cannot be depreciated under MACRS are property you elect to exclude because it is to be depreciated under a method that is not based on a term of years (units-of-production method) and intangible property. Previous depreciation methods have required estimates of useful life (N) and SV at the end of useful life (SV_N). Under MACRS, however, SV_N is defined to be zero, and useful life estimates are not used directly in calculating depreciation amounts.

> MACRS consists of two systems for computing depreciation deductions. The main system is called the *General Depreciation System (GDS)*, and the second system is called the *Alternative Depreciation System (ADS)*.

In general, ADS provides a longer recovery period and uses only the SL method of depreciation. Property that is placed in any tax-exempt use and property used predominantly outside the United States are examples of assets that must be depreciated under ADS. Any property that qualifies under GDS, however, can be depreciated under ADS, if elected.

When an asset is depreciated under MACRS, the following information is needed before depreciation deductions can be calculated:

1. The cost basis (B)
2. The date the property was placed in service
3. The property class and recovery period
4. The MACRS depreciation method to be used (GDS or ADS)
5. The time convention that applies (half year)

The first two items were discussed in Section 7.2. Items 3 through 5 are discussed in the sections that follow.

7.4.1 Property Class and Recovery Period

Under MACRS, tangible depreciable property is categorized (organized) into asset classes. The property in each asset class is then assigned *a class life, GDS recovery period (and property class), and ADS recovery period*. For our use, a partial listing of depreciable assets used in business is provided in Table 7-2. The types of depreciable property grouped together are identified in the second column. Then the class life, GDS recovery period (and property class), and ADS recovery period (all in years) for these assets are listed in the remaining three columns.

Under the GDS, the basic information about property classes and recovery periods is as follows:

1. Most tangible personal property is assigned to one of six *personal property classes* (3-, 5-, 7-, 10-, 15-, and 20-year property). The personal property class (in years) is the same as the *GDS recovery period*. Any depreciable personal property that

TABLE 7-2 MACRS Class Lives and Recovery Periods[a]

Asset Class	Description of Assets	Class Life	Recovery Period GDS[b]	Recovery Period ADS
00.11	Office furniture and equipment	10	7	10
00.12	Information systems, including computers	6	5	5
00.22	Automobiles, taxis	3	5	5
00.23	Buses	9	5	9
00.241	Light general purpose trucks	4	5	5
00.242	Heavy general purpose trucks	6	5	6
00.26	Tractor units for use over the road	4	3	4
01.1	Agriculture	10	7	10
10.0	Mining	10	7	10
13.2	Production of petroleum and natural gas	14	7	14
13.3	Petroleum refining	16	10	16
15.0	Construction	6	5	6
22.3	Manufacture of carpets	9	5	9
24.4	Manufacture of wood products and furniture	10	7	10
28.0	Manufacture of chemicals and allied products	9.5	5	9.5
30.1	Manufacture of rubber products	14	7	14
32.2	Manufacture of cement	20	15	20
34.0	Manufacture of fabricated metal products	12	7	12
36.0	Manufacture of electronic components, products, and systems	6	5	6
37.11	Manufacture of motor vehicles	12	7	12
37.2	Manufacture of aerospace products	10	7	10
48.12	Telephone central office equipment	18	10	18
49.13	Electric utility steam production plant	28	20	28
49.21	Gas utility distribution facilities	35	20	35
79.0	Recreation	10	7	10

[a] Partial listing abstracted from *How to Depreciate Property*, IRS Publication 946, Tables B-1 and B-2, 2006.
[b] Also the *GDS property class*.

does not "fit" into one of the defined asset classes is depreciated as being in the seven-year property class.

2. Real property is assigned to two *real property classes*: nonresidential real property and residential rental property.

3. The *GDS recovery period* is 39 years for nonresidential real property (31.5 years if placed in service before May 13, 1993) and 27.5 years for residential real property.

The following is a summary of basic information for the ADS:

1. For tangible personal property, the *ADS recovery period* is shown in the last column on the right of Table 7-2 (and is normally the same as the class life of the property; there are exceptions such as those shown in asset classes 00.12 and 00.22).

2. Any tangible personal property that does not fit into one of the asset classes is depreciated under a 12-year ADS recovery period.

3. The *ADS recovery period* for nonresidential real property is 40 years.

The use of these rules under the MACRS is discussed further in the next section.

7.4.2 Depreciation Methods, Time Convention, and Recovery Rates

The primary methods used under MACRS for calculating the depreciation deductions over the recovery period of an asset are summarized as follows:

1. GDS 3-, 5-, 7-, and 10-year personal property classes: The 200% DB method, which switches to the SL method when that method provides a greater deduction. The DB method with switchover to SL was illustrated in Section 7.3.3.

2. GDS 15- and 20-year personal property classes: The 150% DB method, which switches to the SL method when that method provides a greater deduction.

3. GDS nonresidential real and residential rental property classes: The SL method over the fixed GDS recovery periods.

4. ADS: The SL method for both personal and real property over the fixed ADS recovery periods.

A *half-year time convention* is used in MACRS depreciation calculations for tangible personal property. This means that all assets placed in service during the year are treated as if use began in the middle of the year, and one-half year of depreciation is allowed. When an asset is disposed of, the half-year convention is also allowed. *If the asset is disposed of before the full recovery period is used, then only half of the normal depreciation deduction can be taken for that year.*

The GDS *recovery rates* (r_k) for the six personal property classes that we will use in our depreciation calculations are listed in Table 7-3. The GDS personal property rates in Table 7-3 include the half-year convention, as well as switchover from the DB method to the SL method when that method provides a greater deduction. Note that, if an asset is disposed of in year $N + 1$, the final BV of the asset will be zero. Furthermore, there are $N + 1$ recovery rates shown for each GDS property class for a property class of N years.

The depreciation deduction (d_k) for an asset under MACRS (GDS) is computed with

$$d_k = r_k \cdot B; \ 1 \leq k \leq N + 1, \tag{7-10}$$

where r_k = recovery rate for year k from Table 7-3.

The information in Table 7-4 provides a summary of the principal features of GDS under MACRS. Included are some selected special rules about depreciable assets. A flow diagram for computing depreciation deductions under MACRS is shown in Figure 7-1. As indicated in the figure, an important choice is whether the main GDS is to be used for an asset or whether ADS is elected instead (or required). Normally, the choice would be to use the GDS for calculating the depreciation deductions.

TABLE 7-3 **GDS Recovery Rates (r_k) for the Six Personal Property Classes**

Year	Recovery Period (and Property Class)					
	3-year[a]	5-year[a]	7-year[a]	10-year[a]	15-year[b]	20-year[b]
1	0.3333	0.2000	0.1429	0.1000	0.0500	0.0375
2	0.4445	0.3200	0.2449	0.1800	0.0950	0.0722
3	0.1481	0.1920	0.1749	0.1440	0.0855	0.0668
4	0.0741	0.1152	0.1249	0.1152	0.0770	0.0618
5		0.1152	0.0893	0.0922	0.0693	0.0571
6		0.0576	0.0892	0.0737	0.0623	0.0528
7			0.0893	0.0655	0.0590	0.0489
8			0.0446	0.0655	0.0590	0.0452
9				0.0656	0.0591	0.0447
10				0.0655	0.0590	0.0447
11				0.0328	0.0591	0.0446
12					0.0590	0.0446
13					0.0591	0.0446
14					0.0590	0.0446
15					0.0591	0.0446
16					0.0295	0.0446
17						0.0446
18						0.0446
19						0.0446
20						0.0446
21						0.0223

Source: IRS Publication 534. *Depreciation*. Washington, D.C.: U.S. Government Printing Office, for 2006 tax returns.

[a]These rates are determined by applying the 200% DB method (with switchover to the SL method) to the recovery period with the half-year convention applied to the first and last years. Rates for each period must sum to 1.0000.

[b]These rates are determined with the 150% DB method instead of the 200% DB method (with switchover to the SL method) and are rounded off to four decimal places.

EXAMPLE 7-4 | **MACRS Depreciation with GDS**

A firm purchased and placed in service a new piece of semiconductor manufacturing equipment. The cost basis for the equipment is $100,000. Determine

(a) the depreciation charge permissible in the fourth year,

(b) the BV at the end of the fourth year,

(c) the cumulative depreciation through the third year,

(d) the BV at the end of the fifth year if the equipment is disposed of at that time.

TABLE 7-4	MACRS (GDS) Property Classes and Primary Methods for Calculating Depreciation Deductions	
GDS Property Class and Depreciation Method	**Class Life (Useful Life)**	**Special Rules**
3-year, 200% DB with switchover to SL	Four years or less	Includes some race horses, and tractor units for over-the-road use.
5-year, 200% DB with switchover to SL	More than 4 years to less than 10	Includes cars and light trucks, semiconductor manufacturing equipment, qualified technological equipment, computer-based central office switching equipment, some renewable and biomass power facilities, and research and development property.
7-year, 200% DB with switchover to SL	10 years to less than 16	Includes single-purpose agricultural and horticultural structures and railroad track. Includes office furniture and fixtures, and property not assigned to a property class.
10-year, 200% DB with switchover to SL	16 years to less than 20	Includes vessels, barges, tugs and similar water transportation equipment.
15-year, 150% DB with switchover to SL	20 years to less than 25	Includes sewage treatment plants, telephone distribution plants, and equipment for two-way voice and data communication.
20-year, 150% DB with switchover to SL	25 years or more	Excludes real property of 27.5 years or more. Includes municipal sewers.
27.5 year, SL	N/A	Residential rental property.
39-year, SL	N/A	Nonresidential real property.

Source: Arthur Andersen & Company, *Tax Reform 1986: Analysis and Planning*, Chicago, 1986, p. 112. Reproduced with permission of Arthur Andersen & Co.

Solution

From Table 7-2, it may be seen that the semiconductor (electronic) manufacturing equipment has a class life of six years and a GDS recovery period of five years. The recovery rates that apply are given in Table 7-3.

(a) The depreciation deduction, or cost-recovery allowance, that is allowable in year four (d_4) is 0.1152 ($100,000) = $11,520.

Figure 7-1
Flow Diagram for
Computing
Depreciation
Deductions under
MACRS

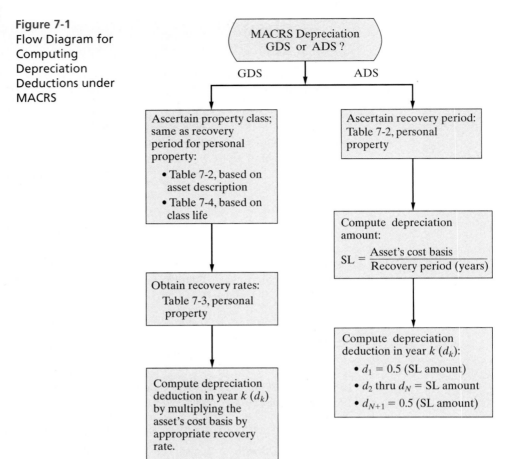

(b) The BV at the end of year four (BV_4) is the cost basis less depreciation charges in years one through four:

$$BV_4 = \$100{,}000 - \$100{,}000(0.20 + 0.32 + 0.192 + 0.1152)$$

$$= \$17{,}280.$$

(c) Accumulated depreciation through year three, d_3^*, is the sum of depreciation amounts in years one through three:

$$d_3^* = d_1 + d_2 + d_3$$

$$= \$100{,}000(0.20 + 0.32 + 0.192)$$

$$= \$71{,}200.$$

(d) The depreciation deduction in year five can only be $(0.5)(0.1152)(\$100{,}000) = \$5{,}760$ when the equipment is disposed of prior to year six. Thus, the BV at the end of year five is $BV_4 - \$5{,}760 = \$11{,}520$.

| EXAMPLE 7-5 | **MACRS with a Trade-in** |

In May 2005, your company traded in a computer and peripheral equipment, used in its business, that had a BV at that time of $25,000. A new, faster computer system having a fair MV of $400,000 was acquired. Because the vendor accepted the older computer as a trade-in, a deal was agreed to whereby your company would pay $325,000 cash for the new computer system.

(a) What is the GDS property class of the new computer system?

(b) How much depreciation can be deducted each year based on this class life? (Refer to Figure 7-1.)

Solution

(a) The new computer is in Asset Class 00.12 and has a class life of six years. (See Table 7-2.) Hence, its GDS property class and recovery period are five years.

(b) The cost basis for this property is $350,000, which is the sum of the $325,000 cash price of the computer and the $25,000 BV remaining on the trade-in. (In this case, the trade-in was treated as a nontaxable transaction.)

MACRS (GDS) rates that apply to the $350,000 cost basis are found in Table 7-3. An allowance (half-year) is built into the year-one rate, so it does not matter that the computer was purchased in May 2005 instead of, say, November 2005. The depreciation deductions (d_k) for 2005 through 2010 are computed using Equation (7-10) and Table 7-3.

Property	Date Placed in Service	Cost Basis	Class Life	MACRS (GDS) Recovery Period
Computer System	May 2005	$350,000	6 years	5 years

Year	Depreciation Deductions
2005	$0.20 \times \$350,000 = \$70,000$
2006	$0.32 \times 350,000 = 112,000$
2007	$0.192 \times 350,000 = 67,200$
2008	$0.1152 \times 350,000 = 40,320$
2009	$0.1152 \times 350,000 = 40,320$
2010	$0.0576 \times 350,000 = 20,160$
	Total $350,000

From Example 7-5, we can conclude that Equation 7-11 is true from the buyer's viewpoint when property of the same type and class is exchanged:

$$\text{Basis} = \text{Actual cash cost} + \text{BV of the trade-in.} \qquad (7\text{-}11)$$

To illustrate Equation (7-11), suppose your company has operated an optical character recognition (OCR) scanner for two years. Its BV now is $35,000, and its fair MV is $45,000. The company is considering a new OCR scanner that costs $105,000. Ordinarily, you would trade the old scanner for the new one and pay the dealer $60,000. The basis ($B$) for depreciation is then $60,000 + $35,000 = $95,000.*

EXAMPLE 7-6	**MACRS Depreciation with ADS**

A large manufacturer of sheet metal products in the Midwest purchased and placed in service a new, modern, computer-controlled flexible manufacturing system for $3.0 million. Because this company would not be profitable until the new technology had been in place for several years, it elected to utilize the ADS under MACRS in computing its depreciation deductions. Thus, the company could slow down its depreciation allowances in hopes of postponing its income tax advantages until it became a profitable concern. What depreciation deductions can be claimed for the new system?

Solution

From Table 7-2, the ADS recovery period for a manufacturer of fabricated metal products is 12 years. Under ADS, the SL method with no SV is applied to the 12-year recovery period, using the half-year convention. Consequently, the depreciation in year one would be

$$\frac{1}{2} \left(\frac{\$3,000,000}{12} \right) = \$125,000.$$

Depreciation deductions in years 2 through 12 would be $250,000 each year, and depreciation in year 13 would be $125,000. Notice how the half-year convention extends depreciation deductions over 13 years ($N + 1$).

7.5 A Comprehensive Depreciation Example

We now consider an asset for which depreciation is computed by the historical and MACRS (GDS) methods previously discussed. Be careful to observe the differences in the mechanics of each method, as well as the differences in the annual depreciation amounts themselves. Also, we compare the present worths (PWs) at $k = 0$ of selected depreciation methods when MARR = 10% per year. As we shall see later in this chapter, depreciation methods that result in larger PWs (of the depreciation amounts) are preferred by a firm that wants to reduce the present worth of its *income taxes paid* to the government.

---

* The *exchange price* of the OCR scanner is $105,000 − $45,000(= actual cash cost). Equation (7-11) prevents exaggerated cost bases to be claimed for new assets having large "sticker prices" compared with their exchange price.

| EXAMPLE 7-7 | **Comparison of Depreciation Methods** |

The La Salle Bus Company has decided to purchase a new bus for $85,000 with a trade-in of their old bus. The old bus has a BV of $10,000 at the time of the trade-in. The new bus will be kept for 10 years before being sold. Its estimated SV at that time is expected to be $5,000.

First, we must calculate the cost basis. The basis is the original purchase price of the bus plus the BV of the old bus that was traded in [Equation (7-11)]. Thus, the basis is $85,000 + $10,000, or $95,000. We need to look at Table 7-2 and find buses, which are asset class 00.23. Hence, we find that buses have a nine-year class (useful) life, over which we depreciate the bus with historical methods discussed in Section 7.3, and a five-year GDS recovery period.

Solution: SL Method

For the SL method, we use the class life of 9 years, even though the bus will be kept for 10 years. By using Equations (7-2) and (7-4), we obtain the following information:

$$d_k = \frac{\$95,000 - \$5,000}{9 \text{ years}} = \$10,000, \quad \text{for } k = 1 \text{ to } 9.$$

| | SL Method | |
EOY k	d_k	BV_k
0	—	$95,000
1	$10,000	85,000
2	10,000	75,000
3	10,000	65,000
4	10,000	55,000
5	10,000	45,000
6	10,000	35,000
7	10,000	25,000
8	10,000	15,000
9	10,000	5,000

Notice that no depreciation was taken after year nine because the class life was only nine years. Also, the final BV was the estimated SV, and the BV will remain at $5,000 until the bus is sold.

Solution: DB Method

To demonstrate this method, we will use the 200% DB equations. With Equations (7-6) and (7-8), we calculate the following:

$$R = 2/9 = 0.2222;$$

$$d_1 = \$95,000(0.2222) = \$21,111;$$

$$d_5 = \$95{,}000(1 - 0.2222)^4(0.2222) = \$7{,}726;$$
$$BV_5 = \$95{,}000(1 - 0.2222)^5 = \$27{,}040.$$

	200% DB Method	
EOY k	d_k	BV_k
0	—	$95,000
1	$21,111	73,889
2	16,420	57,469
3	12,771	44,698
4	9,932	34,765
5	7,726	27,040
6	6,009	21,031
7	4,674	16,357
8	3,635	12,722
9	2,827	9,895

Solution: DB with Switchover to SL Depreciation

To illustrate the mechanics of Table 7-1 for this example, we first specify that the bus will be depreciated by the 200% DB method ($R = 2/N$). Because DB methods never reach a zero BV, suppose that we further specify that a switchover to SL depreciation will be made to ensure a BV of $5,000 at the end of the vehicle's nine-year class life.

EOY k	Beginning-of-Year BV	200% DB Method	SL Method ($BV_9 = \$5{,}000$)	Depreciation Amount Selected
1	$95,000	$21,111	$10,000	$21,111
2	73,889	16,420	8,611	16,420
3	57,469	12,771	7,496	12,771
4	44,698	9,933	6,616	9,933
5	34,765	7,726	5,953	7,726
6	27,040	6,009	5,510	6,009
7	21,031	4,674	5,344	5,344[a]
8	15,687	3,635	5,344	5,344
9	10,344	2,827	5,344	5,344

[a] Switchover occurs in year seven.

Solution: MACRS (GDS) with Half-Year Convention

To demonstrate the GDS with the half-year convention, we will change the La Salle bus problem so that the bus is now sold in year five for Part (a) and in year six for Part (b).

(a) Selling bus in year five:

EOY k	Factor	d_k	BV_k
0	—	—	$95,000
1	0.2000	$19,000	76,000
2	0.3200	30,400	45,600
3	0.1920	18,240	27,360
4	0.1152	10,944	16,416
5	0.0576	5,472	10,944

(b) Selling bus in year six:

EOY k	Factor	d_k	BV_k
0	—	—	$95,000
1	0.2000	$19,000	76,000
2	0.3200	30,400	45,600
3	0.1920	18,240	27,360
4	0.1152	10,944	16,416
5	0.1152	10,944	5,472
6	0.0576	5,472	0

Notice that, when we sold the bus in year five before the recovery period had ended, we took only half of the normal depreciation. The other years (years one through four) were not changed. When the bus was sold in year six, at the end of the recovery period, we did not divide the last year amount by two.

Selected methods of depreciation, illustrated in Example 7-7, are compared in Figure 7-2. In addition, the PW (10%) of each method is shown in Figure 7-2. Because large PWs of depreciation deductions are generally viewed as desirable, it is clear that the MACRS method is very attractive to profitable companies.

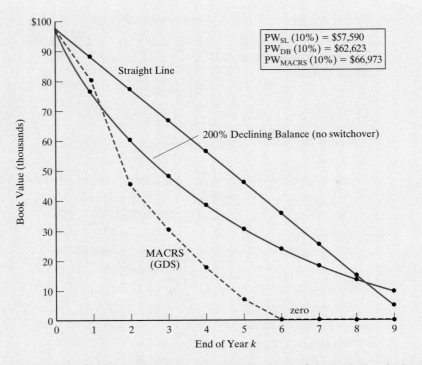

Figure 7-2   BV Comparisons for Selected Methods of Depreciation in Example 7-7 (*Note:* The bus is assumed to be sold in year six for the MACRS-GDS method.)

7.6   Introduction to Income Taxes

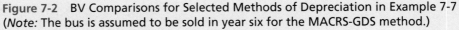

Up to this point, there has been no consideration of income taxes in our discussion of engineering economy, except for the influence of depreciation and other types of deductions. By not complicating our studies with income-tax effects, we have placed primary emphasis on basic engineering economy principles and methodology. There, however, is a wide variety of capital investment problems in which income taxes do affect the choice among alternatives, and after-tax studies are essential.*

In the remainder of this chapter, we shall be concerned with how income taxes affect a project's estimated cash flows. Income taxes resulting from the profitable operation of a firm are usually taken into account in evaluating engineering projects. The reason is quite simple: Income taxes associated with a proposed project may represent a major cash outflow that should be considered together with other cash inflows and outflows in assessing the overall economic profitability of that project.

There are other taxes discussed in Section 7.6.1 that are not directly associated with the income-producing capability of a new project, but they are usually

---

* Some U.S. firms do not pay income taxes, but this is not a valid reason for ignoring income taxes. We assume that most firms have a positive taxable income in their overall operations.

negligible when compared with federal and state income taxes. *When other types of taxes are included in engineering economy studies, they are normally deducted from revenue, as any other operating expense would be, in determining the before-tax cash flow (BTCF) that we considered in Chapters 5 and 6.*

The mystery behind the sometimes complex computation of income taxes is reduced when we recognize that income taxes paid are just another type of expense, and income taxes saved (through depreciation, expenses, and direct tax credits) are identical to other kinds of reduced expenses (e.g., savings in maintenance expenses).

The basic concepts underlying federal and state income tax laws and regulations that apply to most economic analyses of capital investments generally can be understood and applied without difficulty. This discussion of income taxes is not intended to be a comprehensive treatment of the subject. Rather, we utilize some of the more important provisions of the federal Tax Reform Act of 1986 (TRA 86), followed by illustrations of a general procedure for computing the net ATCF for an engineering project and conducting after-tax economic analyses. Where appropriate, important changes to TRA 86 enacted by the Omnibus Budget Reconciliation Act of 1993 (OBRA 93) and the Taxpayer Relief Act of 1997 are also included in this chapter.

7.6.1 Distinctions between Different Types of Taxes

Before discussing the consequences of income taxes in engineering economy studies, we need to distinguish between income taxes and several other types of taxes:

1. *Income taxes* are assessed as a function of gross revenues minus allowable deductions. They are levied by the federal, most state, and occasionally municipal governments.

2. *Property taxes* are assessed as a function of the value of property owned, such as land, buildings, equipment, and so on, and the applicable tax rates. Hence, they are independent of the income or profit of a firm. They are levied by municipal, county, or state governments.

3. *Sales taxes* are assessed on the basis of purchases of goods or services and are thus independent of gross income or profits. They are normally levied by state, municipal, or county governments. Sales taxes are relevant in engineering economy studies only to the extent that they add to the cost of items purchased.

4. *Excise taxes* are federal taxes assessed as a function of the sale of certain goods or services often considered nonnecessities and are hence independent of the income or profit of a business. Although they are usually charged to the manufacturer or original provider of the goods or services, a portion of the cost is passed on to the purchaser.

7.6.2 The Before-Tax and After-Tax Minimum Attractive Rates of Return

In the preceding chapters, we have treated income taxes as if they are not applicable, or we have taken them into account, in general, by using a before-tax MARR, which is larger than the after-tax MARR. An approximation of the before-tax MARR requirement, which includes the effect of income taxes, for studies involving only BTCF can be obtained from the following relationship:

(Before-tax MARR)[(1 − Effective income tax rate)] ≅ After-tax MARR.

Thus,

$$\text{Before-tax MARR} \approx \frac{\text{After-tax MARR}}{(1 - \text{Effective income tax rate})}. \tag{7-12}$$

Determining the effective income tax rate for a firm is discussed in Section 7.7.

This approximation is exact if the asset is nondepreciable and there are no gains or losses on disposal, tax credits, or other types of deductions involved. Otherwise, these factors affect the amount and timing of income tax payments, and some degree of error is introduced into the relationship in Equation (7-12).

7.6.3   The Interest Rate to Use in After-Tax Studies

Any practical definition of a MARR should serve as a guide for investment policy to attain the goals toward which a firm is, or will be, striving. One obvious goal is to meet shareholder expectations. Clearly, long-term goals and capital structure will be changing as a firm matures. In this regard, it is widely accepted that in after-tax engineering economy studies, the MARR should be *at least* the tax-adjusted weighted average cost of capital (WACC):

$$\text{WACC} = \lambda(1 - t)i_b + (1 - \lambda)e_a. \tag{7-13}$$

Here, λ = fraction of a firm's pool of capital that is borrowed from lenders,
   t = effective income tax rate as a decimal,
   i_b = before-tax interest paid on borrowed capital,
   e_a = after-tax cost of equity capital.

The adjustment on the borrowed capital component of WACC, $(1 - t)i_b$, accounts for the fact that interest on borrowed capital is tax deductible.

Often a value higher than the WACC is assigned to the after-tax MARR to reflect the opportunity cost of capital, perceived risk and uncertainty of projects being evaluated, and policy issues such as organizational growth and shareholder satisfaction. We are now in a position to understand how a firm's after-tax MARR is determined and why its value may change substantially over time. To summarize, suppose a firms' WACC is 10% per year. The after-tax MARR may be set at 15% per year to better reflect the business opportunities that are available for capital investment. If the firm's effective income tax rate is 40%, the approximate value of the before-tax MARR is 15%/(1 − 0.4) = 25% per year.

7.6.4   Taxable Income of Corporations (Business Firms)

At the end of each tax year, a corporation must calculate its net (i.e., taxable) before-tax income or loss. Several steps are involved in this process, beginning with the calculation of *gross income*. The corporation may deduct from gross income all ordinary and necessary operating expenses to conduct the business *except capital investments*. Deductions for depreciation are permitted each tax period as a means of consistently and systematically recovering capital investment. Consequently, allowable expenses and deductions may be used to determine taxable income:

$$\text{Taxable income} = \text{Gross income} - \text{All expenses except capital investments}$$
$$- \text{Depreciation deductions.} \tag{7-14}$$

EXAMPLE 7-8   **Determination of Taxable Income**

A company generates $1,500,000 of gross income during its tax year and incurs operating expenses of $800,000. Property taxes on business assets amount to $48,000. The total depreciation deductions for the tax year equal $114,000. What is the taxable income of this firm?

Solution

Based on Equation (7-14), this company's taxable income for the tax year would be

$$\$1,500,000 - \$800,000 - \$48,000 - \$114,000 = \$538,000.$$

7.7 The Effective (Marginal) Corporate Income Tax Rate

The federal corporate income tax rate structure in 2006 is shown in Table 7-5. Depending on the taxable income bracket that a firm is in for a tax year, the marginal federal rate can vary from 15% to a maximum of 39%. Note, however, that the weighted average tax rate at taxable income = $335,000 is 34%, and the weighted average tax rate at taxable income = $18,333,333 is 35%. Therefore, if a corporation has a taxable income for a tax year *greater than* $18,333,333, federal taxes are computed by using a flat rate of 35%.

TABLE 7-5   Corporate Federal Income Tax Rates (2006)

If Taxable Income Is:		The Tax Is:	
Over	But Not Over		of the Amount Over
0	$50,000	15%	0
$50,000	75,000	$7,500 + 25%	$50,000
75,000	100,000	13,750 + 34%	75,000
100,000	335,000	22,250 + 39%	100,000
335,000	10,000,000	113,900 + 34%	335,000
10,000,000	15,000,000	3,400,000 + 35%	10,000,000
15,000,000	18,333,333	5,150,000 + 38%	15,000,000
18,333,333	....	35%	0

Source: Tax Information on Corporations, IRS Publication 542, 2006.

EXAMPLE 7-9	Calculating Income Taxes

Suppose that a firm for a tax year has a gross income of $5,270,000, expenses (excluding capital) of $2,927,500, and depreciation deductions of $1,874,300. What would be its taxable income and federal income tax for the tax year, based on Equation (7-14) and Table 7-5?

Solution

Taxable income = Gross income − Expenses − Depreciation deductions

$$= \$5,270,000 - \$2,927,500 - \$1,874,300$$

$$= \$468,200$$

Income tax = 15% of first $50,000	$7,500
+ 25% of the next $25,000	6,250
+ 34% of the next $25,000	8,500
+ 39% of the next $235,000	91,650
+ 34% of the next $133,200	45,288
Total	$159,188

The total tax liability in this case is $159,188. As an added note, we could have used a flat rate of 34% in this example because the federal weighted average tax rate at taxable income = $335,000 is 34%. The remaining $133,200 of taxable income above this amount is in a 34% tax bracket (Table 7-5). So we have 0.34($468,200) = $159,188.

Although the tax laws and regulations of most of the states (and some municipalities) with income taxes have the same basic features as the federal laws and regulations, there is significant variation in income tax rates. State income taxes are in most cases much less than federal taxes and often can be closely approximated as ranging from 6% to 12% of taxable income. No attempt will be made here to discuss the details of state income taxes. To illustrate the calculation of an effective income tax rate (t) for a large corporation based on the consideration of both federal and state income taxes, however, assume that the applicable federal income tax rate is 35% and the state income tax rate is 8%. Further assume the common case in which taxable income is computed the same way for both types of taxes, except that state income taxes are deductible from taxable income for federal tax purposes, but federal income taxes are not deductible from taxable income for state tax purposes. Based on these assumptions, the general expression for the effective income tax rate is

$$t = \text{State rate} + \text{Federal rate}(1 - \text{State rate}), \text{ or} \qquad (7\text{-}15)$$

$$t = \text{Federal rate} + (1 - \text{Federal rate})(\text{State rate}). \qquad (7\text{-}16)$$

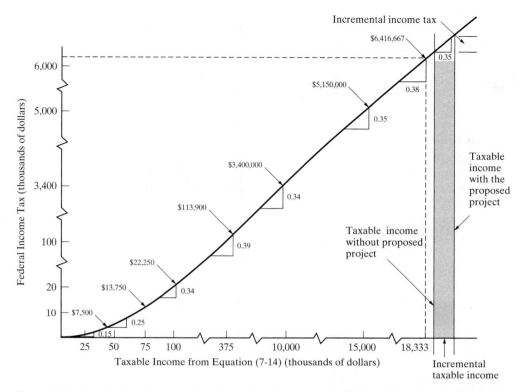

Figure 7-3 The Federal Income Tax Rates for Corporations (Table 7-5) with Incremental Income Tax for a Proposed Project (assumes, in this case, corporate taxable income without project > $18,333,333)

In this example, the effective income tax rate for the corporation would be

$$t = 0.08 + 0.35(1 - 0.08) = 0.402, \text{ or approximately } 40\%.$$

In this chapter, we will often use an effective corporate income tax rate of *approximately* 40% as a representative value that includes state income taxes.

The effective income tax rate on *increments* of taxable income is of importance in engineering economy studies. This concept is illustrated in Figure 7-3, which plots the federal income tax rates and brackets listed in Table 7-5 and shows the added (incremental) taxable income and federal income taxes that would result from a proposed engineering project (shaded in Figure 7-3). In this case, the corporation is assumed to have a taxable income greater than $18,333,333 for their tax year. The same concept, however, applies to a smaller firm with less taxable income for its tax year, which is illustrated in Example 7-10.

EXAMPLE 7-10	**Project-Based Income Taxes**

A small corporation is expecting an annual taxable income of $45,000 for its tax year. It is considering an additional capital investment of $100,000 in an engineering project, which is expected to create an added annual net cash flow (revenues minus expenses) of $35,000 and an added annual depreciation deduction of $20,000. What is the corporation's federal income tax liability

(a) without the added capital investment,

(b) with the added capital investment?

Solution

(a)

Income Taxes	Rate	Amount
On first $45,000	15%	$6,750
Total		$6,750

(b)

Taxable Income		
Before added investment		$45,000
+ added net cash flow		+35,000
− depreciation deduction		−20,000
Taxable Income		$60,000

Income Taxes on $60,000	Rate	Amount
On first $50,000	15%	$7,500
On next $10,000	25%	2,500
Total		$10,000

The increased income tax liability from the investment is $10,000 − $6,750 = $3,250.

As an added note, the change in tax liability can usually be determined more readily by an incremental approach. For instance, this example involved changing the taxable income from $45,000 to $60,000 as a result of the new investment. Thus, the change in income taxes for the tax year could be calculated as follows:

$$\text{First } \$50,000 - \$45,000 = \$5,000 \text{ at } 15\% = \quad \$750$$
$$\text{Next } \$60,000 - \$50,000 = \$10,000 \text{ at } 25\% = \underline{\ 2,500}$$
$$\text{Total} \quad \$3,250$$

The average federal income tax rate on the additional $35,000 − $20,000 = $15,000 of taxable income is calculated as ($3,250/$15,000) = 0.2167, or 21.67%.

7.8   Gain (Loss) on the Disposal of an Asset

When a *depreciable asset* (tangible personal or real property, Section 7.2) is sold, the MV is seldom equal to its BV [Equation (7-1)]. In general, the gain (loss) on sale of depreciable property is the fair market value minus its book value at that time. That is,

$$[\text{Gain (loss) on disposal}]_N = MV_N - BV_N. \tag{7-17}$$

When the sale results in a gain, it is often referred to as *depreciation recapture*. The tax rate for the gain (loss) on disposal of depreciable personal property is usually the same as for ordinary income or loss, which is the effective income tax rate, t.

When a *capital asset* is sold or exchanged, the gain (loss) is referred to as a *capital gain (loss)*. Examples of capital assets are stocks, bonds, gold, silver, and other metals, as well as real property such as a home. Because engineering economic analysis seldom involves an actual capital gain (loss), the more complex details of this situation are not discussed further.

EXAMPLE 7-11	Tax Consequences of Selling an Asset

A corporation sold a piece of equipment during the current tax year for $78,600. The accounting records show that its cost basis, B, is $190,000 and the accumulated depreciation is $139,200. Assume that the effective income tax rate as a decimal is 0.40 (40%). Based on this information, what is

(a) the gain (loss) on disposal,

(b) the tax liability (or credit) resulting from this sale,

(c) the tax liability (or credit) if the accumulated depreciation was $92,400 instead of $139,200?

Solution

(a) The BV at the time of sale is $190,000 − $139,200 = $50,800. Therefore, the gain on disposal is $78,600 − $50,800 = $27,800.

(b) The tax owed on this gain is −0.40($27,800) = −$11,120.

(c) With $d_k^* = \$92,400$, the BV at the time of sale is $190,000 − $92,400 = $97,600. The loss is $78,600 − $97,600 = −$19,000. The tax credit resulting from this loss on disposal is −0.40(−$19,000) = $7,600.

7.9 General Procedure for Making After-Tax Economic Analyses

After-tax economic analyses utilize the same profitability measures as do before-tax analyses. The only difference is that ATCFs are used in place of before-tax cash flows (BTCFs) by including expenses (or savings) to income taxes and then making equivalent worth calculations *using an after-tax MARR*. The tax rates and governing regulations may be complex and subject to changes, but, once those rates and regulations have been translated into their effect on ATCFs, the remainder of the after-tax analysis is relatively straightforward. To formalize the procedure, let

R_k = revenues (and savings) from the project (this is the cash inflow from the project during period k);

E_k = cash outflows during year k for deductible expenses;

d_k = sum of all noncash, or book, costs during year k, such as depreciation;

t = effective income tax rate on *ordinary* income (federal, state, and other), with t assumed to remain constant during the study period;

T_k = income tax consequences during year k;

ATCF_k = ATCF from the project during year k.

Because the taxable income is $(R_k - E_k - d_k)$, the *ordinary income tax consequences* during year k are computed with Equation (7-18):

$$T_k = -t(R_k - E_k - d_k). \tag{7-18}$$

Therefore, when $R_k > (E_k + d_k)$, a tax liability (i.e., negative cash flow) occurs. When $R_k < (E_k + d_k)$, a decrease in the tax amount (a credit) occurs.

In the economic analyses of engineering projects, ATCFs in year k can be computed in terms of BTCF_k (i.e., BTCF in year k):

$$\text{BTCF}_k = R_k - E_k. \tag{7-19}$$

Thus,*

$$\text{ATCF}_k = \text{BTCF}_k + T_k \tag{7-20}$$

$$= (R_k - E_k) - t(R_k - E_k - d_k)$$

$$= (1 - t)(R_k - E_k) + td_k. \tag{7-21}$$

Equation (7-21) shows that after income taxes, a revenue becomes $(1 - t)R_k$ and an expense becomes $(1 - t)E_k$. Note that the ATCF attributable to depreciation (a tax savings) is td_k in year k.

Tabular headings to facilitate the computation of ATCFs using Equations (7-18) and (7-21) are as follows:

Year	(A) BTCF	(B) Depreciation	(C) = (A) − (B) Taxable Income	(D) = −t(C) Cash Flow for Income Taxes	(E) = (A) + (D) ATCF
k	$R_k - E_k$	d_k	$R_k - E_k - d_k$	$-t(R_k - E_k - d_k)$	$(1 - t)(R_k - E_k) + td_k$

Column A consists of the same information used in before-tax analyses, namely the cash revenues (or savings) less the deductible expenses. Column B contains depreciation that can be claimed for tax purposes. Column C is the taxable income or the amount subject to income taxes. Column D contains the income taxes paid (or saved). Finally, column E shows the ATCFs to be used directly in after-tax economic analyses.

A note of caution concerning the definition of BTCFs (and ATCFs) for projects is in order at this point. BTCF is defined to be annual revenue (or savings) attributable to a project minus its annual cash expenses. These expenses should *not* include interest and other financial cash flows. The reason is that *project* cash flows should be analyzed separately from financial cash flows. Including interest expense with project cash flows is incorrect when the firm's pool of capital is being used to

---

* In Figure 7-4, we use $-t$ in column D, so algebraic subtraction of income taxes in Equation (7-20) is accomplished.

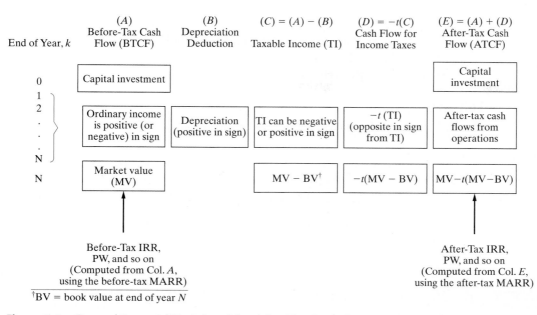

†BV = book value at end of year N

Figure 7-4 General Format (Worksheet) for After-Tax Analysis; Determining the ATCF

undertake an engineering project. Why? A firm's pool of capital consists of debt capital and equity capital. Because the MARR typically uses the weighted average cost of capital as its lower limit, discounting at the MARR for investments from the pool of capital takes account of the cost of debt capital (interest). Thus, there is no need to subtract interest expense in determining BTCFs—to do so would amount to double counting the interest expense associated with debt capital.

A summary of the process of determining ATCF during each year of an N-year study period is provided in Figure 7-4. The format of Figure 7-4 is used extensively throughout the remainder of this chapter, and it provides a convenient way to organize data in after-tax studies.

The column headings of Figure 7-4 indicate the arithmetic operations for computing columns C, D, and E when $k = 1, 2, \ldots, N$. When $k = 0$, capital investments are usually involved, and their tax treatment (if any) is illustrated in the examples that follow. The table should be used with the conventions of $+$ for cash inflow or savings and $-$ for cash outflow or opportunity forgone.

EXAMPLE 7-12 **PW of MACRS Depreciation Amounts**

Suppose that an asset with a cost basis of $100,000 and an ADS recovery period of five years is being depreciated under the *Alternate Depreciation System (ADS) of MACRS*, as follows:

Year	1	2	3	4	5	6
Depreciation deduction	$10,000	$20,000	$20,000	$20,000	$20,000	$10,000

If the firm's effective income tax rate remains constant at 40% during this six-year period, what is the PW of after-tax savings resulting from depreciation when MARR = 10% per year (after taxes)?

Solution

The PW of tax credits (savings) because of this depreciation schedule is

$$PW(10\%) = \sum_{k=1}^{6} 0.4d_k (1.10)^{-k} = \$4,000(0.9091) + \$8,000(0.8264)$$

$$+ \cdots + \$4,000(0.5645) = \$28,948.$$

EXAMPLE 7-13	**After-Tax PW of an Asset**

The asset in Example 7-12 is expected to produce net cash inflows (net revenues) of \$30,000 per year during the six-year period, and its terminal MV is negligible. If the effective income tax rate is 40%, how much can a firm afford to spend for this asset and still earn the MARR? What is the meaning of any excess in affordable amount over the \$100,000 cost basis given in Example 7-12? [Equals the after tax PW.]

Solution

After income taxes, the PW of net revenues is $(1 - 0.4)(\$30,000)(P/A, 10\%, 6) = \$18,000(4.3553) = \$78,395$. After adding to this the PW of tax savings computed in Example 7-12, the affordable amount is \$107,343. Because the capital investment is \$100,000, the net PW equals \$7,343. This same result can be obtained by using the general format (worksheet) of Figure 7-4:

EOY	(A) BTCF	(B) Depreciation Deduction	(C) = (A) − (B) Taxable Income	(D) = − 0.4(C) Income Taxes	(E) = (A) + (D) ATCF
0	−\$100,000				−\$100,000
1	30,000	\$10,000	\$20,000	−\$8,000	22,000
2	30,000	20,000	10,000	−4,000	26,000
3	30,000	20,000	10,000	−4,000	26,000
4	30,000	20,000	10,000	−4,000	26,000
5	30,000	20,000	10,000	−4,000	26,000
6	30,000	10,000	20,000	−8,000	22,000

PW(10%) of ATCF = \$7,343

EXAMPLE 7-14	**PennCo Electric Utility Company**

Here we refer back to the chapter opener where the question of the after-tax cost of electricity production was raised. For the PennCo plant, the annual fuel expense[†] will be $191,346,432 per year. The operating and maintenance expense will be $112,128,000 per year, and the carbon tax will be $73,805,052 per year. The total annual expense of operating the plant will therefore be $377,279,484. Armed with these cost estimates, we can calculate the after-tax cost of the plant by using the template of Figure 7-4 (all numbers are millions of dollars).

EOY	BTCF	Depreciation	Taxable Income	Cash Flow for Income Taxes	ATCF
0	−$1,120.000	—	—	—	−$1,120.000
1 − 30	−377.279	37.333*	−414.613	165.845	−211.434

*$d_k = \$1,120/30 = \$37.333$

The annual worth of the PennCo plant, assuming inflation is negligible, is −$1,120 $(A/P, 10\%, 30)$ − $211.434, which equals −$330.243 million. We can calculate the after-tax cost of a kilowatt-hour by dividing this number ($330,243,000) by the electrical output of 5,606,400,000 kWh. It is close to $0.06 per kWh.

[†] The details of the annual expense calculations are left as a student exercise in Problem 7-60.

7.10 Illustration of Computations of ATCFs

The following problems (Examples 7-15 through 7-19) illustrate the computation of ATCFs, as well as many common situations that affect income taxes. All problems include the assumption that income tax expenses (or savings) occur at the same time (year) as the revenue or expense that gives rise to the taxes. For purposes of comparing the effects of various situations, the after-tax IRR or PW is computed for each example. We can observe from the results of Examples 7-15, 7-16, and 7-17 that the faster (i.e., earlier) the depreciation deduction is, the more favorable the after-tax IRR and PW will become.

EXAMPLE 7-15	**Computing After-Tax PW and IRR**

Certain new machinery, when placed in service, is estimated to cost $180,000. It is expected to *reduce* net annual operating expenses by $36,000 per year for 10 years and to have a $30,000 MV at the end of the 10th year.

(a) Develop the ATCFs and BTCFs.

(b) Calculate the before-tax and after-tax IRR. Assume that the firm is in the federal taxable income bracket of $335,000 to $10,000,000 and that the state income tax rate is 6%. State income taxes are deductible from federal taxable income. This machinery is in the MACRS (GDS) five-year property class.

(c) Calculate the after-tax PW when the *after-tax* MARR = 10% per year.

In this example, the study period is 10 years, but the property class of the machinery is 5 years. Solve by hand and by spreadsheet.

Solution by Hand

(a) Table 7-6 applies the format illustrated in Figure 7-4 to calculate the BTCF and ATCF for this example. In column D, the effective income tax rate is very close to 0.38 [from Equation (7-15)] based on the information just provided.

(b) The before-tax IRR is computed from column A:

$$0 = -\$180,000 + \$36,000(P/A, i'\%, 10) + \$30,000(P/F, i'\%, 10).$$

By trial and error, we find that $i' = 16.1\%$.

The entry in the last year is shown to be $30,000 because the machinery will have this estimated MV. The asset, however, was depreciated to zero with the GDS method. Therefore, when the machine is sold at the end of year 10, there will be $30,000 of *recaptured depreciation*, or gain on disposal [Equation (7-16)], which is taxed at the effective income tax rate of 38%. This tax entry is shown in column D (EOY 10).

By trial and error, the after-tax IRR for Example 7-15 is found to be 12.4%.

(c) When MARR = 10% per year is inserted into the PW equation at the bottom of Table 7-6, it can be determined that the after-tax PW of this investment is $17,208.

Spreadsheet Solution

Figure 7-5 displays the spreadsheet solution for Example 7-15. This spreadsheet uses the form given in Figure 7-4 to compute the ATCFs. The spreadsheet also illustrates the use of the VDB (variable declining balance) function to compute MACRS (GDS) depreciation amounts.

Cell B9 contains the cost basis, cells B10:B19 contain the BTCFs, and the year 10 MV is given in cell B20. The VDB function is used to determine the MACRS (GDS) depreciation amounts in column C. Cell B7 contains the DB percentage used for a five-year property class (see Table 7-4).

Note that in Figure 7-5, there are two row entries for the last year of the study period (year 10). The first entry accounts for expected revenues less expenses, while the second entry accounts for the disposal of the asset. To use the NPV and IRR financial functions, these values must be combined into a net cash flow for the year. This is accomplished by the adjusted ATCF and adjusted BTCF columns in the spreadsheet.

TABLE 7-6 ATCF Analysis of Example 7-15

End of Year, k	(A) BTCF	(B) Cost Basis	×	GDS Recovery Rate	=	Deduction	(C) = (A)−(B) Taxable Income	(D) = −0.38(C) Cash Flow for Income Taxes	(E) = (A)+(D) ATCF
				Depreciation Deduction					
0	−$180,000	—		—					−$180,000
1	36,000	$180,000	×	0.2000	=	$36,000	0	0	36,000
2	36,000	180,000	×	0.3200	=	57,600	−21,600	+8,208	44,208
3	36,000	180,000	×	0.1920	=	34,560	1,440	−547	35,453
4	36,000	180,000	×	0.1152	=	20,736	15,264	−5,800	30,200
5	36,000	180,000	×	0.1152	=	20,736	15,264	−5,800	30,200
6	36,000	180,000	×	0.0576	=	10,368	25,632	−9,740	26,260
7–10	36,000	0				0	36,000	−13,680	22,320
10	30,000						30,000[a]	−11,400[b]	18,600
Total	$210,000							Total	$130,201
								PW (10%) =	$17,208

[a] Depreciation recapture = $MV_{10} - BV_{10} = \$30,000 - 0 = \$30,000$ (gain on disposal).

[b] Tax on depreciation recapture = $\$30,000(0.38) = \$11,400$.

After-tax IRR: Set PW of column E = 0 and solve for i' in the following equation:

$$0 = -\$180,000 + \$36,000(P/F, i', 1) + \$44,208(P/F, i', 2) + \$35,453(P/F, i', 3) + \$30,200(P/F, i', 4) + \$30,200(P/F, i', 5) + \$26,260(P/F, i', 6)$$
$$+ \$22,320(P/A, i', 4)(P/F, i', 6) + \$18,600(P/F, i', 10); \quad \text{IRR} = 12.4\%.$$

	B		D	E
After-tax MARR =	10%		Capital Investment =	$ 180,000
Class Life =	5		Market Value =	$ 30,000
State tax rate =	6%		Annual Savings =	$ 36,000
Federal tax rate =	34%		Useful Life =	10
effective tax rate =	38.00%			
DB rate =	200%			

EOY	BTCF	Depreciation Deduction	Taxable Income	Cash Flow for Income Taxes	ATCF	Adjusted ATCF	Adjusted BTCF
0	$ (180,000)				$ (180,000)	$ (180,000)	$ (180,000)
1	$ 36,000	$ 36,000	$ -	$ -	$ 36,000	$ 36,000	$ 36,000
2	$ 36,000	$ 57,600	$ (21,600)	$ 8,208	$ 44,208	$ 44,208	$ 36,000
3	$ 36,000	$ 34,560	$ 1,440	$ (547)	$ 35,453	$ 35,453	$ 36,000
4	$ 36,000	$ 20,736	$ 15,264	$ (5,800)	$ 30,200	$ 30,200	$ 36,000
5	$ 36,000	$ 20,736	$ 15,264	$ (5,800)	$ 30,200	$ 30,200	$ 36,000
6	$ 36,000	$ 10,368	$ 25,632	$ (9,740)	$ 26,260	$ 26,260	$ 36,000
7	$ 36,000	$ -	$ 36,000	$ (13,680)	$ 22,320	$ 22,320	$ 36,000
8	$ 36,000	$ -	$ 36,000	$ (13,680)	$ 22,320	$ 22,320	$ 36,000
9	$ 36,000	$ -	$ 36,000	$ (13,680)	$ 22,320	$ 22,320	$ 36,000
10	$ 36,000		$ 36,000	$ (13,680)	$ 22,320	$ 22,320	$ 66,000
10	$ 30,000		$ 30,000	$ (11,400)	$ 18,600	$ 40,920	
				PW =	$ 17,209		$ 52,771
				AW =	$ 2,801		$ 8,588
				IRR =	12.4%		16.1%

Annotations:

= ROUND (B3 + B4*(1 – B3), 2)

= IF(B2 >= 15,1.5, 2)

= –E1

= $E$3

= E2

= B10 – C10

= –$B$6 * D10

= B9 + E9

= F9

= B9

= F19 + F20

= B19 + B20

= IRR(G9: G19, $B$1)

= NPV($B$1, G10:G19) + G9

= PMT($B$1, $E$4, – G22)

= IF(A10=$B$2+1, 0.5*C9, VDB($E$1, 0, $B$2, MAX(0, A10-1.5), A10-0.5, $B$7, FALSE))

Figure 7-5   Spreadsheet Solution, Example 7-15

In Example 7-16, we demonstrate the impact of a longer depreciation schedule on the after-tax profitability of an investment.

EXAMPLE 7-16	**Impact of a Longer Depreciation Schedule on After-Tax PW and IRR**

Suppose the machinery in Example 7-15 had been classified in the 10-year MACRS (GDS) property class. Calculate the new after-tax PW and after-tax IRR. Why are these results different than the results of Example 7-15?

Solution

If the machinery in Example 7-15 had been classified in the 10-year MACRS (GDS) property class instead of the five-year property class, depreciation deductions would be slowed down in the early years of the study period and shifted into later years, as shown in Table 7-7. Compared with entries in Table 7-6, entries in columns C, D, and E of Table 7-7 are less favorable, in the sense that a fair amount of ATCF is deferred until later years, producing a lower after-tax PW and IRR. For instance, the PW is reduced from $17,208 in Table 7-6 to $9,136 in Table 7-7. The basic difference between Table 7-6 and Table 7-7 is the *timing of the ATCF*, which is a function of the timing and magnitude of the depreciation deductions. In fact, the curious reader can confirm that the sums of entries in columns A through E of Tables 7-6 and 7-7 are nearly the same (except for the half-year of depreciation only in year 10 of Table 7-7). The timing of cash flows does, of course, make a difference!

Depreciation does not affect BTCF. Fast (accelerated) depreciation produces a larger PW of tax savings than does the same amount of depreciation claimed later in an asset's life.

A minor complication is introduced in ATCF analyses when the study period is shorter than an asset's MACRS recovery period (e.g., for a five-year recovery period, the study period is five years or less). In such cases, we shall assume throughout this book that the asset is sold for its MV in the last year of the study period. Due to the half-year convention, only one-half of the normal MACRS depreciation can be claimed in the year of disposal or end of the study period, so there will usually be a difference between an asset's BV and its MV. Resulting income tax adjustments will be made at the time of the sale (see the last row in Table 7-7) unless the asset in question is not sold but instead kept for standby service. In such a case, depreciation deductions usually continue through the end of the asset's MACRS recovery period. Our assumption of project termination at the end of the study period makes good economic sense, as illustrated in Example 7-17.

TABLE 7-7 ATCF Analysis of Example 7-16 (Reworked Example 7-15 with Machinery in the 10-Year MACRS (GDS) Property Class)

End of Year, k	(A) BTCF	(B) Cost Basis	×	GDS Recovery Rate	=	Deduction	(C) = (A) − (B) Taxable Income	(D) = −0.38(C) Cash Flow for Income Taxes	(E) = (A) + (D) ATCF
0	−$180,000	—		—		—			−$180,000
1	36,000	$180,000	×	0.1000	=	$18,000	$18,000	−$6,840	29,160
2	36,000	180,000	×	0.1800	=	32,400	3,600	−1,368	34,632
3	36,000	180,000	×	0.1440	=	25,920	10,080	−3,830	32,170
4	36,000	180,000	×	0.1152	=	20,736	15,264	−5,800	30,200
5	36,000	180,000	×	0.0922	=	16,596	19,404	−7,374	28,626
6	36,000	180,000	×	0.0737	=	13,266	22,734	−8,639	27,361
7	36,000	180,000	×	0.0655	=	11,790	24,210	−9,200	26,800
8	36,000	180,000	×	0.0655	=	11,790	24,210	−9,200	26,800
9	36,000	180,000	×	0.0656	=	11,808	24,192	−9,193	26,807
10	36,000	180,000	×	0.0655/2	=	5,895	30,105	−11,440	24,560
10	30,000						18,201[a]	−6,916	23,084

Total $130,196
PW (10%) ≈ $9,136
IRR = 11.2%

[a] Gain on disposal = $MV_{10} - BV_{10} = \$30,000 - \left(\frac{0.0655}{2} + 0.0328\right)(\$180,000) = \$18,201$.

EXAMPLE 7-17	Study Period < MACRS Recovery Period

A highly specialized piece of equipment has a first cost of $50,000. If this equipment is purchased, it will be used to produce income (through rental) of $20,000 per year for only four years. At the end of year four, the equipment will be sold for a negligible amount. Estimated annual expenses for upkeep are $3,000 during each of the four years. The MACRS (GDS) recovery period for the equipment is seven years, and the firm's effective income-tax rate is 40%.

(a) If the after-tax MARR is 7% per year, should the equipment be purchased?

(b) Rework the problem, assuming that the equipment is placed on standby status such that depreciation is taken over the full MACRS recovery period.

Solution

(a)

End of Year, k	(A) BTCF	(B) Depreciation Deduction	(C) = (A) − (B) Taxable Income	(D) = −0.4(C) Cash Flow for Income Taxes	(E) = (A) + (D) ATCF
0	−$50,000				−$50,000
1	17,000	$7,145	$9,855	−$3,942	13,058
2	17,000	12,245	4,755	−1,902	15,098
3	17,000	8,745	8,255	−3,302	13,698
4	17,000	3,123[a]	13,877	−5,551	11,449
4	0		−18,742[b]	7,497	7,497

[a] Half-year convention applies with disposal in year four.
[b] Remaining BV.

PW(7%) = $1,026. Because the PW > 0, the equipment should be purchased.

(b)

End of Year, k	(A) BTCF	(B) Depreciation Deduction	(C) Taxable Income	(D) Cash Flow for Income Taxes	(E) ATCF
0	−$50,000				−$50,000
1	17,000	$7,145	$9,855	−$3,942	13,058
2	17,000	12,245	4,755	−1,902	15,098
3	17,000	8,745	8,255	−3,302	13,698
4	17,000	6,245	10,755	−4,302	12,698
5	0	4,465	−4,465	1,786	1,786
6	0	4,460	−4,460	1,784	1,784
7	0	4,465	−4,465	1,786	1,786
8	0	2,230	−2,230	892	892
8	0				0

PW(7%) = $353, so the equipment should be purchased.

The PW is $673 higher in Part (a), which equals the PW of deferred depreciation deductions in Part (b). A firm would select the situation in Part (a) if it had a choice.

An illustration of determining ATCFs for a somewhat more complex, though realistic, capital investment opportunity is provided in Example 7-18.

EXAMPLE 7-18   **After-Tax Analysis of an Integrated Circuit Production Line**

The Ajax Semiconductor Company is attempting to evaluate the profitability of adding another integrated circuit production line to its present operations. The company would need to purchase two or more acres of land for $275,000 (total). The facility would cost $60,000,000 and have no net MV at the end of five years. The facility could be depreciated using a GDS recovery period of five years. An increment of working capital would be required, and its estimated amount is $10,000,000. Gross income is expected to increase by $30,000,000 per year for five years, and operating expenses are estimated to be $8,000,000 per year for five years. The firm's effective income tax rate is 40%.

(a) Set up a table and determine the ATCF for this project.
(b) Is the investment worthwhile when the after-tax MARR is 12% per year?

Solution

TABLE 7-8   After-Tax Analysis of Example 7-18

End of Year, k	(A) BTCF	(B) Depreciation Deduction	(C) = (A)−(B) Taxable Income	(D) = −0.4(C) Cash Flow for Income Taxes	(E) = (A) + (D) ATCF
0	−$60,000,000 −10,000,000 −275,000				−$70,275,000
1	22,000,000	$12,000,000	$10,000,000	−$4,000,000	18,000,000
2	22,000,000	19,200,000	2,800,000	−1,120,000	20,880,000
3	22,000,000	11,520,000	10,480,000	−4,192,000	17,808,000
4	22,000,000	6,912,000	15,088,000	−6,035,200	15,964,800
5	22,000,000	3,456,000	18,544,000	−7,417,600	14,582,400
5	10,275,000[a]		−6,912,000[b]	2,764,800[b]	13,039,800

[a] MV of working capital and land.
[b] Because BV_5 of the production facility is $6,912,000 and net $MV_5 = 0$, a loss on disposal would be taken at EOY 5.

(a) The format recommended in Figure 7-4 is followed in Table 7-8 to obtain ATCFs in years zero through five. Acquisitions of land, as well as additional working capital, are treated as nondepreciable capital investments whose MVs at the end of year five are estimated to equal their first costs. (In economic evaluations, it is customary to assume that land and working capital do not inflate in value during the study period because they are "nonwasting" assets.) By using a variation of Equation (7-21), we are able to compute ATCF in year three (for example) to be

$$ATCF_3 = (\$30,000,000 - \$8,000,000 - \$11,520,000)(1 - 0.40) + \$11,520,000$$
$$= \$17,808,000.$$

(b) The depreciable property in Example 7-18 ($60,000,000) will be disposed of for $0 at the end of year five, and a loss on disposal of $6,912,000 will be claimed at the end of year five. Only a half-year of depreciation ($3,456,000) can be claimed as a deduction in year five, and the BV is $6,912,900 at the end of year five. Because the selling price (MV) is zero, the loss on disposal equals our BV of $6,912,000. As seen from Figure 7-4, a tax credit of 0.40($6,912,000) = $2,764,800 is created at the end of year five. The after-tax IRR is obtained from entries in column E of Table 7-8 and is found to be 12.5%. The after-tax PW equals $936,715 at MARR = 12% per year. Based on economic considerations, this integrated circuit production line should be recommended because it appears to be quite attractive.

In the next example, the after-tax comparison of mutually exclusive alternatives involving only costs is illustrated.

EXAMPLE 7-19   After-Tax Comparison of Purchase versus Leasing Alternatives

An engineering consulting firm can purchase a fully configured Computer-Aided Design (CAD) workstation for $20,000. It is estimated that the useful life of the workstation is seven years, and its MV in seven years should be $2,000. Operating expenses are estimated to be $40 per eight-hour workday, and maintenance will be performed under contract for $8,000 per year. The MACRS (GDS) property class is five years, and the effective income tax rate is 40%.

As an alternative, sufficient computer time can be leased from a service company at an annual cost of $20,000. If the after-tax MARR is 10% per year, how many workdays per year must the workstation be needed in order to justify *leasing* it?

Solution

This example involves an after-tax evaluation of purchasing depreciable property versus leasing it. We are to determine how much the workstation must be utilized so that the lease option is a good economic choice. A *key* assumption is that the cost of engineering design time (i.e., operator time) is unaffected by whether the workstation is purchased or leased. Variable operations expenses associated with ownership result from the purchase of supplies, utilities, and so on. Hardware and software maintenance cost is contractually fixed at $8,000 per year. It is further assumed that the maximum number of working days per year is 250.

Lease fees are treated as an annual expense, and the consulting firm (the lessee) may *not* claim depreciation of the equipment to be an additional expense. (The leasing company presumably has included the cost of depreciation in its fee.) Determination of ATCF for the lease option is relatively straightforward and is not affected by how much the workstation is utilized:

$$(\text{After-tax expense of the lease})_k = -\$20,000(1 - 0.40) = -\$12,000; \quad k = 1, \ldots, 7.$$

ATCFs for the purchase option involve expenses that are fixed (not a function of equipment utilization) in addition to expenses that vary with equipment usage. If we let X equal the number of working days per year that the equipment is utilized, the variable cost per year of operating the workstation is $\$40X$. The after-tax analysis of the purchase alternative is shown in Table 7-9.

The after-tax annual worth (AW) of purchasing the workstation is

$$AW(10\%) = -\$20,000(A/P, 10\%, 7) - \$24X - [\$3,200(P/F, 10\%, 1) + \cdots$$
$$+ \$4,800(P/F, 10\%, 7)](A/P, 10\%, 7) + \$1,200(A/F, 10\%, 7)$$
$$= -\$24X - \$7,511.$$

TABLE 7-9 After-Tax Analysis of Purchase Alternative (Example 7-19)

End of Year, k	(A) BTCF	(B) Depreciation Deduction[a]	(C) = (A) − (B) Taxable Income	(D) = −t(C) Cash Flow for Income Taxes	(E) = (A) + (D) ATCF
0	−$20,000				−$20,000
1	−40X − 8,000	$4,000	−$40X − $12,000	$16X + $4,800	−24X − 3,200
2	−40X − 8,000	6,400	−40X − 14,400	16X + 5,760	−24X − 2,240
3	−40X − 8,000	3,840	−40X − 11,840	16X + 4,736	−24X − 3,264
4	−40X − 8,000	2,304	−40X − 10,304	16X + 4,122	−24X − 3,878
5	−40X − 8,000	2,304	−40X − 10,304	16X + 4,122	−24X − 3,878
6	−40X − 8,000	1,152	−40X − 9,152	16X + 3,661	−24X − 4,339
7	−40X − 8,000	0	−40X − 8,000	16X + 3,200	−24X − 4,800
7	2,000		2,000	−800	1,200

[a] Depreciation deduction$_k$ = $20,000 × (GDS recovery rate).

Figure 7-6
Summary of
Example 7-19

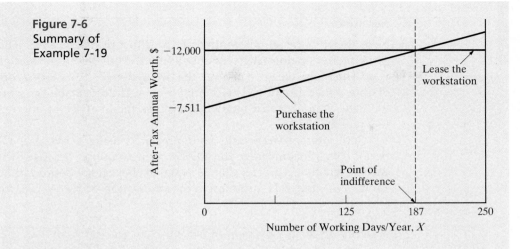

To solve for X, we equate the after-tax annual worth of both alternatives:

$$-\$12,000 = -\$24X - \$7,511.$$

Thus, $X = 187$ days per year. Therefore, if the firm expects to utilize the CAD workstation in its business *more than* 187 days per year, the equipment should be leased. The graphic summary of Example 7-19 shown in Figure 7-6 provides the rationale for this recommendation. The importance of the workstation's estimated utilization, in workdays per year, is now quite apparent.

EXAMPLE 7-20   **After-Tax Analysis of Alternatives with Unequal Lives**

A firm must decide between two system designs, S1 and S2, whose estimated cash flows are shown in the following table. The effective income tax rate is 40% and MACRS (GDS) depreciation is used. Both designs have a GDS recovery period of five years. If the after-tax desired return on investment is 10% per year, which design should be chosen?

	Design	
	S1	S2
Capital investment	$100,000	$200,000
Useful life (years)	7	6
MV at end of useful life	$30,000	$60,000
Annual revenues less expenses	$20,000	$40,000

Solution

Note that the design alternatives have different useful lives. The same basic principles of engineering economy apply to both before-tax and after-tax analyses. Therefore, we must analyze the two system designs over a common period of time. As we discovered in Chapter 6, using the repeatability assumption along with the annual worth method simplifies the analysis of alternatives having unequal lives.

Both alternatives would be depreciated using a five-year GDS recovery period. No adjustments to the GDS rates are required because the useful life of each alternative is greater than or equal to six years of depreciation deductions. Tables 7-10 and 7-11 summarize the calculation of the ATCFs for the design alternatives.

TABLE 7-10 After-Tax Analysis of Design S1, Example 7-20

End of Year, k	(A) BTCF	(B) Depreciation Deduction	(C) = (A) − (B) Taxable Income	(D) = − $t(C)$ Cash Flow for Income Taxes	(E) = (A) + (D) ATCF	PW(10%)
0	−$100,000				−$100,000	−$100,000
1	20,000	$20,000	$0	$0	20,000	18,182
2	20,000	32,000	−12,000	4,800	24,800	20,495
3	20,000	19,200	800	−320	19,680	14,786
4	20,000	11,520	8,480	−3,392	16,608	11,343
5	20,000	11,520	8,480	−3,392	16,608	10,312
6	20,000	5,760	14,240	−5,696	14,304	8,075
7	20,000	0	20,000	−8,000	12,000	6,158
7	30,000		30,000	−12,000	18,000	9,238

$$\text{PW}_{S1}(10\%) = -\$1,411$$

TABLE 7-11 After-Tax Analysis of Design S2, Example 7-20

End of Year, k	(A) BTCF	(B) Depreciation Deduction	(C) = (A) − (B) Taxable Income	(D) = −$t(C)$ Cash Flow for Income Taxes	(E) = (A) + (D) ATCF	PW(10%)
0	−$200,000				−$200,000	−$200,000
1	40,000	$40,000	$0	$0	40,000	36,364
2	40,000	64,000	−24,000	9,600	49,600	40,989
3	40,000	38,400	1,600	−640	39,360	29,571
4	40,000	23,040	16,960	−6,784	33,216	22,687
5	40,000	23,040	16,960	−6,784	33,216	20,624
6	40,000	11,520	28,480	−11,392	28,608	16,149
6	50,000		50,000	−20,000	30,000	16,935

$$\text{PW}_{S2}(10\%) = -\$16,681$$

We can't directly compare the PW of the after-tax cash flows because of the difference in the lives of the alternatives. We can, however, directly compare the AWs of the ATCFs by using the repeatability assumption from Chapter 6.

$$AW_{S1}(10\%) = PW_{S1}(A/P, 10\%, 7) = -\$1,411(0.2054) = -\$290$$

$$AW_{S2}(10\%) = PW_{S2}(A/P, 10\%, 6) = -\$16,681(0.2296) = -\$3,830$$

Based on an after-tax annual worth analysis, Design S1 is preferred since it has the greater (less negative) AW. Neither design however makes money, so if a system is not required, don't recommend either one.

7.11 Economic Value Added

This section discusses an economic measure for estimating the wealth creation potential of capital investments that is experiencing increased attention and use. The measure, called economic value added (EVA),[*] can be determined from some of the data available in an after-tax analysis of cash flows generated by a capital investment. Through retroactive analysis of a firm's common stock valuation, it has been established that some companies experience a statistically significant relationship between the EVA metric and the historical value of their common stock.[†] For our purposes, EVA can also be used to estimate the profit-earning *potential* of proposed capital investments in engineering projects.

Simply stated, EVA is the difference between the company's adjusted net operating profit after taxes (NOPAT) in a particular year and its after-tax cost of capital during that year. Another way to characterize EVA is "the spread between the return on the capital and the cost of the capital."[‡] On a project-by-project basis (i.e., for discrete investments), the EVA metric can be used to gauge the wealth creation opportunity of proposed capital expenditures. We now define annual EVA as

$$EVA_k = (\text{Net operating profit after taxes})_k$$
$$- (\text{Cost of capital used to produce profit})_k$$
$$= NOPAT_k - i \cdot BV_{k-1}, \qquad\qquad (7\text{-}22)$$

where

$$k = \text{an index for the year in question}(1 \leq k \leq N);$$

$$i = \text{after-tax MARR based on a firm's cost of capital},$$

$$BV_{k-1} = \text{beginning-of-year book value};$$

$$N = \text{the study (analysis) period in years}.$$

---

[*] EVA is a registered trademark of Stern Stewart & Company, New York City, NY.
[†] See J. L. Dodd and S. Chen, "EVA: A New Panacea?" *B & E Review*, 42 (July–September 1996): 26–28, and W. Freedman, "How Do You Add Up?" *Chemical Week*, October 9, 1996, pp. 31–34.
[‡] S. Tully, "The Real Key To Creating Wealth," *Fortune*, September 30, 1993, p. 38ff.

NOPAT_k can be determined from Figure 7-4. It is simply the algebraic addition of the entry in Column C and the entry in Column D.

$$\begin{aligned} \text{NOPAT}_k &= \text{Taxable income} + \text{Cash flow for income taxes} \\ &= (R_k + E_k - d_k) + (-t)(R_k + E_k - d_k) \\ &= (R_k + E_k - d_k)(1 - t) \end{aligned} \qquad (7\text{-}23)$$

Substituting Equation (7-21) into Equation (7-23), we can see the relationship between ATCF_k and NOPAT_k to be

$$\text{NOPAT}_k = \text{ATCF}_k - d_k. \qquad (7\text{-}24)$$

Equation (7-23) and Figure 7-4 are demonstrated in Example 7-21 to determine the ATCF amounts, after-tax AW, and the EVA amounts related to a capital investment.

EXAMPLE 7-21   **EVA**

Consider the following proposed capital investment in an engineering project and determine its

(a) year-by-year ATCF,
(b) after-tax AW,
(c) annual equivalent EVA.

Proposed capital investment	= $84,000
Salvage value (end of year four)	= $0
Annual expenses per year	= $30,000
Gross revenues per year	= $70,000
Depreciation method	= Straight Line
Useful life	= four years
Effective income tax rate (t)	= 50%
After-tax MARR (i)	= 12% per year

Solution

(a) Year-by-year ATCF amounts are shown in the following table:

EOY	BTCF	Depreciation	Taxable Income	Income Taxes	ATCF
0	− $84,000	—	—	—	−$84,000
1	70,000 − 30,000	$21,000	$19,000	−$9,500	30,500
2	70,000 − 30,000	21,000	19,000	−9,500	30,500
3	70,000 − 30,000	21,000	19,000	−9,500	30,500
4	70,000 − 30,000	21,000	19,000	−9,500	30,500

(b) The annual equivalent worth of the ATCFs equals $-\$84,000(A/P, 12\%, 4) + \$30,500 = \$2,844.$

(c) The EVA in year k equals $\text{NOPAT}_k - 0.12\,\text{BV}_{k-1}$ [Equation (7-22)]. The year-by-year EVA amounts and the annual equivalent worth of EVA ($\$2,844$) are shown in the next table. Hence, the after-tax AW and the annual equivalent worth of EVA of the project are *identical*.

EOY$_k$	NOPAT	EVA = NOPAT $- i \cdot$ BV$_{k-1}$
1	$\$19,000 - \$9,500 = \$9,500$	$\$9,500 - 0.12(\$84,000) = -\$580$
2	$= \$9,500$	$\$9,500 - 0.12(\$63,000) = \$1,940$
3	$= \$9,500$	$\$9,500 - 0.12(\$42,000) = \$4,460$
4	$= \$9,500$	$\$9,500 - 0.12(\$21,000) = \$6,980$

Annual equivalent EVA $= [-\$580(P/F, 12\%, 1) + \$1,940(P/F, 12\%, 2) + \$4,460(P/F, 12\%, 3) + \$6,980(P/F, 12\%, 4)](A/P, 12\%, 4) = \$2,844.$

In Example 7-21, it was shown that the after-tax AW (12%) of the proposed engineering project is identical to the annual equivalent EVA at the same interest rate. Therefore, the annual equivalent EVA is simply the annual worth, at the after-tax MARR, of a project's ATCFs. This straightforward relationship is also valid when accelerated depreciation methods (such as MACRS) are used in the analysis of a proposed project. The reader is referred to Problems 7-47, 7-48, and 7-49 at the end of the chapter for EVA exercises.

7.12 Summary

In this chapter, we have presented important aspects of federal legislation relating to depreciation and income taxes. It is essential to understand these topics so that correct after-tax engineering economy evaluations of proposed projects may be conducted. Depreciation and income taxes are also integral parts of subsequent chapters in this book.

In this chapter, many concepts regarding current federal income tax laws were described. For example, topics such as taxable income, effective income tax rates, taxation of ordinary income, and gains and losses on disposal of assets were explained. A general format for pulling together and organizing all these apparently diverse subjects was presented in Figure 7-4. This format offers the student or practicing engineer a means of collecting, on one worksheet, information that is required for determining ATCFs and properly evaluating the after-tax economic worth of a proposed capital investment. Figure 7-4 was then employed in numerous examples. The student's challenge is now to use this worksheet in organizing data presented in problem exercises at the end of this and subsequent chapters and to answer questions regarding the after-tax profitability.

Problems

The number in parentheses ( ) that follows each problem refers to the section from which the problem is taken.

7-1. How are depreciation deductions different from other production or service expenses such as labor, material, and electricity? (7.2)

7-2. What conditions must a property satisfy to be considered depreciable? (7.2)

7-3. Explain the difference between real and personal property. (7.2)

7-4. Explain how the cost basis of depreciable property is determined. (7.2)

7-5. Why would a business elect, under MACRS, to use the ADS rather than the GDS? (7.4)

7-6. The "Big-Deal" Company has purchased new furniture for their offices at a retail price of $100,000. An additional $20,000 has been charged for insurance, shipping, and handling. The company expects to use the furniture for 10 years (useful life = 10 years) and then sell it at a salvage (market) value of $10,000. Use the SL method for depreciation to answer these questions. (7.3)

a. What is the depreciation during the second year?

b. What is the BV of the asset at the end of the first year?

c. What is the BV of the asset after 10 years?

7-7. Cisco Systems is purchasing a new bar code–scanning device for its service center in San Francisco. The table that follows lists the relevant cost items for this purchase. The operating expenses for the new system are $10,000 per year, and the useful life of the system is expected to be five years. The SV for depreciation purposes is equal to 25% of the hardware cost. (7.3)

Cost Item	Cost
Hardware	$160,000
Training	$15,000
Installation	$15,000

a. What is the BV of the device at the end of year three if the SL depreciation method is used?

b. Suppose that after depreciating the device for two years with the SL method, the firm decides to switch to the double declining balance depreciation method for the remainder of the device's life (the remaining three years). What is the device's BV at the end of four years?

7-8. An asset for drilling was purchased and placed in service by a petroleum production company. Its cost basis is $60,000, and it has an estimated MV of $12,000 at the end of an estimated useful life of 14 years. Compute the depreciation amount in the third year and the BV at the end of the fifth year of life by each of these methods: (7.3, 7.4)

a. The SL method.

b. The 200% DB method with switchover to SL.

c. The GDS.

d. The ADS.

7-9. What is the depreciation deduction, using each of the following methods, for the second year for an asset that costs $35,000 and has an estimated MV of $7,000 at the end of its seven-year useful life? Assume its MACRS class life is also seven years. (a) 200% DB, (b) GDS (MACRS), and (c) ADS (MACRS). (7.3, 7.4)

7-10. Your company just purchased office furniture (asset class 00.11) for $100,000 and placed it in service on August 13, 2007. The cost basis for the furniture is $100,000, and it will be depreciated with the GDS using half-year convention. The expected salvage (market) value of the furniture is $5,000 in 2015. Determine the recovery period for the furniture and its depreciation deductions over the recovery period. (7.4)

7-11. Your company has purchased a large new truck-tractor for over-the-road use (asset class 00.26). It has a cost basis of $180,000. With additional options costing $15,000, the cost basis for depreciation purposes is $195,000. Its MV at the end of five years is estimated as $40,000. Assume it will be depreciated under the GDS: (7.4)

a. What is the cumulative depreciation through the end of year three?

b. What is the MACRS depreciation in the fourth year?

c. What is the BV at the end of year two?

7-12. A construction company is considering changing its depreciation from the MACRS method to the historical SL method for a general purpose hauling truck. The cost basis of the truck is $100,000, and the expected salvage value for depreciation purposes is $8,000. The company will use the truck for eight years and will depreciate it over this period of time with the SL method. What is the difference in the amount of depreciation that would be claimed in year five (i.e., MACRS versus SL)? (7.3, 7.4)

7-13. A piece of construction equipment (asset class 15.0) was purchased by the Jones Construction Company. The cost basis was $300,000.

a. Determine the GDS and ADS depreciation deductions for this property. (7.4)

b. Compute the difference in PW of the two sets of depreciation deductions in Part (a) if $i = 12\%$ per year. (7.5)

7-14. During its current tax year (year one), a pharmaceutical company purchased a mixing tank that had a fair market price of $120,000. It replaced an older, smaller mixing tank that had a BV of $15,000. Because a special promotion was underway, the old tank was used as a trade-in for the new one, and the cash price (including delivery and installation) was set at $99,500. The MACRS class life for the new mixing tank is 9.5 years. (7.4, 7.3)

a. Under the GDS, what is the depreciation deduction in year three?

b. Under the GDS, what is the BV at the end of year four?

c. If 200% DB depreciation had been applied to this problem, what would be the cumulative depreciation through the end of year four?

7-15. A manufacturer of aerospace products purchased three flexible assembly cells for $500,000 each. Delivery and insurance charges were $35,000, and installation of the cells cost another $50,000. (7.4, 7.8)

a. Determine the cost basis of the three cells.

b. What is the class life of the cells?

c. What is the MACRS depreciation in year five?

d. If the cells are sold to another company for $120,000 each at the end of year six, how much is the recaptured depreciation?

7-16. A special-purpose machine is to be depreciated as a linear function of use (units-of-production method). It costs $25,000 and is expected to produce 100,000 units and then be sold for $5,000. Up to the end of the third year, it had produced 60,000 units, and during the fourth year it produced 10,000 units. What is the depreciation deduction for the fourth year and the BV at the end of the fourth year? (7.3)

7-17. A concrete and rock crusher for demolition work has been purchased for $60,000, and it has an estimated SV of $10,000 at the end of its five-year life. Engineers have estimated that the following units of production (in m^3 of crushed material) will be contracted over the next five years.

EOY	1	2	3	4	5
m^3	16,000	24,000	36,000	16,000	8,000

Using the units of production depreciation method, what is the depreciation allowance in year three, and what is the BV at the end of year two? (7.3)

7-18. Consider a firm that had a taxable income of $90,000, and total gross revenues of $220,000 in the current tax year. Based on this information, answer these questions: (7.6, 7.9)

a. How much federal income tax was paid for the tax year?

b. What was the total amount of deductible expenses (e.g., materials, labor, fuel) and depreciation deductions claimed in the tax year?

7-19. Suppose state income taxes and local income taxes are treated as expenses for purposes of calculating federal taxable income and hence federal income taxes. Determine the effective income tax rate when the federal income tax rate is 35%, the state income tax rate is 6%, and the local income tax rate is 1%. (7.7)

7-20. If the incremental federal income tax rate is 34% and the incremental state income tax rate is 6%, what is the effective combined income tax rate (t)? If state income taxes are 12% of taxable income, what now is the value of t? (7.7)

7-21. The before-tax MARR for a particular firm is 18% per year. The state income tax rate is 5%, and the federal income tax rate is 39%. State income taxes are deductible from federal taxable income. What is this firm's after-tax MARR? (7.7)

7-22. A $125,000 tractor-trailer is being depreciated by the SL method over five years to a final BV of zero. Half-year convention does not apply to this asset. After three years, the rig is sold for (a) $70,000 or (b) $20,000. If the effective income tax rate is 40%, what is the net cash inflow from the sale for situation (a) and situation (b)? (7.8)

7-23. An injection molding machine can be purchased and installed for $90,000. It is in the seven-year GDS property class and is expected to be kept in service for eight years. It is believed that $10,000 can be obtained when the machine is disposed of at the end of year eight. The net annual *value added* (i.e., revenues less expenses) that can be attributed to this machine is constant over eight years and amounts to $15,000. An effective income tax rate of 40% is used by the company, and the after-tax MARR equals 15% per year. (7.4, 7.9)

a. What is the approximate value of the company's before-tax MARR?

b. Determine the GDS depreciation amounts in years one through eight.

c. What is the taxable income at the end of year eight that is related to capital investment?

d. Set up a table and calculate the ATCF for this machine.

e. Should a recommendation be made to purchase the machine?

7-24. A firm is considering investing $10 million in equipment that is expected to have a useful life of four years and is expected to reduce (save) the firm's labor costs by $4 million per year. Assume the firm pays a 40% income tax rate on its taxable income and uses the SL depreciation method. (7.9)

a. What is the ATCF from the investment in years one through four?

b. If the firm's after-tax hurdle rate is 15% per year, what are the investment's PW and IRR?

7-25. An office supply company has purchased a light-duty delivery truck for $15,000. The truck will be depreciated under the MACRS with a property class of five years. The truck's MV is expected to decrease by $2,500 per year. It is anticipated that the purchase of the

truck will increase the company's revenue by $10,000 annually, while the associated operating expenses are expected to be $3,000 annually. The company has an effective income tax rate of 40%, and its after-tax MARR is 15% per year. If the company plans to keep the truck only two years, what would be the equivalent after-tax annual worth of this investment? (7.9)

7-26. An assembly operation at a software company currently requires $100,000 per year in labor costs. A robot can be purchased and installed to automate this operation, and the robot will cost $200,000 with no MV at the end of its 10-year life. The robot, if acquired, will be depreciated using SL depreciation to a terminal BV of zero after 10 years. Maintenance and operation expenses of the robot are estimated to be $64,000 per year. The company has an effective income tax rate of 40%. Invested capital must earn at least 8% after income taxes are taken into account. (7.9)

a. Use the IRR method to determine if the robot is a justifiable investment.

b. If MACRS (seven-year recovery period) had been used in Part (a), would the after-tax IRR be lower or higher than your answer to Part (a)?

7-27. Liberty Airways is considering an investment of $800,000 in ticket purchasing kiosks at selected airports. The kiosks (hardware and software) have an expected life of four years. Extra ticket sales are expected to be 60,000 per year at a discount price of $40 per ticket. Fixed costs, excluding depreciation of the equipment, are $400,000 per year, and variable costs are $24 per ticket. The kiosks will be depreciated over four years, using the SL method with a zero salvage value. The onetime commitment of working capital is expected to be 1/12 of annual sales dollars. The after-tax MARR is 15% per year, and the company pays income tax at the rate of 34%. What's the after-tax PW of this proposed investment? Should the investment be made? (7.9)

7-28. A new municipal refuse-collection truck can be purchased for $84,000. Its expected useful life is six years, at which time the MV and BV will be zero. BTCF will be approximately $18,000 per year over the six-year useful life of the truck. Use SL depreciation, an effective income tax rate of 40%, and an after-tax MARR of 12% to determine the present worth of the investment. What is the after-tax IRR? Is this truck a sound investment? Explain your answer. (7.9)

7-29. Your company has just signed a three-year nonrenewable contract with the city of New Orleans for earthmoving work. You are investigating the purchase of heavy construction equipment for this job. The equipment costs $200,000 and qualifies for five-year MACRS depreciation. At the end of the three-year contract, you expect to be able to sell the equipment for $70,000. If the projected operating expense for the equipment is $65,000 per year, what is the after-tax equivalent uniform annual cost (EUAC) of owning and operating this equipment? The effective income tax rate is 40%, and the after-tax MARR is 12% per year. (7.9)

7-30. The Greentree Lumber Company is attempting to evaluate the profitability of adding another cutting line to its present sawmill operations. They would need to purchase two more acres of land for $30,000 (total). The equipment would cost $130,000 and could be depreciated over a five-year recovery period with the MACRS method. Gross revenue is expected to increase by $50,000 per year for five years, and operating expenses will be $15,000 annually for five years. It is expected that this cutting line will be closed down after five years. The firm's effective income-tax rate is 50%. If the company's after-tax MARR is 5% per year, is this a profitable investment? (7.9)

7-31. A drug store is looking into the possibility of installing a "24/7" automated prescription refill system to increase its projected revenues by $20,000 per year over the next five years. Annual expenses to maintain the system are expected to be $5,000. The system will have no market value at the end of its five-year life, and it will be depreciated by the SL method. The store's effective income tax rate is 40%, and the after-tax MARR is 12% per year. What is the maximum amount that is justified for the purchase of this prescription refill system? (7.9)

7-32. Your company is considering the introduction of a new product line. The initial investment required for this project is $500,000, and annual maintenance costs are anticipated to be $35,000. Annual operating cost will be directly in proportion to the level of production at $7.50 per unit, and each unit of product can be sold for $50.00. If the project has a life of five years, what is the minimum annual production level for which this project is economically viable? Work this problem on an after-tax basis. Assume five-year SL depreciation ($SV_5 = 0$), an effective income tax rate of 40%, and an after-tax MARR of 10% per year. (7.9)

7-33. Your company has purchased equipment (for $50,000) that will reduce materials and labor costs by

$14,000 each year for N years. After N years, there will be no further need for the machine, and because the machine is specially designed, it will have no MV at any time. The IRS, however, has ruled that you must depreciate the equipment on a SL basis with a tax life of five years. If the effective income-tax rate is 40%, what is the minimum number of years your firm must operate the equipment to earn 10% per year after taxes on its investment? (7.9)

7-34. Refer to Problem 6-79. The alternatives all have a MACRS (GDS) property class of three years. If the effective income tax rate is 40% and the after-tax MARR $= (1 - 0.4)(12\%) = 7.2\%$ per year, which alternative should be recommended? Is this the same recommendation you made when the alternatives were analyzed on a before-tax basis? (7.10)

7-35. The following information is for a proposed project that will provide the capability to produce a specialized product estimated to have a short market (sales) life:

- Capital investment is $1,000,000. (This includes land and working capital.)
- The cost of depreciable property, *which is part* of the $1,000,000 total estimated project cost, is $420,000.
- Assume, for simplicity, that the depreciable property is in the MACRS (GDS) three-year property class.
- The analysis period is three years.
- Annual operating and maintenance expenses are $636,000 in the first year, and they increase at the rate of 6% per year (i.e., $\bar{f} = 6\%$) thereafter. (See geometric gradient, Chapter 4.)
- Estimated MV of depreciable property from the project at the end of three years is $280,000.
- Federal income tax rate $= 34\%$; state income tax rate $= 4\%$.
- MARR (after taxes) is 10% per year.

Based on an after-tax analysis using the PW method, what minimum amount of equivalent uniform annual revenue is required to justify the project economically? (7.9, 7.10)

7-36. In a chlorine-fluxing installation in a large aluminum company, engineers are considering the replacement of existing plastic pipe fittings with more expensive, but longer lived, copper fittings. The following table gives a comparison of the capital investments, lives, salvage values, and so on of the two mutually exclusive alternatives under consideration:

	Plastic	Copper
Capital investment	$5,000	$10,000
Useful (class) life	5 years	10 years
Salvage value for depreciation purposes	$1,000(= SV_5)	$5,000(= SV_{10})
Annual expenses	$300	$100
Market value at end of useful life	$0	$0

Depreciation amounts are calculated with the SL method. Assume an income-tax rate of 40% and a MARR after-taxes of 12% per year. Which pipe fitting would you select and why? Carefully list all assumptions that you make in performing the analysis. (7.9, 7.10)

7-37. An industrial coal-fired boiler for process steam is equipped with a 10-year-old electrostatic precipitator (ESP). Changes in coal quality have caused stack emissions to be in noncompliance with federal standards for particulates. Two mutually exclusive alternatives have been proposed to rectify this problem (doing nothing is not an option).

	New Baghouse	New ESP
Capital investment	$1,140,000	$992,500
Annual operating expenses	115,500	73,200

The life of both alternatives is 10 years, the effective income tax rate is 40%, and the after-tax MARR is 9% per year. Both alternatives qualify as seven-year MACRS (GDS) properties. Make a recommendation regarding which alternative to select based on an after-tax analysis. (7.10)

7-38. Storage tanks to hold a highly corrosive chemical are currently made of material Z26. The capital investment in a tank is $30,000, and its useful life is eight years. Your company manufactures electronic components and uses the ADS under MACRS to calculate depreciation deductions for these tanks. The net MV of the tanks at the end of their useful life is zero. When a tank is four years old, it must be relined at a cost of $10,000. This cost is not depreciated and can be claimed as an expense during year four.

Instead of purchasing the tanks, they can be leased. A contract for up to 20 years of storage tank service can be written with the Rent-All Company. If your firm's after-tax MARR is 12% per year, what is the greatest annual amount that you can afford to pay for tank leasing without causing purchasing to be the more economical alternative? Your firm's effective income tax rate is 40%. State any assumptions you make. (7.4, 7.9)

7-39. Two fixtures are being considered for a particular job in a manufacturing firm. The pertinent data for their comparison are summarized in Table P7-39.

The effective federal and state income-tax rate is 50%. Depreciation recapture is also taxed at 50%. If the after-tax MARR is 8% per year, which of the two fixtures should be recommended? State any important assumptions you make in your analysis. (7.9)

7-40. Individual industries will use energy as efficiently as it is economical to do so, and there are several incentives to improve the efficiency of energy consumption. To illustrate, consider the selection of a new water pump. The pump is to operate 800 hours per year. Pump A costs $2,000, has an overall efficiency of 82.06%, and it delivers 11 hp. The other available alternative, pump B, costs $1,000, has an overall efficiency of 45.13%, and delivers 12.1 hp. Both pumps have a useful life of five years and will be sold at that time. (Remember 1 hp = 0.746 kW.)

TABLE P7-39 Table for Problem 7-39		
	Fixture X	Fixture Y
Capital investment	$30,000	$40,000
Annual operating expenses	$3,000	$2,500
Useful life	6 years	8 years
Market value	$6,000	$4,000
Depreciation method	SL to zero book value over 5 years	MACRS (GDS) with 5-year recovery period

Pump A will use SL depreciation over five years with an estimated SV of zero. Pump B will use the MACRS depreciation method with a class life of three years. After five years, pump A has an actual market value of $400, and pump B has an actual market value of $200.

Using the IRR method on the after-tax cash flows and a before-tax MARR of 16.67%, is the incremental investment in pump A economically justifiable? The effective income tax rate is 40%. The cost of electricity is $0.05/kWh, and the pumps are subject to a study period of five years. (7.10)

7-41. Two alternative machines will produce the same product, but one is capable of higher-quality work, which can be expected to return greater revenue. The following are relevant data:

	Machine A	Machine B
Capital investment	$20,000	$30,000
Life	12 years	8 years
Terminal BV (and MV)	$4,000	$0
Annual receipts	$150,000	$188,000
Annual expenses	$138,000	$170,000

Determine which is the better alternative, assuming "repeatability" and using SL depreciation, an income-tax rate of 40%, and an after-tax MARR of 10%. (7.9)

7-42. A firm must decide between two silicon layer chip designs from Intel. Their effective income tax rate is 40%, and MACRS depreciation is used. If the desired after-tax return on investment is 10% per year, which design should be chosen? State your assumptions. (7.10)

	Design A	Design B
Capital investment	$1,000,000	$2,000,000
MV at end of useful life	$1,000,000	$1,100,000
Annual revenues less expenses	$200,000	$400,000
MACRS property class	5 years	5 years
Useful life	7 years	6 years

7-43. Alternative Methods I and II are proposed for a security operation. The following is comparative information:

	Method I	Method II
Initial investment	$10,000	$40,000
Useful (ADR) life	5 years	10 years
Terminal market value	$1,000	$5,000
Annual expenses		
Labor	$12,000	$4,000
Power	$250	$300
Rent	$1,000	$500
Maintenance	$500	$200
Property taxes and insurance	$400	$2,000
Total annual expenses	$14,150	$7,000

Determine which is the better alternative based on an after-tax annual cost analysis with an effective income tax of 40% and an after-tax MARR of 12%, assuming the following methods of depreciation: (7.9)

a. SL

b. MACRS

7-44. A biotech company has an effective income tax rate of 40%. Recaptured depreciation is also taxed at the rate of 40%. The company must choose one of the following mutually exclusive cryogenic freezers for its tissue samples. The after-tax MARR is 12% per year. Which freezer should be selected based on after-tax present worth? (7.10)

	Freezer 1	Freezer 2
Capital investment	$11,000	$33,000
Annual benefit	$3,000	$9,000
Depreciation method	SL	MACRS
Depreciable life	3 years	3 years
IRS approved SV for depreciation	$2,000	$0
Useful life	5 years	5 years
Actual MV at end of useful life	$2,000	$2,000

7-45. A manufacturing process can be designed for varying degrees of automation. The following is relevant cost information:

Degree	First Cost	Annual Labor Expense	Annual Power and Maintenance Expense
A	$10,000	$9,000	$500
B	14,000	7,500	800
C	20,000	5,000	1,000
D	30,000	3,000	1,500

Determine which is best by after-tax analysis using an income-tax rate of 40%, an after-tax MARR of 15%, and SL depreciation. Assume that each has a life of five years and no BV or MV. (7.9)

7-46. Allen International, Inc., manufactures chemicals. It needs to acquire a new piece of production equipment to work on production for a large order that Allen has received. The order is for a period of three years, and at the end of that time the machine would be sold.

Allen has received two supplier quotations, both of which will provide the required service. Quotation I has a first cost of $180,000 and an estimated salvage value of $50,000 at the end of three years. Its cost for operation and maintenance is estimated at $28,000 per year. Quotation II has a first cost of $200,000 and an estimated salvage value of $60,000 at the end of three years. Its cost for operation and maintenance is estimated at $17,000 per year. The company pays income tax at a rate of 40% on ordinary income and 28% on depreciation recovery. The machine will be depreciated using MACRS-GDS (asset class 28.0). Allen uses an after-tax MARR of 12% for economic analysis, and it plans to accept whichever of these two quotations costs less. (7.10)

To perform an after-tax analysis to determine which of these machines should be acquired, you must

a. State the study period you are using.
b. Show all numbers necessary to support your conclusions.
c. State what the company should do.

7-47. AMT, Inc., is considering the purchase of a digital camera for maintenance of design specifications by feeding digital pictures directly into an engineering workstation where computer-aided design files can be superimposed over the digital pictures. Differences between the two images can be noted, and corrections, as appropriate, can then be made by design engineers. (7.12)

a. You have been asked by management to determine the PW of the EVA of this equipment, assuming the following estimates: capital investment = $345,000; market value at end of year six = $120,000; annual revenues = $120,000; annual expenses = $8,000; equipment life = 6 years; effective income tax rate = 50%; and after-tax MARR = 10% per year. MACRS depreciation will be used with a five-year recovery period.

b. Compute the PW of the equipment's ATCFs. Is your answer in Part (a) the same as your answer in Part (b)?

7-48. Refer to Example 7-15. Show that the PW of the annual EVA amounts by the new machinery is the same as the PW of the ATCF amounts ($17,208) given in Table 7-6. (7.10, 7.11)

7-49. Rework Example 7-21 using the MACRS depreciation method (assume three-year property class) instead of the SL depreciation method. (7.11)

7-50. *Extended Learning Exercise*
You have the option to purchase or lease a five-axis horizontal machining center. Any revenues generated from the operation of the machine will be the same whether it is leased or purchased. Considering the information given, should you lease or purchase the machine? Conduct after-tax analyses of both options. The effective income tax rate is 40%, the evaluation period is five years, and the MARR is 10% per year. NOTES: (1) Under the Lease Option, maintenance costs are included in the annual leasing cost. (2) Leasing costs are paid at the beginning of each year and are tax deductible. (3) Depreciation deductions cannot be taken on leased equipment. (4) Deposits are not tax deductible, and refunds of deposits are not taxable; however, owing to the difference in timing between payment and refund, they must be considered in your analysis. (7.10)

Leasing Option
Annual leasing cost: $55,000
Deposit (paid at EOY zero, refunded at EOY five): $75,000
Purchasing Option
Purchase price: $350,000 *capital to be borrowed at i = 8%, equal annual payments (Principal + Interest) for three years*
Depreciation: three year, MACRS
Annual maintenance cost: $20,000
Resale value at EOY five: $150,000

7-51. Suppose that you invest $200 per month (before taxes) for 30 years (360 payments) and the annual interest rate (APR) is 8%, compounded monthly. If your income tax bracket is 28%, what lump sum, after-tax distribution can be taken at the end of 30 years? (7.7)

7-52. A $5,000 balance in a tax-deferred savings plan will grow to $50,313.50 in 30 years at an 8% per year interest rate. What would be the future worth if the $5,000 had been subject to a 28% income tax rate? (7.7)

7-53. Determine the after-tax yield (i.e., IRR on the ATCF) obtained by an individual who purchases a $10,000, 10-year, 10% nominal interest rate bond. The following information is given: (7.7)

- Interest is paid semi-annually, and the bond was bought after the fifth payment had just been received by the previous owner.

- The purchase price for the bond was $9,000.

- All revenues (including capital gains) are taxed at an income rate of 28%.

- The bond is held to maturity.

7-54. A 529-state-approved Individual Retirement Account (IRA) permits parents to invest tax-free dollars into their children's college education fund (this money may only be used for educational expenses). Another popular plan, the Roth IRA, requires after-tax dollars to be invested in a savings fund that may (or may not) be used for paying future college expenses. Both plans are tax free when the money is eventually withdrawn to assist with college expenses. Clearly, the 529 IRA plan is a better way to save for college expenses than the Roth IRA. Quantify "better" when the marginal income tax rate is 28% and $10,000 each year is invested in a mutual fund earning 8% per year for 10 years. *Note:*

The estimated cost of a college education 10 years from now is $110,000. (7.7)

7-55. A Roth IRA enables an individual to invest after-tax dollars during the accumulation phase of a retirement plan. The money is then income tax free when it is withdrawn during retirement. A tax-deductible IRA, on the other hand, provides an up-front tax deduction for the annual contribution, but it then requires income taxes to be paid on all future distributions. A basic assumption as to which plan is more beneficial concerns the current income tax rates versus their projected rates in the future.

To illustrate, suppose that $2,000 is available to invest at the end of each year for 30 years. The income tax rate now and into the foreseeable future is 28%, so $2,000(1 − 0.28) = $1,440 is invested annually into the Roth IRA. However, $2,000 per year can be invested into a tax-deductible IRA. Money invested under either plan will be deposited into a mutual fund earning 8% per year, and all accumulated money will be withdrawn as a lumpsum at the end of year 30. (7.7)

a. Which plan is better if future distributions of the traditional (tax-free) IRA are taxed at an income tax rate of 28%?

b. Which plan is better if the future income tax rate at retirement (end of year 30) is 30%?

7-56. Some experts believe that a new car can lose as much as 25%–30% of its value within a week after the car leaves the dealership! Visit a local automobile dealership (or Web site), and verify that this loss in value is true (or not true). Compare your findings with those of other members of your class. (7.2)

Spreadsheet Exercises

7-57. A bowling alley costs $500,000 and has a useful life of 10 years. Its estimated MV at the end of year 10 is $20,000. Create a spreadsheet that calculates the depreciation for years 1–10 using (i) the SL method, (ii) the 200% DB method, and (iii) the MACRS method (GDS class life = 10 years). For each method, compute the PW of the depreciation deductions (at EOY 0). The MARR is 10% per year. If a large PW is desirable, what do you conclude regarding which method is preferred? (7.5)

7-58. Create a spreadsheet to solve Problem 7-24. What would the MV of the equipment have to be (at the

end of year four) for the firm to consider making this investment? (7.9)

7-59. Interest on municipal bonds is usually exempt from federal income taxes. The interest rate on these bonds is therefore an *after-tax* rate of return (ROR). Other types of bonds (e.g., corporate bonds) pay interest that is taxable for federal income tax purposes. Thus, the before-tax ROR on such bonds is typically higher than the ROR on municipal bonds. Develop a spreadsheet that contains the before-tax ROR (on taxable bonds) that are equivalent to after-tax RORs of 4%, 5%, and 6% for income tax rates of 15%, 28%, and 35%. (7.6)

7-60. Refer to the chapter opener and Example 7-14. As an alternative to the coal-fired plant, PennCo could construct an 800 MW natural gas–fired plant. This plant would require the same initial investment of $1.12 billion dollars to be depreciated over its 30-year life using the SL method with $SV_{30} = 0$. The capacity factor estimate of the plant would still be 80%, but the efficiency of the natural gas–fired plant would be 40%. The annual operating and maintenance expense is expected to be $0.01 per kWh. The cost of natural gas is $8.00 per million Btu and the carbon dioxide tax is $15 per metric ton. Natural gas emits 55 metric tons of carbon dioxide per billion Btu produced. The effective income tax rate is 40%, and the after-tax MARR is 10% per year. Based on the after-tax cost of electricity, create a spreadsheet to determine whether PennCo should construct a natural gas–fired or coal-fired plant. *Note:* 1 kWh = 3,413 Btu. (7.9)

FE Practice Problems

Acme Manufacturing is considering the purchase of a new punch press. This press will cost $150,000. This asset will be depreciated using an MACRS (GDS) recovery period of seven years. Use this information to solve Problems 7-61 to 7-63.

7-61. The depreciation amount in the third year is

(a) $26,235   (b) $28,800   (c) $36,735
(d) $48,000

7-62. The BV at the end of the second year is

(a) $112,500   (b) $113,265   (c) $91,830
(d) $76,525

7-63. If the punch press is sold during the fourth year of ownership, the allowable depreciation charge for the fourth year is

(a) $6,698   (b) $9,368   (c) $17,280
(d) $18,735

An oil refinery has decided to purchase some new drilling equipment for $550,000. The equipment will be kept for 10 years before being sold. The estimated SV for depreciation purposes is to be $25,000. Use this information to solve Problems 7-64 to 7-67.

7-64. Using the SL method, the annual depreciation on the equipment is

(a) $50,000   (b) $51,500   (c) $52,500
(d) $55,000

7-65. Using the SL method, the BV at the end of the depreciable life is

(a) $0   (b) $25,000   (c) $35,000
(d) $50,000

7-66. If SL depreciation is used and the equipment is sold for $35,000 at the end of 10 years, the taxable gain on the disposal of the equipment is

(a) $35,000   (b) $25,000   (c) $15,000
(d) $10,000

7-67. If MACRS depreciation is used, the recovery period of the equipment using the GDS guidelines is

(a) 3 years   (b) 5 years   (c) 7 years
(d) 10 years

A wood products company has decided to purchase new logging equipment for $100,000 with a trade-in of its old equipment. The old equipment has a BV of $10,000 at the time of the trade-in. The new equipment will be kept for 10 years before being sold. Its estimated SV at the time is expected to be $5,000. Use this information to solve Problems 7-68 through 7-72.

7-68. The recovery period of the asset, using the GDS guidelines, is

(a) 10 years   (b) 7 years   (c) 5 years
(d) 3 years

7-69. Using the SL method, the depreciation on the equipment over its depreciable life period is

(a) $10,500   (b) $9,500   (c) $8,000
(d) $7,000

7-70. Using the SL method, the BV at the end of the depreciable life is

(a) $11,811   (b) $10,000   (c) $5,000   (d) $0

7-71. Using the MACRS (GDS recovery period), the depreciation charge permissible at year 6 is equal to

(a) $9,812   (b) $6,336   (c) $4,912   (d) $0

7-72. Using the MACRS (GDS recovery period), if the equipment is sold in year five, the BV at the end of year five is equal to

(a) $29,453   (b) $24,541   (c) $12,672
(d) $6,336

7-73. A small pump costs $16,000 and has a life of eight years and a $2,000 SV at that time. If the 200% DB method is used to depreciate the pump, the BV at the end of year four is

(a) $9,000   (b) $8,000   (c) $6,000
(d) $5,000

7-74. Air handling equipment that costs $12,000 has a life of eight years with a $2,000 SV. What is the SL depreciation amount for each year?

(a) $1,500   (b) $1,000   (c) $1,200
(d) $1,250

7-75. The air handling equipment just described is to be depreciated, using the MACRS with a GDS recovery period of seven years. The BV of the equipment at the end of (including) year four is most nearly

(a) $3,749   (b) $3,124   (c) $5,000
(d) $8,251

7-76. If the federal income tax rate is 35% and the state tax rate is 5% (and state taxes are deductible from federal taxes), the effective income tax rate is

(a) 35%   (b) 37.5%   (c) 38.3%
(d) 40%

7-77. If a company's total effective income tax rate is 40% and its state income tax rate is 20%, what is the company's federal income tax rate?

(a) 20%   (b) 25%   (c) 35%
(d) 40%   (e) 52%

7-78. Acme Manufacturing makes their preliminary economic studies using a before-tax MARR of 18%. More detailed studies are performed on an after-tax basis. If their effective tax rate is 40%, the after-tax MARR is

(a) 6%   (b) 7%   (c) 11%   (d) 13%

7-79. Suppose for some year the income of a small company is $110,000; the expenses are $65,000; the depreciation is $25,000; and the effective income tax rate = 40%. For this year, the ATCF is most nearly

(a) −$8,900   (b) $4,700   (c) $13,200
(d) $29,700   (e) $37,000

Your company is contemplating the purchase of a large stamping machine. The machine will cost $180,000. With additional transportation and installation costs of $5,000 and $10,000, respectively, the cost basis for depreciation purposes is $195,000. Its MV at the end of five years is estimated as $40,000. The IRS has assured you that this machine will fall under a three-year MACRS class life category. The justifications for this machine include $40,000 savings per year in labor and $30,000 savings per year in reduced materials. The before-tax MARR is 20% per year, and the effective income tax rate is 40%. Use this information to solve problems 7-80 through 7-83.

7-80. The *total* before-tax cash flow in year five is most nearly (assuming you sell the machine at the end of year five):

(a) $9,000   (b) $40,000   (c) $70,000
(d) $80,000   (e) $110,000

7-81. The taxable income for year three is most nearly

(a) $5,010   (b) $16,450   (c) $28,880
(d) $41,120   (e) $70,000

7-82. The PW of the *after-tax savings* from the machine, *in labor and materials only*, (neglecting the first cost, depreciation, and the salvage value) is most nearly (using the after tax MARR)

(a) $12,000   (b) $95,000   (c) $151,000
(d) $184,000   (e) $193,000

7-83. Assume the stamping machine will now be used for only three years, owing to the company's losing several government contracts. The MV at the end of year three is $50,000. What is the income tax owed at the end of year three owing to depreciation recapture (capital gain)?

(a) $8,444   (b) $14,220   (c) $21,111
(d) $35,550   (e) $20,000

CHAPTER 8

Price Changes and Exchange Rates

Chapter 8 discusses how inflation/deflation is dealt with in engineering economy studies.

The Impact of Inflation on Household Income

According to government statistics, the median household income in the United States in 1953 was $3,700 per year. Using the consumer price index as a measure of inflation, this is equivalent to $30,000 in 2008. This means that it requires $30,000 in 2008 to purchase the same amount of goods and services that $3,700 would have bought you in 1953. After studying this chapter, you will be able to determine such figures as the compounded annual rate of growth in median household income from 1953 to 2008, as well as many more.

Inflation isn't ruining everything. A dime can still be used as a screwdriver.

—James "Doc" Blakely

8.1 Introduction

In earlier chapters, we assumed that prices for goods and services in the market-place remain relatively unchanged over extended periods. Unfortunately, this is not generally a realistic assumption.

> *General price inflation*, which is defined here as an increase in the average price paid for goods and services bringing about a reduction in the purchasing power of the monetary unit, is a business reality that can affect the economic comparison of alternatives.

The history of price changes shows that price inflation is much more common than general price *deflation*, which involves a decrease in the average price for goods and services with an increase in the purchasing power of the monetary unit. The concepts and methodology discussed in this chapter, however, apply to any price changes.

One measure of price changes in our economy (and an estimate of general price inflation or deflation for the average consumer) is the consumer price index (CPI). The CPI is a composite price index that measures average change in the prices paid for food, shelter, medical care, transportation, apparel, and other selected goods and services used by individuals and families. The average annual rate of CPI inflation from 1982 to 1994 was 3.33%. For the period 1994–2004, the annual rate of CPI inflation was only 2.59%.

Another measure of price changes in the economy (and also an estimate of general price inflation or deflation) is the producer price index (PPI). In actuality, a number of different indexes are calculated covering most areas of the U.S. economy. These indexes are composite measures of average changes in the selling prices of items used in the production of goods and services. These different indexes are calculated by stage of production [crude materials (e.g., iron ore), intermediate materials (rolled sheet steel), and finished goods (automobiles)], by standard industrial classification (SIC), and by the census product code extension of the SIC areas. Thus, PPI information is available to meet the needs of most engineering economy studies.

The CPI and PPI indexes are calculated monthly from survey information by the Bureau of Labor Statistics in the U.S. Department of Labor. These indexes are based on current and historical information and may be used, as appropriate, to represent future economic conditions or for short-term forecasting purposes only.

The published annual CPI and PPI end-of-year values can be used to obtain an estimate of general price inflation (or deflation). This is accomplished by computing an annual change rate (%). The annual change rates are calculated as follows:

$$(\text{CPI or PPI annual change rate, } \%)_k = \frac{(\text{Index})_k - (\text{Index})_{k-1}}{(\text{Index})_{k-1}}(100\%).$$

For example, the end-of-year CPI for 2006 was 201.6, and the estimated end-of-year CPI for 2007 is 208.0. The CPI annual change rate for 2007 is then estimated to be

$$\frac{(CPI)_{2007} - (CPI)_{2006}}{(CPI)_{2006}}(100\%) = \frac{208.0 - 201.6}{201.6}(100\%) = 3.17\%.$$

In this chapter, we develop and illustrate proper techniques to account for the effects of price changes caused by inflation and deflation. The case study at the conclusion of the chapter illustrates the analysis of alternatives when price changes are expected to occur.

8.2 Terminology and Basic Concepts

To facilitate the development and discussion of the methodology for including price changes of goods and services in engineering economy studies, we need to define and discuss some terminology and basic concepts. The dollar is used as the monetary unit in this book except when discussing foreign exchange rates.

1. *Actual dollars (A$)* The number of dollars associated with a cash flow (or a noncash-flow amount such as depreciation) as of the time it occurs. For example, people typically anticipate their salaries two years in advance in terms of actual dollars. Sometimes, A$ are referred to as *nominal* dollars, *current* dollars, *then-current* dollars, and *inflated* dollars, and their relative purchasing power is affected by general price inflation or deflation.

2. *Real dollars (R$)* Dollars expressed in terms of the same purchasing power relative to a particular time. For instance, the future unit prices of goods or services that are changing rapidly are often estimated in real dollars (relative to some base year) to provide a consistent means of comparison. Sometimes, R$ are termed *constant* dollars.

3. *General price inflation (or deflation) rate (f)* A measure of the average change in the purchasing power of a dollar during a specified period of time. The general price inflation or deflation rate is defined by a selected, broadly based index of market price changes. In engineering economic analysis, the rate is projected for a future time interval and usually is expressed as an effective annual rate. Many large organizations have their own selected index that reflects the particular business environment in which they operate.

4. *Market (nominal) interest rate (i_m)* The money paid for the use of capital, normally expressed as an annual rate (%) that includes a market adjustment for the anticipated general price inflation rate in the economy. Thus, it is a *market interest rate* and represents the time value change in future actual dollar cash flows that takes into account both the potential real earning power of money and the estimated general price inflation or deflation in the economy.

5. *Real interest rate (i_r)* The money paid for the use of capital, normally expressed as an annual rate (%) that does *not* include a market adjustment for the anticipated general price inflation rate in the economy. It represents the time value change in future real-dollar cash flows based only on the potential real earning power of money. It is sometimes called the *inflation-free* interest rate.

6. *Base time period (b)* The reference or base time period used to define the constant purchasing power of real dollars. Often, in practice, the base time period is designated as the time of the engineering economic analysis or reference time zero (i.e., $b = 0$). However, b can be any designated point in time.

With an understanding of these definitions, we can delineate and illustrate some useful relationships that are important in engineering economy studies.

8.2.1 The Relationship between Actual Dollars and Real Dollars

> The relationship between actual dollars (A$) and real dollars (R$) is defined in terms of the general price inflation (or deflation) rate; that is, it is a function of f.

Actual dollars as of any period (e.g., a year), k, can be converted into real dollars of *constant* market purchasing power as of any base period, b, by the relationship

$$(R\$)_k = (A\$)_k \left(\frac{1}{1+f}\right)^{k-b} = (A\$)_k (P/F, f\%, k - b), \qquad (8\text{-}1)$$

for a given b value. This relationship between actual dollars and real dollars applies to the unit prices, or costs of fixed amounts of individual goods or services, used to develop (estimate) the individual cash flows related to an engineering project. The designation for a specific type of cash flow, j, would be included as

$$(R\$)_{k,j} = (A\$)_{k,j} \left(\frac{1}{1+f}\right)^{k-b} = (A\$)_{k,j} (P/F, f\%, k - b), \qquad (8\text{-}2)$$

for a given b value, where the terms R$_{k,j}$ and A$_{k,j}$ are the unit price, or cost for a fixed amount, of goods or services j in period k in real dollars and actual dollars, respectively.

EXAMPLE 8-1   **Real-Dollar Purchasing Power of Your Salary**

Suppose that your salary is $45,000 in year one, will increase at 4% per year through year four, and is expressed in actual dollars as follows:

End of Year, k	Salary (A$)
1	45,000
2	46,800
3	48,672
4	50,619

If the general price inflation rate (f) is expected to average 6% per year, what is the real-dollar equivalent of these actual-dollar salary amounts? Assume that the base time period is year one ($b = 1$).

Solution

By using Equation (8-2), we see that the real-dollar salary equivalents are readily calculated relative to the base time period, $b = 1$:

Year	Salary (R\$, $b = 1$)
1	$45,000(P/F, 6\%, 0) = 45,000$
2	$46,800(P/F, 6\%, 1) = 44,151$
3	$48,672(P/F, 6\%, 2) = 43,318$
4	$50,619(P/F, 6\%, 3) = 42,500$

In year one (the designated base time period for the analysis), the annual salary in actual dollars remained unchanged when converted to real dollars. *This illustrates an important point: In the base time period (b), the purchasing power of an actual dollar and a real dollar is the same; that is, $R\$_{b,j} = A\$_{b,j}$.* This example also illustrates the results when the actual annual rate of increase in salary (4% in this example) is less than the general price inflation rate (f). As you can see, the actual-dollar salary cash flow shows some increase, but a decrease in the real-dollar salary cash flow occurs (and thus a decrease in total market purchasing power). This is the situation when people say their salary increases have not kept pace with market inflation.

EXAMPLE 8-2   Real-Dollar Equivalent of Actual After-Tax Cash Flow

An engineering project team is analyzing the potential expansion of an existing production facility. Different design alternatives are being considered. The estimated after-tax cash flow (ATCF) in actual dollars for one alternative is shown in column 2 of Table 8-1. If the general price inflation rate (f) is estimated to be 5.2% per year during the eight-year analysis period, what is the real-dollar ATCF that is equivalent to the actual-dollar ATCF? The base time period is year zero ($b = 0$).

Solution

The application of Equation (8-1) is shown in column 3 of Table 8-1. The ATCF in real dollars shown in column 4 has purchasing power in each year equivalent to the original ATCF in actual dollars (column 2).

TABLE 8-1 ATCFs for Example 8-2

(1) End-of-Year, k	(2) ATCF (A\$)	(3) $(P/F, f\%, k-b)$ $= [1/(1.052)^{k-0}]$	(4) ATCF (R\$), $b=0$
0	−172,400	1.0	−172,400
1	−21,000	0.9506	−19,963
2	51,600	0.9036	46,626
3	53,000	0.8589	45,522
4	58,200	0.8165	47,520
5	58,200	0.7761	45,169
6	58,200	0.7377	42,934
7	58,200	0.7013	40,816
8	58,200	0.6666	38,796

EXAMPLE 8-3 Inflation and Household Income

In this example, we return to the question raised in the chapter opener. According to government statistics, the median household income in the United States in 1953 was \$3,700 per year. Based on the CPI, this is equivalent to \$30,000 in 2008 dollars. We can now answer the following questions.

(a) What was the annual compound rate of growth in median household income from 1953 to 2008?

(b) By 1978, median household income was actually \$15,000. Based on your answer in Part (a), what is the equivalent household income in 2008?

(c) Was the actual median household income in 1978 higher than what you would have predicted in Part (a)? If so, give some reasons for this phenomenon.

Solution

(a) The annual compound rate of growth is the interest rate that makes \$3,700 in 1953 equivalent to \$30,000 in 2008.

$$\$30,000 = \$3,700(F/P, i'\%, 55) = \$3,700 (1 + i')^{55}$$

$$i'\% = 3.88\% \text{ per year}$$

(b) Income$_{2008}$ = \$15,000$(F/P, 3.88\%, 30)$ = \$15,000(3.133) = \$46,995
This is higher than \$30,000 because inflation eased.

(c) Income$_{1978}$ = \$3,700$(F/P, 3.88\%, 25)$ = \$9,583

The actual 1978 income of \$15,000 is greater than \$9,583 because more persons from an individual household were in the workforce in 1978 than in 1953. Also, the inflation rate may have exceeded 3.88% on the average during this time period.

8.2.2 The Correct Interest Rate to Use in Engineering Economy Studies

In general, the interest rate that is appropriate for equivalence calculations in engineering economy studies depends on whether actual-dollar or real-dollar cash-flow estimates are used:

Method	If Cash Flows Are in Terms of	Then the Interest Rate to Use Is
A	Actual dollars (A$)	Market interest rate, i_m
B	Real dollars (R$)	Real interest rate, i_r

This table should make intuitive sense as follows: If one is estimating cash flows in terms of actual (inflated) dollars, the market interest rate (interest rate with an inflation/deflation component) is used. Similarly, if one is estimating cash flows in terms of real dollars, the real (inflation-free) interest rate is used. Thus, one can make economic analyses in either the actual- or real-dollar domain with equal validity, provided that the appropriate interest rate is used for equivalence calculations.

It is important to be consistent in using the correct interest rate for the type of analysis (actual or real dollars) being done. The two mistakes commonly made are as follows:

Interest Rate (MARR)	Type of Analysis	
	A$	R$
i_m	√(Correct)	Mistake 1 Bias is against capital investment
i_r	Mistake 2 Bias is toward capital investment	√(Correct)

In Mistake 1, the market interest rate (i_m), which includes an adjustment for the general price inflation rate (f), is used in equivalent-worth calculations for cash flows estimated in real dollars. There is a tendency to develop future cash-flow estimates in terms of dollars with purchasing power at the time of the study (i.e., real dollars with $b = 0$), and then to use the market interest rate in the analysis [a firm's MARR$_m$ is normally a combined (market) interest rate]. The result of Mistake 1 is a bias *against* capital investment. The cash-flow estimates in real dollars for a project are numerically lower than the actual-dollar estimates with equivalent purchasing power (assuming that $f > 0$). Additionally, the i_m value (which is usually greater than the i_r value that should be used) further reduces (understates) the equivalent worth of the results of a proposed capital investment.

In Mistake 2, the cash-flow estimates are in actual dollars, which include the effect of general price inflation (f), but the real interest rate (i_r) is used for equivalent-worth calculations. Since the real interest rate does not include an adjustment for general price inflation, we again have an inconsistency. The effects of this mistake, in contrast to those in Mistake 1, result in a bias *toward* capital investment by overstating the equivalent worth of future cash flows.

8.2.3   The Relationship among i_m, i_r, and f

Equation (8-1) showed that the relationship between an actual-dollar amount and a real-dollar amount of equal purchasing power in period k is a function of the general inflation rate (f). It is desirable to do engineering economy studies in terms of either actual dollars or real dollars. Thus, the relationship between the two dollar domains is important, as well as the relationship among i_m, i_r, and f, so that the equivalent worth of a cash flow is equal in the base time period when either an actual- or a real-dollar analysis is used. The relationship among these three factors is (derivation not shown)

$$1 + i_m = (1 + f)(1 + i_r); \tag{8-3}$$

$$i_m = i_r + f + i_r(f); \tag{8-4}$$

$$i_r = \frac{i_m - f}{1 + f}. \tag{8-5}$$

Similarly, based on Equation (8-5), the internal rate of return (IRR) of a real-dollar cash flow is related to the IRR of an actual-dollar cash flow (with the same purchasing power each period) as follows: $\text{IRR}_r = (\text{IRR}_m - f)/(1 + f)$.

EXAMPLE 8-4   **Real-Dollar Equivalent of an Investment**

Suppose that $1,000 is deposited each year for five years into an equity (common stock) account earning 8% per year. During this period, general inflation is expected to remain at 3% per year. At the end of five years, what is the dollar value of the account in terms of today's purchasing power (i.e., real dollars)?

Solution

Immediately after the fifth deposit, the actual dollar value of the equity account is

$$(A\$)_5 = \$1,000 \ (F/A, \ 8\%, \ 5) = \$5,866.60.$$

The value of the account in today's purchasing power is

$$(R\$)_5 = \$5,866.60(P/F, \ 3\%, \ 5) = \$5,060.53.$$

EXAMPLE 8-5	**Equivalence of Real-Dollar and Actual-Dollar Cash Flows**

In Example 8-1, your salary was projected to increase at the rate of 4% per year, and the general price inflation rate was expected to be 6% per year. Your resulting estimated salary for the four years in actual and real dollars was as follows:

End of Year, k	Salary (A$)	Salary (R$), $b = 1$
1	45,000	45,000
2	46,800	44,151
3	48,672	43,318
4	50,619	42,500

What is the present worth (PW) of the four-year actual- and real-dollar salary cash flows at the end of year one (base year) if your personal MARR_m is 10% per year (i_m)?

Solution

(a) Actual-dollar salary cash flow:

$$PW(10\%)_1 = \$45,000 + \$46,800(P/F, 10\%, 1)$$
$$+ \$48,672(P/F, 10\%, 2) + \$50,619(P/F, 10\%, 3)$$
$$= \$165,798.$$

(b) Real-dollar salary cash flow:

$$i_r = \frac{i_m - f}{1 + f} = \frac{0.10 - 0.06}{1.06} = 0.03774, \text{ or } 3.774\%;$$

$$PW(3.774\%)_1 = \$45,000 + \$44,151 \left(\frac{1}{1.03774}\right)^1$$

$$+ \$43,318 \left(\frac{1}{1.03774}\right)^2 + \$42,500 \left(\frac{1}{1.03774}\right)^3$$

$$= \$165,798.$$

Thus, we obtain the same PW at the end of year one (the base time period) for both the actual-dollar and real-dollar four-year salary cash flows when the appropriate interest rate is used for the equivalence calculations.

8.3 Fixed and Responsive Annuities

Whenever future cash flows are predetermined by contract, as in the case of a bond or a fixed annuity, these amounts do not respond to general price inflation or deflation. In cases where the future amounts are not predetermined, however, they may respond to general price changes. To illustrate the nature of this situation, let us consider two annuities. The first annuity is fixed (unresponsive to general price

TABLE 8-2   Illustration of Fixed and Responsive Annuities with General Price Inflation Rate of 6% per Year

End of Year k	Fixed Annuity		Responsive Annuity	
	In Actual Dollars	In Equivalent Real Dollars[a]	In Actual Dollars	In Equivalent Real Dollars[a]
1	$2,000	$1,887	$2,120	$2,000
2	2,000	1,780	2,247	2,000
3	2,000	1,679	2,382	2,000
4	2,000	1,584	2,525	2,000
5	2,000	1,495	2,676	2,000
6	2,000	1,410	2,837	2,000
7	2,000	1,330	3,007	2,000
8	2,000	1,255	3,188	2,000
9	2,000	1,184	3,379	2,000
10	2,000	1,117	3,582	2,000

[a] See Equation (8-1).

inflation) and yields $2,000 per year in actual dollars for 10 years. The second annuity is of the same duration and yields enough future actual dollars to be equivalent to $2,000 per year in real dollars (purchasing power). Assuming a general price inflation rate of 6% per year, pertinent values for the two annuities over a 10-year period are as shown in Table 8-2.

Thus, when the amounts are constant in actual dollars (unresponsive to general price inflation), their equivalent amounts in real dollars decline over the 10-year interval to $1,117 in the final year. When the future cash-flow amounts are fixed in real dollars (responsive to general price inflation), their equivalent amounts in actual dollars rise to $3,582 by year 10.

Included in engineering economy studies are certain quantities unresponsive to general price inflation. For instance, depreciation amounts, once determined, do not increase (with present accounting practices) to keep pace with general price inflation; lease fees and interest charges typically are contractually fixed for a given period of time. Thus, it is important when doing an actual-dollar analysis to recognize the quantities that are unresponsive to general price inflation, and when doing a real-dollar analysis to convert these A$ quantities to R$ quantities, using Equation (8-2).

If this is not done, not all cash flows will be in the same dollar domain (A$ or R$), and the analysis results will be distorted. Specifically, the equivalent worths of the cash flows for an A$ and an R$ analysis will not be the same in the base year, b, and the A$ IRR and the R$ IRR for the project will not have the proper relationship based on Equation (8-5); that is, $\text{IRR}_r = (\text{IRR}_m - f)/(1 + f)$.

In Example 8-6, we look at a bond (Chapter 5), which is a fixed-income asset, and see how its current value is affected by a period of projected deflation.

EXAMPLE 8-6	Impact of Deflation on the Current Price of a Bond

Suppose that deflation occurs in the U.S. economy and that the CPI (as a measure of f) is expected to decrease an average of 2% per year for the next five years. A bond with a face (par) value of $10,000 and a life of five years (i.e., it will be redeemed in five years) pays an interest (bond) rate of 5% per year. The interest is paid to the owner of the bond once each year. If an investor expects a *real* rate of return of 4% per year, what is the maximum amount that should be paid now for this bond?

Solution

The cash flows over the life of the bond are 0.05($10,000) = $500 per year in interest (actual dollars) for years one through five, plus the redemption of $10,000 (the face value of the bond), also in actual dollars, at the end of year five. To determine the current value of this bond (i.e., the maximum amount an investor should pay for it), these cash flows must be discounted to the present, using the market interest rate. From Equation (8-4), we can compute i_m (where $f = -2\%$ per year) as follows:

$$i_m = i_r + f + i_r(f) = 0.04 - 0.02 - 0.04(0.02)$$

$$= 0.0192, \text{or } 1.92\% \text{ per year.}$$

Therefore, the current market value of the bond is

$$\text{PW} = \$500(P/A, 1.92\%, 5) + \$10,000(P/F, 1.92\%, 5)$$

$$= \$500(4.7244) + \$10,000(0.9093)$$

$$= \$11,455.$$

In general, if the rate used to discount the future cash flows over the life of a bond is *less* than the bond rate (the situation in this example), then the current (market) value will be greater than the bond's face value. Therefore, during periods of deflation, owners of bonds (or other types of fixed-income assets) need to monitor their market values closely because a favorable *sell situation* may occur.

Engineering economy studies that include the effects of price changes caused by inflation or deflation may also include such items as interest charges, depreciation amounts, lease payments, and other contract amounts that are actual-dollar cash flows based on past commitments. They are generally *unresponsive* to further price changes. At the same time, many other types of cash flows (e.g., labor, materials) are *responsive* to market price changes. In Example 8-7, an *after-tax analysis* is presented that shows the correct handling of these different situations.

EXAMPLE 8-7	**After-Tax Analysis: A Mixture of Fixed and Responsive Cash Flows**

The cost of a new and more efficient electrical circuit switching equipment is $180,000. It is estimated (in base year dollars, $b = 0$) that the equipment will reduce current net operating expenses by $36,000 per year (for 10 years) and will have a $30,000 market value at the end of the 10th year. For simplicity, these cash flows are estimated to increase at the general price inflation rate ($f = 8\%$ per year). Due to new computer control features on the equipment, it will be necessary to contract for some maintenance support during the first three years. The maintenance contract will cost $2,800 per year. This equipment will be depreciated under the MACRS (GDS) method, and it is in the five-year property class. The effective income tax rate (t) is 38%; the selected analysis period is 10 years; and the MARR$_m$ (after taxes) is $i_m = 15\%$ per year.

(a) Based on an actual-dollar after-tax analysis, is this capital investment justified?

(b) Develop the ATCF in real dollars.

Solution

(a) The actual-dollar after-tax economic analysis of the new equipment is shown in Table 8-3 (columns 1–7). The capital investment, savings in operating expenses, and market value (in the 10th year) are estimated in actual dollars (column 1), using the general price inflation rate and Equation (8-1). The maintenance contract amounts for the first three years (column 2) are already in actual dollars. (They are unresponsive to further price changes.) The algebraic sum of columns 1 and 2 equals the before-tax cash flow (BTCF) in actual dollars (column 3).

In columns 4, 5, and 6, the depreciation and income-tax calculations are shown. The depreciation deductions in column 4 are based on the MACRS (GDS) method and, of course, are in actual dollars. The entries in columns 5 and 6 are calculated as discussed in Chapter 7. The effective income tax rate (t) is 38% as given. The entries in column 6 are equal to the entries in column 5 multiplied by $-t$. The algebraic sum of columns 3 and 6 equals the ATCF in actual dollars (column 7). The PW of the actual-dollar ATCF, using $i_m = 15\%$ per year, is

$$PW(15\%) = -\$180,000 + \$36,050(P/F, 15\%, 1)$$
$$+ \cdots + \$40,156(P/F, 15\%, 10) = \$33,790.$$

Therefore, the project is economically justified.

(b) Next, Equation (8-1) is used to calculate the ATCF in real dollars from the entries in column 7. The real-dollar ATCF (column 9) shows the estimated economic consequences of the new equipment in dollars that have the constant purchasing power of the base year. The actual-dollar ATCF (column 7) is in dollars that have the purchasing power of the year in which the cost or saving occurs. The comparative information provided by the ATCF in both actual dollars and real dollars is helpful in interpreting the

TABLE 8-3 Example 8-7 When the General Price Inflation Rate Is 8% per Year

End of Year (k)	(1) A$ Cash Flows	(2) Contract (A$)	(3) BTCF (A$)	(4) Depreciation (A$)	(5) Taxable Income	(6) Income Taxes (t = 0.38)	(7) ATCF (A$)	(8) R$ Adjustment $[1/(1+f)]^{k-b}$	(9) ATCF (R$)
0	−$180,000		−$180,000				−$180,000	1.0000	−$180,000
1	38,880[a]	−$2,800	36,080	$36,000	$80	−$30	36,050	0.9259	33,379
2	41,990	−2,800	39,190	57,600	−18,410	+6,996	46,186	0.8573	39,595
3	45,349	−2,800	42,549	34,560	7,989	−3,036	39,513	0.7938	31,366
4	48,978		48,978	20,736	28,242	−10,732	38,246	0.7350	28,111
5	52,895		52,895	20,736	32,159	−12,220	40,675	0.6806	27,683
6	57,128		57,128	10,368	46,760	−17,769	39,359	0.6302	24,804
7	61,697		61,697		61,697	−23,445	38,252	0.5835	22,320
8	66,632		66,632		66,632	−25,320	41,312	0.5403	22,320
9	71,964		71,964		71,964	−27,346	44,618	0.5003	22,320
10	77,720		77,720		77,720	−29,534	48,186	0.4632	22,320
10	64,767[b]		64,767		64,767	−24,611	40,156	0.4632	18,600

[a] $(A\$)_k = \$36,000(1.08)^{k-0}, k = 1,\ldots,10.$
[b] $MV_{10,A\$} = \$30,000(1.08)^{10} = \$64,767.$

results of an economic analysis. Also, as illustrated in this example, the conversion between actual dollars and real dollars can easily be done. The PW of the real-dollar ATCF (column 9), using $i_r = (i_m - f)/(1 + f) = (0.15 - 0.08)/1.08 = 0.06481$, or 6.48%, is

$$PW(6.48\%) = -\$180,000 + \$33,379(P/F, 6.48\%, 1)$$
$$+ \cdots + \$18,600(P/F, 6.48\%, 10)$$
$$= \$33,790.$$

Thus, the PW (equivalent worth in the base year with $b = 0$) of the real-dollar ATCF is the same as the PW calculated previously for the actual-dollar ATCF.

8.4   Spreadsheet Application

The following example illustrates the use of spreadsheets to convert actual dollars to real dollars, and vice versa.

EXAMPLE 8-8	Saving to Meet a Retirement Goal

Sara B. Goode wishes to retire in the year 2022 with personal savings of $500,000 (1997 spending power). Assume that the expected inflation rate in the economy will average 3.75% per year during this period. Sara plans to invest in a 7.5% per year savings account, and her salary is expected to increase by 8.0% per year between 1997 and 2022. Assume that Sara's 1997 salary was $60,000 and that the first deposit took place at the end of 1997. What percent of her yearly salary must Sara put aside for retirement purposes to make her retirement plan a reality?

Solution

This example demonstrates the flexibility of a spreadsheet, even in instances where all of the calculations are based on a piece of information (percentage of salary to be saved) that we do not yet know. If we deal in actual dollars, the cash-flow relationships are straightforward. The formula in cell B7 in Figure 8-1 converts the desired ending balance into actual dollars. The salary is paid at the end of the year, at which point some percentage is placed in a bank account. The interest calculation is based on the cumulative deposits and interest in the account at the beginning of the year, but not on the deposits made at the end of the year. The salary is increased and the cycle repeats.

The spreadsheet model is shown in Figure 8-1. We can enter the formulas for the annual salary increase (column B), the percentage of the salary (cell E1) that goes into savings (column C), and the bank balance at the end of the year (column D) without knowing the percentage of salary being saved.

A slider control (cells F1 and G1) is used in this model to vary the percentage of salary saved value in cell E1. The value in cell E1 can be changed by using the

Formula callouts:

- $= B5*(1 + B4)^{25}$
- $= G1/10000$
- $= B10*\$E\$1$
- $= B10*(1 + \$B\$2)$
- $= C10$
- $= B1$
- $= C11 + D10*(1 + \$B\$3)$
- $= D35 - B7$

	A	B	C	D	E	F	G
1	Starting Salary =	$ 60,000		% Salary to Save Annually	12.44%	◄	►
2	Annual Salary Increase =	8.00%					
3	Savings Interest Rate =	7.50%					
4	Average Inflation Rate =	3.75%					
5	Desired amount in 2022 (R$) =	$ 500,000					
6							
7	Desired Amount in 2022 (A$) =	$ 1,255,084					
8							
9	Year	Salary (A$)	Savings (A$)	Bank Balance (A$)			
10	1997	$ 60,000	$ 7,464	$ 7,464			
11	1998	$ 64,800	$ 8,061	$ 16,085			
12	1999	$ 69,984	$ 8,706	$ 25,997			
13	2000	$ 75,583	$ 9,402	$ 37,350			
14	2001	$ 81,629	$ 10,155	$ 50,305			
15	2002	$ 88,160	$ 10,967	$ 65,045			
16	2003	$ 95,212	$ 11,844	$ 81,768			
17	2004	$ 102,829	$ 12,792	$ 100,693			
18	2005	$ 111,056	$ 13,815	$ 122,060			
19	2006	$ 119,940	$ 14,921	$ 146,135			
20	2007	$ 129,535	$ 16,114	$ 173,210			
21	2008	$ 139,898	$ 17,403	$ 203,604			
22	2009	$ 151,090	$ 18,796	$ 237,670			
23	2010	$ 163,177	$ 20,299	$ 275,794			
24	2011	$ 176,232	$ 21,923	$ 318,402			
25	2012	$ 190,330	$ 23,677	$ 365,959			
26	2013	$ 205,557	$ 25,571	$ 418,977			
27	2014	$ 222,001	$ 27,617	$ 478,018			
28	2015	$ 239,761	$ 29,826	$ 543,695			
29	2016	$ 258,942	$ 32,212	$ 616,685			
30	2017	$ 279,657	$ 34,789	$ 697,725			
31	2018	$ 302,030	$ 37,573	$ 787,627			
32	2019	$ 326,192	$ 40,578	$ 887,278			
33	2020	$ 352,288	$ 43,825	$ 997,648			
34	2021	$ 380,471	$ 47,331	$ 1,119,802			
35	2022	$ 410,909	$ 51,117	$ 1,254,905			
36							
37	Difference between Desired and Actual =			$ (179)			

Figure 8-1 Spreadsheet Solution, Example 8-8

mouse to click on either the left or right arrows. We'll know when we've found the correct percentage when the difference between the desired balance and actual balance (cell D37) is approximately zero. The steps involved in creating a slider control in Excel can be found in Appendix A. (As an alternative to the slider control, we could have manually changed the value in cell E1 or made use of the *solver* feature that most spreadsheet packages include.)

Once the problem has been formulated in a spreadsheet, we can determine the impact of different interest rates, inflation rates, and so on, on the retirement plan, with minimal changes and effort. [Answer: 12.44%]

8.5 Foreign Exchange Rates and Purchasing Power Concepts

Unless you are planning to travel abroad, you probably haven't given much thought to how currency fluctuations can affect purchasing power. With the world economy becoming increasingly interconnected, a dollar rising in price (or a weakening euro) will have an impact on the estimated cash flows associated with foreign investments. Many factors influence the price of a currency due to supply-and-demand considerations, including a country's interest rates, cross-border investing, national debt, inflation, and trade balance. When a currency changes in value because of these factors, holders of that country's currency experience a change in overseas buying power. For example, if one dollar goes from being worth 1.0 euros to 0.8 euros (in this situation the dollar is weakening), 1,000 euros worth of goods from a European company that previously cost a U.S. purchaser $1,000 must now be bought for $1,250. This applies to consumer products as well as capital investments, and the purpose of Section 8.5 is to explore the consequences of exchange rate fluctuations on the profitability of foreign investments.

> Changes in the exchange rate between two currencies over time are analogous to changes in the general inflation rate because the relative purchasing power between the two currencies is changing similar to the relative purchasing power between actual dollar amounts and real dollar amounts.

Assume the following:

i_{us} = rate of return in terms of a market interest rate relative to U.S. dollars.

i_{fm} = rate of return in terms of a market interest rate relative to the currency of a foreign country.

f_e = annual devaluation rate (rate of annual change in the exchange rate) between the currency of a foreign country and the U.S. dollar. In the following relationships, a positive f_e is used when the foreign currency is being devalued relative to the dollar, and a negative f_e is used when the dollar is being devalued relative to the foreign currency.

Then (derivation not shown)

$$1 + i_{us} = \frac{1 + i_{fm}}{1 + f_e},$$

or

$$i_{fm} = i_{us} + f_e + f_e(i_{us}), \qquad (8\text{-}6)$$

and

$$i_{us} = \frac{i_{fm} - f_e}{1 + f_e}. \qquad (8\text{-}7)$$

| EXAMPLE 8-9 | Investing in a Foreign Assembly Plant |

The CMOS Electronics Company is considering a capital investment of 50,000,000 pesos in an assembly plant located in a foreign country. Currency is expressed in pesos, and the exchange rate is now 100 pesos per U.S. dollar.

The country has followed a policy of devaluing its currency against the dollar by 10% per year to build up its export business to the United States. This means that each year the number of pesos exchanged for a dollar increases by 10% ($f_e = 10\%$), so in two years $(1.10)^2(100) = 121$ pesos would be traded for one dollar. Labor is quite inexpensive in this country, so management of CMOS Electronics feels that the proposed plant will produce the following rather attractive ATCF, stated in pesos:

End of Year	0	1	2	3	4	5
ATCF (millions of pesos)	−50	+20	+20	+20	+30	+30

If CMOS Electronics requires a 15% IRR per year, after taxes, in U.S. dollars (i_{us}) on its foreign investments, should this assembly plant be approved? Assume that there are no unusual risks of nationalization of foreign investments in this country.

Solution

To earn a 15% annual rate of return in U.S. dollars, the foreign plant must earn, based on Equation (8-6), $0.15 + 0.10 + 0.15(0.10) = 0.265$, which is 26.5% on its investment in pesos (i_{fm}). As shown next, the PW (at 26.5%) of the ATCF, in pesos, is 9,165,236, and its IRR is 34.6%. Therefore, investment in the plant appears to be economically justified. We can also convert pesos into dollars when evaluating the prospective investment:

End of Year	ATCF (Pesos)	Exchange Rate	ATCF (Dollars)
0	−50,000,000	100 pesos per $1	−500,000
1	20,000,000	110 pesos per $1	181,818
2	20,000,000	121 pesos per $1	165,289
3	20,000,000	133.1 pesos per $1	150,263
4	30,000,000	146.4 pesos per $1	204,918
5	30,000,000	161.1 pesos per $1	186,220
IRR:	34.6%	IRR:	22.4%
PW(26.5%):	9,165,236 pesos	PW(15%):	$91,632

The PW (at 15%) of the ATCF, in dollars, is $91,632, and its IRR is 22.4%. Therefore, the plant again appears to be a good investment, in economic terms. Notice that the two IRRs can be reconciled by using Equation (8-7):

$$i_{us}(\text{IRR in \$}) = \frac{i_{fm}(\text{IRR in pesos}) - 0.10}{1.10}$$

$$= \frac{0.346 - 0.10}{1.10}$$

$$= 0.224, \text{ or } 22.4\%.$$

Remember that devaluation of a foreign currency relative to the U.S. dollar produces less expensive exports to the United States. Thus, devaluation means that the U.S. dollar is stronger relative to the foreign currency. That is, fewer dollars are needed to purchase a fixed amount (barrels, tons, items) of goods and services from the foreign source or, stated differently, more units of the foreign currency are required to purchase U.S. goods. This phenomenon was observed in Example 8-9.

Conversely, when exchange rates for foreign currencies gain strength against the U.S. dollar (f_e has a negative value and the U.S. dollar is weaker relative to the foreign currency), the prices for imported goods and services into the United States rise. In such a situation, U.S. products are less expensive in foreign markets.

In summary, if the average devaluation of currency A is $f_e\%$ per year relative to currency B, then each year it will require $f_e\%$ more of currency A to exchange for the same amount of currency B.

EXAMPLE 8-10 **Estimating Future Exchange Rates**

The monetary (currency) unit of another country, A, has a present exchange rate of 10.7 units per U.S. dollar.

(a) If the average devaluation of currency A in the international market is estimated to be 4.2% per year (for the next five years) relative to the U.S. dollar, what would be the exchange rate three years from now?

(b) If the average devaluation of the U.S. dollar (for the next five years) is estimated to be 3% per year relative to currency A, what would be the exchange rate three years from now?

Solution

(a) $10.7(1.042)^3 = 12.106$ units of A per U.S. dollar.

(b) 10.7 units of $A = 1(1.03)^3$ U.S. dollars,

$$1 \text{ U.S. dollar} = \frac{10.7}{1.09273} = 9.792 \text{ units of } A.$$

EXAMPLE 8-11 **Profitability Analysis of a Foreign Investment**

A U.S. firm is analyzing a potential investment project in another country. The present exchange rate is 425 of their currency units (A) per U.S. dollar. The best estimate is that currency A will be devalued in the international market at an average of 2% per year relative to the U.S. dollar over the next several years. The estimated before-tax net cash flow (in currency A) of the project is as follows:

End of Year	Net Cash Flow (Currency A)
0	−850,000,000
1	270,000,000
2	270,000,000
3	270,000,000
4	270,000,000
5	270,000,000
6	270,000,000
6 (MV)[a]	120,000,000

[a] Estimated market value at the end of year six.

(a) If the MARR value of the U.S. firm (before taxes and based on the U.S. dollar) is 20% per year, is the project economically justified?

(b) Instead, if the U.S. dollar is estimated to be devalued in the international market an average of 2% per year over the next several years, what would be the rate of return based on the U.S. dollar? Is the project economically justified?

Solution

(a)
$$PW(i'\%) = 0 = -850,000,000 + 270,000,00(P/A, \ i'\%, 6)$$

$$+ \ 120,000,000(P/F, \ i'\%, 6).$$

By trial and error (Chapter 5), we determine that $i'\% = i_{fm} = IRR_{fm} = 24.01\%$. Now, using Equation (8-7), we have

$$i_{us} = IRR_{us} = \frac{0.2401 - 0.02}{1.02} = 0.2158, \text{or } 21.58\%.$$

Since this rate of return in terms of the U.S. dollar is greater than the firm's MARR (20% per year), the project is economically justified (but very close to the minimum rate of return required on the investment).

A Second Solution to Part (a)

Using Equation (8-6), we can determine the firm's MARR value in terms of currency A as follows:

$$i_{fm} = MARR_{fm} = 0.20 + 0.02 + 0.02(0.20) = 0.244, \text{or } 22.4\%.$$

Using this $MARR_{fm}$ value, we can calculate the PW(22.4%) of the project's net cash flow (which is estimated in units of currency A); that is, PW(22.4%) = 32,597,000 units of currency A. Since this PW is greater than zero, we confirm that the project is economically justified.

(b) Using Equation (8-7) and the IRR_{fm} value [24.01% calculated in the solution to Part (a)], we have

$$i_{us} = IRR_{us} = \frac{0.2401 - (-0.02)}{1 - 0.02} = 0.2654, \text{or } 26.54\%.$$

Since the rate of return in terms of the U.S. dollar is greater than the required $MARR_{us} = 20\%$, the project is economically justified.

As additional information, notice in the first solution to Part (a), when currency A is estimated to be devalued relative to the U.S. dollar, $IRR_{us} = 21.58\%$; whereas in Part (b), when the U.S. dollar is estimated to be devalued relative to currency A, $IRR_{us} = 26.54\%$. What is the relationship between the different currency devaluations in the two parts of the problem and these results?

Answer: Since the annual earnings to the U.S. firm from the investment are in units of currency A originally, and currency A is being devalued relative to the U.S. dollar in Part (a), these earnings would be exchanged for decreasing annual amounts of U.S. dollars, causing an unfavorable impact on the project's IRR_{us}. In Part (b), however, the U.S. dollar is being devalued relative to currency A, and these annual earnings would be exchanged for increasing annual amounts of U.S. dollars, causing a favorable impact on the project's IRR_{us}.

8.6 CASE STUDY—Selecting Electric Motors to Power an Assembly Line

A large computer manufacturing company is expanding its operations in response to an increase in demand. To meet the new production goals, the company needs to upgrade its assembly line. This upgrade requires additional electric motors that will be able to deliver a total of 1,550 horsepower output.

There are two types of motors available, induction and synchronous, whose specifications are listed in the accompanying table. Both motor types are three-phase, and the quoted voltage is rms line-to-line. If the synchronous motors are operated at a power factor (pf) of 1.0, their efficiency is reduced to 80%, but the local power utility will provide electricity at a discounted rate of $0.025 per kilowatt-hour for the first year.

Specifications	Type of Motor		
	Induction	Synchronous (pf = 0.9 lag)	Synchronous (pf = 1.0)
Rated Horsepower	400	500	500
Efficiency	85%	90%	80%
Line Voltage	440	440	440
Capital Investment	$17,640	$24,500	$24,500
Maintenance Cost (First Year)	$1,000	$1,250	$1,250
Electricity Cost (First Year)	$0.030/kWh	$0.030/kWh	$0.025/kWh

Historical data indicate that maintenance costs increase at the rate of 4% per year and that the cost of electricity increases at the rate of 5% per year. The assembly line must operate two eight-hour shifts per day, six days per week, for 50 weeks per year. An eight-year study period is being used for the project, and the company's market-based cost of capital (MARR$_m$) is 8% per year. Your engineering project team has been tasked with determining what combination of motors should be purchased to provide the additional power requirements of the upgraded assembly line.

Solution

Your project team has decided that the overall problem of motor selection can be broken down into the following subproblems to arrive at a recommendation:

(a) Determine whether it is less expensive to operate a synchronous motor at a power factor of 0.9 or 1.0.

(b) Determine the PW of the cost of operating one motor of each type at rated horsepower for the total period of operational time during the eight-year project lifetime.

(c) Determine the combination of motors that should be purchased. The objective is to minimize the PW of the total cost.

Solution to Subproblem (a)

Because the capital investment and maintenance costs of the synchronous motor are independent of the power factor employed, only the cost of electricity needs to be considered at this stage of the analysis.

The efficiency of the 500-horsepower synchronous motor operating at a power factor of 0.9 is 90%. Therefore, the cost of electricity per operating hour is (recall that 1 hp = 746 watts)

$$(500 \text{ hp}/0.9)(0.746 \text{ kW/hp})(\$0.030/\text{kWh}) = \$12.43/\text{hour}$$

When operated at a power factor (pf) of 1.0, the efficiency of the motor is reduced to 80%. The electric company, however will provide electricity at a discounted rate. The resulting cost of electricity is

$$(500 \text{ hp}/0.8)(0.746 \text{ kW/hp})(\$0.025/\text{kWh}) = \$11.66/\text{hour}.$$

Therefore, it is less expensive to operate the synchronous motor at pf = 1.0. The discounted electricity rate more than compensates for the loss in efficiency. Given this information, operating the synchronous motor at pf = 0.9 is no longer a feasible alternative.

Solution to Subproblem (b)

The total cost of each motor over the study period consists of the capital investment, maintenance costs, and electricity costs. Specifically, for each motor type,

$$PW_{\text{Total costs}} = \text{Capital investment} + PW(\text{maintenance costs})$$
$$+ PW(\text{electricity costs}).$$

Maintenance costs increase 4% annually over the eight-year period and can be modeled as a geometric gradient series. Recall from Chapter 4 that the present equivalent of a geometric gradient series is given by:

$$P = \begin{cases} \dfrac{A_1[1 - (P/F, i\%, N)(F/P, \bar{f}\%, N)]}{i - \bar{f}} & \bar{f} \neq i \\ A_1 N(P/F, i\%, 1) & \bar{f} = i, \end{cases} \qquad (8\text{-}8)$$

where $\bar{f}$ is the rate of increase per period. Using this relationship,

$$PW(\text{maintenance costs}) = \frac{A_1[1 - (P/F, 8\%, 8)(F/P, 4\%, 8)]}{0.08 - 0.04}$$
$$= \frac{A_1}{0.04}[1 - (0.5403)(1.3686)]$$
$$= A_1(6.5136),$$

where $A_1 = \$1,000$ for the induction motor and $A_1 = \$1,250$ for the synchronous motor.

Electricity costs are estimated to increase 5% annually over the eight-year study period and can also be modeled as a geometric gradient series.

$$PW(\text{electricity costs}) = \frac{A_1[1 - (P/F, 8\%, 8)(F/P, 5\%, 8)]}{0.08 - 0.05}$$

$$= \frac{A_1}{0.03}[1 - (0.5403)(1.4775)]$$

$$= A_1(6.7236).$$

Now, the number of hours of motor operation per year is

(8 hours/shift)(2 shifts/day)(6 days/week)(50 weeks/year) = 4,800 hours/year.

Thus, for the induction motor,

$$A_1 = (400 \text{ hp}/0.85)(0.746 \text{ kW/hp})(4,800 \text{ hours/year})(\$0.030/\text{kWh}) = \$50,552$$

and

$$A_1 = (500 \text{ hp}/0.8)(0.746 \text{ kW/hp})(4,800 \text{ hours/year})(\$0.025/\text{kWh}) = \$55,950$$

for the synchronous motor running at pf = 1.0.

We can now compute the PW of total costs for each motor type over the entire eight-year study period.

$$PW_{\text{Total costs}}(\text{induction motor}) = \$17,640 + \$1,000(6.5136) + \$50,552(6.7236)$$

$$= \$17,640 + \$6,514 + \$339,891$$

$$= \$364,045.$$

$$PW_{\text{Total costs}}(\text{synchronous motor}) = \$24,500 + \$1,250(6.5136) + \$55,950(6.7236)$$

$$= \$24,500 + \$8,142 + \$376,185$$

$$= \$408,827.$$

Solution to Subproblem (c)

The last step of the analysis is to determine what combination of motors to use to obtain the required 1,550 horsepower output. The following four options will be considered:

(A) Four induction motors (three at 400 horsepower, one at 350 horsepower)
(B) Four synchronous motors (three at 500 horsepower, one at 50 horsepower)
(C) Three induction motors at 400 horsepower plus one synchronous motor at 350 horsepower
(D) Three synchronous motors at 500 horsepower plus one induction motor at 50 horsepower.

For a motor used at less than rated capacity, the electricity cost is proportional to the fraction of capacity used—all other costs are unaffected.

The PW of total costs for each of the preceding options is computed using the results of subproblem (b):

$$PW_{\text{Total costs}}(A) = (3)(\$364,045) + [\$17,640 + \$6,514 + (350/400)(\$339,891)]$$

$$= \$1,413,694$$

$$PW_{\text{Total costs}}(B) = (3)(\$408,827) + [\$24,500 + \$8,142 + (50/500)(\$376,185)]$$

$$= \$1,296,742$$

$$PW_{\text{Total costs}}(C) = (3)(\$364,045) + [\$24,500 + \$8,142 + (350/500)(\$376,185)]$$

$$= \$1,388,107$$

$$PW_{\text{Total costs}}(D) = (3)(\$408,827) + [\$17,640 + \$6,514 + (50/400)(\$339,891)]$$

$$= \$1,293,121$$

Option (D) has the lowest PW of total costs. Thus, your project team's recommendation is to power the assembly line using three 500-horsepower synchronous motors operated at a power factor of 1.0 and one 400-horsepower induction motor. You also include a note to management that there is additional capacity available should future modifications to the assembly line require more power.

This case study illustrates the inclusion of price changes in the analysis of engineering projects. The cost of goods and services changes over time and needs to be addressed. Sensitivity analyses on key performance and cost parameters can be used to quantify project risk. ▄

8.7 Summary

Price changes caused by inflation or deflation are an economic and business reality that can affect the comparison of alternatives. In fact, since the 1920s, the historical U.S. inflation rate has averaged approximately 4% per year. Much of this chapter has dealt with incorporating price changes into before-tax and after-tax engineering economy studies.

In this regard, it must be ascertained whether cash flows have been estimated in actual dollars or real dollars. The appropriate interest rate to use when discounting or compounding actual-dollar amounts is a nominal, or marketplace, rate, while the corresponding rate to apply in real dollar analysis is the firm's real interest rate.

Engineering economy studies often involve quantities that do not respond to inflation, such as depreciation amounts, interest charges, and lease fees and other amounts established by contract. Identifying these quantities and handling them properly in an analysis are necessary to avoid erroneous economic results. The use of the chapter's basic concepts in dealing with foreign exchange rates has also been demonstrated.

Problems

The number in parentheses ( ) that follows each problem refers to the section from which the problem is taken.

8-1. Consider this statement: "Real gross domestic product (GDP) in the U.S. economy grows at 3.0%." Explain what this means in terms that your younger sister can understand. (8.1)

8-2. Your rich aunt is going to give you an end-of-year gift of $1,000 for each of the next 10 years.

a. If general price inflation is expected to average 6% per year during the next 10 years, what is the equivalent value of these gifts at the present time? The real interest rate is 4% per year.

b. Suppose that your aunt specified that the annual gifts of $1,000 are to be increased by 6% each year to keep pace with inflation. With a real interest rate of 4% per year, what is the current PW of the gifts? (8.2)

8-3. Because of general price inflation in our economy, the purchasing power of the dollar shrinks with the passage of time. If the average general price inflation rate is expected to be 4% per year into the foreseeable future, how many years will it take for the dollar's purchasing power to be one-half of what it is now? (That is, at what future point in time will it take $2 to buy what can be purchased today for $1?) (8.2)

8-4. Health care costs are reportedly rising at an annual rate that is triple the general inflation rate. The current inflation rate is running at 4% per year. In 10 years, how much greater will health care costs be compared to a service/commodity that increases at exactly the 4% general inflation rate? (8.2)

8-5. Which of these situations would you prefer? (8.2)

a. You invest $2,500 in a certificate of deposit that earns an effective interest rate of 8% per year. You plan to leave the money alone for five years, and the general price inflation rate is expected to average 5% per year. Income taxes are ignored.

b. You spend $2,500 on a piece of antique furniture. You believe that in five years the furniture can be sold for $4,000. Assume that the average general price inflation rate is 5% per year. Again, income taxes are ignored.

8-6. Annual expenses for two alternatives have been estimated on different bases as follows:

End of Year	Alternative A Annual Expenses Estimated in Actual Dollars	Alternative B Annual Expenses Estimated in Real Dollars with $b = 0$
1	$120,000	$100,000
2	132,000	110,000
3	148,000	120,000
4	160,000	130,000

If the average general price inflation rate is expected to be 6% per year and the real rate of interest is 9% per year, show which alternative has the least negative equivalent worth in the base period. (8.2)

8-7. The incredible shrinking $50 bill in 1957 was worth $50, but in 2007 it is worth only $6.58. (8.2)

a. What was the compounded average annual inflation rate (loss of purchasing power) during this period of time?

b. Fifty dollars invested in the stock market in 1957 was worth $1,952 in 2007. In view of your answer to Part (a), what was the annual real interest rate earned on this investment?

8-8. A firm desires to determine the most economic equipment-overhauling schedule alternative to provide for service for the next nine years of operation. The firm's real MARR is 8% per year, and the inflation rate is estimated at 7% per year. The following are alternatives with all the costs expressed in real (constant worth) dollars: (8.2)

I. Completely overhaul for $10,000 now.

II. A major overhaul for $7,000 now that can be expected to provide six years of service and then a minor overhaul costing $5,000 at the end of six years.

III. A minor overhaul costing $5,000 now as well as at the end of three years and six years from now.

8-9. A recent college graduate has received the annual salaries shown in the following table over the past four years. During this time, the CPI has performed as indicated. Determine the annual salaries in *year-zero dollars* ($b = 0$), using the CPI as the indicator of general price inflation. (8.2)

End of Year	Salary (A$)	CPI
1	$34,000	2.4%
2	36,200	1.9%
3	38,800	3.3%
4	41,500	3.4%

8-10. Paul works for a government agency in southern California making $70,000 per year. He is now being transferred to a branch office in Tennessee. The salary reduction associated with this transfer is 11%. Paul is not insulted by this reduction in pay and accepts his new location and salary gladly. He researched that the cost of living index in California is 132 whereas the cost of living index in Tennessee is 95. Over the next five years, what is the FW of Paul's extra income/improved life style (through the reduced cost of living) from having made this move? Paul's MARR is 10% per year (i_m). (8.2)

8-11. A large corporation's electricity bill amounts to $400 million. During the next 10 years, electricity usage is expected to increase by 75%, and the estimated electricity bill 10 years hence has been projected to be $920 million. Assuming electricity usage and rates increase at uniform annual rates over the next 10 years, what is the annual rate of inflation of electricity prices expected by the corporation? (8.2)

8-12. Your older brother is concerned more about investment safety than about investment performance. For example, he has invested $100,000 in safe 10-year corporate AAA bonds yielding an average of 6% per year, payable each year. His effective income tax rate is 33%, and inflation will average 3% per year. How much will his $100,000 be worth in 10 years in today's purchasing power after income taxes and inflation are taken into account? (8.2)

8-13. Your younger brother is very concerned about investment performance, but he is willing to tolerate some risk. He decides to invest $100,000 in a mutual fund that will earn 10% per year. The proceeds will be accumulated as a lump sum and paid out at the end of year 10. Based on the data given in Problem 8-12, what is today's purchasing power equivalent of your younger brother's investment? (8.2)

8-14. A commercial building design cost $89/square-foot to construct eight years ago (for an 80,000-square-foot building). This construction cost has increased 5.4% per year since then. Presently, your company

is considering construction of a 125,000-square-foot building of the same design. The cost capacity factor is $X = 0.92$. In addition, it is estimated that working capital will be 5% of construction costs, and that project management, engineering services, and overhead will be 4.2%, 8%, and 31%, respectively, of construction costs. Also, it is estimated that annual expenses in the first year of operation will be $5/square-foot, and these are estimated to increase 5.66% per year thereafter. The future general inflation rate is estimated to be 7.69% per year, and the market-based MARR = 12% per year (i_m). (Chapter 3 and 8.2)

a. What is the estimated capital investment for the 125,000-square-foot building?

b. Based on a before-tax analysis, what is the PW for the first 10 years of ownership of the building?

c. What is the AW for the first 10 years of ownership in real dollars (R$)?

8-15. Twenty years ago your rich uncle invested $10,000 in an aggressive (i.e., risky) mutual fund. Much to your uncle's chagrin, the value of his investment declined by 18% during the first year and then declined another 31% during the second year. But your uncle decided to stick with this mutual fund, reasoning that long-term sustainable growth of the U.S. economy was bound to occur and enhance the value of his mutual fund. Eighteen more years have passed, and your uncle's cumulative return over the 20-year period is a whopping 507%! (8.2)

a. What is the value of the original investment now?

b. If inflation has averaged 6% per year over the past 20 years, what is the spending power equivalent of the answer to Part (a) in terms of real dollars 20 years ago?

c. What is the real compound interest rate earned over the 20-year period?

d. During the past 18 years, what compound annual rate of return (yield) was earned on your uncle's investment?

8-16. The operating budget estimate for an engineering staff for fiscal year 2009 is $1,780,000. The actual budget expenditures of the staff for the previous two fiscal years, as well as estimates for the next two years, are as shown in the table on page 382. These are actual-dollar amounts. Management, however, also wants annual budget amounts for these years, using a constant-dollar perspective. The 2009 fiscal year is to be used for this purpose ($b = 2009$). The estimated annual general price inflation rate is 5.6%. What are the annual constant- (real-) dollar budget amounts? (8.2)

Fiscal Year	Budget Amount (A$)
2007	$1,615,000
2008	1,728,000
2009	1,780,000
2010	1,858,300
2011	1,912,200

8-17. An investor lends $10,000 today, to be repaid in a lump sum at the end of 10 years with interest at 10% ($= i_m$) compounded annually. What is the real rate of return, assuming that the general price inflation rate is 8% annually? (8.2)

8-18. In 2005, the average cost to construct a 150,000-barrel-a-day oil refinery was $2.4 billion (excluding any interest payments). This capital investment is spread uniformly over a five-year construction cycle (i.e., the refinery starts producing products in 2010). Assume end-of-year cash-flow convention and answer the following questions. (Chapter 3 and 8.2)

a. If the cost of capital spent to build the refinery is 10% per year, how much interest will be charged (as a lump sum in 2010) during the construction of this project?

b. If the cost-capacity factor is 0.91, what is the approximate cost to build a 200,000-barrel-a-day refinery in 2005?

c. If inflation of the construction cost of a refinery is 9.2% per year, what is the estimated cost of building a 200,000-barrel-a-day refinery in 2015?

8-19. In January of 1980, a Troy ounce of gold sold for $850 (an all-time high). Over the past 28 years, suppose the CPI has grown at a compounded annual rate of 3.2%. Today a Troy ounce of gold sells for $690. (8.2)

a. In real terms, with 1980 as the reference year, what is today's price of gold per ounce in 1980 purchasing power?

b. If gold increases in value to keep pace with the CPI, how many years will it take to grow to $850 per ounce in today's purchasing power?

c. What was the real interest rate earned from 1980 to 2008 on an ounce of gold?

8-20. Suppose it costs $15,000 per year (in early 2007 dollars) for tuition, room, board, and living expenses to attend a public university in the Southeastern United States. (8.2)

a. If inflation averages 6% per year, what is the total cost of a four-year college education starting in 2020 (i.e., 13 years from now)?

b. Starting in January of 2007, what monthly savings amount must be put aside to pay for this four-year college education? Assume your savings account earns 6% per year, compounded monthly (0.5% per month).

8-21. Suppose natural gas experiences a 1.8% increase per year in real terms over the foreseeable future. The cost of a 1,000 cubic feet of natural gas is now $7.50. (8.2)

a. What will be the cost in real terms of 1,000 cubic feet of natural gas in 22 years?

b. If the general price inflation rate (e.g., the CPI) for the next 22 years is expected to average 3.2% per year, what will a 1,000 cubic feet of natural gas cost in actual dollars 22 years from now?

8-22. Refer to Example 4-8. If inflation is expected to average 3% per year over the next 60 years, your purchasing power will be less than $1,107,706. (8.2)

a. When $f = 3$% per year, what is the purchasing power of this future sum of money in today's dollars when you reach age 80?

b. Repeat Part (a) when inflation averages 2% per year.

8-23. Four hundred pounds of copper are used in a typical 2,200-square-foot newly constructed house. Today's copper price is $3.75 per pound. (8.2)

a. If copper prices are expected to increase 8.5% per year (on average), what is the cost of copper going to be in a new 2,400-square-foot house built 10 years from now? Use a cost-capacity factor of 1.0.

b. To minimize building costs, do you think that another less-expensive material (e.g., aluminum) will be substituted for copper in the electrical wiring in new homes?

8-24. The cost of first-class postage has risen by about 5% per year over the past 30 years. The U.S. Postal Service introduced a one-time forever stamp in 2008 that cost 41 cents for first-class postage (one ounce or less), and it will be valid as first-class postage regardless of all future price increases. Let's say you decided to purchase 1,000 of these stamps for this one-time special rate. Assume 5% inflation and your personal MARR is 10% per year (i_m). Did you make a sound economical decision? (8.2)

8-25. Refer to Problem 5-56. If the government expects inflation to average 3% per year, what is the equivalent uniform annual savings in fuel (jet fuel savings will be $2.5 billion one year from now). MARR remains at 7% per year (i_m).

8-26. With interest rates on the rise, many Americans are wondering what their investment strategy should be. A safe (i.e., a virtually risk-free) and increasingly popular way to keep pace with the cost of living is to purchase inflation-indexed government bonds. These so called I-bonds pay competitive interest rates and increase in value when the CPI rises. These bonds can be held for up to 30 years.

Suppose you purchased an I-bond for $10,000 and held it for 11 years, at which time you received $20,000 for the bond. Inflation has averaged 3% per year during this 11-year period. What real annual rate of return did you earn on your inflation-adjusted I-bond? Is it really competitive? (8.3)

8-27. The AZROC Corporation needs to acquire a computer system for one of its regional engineering offices. The purchase price of the system has been quoted at $50,000, and the system will reduce annual expenses by $18,000 per year in real dollars. Historically, these annual expenses have increased at an average rate of 8% per year, and this is expected to continue into the future. Maintenance services will be contracted for, and their cost per year (in actual dollars) is constant at $3,000. Also, assume $f = 8\%$ per year.

What is the minimum (integer-valued) life of the system such that the new computer can be economically justified? Assume that the computer's market value is zero at all times. The firm's MARR is 25% per year (i_m). Show all calculations. (8.3)

8-28. A hospital has two different medical devices it can purchase to perform a specific task. Both devices will perform an accurate analysis. Device A costs $100,000 initially, whereas device B (the deluxe model) costs $150,000. It has been estimated that the cost of maintenance will be $5,000 for device A and $3,000 for device B in the first year. Management expects these costs to increase 10% per year. The hospital uses a six-year study period, and its effective income tax rate is 50%. Both devices qualify as 5-year MACRS (GDS) property. Which device should the hospital choose if

the after-tax, market-based MARR is 8% per year (i_m)? (8.3)

8-29. The cost of conventional medications for diabetes has risen by an average of 13.1% per year over the past four years. During this same period of time, the cost of a specialty drug for rheumatoid arthritis (RA) has increased 22.5% per year. If these increases continue into the foreseeable future, in how many years will an RA medicine now costing $20 per dose be at least five times as expensive as a diabetes inhaler that now costs $10 per use? (8.3)

8-30. XYZ rapid prototyping (RP) software costs $20,000, lasts one year, and will be expensed (i.e., written off in one year). The cost of the upgrades will increase 10% per year, starting at the beginning of year two. How much can be spent now for an RP software upgrade agreement that lasts three years and must be depreciated with the SL method to zero over three years? MARR is 20% per year (i_m), and the effective income tax rate (t) is 34%. (Chapter 7, and 8.3)

8-31. In mid-2006, a British pound sterling (the monetary unit in the United Kingdom) was worth 1.4 euros (the monetary unit in the European Union). If a U.S. dollar bought 0.55 pound sterling in 2006, what was the exchange rate between the U.S. dollar and the euro? (8.5)

8-32. In a certain foreign country in 2005, the local currency (the 'Real') was pegged to the U.S. dollar at the rate of $1 U.S. = 1 Real. The Real was then devalued over the next five years so that $1 U.S. = 2 Real. A bank in the Northeastern United States bought assets in this country valued at 100 million Real in 2005. Now that it is year 2010, what is the worth of this bank's investment in U.S. dollars? Should the bank sell out of its investment in this foreign country or should it buy more assets? (8.5)

8-33. An international corporation located in Country A is considering a project in the United States. The currency in Country A, say X, has been strengthening relative to the U.S. dollar; specifically, the average devaluation of the U.S. dollar has been 2.6% per year (which is projected to continue). Assume the present exchange rate is 6.4 units of X per U.S. dollar. (a) What is the estimated exchange rate two years from now, and (b) if, instead, currency X was devaluing at the same rate (2.6% per year) relative to the U.S. dollar, what would be the exchange rate three years from now? (8.5)

8-34. A certain commodity cost 0.5 pound sterling in the United Kingdom. In U.S. denominated dollars, this same item can be purchased for 85 cents. The exchange rate is 1 pound sterling = $1.80 U.S. (8.5)

a. Should this commodity be purchased in the United Kingdom or in the United States?

b. If 100,000 items comprise the purchase quantity, how much can be saved in view of your answer to Part (a)?

8-35. A company requires a 26% internal rate of return (before taxes) in U.S. dollars on project investments in foreign countries. (8.5)

a. If the currency of Country A is projected to average an 8% annual devaluation relative to the dollar, what rate of return (in terms of the currency there) would be required for a project?

b. If the dollar is projected to devaluate 6% annually relative to the currency of Country B, what rate of return (in terms of the currency there) would be required for a project?

8-36. Apex, Inc. is a U.S.-based firm that is in the process of buying a medium-sized company in Brazil. Because the Brazilian company sells its services to the global marketplace, revenues are dollar denominated. Its costs, however, are expressed in the Brazilian currency. If Apex anticipates a stronger dollar, which tends to decrease the value of the Brazilian currency, explain why the Brazilian company's purchase price may be relatively insensitive to exchange-rate fluctuations. (8.5)

8-37. A U.S. company is considering a high-technology project in a foreign country. The estimated economic results for the project (after taxes), in the foreign currency (T-marks), is shown in the following table for the seven-year analysis period being used. The company requires an 18% rate of return in U.S. dollars (after taxes) on any investments in this foreign country. (8.5)

End of Year	Cash Flow (T-marks after Taxes)
0	−3,600,000
1	450,000
2	1,500,000
3	1,500,000
4	1,500,000
5	1,500,000
6	1,500,000
7	1,500,000

a. Should the project be approved, based on a PW analysis in U.S. dollars, if the devaluation of the T-mark, relative to the U.S. dollar, is estimated to average 12% per year and the present exchange rate is 20 T-marks per dollar?

b. What is the IRR of the project in T-marks?

c. Based on your answer to (b), what is the IRR in U.S. dollars?

8-38. Sanjay has 100 euros to spend before he flies back to the United States. He wishes to purchase jewelry priced at $160 (U.S.) in a duty-free shop at the airport. Euros can be bought for $1.32 and sold for $1.24. The owner of the duty-free shop tells Sanjay that the jewelry can be purchased for 100 euros plus $40 (U.S.). Is this a good deal for Sanjay? Explain your reasoning. (8.5)

8-39. An automobile manufacturing company in Country X is considering the construction and operation of a large plant on the eastern seaboard of the United States. Their MARR = 20% per year on a before-tax basis. (*This is a market rate relative to their currency in Country X.*) The study period used by the company for this type of investment is 10 years. Additional information is provided as follows:

- The currency in Country X is the Z-Kron.

- It is estimated that the U.S. dollar will become weaker relative to the Z-Kron during the next 10 years. Specifically, the dollar is estimated to be devalued at an average rate of 2.2% per year.

- The present exchange rate is 92 Z-Krons per U.S. dollar.

- The estimated before-tax net cash flow (in U.S. dollars) is as follows:

EOY	Net Cash Flow (U.S. Dollars)
0	−$168,000,000
1	−32,000,000
2	69,000,000
⋮	⋮
10	69,000,000

Based on a before-tax analysis, will this project meet the company's economic decision criterion? (8.5)

8-40. Your company *must* obtain some laser measurement devices for the next six years and is considering leasing. You have been directed to perform an actual-dollar after-tax study of the leasing approach. The pertinent information for the study is as follows:

Lease costs: First year, $80,000; second year, $60,000; third through sixth years, $50,000 per year. Assume that a six-year contract has been offered by the lessor that fixes these costs over the six-year period.

Other costs (not covered under contract): $4,000 in year-zero dollars, and estimated to increase 10% each year.

Effective income tax rate: 40%. (8.6)

a. Develop the actual-dollar ATCF for the leasing alternative.

b. If the real MARR (i_r) after taxes is 5% per year and the annual inflation rate (f) is 9.524% per year, what is the actual-dollar after-tax equivalent annual cost for the leasing alternative?

8-41. A gas-fired heating unit is expected to meet an annual demand for thermal energy of 500 million Btu, and the unit is 80% efficient. Assume that each 1,000 cubic feet of natural gas, if burned at 100% efficiency, can deliver one million Btu. Further suppose that natural gas is now selling for $7.50 per 1,000 cubic feet. What is the PW of fuel cost for this heating unit over a 12-year period if natural gas prices are expected to increase at an average rate of 10% per year? The firm's MARR ($= i_m$) is 18% per year. (8.6)

8-42. A small heat pump, including the duct system, now costs $2,500 to purchase and install. It has a useful life of 15 years and incurs annual maintenance expenses of $100 per year in real (year-zero) dollars over its useful life. A compressor replacement is required at the end of the eighth year at a cost of $500 in real dollars. The annual cost of electricity for the heat pump is $680, based on present prices. Electricity prices are projected to increase at an annual rate of 10%. All other costs are expected to increase at 6%, which is the projected general price inflation rate. The firm's MARR, which includes an allowance for general price inflation, is 15% per year (i_m). No market value is expected from the heat pump at the end of 15 years. (8.6)

a. What is the AW, expressed in actual dollars, of owning and operating the heat pump?

b. What is the AW in real dollars of owning and operating the heat pump?

8-43. A certain engine lathe can be purchased for $150,000 and depreciated over three years to a zero salvage value with the SL method. This machine will produce metal parts that will generate revenues of $80,000 (time zero dollars) per year. It is a policy of the company that the annual revenues will be increased each year to keep pace with the general inflation rate, which is expected to average 5%/year ($f = 0.05$). Labor, materials, and utilities totaling $20,000 (time 0 dollars) per year are all expected to increase at 9% per year. The firm's effective income tax rate is 50%, and its after-tax MARR (i_m) is 26% per year.

Perform an actual-dollar ($A\$$) analysis and determine the annual ATCFs of the preceding investment opportunity. Use a life of three years and work to the nearest dollar. What interest rate would be used for discounting purposes? (Chapter 7 and 8.3, 8.6)

8-44. Your company manufactures circuit boards and other electronic parts for various commercial products. Design changes in part of the product line, which are expected to increase sales, will require changes in the manufacturing operation. The cost basis of new equipment required is $220,000 (MACRS five-year property class). Increased annual revenues, in year zero dollars, are estimated to be $360,000. Increased annual expenses, in year zero dollars, are estimated to be $239,000. The estimated market value of equipment in actual dollars at the end of the six-year analysis period is $40,000. General price inflation is estimated at 4.9% per year; the total increase rate of annual revenues is 2.5%, and for annual expenses it is 5.6%; the after-tax MARR (in market terms) is 10% per year (i_m); and $t = 39\%$. (Chapter 7 and 8.3, 8.6)

a. Based on an after-tax, actual-dollar analysis, what is the maximum amount that your company can afford to spend on the total project (i.e., changing the manufacturing operations)? Use the PW method of analysis.

b. Develop (show) your ATCF in real dollars.

8-45. You have been assigned the task of analyzing whether to purchase or lease some transportation equipment for your company. The analysis period is six years, and the base year is year zero ($b = 0$). Other pertinent information is given in Table P8-45. Also,

a. The contract terms for the lease specify a cost of $300,000 in the first year and $200,000 annually in years two through six (the contract, i.e., these rates, does not cover the annual expense items).

b. The after-tax MARR (*not* including inflation) is 13.208% per year (i_r).

c. The general inflation rate (f) is 6%.

d. The effective income tax rate (t) is 34%.

TABLE P8-45 Table for Problem 8-45

Cash Flow Item	Estimate in Year-0 Dollars		Best Estimate of Price Change (% per year)
	Purchase	Lease	
Capital investment	$600,000	—	—
MV at end of six years	90,000	—	2%
Annual operating, insurance, and other expenses	26,000	$26,000	6
Annual maintenance expenses	32,000	32,000	9

e. Assume the equipment is in the MACRS (GDS) five-year property class.

Which alternative is preferred? (Use an after-tax, actual dollar analysis and the FW criterion.) (Chapter 7 and 8.3, 8.6)

8-46. *Extended Learning Exercise*
Chrysler is considering a cost reduction program with its major suppliers wherein the suppliers must reduce the cost of the components that they furnish to Chrysler *by 5% each year.* (The total amount of savings would increase each year.) The initial costs for setting up the program include the following:

- Program Setup: $12,500,000
- Feasibility Study
  (completed six months ago):   1,500,000
- Supplier Training: 7,500,000

Chrysler is currently paying a total of $85,000,000 per year for components purchased from the vendors who will be involved in this program. (Assume that, if the program is not approved, the annual cost of purchased components will remain constant at $85,000,000 per year.) The program has been designed as a five-year initiative, and Chrysler's MARR for such projects is 12% (i_m). There will be annual operating expenses associated with the program for further training of vendors, updating internal documentation, and so on. Given the projected savings in purchased components, what would be the maximum annual operating expense for this program such that it is marginally justified? (8.3)

8-47. *Extended Learning Exercise* The yuan is the currency of the People's Republic of China (PRC). The yuan has been pegged by the PRC government to the U.S. dollar for the past 10 years at a fixed rate of 8.28 yuan to the dollar. If the presently (May 2005) undervalued yuan were allowed to rise to its true value, Chinese imports would become more expensive

compared to U.S.-made products. One scenario is to allow the yuan to float in value against the U.S. dollar. Assume the yuan increases in value 25% against the dollar so that the dollar is worth 0.75(8.28 yuan) = 6.21 yuan.

If an American firm buys a lot of Chinese-made textile products for 10 yuan apiece, it would cost 10 yuan ($1 U.S./6.21 yuan) = $1.6103 per item with the revalued yuan. The lot of 10,000 items would cost $16,103. Before the revaluing of the yuan, each item would have cost 10 yuan ($1 U.S./8.28 yuan) = $1.2077 per item, or $12,077 for the lot. Discuss the following with your classmates. (8.5)

a. What happens to the cost of PRC goods in European countries when the dollar is strong (still 8.28 yuan per dollar)? What happens when the dollar weakens?

b. Will Americans buy the same amount of Chinese goods when $1 U.S. = 6.21 yuan? Will this fuel inflation in the United States?

c. What happens to the PRC–U.S. trade balance if more expensive Chinese goods are bought at the same level as with the $1 U.S. per 8.28 yuan policy?

8-48. *Extended Learning Exercise*
This case study is a justification of a computer system for, the ABC Company. The following are known data:

- The combined initial hardware and software cost is $80,000.

- Contingency costs have been set at $15,000. These are not necessarily incurred.

- A service contract on the hardware costs $500 per month.

- The effective income tax rate (*t*) is 38%.

- Company management has established a 15% (= i_m) per year hurdle rate (MARR).

In addition, the following assumptions and projections have been made:

- In order to support the system on an ongoing basis, a programmer/analyst will be required. The starting salary (first year) is $28,000, and fringe benefits amount to 30% of the base salary. Salaries are expected to increase by 6% each year thereafter.
- The system is expected to yield a staff savings of three persons (to be reduced through normal attrition) at an average salary of $16,200 per year per person (base salary plus fringes) in year-zero (base year) dollars. It is anticipated that one person will retire during the second year, another in the third year, and a third in the fourth year.

- A 3% reduction in purchased material costs is expected; first year purchases are $1,000,000 in year-zero dollars and are expected to grow at a compounded rate of 10% per year.
- The project life is expected to be six years, and the computer capital investment will be fully depreciated over that time period [MACRS (GDS) five-year property class].

Based on this information, perform an actual-dollar ATCF analysis. Is this investment acceptable based on economic factors alone? (8.3, 8.6)

Spreadsheet Exercises

8-49. Develop a spreadsheet to verify that the following table entries are correct (to the nearest whole percent). (8.2)

	Erosion of Money's Purchasing Power	
Inflation Rate	10 years	25 years
2%	−18%	−39%
4%	−32%	−62%

8-50. Refer to Example 8-8. Sara has decided that saving more than 12% of her annual salary was unrealistic. She is considering postponing her retirement until 2027, but still wishes to have $500,000 (in 1997 spending power) at that time. Determine what percentage of her annual salary Sara will have to save to meet her objective. (8.4)

8-51. Because of tighter safety regulations, an improved air filtration system must be installed at a plant that produces a highly corrosive chemical compound. The capital investment in the system is $260,000 in present-day dollars. The system has a useful life of 10 years and is in the MACRS (GDS) five-year property class. It is expected that the MV of the system at the end of its 10-year life will be $50,000 in present-day dollars. Annual expenses, estimated in present-day dollars, are expected to be $6,000 per year, not including an annual property tax of 4% of the investment cost (*does not inflate*). Assume that the *plant* has a remaining life of 20 years and that replacement costs, annual expenses, and MV increase at 6% per year.

If the effective income tax rate is 40%, set up a spreadsheet to determine the ATCF for the system over a 20-year period. The after-tax market rate of return desired on investment capital is 12% per year (i_m). What is the PW of costs of this system after income taxes have been taken into account? Develop the real-dollar ATCF. (Assume that the annual general price inflation rate is 4.5% over the 20-year period.) (Chapter 7 and 8.3, 8.6)

Case Study Exercises

8-52. Repeat the analysis for a reduced study period of five years. (8.6)

8-53. Suppose the cost of electricity is expected to increase at the rate of 6% per year. How will this new information change your recommendation? Use the original eight-year study period in your analysis. (8.6)

8-54. A re-design of the assembly line would only require 1,300 horsepower output. Determine the best combination of motors given this reduced horsepower requirement. (8.6)

FE Practice Problems

8-55. A machine cost $2,550 on January 1, 2000, and $3,930 on January 1, 2004. The average inflation rate over these four years was 7% per year. What is the true percentage increase in the cost of the machine from 2000 to 2004? (8.2)

(a) 14.95%   (b) 54.12%   (c) 7.00%
(d) 17.58%   (e) 35.11%

8-56. If you want to receive a 7% inflation-free return on your investment and you expect inflation to be 9% per year, what actual interest rate must you earn? (8.2)

(a) 16%   (b) −2%   (c) 1%   (d) 7%   (e) 9%

8-57. A person just turned 21 years old. If inflation is expected to average 2.4% per year for the next 44 years, how much will $1 today be worth when this person retires at age 65?

(a) $0.055   (b) $0.35   (c) $0.56   (d) $0.65

8-58. Two alternatives, A and B, are under consideration. Both have a life of five years. Alternative A needs an initial investment of $17,000 and provides a net revenue of $4,000 per year for five years. Alternative B requires an investment of $19,000 and has an annual net revenue of $5,000. All estimates are in actual dollars. Inflation is expected to be 2% per year for the next five years, and the inflation-free (real) MARR is 9.8% per year. Which alternative should be chosen? (8.2)

(a) Alternative A   (b) Alternative B
(c) Neither alternative

8-59. How much life insurance should Stan Moneymaker buy if he wants to leave enough money to his family so that they get $25,000 per year forever in constant dollars as of the year of his death? The interest rate from the life insurance company is 7% per year, while the inflation rate is expected to be 4% per year. The first payment occurs one year after Stan's death. (8.2)

(a) $866,000   (b) $357,000   (c) $625,000
(d) $841,000

8-60. If the European euro is worth Canadian $1.60 and a Canadian dollar trades for U.S. $0.75, what is the exchange rate from U.S. dollars to European euros? (8.5)

(a) $2.13   (b) $1.44   (c) $1.20   (d) $0.95

8-61. Light, sweet crude oil in the world market is U.S. dollar denominated. The price of oil is rising partially because the dollar is not worth what it used to be (i.e., the dollar is now worth less than last year or last month). If the dollar today is worth 0.75 euros in Europe (i.e., 1 euro = $1.333 U.S.), the value in euros of $67 U.S. per barrel oil is most nearly (8.5)

(a) 40 euros per barrel   (b) 50 euros per barrel
(c) 60 euros per barrel   (d) 80 euros per barrel

CHAPTER 9

Replacement Analysis

The objective of Chapter 9 is to address the question of whether a currently owned asset should be kept in service or immediately replaced.

High-Tech Replacement

Many students purchase laptop computers for use in college. A typical laptop costs $800, followed by annual expenses for software upgrades and maintenance. Suppose that the computer will be kept for a maximum of three years. When is the most economical time to replace the laptop with a new computer if one's personal interest rate is 10% per year? Annual expenses and year-end resale values are expected to be the following:

	Year 1	Year 2	Year 3
Software upgrades and maintenance	$150	$200	$300
Resale value at end of year	$600	$450	$200

In this chapter, you will learn how to answer this question as well as others involving replacement and retirement issues of deteriorating assets.

When firms defer the replacement of equipment, the immediate result is a decline in the production of goods and/or services, which in turn leads to unemployment, and this is often taken as one of the signals that a business recession is under way.

—William T. Morris (1964)

9.1 Introduction

A decision situation often encountered in business firms and government organizations, as well as by individuals, is whether an existing asset should be retired from use, continued in service, or replaced with a new asset. As the pressures of worldwide competition continue to increase, requiring higher quality goods and services, shorter response times, and other changes, this type of decision is occurring more frequently. Thus, the *replacement problem*, as it is commonly called, requires careful engineering economy studies to provide the information needed to make sound decisions that improve the operating efficiency and the competitive position of an enterprise.

Engineering economy studies of replacement situations are performed using the same basic methods as other economic studies involving two or more alternatives. Often the decision is whether to replace an existing (old) asset, descriptively called the *defender*, with a new asset. The one or more alternative replacement (new) assets are then called *challengers*.

The specific decision situation, however, occurs in different forms. Sometimes it may be whether to retire an asset without replacement (*abandonment*) or to retain the asset for backup rather than primary use. Also, the decision may be influenced if changed production requirements can be met by *augmenting* the capacity or capability of the existing asset(s).

9.2 Reasons for Replacement Analysis

The need to evaluate the replacement, retirement, or augmentation of assets results from changes in the economics of their use in an operating environment. Various reasons can underlie these changes, and unfortunately they are sometimes accompanied by unpleasant financial facts. The following are the three major reasons that summarize most of the factors involved:

1. *Physical Impairment (Deterioration)* These are changes that occur in the physical condition of an asset. Normally, continued use (aging) results in the less efficient operation of an asset. Routine maintenance and breakdown repair costs increase, energy use may increase, more operator time may be required, and so forth. Or, some unexpected incident such as an accident occurs that affects the physical condition and the economics of ownership and use of the asset.

2. *Altered Requirements* Capital assets are used to produce goods and services that satisfy human wants. When the demand for a good or service either increases or decreases or the design of a good or service changes, the related asset(s) may have the economics of its use affected.

3. *Technology* The impact of changes in technology varies among different types of assets. For example, the relative efficiency of heavy highway construction equipment is impacted less rapidly by technological changes than automated manufacturing equipment. In general, the costs per unit of production, as well as quality and other factors, are favorably impacted by changes in technology, which result in more frequent replacement of existing assets with new and better challengers.

Reason 2 (altered requirements) and Reason 3 (technology) are sometimes referred to as different categories of *obsolescence*. In any replacement problem, factors from more than one of these three major areas may be involved. Regardless of the specific considerations, the replacement of assets often represents an economic opportunity for the firm.

For the purposes of our discussion of replacement studies, the following is a distinction between various types of lives for typical assets:

Economic life is the period of time (years) that results in the minimum equivalent uniform annual cost (EUAC) of owning and operating an asset.* If we assume good asset management, economic life should coincide with the period of time extending from the date of acquisition to the date of abandonment, demotion in use, or replacement from the primary intended service.

Ownership life is the period between the date of acquisition and the date of disposal by a specific owner. A given asset may have different categories of use by the owner during this period. For example, a car may serve as the primary family car for several years and then serve only for local commuting for several more years.

Physical life is the period between original acquisition and final disposal of an asset over its succession of owners. For example, the car just described may have several owners over its existence.

Useful life is the time period (years) that an asset is kept in productive service (either primary or backup). It is an estimate of how long an asset is expected to be used in a trade or business to produce income.

9.3 Factors that Must Be Considered in Replacement Studies

There are several factors that must be considered in replacement studies. Once a proper perspective has been established regarding these factors, little difficulty should be experienced in making replacement studies. Six factors and related concepts are discussed in this section:

1. Recognition and acceptance of past errors
2. Sunk costs

---

* The AW of a primarily cost cash-flow pattern is sometimes called the EUAC. Because this term is commonly used in the definition of the economic life of an asset, we will often use EUAC in this chapter.

3. Existing asset value and the *outsider viewpoint*
4. Economic life of the proposed replacement asset (challenger)
5. Remaining (economic) life of the old asset (defender)
6. Income tax considerations

9.3.1 Past Estimation Errors

The economic focus in a replacement study is the future. Any *estimation errors* made in a previous study related to the defender are not relevant (unless there are income tax implications). For example, when an asset's book value (BV) is greater than its current market value (MV), the difference frequently has been designated as an estimation error. Such errors also arise when capacity is inadequate, maintenance expenses are higher than anticipated, and so forth.

This implication is unfortunate because in most cases these differences are not the result of errors but of the inability to foresee future conditions better at the time of the original estimates. Acceptance of unfavorable economic realities may be made easier by posing a hypothetical question, What will be the costs of my competitor who has no past errors to consider? In other words, we must decide whether we wish to live in the *past*, with its errors and discrepancies, or to be in a sound competitive position in the *future*. A common reaction is, I can't afford to take the loss in value of the existing asset that will result if the replacement is made. The fact is that the loss already has occurred, whether or not it could be afforded, and it exists whether or not the replacement is made.

9.3.2 The Sunk-Cost Trap

Only present and future cash flows should be considered in replacement studies. Any unamortized values (i.e., unallocated value of an asset's capital investment) of an existing asset under consideration for replacement are strictly the result of *past* decisions—that is, the initial decision to invest in that asset and decisions as to the method and number of years to be used for depreciation purposes. For purposes of this chapter, we define a *sunk cost* to be the difference between an asset's BV and its MV at a particular point in time. Sunk costs have no relevance to the replacement decisions that must be made (*except to the extent that they affect income taxes*). When income tax considerations are involved, we must include the sunk cost in the engineering economy study. Clearly, serious errors can be made in practice when sunk costs are incorrectly handled in replacement studies.

9.3.3 Investment Value of Existing Assets and the Outsider Viewpoint

Recognition of the nonrelevance of BVs and sunk costs leads to the proper viewpoint to use in placing value on existing assets for replacement study purposes. In this chapter, we use the so-called *outsider viewpoint* for approximating the investment worth of an existing asset (defender). In particular, the outsider viewpoint* is the perspective that would be taken by an impartial third party to

---

* The outsider viewpoint is also known as the opportunity cost approach to determining the value of the defender.

establish the fair MV of a used (secondhand) asset. This viewpoint forces the analyst to focus on present and future cash flows in a replacement study, thus avoiding the temptation to dwell on past (sunk) costs.

The *present realizable* MV is the correct capital investment amount to be assigned to an existing asset in replacement studies.* A good way to reason that this is true is to use the *opportunity cost* or *opportunity forgone principle*. That is, if it is decided to keep the existing asset, we are giving up the opportunity to obtain its net realizable MV at that time. Therefore, this situation represents the *opportunity cost* of keeping the defender.

There is one addendum to this rationale: If any new investment expenditure (such as for overhaul) is needed to upgrade the existing asset so that it will be competitive in level of service with the challenger, the extra amount should be added to the present realizable MV to determine the total investment in the existing asset for replacement study purposes.

> When using the outsider viewpoint, the total investment in the defender is the opportunity cost of not selling the existing asset for its current MV, *plus* the cost of upgrading it to be competitive with the best available challenger. (All feasible challengers are to be considered.)

Clearly, the MV of the defender must not also be claimed as a reduction in the challenger's capital investment, because doing so would provide an unfair advantage to the challenger by double counting the defender's selling price.

EXAMPLE 9-1   Investment Cost of the Defender (Existing Asset)

The purchase price of a certain new automobile (challenger) being considered for use in your business is $21,000. Your firm's present automobile (defender) can be sold on the open market for $10,000. The defender was purchased with cash three years ago, and its current BV is $12,000. To make the defender comparable in continued service to the challenger, your firm would need to make some repairs at an estimated cost of $1,500.

Based on this information,

(a) What is the total capital investment in the defender, using the outsider viewpoint?

(b) What is the unamortized value of the defender?

Solution

(a) The total capital investment in the defender (if kept) is its current MV (an opportunity cost) plus the cost of upgrading the car to make it comparable in service to the challenger. Hence, the total capital investment in the defender is $10,000 + $1,500 = $11,500 (from an outsider's viewpoint). This represents a good starting point for estimating the cost of keeping the defender.

---

* In after-tax replacement studies, the before-tax MV is modified by income tax effects related to potential gains (losses) forgone if the defender is kept in service.

(b) The unamortized value of the defender is the book loss (if any) associated with disposing of it. Given that the defender is sold for $10,000, the unamortized value (loss) is $12,000 − $10,000 = $2,000. This is the difference between the current MV and the current BV of the defender. As discussed in Section 9.3.2, this amount represents a sunk cost and has no relevance to the replacement decision, except to the extent that it may impact income taxes (to be discussed in Section 9.9).

9.3.4 Economic Life of the Challenger

The economic life of an asset minimizes the EUAC of owning and operating the asset, and it is often shorter than the useful or physical life. It is essential to know a challenger's economic life, in view of the principle that new and existing assets should be compared over their economic (optimum) lives. Economic data regarding challengers are periodically updated (often annually), and replacement studies are then repeated to ensure an ongoing evaluation of improvement opportunities.

9.3.5 Economic Life of the Defender

As we shall see later in this chapter, the economic life of the defender is often one year. Consequently, care must be taken when comparing the defender asset with a challenger asset, because *different lives* (Chapter 6) are involved in the analysis. We shall see that the defender should be kept longer than its apparent economic life, as long as its *marginal cost* is less than the minimum EUAC of the challenger over its economic life. What *assumptions* are involved when two assets having different apparent economic lives are compared, given that the defender is a nonrepeating asset? These concepts will be discussed in Section 9.7.

9.3.6 The Importance of Income Tax Consequences

The replacement of assets often results in gains or losses from the sale of *depreciable property*, as discussed in Chapter 7. Consequently, to perform an accurate economic analysis in such cases, the studies must be made on an *after-tax basis*. It is evident that the existence of a taxable gain or loss, in connection with replacement, can have a considerable effect on the results of an engineering study. A prospective gain from the disposal of assets can be reduced by as much as 40% or 50%, depending on the effective income tax rate used in a particular study. Hence, the decision to dispose of or retain an existing asset can be influenced by income tax considerations.

9.4 Typical Replacement Problems

The following typical replacement situations are used to illustrate several of the factors that must be considered in replacement studies. These analyses use the outsider viewpoint to determine the investment in the defenders.

| EXAMPLE 9-2 | **Replacement Analysis Using Present Worth (before Taxes)** |

A firm owns a pressure vessel that it is contemplating replacing. The old pressure vessel has annual operating and maintenance expenses of $60,000 per year and it can be kept for five more years, at which time it will have zero MV. It is believed that $30,000 could be obtained for the old pressure vessel if it were sold now.

A new pressure vessel can be purchased for $120,000. The new pressure vessel will have an MV of $50,000 in five years and will have annual operating and maintenance expenses of $30,000 per year. Using a before-tax MARR of 20% per year, determine whether or not the old pressure vessel should be replaced. A study period of five years is appropriate.

Solution

The first step in the analysis is to determine the investment value of the defender (old pressure vessel). Using the outsider viewpoint, the investment value of the defender is $30,000, its present MV. We can now compute the PW (or FW or AW) of each alternative and decide whether the old pressure vessel should be kept in service or replaced immediately.

$$Defender: \text{PW}(20\%) = -\$30,000 - \$60,000(P/A, 20\%, 5)$$

$$= -\$209,436.$$

$$Challenger: \text{PW}(20\%) = -\$120,000 - \$30,000(P/A, 20\%, 5)$$

$$+ \$50,000(P/F, 20\%, 5) = -\$189,623.$$

The PW of the challenger is greater (less negative) than the PW of the defender. Thus, the old pressure vessel should be replaced immediately. (The EUAC of the defender is $70,035 and that of the challenger is $63,410.)

| EXAMPLE 9-3 | **Before-Tax Replacement Analysis Using EUAC** |

The manager of a carpet manufacturing plant became concerned about the operation of a critical pump in one of the processes. After discussing this situation with the supervisor of plant engineering, they decided that a replacement study should be done, and that a nine-year study period would be appropriate for this situation. The company that owns the plant is using a before-tax MARR of 10% per year for its capital investment projects.

The existing pump, Pump *A*, including driving motor with integrated controls, cost $17,000 five years ago. An estimated MV of $750 could be obtained for the pump if it were sold now. Some reliability problems have been experienced with Pump *A*, including annual replacement of the impeller and bearings at a cost of $1,750. Annual operating and maintenance expenses have been averaging $3,250. Annual insurance and property tax expenses are 2% of the initial capital investment. It appears that the pump will provide adequate

TABLE 9-1 Summary of Information for Example 9-3		
MARR (before taxes) = 10% per year		
Existing Pump A *(defender)*		
Capital investment when purchased five years ago		$17,000
Annual expenses:		
Replacement of impeller and bearings	$1,750	
Operating and maintenance	3,250	
Taxes and insurance: $17,000 × 0.02	340	
Total annual expenses		$5,340
Present MV		$750
Estimated market value at the end of nine additional years		$200
Replacement Pump B *(challenger)*		
Capital investment		$16,000
Annual expenses:		
Operating and maintenance	$3,000	
Taxes and insurance: $16,000 × 0.02	320	
Total annual expenses		$3,320
Estimated MV at the end of nine years: $16,000 × 0.20		$3,200

service for another nine years if the present maintenance and repair practice is continued. It is estimated that if this pump is continued in service, its final MV after nine more years will be about $200.

An alternative to keeping the existing pump in service is to sell it immediately and to purchase a replacement pump, Pump B, for $16,000. An estimated MV at the end of the nine-year study period would be 20% of the initial capital investment. Operating and maintenance expenses for the new pump are estimated to be $3,000 per year. Annual taxes and insurance would total 2% of the initial capital investment. The data for Example 9-3 are summarized in Table 9-1.

Based on these data, should the defender (Pump A) be kept [and the challenger (Pump B) not purchased], or should the challenger be purchased now (and the defender sold)? Use a before-tax analysis and the outsider viewpoint in the evaluation.

Solution

In an analysis of the defender and challenger, care must be taken to correctly identify the investment amount in the existing pump. Based on the outsider viewpoint, this would be the current MV of $750; that is, the *opportunity cost* of keeping the defender. Note that the investment amount of Pump A ignores the original purchase price of $17,000. Using the principles discussed thus far, a before-tax analysis of EUAC of Pump A and Pump B can now be made.

The solution of Example 9-3 using EUAC (before taxes) as the decision criterion follows:

Study Period = 9 Years	Keep Old Pump A	Replacement Pump B
EUAC(10%):		
Annual expenses	$5,340	$3,320
Capital recovery cost (Equation 5-5):		
[$750(A/P, 10%, 9) − $200(A/F, 10%, 9)]	115	
[$16,000(A/P, 10%, 9) − $3,200(A/F, 10%, 9)]		2,542
Total EUAC(10%)	$5,455	$5,862

Because Pump A has the smaller EUAC ($5,455 < $5,862), the replacement pump is apparently not justified and the defender should be kept at least one more year. We could also make the analysis using other methods (e.g., PW), and the indicated choice would be the same.

9.5 Determining the Economic Life of a New Asset (Challenger)

Sometimes in practice the useful lives of the defender and the challenger(s) are not known and cannot be reasonably estimated. The time an asset is kept in productive service might be extended indefinitely with adequate maintenance and other actions, or it might be suddenly jeopardized by an external factor such as technological change.

> It is important to know the economic life, minimum EUAC, and total year-by-year (marginal) costs for both the best challenger and the defender *so that they can be compared on the basis of an evaluation of their economic lives and the costs most favorable to each.*

The economic life of an asset was defined in Section 9.2 as the time that results in the minimum EUAC of owning and operating the asset. Also, the economic life is sometimes called the minimum-cost life or optimum replacement interval. For a new asset, its EUAC can be computed if the capital investment, annual expenses, and year-by-year MVs are known or can be estimated. The apparent difficulties of estimating these values in practice may discourage performing the economic life and equivalent cost calculations. Similar difficulties, however, are encountered in most engineering economy studies when estimating the *future* economic consequences of alternative courses of action. Therefore, the estimating problems in replacement analysis are not unique and can be overcome in most application studies (see Chapter 3).

The estimated initial capital investment, as well as the annual expense and MV estimates, may be used to determine the PW through year k of total costs, PW_k.

That is, on a *before-tax* basis,

$$PW_k(i\%) = I - MV_k(P/F, i\%, k) + \sum_{j=1}^{k} E_j(P/F, i\%, j), \qquad (9\text{-}1)$$

which is the sum of the initial capital investment, I, (PW of the initial investment amounts, if any, occur after time zero) adjusted by the PW of the MV at the end of year k, and of the PW of annual expenses (E_j) through year k. The *total marginal cost* for each year k, TC_k, is calculated using Equation (9-1) by finding the increase in the PW of total cost from year $k-1$ to year k and then determining the equivalent worth of this increase at the end of year k. That is, $TC_k = (PW_k - PW_{k-1})(F/P, i\%, k)$. The algebraic simplification of this relationship results in the formula

$$TC_k(i\%) = MV_{k-1} - MV_k + iMV_{k-1} + E_k, \qquad (9\text{-}2)$$

which is the sum of the loss in MV during the year of service, the opportunity cost of capital invested in the asset at the beginning of year k, and the annual expenses incurred in year k (E_k). These total marginal (or year-by-year) costs, according to Equation (9-2), are then used to find the EUAC of each year prior to and including year k. The minimum $EUAC_k$ value during the useful life of the asset identifies its economic life, N_C^*. This procedure is illustrated in Examples 9-4 and 9-5.

EXAMPLE 9-4   **Economic Life of a Challenger (New Asset)**

A new forklift truck will require an investment of $30,000 and is expected to have year-end MVs and annual expenses as shown in columns 2 and 5, respectively, of Table 9-2. If the before-tax MARR is 10% per year, how long should the asset be retained in service? Solve by hand and by spreadsheet.

TABLE 9-2   **Determination of the Economic Life, N^*, of a New Asset (Example 9-4)**

(1) End of Year, k	(2) MV, End of Year k	(3) Loss in Market Value (MV) during Year k	(4) Cost of Capital = 10% of Beginning of Year MV	(5) Annual Expenses (E_k)	(6) [= (3) + (4) + (5)] Total (Marginal) Cost for Year (TC_k)	(7) EUAC[a] through Year k
0	$30,000					
1	22,500	$7,500	$3,000	$3,000	$13,500	$13,500
2	16,875	5,625	2,250	4,500	12,375	12,964
3	12,750	4,125	1,688	7,000	12,813	**12,918** $N_C^* = 3$
4	9,750	3,000	1,275	10,000	14,275	13,211
5	7,125	2,625	975	13,000	16,600	13,766

[a] $EUAC_k = \left[\sum_{j=1}^{k} TC_j(P/F, 10\%, j)\right](A/P, 10\%, k)$

✍ Solution by Hand

The solution to this problem is obtained by completing columns 3, 4, 6 [Equation (9-2)], and 7 of Table 9-2. In the solution, the customary year-end occurrence of all cash flows is assumed. The loss in MV during year k is simply the difference between the beginning-of-year market value, MV_{k-1}, and the end-of-year market value, MV_k. The opportunity cost of capital in year k is 10% of the capital unrecovered (invested in the asset) at the beginning of each year. The values in column 7 are the EUACs that would be incurred each year (1 to k) if the asset were retained in service through year k and then replaced (or retired) at the end of the year. The minimum EUAC occurs at the end of year N_C^*.

It is apparent from the values shown in column 7 that the new forklift truck will have a minimum EUAC if it is kept in service for only three years (i.e., $N_C^* = 3$).

Spreadsheet Solution

Figure 9-1 illustrates the use of a spreadsheet to solve for the economic life of the asset described in Example 9-4. The capital recovery formula in column C uses absolute addressing for the original capital investment and relative addressing to obtain the current year market value. Column E calculates the PW of the

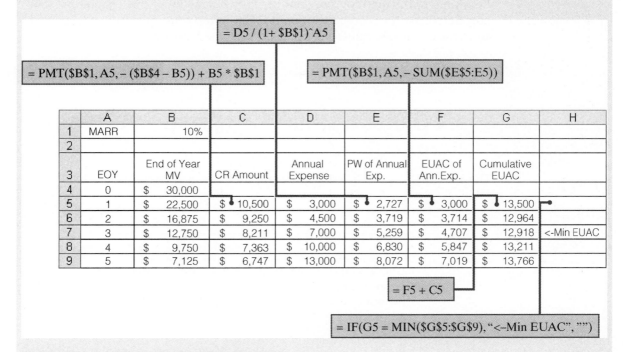

Figure 9-1 Spreadsheet Solution, Example 9-4

annual expense values in column D. Column F determines the EUAC of the annual expenses. The cumulative EUAC column (G) is the sum of columns C and F. The IF function in column H places a label that identifies the minimum EUAC value, which corresponds to the economic life of the asset.

The computational approach in the preceding example, as shown in Table 9-2, was to determine the total marginal cost for each year and then to convert these into an EUAC through year k. The before-tax EUAC for any life can also be calculated using the more familiar capital recovery formula presented in Chapter 5. For example, for a life of two years, the EUAC can be calculated with the help of Equation (5-5), as follows:

$$EUAC_2(10\%) = \$30,000(A/P, 10\%, 2) - \$16,875(A/F, 10\%, 2)$$
$$+ [\$3,000(P/F, 10\%, 1) + \$4,500(P/F, 10\%, 2)]$$
$$\times (A/P, 10\%, 2) = \$12,964.$$

This agrees with the corresponding row in column 7 of Table 9-2.

EXAMPLE 9-5 **Economic Life of a Laptop Computer**

Returning to the question posed at the very beginning of this chapter, we can now determine the economic life of the $800 laptop computer. Annual expenses and year-end resale values are expected to be the following:

	Year 1	Year 2	Year 3
Software upgrades and maintenance	$150	$200	$300
Resale value at end of year	$600	$450	$200

The marginal cost of keeping the computer for each of the three years is as follows:

	Year 1	Year 2	Year 3
Upgrades and maintenance	$150	$200	$300
Depreciation	$200	$150	$250
Interest on capital at 10%	$80	$60	$45
Total marginal cost	$430	$410	$595

$EUAC_1(10\%) = \$430$

$EUAC_2(10\%) = [\$430(P/F, 10\%, 1) + \$410(P/F, 10\%, 2)](A/P, 10\%, 2) = \$420$

$EUAC_3(10\%) = [\$420(P/A, 10\%, 2) + \$595(P/F, 10\%, 3)](A/P, 10\%, 3) = \$473$

Therefore, based upon economics alone, the computer should be kept for two years before being replaced.

Although the differences in the $EUAC_k(10\%)$ values may seem small enough to make this analysis not worth the effort, consider the viewpoint of an international company that supplies its employees with laptop computers. The money saved by replacing the computer after two years (as opposed to keeping it for a third year) becomes significant when you are considering hundreds (or even thousands) of computers company-wide. This also applies to the thousands of college students who need to purchase laptop computers.

9.6 Determining the Economic Life of a Defender

In replacement analyses, we must also determine the economic life (N_D^*) that is most favorable to the defender. This gives us the choice of keeping the defender as long as its EUAC at N_D^* is less than the minimum EUAC of the challenger. When a major outlay for defender alteration or overhaul is needed, the life that will yield the minimum EUAC is likely to be the period that will elapse before the next major alteration or overhaul is needed. Alternatively, *when there is no defender MV now or later (and no outlay for alteration or overhaul) and when the defender's operating expenses are expected to increase annually, the remaining life that will yield the minimum EUAC will be one year.*

When MVs are greater than zero and expected to decline from year to year, it is necessary to calculate the apparent remaining economic life, which is done in the same manner as for a new asset. Using the outsider viewpoint, the investment value of the defender is considered to be its present realizable MV.

Regardless of how the remaining economic life for the defender is determined, a decision to keep the defender does not mean that it should be kept only for this period of time. Indeed, the defender should be kept longer than the apparent economic life as long as its *marginal* cost (total cost for an additional year of service) is less than the minimum EUAC for the best challenger.

This important principle of replacement analysis is illustrated in Example 9-6.

| EXAMPLE 9-6 | **Economic Life of a Defender (Existing Asset)** |

It is desired to determine how much longer a forklift truck should remain in service before it is replaced by the new truck (challenger) for which data were given in Example 9-4 and Table 9-2. The defender in this case is two years old, originally cost $19,500, and has a present realizable MV of $7,500. If kept, its MVs and annual expenses are expected to be as follows:

End of Year, k	MV, End of Year k	Annual Expenses, E_k
1	$6,000	$8,250
2	4,500	9,900
3	3,000	11,700
4	1,500	13,200

Determine the most economical period to keep the defender before replacing it (if at all) with the present challenger of Example 9-4. The before-tax cost of capital (MARR) is 10% per year.

Solution

Table 9-3 shows the calculation of total cost for each year (marginal cost) and the EUAC at the end of each year for the defender based on the format used in Table 9-2. Note that the minimum EUAC of $10,500 corresponds to keeping the defender for one more year. The marginal cost of keeping the truck for the second year is, however, $12,000, which is still less than the minimum EUAC for the challenger (i.e., $12,918, from Example 9-4). The *marginal cost* for keeping the defender the third year and beyond is greater than the $12,918 minimum EUAC for the challenger. Based on the available data shown, it would be most economical to keep the defender for two more years and then to replace it with the challenger.

TABLE 9-3   Determination of the Economic Life N^* of an Existing Asset (Example 9-6)

(1) End of Year, k	(2) MV, End of Year, k	(3) Loss in Market Value (MV) during Year k	(4) Cost of Capital = 10% of Beginning of Year MV	(5) Annual Expenses (E_k)	(6) [= (3) + (4) + (5)] Total (Marginal) Cost for Year, (TC_k)	(7) EUAC[a] through Year k
0	$7,500					
1	6,000	$1,500	$750	$8,250	$10,500	$10,500 $N_D^* = 1$
2	4,500	1,500	600	9,900	12,000	11,214
3	3,000	1,500	450	11,700	13,650	11,950
4	1,500	1,500	300	13,200	15,000	12,607

[a] $\text{EUAC}_k = \left[\sum_{j=1}^{k} TC_j (P/F, 10\%, j) \right] (A/P, 10\%, k)$

Example 9-6 assumes that a comparison is being made with the best challenger alternative available. In this situation, if the defender is retained beyond the point where its marginal costs exceed the minimum EUAC for the challenger, the

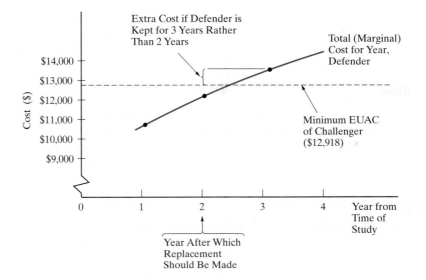

Figure 9-2 Defender versus Challenger Forklift Trucks (Based on Examples 9-4 and 9-6)

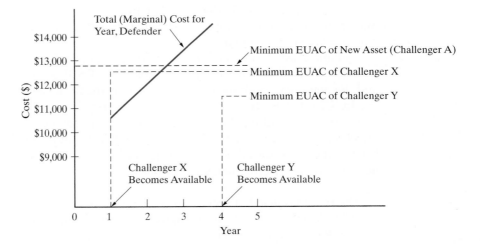

Figure 9-3 Old versus New Asset Costs with Improved Challengers Becoming Available in the Future

difference in costs continues to grow and replacement becomes more urgent. This is illustrated to the right of the intersection in Figure 9-2.

Figure 9-3 illustrates the effect of improved new challengers in the future. If an improved challenger X becomes available before replacement with the new asset of Figure 9-2, then a new replacement study should take place to consider the improved challenger. If there is a possibility of a further-improved challenger Y as of, say, four years later, it may be better still to postpone replacement until that challenger becomes available. Although retention of the old asset beyond its breakeven point with the best available challenger has a cost that may well grow with time, this cost of waiting can, in some instances, be worthwhile if it permits the purchase of an improved asset having economies that offset the cost of waiting. Of course, a decision to postpone a replacement may also "buy time and information." Because technological change tends to be sudden and dramatic rather than uniform and gradual, new challengers with significantly improved features can arise sporadically and can change replacement plans substantially.

When replacement is not signaled by the engineering economy study, more information may become available before the next analysis of the defender. Hence, the next study should include any additional information. *Postponement* generally should mean a postponement of the decision on when to replace, not the decision to postpone replacement until a specified future date.

9.7 Comparisons in which the Defender's Useful Life Differs from that of the Challenger

In Section 9.4, we discussed a typical replacement situation in which the useful lives of the defender and the challenger were known and were the same, as well as equal to the study period. When this situation occurs, any of the equivalence methods, properly applied, may be used.

In the previous two sections (9.5 and 9.6), we discussed the economic lives of a new asset and of a defender and how these results (along with the related cost information) are used in replacement analysis when the useful lives of the assets may or may not be known.

A third situation occurs when the useful lives of the best challenger and the defender are known, or can be estimated, but are not the same. The comparison of the challenger and the defender under these circumstances is the topic of this section.

In Chapter 6, two assumptions used for the economic comparisons of alternatives, including those having different useful lives, were described: (1) *repeatability* and (2) *cotermination*. Under either assumption, the analysis period used is the same for all alternatives in the study. The repeatability assumption, however, involves two main stipulations:

1. The period of needed service for which the alternatives are being compared is either indefinitely long or a length of time equal to a common multiple of the useful lives of the alternatives.
2. What is estimated to happen in the first useful life span will happen in all succeeding life spans, if any, for each alternative.

For replacement analyses, the first of these conditions may be acceptable, but normally the second is not reasonable for the defender. The defender is typically an older and used piece of equipment. An identical replacement, even if it could be found, probably would have an installed cost in excess of the current MV of the defender.

Failure to meet the second stipulation can be circumvented if the period of needed service is assumed to be indefinitely long and *if we recognize that the analysis is really to determine whether* **now** *is the time to replace the defender*. When the defender is replaced, either now or at some future date, it will be by the challenger—the best available replacement.

Figure 9-4 Effect of the Repeatability Assumption Applied to Alternatives in Example 9-6

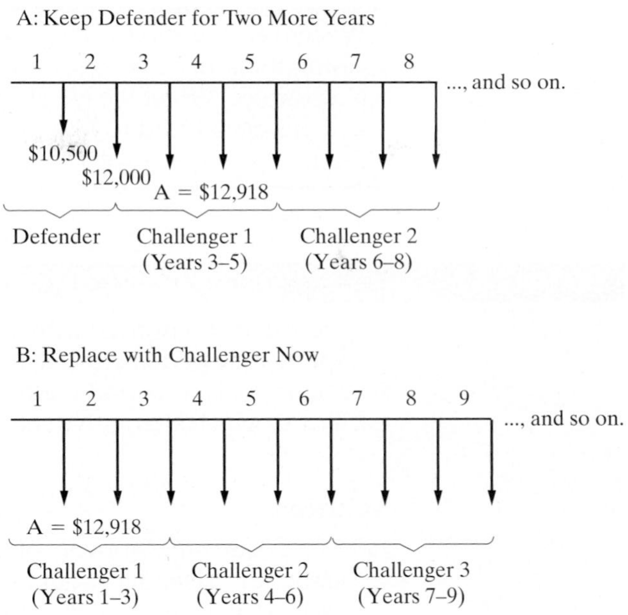

A: Keep Defender for Two More Years

B: Replace with Challenger Now

Example 9-6, involving a before-tax analysis of the defender versus a challenger forklift truck, made implicit use of the *repeatability* assumption. That is, it was assumed that the particular challenger analyzed in Table 9-2 would have a minimum EUAC of $12,918 regardless of when it replaces the defender. Figure 9-4 shows time diagrams of the cost consequences of keeping the defender for two more years versus adopting the challenger now, with the challenger costs to be repeated in the indefinite future. Recall that the economic life of the challenger is three years. *It can be seen in Figure 9-4 that the only difference between the alternatives is in years one and two.*

The repeatability assumption, applied to replacement problems involving assets with different useful or economic lives, often simplifies the economic comparison of the alternatives. For example, the comparison of the PW values of the alternatives in Figure 9-4 *over an infinite analysis period* (refer to the capitalized worth calculation in Chapter 5) will confirm our previous answer to Example 9-6 that Alternative A (keep defender for two more years) is preferred to Alternative B (replace with challenger now). Using a MARR = 10% per year, we have

$$PW_A(10\%) = -\$10,500(P/F, 10\%, 1) - \$12,000(P/F, 10\%, 2)$$

$$- \frac{\$12,918}{0.10}(P/F, 10\%, 2)$$

$$= -\$126,217;$$

$$PW_B(10\%) = -\frac{\$12,918}{0.10} = -\$129,180.$$

The difference $(PW_B - PW_A)$ is $-\$2,963$, which confirms that the additional cost of the challenger over the next two years is not justified and it is best to keep the defender for two more years before replacing it with the challenger.

Whenever the *repeatability* assumption is not applicable, the *coterminated* assumption may be used; it involves using a finite study period for all alternatives. When the effects of price changes and taxes are to be considered in replacement studies, it is recommended that the coterminated assumption be used.

EXAMPLE 9-7   **Replacement Analysis Using the Coterminated Assumption**

Suppose that we are faced with the same replacement problem as in Example 9-6, except that the period of needed service is (a) three years or (b) four years. That is, a finite analysis period under the coterminated assumption is being used. In each case, which alternative should be selected?

Solution

(a) For a planning horizon of three years, we might intuitively think that either the defender should be kept for three years or it should be replaced immediately by the challenger to serve for the next three years. From Table 9-3, the EUAC for the defender for three years is $11,950; and from Table 9-2, the EUAC for the challenger for three years is $12,918. Thus, following this reasoning, the defender would be kept for three years. This, however, is not quite right. Focusing on the *Total (Marginal) Cost for Each Year* columns, we can see that the defender has the lowest cost in the first two years, but in the third year its cost is $13,650; the EUAC of one year of service for the challenger is only $13,500. Hence, it would be more economical to replace the defender after the second year. This conclusion can be confirmed by enumerating all replacement possibilities and their respective costs and then computing the EUAC for each, as will be done for the four-year planning horizon in Part (b).

TABLE 9-4   **Determination of When to Replace the Defender for a Planning Horizon of Four Years [Example 9-7, Part (b)]**

Keep Defender for	Keep Challenger for	Total (Marginal) Costs for Each Year				EUAC at 10% for 4 Years	
		1	2	3	4		
0 years	4 years	$13,500[a]	$12,375[a]	$12,813[a]	$14,275[a]	$13,211	
1	3	10,500	13,500	12,375	12,813	12,225	
2	2	10,500	12,000	13,500	12,375	**12,006**	← Least cost alternative
3	1	10,500	12,000	13,650	13,500	12,284	
4	0	10,500[b]	12,000[b]	13,650[b]	15,000[b]	12,607	

[a] Column 6 of Table 9-2.
[b] Column 6 of Table 9-3.

(b) For a planning horizon of four years, the alternatives and their respective costs for each year and the EUAC of each are given in Table 9-4. Thus, the most economical alternative is to keep the defender for two years and then replace it with the challenger, to be kept for the next two years. The decision to keep the defender for two years happens to be the same as when the repeatability assumption was used, which, of course, would not be true in general.

When a replacement analysis involves a defender that cannot be continued in service because of changes in technology, service requirements, and so on, a choice among two or more new challengers must be made. Under this situation, the repeatability assumption may be a convenient economic modeling approach for comparing the alternatives and making a present decision. Note that, when the defender is not a feasible alternative, the replacement problem is no different than any other analysis involving mutually exclusive alternatives.

9.8 Retirement without Replacement (Abandonment)

Consider a project for which the period of required service is finite and that has *positive* net cash flows following an initial capital investment. Market values, or abandonment values, are estimated for the end of each remaining year in the project's life. In view of an opportunity cost (MARR) of $i\%$ per year, should the project be undertaken? Given that we have decided to implement the project, what is the best year to abandon the project? In other words, what is the economic life of the project?

For this type of problem, the following assumptions are applicable:

1. Once a capital investment has been made, the firm desires to postpone the decision to abandon a project as long as its present equivalent value (PW) is not decreasing.
2. The existing project will be terminated at the best abandonment time and will not be replaced by the firm.

Solving the abandonment problem is similar to determining the economic life of an asset. In abandonment problems, however, annual benefits (cash inflows) are present, but in economic life analysis, costs (cash outflows) are dominant. In both cases, the objective is to increase the overall wealth of the firm by finding the life that maximizes profits or, equivalently, minimizes costs.

EXAMPLE 9-8	When to Retire an Asset (No Replacement)

An $80,000 baling machine for recycled paper was purchased by the XYZ company two years ago. The current MV of the machine is $50,000, and it can be kept in service for seven more years. MARR is 12% per year and the projected net annual receipts (revenues less expenses) and end-of-year market values for the machine are shown below. When is the best time for the company to abandon this project?

	End of Year						
	1	2	3	4	5	6	7
Net annual receipts	$20,000	$20,000	$18,000	$15,000	$12,000	$6,000	$3,000
Market value	40,000	32,000	25,000	20,000	15,000	10,000	5,000

Solution

The PWs that result from deciding now to keep the machine exactly one, two, three, four, five, six, and seven years are as follows:

Keep for one year:

$$PW(12\%) = -\$50,000 + (\$20,000 + \$40,000)(P/F, 12\%, 1) = \$3,571.$$

Keep for two years:

$$PW(12\%) = -\$50,000 + \$20,000(P/F, 12\%, 1) + (\$20,000 + \$32,000)$$
$$\times (P/F, 12\%, 2) = \$9,311.$$

In the same manner, the PW for years three through seven can be computed. The results are as follows.

Keep for three years: $PW(12\%) = \$14,408$

Keep for four years: $PW(12\%) = \$18,856$

Keep for five years: $PW(12\%) = $**$21,466**$ ← best abandonment time

Keep for six years: $PW(12\%) = \$21,061$

Keep for seven years: $PW(12\%) = \$19,614$

As you can see, PW is maximized ($21,466) by retaining the machine for five years. Thus, the best abandonment time for this project would be in five years.

9.9 After-Tax Replacement Studies

As discussed in Chapter 7, income taxes associated with a proposed project may represent a major cash outflow for a firm. Therefore, income taxes should be considered along with all other relevant cash flows when assessing the economic profitability of a project. This fact also holds true for replacement decisions. The replacement of an asset often results in gains or losses from the sale of the existing asset (defender). The income tax consequence resulting from the gain (loss) associated with the sale (or retention) of the defender has the potential to impact the decision to keep the defender or to sell it and purchase the challenger. The remainder of this section is devoted to demonstrating the procedure for performing replacement analyses on an after-tax basis. Note that after-tax replacement analyses require knowledge of the depreciation schedule already in use for the defender, as well as the appropriate depreciation schedule to be used for the challenger.

9.9.1 After-Tax Economic Life

In earlier sections, the economic life of a new asset (Examples 9-4 and 9-5) and the economic life of an existing asset (Example 9-6) were determined on a before-tax basis. An *after-tax* analysis, however, can also be used to determine the economic life of an asset by extending Equation (9-1) to account for income tax effects:

$$PW_k(i\%) = I + \sum_{j=1}^{k} [(1-t)E_j - td_j](P/F, i\%, j) - [(1-t)MV_k + t(BV_k)](P/F, i\%, k).$$

(9-3)

This computation finds the PW of the after-tax cash flows (ATCFs) (expressed as costs) through year k, PW_k, by (1) adding the initial capital investment, I, (PW of investment amounts if any occur after time zero) and the sum of the after-tax PW of annual expenses through year k, including adjustments for annual depreciation amounts (d_j), and then (2) adjusting this total after-tax PW of costs by the after-tax consequences of gain or loss on disposal of the asset at the end of year k. Similar to the previous before-tax analysis using Equation (9-1), Equation (9-3) is used to develop the after-tax total marginal cost for each year k, TC_k. That is, $TC_k = (PW_k - PW_{k-1})(F/P, i\%, k)$. The algebraic simplification of this relationship results in Equation (9-4):

$$TC_k(i\%) = (1-t)(MV_{k-1} - MV_k + iMV_{k-1} + E_k) + i(t)(BV_{k-1}).$$ (9-4)

Equation (9-4) is $(1-t)$ times Equation (9-2) plus interest on the tax adjustment from the BV of the asset at the beginning of year k. A tabular format incorporating Equation (9-4) is used in the solution of the next example to find the economic life of a new asset on an after-tax basis (N_{AT}^*). This same procedure can also be used to find the after-tax economic life of an existing asset.

EXAMPLE 9-9 **After-Tax Economic Life of an Asset**

Find the economic life on an after-tax basis for the new forklift truck (challenger) described in Example 9-4. Assume that the new forklift is depreciated as a MACRS (GDS)* three-year property class asset, the effective income tax rate is 40%, and the after-tax MARR is 6% per year.

Solution

The calculations using Equation (9-4) are shown in Table 9-5. The expected year-by-year MVs and annual expenses are repeated from Example 9-4 in columns 2

---

* In Chapter 7, the GDS (general depreciation system) and ADS (alternative depreciation system) are discussed.

TABLE 9-5   Determination of the After-Tax Economic Life for the Asset Described in Example 9-4

(1) End of Year, k	(2) MV, End of Year k	(3) Loss in MV during Year k	(4) Cost of Capital = 6% of BOY MV in Col. 2	(5) Annual Expenses	(6) Approximate After-Tax Total (Marginal) Cost for Year k $(1-t) \cdot (\text{Col.3} + 4 + 5)$
0	$30,000	0	0	0	0
1	22,500	$7,500	$1,800	$3,000	$7,380
2	16,875	5,625	1,350	4,500	6,885
3	12,750	4,125	1,013	7,000	7,283
4	9,750	3,000	765	10,000	8,259
5	7,125	2,625	585	13,000	9,726

(1) End of Year, k	(7) MACRS BV at End of Year k	(8) Interest on Tax Adjustment = 6% · t · BOY BV in Col. 7	(9) Adjusted After-Tax Total (Marginal) Cost (TC_k) (Col. 6 + Col. 8)	(10) EUAC[a] (After Tax) through Year k
0	$30,000	0	0	0
1	20,001	$720	$8,100	$8,100
2	6,666	480	7,365	7,743
3	2,223	160	7,442	**7,649** $N^*_{\text{AT}} = 3$
4	0	53	8,312	7,800
5	0	0	9,726	8,142

[a] $\text{EUAC}_k = \left[\sum_{j=1}^{k} (\text{Col.9})_j (P/F, 6\%, j) \right] (A/P, 6\%, k)$

and 5, respectively. In column 6, the *sum* of the loss in MV during year k, cost of capital based on the MV at the beginning-of-year (BOY) k, and annual expenses in year k are multiplied by $(1-t)$ to determine an *approximate* after-tax total marginal cost in year k.

The BV amounts at the end of each year, based on the new forklift truck being a MACRS (GDS) three-year property class asset, are shown in column 7. These amounts are then used in column 8 to determine an annual tax adjustment [last term in Equation (9-4)], based on the beginning of year (BOY) book values (BV_{k-1}). This annual tax adjustment is algebraically added to the entry in column 6 to obtain an *adjusted* after-tax total marginal cost in year k, TC_k. The total marginal cost amounts are used in column 10 to calculate, successively, the EUAC_k of retirement of the asset at the end of year k. In this case, the after-tax economic life (N^*_{AT}) is three years, the same result obtained on a before-tax basis in Example 9-4.

It is not uncommon for the before-tax and after-tax economic lives of an asset to be the same (as occurred in Examples 9-4 and 9-9).

9.9.2 After-Tax Investment Value of the Defender

The outsider viewpoint has been used in this chapter to establish a before-tax investment value of an existing asset. Using this viewpoint, the present realizable MV of the defender is the appropriate before-tax investment value. This value (although not an actual cash flow) represents the opportunity cost of keeping the defender. In determining the after-tax investment value, we must also include the opportunity cost of gains (losses) [Chapter 7, Section 7.8] not realized if the defender is kept.

Consider, for example, a printing machine that was purchased three years ago for $30,000. It has a present MV of $5,000 and a current BV of $8,640. If the printing machine were sold now, the company would experience a loss on disposal of $5,000 − $8,640 = −$3,640. Assuming a 40% effective income tax rate, this loss would translate into a (−0.40)(−$3,640) = $1,456 tax savings. Thus, if it is decided to keep the printing machine, not only would the company be giving up the opportunity to obtain the $5,000 MV, it would also be giving up the opportunity to obtain the $1,456 tax credit that would result from selling the printing machine at a price less than its current BV. Thus, the total after-tax investment value of the existing printing machine is $5,000 + $1,456 = $6,456.

The computation of the after-tax investment value of an existing asset is quite straightforward. Using the general format for computing ATCFs presented previously in Figure 7-4, we would have the following entries if the defender were sold now (year zero). Note that MV_0 and BV_0 represent the MV and BV, respectively, of the defender at the time of the analysis.

End of Year, k	BTCF	Depreciation	Taxable Income	Cash Flow for Income Taxes	ATCF (if defender is sold)
0	MV_0	None	$MV_0 - BV_0$	$-t(MV_0 - BV_0)$	$MV_0 - t(MV_0 - BV_0)$

Now, if it was decided to keep the asset, the preceding entries become the opportunity costs associated with keeping the defender. The appropriate year-zero entries for analyzing the after-tax consequences of keeping the defender are shown in Figure 9-5. Note that the entries in Figure 9-5 are simply the same values shown in the preceding table, only reversed in sign to account for the change in perspective (keep versus sell).

End of Year, k	(A) BTCF	(B) Depreciation	(C) Taxable Income	(D) = −t(C) Cash Flow for Income Taxes	(E) = (A) + (D) ATCF (if defender is kept)
0	$-MV_0$	None	$-(MV_0 - BV_0)$	$-t[-(MV_0 - BV_0)]$ $= t(MV_0 - BV_0)$	$-MV_0 + t(MV_0 - BV_0)$

Figure 9-5   General Procedure for Computing the After-Tax Investment Value of a Defender

EXAMPLE 9-10   **After-Tax Investment Cost of a Defender: BV > MV**

An existing asset being considered for replacement has a current MV of $12,000 and a current BV of $18,000. Determine the after-tax investment value of the existing asset (if kept) using the outsider viewpoint and an effective income tax rate of 34%.

Solution

Given that $MV_0 = \$12,000$, $BV_0 = \$18,000$, and $t = 0.34$, we can easily compute the ATCF associated with keeping the existing asset by using the format of Figure 9-5:

End of Year, k	BTCF	Depreciation	Taxable Income	Cash Flow for Income Taxes	ATCF
0	−$12,000	None	−($12,000 − $18,000) = $6,000	(−0.34)($6,000) = −$2,040	−$12,000 − $2,040 = −$14,040

The appropriate after-tax investment value for the existing asset is $14,040. Note that this is higher than the before-tax investment value of $12,000, owing to the tax credit given up by *not* selling the existing machine at a loss.

EXAMPLE 9-11   **After-Tax Investment Cost of a Defender: BV < MV**

An engineering consulting firm is considering the replacement of its CAD workstation. The workstation was purchased four years ago for $20,000. Depreciation deductions have followed the MACRS (GDS) five-year property class schedule. The workstation can be sold now for $4,000. Assuming the effective income tax rate is 40%, compute the after-tax investment value of the CAD workstation if it is kept.

Solution

To compute the ATCF associated with keeping the defender, we must first compute the current BV, BV_0. The workstation has been depreciated for four years under the MACRS (GDS) system with a five-year property class. Thus,

$$BV_0 = \$20,000(1 - 0.2 - 0.32 - 0.192 - 0.1152) = \$3,456.^{*}$$

Using the format presented in Figure 9-5, we find that the ATCF associated with keeping the defender can be computed as follows:

---

* Current tax law dictates that gains and losses be taxed as ordinary income. As a result, it is not necessary to explicitly account for the MACRS half-year convention when computing the *if sold* BV (the increase in taxable income because of a higher BV is offset by the halfyear of depreciation that could be claimed if the defender is kept). This allows us to simplify the procedure for computing the after-tax investment value of the defender.

End of Year, k	BTCF	Depreciation	Taxable Income	Cash Flow for Income Taxes	ATCF
0	−$4,000	None	−($4,000 − $3,456) = −$544	(−0.4)(−$544) =$218	−$4,000 + $218 = −$3,782

The after-tax investment value of keeping the existing CAD workstation is $3,782. Note that, in the case where MV_0 is higher than BV_0, the after-tax investment value is lower than the before-tax investment value. This is because the gain on disposal (and resulting tax liability) does not occur at this time if the defender is retained.

9.9.3 Illustrative After-Tax Replacement Analyses

The following examples represent typical after-tax replacement analyses. They illustrate the appropriate method for including the effect of income taxes, as well as several of the factors that must be considered in general replacement studies.

EXAMPLE 9-12

After-Tax EUAC Analysis (Restatement of Example 9-3 with Tax Information)

The manager of a carpet manufacturing plant became concerned about the operation of a critical pump in one of the processes. After discussing this situation with the supervisor of plant engineering, they decided that a replacement study should be done and that a nine-year study period would be appropriate for this situation. The company that owns the plant is using an after-tax MARR of 6% per year for its capital investment projects. The effective income tax rate is 40%.

The existing pump, Pump A, including driving motor with integrated controls, cost $17,000 five years ago. The accounting records show the depreciation schedule to be following that of an asset with a MACRS (ADS) recovery period of nine years. Some reliability problems have been experienced with Pump A, including annual replacement of the impeller and bearings at a cost of $1,750. Annual expenses have been averaging $3,250. Annual insurance and property tax expenses are 2% of the initial capital investment. It appears that the pump will provide adequate service for another nine years if the present maintenance and repair practice is continued. An estimated MV of $750 could be obtained for the pump if it is sold now. It is estimated that, if this pump is continued in service, its final MV after nine more years will be about $200.

An alternative to keeping the existing pump in service is to sell it immediately and to purchase a replacement pump, Pump B, for $16,000. A nine-year class life (MACRS five-year property class) would be applicable to the new pump under the GDS. An estimated MV at the end of the nine-year study period would be 20% of the initial capital investment. Operating and maintenance expenses for the new pump are estimated to be $3,000 per year.

TABLE 9-6	Summary of Information for Example 9-12

MARR (after taxes) = 6% per year
Effective income tax rate = 40%

Existing Pump A (defender)	
MACRS (ADS) recovery period	9 years
Capital investment when purchased five years ago	$17,000
Total annual expenses	$5,340
Present MV	$750
Estimated market value at the end of nine additional years	$200

Replacement Pump B (challenger)	
MACRS (GDS) property class	5 years
Capital investment	$16,000
Total annual expenses	$3,320
Estimated MV at the end of nine years	$3,200

Annual taxes and insurance would total 2% of the initial capital investment. The data for Example 9-12 are summarized in Table 9-6.

Based on these data, should the defender (Pump *A*) be kept [and the challenger (Pump *B*) not purchased], or should the challenger be purchased now (and the defender sold)? Use an after-tax analysis and the outsider viewpoint in the evaluation.

Solution

The after-tax computations for keeping the defender (Pump *A*) and not purchasing the challenger (Pump *B*) are shown in Table 9-7. *Year zero of the analysis period* is at the *end of the current (fifth) year of service* of the defender. The

TABLE 9-7	ATCF Computations for the Defender (Existing Pump *A*) in Example 9-12

End of Year, k	(A) BTCF^a	(B) MACRS (ADS) Depreciation	(C) = (A) − (B) Taxable Income	(D) = −0.4(C) Income Taxes at 40%	(E) = (A) + (D) ATCF
0	−$750	None	$7,750	−$3,100	−$3,850
1–4	−5,340	$1,889	−7,229	2,892	−2,448
5	−5,340	944	−6,284	2,514	−2,826
6–9	−5,340	0	−5,340	2,136	−3,204
9	200		200^b	−80	120

^{a} Before-tax cash flow (BTCF).
^{b} Gain on disposal (taxable at the 40% rate).

year-zero entries of Table 9-7 are computed using the general format presented in Figure 9-5 and are further explained in the following:

1. BTCF (−$750): The same amount used in the before-tax analysis of Example 9-3. This amount is based on the outsider viewpoint and is the opportunity cost of keeping the defender instead of replacing it (and selling it for the estimated present MV of $750).

2. Taxable income ($7,750): This amount is the result of an increase in taxable income of $7,750 due to the tax consequences of keeping the defender instead of selling it. Specifically, *if we sold the defender now*, the loss on disposal would be as follows:

$$\text{Gain or loss on disposal (if sold now)} = MV_0 - BV_0,$$

$$BV_0 = \$17,000[1 - 0.0556 - 4(0.1111)] = \$8,500,$$

$$\text{Loss on disposal (if sold now)} = \$750 - \$8,500 = -\$7,750.$$

But *since we are keeping the defender (Pump A) in this alternative,* we have the reverse effect on taxable income, an increase of $7,750 due to an opportunity forgone.

3. Cash flow for income taxes (−$3,100): The increase in taxable income because of the tax consequences of keeping the defender results in an increased tax liability (or tax credit forgone) of $-0.4(\$7,750) = -\$3,100$.

4. ATCF (−$3,850): The total after-tax investment value of the defender is the result of two factors: the present MV ($750) and the tax credit ($3,100) forgone by keeping the existing Pump A. Therefore, the ATCF representing the investment in the defender (based on the outsider viewpoint) is $-\$750 - \$3,100 = -\$3,850$.

The remainder of the ATCF computations over the nine-year analysis period for the alternative of keeping the defender are shown in Table 9-7. The after-tax computations for the alternative of purchasing the challenger (Pump *B*) are shown in Table 9-8.

The next step in an after-tax replacement study involves equivalence calculations using an after-tax MARR. The following is the after-tax EUAC analysis for Example 9-12:

$$\begin{aligned}
\text{EUAC(6\%) of Pump } A(\text{defender}) = {} & \$3,850(A/P, 6\%, 9) \\
& + \$2,448(P/A, 6\%, 4)(A/P, 6\%, 9) \\
& + [\$2,826(F/P, 6\%, 4) \\
& + \$3,204(F/A, 6\%, 4) - \$120] \\
& \times (A/F, 6\%, 9) \\
= {} & \$3,332;
\end{aligned}$$

TABLE 9-8   ATCF Computations for the Challenger (Replacement Pump *B*) in Example 9-12

End of Year, *k*	(A) BTCF	(B) MACRS (GDS) Depreciation	(C) = (A) − (B) Taxable Income	(D) = −0.4 (C) Income Taxes at 40%	(E) = (A) + (D) ATCF
0	−$16,000	None			−$16,000
1	−3,320	$3,200	−$6,520	$2,608	−712
2	−3,320	5,120	−8,440	3,376	56
3	−3,320	3,072	−6,392	2,557	−763
4	−3,320	1,843	−5,163	2,065	−1,255
5	−3,320	1,843	−5,163	2,065	−1,255
6	−3,320	922	−4,242	1,697	−1,623
7–9	−3,320	0	−3,320	1,328	−1,992
9	3,200		3,200[a]	−1,280	1,920

[a] Gain on disposal (taxable at the 40% rate).

$$\text{EUAC}(6\%)\text{ of Pump }B\text{(challenger)} = \$16,000(A/P, 6\%, 9)$$
$$+ [\$712(P/F, 6\%, 1) - \$56(P/F, 6\%, 2)$$
$$+ \$763(P/F, 6\%, 3)$$
$$+ \cdots + \$1,992(P/F, 6\%, 9)](A/P, 6\%, 9)$$
$$- \$1,920(A/F, 6\%, 9)$$
$$= \$3,375.$$

Because the EUACs of both pumps are very close, other considerations, such as the improved reliability of the new pump, could detract from the slight economic preference for Pump *A*. The after-tax annual costs of both alternatives are considerably less than their before-tax annual costs.

The after-tax analysis does not *reverse* the results of the before-tax analysis for this problem (see Example 9-3). Due to income tax considerations, however, identical before-tax and after-tax recommendations should not necessarily be expected.

The next example involves the determination of the economic life of a defender on an after-tax basis and the use of after-tax marginal costs to determine the most economical time to replace the defender.

EXAMPLE 9-13   Another After-Tax Replacement Problem

The Hokie Metal Stamping Company is considering the replacement of a spray system. The new improved system will cost $60,000 for installation and will have an estimated economic life of 12 years. The MV of the new system at the end of

12 years is expected to be $6,000. Further, it is estimated that annual operating and maintenance expenses will average $32,000 per year for the new system and that straight-line depreciation (with a $6,000 terminal MV) will be used.

The present system has a remaining useful life of three years. It has a current BV of $12,000 and a present realizable MV of $8,000. The estimated operating expenses, MVs, and BVs of the present system for the next three years are as follows:

Year	Market Value at End of Year	Book Value at End of Year	Operating Expenses during Year
1	$6,000	$9,000	$40,000
2	5,000	6,000	50,000
3	4,000	3,000	60,000

A spray system will be needed for as long as the company remains in business (which it hopes is a long, long time). Perform an after-tax analysis to determine the most economical period to keep the defender *before* replacing it with the new system. The after-tax MARR is 15% per year, and the effective income tax rate is 50%.

Solution

This analysis begins with the determination of the after-tax economic life of the present system (the economic life of the challenger was given to be 12 years). Table 9-9 shows the calculations of the year-by-year after-tax marginal costs [Equation (9-4)] of the defender and the corresponding EUAC. It can be seen in column 10 that the economic life of the defender is one year.

Table 9-10 contains the ATCF calculations for the challenger. The ATCFs are used to compute the after-tax EUAC of the challenger as follows:

$$\text{EUAC} (15\%) = \$60,000(A/P, 15\%, 12) + \$13,750 - \$6,000(A/F, 15\%, 12)$$

$$= \$24,613.$$

Comparing just the EUACs of the defender with the EUAC of the challenger, you may be tempted to conclude that the old system should be kept at least one more year, perhaps even two more years. In this situation, however, you should examine marginal costs. The valid economic criterion when operating expenses are increasing over time is to keep the old system as long as the marginal cost of an additional year of service is less than the EUAC of the new system. The marginal cost of keeping the old system for the first year is $22,500. This $22,500 is less than the $24,613 EUAC of the new system, thus justifying keeping the old system for the first year. The marginal cost of keeping the old system for the second year is $26,625. This is greater than the $24,613 average annual cost of the new system, thus indicating that the old system should not be kept the second year but rather that it be replaced at the end of the first year.

TABLE 9-9 Determination of the After-Tax Economic Life for the Defender Described in Example 9-13

(1) End of Year, k	(2) MV, End of Year k	(3) Loss in MV during Year k	(4) Cost of Capital = 15% of BOY MV in Col.2	(5) Annual Expenses	(6) Approximate After-Tax Total (Marginal) Cost for Year k $(1-t) \cdot (\text{Col.3} + 4 + 5)$
0	$8,000	0	0	0	0
1	6,000	$2,000	$1,200	$40,000	$21,600
2	5,000	1,000	900	50,000	25,950
3	4,000	1,000	750	60,000	30,875

(1) End of Year, k	(7) BV at End of Year k	(8) Interest on Tax Adjustment = $15\% \cdot t \cdot \text{BOY} \cdot \text{BV}$ In Col. 7	(9) Adjusted After-Tax Total (Marginal) Cost (TC_k) (Col. 6 + Col. 8)	(10) $EUAC^a$ (After-Tax) through Year k
0	$12,000	0	0	0
1	9,000	$900	$22,500	$22,500 $N_{AT}^* = 1$
2	6,000	675	26,625	24,418
3	3,000	450	31,325	26,408

[a] $\text{EUAC}_k = [\sum_{j=1}^{k}(\text{Col.9})_j \cdot (P/F, 15\%, j)](A/P, 15\%, k)$.

TABLE 9-10 ATCF Computations for the Challenger in Example 9-13

End of Year, k	(A) BTCF	(B) Straight-line Depreciation	(C) = (A) − (B) Taxable Income	(D) = −0.4(C) Income Taxes at 40%	(E) = (A) + (D) ATCF
0	−$60,000	None			−$60,000
1–12	−32,000	$4,500^a$	−$36,500	$18,250	−13,750
12	6,000		0^b	0	6,000

[a] Straight-line depreciation amount = ($60,000 − $6,000)/12 = $4,500.
[b] $BV_{12} = $60,000 − (12)($4,500) = $6,000; MV_{12} − BV_{12} = 0$.

9.10 CASE STUDY—Replacement of a Hospital's Emergency Electrical Supply System

Sometimes in engineering practice a replacement analysis involves an existing asset that cannot meet future service requirements without *augmentation* of its capabilities. When this is the case, the defender with increased capability should be competitive with the best available challenger. The analysis of this situation is included in the following case study.

The emergency electrical supply system of a hospital owned by a medical service corporation is presently supported by an 80-kW diesel-powered electrical generator that was put into service five years ago [capital investment = $210,000; MACRS (GDS) seven-year property class]. An engineering firm is designing modifications to the electrical and mechanical systems of the hospital as part of an expansion project. The redesigned emergency electrical supply system will require 120 kW of generating capacity to serve the increased demand. Two preliminary designs for the system are being considered. The first involves the augmentation of the existing 80-kW generator with a new 40-kW diesel-powered unit (GDS seven-year property class). This alternative represents the augmented defender. The second design includes replacement of the existing generator with the best available alternative, a new turbine-powered unit with 120 kW of generating capacity (the challenger). Both alternatives will provide the same level of service to the operation of the emergency electrical supply system.

The challenger, if selected, will be leased by the hospital for a 10-year period. At that time, the lease contract would be renegotiated either for the original piece of equipment or for a replacement generator with the same capacity. The following additional estimates have been generated for use in the replacement analysis.

| | Alternative | | |
| | Defender | | |
	80 kW	40 kW	Challenger
Capital investment	$90,000[a]	$140,000	$10,000[b]
Annual lease amount	0	0	$39,200
Operating hours per year	260	260	260
Annual expenses (year zero $):			
Operating and maintenance (O&M)			
expense per hour	$80	$35	$85
Other expenses	$3,200	$1,000	$2,400
Useful life	10 years	15 years	15 years

[a] Opportunity cost based on present MV of the defender (outsider viewpoint).

[b] Deposit required by the terms of the contract to lease the challenger. It is refundable at the end of the study period.

The annual lease amount for the challenger will not change over the 10-year contract period. The operating and maintenance expense per hour of operation and the other annual expense amounts for both alternatives are estimated in year-zero dollars and are expected to increase at the rate of 4% per year (assume base year, b, is year zero; see Chapter 8 for dealing with price changes).

The present estimated MV of the 80-kW generator is $90,000, and its estimated MV at the end of an additional 10 years, in year-zero dollars, is $30,000. The estimated MV of the new 40-kW generator, 10 years from now in year-zero dollars, is $38,000. Both future market values are estimated to increase at the rate of 2% per year.

The corporation's after-tax, market-based MARR (i_m) is 12% per year, and its effective income tax rate is 40%. A 10-year planning horizon (study period)

Formula annotations:

- = IF(A12 = $B$3 − $E$5 + 1, 0.5*C11, VDB($B$5, 0, $B$3, MAX(0, $E$5 + A12 − 1.5), (A12 + $E$5) − 0.5, $B$4, FALSE))
- = IF(A12 = $B$3 + 1, 0.5*D11, VDB($F$2, 0, $B$3, MAX(0, A12 − 1.5), A12 − 0.5, $B$4, FALSE))
- = −(E2 − SUM(C12:C14))
- = G11
- = B11 + F11
- = G21 + G22
- = NPV(B2, H12:H21) + H11
- = −E11*$B$1
- = B12 − C12 − D12
- = (E6 + F6)*(1 +B7)^A22
- = IF(B3> = 15, 1.5, 2)
- = −(E2 + F2)
- = −($B$8*($E$3 + $F$3) + ($E$4 + $F$4))*(1 + $B$6)^A12

Parameters:

	A	B
1	Effective Tax Rate =	40%
2	After tax MARR =	12%
3	Class Life =	7
4	DB Rate =	200%
5	80 kW Cost Basis =	$ 210,000
6	Annual Expense Rate of Increase / yr. =	4%
7	MV Rate of increase / yr. =	2%
8	Operating Hours / yr. =	260

	D	E (80 kW)	F (40 kW)
1	Capital Investment	$ 90,000	$ 140,000
2	Annual O & M Expenses / hr.	$ 80	$ 35
3	Other Annual Expenses	$ 3,200	$ 1,000
4	Depreciation year at study start	5	0
5	MV at end of life	$ 30,000	$ 38,000

NOTE: All estimates expressed in year 0 dollars

Year	BTCF	80 kW d(k)	40 kW d(k)	Taxable Income	Cash Flow for Income Taxes	ATCF	Adjusted ATCF
0	$ (230,000)			$ (43,145)	$ 17,258	$ (212,742)	$ (212,742)
1	$ (35,464)	$ 18,742	$ 20,000	$ (74,206)	$ 29,682	$ (5,782)	$ (5,782)
2	$ (36,883)	$ 18,742	$ 34,286	$ (89,910)	$ 35,964	$ (918)	$ (918)
3	$ (38,358)	$ 9,371	$ 24,490	$ (72,219)	$ 28,888	$ (9,470)	$ (9,470)
4	$ (39,892)		$ 17,493	$ (57,385)	$ 22,954	$ (16,938)	$ (16,938)
5	$ (41,488)		$ 12,495	$ (53,983)	$ 21,593	$ (19,895)	$ (19,895)
6	$ (43,147)		$ 12,495	$ (55,642)	$ 22,257	$ (20,891)	$ (20,891)
7	$ (44,873)		$ 12,495	$ (57,368)	$ 22,947	$ (21,926)	$ (21,926)
8	$ (46,668)		$ 6,247	$ (52,916)	$ 21,166	$ (25,502)	$ (25,502)
9	$ (48,535)			$ (48,535)	$ 19,414	$ (29,121)	$ (29,121)
10	$ (50,476)			$ (50,476)	$ 20,191	$ (30,286)	$ 19,449
10	$ 82,892			$ 82,892	$ (33,157)	49,735	

PW = $ (282,472)

Figure 9-6 Spreadsheet Analysis of the Defender (Existing Generator with Augmentation)

is considered appropriate for this decision situation. (Note that, with income tax considerations and price changes in the analysis, a study period based on the coterminated assumption is being used.)

Based on an after-tax, actual-dollar analysis, which alternative (augmentation of the defender or lease of the challenger) should be selected as part of the design of the modified emergency electrical power system?

Solution

The after-tax analysis of the first alternative (defender), keeping the existing 80-kW generator and augmenting its capacity with a new 40-kW generator, is shown in Figure 9-6. The initial $230,000 before-tax capital investment amount (cell B11) is the sum of (1) the present $90,000 MV of the existing 80-kW generator, which is an opportunity cost based on an outsider viewpoint and (2) the $140,000 capital investment for the new 40-kW generator. The −$43,145 of taxable income at time zero (cell E11) is because of the gain on disposal, *which is not incurred* when the 80-kW generator is kept instead of sold. The after-tax PW of keeping the defender and augmenting its capacity is −$282,472.

A spreadsheet analysis of the leasing alternative (challenger) is shown in Figure 9-7. Under the contract terms for leasing the challenger, there is an initial $10,000 deposit, which is fully refundable at the end of the 10-year period. There are no tax consequences associated with the deposit transaction. The annual before-tax cash flow (BTCF) for the challenger is the sum of (1) the annual lease amount, which stays constant over the 10-year contract period, and (2) the annual operating and maintenance and other expenses, which increase at the rate of 4% per year. For example, the BTCF for the challenger in year one is −$39,200 − [$85(260) + $2,400](1.04) = −$64,680. These annual BTCF amounts for years one through ten are fully deductible from taxable income by the corporation, and they are also the taxable income amounts for the alternative (the corporation cannot claim any depreciation on the challenger because it does not own the equipment). Hence, the after-tax PW of selecting the challenger, assuming it is leased under these contract terms, is −$239,695.

Based on an after-tax analysis, the challenger or a new turbine-powered unit with 120 kW of generating capacity is economically preferable for use in the emergency electrical supply system because its PW has the least negative value. This case study also illustrates the importance of a life cycle costing (LCC) approach to the economic evaluation of engineering alternatives. In the foregoing case, not only is the initial capital investment considered but annual operating and maintenance costs for each alternative as well. Leasing the 120-kW generator over a 10-year period is economically preferable to the other alternatives in this case.   ■

9.11 Summary

In sum, there are several important points to keep in mind when conducting a replacement or retirement study:

The MV of the defender must *not* also be deducted from the purchase price of the challenger *when using the outsider viewpoint* to analyze a replacement problem.

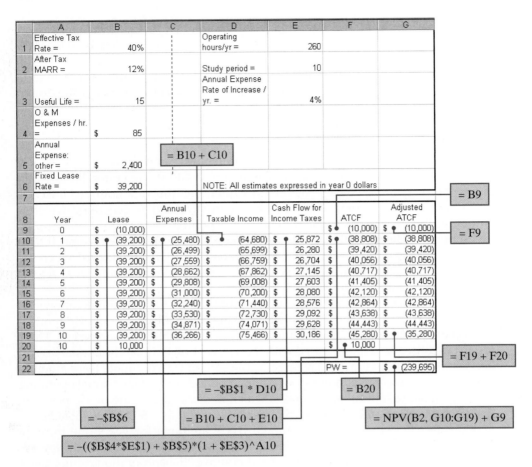

Figure 9-7   Spreadsheet Analysis of the Challenger (Leasing a New Turbine-Powered Unit)

This error double-counts the defender's MV and biases the comparison toward the challenger.

A sunk cost (i.e., MV − BV < 0) associated with keeping the defender must *not* be added to the purchase price of the best available challenger. This error results in an incorrect penalty that biases the analysis in favor of retaining the defender.

In Section 9.6, we observed that the economic life of the defender is often one year, which is generally true if annual expenses are high relative to the defender's investment cost when using the outsider viewpoint. Hence, the marginal cost of the defender should be compared with the EUAC at the *economic life* of the challenger to answer the fundamental question, Should the defender be kept for one or more years or disposed of now? The typical pattern of the EUAC for a defender and a challenger is illustrated in Figure 9-8.

The income tax effects of replacement decisions should not be ignored. The forgone income tax credits associated with keeping the defender may swing the economic preference away from the defender, thus making the challenger the better choice.

The *best available* challenger(s) must be determined. Failure to do so represents unacceptable engineering practice.

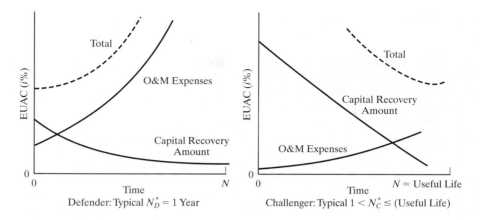

Figure 9-8 Typical Pattern of the EUAC for a Defender and a Challenger

Any excess capacity, reliability, flexibility, safety, and so on of the challenger may have value to the owner and should be claimed as a dollar benefit if a dollar estimate can be placed on it. Otherwise, this value would be treated as a *nonmonetary* benefit. (Chapter 14 describes methods for including nonmonetary benefits in the analysis of alternatives.)

Problems

The number in parentheses () that follows each problem refers to the section from which the problem is taken.

9-1. An industrial lift truck has been in service for several years, and management is contemplating replacing it. A planning horizon of five years is to be used in the replacement study. The old lift truck (defender) has a current MV of $7,000. If the defender is retained, it is anticipated to have annual operating and maintenance costs of $8,000. It will have a zero MV at the end of five additional years of service. The new lift truck (challenger) will cost $22,000 and will have operating and maintenance costs of $5,100. It will have an MV of $9,000 at the end of the planning horizon. Determine the preferred alternative, using a PW comparison and a minimum attractive rate of return (before taxes) of 20% per year. (9.4)

9-2. Stan Moneymaker just graduated from college and intends to travel across the United States for a year, selling the landscape photographs he collected while in college. Stan presently owns an older-model van that he'll possibly use for this trip. But the van is very fuel inefficient and gets only 10 miles per gallon of gasoline. For 20,000 miles of travel and gasoline costing $3.00 per gallon, Stan's analysis of his annual transportation expenses is the following:

$$\text{Loss in value of the van during the next year} = \$1,000$$

$$\text{Fuel}[(20,000 \text{ miles} \div 10 \text{ mpg})(\$3.00/\text{gallon})] = \$6,000$$

$$\text{Maintenance and miscellaneous expenses} = \$1,000$$

$$\text{Insurance} = \underline{\$700}$$

$$\text{Total} = \$8,700$$

As an alternative to driving his old van, Stan is considering the purchase of a newer-model truck for a net price (sticker price minus trade-in) of $17,000. The fuel economy would be much improved at 20 miles per gallon of gasoline, and its road worthiness (i.e., reliability) would probably be better. Stan's analysis of annual transportation expenses with the newer truck is itemized below.

$$\text{Loss in value of the truck during the year} = \$3,500$$

$$\text{Fuel}[(20,000 \text{ miles} \div 20 \text{ mpg})(\$3.00/\text{gallon})] = \$3,000$$

$$\text{Maintenance and miscellaneous expenses} = \$800$$

$$\text{Insurance and finance charges on new truck} = \underline{\$1,900}$$

$$\text{Total} = \$9,200$$

What would you advise Stan to do—keep the older van or buy the replacement truck? What other major expenses or considerations should be included in the analysis? (9.4)

9-3. The Ajax Corporation has an overhead crane that has an estimated remaining life of 10 years. The crane can be sold now for $8,000. If the crane is kept in service, it must be overhauled immediately at a cost of $4,000. Operating and maintenance costs will be $3,000 per year after the crane is overhauled. The overhauled crane will have zero MV at the end of the 10-year study period. A new crane will cost $18,000, will last for 10 years, and will have a $4,000 MV at that time. Operating and maintenance costs are $1,000 per year for the new crane. The company uses a before-tax interest rate of 10% per year in evaluating investment alternatives. Should the company replace the old crane? (9.4)

9-4. A special purpose NASA fuel cell requires an investment of $80,000 and has no MV at any time. Operating expenses in year k are given by $C_k = \$10,000 + \$6,000\ (k - 1)$. Obsolescence is reflected in increased operating expenses and is equivalent to $4,000 per year—this expense has not been included in the equation for C_k. (9.5)

a. Determine the economic life of the fuel cell if $i = 0\%$.

b. Repeat Part (a) when the $4,000 per year is ignored. What can you conclude about the influence of an expense that is constant over time on the economic life?

9-5. In a replacement analysis for a vacuum seal on a spacecraft, the following data are known about the challenger: the initial investment is $12,000; there is no annual maintenance cost for the first three years, however, it will be $2,000 in each of years four and five, and then $4,500 in the sixth year and increasing by $2,500 each year thereafter. The salvage value is $0 at all times, and MARR is 10% per year. What is the economic life of this challenger? (9.5)

9-6. A steam generation system at a biomass-fueled power plant uses an electrostatic precipitator (ESP) to clean its gaseous effluents. The power plant has consistently made use of the same type of ESP over the past several years. The installed cost of a new ESP has been relatively constant at $80,000. Records of operation and maintenance expenses indicate the following average expenses per year as a function of the age of the ESP. The MVs of the ESP are also reasonably well known as a function of age.

Year	1	2	3	4	5
O&M expense	$30,000	$30,000	$35,000	$40,000	$45,000
Market value	60,000	50,000	40,000	25,000	12,500

Determine the best time to replace the ESP if the MARR is 15% per year. (9.5)

9-7. A high-speed electronic assembly machine was purchased two years ago for $50,000. At the present time, it can be sold for $25,000 and replaced by a newer model having a purchase price of $42,500; or it can be kept in service for a maximum of one more year. The new assembly machine, if purchased, has a useful life of not more than two years. The projected resale values and operating and maintenance costs for the challenger and the defender are shown in the accompanying table on a year-by-year basis. If the before-tax MARR is 15%, when should the old assembly machine be replaced? Use the following data table for your analysis. (9.6, 9.7)

	Challenger		Defender	
Year	Market Value	O&M Costs	Market Value	O&M Costs
0	$42,500	—	$25,000	—
1	31,000	$10,000	17,000	$14,000
2	25,000	12,500	—	—

9-8. A city water and waste-water department has a four-year-old sludge pump that was initially purchased for $65,000. This pump can be kept in service for an additional four years, or it can be sold for $35,000 and replaced by a new pump. The purchase price of the replacement pump is $50,000. The projected MVs and operating and maintenance costs over the four-year planning horizon are shown in the table that follows. Assuming the MARR is 10% (a) determine the *economic life* of the challenger, and (b) determine when the defender should be replaced. (9.5, 9.6)

	Defender		Challenger	
Year	MV at EOY	O&M Cost	MV at EOY	O&M Cost
1	$25,000	$18,500	$40,000	$13,000
2	21,000	21,000	32,000	15,500
3	17,000	23,500	24,000	18,000
4	13,000	26,000	16,000	20,500

9-9. The replacement of a planing machine is being considered by the Reardorn Furniture Company. (There is an indefinite future need for this type of machine.) The best challenger will cost $30,000 for installation and will have an estimated economic life of 12 years and a $2,000 MV at that time. It is estimated that annual expenses will average $16,000 per year. The defender has a present BV of $6,000 and a present MV of $4,000. Data for the defender for the next three years are as follows:

Year	MV at End of Year	BV at End of Year	Expenses during the Year
1	$3,000	$4,500	$20,000
2	2,500	3,000	25,000
3	2,000	1,500	30,000

Using a before-tax interest rate of 15% per year, make a comparison to determine whether it is economical to make the replacement now. (9.6, 9.7)

9-10. An existing robot is used in a commercial material laboratory to handle ceramic samples in the high-temperature environment that is part of several test procedures. Due to changing customer needs, the robot will not meet future service requirements unless it is upgraded at a cost of $2,000. Because of this situation, a new advanced technology robot has been selected for potential replacement of the existing robot. The accompanying estimates have been developed from information provided by some current users of the new robot and data obtained from the manufacturer. The firm's before-tax MARR is 25% per year. Based on this information, should the existing robot be replaced? Assume that a robot will be needed for an indefinite period. (9.4, 9.7)

Defender	
Current MV	$38,200
Upgrade cost (year 0)	$2,000
Annual expenses	$1,400 in year one, and increasing at the rate of 8% per year thereafter
Useful life (years)	6
MV at end of useful life	−$1,500

Challenger	
Purchase price	$51,000
Installation cost	$5,500
Annual expenses	$1,000 in year one, and increasing by $150 per year thereafter
Useful life (years)	10
MV at end of useful life	$7,000

9-11. A steel pedestrian overpass must either be reinforced or replaced. Reinforcement would cost $22,000 and would make the overpass adequate for an additional five years of service. If the overpass is torn down now, the scrap value of the steel would exceed the removal cost by $14,000. If it is reinforced, it is estimated that its net salvage (market) value would be $16,000 at the time it is retired from service. A new prestressed concrete overpass would cost $140,000 and would meet the foreseeable requirements of the next 40 years. Such a design would have no net scrap or MV. It is estimated that the annual expenses of the reinforced overpass would exceed those of the concrete overpass by $3,200. Assume that money costs the state 10% per year and that the state pays no taxes. What would you recommend? (9.4, 9.7)

9-12. The corporate average fuel economy (CAFE) standard for mileage is currently 27.5 miles per gallon of gasoline (the defender) for passenger cars. To conserve fuel and reduce air pollution, suppose the U.S. Congress sets the CAFE standard at 36 miles per gallon (the challenger) in 2010. An auto will emit on average 0.9 pounds of carbon dioxide (CO_2) per mile driven at 27.5 miles per gallon, and it will emit 0.8 pounds of CO_2 per mile driven at 36 miles per gallon. (9.4)

a. How much fuel and carbon dioxide would be saved over the lifetime of a passenger car with the new standard? Assume a car will be driven 99,000 miles over its lifetime.

b. If CO_2 costs $0.02/lb to capture and sequester, what penalty does this place on the defender? Should this penalty affect the CAFE replacement analysis?

9-13. Use the PW method to select the better of the following alternatives:

Annual Expenses	Defender: Alternative A	Challenger: Alternative B
Labor	$300,000	$250,000
Material	250,000	100,000
Insurance and property taxes	4% of initial capital investment	None
Maintenance	$8,000	None
Leasing cost	None	$100,000

Assume that the defender was installed five years ago. The MARR is 10% per year. (9.7)

Definition of alternatives:

A: Retain an already owned machine (defender) in service for eight more years.

B: Sell the defender and lease a new one (challenger) for eight years.

Alternative A (additional information):

Cost of defender five years ago = $500,000
BV now = $111,550
Estimated MV eight years from now = $50,000
Present MV = $150,000

9-14. Four years ago, the Attaboy Lawn Mower Company purchased a piece of equipment. Because of increasing maintenance costs for this equipment, a new piece of machinery is being considered for the assembly line. The cost characteristics of the defender (present equipment) and the challenger are shown below:

Defender	Challenger
Original cost = $9,000	Purchase cost = $11,000
Maintenance = $500 in year one (four years ago) increasing by a uniform gradient of $100 per year thereafter	Maintenance = $150 per year
MV at end of life = 0	MV at end of life = $3,000
Original estimated life = nine years	Estimated life = five years

Suppose a $6,000 MV is available now for the defender. Perform a before-tax analysis, using a before-tax MARR of 15%, to determine which alternative to select. Be sure to state all important assumptions you make, and utilize a uniform gradient in your analysis of the defender. (9.4)

9-15. A small high-speed commercial centrifuge has the following net cash flows and abandonment values over its useful life (see Table P9-15, page 427).

The firm's MARR is 10% per year. Determine the optimal time for the centrifuge to be abandoned if its current MV is $7,500 and it won't be used for more than five years. (9.8)

9-16. Consider a piece of equipment that initially cost $8,000 and has these estimated annual expenses and MV:

End of Year, k	Annual Expenses	MV at End of Year
1	$3,000	$4,700
2	3,000	3,200
3	3,500	2,200
4	4,000	1,450
5	4,500	950
6	5,250	600
7	6,250	300
8	7,750	0

If the after-tax MARR is 7% per year, determine the after-tax economic life of this equipment. MACRS (GDS) depreciation is being used (five-year property class). The effective income tax rate is 40%. (9.9)

9-17. A current asset (defender) is being evaluated for potential replacement. It was purchased four years ago at a cost of $62,000. It has been depreciated as a MACRS (GDS) five-year property-class asset. The present MV of the defender is $12,000. Its remaining useful life is estimated to be four years, but it will require additional repair work now (a one-time $4,000 expense) to provide continuing service equivalent to the challenger. The current effective income tax rate is 39%, and the after-tax MARR = 15% per year. Based on an outsider viewpoint, what is the after-tax initial investment in the defender if it is kept (*not* replaced now)? (9.9)

9-18. The PW of the ATCFs through year k, PW_k, for a defender (three-year remaining useful life) and a challenger (five-year useful life) are given in the following table:

TABLE P9-15   Cash Flows and Abandonment Values for Problem 9-15

	End of Year				
	1	2	3	4	5
Annual revenues less expenses	$2,000	$2,000	$2,000	$2,000	$2,000
Abandonment value of machine[a]	$6,200	$5,200	$4,000	$2,200	0

[a] Estimated MV.

	PW of ATCF through Year k, PW_k	
Year	Defender	Challenger
1	−$14,020	−$18,630
2	−28,100	−34,575
3	−43,075	−48,130
4		−65,320
5		−77,910

Assume the after-tax MARR is 12% per year. On the basis of this information, answer the following questions:

a. What are the economic life and the related minimum EUAC when $k = N_{AT}^*$, for both the defender and the challenger? (9.5, 9.6)

b. When should the challenger (based on the present analysis) replace the defender? Why? (9.6, 9.7)

c. What assumption(s) have you made in answering Part (b)?

9-19. In 2007, a certain pharmaceutical company has some existing semiautomated production equipment that they are considering replacing. This equipment has a present MV of $57,000 and a BV of $30,000. It has five more years of straight-line depreciation available (if kept) of $6,000 per year, at which time its BV would be $0. The estimated MV of the equipment five years from now (in year-zero dollars) is $18,500. The MV rate of increase on this type of equipment has been averaging 3.2% per year. The total operating and maintenance and other related expenses are averaging $27,000 (A$) per year.

New automated replacement equipment would be leased. Estimated operating and maintenance and related company expenses for the new equipment

are $12,200 per year. The annual leasing cost would be $24,300. The after-tax MARR (with an inflation component) is 9% per year (i_m); $t = 40\%$; and the analysis period is five years. Based on an after-tax, A$ analysis, should the replacement be made? Base your answer on the actual IRR of the incremental cash flow. (9.9)

9-20. It is being decided whether or not to replace an existing piece of equipment with a newer, more productive one that costs $80,000 and has an estimated MV of $20,000 at the end of its useful life of six years. Installation charges for the new equipment will amount to $3,000; this is not added to the capital investment but will be an expensed item during the first year of operation. MACRS (GDS) depreciation (five-year property class) will be used. The new equipment will reduce direct costs (labor, maintenance, rework, etc.) by $10,000 in the first year, and this amount is expected to increase by $500 each year thereafter during its six-year life. It is also known that the BV of the fully depreciated old machine is $10,000 but that its present fair MV is $14,000. The MV of the old machine will be zero in six years. The effective income tax rate is 40%. (9.9)

a. Determine the prospective after-tax *incremental* cash flow associated with the new equipment if it is believed that the existing machine could perform satisfactorily for six more years.

b. Assume that the after-tax MARR is 12% per year. Based on the ERR method, should you replace the defender with the challenger? Assume $\epsilon = $ MARR.

9-21. Five years ago a chemical plant invested $90,000 in a pumping station on a nearby river to provide the water required for their production process. Straight-line depreciation is employed for tax purposes, using a 30-year life and zero salvage value. During the past five years, operating costs have been as follows:

Maintenance	$2,000 per year
Power and labor	4,500 per year
Taxes and insurance	1,000 per year
Total	$7,500 per year

A nearby city has offered to purchase the pumping station for $75,000 as it is in the process of developing a city water distribution system. The city is willing to sign a 10-year contract with the plant to provide the plant with its required volume of water for a price of $10,000 per year. Plant officials estimate that the salvage value of the pumping station 10 years hence will be about $30,000. The before-tax MARR is 10% per year. An effective income tax rate of 50% is to be assumed. Compare the annual cost (after taxes) of the defender (keep the pumping station) to the challenger (sell station to the city) and help plant management determine the more economical course of action. (9.9)

9-22. Five years ago, an airline installed a baggage conveyor system in a terminal, knowing that within a few years it would have to be moved. The original cost of the installation was $120,000, and through accelerated depreciation methods the company has been able to depreciate the entire cost. It now finds that it will cost $40,000 to move and upgrade the conveyor. This additional cost would be recovered over the next six years [the MACRS (ADS) with half-year convention and a five-year recovery period would be used], which the airline believes is a good estimate of the remaining useful life of the system if moved. As an alternative, the airline finds that it can purchase a somewhat more efficient conveyor system for an installed cost of $120,000. The new system would result in an estimated reduction in annual expenses of $6,000 in year-zero dollars. Annual expenses are expected to increase by 6% per year. The new system is in the five-year MACRS (GDS) property class, and its estimated MV six years from now, in year-zero dollars, is 50% of its installed cost. This MV is estimated to increase by 3% per year. A small airline company, which will occupy the present space, has offered to buy the old conveyor for $90,000.

Annual property taxes and insurance on the present equipment have been $1,500, but it is estimated that they would increase to $1,800 if the equipment is moved and upgraded. For the new system, it is estimated that these costs would be about $2,750 per year. All other expenses would be about equal for the two alternatives. The company is in the 40% income tax bracket. It wishes to

obtain at least 10% per year, after taxes, on any invested capital. What would you recommend? (9.9, 9.10)

9-23. A manufacturing company has some existing semiautomatic production equipment that it is considering replacing. This equipment has a present MV of $57,000 and a BV of $27,000. It has five more years of depreciation available under MACRS (ADS) of $6,000 per year for four years and $3,000 in year five. (The original recovery period was nine years.) The estimated MV of the equipment five years from now is $18,500. The total annual operating and maintenance expenses are averaging $27,000 per year.

New automated replacement equipment would then be leased. Estimated annual operating expenses for the new equipment are $12,200 per year. The *annual* leasing costs would be $24,300. The MARR (after taxes) is 9% per year, $t = 40\%$, and the analysis period is five years. (*Remember*: The owner claims depreciation, and the leasing cost is an operating expense.)

Based on an after-tax analysis, should the new equipment be leased? Base your answer on the IRR of the incremental cash flow. (9.9, 9.10)

9-24. Five years ago a company in New Jersey installed a diesel-electric unit costing $50,000 at a remote site because no dependable electric power was available from a public utility. The company has computed depreciation by the straight-line method with a useful life of 10 years and a zero salvage value. Annual operation and maintenance expenses are $16,000, and property taxes and insurance cost another $3,000 per year.

Dependable electric service is now available at an estimated annual cost of $30,000. The company in New Jersey wishes to know whether it would be more economical to dispose of the diesel-electric unit now, when it can be sold for $35,000, or to wait five years when the unit would have to be replaced anyway (with no MV). The company has an effective income tax rate of 50% and tries to limit its capital expenditures to opportunities that will earn at least 15% per year after income taxes. What would you recommend? (9.9)

9-25. *Extended Learning Exercise* There are two customers requiring three-phase electrical service, one existing at location *A* and a new customer at location *B*. The load at location *A* is known to be 110 KVA, and at location *B* it is contracted to be 280 KVA. Both loads are expected to remain constant indefinitely into the future. Already in service at *A* are three 100-KVA transformers that were installed some years ago when the load was

TABLE P9-25	Table for Problem 9-25	
	Existing and New Transformers	
	Three 37.5-KVA	Three 100-KVA
Capital Investment:		
Equipment	$900	$2,100
Installation	$340	$475
Property tax	2% of capital investment	2% of capital investment
Removal cost	$100	$110
Market value	$100	$110
Useful life (years)	30	30

much greater. Thus, the alternatives are as follows:

Alternative A: Install three 100-KVA transformers (new) at B now and replace those at A with three 37.5-KVA transformers only when the existing ones must be retired.

Alternative B: Remove the three 100-KVA transformers now at A and relocate them at B. Then install three 37.5-KVA transformers (new) at A.

Data for both alternatives are provided in Table P9-25. The existing transformers have 10 years of life remaining. Suppose that the before-tax MARR = 8% per year. Recommend which action to follow after calculating an appropriate criterion for comparing these alternatives. List all assumptions necessary and ignore income taxes. (9.7)

9-26. *Extended Learning Exercise* A truck was purchased four years ago for $65,000 to move raw materials and finished goods between a production facility and four remote warehouses. This truck (the defender) can be sold at the present time for $40,000 and replaced by a new truck (the challenger) with a purchase price of $70,000.

a. Given the MVs and operating and maintenance costs that follow, what is the economic life of the challenger if MARR = 10%? Note: *This is a*

before-tax analysis that does not require any calculations involving the defender. (9.5)

Defender			Challenger		
EOY	Market Value	O&M Costs	EOY	Market Value	O&M Costs
1	$30,000	$8,500	1	$56,000	$5,500
2	20,000	10,500	2	44,000	6,800
3	12,000	14,000	3	34,000	7,400
4	4,000	16,000	4	22,000	9,700

b. Suppose that the defender was set up on a depreciation schedule with a five-year MACRS class life at the time of its purchase (four years ago). The defender can be sold now for $40,000, or a rebuilt engine and transmission can be purchased and installed at a cost of $12,000 (capital investment with three-year depreciable life, straight line, salvage value = 0). If the defender is kept in service, assume that it will have operating and maintenance costs as shown in Part (a) and a MV of $0 at the end of four years. Determine the ATCFs for the defender. (9.9)

Spreadsheet Exercises

9-27. Refer to Example 9-4. By how much would the MARR have to change before the economic life decreases by one year? How about to increase the economic life by one year? (9.5)

9-28. Create a spreadsheet to solve Problem 9-9. What percent reduction in the annual expenses of the defender would result in a delay of its replacement by the challenger? (9.4, 9.7)

Case Study Exercises

9-29. Determine how much the annual lease amount of the challenger can increase before the defender becomes the preferred alternative. (9.10)

9-30. Suppose the study period is reduced to five years. Will this change the replacement decision? (9.10)

9-31. Before committing to a 10-year lease, it was decided to consider one more alternative. Instead of simply augmenting the existing generator with a new 40-kW unit, this alternative calls for replacing the 80-kW generator with two 40-kW units (for a total of three 40-kW units). The company selling the 40-kW units will reduce the purchase price to $120,000 per generator if three generators are bought. How does this new challenger compare to the augmented defender and the original challenger? (9.10)

FE Practice Problems

9-32. Machine A was purchased last year for $20,000 and had an estimated MV of $2,000 at the end of its six-year life. Annual operating costs are $2,000. The machine will perform satisfactorily over the next five years. A salesman for another company is offering a replacement, Machine B, for $14,000, with an MV of $1,400 after five years. Annual operating costs for Machine B will only be $1,400. A trade-in allowance of $10,400 has been offered for Machine A. If the before-tax MARR is 12% per year, should you buy the new machine? (9.4)

(a) No, continue with Machine A.
(b) Yes, purchase Machine B.

9-33. A company is considering replacing a machine that was bought six years ago for $50,000. The machine, however, can be repaired and its life extended by five more years. If the current machine is replaced, the new machine will cost $44,000 and will reduce the operating expenses by $6,000 per year. The seller of the new machine has offered a trade-in allowance of $15,000 for the old machine. If MARR is 12% per year before taxes, how much can the company spend to repair the existing machine? Choose the closest answer. (9.4)

(a) $22,371   (b) $50,628   (c) $7,371   (d) −$1,000

Machine A was purchased three years ago for $10,000 and had an estimated MV of $1,000 at the end of its 10-year life. Annual operating costs are $1,000. The machine will perform satisfactorily for the next seven years. A salesman for another company is offering machine B for $50,000 with an MV of $5,000 after 10 years. Annual operating costs will be $600. Machine A could be sold now for $7,000, and MARR is 12% per year. Use this information to answer Problems 9-34 and 9-35.

9-34. Using the outsider viewpoint, what is the EUAC of continuing to use Machine A? Choose the closest answer. (9.6)

(a) $1,000   (b) $2,182   (c) $2,713
(d) $901   (e) $2,435

9-35. Using the outsider viewpoint, what is the EUAC of buying Machine B? Choose the closest answer. (9.5)

(a) $8,565   (b) $11,361   (c) $9,750
(d) $9,165   (e) $900

9-36. A corporation purchased a machine for $60,000 five years ago. It had an estimated life of 10 years and an estimated salvage value of $9,000. The current BV of this machine is $34,500. If the current MV of the machine is $40,500 and the effective income tax rate is 29%, what is the after-tax investment value of the machine? Use the outsider viewpoint. (9.9)

(a) $28,755   (b) $40,500   (c) $38,760
(d) $37,455   (e) $36,759

CHAPTER 10

Evaluating Projects with the Benefit–Cost Ratio Method

The objective of Chapter 10 is to demonstrate the use of the benefit–cost (B–C) ratio for the evaluation of public projects.

Constructing a Bypass to Relieve Traffic Congestion

Two heavily traveled interstate highways currently intersect in the middle of a major metropolitan area. A 25-mile, four-lane bypass is being considered to connect the busy interstate highways at a point outside of the metropolitan area. The projected construction cost of the bypass is $20 million. Annual maintenance of the roadway is expected to be $500,000. Major monetary benefits of reduced time delays due to traffic congestion, improved traveler safety (fewer traffic accidents), and expanded opportunities for commercial businesses are anticipated to be around $2 million per year.

This bypass would be state owned and maintained and is therefore considered a public project. In this chapter, you will learn how public projects such as this are evaluated using a benefit–cost ratio.

> ... the Federal Government should improve or participate in the improvement of navigable
> waters or their tributaries, including watersheds thereof, for flood-control purposes if the
> benefits to whomsoever they may accrue are in excess of the estimated costs ...
>
> —Flood Control Act (1936)

10.1 Introduction

Public projects are those authorized, financed, and operated by federal, state, or local governmental agencies. Such public works are numerous, and although they may be of any size, they are frequently much larger than private ventures. Since they require the expenditure of capital, such projects are subject to the principles of engineering economy with respect to their design, acquisition, and operation. Because they are public projects, however, a number of important special factors exist that are not ordinarily found in privately financed and operated businesses. The differences between public and private projects are listed in Table 10-1.

TABLE 10-1   Some Basic Differences between Privately Owned and Publicly Owned Projects

	Private	Public
Purpose	Provide goods or services at a profit; maximize profit or minimize cost	Protect health; protect lives and property; provide services (at no profit); provide jobs
Sources of capital	Private investors and lenders	Taxation; private lenders
Method of financing	Individual ownership; partnerships; corporations	Direct payment of taxes; loans without interest; loans at low interest; self-liquidating bonds; indirect subsidies; guarantee of private loans
Multiple purposes	Moderate	Common (e.g., reservoir project for flood control, electrical power generation, irrigation, recreation, education)
Project life	Usually relatively short (5 to 10 years)	Usually relatively long (20 to 60 years)
Relationship of suppliers of capital to project	Direct	Indirect, or none
Nature of benefits	Monetary or relatively easy to equate to monetary terms	Often nonmonetary, difficult to quantify, difficult to equate to monetary terms
Beneficiaries of project	Primarily, entity undertaking project	General public
Conflict of purposes	Moderate	Quite common (dam for flood control versus environmental preservation)
Conflict of interests	Moderate	Very common (between agencies)
Effect of politics	Little to moderate	Frequent factors; short-term tenure for decision makers; pressure groups; financial and residential restrictions; etc.
Measurement of efficiency	Rate of return on capital	Very difficult; no direct comparison with private projects

As a consequence of these differences, it is often difficult to make engineering economy studies and investment decisions for public-works projects in exactly the same manner as for privately owned projects. Different decision criteria are often used, which creates problems for the public (who pays the bill), for those who must make the decisions, and for those who must manage public-works projects.

The benefit–cost ratio method, which is normally used for the evaluation of public projects, has its roots in federal legislation. Specifically, the Flood Control Act of 1936 requires that, for a federally financed project to be justified, its benefits must be in excess of its costs. In general terms, B–C analysis is a systematic method of assessing the desirability of government projects or policies when it is important to take a long-term view of future effects and a broad view of possible side effects. In meeting the requirements of this mandate, the B–C ratio method evolved into the calculation of a ratio of project benefits to project costs. Rather than allowing the analyst to apply criteria more commonly used for evaluating private projects (IRR, PW, and so on), most governmental agencies require the use of the B–C method.

10.2 Perspective and Terminology for Analyzing Public Projects

Before applying the B–C ratio method to evaluate a public project, the appropriate perspective (Chapter 1, Principle 3) must be established. In conducting an engineering economic analysis of any project, whether it is a public or a private undertaking, the proper perspective is to consider the *net* benefits to the *owners* of the enterprise considering the project. This process requires that the question of who owns the project be addressed. Consider, for example, a project involving the expansion of a section of I-80 from four to six lanes. Because the project is paid for primarily with federal funds channeled through the Department of Transportation, we might be inclined to say that the federal government is the owner. These funds, however, originated from tax dollars—thus, the true owners of the project are the taxpayers.

As mentioned previously, the B–C method requires that a ratio of benefits to costs be calculated. Project *benefits* are defined as the favorable consequences of the project to the public, but project *costs* represent the monetary disbursement(s) required of the government. It is entirely possible, however, for a project to have unfavorable consequences to the public. Considering again the widening of I-80, some of the *owners* of the project—farmers along the interstate—would lose a portion of their arable land, along with a portion of their annual revenues. Because this negative financial consequence is borne by (a segment of) the public, it cannot be classified as either a benefit or a cost. The term *disbenefits* is generally used to represent the negative consequences of a project to the public.

EXAMPLE 10-1	Benefits and Costs of a Convention Center and Sports Complex

A new convention center and sports complex has been proposed to the Gotham City Council. This public-sector project, if approved, will be financed through the issue of municipal bonds. The facility will be located in the City Park near downtown Gotham City, in a wooded area, which includes a bike path, a nature trail, and a pond. Because the city already owns the park, no purchase of land is necessary. List separately the project's *benefits, costs,* and any *disbenefits.*

Solution

BENEFITS	Improvement of the image of the downtown area of Gotham City
	Potential to attract conferences and conventions to Gotham City
	Potential to attract professional sports franchises to Gotham City
	Revenues from rental of the facility
	Increased revenues for downtown merchants of Gotham City
	Use of facility for civic events
COSTS	Architectural design of the facility
	Construction of the facility
	Design and construction of parking garage adjacent to the facility
	Operating and maintenance costs of the facility
	Insurance costs of the facility
DISBENEFITS	Loss of use of a portion of the City Park to Gotham City residents, including the bike path, the nature trail, and the pond
	Loss of wildlife habitat in urban area

10.3 Self-Liquidating Projects

The term *self-liquidating project* is applied to a governmental project that is expected to earn direct revenue sufficient to repay its cost in a specified period of time. Most of these projects provide utility services—for example, the fresh water, electric power, irrigation water, and sewage disposal provided by a hydroelectric dam. Other examples of self-liquidating projects include toll bridges and highways.

As a rule, self-liquidating projects are expected to earn direct revenues that offset their costs, but they are not expected to earn profits or pay income taxes. Although they also do not pay property taxes, in some cases in-lieu payments are made to state, county, or municipal governments in place of the property or franchise taxes that would have been paid had the project been under private ownership. For example, the U.S. government agreed to pay the states of Arizona and Nevada $300,000 each annually for 50 years in lieu of taxes that would have accrued if Hoover Dam had been privately constructed and operated.

10.4 Multiple-Purpose Projects

An important characteristic of public-sector projects is that many such projects have multiple purposes or objectives. One example of this would be the construction of a dam to create a reservoir on a river. (See Figure 10-1.) This project would have multiple purposes: (1) assist in flood control, (2) provide water for irrigation,

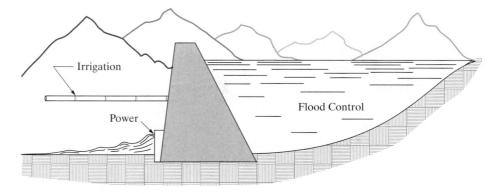

Figure 10-1   Schematic Representation of a Multiple-Purpose Project Involving Flood Control, Irrigation, and Power

(3) generate electric power, (4) provide recreational facilities, and (5) provide drinking water. Developing such a project to meet more than one objective ensures that greater overall economy can be achieved. Because the construction of a dam involves very large sums of capital and the use of a valuable natural resource—a river—it is likely that the project could not be justified unless it served multiple purposes. This type of situation is generally desirable, but, at the same time, it creates economic and managerial problems due to the overlapping utilization of facilities and the possibility of a conflict of interest between the several purposes and the agencies involved.

The basic problems that often arise in evaluating public projects can be illustrated by returning to the dam shown in Figure 10-1. The project under consideration is to be built in the semiarid central portion of California, primarily to provide control against spring flooding resulting from the melting snow in the Sierra Nevadas. If a portion of the water impounded behind the dam could be diverted onto the adjoining land below the dam, the irrigation water would greatly increase the productivity, and thus the value, of that land, which would result in an increase in the nation's resources. So the objectives of the project should be expanded to include both flood control and irrigation.

The existence of a dam with a high water level on one side and a much lower level on the other side also suggests that some of the nation's resources will be wasted unless a portion of the water is diverted to run through turbines, generating electric power. This electricity can be sold to customers in the areas surrounding the reservoir, giving the project the third purpose of generating electric power.

In this semiarid region, the creation of a large reservoir behind the dam would provide valuable facilities for hunting, fishing, boating, swimming, and camping. Thus, the project has the fourth purpose of providing recreation facilities. A fifth purpose would be the provision of a steady, reliable supply of drinking water.

Each of the above-mentioned objectives of the project has desirable economic and social value, so what started out as a single-purpose project now has five purposes. The failure to meet all five objectives would mean that valuable national resources are being wasted. On the other hand, there are certain *disbenefits* to the public that must also be considered. Most apparent of these is the loss of farm land above the dam in the area covered by the reservoir. Other disbenefits might include (1) the loss of a white-water recreational area enjoyed by canoeing, kayaking, and rafting enthusiasts, (2) the loss of annual deposits of fertile soil in the river basin below the dam due to spring flooding, and (3) the negative ecological impact of obstructing the flow of the river.

If the project is built to serve five purposes, the fact that one dam will serve all of them leads to at least three basic problems. The *first* of these is the allocation of the cost of the dam to each of its intended purposes. Suppose, for example, that the estimated costs of the project are $35,000,000. This figure includes costs incurred for the purchase and preparation of land to be covered by water above the dam site; the actual construction of the dam, the irrigation system, the power generation plant, and the purification and pumping stations for drinking water; and the design and development of recreational facilities. The allocation of some of these costs to specific purposes is obvious (e.g., the cost of constructing the irrigation system), but what portion of the cost of purchasing and preparing the land should be assigned to flood control? What amounts should be assigned to irrigation, power generation, drinking water, and recreation?

The *second* basic problem is the conflict of interest among the several purposes of the project. Consider the decision as to the water level to be maintained behind the dam. In meeting the first purpose—flood control—the reservoir should be maintained at a near-empty level to provide the greatest storage capacity during the months of the spring thaw. This lower level would be in direct conflict with the purpose of power generation, which could be maximized by maintaining as high a level as possible behind the dam at all times. Further, maximizing the recreational benefits would suggest that a constant water level be maintained throughout the year. Thus, conflicts of interest arise among the multiple purposes, and compromising decisions must be made. These decisions ultimately affect the magnitude of benefits resulting from the project.

A *third* problem with multiple-purpose public projects is political sensitivity. Because each of the various purposes, or even the project itself, is likely to be desired or opposed by some segment of the public and by various interest groups that may be affected, inevitably such projects frequently become political issues.* This conflict often has an effect on cost allocations and thus on the overall economy of these projects.

The net result of these three factors is that the cost allocations made in multiple-purpose public-sector projects tend to be arbitrary. As a consequence, production and selling costs of the services provided also are arbitrary. Because of this, these costs cannot be used as valid yardsticks with which similar private-sector projects can be compared to determine the relative efficiencies of public and private ownership.

---

* The construction of the Tellico Dam on the Little Tennessee River was considerably delayed due to two important *disbenefits:* (1) concerns over the impact of the project on the environment of a small fish, the snail darter, and (2) the flooding of burial grounds considered sacred by the Cherokee Nation.

10.5 Difficulties in Evaluating Public-Sector Projects

With all of the difficulties that have been cited in evaluating public-sector projects, we may wonder whether engineering economy studies of such projects should be attempted. In most cases, economy studies cannot be made in as complete, comprehensive, and satisfactory a manner as in the case of studies of privately financed projects. In the private sector, the *costs* are borne by the firm undertaking the project, and the *benefits* are the favorable outcomes of the project accrued by the firm. Any costs and benefits that occur outside of the firm are generally ignored in evaluations unless it is anticipated that those external factors will indirectly impact the firm. But the opposite is true in the case of public-sector projects. In the wording of the Flood Control Act of 1936, "if the benefits to whomsoever they may accrue are in excess of the estimated costs," all of the potential benefits of a public project are relevant and should be considered. Simply enumerating all of the benefits for a large-scale public project is a formidable task! Further, the monetary value of these benefits to all of the affected segments of the public must somehow be estimated. Regardless, decisions about the investment of capital in public projects must be made by elected or appointed officials, by managers, or by the general public in the form of referendums. Because of the magnitude of capital and the long-term consequences associated with many of these projects, following a systematic approach for evaluating their worthiness is vital.

There are a number of difficulties inherent in public projects that must be considered in conducting engineering economy studies and making economic decisions regarding those projects. Some of these are as follows:

1. There is no profit standard to be used as a measure of financial effectiveness. Most public projects are intended to be nonprofit.

2. The monetary impact of many of the benefits of public projects is difficult to quantify.

3. There may be little or no connection between the project and the public, which is the owner of the project.

4. There is often strong political influence whenever public funds are used. When decisions regarding public projects are made by elected officials who will soon be seeking reelection, *the immediate benefits and costs are stressed, often with little or no consideration for the more important long-term consequences.*

5. Public projects are usually much more subject to legal restrictions than are private projects. For example, the area of operations for a municipally owned power company may be restricted such that power can be sold only within the city limits, regardless of whether a market for any excess capacity exists outside the city.

6. The ability of governmental bodies to obtain capital is much more restricted than that of private enterprises.

7. The appropriate interest rate for discounting the benefits and costs of public projects is often controversially and politically sensitive. Clearly, lower interest rates favor long-term projects having major social or monetary benefits in the future. Higher interest rates promote a short-term outlook whereby decisions are based mostly on initial investments and immediate benefits.

A discussion of several viewpoints and considerations that are often used to establish an appropriate interest rate for public projects is included in the next section.

10.6 What Interest Rate Should Be Used for Public Projects?

When public-sector projects are evaluated, interest rates* play the same role of accounting for the time value of money as in the evaluation of projects in the private sector. The rationale for the use of interest rates, however, is somewhat different. The choice of an interest rate in the private sector is intended to lead directly to a selection of projects to maximize profit or minimize cost. In the public sector, on the other hand, projects are not usually intended as profit-making ventures. Instead, the goal is the *maximization of social benefits*, assuming that these have been appropriately measured. The choice of an interest rate in the public sector is intended to determine how available funds should best be allocated among competing projects to achieve social goals. The relative differences in magnitude of interest rates between governmental agencies, regulated monopolies, and private enterprises are illustrated in Figure 10-2.

Three main considerations bear on what interest rate to use in engineering economy studies of public-sector projects:

1. The interest rate on borrowed capital
2. The opportunity cost of capital to the governmental agency
3. The opportunity cost of capital to the taxpayers

As a general rule, it is appropriate to use the interest rate on borrowed capital as the interest rate for cases in which money is borrowed specifically for the project(s) under consideration. For example, if municipal bonds are issued specifically for

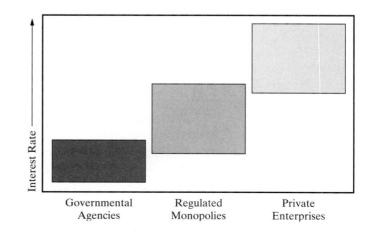

Figure 10-2   Relative Differences in Interest Rates for Governmental Agencies, Regulated Monopolies, and Private Enterprise

* The term *discount rate* is often used instead of interest rate (or MARR) in government literature on B–C analysis.

the financing of a new school, the effective interest rate on those bonds should be the interest rate.

For public-sector projects, the opportunity cost of capital to a governmental agency encompasses the annual rate of *benefit* to either the constituency served by that agency or the composite of taxpayers who will eventually pay for the project. If projects are selected such that the estimated *return* (in terms of benefits) on all accepted projects is higher than that on any of the rejected projects, then the interest rate used in economic analyses is that associated with the best opportunity forgone. If this process is done for all projects and investment capital available within a governmental agency, the result is *an opportunity cost of capital for that governmental agency.* A strong argument against this philosophy, however, is that the different funding levels of the various agencies and the different nature of projects under the direction of each agency would result in different interest rates for each of the agencies, even though they all share a common primary source of funds—taxation of the public.

The third consideration—the opportunity cost of capital to the taxpayers—is based on the philosophy that all government spending takes potential investment capital away from the taxpayers. The taxpayers' opportunity cost is generally greater than either the cost of borrowed capital or the opportunity cost to governmental agencies, and there is a compelling argument for applying the largest of these three rates as the interest rate for evaluating public projects; it is not economically sound to take money away from a taxpayer to invest in a government project yielding benefits at a rate less than what could have been earned by that taxpayer.

This argument was supported by a federal government directive issued in 1997—and still in force—by the Office of Management and Budget (OMB).* According to this directive, a 7% interest rate should be used in economic evaluations for a wide range of federal projects, with certain exceptions (e.g., a lower rate can be applied in evaluating water resource projects). This 7%, it can be argued, is at least a rough approximation of the real-dollar return that taxpayers could earn from the use of that money for private investment. This corresponds to an approximate nominal interest rate of 10% per year.

One additional theory on establishing interest rates for federal projects advocates that the *social discount rate* used in such analyses should be the market-determined *risk-free* rate for private investments.[†] According to this theory, a nominal interest rate in the order of 3–4% per year should be used.

The preceding discussion focuses on the *considerations* that should play a role in establishing an interest rate for public projects. As in the case of the private sector, there is no simple formula for determining the appropriate interest rate for public projects. With the exception of projects falling under the 1997 OMB directive, setting the interest rate is ultimately a policy decision at the discretion of the governmental agency conducting the analysis.

---

* Office of Management and Budget, "Guidelines and Discount Rates for Benefit–Costs Analysis of Federal Programs," *OMB Circular No. A-94 (revised)*, February 21, 1997. The OMB home page is http://www.whitehouse.gov/omb.
[†] K. J. Arrow and R. C. Lind, "Uncertainty and the Evaluation of Public Investment Decisions," *American Economic Review*, 60 (June 1970): 364–378.

10.7 The Benefit–Cost Ratio Method

As the name implies, the B–C ratio method involves the calculation of a ratio of benefits to costs. Whether evaluating a project in the private sector or in the public sector, the time value of money must be considered to account for the timing of cash flows (or benefits) occurring after the inception of the project. Thus, the B–C ratio is actually a ratio of *discounted benefits* to *discounted costs.*

Any method for formally evaluating projects in the public sector must consider the worthiness of allocating resources to achieve social goals. For over 70 years, the B–C ratio method has been the accepted procedure for making go/no-go decisions on independent projects and for comparing mutually exclusive projects in the public sector, even though the other methods discussed in Chapter 5 (PW, AW, IRR, etc.) will lead to identical recommendations, *assuming all these procedures are properly applied.*

Accordingly, the purpose of this section is to describe and illustrate the mechanics of the B–C ratio method for evaluating projects. Two different B–C ratios will be presented because they are used in practice by various government agencies and municipalities. Both ratios lead to the *identical choice* of which project is best when comparing mutually exclusive alternatives.

The B–C ratio is defined as the ratio of the equivalent worth of benefits to the equivalent worth of costs. The equivalent-worth measure applied can be present worth, annual worth, or future worth, but customarily, either PW or AW is used. An interest rate for public projects, as discussed in the previous section, is used in the equivalent-worth calculations. The B–C ratio is also known as the *savings-investment ratio (SIR)* by some governmental agencies.

Several different formulations of the B–C ratio have been developed. Two of the more commonly used formulations are presented in this section, illustrating the use of both present worth and annual worth.

Conventional B–C ratio with PW:

$$\text{B--C} = \frac{\text{PW(benefits of the proposed project)}}{\text{PW(total costs of the proposed project)}}$$

$$= \frac{\text{PW}(B)}{I - \text{PW(MV)} + \text{PW(O\&M)}} \tag{10-1}$$

where $\text{PW}(\cdot)$ = present worth of $(\cdot)$;
  B = benefits of the proposed project;
  I = initial investment in the proposed project;
  MV = market value at the end of useful life;
  $\text{O\&M}$ = operating and maintenance costs of the proposed project.

Modified B–C ratio with PW:

$$\text{B--C} = \frac{\text{PW}(B) - \text{PW(O\&M)}}{I - \text{PW(MV)}}. \tag{10-2}$$

The numerator of the modified B–C ratio expresses the equivalent worth of the benefits minus the equivalent worth of the operating and maintenance costs, and the denominator includes only the initial investment costs (less any market value). A project is acceptable when the B–C ratio, as defined in either Equation (10-1) or (10-2), is greater than or equal to 1.0.

Equations (10-1) and (10-2) can be rewritten in terms of equivalent annual worth, as follows:

Conventional B–C ratio with AW:

$$B\text{–}C = \frac{AW(\text{benefits of the proposed project})}{AW(\text{total costs of the proposed project})}$$

$$= \frac{AW(B)}{CR + AW(O\&M)}, \qquad (10\text{-}3)$$

where $AW(\cdot)$ = annual worth of $(\cdot)$;

  B = benefits of the proposed project;

  CR = capital-recovery amount (i.e., the equivalent annual cost of the initial investment, I, including an allowance for market, or salvage value, if any);

  $O\&M$ = operating and maintenance costs of the proposed project.

Modified B–C ratio with AW:

$$B\text{–}C = \frac{AW(B) - AW(O\&M)}{CR}. \qquad (10\text{-}4)$$

The resulting B–C ratios for all the previous formulations will give identical results in determining the acceptability of a project (i.e., either B–C $\geq$ 1.0 or B–C $<$ 1.0). The conventional B–C ratio will give identical numerical results for both PW and AW formulations; similarly, the modified B–C ratio gives identical numerical results whether PW or AW is used. Although the magnitude of the B–C ratio will differ between conventional and modified B–C ratios, go/no-go decisions are not affected by the choice of approach, as shown in Example 10-2.

EXAMPLE 10-2   **Equivalence of the B–C Ratio Formulations**

The city of Columbia is considering extending the runways of its municipal airport so that commercial jets can use the facility. The land necessary for the runway extension is currently a farmland that can be purchased for $350,000. Construction costs for the runway extension are projected to be $600,000, and the additional annual maintenance costs for the extension are estimated to be $22,500. If the runways are extended, a small terminal will be constructed at a cost of $250,000. The annual operating and maintenance costs for the terminal are estimated at $75,000. Finally, the projected increase in flights will require the addition of two air traffic controllers at an annual cost of $100,000. Annual *benefits* of the runway extension have been estimated as follows:

$325,000	Rental receipts from airlines leasing space at the facility
$65,000	Airport tax charged to passengers
$50,000	Convenience benefit for residents of Columbia
$50,000	Additional tourism dollars for Columbia

Apply the B–C ratio method with a study period of 20 years and a MARR of 10% per year to determine whether the runways at Columbia Municipal Airport should be extended.

Solution

Conventional B–C: *Equation (10-1)*	B–C = PW(B)/[I − PW(MV) + PW(O&M)] B–C = $490,000 (P/A, 10%, 20)/[$1,200,000 + $197,500 (P/A, 10%, 20)] **B–C = 1.448 > 1; extend runways.**
Modified B–C: *Equation (10-2)*	B–C = [PW(B) − PW(O&M)]/[I − PW(MV)] B–C = [$490,000 (P/A, 10%, 20) − $197,500 (P/A, 10%, 20)]/$1,200,000 **B–C = 2.075 > 1; extend runways.**
Conventional B–C: *Equation (10-3)*	B–C = AW(B)/[CR + AW(O&M)] B–C = $490,000/[$1,200,000 (A/P, 10%, 20) + $197,500] **B–C = 1.448 > 1; extend runways.**
Modified B–C: *Equation (10-4)*	B–C = [AW(B) − AW(O&M)]/CR B–C = [$490,000 − $197,500]/[$1,200,000 (A/P, 10%, 20)] **B–C = 2.075 > 1; extend runways.**

As can be seen in the preceding example, the difference between conventional and modified B–C ratios is essentially due to subtracting the equivalent-worth measure of operating and maintenance costs from both the numerator and the denominator of the B–C ratio. In order for the B–C ratio to be greater than 1.0, the numerator must be greater than the denominator. Similarly, the numerator must be less than the denominator for the B–C ratio to be less than 1.0. Subtracting a constant (the equivalent worth of operating and maintenance costs) from both numerator and denominator does not alter the *relative* magnitudes of the numerator and denominator. Thus, project acceptability is not affected by the choice of conventional versus modified B–C ratio. This information is stated mathematically as follows for the case of B–C > 1.0:

Let N = the numerator of the conventional B–C ratio,
D = the denominator of the conventional B–C ratio,
O&M = the equivalent worth of operating and maintenance costs.

If B–C $= \dfrac{N}{D} > 1.0$, then $N > D$.

If $N > D$, and $[N - \text{O\&M}] > [D - \text{O\&M}]$, then $\dfrac{N - \text{O\&M}}{D - \text{O\&M}} > 1.0$.

Note that $\dfrac{N - \text{O\&M}}{D - \text{O\&M}}$ is the modified B–C ratio; thus, if conventional B–C > 1.0, then modified B–C > 1.0

EXAMPLE 10-3	**B–C Ratio of the Proposed Bypass**

In this example, we will evaluate the bypass described in the beginning of the chapter. The construction cost of the bypass is $20 million, and $500,000 would be required each year for annual maintenance. The annual benefits to the public have been estimated to be $2 million. If the study period is 50 years and the state's interest rate is 8% per year, should the bypass be constructed? What impact does a social interest rate of 4% per year have on the B–C ratio of the project?

Solution

At an interest rate of 8% per year, the conventional B–C ratio of the proposed bypass is

$$\text{B–C} = \frac{\$2,000,000}{\$20,000,000\,(A/P,\ 8\%,\ 50) + \$500,000} = 0.94.$$

Because this ratio is less than one, the bypass is not economically acceptable at 8% interest.

If a social interest rate of 4% per year was used, the B–C ratio would be 1.40 and the bypass would be acceptable.

Two additional issues of concern are the treatment of *disbenefits* in B–C analyses and the decision as to whether certain cash flow items should be treated as *additional benefits* or as *reduced costs*. The first concern arises whenever disbenefits are formally defined in a B–C evaluation of a public-sector project. An example of the second concern would be a public-sector project proposing to replace an existing asset having high annual operating and maintenance costs with a new asset having lower operating and manual costs. As will be seen in Sections 10.7.1 and 10.7.2, the final recommendation on a project is not altered by either the approach to incorporating disbenefits or the classification of an item as a reduced cost or an additional benefit.

10.7.1 Disbenefits in the B–C Ratio

In a previous section, disbenefits were defined as negative consequences to the public resulting from the implementation of a public-sector project. The traditional approach for incorporating disbenefits into a B–C analysis is to reduce benefits by the amount of disbenefits (i.e., to subtract disbenefits from benefits in the numerator of the B–C ratio). Alternatively, the disbenefits could be treated as costs (i.e., add disbenefits to costs in the denominator). Equations (10-5) and (10-6) illustrate the two approaches for incorporating disbenefits in the conventional B–C ratio, with benefits, costs, and disbenefits in terms of equivalent AW. (Similar equations could also be developed for the modified B–C ratio or for PW as the measure of equivalent worth.) Again, the *magnitude* of the B–C ratio will be different depending upon which approach is used to incorporate disbenefits, but project acceptability—that is, whether the B–C ratio is >, <, or = 1.0—will not be affected, as shown in Example 10-4.

Conventional B–C ratio with AW, *benefits* reduced by amount of *disbenefits*:

$$\text{B–C} = \frac{\text{AW(benefits)} - \text{AW(disbenefits)}}{\text{AW(costs)}} = \frac{\text{AW}(B) - \text{AW}(D)}{\text{CR} + \text{AW(O\&M)}}. \quad (10\text{-}5)$$

Here, $\text{AW}(\cdot)$ = annual worth of $(\cdot)$;
B = benefits of the proposed project;
D = disbenefits of the proposed project;
CR = capital recovery amount (i.e., the equivalent annual cost of the initial investment, I, including an allowance for market value, if any);
$\text{O\&M}$ = operating and maintenance costs of the proposed project.

Conventional B–C ratio with AW, *costs* increased by amount of *disbenefits*:

$$\text{B–C} = \frac{\text{AW(benefits)}}{\text{AW(costs)} + \text{AW(disbenefits)}} = \frac{\text{AW}(B)}{\text{CR} + \text{AW(O\&M)} + \text{AW}(D)}. \quad (10\text{-}6)$$

EXAMPLE 10-4	Including Disbenefits in a B–C Analysis

Refer back to Example 10-2. In addition to the benefits and costs, suppose that there are disbenefits associated with the runway extension project. Specifically, the increased noise level from commercial jet traffic will be a serious nuisance to homeowners living along the approach path to the Columbia Municipal Airport. The annual disbenefit to citizens of Columbia caused by this noise pollution is estimated to be $100,000. Given this additional information, reapply the conventional B–C ratio, with equivalent annual worth, to determine whether this disbenefit affects your recommendation on the desirability of this project.

Solution

Disbenefits reduce benefits, Equation (10-5)	B–C = [AW(B) − AW(D)]/[CR + AW(O&M)] B–C = [$490,000 − $100,000]/[$1,200,000 $(A/P, 10\%, 20)$ + $197,500] **B–C = 1.152 > 1; extend runways.**
Disbenefits treated as additional costs, Equation (10-6)	B–C = AW(B)/[CR + AW(O&M) + AW(D)] B–C = $490,000/[$1,200,000 $(A/P, 10\%, 20)$ + $197,500 + $100,000] **B–C = 1.118 > 1; extend runways.**

As in the case of conventional and modified B–C ratios, the treatment of disbenefits may affect the magnitude of the B–C ratio, but it has no effect on project desirability in go/no-go decisions. It is left to the reader to develop a mathematical rationale for this, similar to that included in the discussion of conventional versus modified B–C ratios.

10.7.2 Added Benefits versus Reduced Costs in B–C Analyses

The analyst often needs to classify certain cash flows as either added benefits or reduced costs in calculating a B–C ratio. The questions arise, How critical is the proper assignment of a particular cash flow as an added benefit or a reduced cost? Is the outcome of the analysis affected by classifying a reduced cost as a benefit? *An arbitrary decision as to the classification of a benefit or a cost has no impact on project acceptability.* The mathematical rationale for this information is presented next and in Example 10-5.

Let B = the equivalent annual worth of project benefits,
C = the equivalent annual worth of project costs,
X = the equivalent annual worth of a cash flow (either an added benefit or a reduced cost) not included in either B or C.

If X is classified as an added benefit, *then* $\text{B–C} = \dfrac{B + X}{C}$. Alternatively,

if X is classified as a reduced cost, *then* $\text{B–C} = \dfrac{B}{C - X}$.

Assuming that the project is acceptable, that is, B–C ≥ 1.0,

$$\frac{B + X}{C} \geq 1.0, \text{ which indicates that } B + X \geq C, \text{ and}$$

$$\frac{B}{C - X} \geq 1.0, \text{ which indicates that } B \geq C - X,$$

which can be restated as $B + X \geq C$.

EXAMPLE 10-5	Added Benefit versus Reduced Cost in a Bridge Widening Project

A project is being considered by the Tennessee Department of Transportation to replace an aging bridge across the Cumberland River on a state highway. The existing two-lane bridge is expensive to maintain and creates a traffic bottleneck because the state highway is four lanes wide on either side of the bridge. The new bridge can be constructed at a cost of $300,000, and estimated annual maintenance costs are $10,000. The existing bridge has annual maintenance costs of $18,500. The annual benefit of the new four-lane bridge to motorists, due to the removal of the traffic bottleneck, has been estimated to be $25,000. Conduct a B–C analysis, using a MARR of 8% and a study period of 25 years, to determine whether the new bridge should be constructed.

Solution

Treating the reduction in annual maintenance costs as a *reduced cost*:

$$\text{B–C} = \$25,000/[\$300,000(A/P, 8\%, 25) - (\$18,500 - \$10,000)]$$

$$\text{B–C} = 1.275 > 1; \text{ construct new bridge}.$$

Treating the reduction in annual maintenance costs as an *increased benefit*:

$$\text{B–C} = [\$25,000 + (\$18,500 - \$10,000)]/[\$300,000(A/P, 8\%, 25)]$$

$$\text{B–C} = 1.192 > 1; \text{ construct new bridge}.$$

Therefore, the decision to classify a cash flow item as an additional benefit or as a reduced cost will affect the magnitude of the calculated B–C ratio, but it will have no effect on project acceptability.

10.8 Evaluating Independent Projects by B–C Ratios

Independent projects are categorized as groupings of projects for which the choice to select any particular project in the group is *independent* of choices regarding any and all other projects within the group. Thus, it is permissible to select none of the projects, any combination of projects, or all of the projects from an independent group. (Note that this does not hold true under conditions of *capital rationing*. Methods of evaluating otherwise independent projects under capital rationing are discussed in Chapter 13.) Because any or all projects from an independent set can be selected, formal comparisons of independent projects are unnecessary. The issue of whether one project is *better* than another is unimportant if those projects are independent; the only criterion for selecting each of those projects is whether their respective B–C ratios are equal to or greater than 1.0.

A typical example of an economy study of a federal project—using the conventional B–C-ratio method—is the study of a flood control and power project on the White River in Missouri and Arkansas. Considerable flooding and consequent damage had occurred along certain portions of this river, as shown in Table 10-2. In addition, the uncontrolled water flow increased flood conditions on the lower Mississippi River. In this case, there were independent options of building a reservoir *and/or* a channel improvement to alleviate the problem. The cost and benefit summaries for the Table Rock reservoir and the Bull Shoals channel improvement are shown in Table 10-3. The fact that the Bull Shoals channel improvement project has the higher B–C ratio is irrelevant; both options are acceptable because their B–C ratios are greater than one.

TABLE 10-2 Annual Loss as a Result of Floods on Three Stretches of the White River

Item	Annual Value of Loss	Annual Loss per Acre of Improved Land in Floodplain	Annual Loss per Acre for Total Area in Floodplain
Crops	$1,951,714	$6.04	$1.55
Farm (other than crops)	215,561	0.67	0.17
Railroads and highways	119,800	0.37	0.09
Levees[a]	87,234	0.27	0.07
Other losses	168,326	0.52	0.13
Total	**$2,542,635**	**$7.87**	**$2.01**

[a] Expenditures by the United States for levee repairs and high-water maintenance.

TABLE 10-3 Estimated Costs, Annual Charges, and Annual Benefits for the Table Rock Reservoir and Bull Shoals Channel Improvement Projects

Item	Table Rock Reservoir	Bull Shoals Channel Improvement
Cost of dam and appurtenances, and reservoir:		
Dam, including reservoir-clearing, camp, access railroads and highways, and foundation exploration and treatment	$20,447,000	$25,240,000
Powerhouse and equipment	6,700,000	6,650,000
Power-transmission facilities to existing load-distribution centers	3,400,000	4,387,000
Land	1,200,000	1,470,000
Highway relocations	2,700,000	140,000
Cemetery relocations	40,000	18,000
Damage to villages	6,000	94,500
Damage to miscellaneous structures	7,000	500
Total Construction Cost (estimated appropriation of public funds necessary for the execution of the project)	$34,500,000	$38,000,000
Federal investment:		
Total construction cost	$34,500,000	$38,000,000
Interest during construction	$1,811,300	1,995,000
Total	$36,311,300	$39,995,000
Present value of federal properties	1,200	300
Total Federal Investment	$36,312,500	$39,995,300
Total Annual Costs	**$1,642,200**	**$1,815,100**

(*Continued*)

TABLE 10-3	Estimated Costs, Annual Charges, and Annual Benefits for the Table Rock Reservoir and Bull Shoals Channel Improvement Projects (*continued*)		
Item		Table Rock Reservoir	Bull Shoals Channel Improvement
Annual benefits:			
Prevented direct flood losses in White River basin:			
Present conditions		60,100	266,900
Future developments		19,000	84,200
Prevented indirect flood losses owing to floods in White River basin		19,800	87,800
Enhancement in property values in White River valley		7,700	34,000
Prevented flood losses on Mississippi River		220,000	980,000
Annual Flood Benefits		326,000	1,452,900
Power value		1,415,600	1,403,400
Total Annual Benefits		$1,742,200	$2,856,300
Conventional B–C Ratio = Total Annual Benefits ÷ Annual Costs		1.06	1.57

Several interesting facts may be noted concerning this study. *First*, there was no attempt to allocate the cost of the projects between flood control and power production. *Second*, very large portions of the flood-control benefits were shown to be in connection with the Mississippi River and are not indicated in Table 10-2; these were not detailed in the main body of the report but were shown in an appendix. Only a moderate decrease in the value of these benefits would have changed the B–C ratio considerably. *Third*, without the combination of flood-control and power-generation objectives, neither project would have been economical for either purpose. These facts point to the advantages of multiple purposes for making flood-control projects economically feasible and to the necessity for careful enumeration and evaluation of the prospective benefits of a public-sector project.

10.9   Comparison of Mutually Exclusive Projects by B–C Ratios

Recall that a group of *mutually exclusive projects* was defined as *a group of projects from which, at most, one project may be selected*. When using an equivalent-worth method to select from among a set of mutually exclusive alternatives (MEAs), the best alternative can be selected by maximizing the PW (or AW, or FW). Because the B–C method provides a *ratio* of benefits to costs rather than a direct measure of each project's *profit potential*, selecting the project that maximizes the B–C ratio does not guarantee that the best project is selected. This phenomenon is illustrated in Example 10-6. As with the rate-of-return procedures in Chapter 6, an evaluation of mutually exclusive alternatives by the B–C ratio requires that an *incremental* B–C analysis be conducted.

EXAMPLE 10-6	Inconsistent Ranking Problem When B–C Ratios Are Inappropriately Compared

The required investments, annual operating and maintenance costs, and annual benefits for two mutually exclusive alternative projects are shown subsequently. Both conventional and modified B–C ratios are included for each project. Note that Project A has the greater *conventional* B–C, but Project B has the greater *modified* B–C. Given this information, which project should be selected?

	Project A	Project B	
Capital investment	$110,000	$135,000	MARR = 10% per year
Annual O&M cost	12,500	45,000	Study period = 20 years
Annual benefit	37,500	80,000	
Conventional B–C	**1.475**	1.314	
Modified B–C	1.934	**2.206**	

Solution

The B–C analysis has been conducted improperly. Although each of the B–C ratios shown is *numerically correct*, a comparison of mutually exclusive alternatives requires that an incremental analysis be conducted.

When comparing mutually exclusive alternatives with the B–C ratio method, *they are first ranked in order of increasing total equivalent worth of costs*. This rank-ordering will be identical whether the ranking is based on PW, AW, or FW of costs. The do-nothing alternative is selected as a baseline alternative. The B–C ratio is then calculated for the alternative having the lowest equivalent cost. If the B–C ratio for this alternative is equal to or greater than 1.0, then that alternative becomes the new baseline; otherwise, do-nothing remains as the baseline. The next least equivalent cost alternative is then selected, and the difference (Δ) in the respective benefits and costs of this alternative and the baseline is used to calculate an incremental B–C ratio ($\Delta B/\Delta C$). If that ratio is equal to or greater than 1.0, then the higher equivalent cost alternative becomes the new baseline; otherwise, the last baseline alternative is maintained. Incremental B–C ratios are determined for each successively higher equivalent cost alternative until the last alternative has been compared. The flowchart of this procedure is included as Figure 10-3, and the procedure is illustrated in Example 10-7.

EXAMPLE 10-7	Incremental B–C Analysis of Mutually Exclusive Projects

Three mutually exclusive alternative public-works projects are currently under consideration. Their respective costs and benefits are included in the table that follows. Each of the projects has a useful life of 50 years, and MARR is 10% per

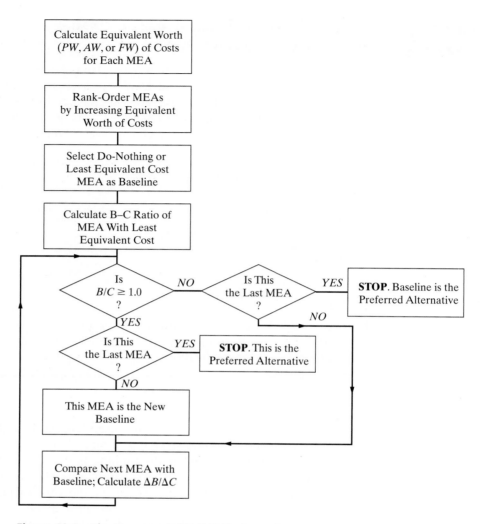

Figure 10-3 The Incremental B–C Ratio Procedure

year. Which, if any, of these projects should be selected? Solve by hand and by spreadsheet.

	A	B	C
Capital investment	$8,500,000	$10,000,000	$12,000,000
Annual operating and maintenance costs	750,000	725,000	700,000
Market value	1,250,000	1,750,000	2,000,000
Annual benefit	2,150,000	2,265,000	2,500,000

Solution by Hand

$$PW(\text{Costs}, A) = \$8,500,000 + \$750,000(P/A, 10\%, 50)$$
$$-\$1,250,000(P/F, 10\%, 50) = \$15,925,463,$$
$$PW(\text{Costs}, B) = \$10,000,000 + \$725,000(P/A, 10\%, 50)$$
$$-\$1,750,000(P/F, 10\%, 50) = \$17,173,333,$$
$$PW(\text{Costs}, C) = \$12,000,000 + \$700,000(P/A, 10\%, 50)$$
$$-\$2,000,000(P/F, 10\%, 50) = \$18,923,333,$$
$$PW(\text{Benefit}, A) = \$2,150,000(P/A, 10\%, 50) = \$21,316,851,$$
$$PW(\text{Benefit}, B) = \$2,265,000(P/A, 10\%, 50) = \$22,457,055,$$
$$PW(\text{Benefit}, C) = \$2,750,000(P/A, 10\%, 50) = \$24,787,036.$$
$$B\text{–}C(A) = \$21,316,851/\$15,925,463$$
$$= 1.3385 > 1.0$$

Therefore, Project A is acceptable.

$$\Delta B/\Delta C \text{ of } (B - A) = (\$22,457,055 - \$21,316,851)/(\$17,173,333 - \$15,925,463)$$
$$= 0.9137 < 1.0.$$

Therefore, increment required for Project B is not acceptable.

$$\Delta B/\Delta C \text{ of } (C - A) = (\$24,787,036 - \$21,316,851)/(\$18,923,333 - \$15,925,463)$$
$$= 1.1576 > 1.0.$$

Therefore, increment required for Project C is acceptable.

Decision: *Recommend Project C.*

Spreadsheet Solution

A spreadsheet analysis for this example is shown in Figure 10-4. For each mutually exclusive project, the PW of total benefits (cells B11:B13) and the PW of total costs (cells C11:C13) are calculated. Note that annual operating and maintenance costs are included in the total cost calculation in accordance with the conventional B–C ratio formulation.

Since the projects are mutually exclusive, they must be ranked from smallest to largest according to the equivalent worth of costs (do nothing → A → B → C). The B–C ratio for Project A is calculated to be 1.34 (row 17), and Project A replaces do-nothing as the baseline alternative.

Project B is now compared with Project A (row 18). The ratio of incremental benefits to incremental costs is less than 1.0, so Project A remains the baseline alternative. Finally, in row 19, the incremental benefits and incremental costs

$$= B5 + PV(\$B\$2, \$B\$1, -C5) - E5 / (1 + \$B\$2)^\wedge\$B\$1$$

	A	B	C	D	E
1	Useful Life =	50			
2	MARR =	10%			
3					
4	Project	Capital Investment	Ann. Maint Cost	Annual Benefit	Market Value
5	A	$8,500,000	$750,000	$ 2,150,000	$ 1,250,000
6	B	$10,000,000	$725,000	$ 2,265,000	$ 1,750,000
7	C	$12,000,000	$700,000	$ 2,500,000	$ 2,000,000
8					
9		Present Worth			
10	Project	Benefits	Costs		
11	A	$ 21,316,851	$ 15,925,463		
12	B	$ 22,457,055	$ 17,173,333		
13	C	$ 24,787,036	$ 18,923,333		
14					
15	Present Worth Formulation				
16		Inc. Benefit	Inc. Cost	Inc. B/C	
17	0 -> A	$ 21,316,851	$ 15,925,463	1.34	
18	A -> B	$ 1,140,204	$ 1,247,870	0.91	
19	A -> C	$ 3,470,185	$ 2,997,870	1.16	

$$= PV(\$B\$2, \$B\$1, -D5)$$

$$= B11$$

$$= B12 - B11$$

$$= B13 - B11$$

$$= B17 / C17$$

Figure 10-4   Spreadsheet Solution, Example 10-7

associated with selecting Project C instead of Project A are used to calculate an incremental B–C ratio of 1.16. Since this ratio is greater than one, the increment required for Project C is acceptable, and thus Project C becomes the recommended project.

It is not uncommon for some of the projects in a set of mutually exclusive public-works projects to have different lives. Recall from Chapter 6 that the AW criterion can be used to select from among alternatives with different lives as long as the assumption of *repeatability* is valid. Similarly, if a mutually exclusive set of public-works projects includes projects with varying useful lives, it may be possible to conduct an incremental B–C analysis by using the AW of benefits and costs of the various projects. This analysis is illustrated in Example 10-8.

EXAMPLE 10-8   B–C Analysis with Unequal Project Lives

Two mutually exclusive alternative public-works projects are under consideration. Their respective costs and benefits are included in the table that follows. Project I has an anticipated life of 35 years, and the useful life of Project II has been estimated to be 25 years. If the MARR is 9% per year, which, if either, of these projects should be selected? The effect of inflation is negligible.

	Project I	Project II
Capital investment	$750,000	$625,000
Annual operating and maintenance	120,000	110,000
Annual benefit	245,000	230,000
Useful life of project (years)	35	25

Solution

$$\text{AW(Costs, I)} = \$750,000(A/P, 9\%, 35) + \$120,000 = \$190,977,$$

$$\text{AW(Costs, II)} = \$625,000(A/P, 9\%, 25) + \$110,000 = \$173,629,$$

$$\text{B–C(II)} = \$230,000/\$173,629 = 1.3247 > 1.0.$$

Therefore, Project II is acceptable.

$$\Delta B/\Delta C \text{ of (I–II)} = (\$245,000 - \$230,000)/(\$190,977 - \$173,629)$$
$$= 0.8647 < 1.0.$$

Therefore, increment required for Project I is not acceptable.

Decision: *Project II should be selected.*

10.10 Criticisms and Shortcomings of the Benefit–Cost Ratio Method*

Although the B–C ratio method is firmly entrenched as *the* procedure used by most governmental agencies for the evaluation of public-sector projects, it has been widely criticized over the years. Among the criticisms are that (1) it is often used as a tool for after-the-fact justifications rather than for project evaluation, (2) serious distributional inequities (i.e., one group reaps the benefits while another incurs the costs) may not be accounted for in B–C studies, and (3) qualitative information is often ignored in B–C studies.

Regarding the first criticism, the B–C ratio is considered by many to be a method of using the numbers to support the views and interests of the group paying for

---

* J. T. Campen, *Benefit, Cost, and Beyond: The Political Economy of Benefit–Cost Analysis* (Cambridge, MA: Ballinger, 1986). All quotes in Section 10.10 are taken from this source.

the analysis. A congressional subcommittee once supported this critical view with the following conclusion:

> . . . the most significant factor in evaluating a B–C study is the name of the sponsor. B–C studies generally are formulated after basic positions on an issue are taken by the respective parties. The results of competing studies predictably reflect the respective positions of the parties on the issue (p. 55 of Campen).

One illustration of such a flawed B–C study is the Bureau of Reclamation's 1967 analysis supporting the proposed Nebraska Mid-State Project. The purpose of this project was to divert water from the Platte River to irrigate cropland, and the computed B–C ratio for this project was 1.24, indicating that the project should be undertaken. This favorable ratio was based, in part, on the following fallacies (p. 53):

1. An unrealistically low interest rate of only 3.125% per year was used.
2. The analysis used a project life of 100 years rather than the generally more accepted and supportable assumption of a 50-year life.
3. The analysis counted wildlife and fish benefits among the project's benefits, although in reality, the project would have been devastating to wildlife. During a 30-year history of recorded water flow (1931–1960), the proposed water diversion would have left a substantial stretch of the Platte River dry for over half of that time, destroying fish and waterfowl habitats on 150 miles of the river. This destruction of wildlife habitats would have adversely affected the populations of endangered species, including the bald eagle, the whooping crane, and the sandhill crane.
4. The benefits of increased farm output were based on government support prices that would have required a substantial federal subsidy.

Campen states that "the common core of these criticisms is not so much the fact that B–C analysis is used to justify particular positions but that it is presented as a scientific unbiased method of analysis" (pp. 52–53). In order for an analysis to be fair and reliable, it must be based on an accurate and realistic assessment of all the relevant benefits and costs. Thus, the analysis must be performed either by an unbiased group or by a group that includes representatives from each of the interest groups involved. When the Nebraska Mid-State Project was reevaluated by an impartial third party, for example, a more realistic B–C ratio of only 0.23 was determined. Unfortunately, analyses of public-sector projects are sometimes performed by parties who have already formed strong opinions on the worth of these projects.

Another shortcoming of the B–C ratio method is that benefits and costs effectively cancel each other out without regard for who reaps the benefits or who incurs the costs, which does not cause significant difficulties in the private sector where the benefits accrue to and the costs are borne by the owners of the firm. Recall that in the public sector, however, the benefits to whomsoever they accrue must be considered, which may result in serious distributional inequities in B–C studies. Consider, for example, the inequity shown in the following situation:

> Consider a proposal to raise property taxes by 50% on all properties with odd-numbered street addresses and simultaneously to lower property taxes by 50% on all properties with even-numbered addresses. A conventional B–C analysis of this proposal would conclude that its net benefits were approximately zero, and an analysis of its impact on the overall level of inequality in the size distribution of income would also show no significant effects. Nevertheless, such a proposal would be generally, and rightly, condemned as consisting of an arbitrary and unfair redistribution of income (p. 56 of Campen).

Another, more realistic, example of adverse distributional consequences would involve a project to construct a new chemical plant in *Town A*. The chemical plant would employ hundreds of workers in an economically depressed area, but, according to one group of concerned citizens, it would also release hazardous byproducts that could pollute the groundwater and a nearby stream that provides most of the drinking water for *Town B*. Thus, the benefits of this project would be the addition of jobs and a boost to the local economy for *Town A*; but *Town B* would have increased water-treatment costs, and residents may be subject to long-term health hazards and increased medical bills. Unfortunately, a B–C-ratio analysis would show only the net monetary effect of the project, without regard to the serious distributional inequities.

Although these criticisms have often been leveled at the B–C method itself, problems in the use (and abuse) of the B–C procedure are largely due to the inherent difficulties in evaluating public projects (see Section 10.5) and the manner in which the procedure is applied. Note that these same criticisms might be for a poorly conducted analysis that relied on an equivalent worth or rate-of-return method.

10.11   CASE STUDY—Improving a Railroad Crossing

Traffic congestion and vehicle safety are significant concerns in most major cities in the Northeast United States. A major metropolitan city in New Jersey is considering the elimination of a railroad grade crossing by building an overpass. Traffic engineers estimated that approximately 2,000 vehicles per day are delayed at an average of two minutes each due to trains at the grade crossing. Trucks comprise 40% of the vehicles, and the opportunity cost of their delay is assumed to average $20 per truck-hour. The other vehicles are cars having an assumed average opportunity cost of $4 per car-hour. It is also estimated that the new overpass will save the city approximately $4,000 per year in expenses directly due to accidents.

The traffic engineers determined that the overpass would cost $1,000,000 and is estimated to have a useful life of 40 years and a $100,000 salvage value. Annual maintenance costs of the overpass would cost the city $5,000 more than the maintenance costs of the existing grade crossing. The installation of the overpass will save the railroad an annual expense of $30,000 for lawsuits and maintenance of crossing guards.

Since this is a public project, there are special considerations and a complete and comprehensive engineering economy study is more challenging than in the case of privately financed projects. For example, in the private sector, costs are accrued by the firm undertaking the project, and benefits are the favorable outcomes achieved by the firm. Typically, any costs and benefits that are external to the firm are ignored in economic evaluations unless those external costs and benefits indirectly affect the firm. With economic evaluations of public projects, however, the opposite is true. As in the case of improving the railroad crossing, there are multiple purposes or objectives to consider. The true owners of the project are the taxpayers! The monetary impacts of the diverse benefits are oftentimes hard to quantify, and there may be special political or legal issues to consider.

In this case study, the city council is now in the process of considering the merits of the engineering proposal to improve the railroad crossing. The city council is considering the following questions in its deliberations:

- Should the overpass be built by the city if it is to be the owner and the opportunity cost of the city's capital is 8% per year?

- How much should the railroad reasonably be asked to contribute toward construction of the bridge if its opportunity cost of capital is assumed to be 15% per year?

Solution

The city uses the conventional B–C ratio with AW for its analyses of public projects. The annual benefits of the overpass are comprised of the time savings for vehicles (whose drivers are "members" of the city) and the reduction in accident expenses. The city's cost engineer makes the following estimates.

Annual Benefits:

$$\text{Cars} = \left(\frac{2,000 \times 0.6 \text{ vehicles}}{\text{day}}\right)\left(\frac{365 \text{ days}}{\text{year}}\right)\left(\frac{\text{hour}}{60 \text{ minutes}}\right)\left(\frac{\$4.00}{\text{car} - \text{hour}}\right) = \$58,400$$

$$\text{Trucks} = \left(\frac{2,000 \times 0.4 \text{ vehicles}}{\text{day}}\right)\left(\frac{365 \text{ days}}{\text{year}}\right)\left(\frac{\text{hour}}{60 \text{ minutes}}\right)\left(\frac{\$20.00}{\text{truck} - \text{hour}}\right)$$

$$= \$194,667$$

Annual savings $= \$4,000$

Total annual benefits $= \$58,400 + \$194,667 + \$4,000 = \$257,067$

Notice that the estimated $30,000 annual expense savings for lawsuits and maintenance of the crossing guard is not included in the annual benefits calculation. This savings will be experienced by the owners of the railroad, not by the city.

The costs of the overpass to the city are the construction of the overpass (less its salvage value) and the increased maintenance costs. The cost engineer makes the following estimates.

Annual Costs:

Capital recovery $= \$1,000,000(A/P, 8\%, 40) - \$100,000(A/F, 8\%, 40) = \$83,474$

Increased maintenance $= \$5,000$

Total annual costs $= \$83,474 + \$5,000 = \$88,474$

Based on these estimates, the B–C ratio of the proposed overpass is

$$\text{B–C Ratio} = \frac{\text{Annual benefits}}{\text{Annual costs}} = \frac{\$257,067}{\$88,474} = 2.91.$$

The cost engineer recommends to the city council that the new overpass be built since the B–C ratio is greater than 1.0.

The cost engineer also advises the council that since the railroad company stands to directly benefit from the replacement of the existing grade crossing by the overpass, it would not be unreasonable for the city to request a contribution to the construction cost of the overpass. Given the railroad company's cost of capital of 15% per year and an estimated annual savings of \$30,000, the cost engineer calculates that the overpass is worth

$$\text{PW}(15\%) = \$30,000(P/A, 15\%, 40) = \$199,254$$

to the railroad. Any amount contributed by the railroad company would serve to reduce the denominator of the B–C ratio, thereby increasing the value of the ratio. The B–C ratio was computed to be 2.91 without any contribution by the railroad. Therefore, the cost engineer concludes that the city should plan on constructing the overpass regardless of whether or not the railroad company can be persuaded to contribute financially to the project.

This case illustrates some of the special issues associated with providing economic evaluations of public-sector projects. While the B–C ratio is a useful method for evaluating projected financial performance of a public-sector project, the quantification of benefits and costs may prove difficult. As the case illustrates, typically, surrogate or proxy measures are used in the estimation of benefits and costs. Also, it is especially important to remember the perspectives of the owners in the evaluation of public projects—not necessarily the elected city council members, but the taxpayers!

10.12 Summary

From the discussion and examples of public projects presented in this chapter, it is apparent that, because of the methods of financing, the absence of tax and profit requirements, and political and social factors, the criteria used in evaluating privately financed projects frequently cannot be applied to public works. Nor should public projects be used as yardsticks with which to compare private projects. Nevertheless, whenever possible, public works should be justified on an economic basis to ensure that the public obtains the maximum return from the tax money that is spent. Whether an engineer is working on such projects, is called upon to serve as a consultant, or assists in the conduct of the B–C analysis, he or she is bound by professional ethics to do his or her utmost to see that the projects and the associated analyses are carried out in the best possible manner and within the limitations of the legislation enacted for their authorization.

The B–C ratio has remained a popular method for evaluating the financial performance of public projects. Both the conventional and modified B–C ratio methods have been explained and illustrated for the case of *independent* and *mutually exclusive* projects. A final note of caution: The best project among a *mutually exclusive* set of projects is not necessarily the one that maximizes the B–C ratio. In this chapter, we have seen that an incremental analysis approach to evaluating benefits and costs is necessary to ensure the correct choice.

Problems

The number in parentheses () that follows each problem refers to the section from which the problem is taken.

10-1. A sewage containment project in Louisiana is expected to require an initial investment of $3 million and annual maintenance expenses of $57,000. The benefits to the public are valued at $460,000 per year. This project can be assumed to have an infinite life. If MARR is 10% per year, determine whether the project is economically attractive using the B–C ratio measure of merit. (10.7)

10-2. An environmentally friendly 2,800-square-foot green home (99% air tight) costs about 8% more to construct than a same-sized conventional home. Most green homes can save 15% per year on energy expenses to heat and cool the living space. For a $250,000 conventional home with a heating and cooling bill of $3,000 per year, how much would have to be saved in energy expenses per year to justify this home (i.e., B–C ratio greater than or equal to one)? The discount rate is 10% per year, and the expected life of the home is 30 years. (10.7)

10-3. A proposal has been made for improving the downtown area of a small town. The plan calls for banning vehicular traffic on the main street and turning this street into a pedestrian mall with tree plantings and other beautification features. This plan will involve actual costs of $7,000,000 and, according to its proponents, the plan will produce benefits and disbenefits to the town as follows:

Benefits:
Increased sales tax revenue $400,000 per year
Increased real estate property taxes $300,000 per year
Benefits due to decreased air pollution $50,000 per year
Quality of life improvements to users $50,000 per year

Disbenefits:
Increased maintenance $150,000 per year

a. Compute the B–C ratio of this plan based on a MARR of 8% per year and an infinite life for the project. (10.7)

b. How does the B–C ratio change for a 20-year project life and a MARR of 10% per year? (10.7)

10-4. A retrofitted space-heating system is being considered for a small office building. The system can be purchased and installed for $120,000, and it will save an estimated 300,000 kilowatt-hours (kWh) of electric power each year over a six-year period. A kilowatt-hour of electricity costs $0.10, and the company uses a MARR of 15% per year in its economic evaluations of refurbished systems. The market value of the system will be $8,000 at the end of six years, and additional annual operating and maintenance expenses are negligible. Use the benefit–cost method to make a recommendation. (10.7)

10-5. In Problem 10-4, what is the benefit–cost ratio of the project if the general inflation rate is 4% per year and the market value is negligible? The market interest rate (i_m) is 18% per year, and the annual savings are expressed in year-zero dollars. (10.7, Chapter 8)

10-6. A tunnel through a mountain is being considered as a replacement for an existing stretch of highway in southeastern Kentucky. The existing road is a steep, narrow, winding two-lane highway that has been the site of numerous fatal accidents, with an average of 2.05 fatalities and 3.35 serious injuries per year. It has been projected that the tunnel will significantly reduce the frequency of accidents, with estimates of not more than 0.15 fatalities and 0.35 serious injuries per year. Initial capital investment requirements, including land acquisition, tunnel excavation, lighting, roadbed preparation, and so on, have been estimated to be $45,000,000. Annualized upkeep costs for the tunnel will be significantly less than for the existing highway, resulting in an annual savings of $85,000. For purposes of this analysis, a *value per life saved* of $1,000,000 will be applied, along with an estimate of $750,000 for medical costs, disability, and so on, per serious injury. (10.7)

a. Apply the B–C ratio method, with an anticipated life of the tunnel project of 50 years and an interest rate of 8% per year, to determine whether the tunnel should be constructed.

b. Assuming that the cost per serious injury is unchanged, determine the value per life saved at which the tunnel project is marginally justified (i.e., B–C = 1.0).

10-7. A toll bridge across the Mississippi River is being considered as a replacement for the current I-40 bridge linking Tennessee to Arkansas. Because this bridge, if approved, will become a part of the U.S. Interstate Highway system, the B–C ratio method must be applied in the evaluation. Investment costs of the structure are estimated to be $17,500,000, and $325,000 per year in operating and maintenance costs are anticipated. In addition, the bridge must be resurfaced every fifth year of its 30-year projected life at a cost of $1,250,000 per occurrence (no resurfacing cost in year 30). Revenues generated from the toll are anticipated to be $2,500,000 in its first year of operation, with a projected annual rate of increase of 2.25% per year due to the anticipated annual increase in traffic across the bridge. Assuming zero market (salvage) value for the bridge at the end of 30 years and a MARR of 10% per year, should the toll bridge be constructed? (10.7)

10-8. Refer back to Problem 10-7. Suppose that the toll bridge can be redesigned such that it will have a (virtually) infinite life. MARR remains at 10% per year. Revised costs and revenues (benefits) are given as follows: (10.7, 10.9)

Capital investment: $22,500,000

Annual operating and maintenance costs: $250,000

Resurface cost every seventh year: $1,000,000

Structural repair cost, every 20th year: $1,750,000

Revenues (treated as constant—no rate of increase): $3,000,000

a. What is the capitalized worth of the bridge?

b. Determine the B–C ratio of the bridge over an infinite time horizon.

c. Should the initial design (*Problem 10-7*) or the new design be selected?

10-9. Five independent projects are available for funding by a certain public agency. The following tabulation shows the equivalent annual benefits and costs for each: (10.8)

Project	Annual Benefits	Annual Costs
A	$1,800,000	$2,000,000
B	5,600,000	4,200,000
C	8,400,000	6,800,000
D	2,600,000	2,800,000
E	6,600,000	5,400,000

a. Assume that the projects are of the type for which the benefits can be determined with considerable certainty and that the agency is willing to invest money as long as the B–C ratio is at least one. Which alternatives should be selected for funding?

b. What is the rank-ordering of projects from best to worst?

c. If the projects involved intangible benefits that required considerable judgment in assigning their values, would your recommendation be affected?

10-10. In the development of a publicly owned, commercial waterfront area, three possible independent plans are being considered. Their costs and estimated benefits are as follows: (10.8)

	PW ($000s)	
Plan	Costs	Benefits
A	$123,000	$139,000
B	135,000	150,000
C	99,000	114,000

a. Which plan(s) should be adopted, if any, if the controlling board wishes to invest any amount required, provided that the B–C ratio on the required investment is at least 1.0?

b. Suppose that 10% of the costs of each plan are reclassified as disbenefits. What percentage change in the B–C ratio of each plan results from the reclassification?

c. Comment on why the rank-orderings in (a) are unaffected by the change in (b).

10-11. Five mutually exclusive alternatives are being considered for providing a sewage-treatment facility.

The annual equivalent costs and estimated benefits of the alternatives are as follows:

Alternative	Annual Equivalent (in thousands)	
	Cost	Benefits
A	$1,050	$1,110
B	900	810
C	1,230	1,390
D	1,350	1,500
E	990	1,140

Which plan, if any, should be adopted if the Sewage Authority wishes to invest if, and only if, the B–C ratio is at least 1.0.? (10.9)

10-12. A town in northern Colorado is planning on investing in a water purification system. Three mutually exclusive systems have been proposed, and their capital investment costs and net annual benefits are the following (salvage values are negligible).

EOY	System		
	A	B	C
0	−$160,000	−$245,000	−$200,000
1	80,000	120,000	70,000
2	70,000	100,000	70,000
3	60,000	80,000	70,000
4	50,000	60,000	70,000

If the town's MARR is 10% per year, use the B–C ratio method to determine which system is best. (10.9)

10-13. A nonprofit government corporation is considering two alternatives for generating power:

Alternative A. Build a coal-powered generating facility at a cost of $20,000,000. Annual power sales are expected to be $1,000,000 per year. Annual operating and maintenance costs are $200,000 per year. A benefit of this alternative is that it is expected to attract new industry, worth $500,000 per year, to the region.

Alternative B. Build a hydroelectric generating facility. The capital investment, power sales, and operating costs are $30,000,000, $800,000, and $100,000 per year, respectively. Annual benefits of this alternative are as follows:

Flood-control savings	$600,000
Irrigation	$200,000
Recreation	$100,000
Ability to attract new industry	$400,000

The useful life of both alternatives is 50 years. Using an interest rate of 5%, determine which alternative (if either) should be selected according to the conventional B–C-ratio method. (10.9)

10-14. A new storm drainage system must be constructed right away to reduce periodic flooding that occurs in a city that is in a valley. Five mutually exclusive designs have been proposed, and their present worth (in thousands of dollars) of costs and benefits are the following. (10.9)

	System				
	1	2	3	4	5
PW of costs	$1,000	$4,000	$4,000	$10,000	$12,000
PW of benefits	8,000	8,000	14,000	16,000	24,000

a. Which system has the greatest B–C ratio?

b. Which system has the largest incremental B–C ratio (based on differences between alternatives)?

c. Which system should be chosen?

10-15. Refer to Example 10-7. At a public hearing, someone argued that the market values cannot be adequately estimated 50 years into the future. How is the decision changed if the market values are treated as $0 for all three alternatives? (10.9)

10-16. Consider the mutually exclusive alternatives in Table P10-16. Which alternative would be chosen according to these decision criteria?

a. Maximum benefit

b. Minimum cost

c. Maximum benefits minus costs

d. Largest investment having an incremental B–C ratio larger than one

e. Largest B–C ratio

Which project should be chosen? (10.9)

TABLE P10-16 Mutually Exclusive Alternatives for Problem 10-16

Alternative	Equivalent Annual Cost of Project	Expected Annual Flood Damage	Annual Benefits
I. No flood control	0	$100,000	0
II. Construct levees	$30,000	80,000	$112,000
III. Build small dam	$100,000	5,000	110,000

TABLE P10-18 Costs and Benefits for Problem 10-18*

		Costs		Benefits		
Option	Dam Sites	Construction	Annual Maintenance	Annual Flood	Annual Fire	Annual Recreation
A	1	$3,120,000	$52,000	$520,000	$52,000	$78,000
B	1 and 2	3,900,000	91,000	630,000	104,000	78,000
C	1, 2 and 3	7,020,000	130,000	728,000	156,000	156,000
D	1, 2, 3 and 4	9,100,000	156,000	780,000	182,000	182,000

* Fashioned after a problem from James L. Riggs, *Engineering Economics* (New York: McGraw-Hill, 1977), 432–434.

10-17. Four mutually exclusive projects are being considered for a new 2-mile jogging track. The life of the track is expected to be 80 years, and the sponsoring agency's MARR is 12% per year. Annual benefits to the public have been estimated by an advisory committee and are shown below. Use the B–C method (incrementally) to select the best jogging track. (10.9)

	Alternative			
	A	B	C	D
Initial cost	$62,000	$52,000	$150,000	$55,000
Annual benefits	$10,000	$8,000	$20,000	$9,000
B–C ratio	1.34	1.28	1.11	1.36

10-18. A river that passes through private lands is formed from four branches of water that flow from a national forest. Some flooding occurs each year, and a major flood generally occurs every few years. If small earthen dams are placed on each of the four branches, the chances of major flooding would be practically eliminated. Construction of one or more dams would reduce the amount of flooding by varying degrees.

Other potential benefits from the dams are the re-duction of damages to fire and logging roads in the forest, the value of the dammed water for protection against fires, and the increased recreational use. Table P10-18 contains the estimated benefits and costs associated with building one or more dams.

The equation used to calculate B–C ratios is as follows:

$$\text{B–C ratio} = \frac{\text{Annual flood and fire savings} + \text{Recreational benefits}}{\text{Equivalent annual construction costs} + \text{Maintenance}}$$

Benefits and costs are to be compared by using the AW method with an interest rate of 8%; useful lives of the dams are 100 years.

a. Which of the four options would you recommend? Explain why.

b. If fire benefits are reclassified as reduced costs, would the choice in Part (a) be affected? Show your work.

10-19. A state-sponsored Forest Management Bureau is evaluating alternative routes for a new road into a formerly inaccessible region. Three mutually exclusive plans for routing the road provide different benefits, as indicated in Table P10-19 (page 462). The roads are assumed to have an economic life of 50 years, and MARR is 8% per year. Which route should be selected according to the B–C ratio method? (10.9)

TABLE P10-19   Mutually Exclusive Plans for Problem 10-19

Route	Construction Costs	Annual Maintenance Cost	Annual Savings in Fire Damage	Annual Recreational Benefit	Annual Timber Access Benefit
A	$185,000	$2,000	$5,000	$3,000	$500
B	220,000	3,000	7,000	6,500	1,500
C	290,000	4,000	12,000	6,000	2,800

10-20. The city of Oak Ridge is evaluating three mutually exclusive landscaping plans for refurbishing a public greenway. Benefits to the community have been estimated by a landscaping committee, and the costs of planting trees and shrubbery, as well as maintaining the greenway, are summarized below. The city's discount rate is 8% per year, and the planning horizon is 10 years. (10.9)

	Landscaping Plan		
	A	B	C
Initial planting cost	$75,000	$50,000	$65,000
Annual maintenance expense	4,000	5,000	4,700
Annual community benefits	20,000	18,000	20,000

a. Use the B–C ratio method to recommend the best plan when annual maintenance expenses offset annual benefits in the numerator.

b. Repeat Part (a) when annual maintenance expenses add to total costs in the denominator. Which plan is best?

c. Should the recommendations in Part (a) and (b) be the same? Why or why not?

10-21. An area on the Colorado River is subject to periodic flood damage that occurs, on the average, every two years and results in a $2,000,000 loss. It has been proposed that the river channel be straightened and deepened, at a cost of $2,500,000, to reduce the probable damage to not over $1,600,000 for each occurrence during a period of 20 years before it would have to be deepened again. This procedure would also involve annual expenditures of $80,000 for minimal maintenance. One legislator in the area has proposed that a better solution would be to construct a flood-control dam at a cost of $8,500,000, which would last indefinitely, with annual maintenance costs of not over $50,000. He estimates that this project would reduce the probable annual flood damage to not over $450,000. In addition, this solution would provide a substantial amount of irrigation water that would produce annual revenue of $175,000 and recreational facilities, which he estimates would be worth at least $45,000 per year to the adjacent populace. A second legislator believes that the dam should be built and that the river channel also should be straightened and deepened, noting that the total cost of $11,000,000 would reduce the probable annual flood loss to not over $350,000 while providing the same irrigation and recreational benefits. If the state's capital is worth 10%, determine the B–C ratios and the incremental B–C ratio. Recommend which alternative should be adopted. (10.9)

10-22. Ten years ago, the port of Secoma built a new pier containing a large amount of steel work, at a cost of $300,000, estimating that it would have a life of 50 years. The annual maintenance cost, much of it for painting and repair caused by environmental damage, has turned out to be unexpectedly high, averaging $27,000. The port manager has proposed to the port commission that this pier be replaced immediately with a reinforced concrete pier at a construction cost of $600,000. He assures them that this pier will have a life of at least 50 years, with annual maintenance costs of not over $2,000. He presents the information in Table P10-22 as justification for the replacement, having determined that the net market value of the existing pier is $40,000.

He has stated that, because the port earns a net profit of over $3,000,000 per year, the project could be financed out of annual earnings. Thus, there would be no interest cost, and an annual savings of $19,000 would be obtained by making the replacement. (10.9)

a. Comment on the port manager's analysis.

b. Make your own analysis and recommendation regarding the proposal.

10-23. You have been requested to recommend one of the mutually exclusive industrial sanitation control

TABLE P10-22 Pier Replacement Cost for Problem 10-22

Annual Cost of Present Pier		Annual Cost of Proposed Pier	
Depreciation ($300,000/50)	$6,000	Depreciation ($600,000/50)	$12,000
Maintenance cost	27,000	Maintenance cost	2,000
Total	$33,000	Total	$14,000

TABLE P10-25 Bridge Design Information for Problem 10-25

	Bridge Design		
	A	B	C
Capital investment	$17,000,000	$14,000,000	$12,500,000
Annual maintenance cost*	12,000	17,500	20,000
Resurface (every fifth year)*	–	40,000	40,000
Resurface (every seventh year)*	40,000	–	–
Bridge replacement cost	3,000,000	3,500,000	3,750,000
Annual benefit	2,150,000	1,900,000	1,750,000
Useful life of bridge (years)**	35	25	25

* Cost not incurred in last year of bridge's useful life.

** Applies to roadbed only; structural portion of bridge has indefinite useful life.

systems that are given below. If MARR is 15% per year, which system would you select? Use the B–C method. (10.9)

	Alternative	
	Gravity-fed	Vacuum-led
Capital investment	$24,500	$37,900
Annual receipts less expenses	8,000	8,000
Life in years	5	10

10-24. In the aftermath of Hurricane Thelma, the U.S. Army Corps of Engineers is considering two alternative approaches to protect a freshwater wetland from the encroaching seawater during high tides. The first alternative, the construction of a 5-mile long, 20-foot-high levee, would have an investment cost of $25,000,000 with annual upkeep costs estimated at $725,000. A new roadway along the top of the levee would provide two major benefits: (1) improved recreational access for fishermen and (2) reduction of the driving distance between the towns at opposite ends of the proposed levee by 11 miles. The annual benefit for the levee has been estimated at $1,500,000. The second alternative, a channel-dredging operation, would have

an investment cost of $15,000,000. The annual cost of maintaining the channel is estimated at $375,000. There are no documented benefits for the channel-dredging project. Using a MARR of 8% and assuming a 25-year life for either alternative, apply the incremental B–C ratio ($\Delta B/\Delta C$) method to determine which alternative should be chosen. (Note: The null alternative, Do nothing, is not a viable alternative.) (10.9)

10-25. *Extended Learning Exercise* The Fox River is bordered on the east by Illinois Route 25 and on the west by Illinois Route 31. Along one stretch of the river, there is a distance of 16 miles between adjacent crossings. An additional crossing in this area has been proposed, and three alternative bridge designs are under consideration. Two of the designs have 25-year useful lives, and the third has a useful life of 35 years. Each bridge must be resurfaced periodically, and the roadbed of each bridge will be replaced at the end of its useful life, at a cost significantly less than initial construction costs. The annual benefits of each design differ on the basis of disruption to normal traffic flow along Routes 25 and 31. Given the information in Table P10-25, use the B–C ratio method to determine which bridge design should be selected. Assume that the selected design will be used indefinitely, and use a MARR of 10% per year. (10.9)

FE Practice Problems

The city council of Morristown is considering the purchase of one new fire truck. The options are Truck X and Truck Y. The appropriate financial data are as follows:

	Truck X	Truck Y
Capital investment	$50,000	$64,000
Maintenance cost per year	$6,000	$5,000
Useful life	6 years	6 years
Reduction in fire damage per year	$20,000	$22,000

The purchase is to be financed by money borrowed at 12% per year. Use this information to answer problems 10-26 and 10-27.

10-26. What is the conventional B–C ratio for Truck X? (10.7)

(a) 1.41   (b) 0.87   (c) 1.64   (d) 1.10   (e) 1.15

10-27. Which fire truck should be purchased? (10.9)

(a) Neither Truck X nor Truck Y

(b) Truck X

(c) Truck Y

(d) Both Truck X and Truck Y

10-28. A state government is considering construction of a flood control dike having a life span of 15 years. History indicates that a flood occurs every five years and causes $600,000 in damages, on average. If the state uses a MARR of 12% per year and expects every public-works project to have a B–C ratio of at least 1.0, what is the maximum investment that will be allowed for the dike? Assume that the flood occurs in the middle of each five-year period. (10.7)

(a) $1,441,000   (b) $643,000   (c) $843,000
(d) $4,087,000   (e) $1,800,000

A flood control project with a life of 16 years will require an investment of $60,000 and annual maintenance costs of $5,000. The project will provide no benefits for the first two years but will save $24,000 per year in flood damage starting in the third year. The appropriate MARR is 12% per year. Use this information to answer problems 10-29 and 10-30. Select the closest answer. (10.7)

10-29. What is the conventional B–C ratio for the flood control project?

(a) 1.53   (b) 1.33   (c) 1.76   (d) 2.20   (e) 4.80

10-30. What is the modified B–C ratio for the flood control project?

(a) 1.53   (b) 1.33   (c) 1.76   (d) 2.20   (e) 4.80

CHAPTER 11

Breakeven and Sensitivity Analysis

Our aim in Chapter 11 is to illustrate breakeven and sensitivity methods for investigating variability in outcomes of engineering projects.

Renting versus Purchasing a Home

A common situation faced by young families is whether to rent or buy a home. Many of the decision factors are speculative, such as the future resale value of the house and how long the family will be in a particular area. For example, what will be the resale value of a home currently valued at $150,000 five years from now? If you plan to stay in your current location for only three years, should you rent or purchase a home? In this chapter, you will learn how to evaluate decision problems such as this by examining the sensitivity of the decision to changes in the estimates of selected factor estimates.

Predictions are usually difficult, especially about the future.

—Yogi Berra

11.1 Introduction

There is much common wisdom relating to uncertainty in Yogi Berra's observation about the future! In previous chapters, we presented specific assumptions concerning revenues, costs, and other quantities important to an engineering economy study. It was assumed that a high degree of confidence could be placed in all estimated values. That degree of confidence is sometimes called *assumed certainty*. Decisions made solely on the basis of this kind of analysis are sometimes called *decisions under certainty*. The term is rather misleading in that there rarely is a case in which the best of estimates of quantities can be assumed as certain.

In virtually all situations, there is doubt as to the ultimate economic results that will be obtained from an engineering project. The motivation for breakeven and sensitivity analysis is to establish the bounds of error in our estimates such that another alternative being considered may become a better choice than the one we recommended under assumed certainty. Thus, this chapter deals with nonprobabilistic techniques for dealing with risk and uncertainty.

Breakeven analysis determines the value of a critical factor at which economic trade-offs are balanced. It is also helpful in some situations to investigate how sensitive a project's economics are to variations in its estimates of life, interest rate, initial capital investment, and so on. *Sensitivity*, in general, means the relative magnitude of change in the measure of merit (such as present worth [PW] or internal rate of return [IRR]) caused by one or more changes in estimated factor values. The specific factors of concern will vary with each project, but one or more of them will normally need to be further analyzed before the best decision can be made. Simply stated, engineering economy studies focus on the future, and lack of knowledge about the estimated economic results cannot be avoided.

In engineering economy studies, breakeven and sensitivity analysis are general methodologies, readily available, that provide information about the potential impact on equivalent worth due to variability in selected factor estimates. Their routine use is fundamental to developing economic information useful in the decision process.

11.2 Breakeven Analysis

When the selection between two engineering project alternatives (or outcomes) is heavily dependent on a single factor, we can solve for the value of that factor at which the conclusion is a standoff. That value is known as the *breakeven point*, that is, the value at which we are indifferent between the two alternatives. (The use of breakeven points with respect to production and sales volumes was discussed in Chapter 2.) Then, if the best estimate of the actual outcome of the common factor is higher or lower than the breakeven point, and assumed certain, the best alternative becomes apparent.

In mathematical terms, we have

$$EW_A = f_1(y) \quad \text{and} \quad EW_B = f_2(y),$$

where EW_A = an equivalent worth (PW, annual worth [AW], or future worth [FW]) calculation for the net cash flow of Alternative A;

 EW_B = the same equivalent-worth calculation for the net cash flow of Alternative B;

 y = a common factor of interest affecting the equivalent-worth values of Alternative A and Alternative B.

Therefore, the breakeven point between Alternative A and Alternative B is the value of factor y for which the two equivalent-worth values are equal. That is, $EW_A = EW_B$, or $f_1(y) = f_2(y)$, which may be solved for y.

Similarly, when the economic acceptability of an engineering project depends upon the value of a single factor, say z, mathematically we can set an equivalent worth of the project's net cash flow for the analysis period equal to zero [EW $= f(z) = 0$] and solve for the breakeven value of z. That is, the breakeven value of z is the value of z at which we would be indifferent (economically) between accepting and rejecting the project. Then, if the best estimate of the value of z is higher or lower than the breakeven point value, and assumed certain, the economic acceptability of the project is known.

The following are examples of common factors for which breakeven analyses might provide useful insights into the decision problem:

1. *Annual revenue and expenses.* Solve for the annual revenue required to equal (breakeven with) annual expenses. Breakeven annual expenses of an alternative can also be determined in a pairwise comparison when revenues are identical for both alternatives being considered.

2. *Rate of return.* Solve for the rate of return on the increment of invested capital at which two given alternatives are equally desirable.

3. *Market (or salvage) value.* Solve for the future resale value that would result in indifference as to preference for an alternative.

4. *Equipment life.* Solve for the useful life required for an engineering project to be economically justified.

5. *Capacity utilization.* Solve for the hours of utilization per year, for example, at which an alternative is justified or at which two alternatives are equally desirable.

The usual breakeven problem involving two alternatives can be most easily approached mathematically by equating an equivalent worth of the two alternatives expressed as a function of the factor of interest. Using the same approach for the economic acceptability of an engineering project, we can mathematically equate an equivalent worth of the project to zero as a function of the factor of concern. In breakeven studies, project lives may or may not be equal, so care should be taken to determine whether the coterminated or repeatability assumption best fits the situation.

The following examples illustrate both mathematical and graphical solutions to typical breakeven problems involving one or more engineering projects.

| EXAMPLE 11-1 | Single Project Breakeven Analysis: Solar Panels |

A homeowner is considering whether to invest in solar panels. This homeowner has gathered the following information for her town:

- The average cost for residential energy is $0.124/kWh.
- The government subsidizes 70% of whatever is invested by the household.
- Each square foot of panel produces 19.13 kWh every year.
- The cost of installation per square foot of solar panel is $93.91.

The house has 350 square feet of roof space. What is the breakeven point in years for the installation of solar panels? Assume the homeowner's interest rate is 6% per year.

Solution

The breakeven point is where the annual savings from the solar panels offsets the initial investment. The initial investment is calculated as follows.

$$350 \text{ square feet of roof space} \times \$93.91/\text{square foot} = \$32,869$$

With a 70% government subsidy, the purchase price is $(0.3)(\$32,869) = \$9,861$. To find the breakeven point in years (X), we set the purchase price equal to the potential energy savings.

$\$9,861 = (350 \text{ square feet})(19.13 \text{ kWh/sq ft-year})(\$0.124/\text{kWh})(P/A, 6\%, X)$

$\$9,861 = (\$830.24)(P/A, 6\%, X)$

$11.87 = (P/A, 6\%, X)$

From Table C-9 $(i = 6\%)$, we see that $21 \leq X \leq 22$ years. It would take approximately 22 years of energy savings to offset the installation cost of the solar panels. A graph of this breakeven situation is shown in Figure 11-1.

Figure 11-1 Graphical Plot of the Breakeven Point for the Solar Panel Investment

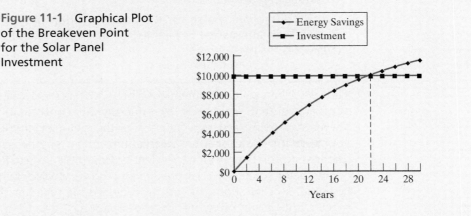

EXAMPLE 11-2   **Two Alternative Breakeven Analysis: Hybrid Vehicles**

Gas-electric (so-called hybrid) vehicles save on gasoline consumption by shutting off the vehicle's engine while idling, giving the vehicle a boost of electric power during acceleration, and capturing electrical energy while braking. In addition to environmental benefits, the primary monetary benefit to the owner is reduced fuel cost as a result of improved gas mileage. The trade-off, however, is that the purchase price of the hybrid vehicle is higher than that of a standard gasoline-only fueled vehicle.

Consider a hybrid vehicle with a sticker price of $31,500. This vehicle will average 30 miles per gallon of gasoline. A tax credit* of $1,500 for the hybrid vehicle effectively reduces its sticker price to $30,000. A comparably equipped gasoline-only vehicle will cost $28,000 and will average 25 miles per gallon of gasoline. Assuming an interest rate of 3% per year and a study period of five years, find the breakeven cost of gasoline ($/gal) if the vehicle will be driven 18,000 miles each year.

Solution

To find the cost of gasoline that makes you indifferent between the two vehicles, you need to develop an equivalent-worth equation in terms of the factor of interest—the cost of gasoline. In this simple example, we are only considering the sticker price (investment cost) and fuel cost (annual expenses). By default we are assuming that the maintenance of the two vehicles will be the same as will the future resale value.

In this example, we develop equivalent uniform annual cost (EUAC) expressions for each of the vehicles. Letting X = cost of gasoline, we get the following EUAC equations.

Hybrid: $EUAC_H(3\%) = (\$31{,}500 - \$1{,}500)(A/P, 3\%, 5)$

$$+ (\$X/\text{gal}) \left[\frac{18{,}000 \text{ mi/year}}{30 \text{ mi/gal}} \right]$$

Gas-only: $EUAC_G(3\%) = \$28{,}000(A/P, 3\%, 5) + (\$X/\text{gal}) \left[\dfrac{18{,}000 \text{ mi/year}}{25 \text{ mi/gal}} \right]$

Setting $EUAC_H(3\%) = EUAC_G(3\%)$ and solving for X, we find the breakeven cost of gasoline to be

$$X = \$3.64/\text{gal}.$$

Figure 11-2 shows the breakeven chart for these vehicles as a function of the cost of gasoline. If our best estimate of the average cost of gasoline over the next five years is less than $3.64 per gallon, then purchasing a traditional gasoline-only vehicle is more economical. The hybrid would be the vehicle of choice if the cost of gasoline is projected to be higher than $3.64 per gallon.

---

* The Energy Policy Act of 2005 provides for certain tax credits for purchasers of hybrid and other alternatively fueled vehicles. Very specific requirements must be met. The $1,500 used in the example is for demonstration purposes only.

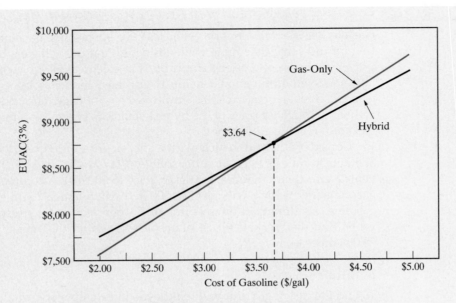

Figure 11-2 Breakeven Chart for Hybrid versus Gas-Only Vehicle

Comment

In this example, we only considered the easily measured monetary benefit of the hybrid vehicle. Other important factors that may influence this decision are reduced emissions and more efficient use of a scarce resource.

EXAMPLE 11-3 **Three Alternative Breakeven Analysis: Hours of Operation**

Suppose that there are three alternative electric motors capable of providing 100 horsepower (hp) output. These motors differ in their purchase price, maintenance costs, and efficiency. The efficiency of the motor impacts the energy cost of the motor. Specific data are listed below.

	Alpha	Beta	Gamma
Purchase price	$10,000	$15,000	$20,000
Annual maintenance	$800	$500	$250
Efficiency	74%	86%	92%

If we look at the annual cost of owning and operating a motor, a factor of interest is the hours of operation, which drives the energy cost of the motor. Over what range of hours operated per year (at full load) is the Beta motor the best choice? How many hours would the motor have to be operated each year before the Gamma motor is the most economical? Use a spreadsheet to answer these questions. Assume that the minimum attractive rate of return (MARR) = 15% per year, the study period is 10 years, and none of the motors will have a market value after 10 years.

Spreadsheet Solution

Because we are dealing with costs only in this example (revenues are assumed to be equal), we use the EUAC metric to solve for the breakeven points between the motors. The basic equation used to develop the spreadsheet model is

$$\text{EUAC}(15\%) = \text{Purchase price } (A/P, 15\%, 10) + \text{Annual maintenance}$$

$$+ \text{Annual energy cost.}$$

The energy cost portion of the equation can be expressed in terms of the hours of operation.

$$\text{Energy cost} = (\text{Operating hours/year})(\$0.05/\text{kWh}) \left[\frac{(100 \text{ hp})(0.746 \text{ kW/hp})}{\text{motor efficiency}} \right]$$

For instance, the annual energy expense of the Alpha motor is

$$(X \text{ hours/year})(\$0.05/\text{kWh}) \left[\frac{(100 \text{ hp})(0.746 \text{ kW/hp})}{0.74} \right] = \$5.04 \ X$$

where X is the annual hours of operation. We can now solve for the breakeven point between each pair of motors.

Figure 11-3 shows the spreadsheet model. The Goal Seek function was used to solve for the individual breakeven points between motors (cells B18, B21, and B24). In Goal Seek (under the Tools menu), the spreadsheet software changes the value of a single cell (hours of operation in this example) to find the value that sets a different cell (difference in EUAC) equal to a target amount ($0).

Goal Seek	
Set cell:	E18
To value:	0
By changing cell:	B18

The breakeven graph is shown in Figure 11-4. The numbers at the bottom of Figure 11-3 (cells A27:D38) are used to generate the breakeven chart. Although we solved for the exact values of the breakeven points, the graph is useful in reminding us which motor is preferred over which range of operating hours. In summary, the Alpha motor is preferred if annual operating hours are less than 1,694; the Gamma motor would be selected if annual operating hours are greater than 4,389; otherwise the Beta motor is the most economical choice.

	A	B	C	D	E
1	MARR	15%			
2	Electricity ($/kWh)	$ 0.05			
3	Useful Life (years)	5			
4					
5		**Alpha**	**Beta**	**Gamma**	
6	Purchase Price	$10,000	$15,000	$20,000	
7	Annual Maintenance	$800	$500	$250	
8	Efficiency	74%	86%	92%	
9					
10		**Alpha**	**Beta**	**Gamma**	
11	CR Amount	$2,983	$4,475	$5,966	
12	Annual Maintenance	$800	$500	$250	
13	Sub-total	$3,783	$4,975	$6,216	
14					
15	Energy Expense ($/hr)	5.04	4.34	4.05	
16					
17	Alpha - Beta	**Hours**	**Alpha**	**Beta**	**Difference**
18	Breakeven Point	1694	$12,323	$12,323	$0
19					
20	Alpha - Gamma	**Hours**	**Alpha**	**Gamma**	**Difference**
21	Breakeven Point	2467	$16,219	$16,219	$0
22					
23	Beta - Gamma	**Hours**	**Beta**	**Gamma**	**Difference**
24	Breakeven Point	4389	$24,012	$24,012	$0
25					
26					
27		**Hours**	**Alpha**	**Beta**	**Gamma**
28		1500	$11,344	$11,481	$12,298
29		2000	$13,864	$13,649	$14,325
30		2500	$16,385	$15,818	$16,352
31		3000	$18,905	$17,986	$18,379
32		3500	$21,425	$20,155	$20,407
33		4000	$23,945	$22,324	$22,434
34		4500	$26,466	$24,492	$24,461
35		5000	$28,986	$26,661	$26,488
36		5500	$31,506	$28,829	$28,515
37		6000	$34,026	$30,998	$30,542
38		6500	$36,547	$33,167	$32,570

Figure 11-3   Spreadsheet Solution to Example 11-3

Figure 11-4   Graphical Plot of Breakeven Analysis of Example 11-3

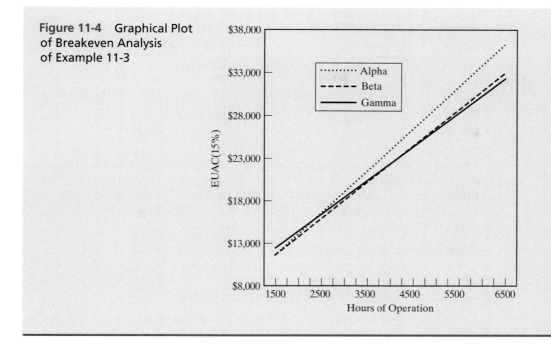

11.3 Sensitivity Analysis

Sensitivity analysis is used to explore what happens to a project's profitability when the estimated value of study factors are changed. For example, if annual expenses turn out to be 10% higher than expected, will the project still be acceptable? Sometimes sensitivity is specifically defined to mean the percent change in one or more factors that will reverse a decision among project alternatives or reverse a decision about the economic acceptability of a single project. This percent change is called sensitivity with respect to *decision reversal*. Example 11-4 shows the calculation of decision reversal points.

Another useful sensitivity tool is the spiderplot. This approach makes explicit the impact of variability in the estimates of each factor of concern on the economic measure of merit. Example 11-5 demonstrates this technique by plotting the results of changes in the estimates of several factors, separately, on the PW of an engineering project. Spreadsheet applications provide an excellent capability to answer *what-if* questions and are useful for generating the spiderplot.

| EXAMPLE 11-4 | Decision Reversal |

Consider a proposal to enhance the vision system used by a postal service to sort mail. The new system is estimated to cost $1.3 million and will incur an additional $100,000 per year in maintenance costs. The system will produce annual savings of $500,000 each year (primarily by decreasing the percentage of misdirected mail and reducing the amount of mail that must be sorted manually).

The MARR is 12% per year, and the study period is five years at which time the system will be technologically obsolete (worthless). The PW of this proposal is

$$PW(12\%) = -\$1,300,000 + (\$500,000 - \$100,000)(P/A, 12\%, 5) = \$141,910.$$

Determine how sensitive the decision to invest in the system is to the estimates of investment cost and annual savings.

Solution

Our initial appraisal of the project shows it to be a profitable venture. Now let's look at what happens if we are wrong in our estimates. Essentially, we need to find the breakeven investment cost and the breakeven annual savings. Let x be the percent change in investment cost that would cause us to reverse our decision. Then

$$PW(12\%) = 0 = -\$1,300,000(1 + x) + (\$500,000 - \$100,000)(P/A, 12\%, 5)$$

$$x = +10.9\%.$$

Similarly, let y be the percent change in annual savings.

$$PW(12\%) = 0 = -\$1,300,000 + [\$500,000(1 + y) - \$100,000](P/A, 12\%, 5)$$

$$y = -7.9\%$$

If the investment cost increases by more than 10.9%, the new vision system would no longer be acceptable. Likewise, if the estimate of annual savings is lower by more than 7.9% of its most likely value, the project would be a no-go. The responsible engineer now needs to take a hard look at how the original estimates were made and decide whether or not more detailed estimates are required.

Note

When examining the impact of changing one estimate value, all other factor values are held at their original amounts.

EXAMPLE 11-5 **Spiderplot for the Proposed Vision System**

We will now further explore the sensitivity of the proposed vision system by creating a spiderplot. The best (most likely) estimates for the vision system described in Example 11-4 are listed below.

Capital investment, I	$1,300,000
Annual savings, A	500,000
Annual expenses, E	100,000
Useful life, N	5 years

We are interested in investigating the sensitivity of the PW of the system over a range of ±40% changes in all of the above estimates. Recall that MARR = 12% per year.

Spreadsheet Solution

Figure 11-5(a) shows the table of PW values as each factor of the PW calculations is varied over a range of ±40% from the most likely estimate. Each column has a unique formula that refers to the factors located in the range C2:C5 to determine PW. The particular factor of interest—for example, the capital investment in column B—is multiplied by the factor (1 + percent change as a decimal) to create the table. You can verify your formulas by noting that all columns are equal at the most likely value (percent change = 0). The cell formulas in the range B9:E9 (listed below) are simply copied down the columns to complete the spreadsheet.

Cell	Contents
B9	$= -\$C\$2\,^*(1 + A9) + PV(\$C\$6, \$C\$5, -(\$C\$3 - \$C\$4))$
C9	$= -\$C\$2 + PV(\$C\$6, \$C\$5, -(\$C\$3\,^*(1 + A9) - \$C\$4))$
D9	$= -\$C\$2 + PV(\$C\$6, \$C\$5, -(\$C\$3 - \$C\$4\,^*(1 + A9)))$
E9	$= -\$C\$2 + PV(\$C\$6, \$C\$5\,^*(1 + A9), -(\$C\$3 - \$C\$4))$

The spiderplot in Figure 11-5(b) shows the sensitivity of the PW to percent deviation changes in each factor's best estimate. The other factors are assumed to remain at their most likely values. The PW of this project based on the best estimates of the factors is

$$PW(12\%) = -\$1,300,000 + (\$500,000 - \$100,000)(P/A, 12\%, 5) = \$141,910.$$

	A	B	C	D	E
1	Most Likely Values				
2	Capital Investment		$1,300,000		
3	Annual Savings		$500,000		
4	Annual Maintenance		$100,000		
5	Useful Life (years)		5		
6	MARR		12%		
7					
8	% Change in Factor	Capital Investment	Annual Savings	Annual Maintenance	Useful Life
9	−40%	$661,910	−$579,045	$286,102	−$339,267
10	−30%	$531,910	−$398,806	$250,054	−$208,564
11	−20%	$401,910	−$218,567	$214,006	−$85,060
12	−10%	$271,910	−$38,328	$177,958	$31,640
13	0%	$141,910	$141,910	$141,910	$141,910
14	10%	$11,910	$322,149	$105,863	$246,107
15	20%	−$118,090	$502,388	$69,815	$344,563
16	30%	−$248,090	$682,627	$33,767	$437,595
17	40%	−$378,090	$862,866	−$2,281	$525,503

(a) Table of PW Values for Varied Factor Values

Figure 11-5 Spreadsheet Solution for Example 11-5

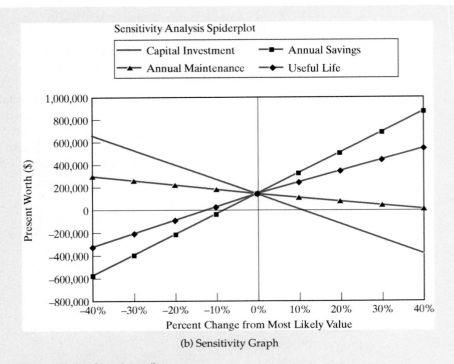

(b) Sensitivity Graph

Figure 11-5 *(continued)*

This value of the highly favorable PW occurs at the common intersection point of the percent deviation graphs for the four separate project factors. The relative degree *of sensitivity of the PW to each factor is indicated by the slope of the curves* (the steeper the slope of a curve the more sensitive the PW is to the factor). Also, the intersection of each curve with the abscissa (PW = 0) shows the decision reversal point—the percent change from each factor's most likely value at which the PW is zero. Based on the spiderplot, we see that the PW is insensitive to annual maintenance expenses but quite sensitive to changes in the capital investment, annual savings, and useful life. Such an analytical tool as the spiderplot can assist with the insightful exploration of the variable aspects of an engineering economy study.

For additional information, consider using the sensitivity graph technique to compare two or more mutually exclusive project alternatives. If only two alternatives are being compared, a spiderplot based on the incremental cash flow between the alternatives can be used to aid in the selection of the preferred alternative. Extending this approach to three alternatives, two sequential paired comparisons can be used to help select the preferred alternative. Another approach is to plot (overlay) in the same figure a sensitivity graph for each alternative. Obviously, if this later approach is used for the comparison of more than, say, three alternatives (with two or three factors each), interpreting the results may become a problem.

EXAMPLE 11-6	Rent or Purchase a Home?

In this example, we return to the decision problem discussed in the chapter opener: Is it more economical to purchase or rent a home? Evaluate the economics of renting versus buying a $150,000 home and living in it for five years. Use the data below in your analysis.*

Rental Option: Rent is $1,200 per month for the first year. The monthly rental fee will increase by $25 for each subsequent year of renting. There is also a $1,200 deposit payable when the lease is signed and refundable when the house is left in good condition. Renter's insurance is $35 per month.

Purchase Option: A $30,000 down payment is made, so $120,000 will be financed with a 30-year mortgage having a 7% annual interest rate. Additional closing costs of $2,000 are paid at the time of purchase. Property taxes and homeowners' insurance total $200 per month, and maintenance is expected to average $50 per month. The resale value of the home after five years is anticipated to be $160,000. The commission paid to the realtor at the time of the sale is expected to be 7% of the selling price.

If your personal interest rate is 10% per year (compounded monthly), is it more economical to rent or purchase this home? Examine the sensitivity of the decision to changes in the resale value of the home, the mortgage rate, and the length of ownership.

Solution

Figure 11-6 shows a spreadsheet solution for this example. For the stated problem data, it is more economical to purchase this home than to rent it (PW of ownership costs < PW of rental costs). The formulas used to compute the values in the highlighted cells are shown below.

Cell	Contents
F8	= C8 + PV(C5/12, C4 * 12,, C8)
F9	= −PV(C5/12, C4 * 12, C9)
F10	= (C10 * 12/C5) * (−PV(C5, C4, 1) + PV(C5, C4,, C4))
F11	= −PV(C5/12, C4 * 12, C11)
F12	= SUM(F8:F11)
F15	= C15
F16	= −PV(C5/12, C4 * 12, C26)
F17	= C16 * C17
F20	= −PV(C5/12, C4 * 12, C20)
F21	= −PV(C5/12, C4 * 12, C21)
F22	= PV(C5/12, C4 * 12,, (C22 − C27))
F23	= −PV(C5/12, C4 * 12,, C23 * C22)
F24	= SUM(F15:F23)
C26	= −PMT(C18/12, C19 * 12, (C16 * (1− C17)))
C27	= −FV(C18/12, C4 * 12,, (C16 * (1− C17))) + FV(C18/12, C4 * 12, C26)
E27	= IF(F12 < F24, "Rent Home", "Purchase Home")
F27	= ABS(F12 − F24)

---

* The data used in this example are for demonstration purposes only. Actual data will be location specific.

	A	B	C	D	E	F	G
1	Purchase or Rent Analysis						
2							
3	Inputs				Present Worth Analysis		
4		Length of ownership/renting (years)	5		Note: cash inflows are negative values; cash outflows are positive values		
5		Personal interest rate (percent)	10.0%				
6							
7		Renting			Renting	Present Worth	
8		Deposit (returned at departure)	$1,200		Deposit	$470.65	
9		Monthly payments (first year)	$1,200		Monthly payment (base)	$56,478.44	
10		Monthly rent change each year	$25		Monthly rent change each year	$2,058.54	
11		Renters insurance (monthly)	$35		Renters insurance	$1,647.29	
12					Renting total cost	$60,654.92	
13							
14		Purchase			Purchase	Present Worth	
15		Application/closing fees	$2,000		Application/closing fees	$2,000.00	
16		Purchase price	$150,000		Monthly mortgage payments	$37,575.25	
17		Down payment percentage	20.0%		Down payment	$30,000.00	
18		Loan interest rate (percent)	7.0%				
19		Loan length (years)	30				
20		Monthly insurance and tax	$200		Monthly insurance and tax	$9,413.07	
21		Monthly maintenance	$50		Monthly maintenance	$2,353.27	
22		Future selling price	$160,000		Equity in house	($28,591.65)	
23		Sales commission/fees (percent)	7.0%		Sales commission/fees	$6,807.23	
24					Owning total cost	$59,557.18	
25	Calculations						
26		Monthly mortgage payment	$798.36		Present worth advantage:		
27		Mortgage balance at time of sale	$112,957.91		Purchase Home	$1,097.75	

Figure 11-6   Spreadsheet Solution for Example 11-6

Now we will examine the sensitivity of this decision to changes in the most likely estimates of the future resale value of the home, the mortgage rate, and the length of ownership. Figure 11-7 is a spiderplot of the incremental PW. The difference in PW of the purchase versus rent alternatives is plotted against the percent change in the factors of interest. The slope of the lines tells us that the decision to purchase rather than rent is most sensitive to changes in the future selling price of the home. The least sensitive of these three factors is the length of ownership. The Goal Seek function of Excel returns the following breakeven values for these factors. Note that all other factors are held at their most likely values when determining the breakeven point of an individual factor.

Factor	Breakeven Value	Percent Change from Most Likely
Future selling price	$158,058	−1.21%
Mortgage rate	7.2%	+2.86%
Length of ownership	3.95 years	−21.00%

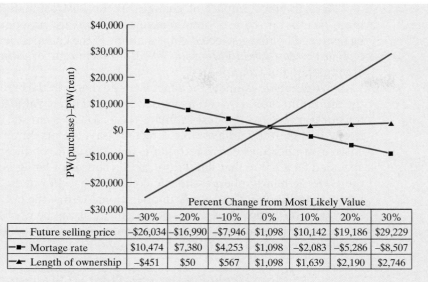

	−30%	−20%	−10%	0%	10%	20%	30%
Future selling price	−$26,034	−$16,990	−$7,946	$1,098	$10,142	$19,186	$29,229
Mortage rate	$10,474	$7,380	$4,253	$1,098	−$2,083	−$5,286	−$8,507
Length of ownership	−$451	$50	$567	$1,098	$1,639	$2,190	$2,746

Figure 11-7   Sensitivity Graph for Example 11-6

In reality, there are more factors to consider in the decision to rent or purchase a home. Many other mortgage options exist, such as different repayment periods (15 or 20 years) and the option of paying points to reduce the interest rate. Also, interest paid on home loans is tax deductible. These factors are considered in Problem 11-23. The following Web sites offer more comprehensive calculators for the rent versus purchase decision.

http://calculators.interest.com/index.asp
http://www.mortgage-net.com/calculators

11.4 Multiple Factor Sensitivity Analysis

In the previous section, we looked at the impact of changing a single factor estimate on the equivalent worth of a project (all other values were held constant). We are often concerned about the *combined* effects of changes in two or more project factors. In this situation, the following approach can be used to develop additional information.

1. Develop a sensitivity graph (spiderplot) for the project as discussed in Section 11.3. For the most sensitive factors, try to develop improved estimates and reduce the range of variability before proceeding further with the analysis.

2. Select the most sensitive project factors based on the information in the sensitivity graph. Analyze the combined effects of these factors on the project's economic measure of merit by determining the impact of selected combinations of three or more factors. (These combinations are sometimes called *scenarios*.)

The combined impact of changes in the best estimate values for three or more factors on the economic measure of merit for an engineering project can be analyzed by using selected combinations of the changes. This approach and the *Optimistic-Most Likely-Pessimistic (O-ML-P)* technique of estimating factor values are illustrated in Example 11-7.

An optimistic estimate for a factor is one that is in the favorable direction (say, the minimum capital investment cost). The most likely value for a factor is defined for our purposes as the best estimate value. The pessimistic estimate for a factor is one that is in the unfavorable direction (say, the maximum capital investment cost). In applications of this technique, the optimistic condition for a factor is often specified as a value that has 19 out of 20 chances of being better than the actual outcome. Similarly, the pessimistic condition has 19 out of 20 chances of being worse than the actual outcome. In operational terms, the optimistic condition for a factor is the value when things occur as well as can be reasonably expected, and the pessimistic estimate is the value when things occur as detrimentally as can be reasonably expected.

| EXAMPLE 11-7 | **Optimistic-Most Likely-Pessimistic (O-ML-P) Scenarios** |

Consider a proposed ultrasound inspection device for which the optimistic, and most likely, pessimistic estimates are given in Table 11-1. MARR is 8% per year. Also shown at the end of Table 11-1 are the AWs for all three estimation conditions. Based on this information, analyze the combined effects of uncertainty in the factors on the AW value.

Solution

Step 1: Before proceeding further with the solution, we need to evaluate the two extreme values of the AW. As shown at the end of Table 11-1, the AW for the optimistic estimates is very favorable ($73,995), whereas the AW for the pessimistic estimates is quite unfavorable (−$33,100). If both extreme AW values were positive, we would make a *go* decision with respect to the device without further analysis, because no combination of factor values based on the estimates will result in AW less than zero. By similar reasoning, if all AW values were negative, a *no-go* decision would be made regarding the device. In this example,

TABLE 11-1 **Optimistic, Most Likely, and Pessimistic Estimates and AWs for Proposed Ultrasound Device (Example 11-7)**

	Estimation Condition		
	Optimistic (O)	Most Likely (ML)	Pessimistic (P)
Capital investment, I	$150,000	$150,000	$150,000
Useful life, N	18 years	10 years	8 years
Market value, MV	0	0	0
Annual revenues, R	$110,000	$70,000	$50,000
Annual expenses, E	20,000	43,000	57,000
AW (8%):	+$73,995	+$4,650	−$33,100

however, the decision is sensitive to other combinations of outcomes, and we proceed to Steps 2 and 3.

Step 2: A sensitivity graph (spiderplot) for this situation is needed to show explicitly the sensitivity of the AW to the three factors of concern: useful life, N; annual revenues, R; and annual expenses, E. The spiderplot is shown in Figure 11-8. The curves for N, R, and E plot percent deviation changes from the most likely (best) estimate, over the range of values defined by the optimistic and pessimistic estimates for each factor, versus AW. As additional information, a curve is also shown for MARR versus AW. Based on the spiderplot, AW for the proposed ultrasound device appears very sensitive to annual revenues and quite sensitive to annual expenses and reductions in useful life. Even if we were to significantly change the MARR (8%), however, it would have little impact on the AW.

Step 3: The various combinations of the optimistic, most likely, and pessimistic factor values (outcomes) for annual revenues, useful life, and annual expenses need to be analyzed for their combined impacts on the AW. The results for these 27 ($3 \times 3 \times 3$) combinations are shown in Table 11-2.

Making the AW Results Easier to Interpret

Since the AW values in Table 11-2 result from estimates subject to differing degrees of variation, little information of value would be lost if the numbers were rounded to the nearest thousand dollars. Further, suppose that management is most interested in the number of combinations of outcomes in which AW is,

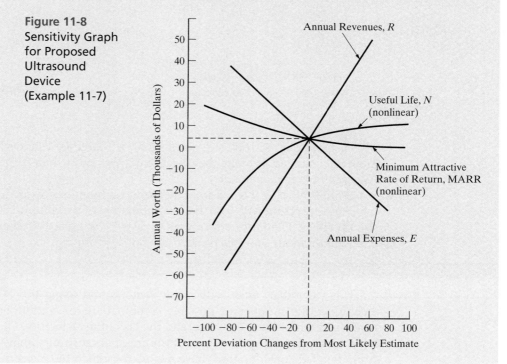

Figure 11-8
Sensitivity Graph for Proposed Ultrasound Device (Example 11-7)

Annual Revenues, R

Useful Life, N (nonlinear)

Minimum Attractive Rate of Return, MARR (nonlinear)

Annual Expenses, E

Annual Worth (Thousands of Dollars)

Percent Deviation Changes from Most Likely Estimate

TABLE 11-2 AWs ($) for All Combinations of Estimated Outcomes[a] for Annual Revenues, Annual Expenses, and Useful Life: Proposed Ultrasound Device (Example 11-7)

Annual Revenues, R	Annual Expenses, E								
	O			ML			P		
	Useful Life, N			Useful Life, N			Useful Life, N		
	O	ML	P	O	ML	P	O	ML	P
O	73,995	67,650	63,900	50,995	44,650	40,900	36,995	30,650	26,900
ML	34,000	27,650	23,900	10,995	4,650	900	−3,005	−9,350	−13,100
P	14,000	7,650	3,900	−9,005	−15,350	−19,100	−23,005	−29,350	−33,100

[a] Estimates: O, optimistic; ML, most likely; P, pessimistic.

TABLE 11-3 Results in Table 11-2 Made Easier to Interpret (AWs in $000s)[a, b]

Annual Revenues, R	Annual Expenses, E								
	O			ML			P		
	Useful Life, N			Useful Life, N			Useful Life, N		
	O	ML	P	O	ML	P	O	ML	P
O	74	68	64	51	45	41	37	31	27
ML	34	28	24	11	5	1	−3	−9	−13
P	14	8	4	−9	−15	−19	−23	−29	−33

[a] Estimates: O, optimistic; ML, most likely; P, pessimistic.
[b] Boxed entries, AW>$50,000 (4 out of 27 combinations); underscored entries, AW<$0 (9 out of 27 combinations).

say, (1) more than $50,000 and (2) less than $0. Table 11-3 shows how Table 11-2 might be changed to make it easier to interpret and use in communicating the AW results to management.

From Table 11-3, it is apparent that four combinations result in AW > $50,000, while nine produce AW < $0. Each combination of conditions is not necessarily equally likely. Therefore, statements such as "There are 9 chances out of 27 that we will lose money on this project" are not appropriate.

It is clear that, even with a few factors, and using the O-ML-P estimating technique, the number of possible combinations of conditions in a sensitivity analysis can become quite large, and the task of investigating all of them might be quite time-consuming. One goal of progressive sensitivity analysis is to eliminate from detailed consideration those factors for which the measure of merit is quite insensitive, highlighting the conditions for other factors to be studied further in accordance with the degree of sensitivity of each. Thus, the number of combinations of conditions included in the analysis can perhaps be kept to a manageable size.

11.5 CASE STUDY—Analysis of a Business Venture

A small group of investors is considering starting a small premixed-concrete plant in a rapidly developing suburban area about 15 miles from a large city. The group believes that there will be a good market for premixed concrete in this area for at least the next 10 years and that if they establish such a local plant, it will be unlikely that another local plant would be established. Existing plants in the adjacent large city would, of course, continue to serve this new area. The investors believe that the plant could operate at about 75% of capacity 250 days per year, because it is located in an area where the weather is mild throughout the year.

The plant will cost $100,000 and will have a maximum capacity of 72 cubic yards of concrete per day. Its market value at the end of 10 years is estimated to be $20,000, which is the value of the land. To deliver the concrete, four secondhand trucks would be acquired, costing $8,000 each, having an estimated life of five years and a market value of $500 each at the end of that time. In addition to the four truck drivers, who would be paid $50.00 per day each, four people would be required to operate the plant and office at a total cost of $175.00 per day. Annual operating and maintenance expenses are estimated for the plant and office at $7,000 and for each truck at $2,250, both in view of 75% capacity utilization. Raw material costs are estimated to be $27.00 per cubic yard of concrete. Payroll taxes, vacations, and other fringe benefits would amount to 25% of the annual payroll. Annual taxes and insurance on each truck would be $500, and taxes and insurance on the plant would be $1,000 per year. The investors would not contribute any labor to the business, but a manager would be employed at an annual salary of $20,000.

Delivered, premixed concrete is currently selling for an average of $45 per cubic yard. A useful plant life of 10 years is expected, and capital invested elsewhere by these investors is earning about 15% per year before income taxes. It is desired to find the AW for the expected conditions described and to perform sensitivity analyses for certain factors.

Annual Worth of Proposed Plant at Expected Conditions

Annual revenue: $72 \times 250 \times \$45 \times 0.75 = \$607,500.$

Annual expenses:

1. Capital recovery amount

   Plant: $\$100,000(A/P, 15\%, 10)$
   $-\$20,000 (A/F, 15\% \ 10)$ = $18,940

   Trucks: $4[\$8,000(A/P, 15\%, 5)$
   $-\$500(A/F, 15\%, 5)]$ = $9,250

   $28,190

2. Labor:

   Plant and office: $\$175 \times 250$ = 43,750

   Truck drivers: $4 \times \$50 \times 250$ = 50,000

   Manager = 20,000

   113,750

3.	Payroll taxes, fringe benefits, and so on:			
	$113,750 × 0.25	=		28,438
4.	Taxes and insurance:			
	Plant	=	1,000	
	Trucks: $500 × 4	=	2,000	
				3,000
5.	Operations and maintenance at 75% capacity:			
	Plant and office	=	7,000	
	Trucks: $2,250 × 4	=	9,000	
				16,000
6.	Materials: 72 × 0.75 × 250 × $27.00			364,500
			Total expenses	$553,878

The net AW for these most likely (best) estimates is $607,500 − $553,878 = $53,622. Apparently, the project is an attractive investment opportunity.

In this case study, there are three factors that are of great importance and that must be estimated: *capacity utilization,* the *selling price of the product,* and the *useful life of the plant.* A fourth factor—raw material costs—is important, but any significant change in this factor would probably have an equal impact on competitors and probably would be reflected in a corresponding change in the selling price of mixed concrete. The other cost elements should be determinable with considerable accuracy. Therefore, we need to investigate the effect of variations in the plant utilization, selling price, and useful life. Sensitivity analysis is needed in this situation.

11.5.1 Sensitivity to Capacity Utilization

As a first step, we will determine how expenses would vary, if at all, as capacity utilization is varied. In this case, it is probable that the annual expense items listed under groups 1, 2, 3, and 4 in the previous tabulation would be virtually unaffected if capacity utilization should vary over a quite wide range—from 50 to 90%, for example. To meet peak demands, the same amount of plant, trucks, and personnel probably would be required. Operations and maintenance expenses (group 5) would be affected somewhat. For this factor, we must try to determine what the variation would be or make a reasonable assumption as to the probable variation. For this case, it will be *assumed* that one-half of these expenses would be fixed and the other half would vary with capacity utilization by a straightline relationship. Certain other factors, such as the cost of materials in this case, will vary in direct proportion to capacity utilization.

Using these assumptions, Table 11-4 shows how the revenue, expenses, and net AW would change with different capacity utilizations. It will be noted that AW is moderately sensitive to capacity utilization. The plant could be operated at a little less than 65% of capacity, instead of the assumed 75%, and still produce an AW greater than zero. Also, quite clearly, if they can operate above the assumed 75% of capacity, AW would be very good. This type of analysis gives the analyst a good idea of how much leeway the company has in capacity utilization and still have an acceptable venture.

TABLE 11-4 AW at $i = 15\%$ per Year for the Premixed-Concrete Plant for Various Capacity Utilizations (Average Selling Price Is $45 per Cubic Yard)

	50% Capacity	65% Capacity	90% Capacity
Annual revenue	$405,000	$526,500	$729,000
Annual expenses:			
Capital recovery	28,190	28,190	28,190
Labor	113,750	113,750	113,750
Payroll taxes and similar items	28,438	28,438	28,438
Taxes and insurance	3,000	3,000	3,000
Operations and maintenance[a]	13,715	15,086	17,372
Materials	243,000	315,900	437,400
Total expenses	$430,093	$504,364	$628,150
AW (15%)	−$25,093	+$22,136	+$100,850

[a] Let x = annual operations and maintenance expenses and assume that 50% of the cost varies directly with capacity utilization. At 75% capacity utilization, $x/2 + (x/2)(0.75) = \$16,000$, so that $x = \$18,286$ at 100% capacity utilization. Therefore, at 50% utilization, the operations and maintenance expense would be $\$9,143 + 0.5(\$9,143) = \$13,715$.

TABLE 11-5 Effect of Various Selling Prices on the AW for the Premixed-Concrete Plant Operating at 75% of Capacity

	Selling Price			
	$45.00	$43.65(3%)[a]	$42.75(5%)[a]	$40.50(10%)[a]
Annual revenue	$607,500	$589,275	$577,125	$546,750
Annual expenses	553,878	553,878	553,878	553,878
AW(15%)	$53,622	$35,397	$23,247	−$7,128

[a] Percentage values shown in parentheses are reductions in price below $45.

11.5.2 Sensitivity to Selling Price

Examination of the sensitivity of the project to the selling price of the concrete reveals the situation shown in Table 11-5. The values in this table assume that the plant would operate at 75% of capacity; the expenses would thus remain constant, with only the selling price varying. Here it will be noted that the project is quite sensitive to price. A decrease in price of 10% would reduce the IRR to less than 15% (i.e., AW < 0). Because a decrease of 10% is not very large, the investors would want to make a thorough study of the price structure of concrete in the area of the proposed plant, particularly with respect to the possible effect of the increased competition that the new plant would create. If such a study reveals price instability in the market for concrete, the plant could be a risky investment.

11.5.3 Sensitivity to Useful Life

The effect of the third factor, assumed useful life of the plant, can be investigated readily. If a life of five years were assumed for the plant, instead of the assumed value of 10 years, the only factor in the study that would be changed would be the cost of capital recovery. If the market value is assumed to remain constant, the capital recovery amount over a five-year period is

$$\$100{,}000(A/P, 15\%, 5) - \$20{,}000(A/F, 15\%, 5) = \$26{,}866 \text{ per year,}$$

which is \$7,926 more expensive than the initial value of \$18,940. In this case, AW would be reduced to \$45,696—a decline of 14.8%. Hence, a 50% reduction in useful life causes only a 14.8% reduction in AW. Clearly, the venture is rather insensitive to the assumed useful life of the plant.

With the added information supplied by the sensitivity analyses that have just been described, those who make the investment decision concerning the proposed concrete plant would be in a much better position than if they had only the initial study results, based on an assumed utilization of 75% of capacity, available to them. ———————■

11.6 Summary

Engineering economy involves decision making among competing uses of scarce capital resources. The consequences of these decisions usually extend far into the future. In this chapter, we have used techniques to deal with the realization that the consequences or outcomes (cash flows, useful lives, etc.) of engineering projects can never be known with absolute certainty.

Several of the most commonly applied and useful procedures for dealing with the variability of estimates in engineering economy studies have been presented in this chapter. Breakeven analysis determines the value of a key common factor, such as utilization of capacity, at which the economic desirability of two alternatives is equal or a project is economically justified. This breakeven point is then compared to an independent estimate of the factor's most likely (best estimate) value to assist with the selection between alternatives or to make a decision about a project. The sensitivity graph technique makes explicit the impact of variations in the estimates of each project factor of concern on the economic measure of merit, and it is a valuable analysis tool. The technique discussed in Section 11.4 for evaluating the combined impact of changes in two or more factors is important when additional information is needed to assist decision making.

Regrettably, there is no quick and easy answer to the question, How should variability best be considered in an engineering economic analysis? Generally, simple procedures (e.g., sensitivity analysis) allow reasonable discrimination among alternatives to be made or the acceptability of a project to be determined on the basis of the variations present, and they are relatively inexpensive to apply. So Yogi Berra is right—prediction is usually difficult, especially about the future!

Problems

The number in parentheses () that follows each problem refers to the section from which the problem is taken.

11-1. An electronics firm is planning to manufacture a new handheld gaming device for the preteen market. The following data have been estimated for the product.

Sales price	$12.50 per unit
Equipment cost	$200,000
Overhead cost	$50,000 per year
Operating and maintenance cost	$25 per operating hour
Production time per 1,000 units	100 hours
Study period (planning horizon)	5 years
MARR	15% per year

Assuming a negligible market (salvage) value for the equipment at the end of five years, determine the breakeven annual sales volume for this product. (11.2)

11-2. Refer to Example 11-2. Assuming gasoline costs $3.50 per gallon, find the breakeven mileage per year between the hybrid vehicle and the gas-only vehicle. All other factors remain the same. (11.2)

11-3. The Universal Postal Service is considering the possibility of fixing wind deflectors on the tops of 500 of their long-haul tractors. Three types of deflectors, with the following characteristics, are being considered (MARR =10% per year):

	Windshear	Blowby	Air-vantage
Capital investment	$1,000	$400	$1,200
Drag reduction	20%	10%	25%
Maintenance per year	$10	$5	$5
Useful life	10 years	10 years	5 years

If 5% in drag reduction means 2% in fuel savings per mile, how many miles do the tractors have to be driven per year before the windshear deflector is favored over the other deflectors? Over what range of miles driven per year is air-vantage the best choice? (*Note:* Fuel cost is expected to be $3.00 per gallon, and average fuel consumption is 5 miles per gallon without the deflectors.) State any assumptions you make. (11.2)

11-4. In planning a small two-story office building, the architect has submitted two designs. The first provides foundation and structural details so that two additional stories can be added to the required initial two stories at a later date and without modifications to the original structure. This building would cost $1,400,000. The second design, without such provisions, would cost only $1,250,000. If the first plan is adopted, an additional two stories could be added at a later date at a cost of $850,000. If the second plan is adopted, however, considerable strengthening and reconstruction would be required, which would add $300,000 to the cost of a two-story addition. Assuming that the building is expected to be needed for 75 years, by what time would the additional two stories have to be built to make the adoption of the first design justified? MARR is 10% per year. (11.2)

11-5. Consider these two alternatives. (11.2)

	Alternative 1	Alternative 2
Capital investment	$4,500	$6,000
Annual revenues	$1,600	$1,850
Annual expenses	$400	$500
Estimated market value	$800	$1,200
Useful life	8 years	10 years

a. Suppose that the capital investment of Alternative 1 is known with certainty. By how much would the estimate of capital investment for Alternative 2 have to vary so that the *initial* decision based on these data would be reversed? The annual MARR is 15% per year.

b. Determine the life of Alternative 1 for which the AWs are equal.

11-6. An area can be irrigated by pumping water from a nearby river. Two competing installations are being considered.

	Pump A 6 in. System	Pump B 8 in. System
Operating load on motor	15 hp	10 hp
Efficiency of pump motor	0.60	0.75
Cost of installation	$2,410	$4,820
Market value	$80	$0
Useful life	8 years	8 years

The MARR is 12% per year and electric power for the pumps costs $0.06 per kWh. Recall that 1 horsepower (hp) equals 0.746 kilowatts. (11.2, 11.3)

a. At what level of operation (hours per year) would you be indifferent between the two pumping systems? If the pumping system is expected to operate 2,000 hours per year, which system should be recommended?

b. Perform a sensitivity analysis on the efficiency of Pump A. Over what range of pumping efficiency is Pump A preferred to Pump B? Assume 2,000 hours of operation per year, and draw a graph to illustrate your answer.

11-7. Consider the following two investment alternatives. Determine the range of investment costs for Alternative B (i.e., min. value $< X <$ max. value) that will convince an investor to select Alternative B. MARR = 10% per year, and other relevant data are shown in the following table. State clearly any assumptions that are necessary to support your answer. (11.2)

	Alt. A	Alt. B
Capital investment	$5,000	$ X
Net annual receipts	$1,500	$1,400
Market value	$1,900	$4,000
Useful life	5 years	7 years

11-8. The Ford Motor Company is considering three mutually exclusive electronic stability control systems for protection against rollover of its automobiles. The investment (study) period is four years, and MARR is 12% per year. Data for the fixture costs of the systems are as follows.

	Alternatives		
	A	B	C
Capital investment	$12,000	$15,800	$8,000
Annual savings	$4,000	$5,200	$3,000
MV (after 4 years)	$3,000	$3,500	$1,500
IRR	19.2%	18%	23%

Plot the AW of each alternative against MARR as the MARR varies across this range: 4%, 8%, 12%, 16%, and 20%. What can you generalize about the range of the MARR for which each alternative is preferred? (11.2)

11-9. Your company operates a fleet of light trucks that are used to provide contract delivery services. As the engineering and technical manager, you are analyzing the purchase of 55 new trucks as an addition to the fleet. These trucks would be used for a new contract the sales staff is trying to obtain. If purchased, the trucks would cost $21,200 each; estimated use is 20,000 miles per year per truck; estimated operation and maintenance and other related expenses (year-zero dollars) are $0.45 per mile, which is forecasted to increase at the rate of 5% per year; and the trucks are MACRS (GDS) three-year property class assets. The analysis period is four years; $t = 38\%$; MARR = 15% per year (after taxes; includes an inflation component); and the estimated MV at the end of four years (in year-zero dollars) is 35% of the purchase price of the vehicles. This estimate is expected to increase at the rate of 2% per year.

Based on an after-tax, actual-dollar analysis, what is the annual revenue required by your company from the contract to justify these expenditures before any profit is considered? This calculated amount for annual revenue is the breakeven point between purchasing the trucks and which other alternative? (11.2, Chapters 7 and 8)

11-10. A nationwide motel chain is considering locating a new motel in Bigtown, USA. The cost of building a 150-room motel (excluding furnishings) is $5 million. The firm uses a 15-year planning horizon to evaluate investments of this type. The furnishings for this motel must be replaced every five years at an estimated cost of $1,875,000 (at $k = 0, 5$, and 10). The old furnishings have no market value. Annual operating and maintenance expenses for the facility are estimated to be $125,000. The market value of the motel after 15 years is estimated to be 20% of the original building cost.

Rooms at the motel are projected to be rented at an average rate of $45 per night. On the average, the motel will rent 60% of its rooms each night. Assume the motel will be open 365 days per year. MARR is 10% per year. (11.3)

a. Using an annual-worth measure of merit, is the project economically attractive?

b. Investigate sensitivity to decision reversal for the following three factors: (1) capital investment, (2) MARR, and (3) occupancy rate (average percent of rooms rented per night). To which of these factors is the decision most sensitive?

c. Graphically investigate the sensitivity of the AW to changes in the above three factors. Investigate changes over the interval ±40%. On your graph, use percent change as the x-axis and AW as the y-axis.

11-11. A large city in the mid-West needs to acquire a street-cleaning machine to keep its roads looking nice year round. A used cleaning vehicle will cost $75,000 and have a $20,000 market (salvage) value at the end of its five-year life. A new system with advanced features will cost $150,000 and have a $50,000 market value at the end of its five-year life. The new system is expected to reduce labor hours compared with the used system. Current street-cleaning activity requires the used system to operate 8 hours per day for 20 days per month. Labor costs $40 per hour (including fringe benefits), and MARR is 12% per year compounded monthly. (11.2)

a. Find the breakeven percent reduction in labor hours for the new system.

b. If the new system is expected to be able to reduce labor hours by 20% compared with the used system, which machine should the city purchase?

11-12. Refer to Problem 11-11. The best estimate for reduction of labor hours for the new system is 20% (compared with the used system). Investigate how sensitive the decision is to (a) changes in the MARR, (b) changes in the market value of the new system, and (c) the productivity improvement of the new system. Graph your results. *Hint: Think incrementally!* (11.3)

11-13. A group of private investors borrowed $30 million to build 300 new luxury apartments near a large university. The money was borrowed at 6% annual interest, and the loan is to be repaid in equal annual amounts over a 40-year period. Annual operating, maintenance, and insurance expenses are estimated to be $4,000 per apartment. This expense will be incurred even if an apartment is vacant. The rental fee for each apartment will be $12,000 per year, and the worst-case occupancy rate is projected to be 80%. Investigate the sensitivity of annual profit (or loss) to (a) changes in the occupancy rate and (b) changes in the annual rental fee. (11.3)

11-14. The managers of a company are considering an investment with the following estimated cash flows. MARR is 15% per year.

Capital investment	$30,000
Annual revenues	$20,000
Annual expenses	$5,000
Market value	$1,000
Useful life	5 years

The company is inclined to make the investment; however, the managers are nervous because all of the cash flows and the useful life are approximate values. The capital investment is known to be within ±5%. Annual expenses are known to be within ±10%. The annual revenue, market value, and useful life estimates are known to be within ±20%. (11.3)

a. Analyze the sensitivity of PW to changes in each estimate individually. Based on your results, make a recommendation regarding whether or not they should proceed with this project. Graph your results for presentation to management.

b. The company can perform market research and/or collect more data to improve the accuracy of these estimates. Rank these variables by ordering them in accordance with the need for more accurate estimates (from highest need to lowest need).

11-15. An industrial machine costing $10,000 will produce net cash savings of $4,000 per year. The machine has a five-year useful life but must be returned to the factory for major repairs after three years of operation. These repairs cost $5,000. The company's MARR is 10% per year. What IRR will be earned on the purchase of this machine? Analyze the sensitivity of IRR to ±$2,000 changes in the repair cost. (11.3)

11-16. You have decided to purchase a new automobile with a hybrid-fueled engine and a six-speed transmission. After the trade-in of your present car, the purchase price of the new automobile is $30,000. This balance can be financed by the auto dealer at 2.9% APR (compounded monthly) and repaid over 48 monthly payments. Alternatively, you can get an instant rebate on the purchase price if you finance the loan balance at an APR of 8.9% (compounded monthly) over 48 months. Determine the amount of the instant rebate that would make you indifferent between the financing plans. (11.2)

11-17. Two traffic signal systems are being considered for an intersection. One system costs $32,000 for installation and has an efficiency rating of 78%, requires 28 kW power (output), incurs a user cost of $0.24 per vehicle, and has a life of 10 years. A second system costs $45,000 to install, has an efficiency rating of 90%, requires 34 kW power (output), has a user cost of $0.22 per vehicle, and has a life of 15 years. Annual maintenance costs are $75 and $100, respectively. MARR = 10% per year. How many vehicles must use the intersection to justify the second system when electricity costs $0.08/kWh? (11.2)

11-18. An office building is considering converting from a coal-burning furnace to one that burns either fuel oil or natural gas. The cost of converting to fuel oil is estimated to be $80,000 initially; annual operating expenses are estimated to be $4,000 less than that experienced when using the coal furnace. Approximately 140,000 Btus are produced per gallon of fuel oil; fuel oil is anticipated to cost $2.20 per gallon.

The cost of converting to natural gas is estimated to be $60,000 initially; additionally, annual operating and maintenance expenses are estimated to be $6,000 less than that for the coal-burning furnace. Approximately 1,000 Btus are produced per cubic foot of natural gas; it is estimated that natural gas will cost $0.04 per cubic foot.

A planning horizon of 20 years is to be used. Zero market values and a MARR of 10% per year are appropriate. Perform a sensitivity analysis for the annual Btu requirement for the heating system. (*Hint:* First calculate the breakeven number of Btus (in thousands). Then determine AWs if Btu requirement varies over ±30% of the breakeven amount.) (11.3)

11-19. Suppose that, for a certain potential investment project, the optimistic, most likely, and pessimistic estimates are as shown in the accompanying table. (11.4)

	Optimistic	Most Likely	Pessimistic
Capital investment	$90,000	$100,000	$120,000
Useful life	12 years	10 years	6 years
Market value	$30,000	$20,000	$0
Net annual cash flow	$35,000	$30,000	$20,000
MARR (per year)	10%	10%	10%

a. What is the AW for each of the three estimation conditions?

b. It is thought that the most critical factors are useful life and net annual cash flow. Develop a table showing the net AW for all combinations of the estimates for these two factors, assuming all other factors to be at their most likely values.

11-20. A bridge is to be constructed now as part of a new road. Engineers have determined that traffic density on the new road will justify a two-lane road and a bridge at the present time. Because of uncertainty regarding future use of the road, the time at which an extra two lanes will be required is currently being studied.

The two-lane bridge will cost $200,000 and the four-lane bridge, if built initially, will cost $350,000. The future cost of widening a two-lane bridge to four lanes will be an extra $200,000 plus $25,000 for every year that widening is delayed. The MARR used by the highway department is 12% per year. The following estimates have been made of the times at which the four-lane bridge will be required:

Pessimistic estimate	4 years
Most likely estimate	5 years
Optimistic estimate	7 years

In view of these estimates, what would you recommend? What difficulty, if any, do you have in interpreting your results? List some advantages and disadvantages of this method of preparing estimates. (11.4)

11-21. An aerodynamic three-wheeled automobile (the Dart) runs on compressed natural gas stored in two cylinders in the rear of the vehicle. The $13,000 Dart can cruise at speeds up to 80 miles per hour, and it can travel 100 miles per gallon of fuel. Another two-seater automobile costs $10,000 and averages 50 miles per gallon of compressed natural gas. If fuel costs $8.00 per gallon and MARR is 10% per year, over what range of annual miles driven is the Dart more economical? Assume a useful life of five years for both cars. (11.2)

11-22. *Extended Learning Exercise*
Consider these two alternatives for solid-waste removal (11.3, Chapter 7):

Alternative A: Build a solid-waste processing facility. Financial variables are as follows:

Capital investment	$108 million in 2008 (commercial operation starts in 2008)
Expected life of facility	20 years
Annual operating expenses	$3.46 million
Estimated market value	40% of initial capital cost at all times

Alternative B: Contract with vendors for solid-waste disposal after intermediate recovery. Financial variables are as follows:

Capital investment	$17 million in 2008 (This is for *intermediate* recovery from the solid-waste stream.)
Expected contract period	20 years
Annual operating expenses	$2.10 million
Repair costs to intermediate recovery system every five years	$3.0 million
Annual fee to vendors	$10.3 million
Estimated market value at all times	$0

Related Data:

MACRS (GDS) property class	15 yr (Chapter 7)
Study period	20 yr
Effective income tax rate	40%
Company MARR (after-tax)	10% per year
Inflation rate	0% (ignore inflation)

a. How much more expensive (in terms of capital investment only) could Alternative B be in order to breakeven with Alternative A?

b. How sensitive is the after-tax PW of Alternative B to cotermination of both alternatives at the end of *year 10*?

c. Is the initial decision to adopt Alternative B in Part (a) reversed if our company's annual operating expenses for Alternative B ($2.10 million per year) unexpectedly double? Explain why (or why not).

d. Use a computer spreadsheet available to you to solve this problem.

11-23. *Extended Learning Exercise* Construct the spreadsheet used in Example 11-6. Use this spreadsheet as a starting point to answer the following questions. Be sure to clearly state your assumptions. (11.3)

a. How sensitive is the decision to the amount of the down payment for the purchase option? If a down payment of less than 20% is used, you will need to account for monthly mortgage insurance (about $56/month).

b. How sensitive is the decision to the type of mortgage acquired? You can look at different repayment periods (15 or 20 years) as well as the impact of points paid on the loan. One point is equal to 1% of the loan value and is paid as part of the closing costs.

c. Interest paid on a mortgage is tax deductible if you itemize deductions on your personal income tax return. Discuss how you would incorporate this and what impact it would have on the decision to purchase versus rent a home.

Spreadsheet Exercises

11-24. We know the standard means of cutting the high cost of driving our automobiles—slow down your speed, no *jack rabbit* starts, inflate tires properly, clean air filters regularly, and so on. Another way to reduce the cost of driving is to join every big-rigger in the United States and half of Europe—go diesel. Diesels are inherently more efficient than gasoline engines because they deliver a third better fuel mileage. They also offer a reliable high-torque engine!

The initial extra sticker price of a diesel-fueled car is about $1,200. A gallon of diesel fuel costs about the same as a gallon of gasoline. Assume your gasoline-fueled car averages 24–27 miles per gallon. You drive an average of 20,000 miles per year, and your personal MARR is in the 12–15% per year ballpark. Develop a spreadsheet

to compare the economics of driving 100,000 miles in your car fueled by gasoline versus diesel. Explore various *what if* questions that you have regarding this comparison. (11.3)

11-25. It is desired to determine the most economical thickness of insulation for a large cold-storage room. Insulation is expected to cost $150 per 1,000 square feet of wall area per inch of thickness installed, and to require annual property taxes and insurance of 5% of the capital investment. It is expected to have $0 market value after a 20-year life. The following are estimates of the heat loss per 1,000 square feet of wall area for several thicknesses:

Insulation, in.	Heat loss, Btu per hr
3	4,400
4	3,400
5	2,800
6	2,400
7	2,000
8	1,800

The cost of heat removal (loss) is estimated at $0.04 per 1,000 Btu. The MARR is 20% per year. Assuming continuous operation throughout the year, develop a spreadsheet to analyze the sensitivity of the optimal thickness to errors in estimating the cost of heat removal. Use the AW technique. (11.3)

11-26. Suppose that, for an engineering project, the optimistic, most likely, and pessimistic estimates are as shown in the accompanying table.

	Optimistic	Most Likely	Pessimistic
Capital investment	$90,000	$100,000	$120,000
Useful life	12 years	10 years	6 years
Market value	$30,000	$20,000	$0
Net annual cash flow	$35,000	$30,000	$20,000
MARR (per year)	11%	11%	11%

Develop a spreadsheet to determine the AW for each of the three estimate conditions. It is thought that the most critical elements are useful life and net annual cash flow. Include in your spreadsheet a table showing the AW for all combinations of the estimates for these two factors, assuming that all other factors remain at their most likely values. (11.4)

11-27. The world's largest carpet maker has just completed a feasibility study of what to do with the 16,000 tons of overruns, rejects, and remnants it produces every year. The company's CEO launched the feasibility study by asking, Why pay someone to dig coal out of the ground and then pay someone else to put our waste into a landfill? Why not just burn our own waste? The company is proposing to build a $10-million power plant to burn its waste as fuel, thereby saving $2.8 million a year in coal purchases. Company engineers have determined that the waste-burning plant will be environmentally sound, and after its four-year study period the plant can be sold to a local electric utility for $5 million.

The firm's MARR is 15%. Develop a spreadsheet to investigate the sensitivity of the proposal's PW to changes in the estimates of (a) initial investment amount, (b) annual savings of fuel cost, and (c) future resale value of the plant. To which factor is the decision most sensitive? Include in your solution the specific values of these factors that would cause decision reversal. Use the charting feature to create a spiderplot. (11.3)

Case Study Exercises

11-28. Develop an appropriate sensitivity graph (spiderplot) for this analysis. Include any additional values of the factors you consider necessary. Also, include raw material costs as an additional factor in the sensitivity graph, under the assumption that all competitors may not respond to changes in these costs the same way. (11.5)

11-29. For the three most sensitive factors, further analyze the combined effect of their outcomes on the AW. You must determine how to best formulate the combinations of factor outcomes—Example 11-7 illustrated one approach with O-ML-P estimating; however, you can develop three or four scenarios with selected changes in specific factors. (11.5)

FE Practice Problems

11-30. A highway bridge is being considered for replacement. The new bridge would cost $X and would last for 20 years. Annual maintenance costs for the new bridge are estimated to be $24,000. People will be charged a toll of $0.25 per car to use the new bridge. Annual car traffic is estimated at 400,000 cars. The cost of collecting the toll consists of annual salaries for five collectors at $10,000 per collector. The existing bridge can be refurbished for $1,600,000 and would need to be replaced in 20 years. There would be additional refurbishing costs of $70,000 every five years and regular annual maintenance costs of $20,000 for the existing bridge. There would be no toll to use the refurbished bridge. If MARR is 12% per year, what is the maximum acceptable cost (X) of the new bridge? (11.2)

(a) $1,943,594   (b) $2,018,641   (c) $1,652,425
(d) $1,570,122   (e) $2,156,209

11-31. An analysis of accidents in a rural state indicates that widening a highway from 30 ft to 40 ft will decrease the annual accident rate from 1,250 to 710 per million vehicle-miles. Calculate the average daily number of vehicles that should use the highway to justify widening on the basis of the following estimates: (i) the average loss per accident is $1,200; (ii) the cost of widening is $117,000 per mile; (iii) the useful life of the widened road is 25 years; (iv) annual maintenance costs are 3% of the capital investment; and (v) MARR is 12% per year. (11.2)

(a) 78   (b) 63   (c) 34   (d) 59   (e) 27

A supermarket chain buys loaves of bread from its supplier at $0.50 per loaf. The chain is considering two options to bake its own bread.

	Machine A	Machine B
Capital investment	$8,000	$16,000
Useful life (years)	7	7
Annual fixed cost	$2,000	$4,000
Variable cost per loaf	$0.26	$0.16

Neither machine has a market value at the end of seven years, and MARR is 12% per year. Use this information to answer Problems 11-32, 11-33, and 11-34. Select the closest answer. (11.2)

11-32. What is the minimum number of loaves that must be sold per year to justify installing Machine A instead of buying the loaves from the supplier?

(a) 7,506   (b) 22,076   (c) 37,529   (d) 75,059
(e) 15,637

11-33. What is the minimum number of loaves that must be sold per year to justify installing Machine B instead of buying the loaves from the supplier?

(a) 37,529   (b) 15,637   (c) 22,076   (d) 75,059
(e) 7,506

11-34. If the demand for bread at this supermarket is 35,000 loaves per year, what strategy should be adopted for acquiring bread? Both Machine A and Machine B are capable of meeting annual demand.

(a) Continue buying from the supplier.
(b) Install Machine A.
(c) Install Machine B.
(d) Install both Machine A and Machine B.

Answer Problems 11-35 through 11-39 on the basis of the figure on page 494 and the most likely estimates given as follows:

MARR	12% per year
Useful Life	5 years
Initial investment	$5,000
Receipts–Expenses ($R - E$)	$1,500/year

11-35. If the initial investment is increased by more than 9%, the project is profitable.

(a) True   (b) False

11-36. If the profit ($R - E$) is decreased by 5%, this project is not profitable.

(a) True   (b) False

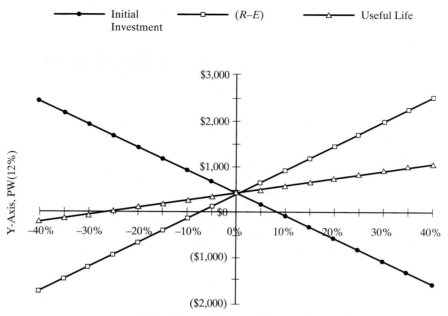

X-Axis (Percent Change in Parameter)

11-37. Variations in the profit and useful life of the project are inversely related.

(a) True   (b) False

11-38. This project (based upon the most likely estimates) is profitable.

(a) True   (b) False

11-39. An initial investment of $5,500 is profitable.

(a) True   (b) False

CHAPTER 12

Probabilistic Risk Analysis

The aim of Chapter 12 is to discuss and illustrate several probabilistic methods that are useful in analyzing risk and uncertainty associated with engineering economy studies.

The Risks of Global Warming*

Risk is a condition where there is a possibility of adverse deviation from a desired and expected outcome. The risks of global climate change caused by carbon dioxide and other greenhouse gases include heightened regulation, revenue loss, and increased physical property impairment. Opportunities for mitigating the risks associated with climate change are numerous: increased efficiency of energy production and use, improved agricultural practices, and carbon capture and sequestration are just a few of the choices we have. In this chapter, you will learn how various probabilistic techniques can be used to assess the risks of engineering projects such as those that mitigate the effects of global warming.

---

* *Financial Times*, October 31, 2006, 1–2.

Use every possible means . . . to come with certainty at the enemy's strength, situation and movements—without this we wander in a wilderness of uncertainties.

—George Washington (1776)

12.1 Introduction

In previous chapters, we stated specific assumptions concerning applicable revenues, costs, and other quantities important to an engineering economy analysis. It was assumed that a high degree of confidence could be placed in all estimated values. That degree of confidence is sometimes called *assumed certainty*. Decisions made solely on the basis of this kind of analysis are sometimes called *decisions under certainty*. The term is rather misleading in that there rarely is a case in which the best of estimates of quantities can be assumed as certain.

We now consider the more realistic situation in which estimated future quantities are uncertain and project outcomes are risky. The motivation for dealing with risk and uncertainty is to establish the bounds of error such that another alternative being considered may turn out to be a better choice than the one we recommended under assumed certainty.

Both *risk* and *uncertainty* in decision-making activities are caused by lack of precise knowledge regarding future business conditions, technological developments, synergies among funded projects, and so on. *Decisions under risk* are decisions in which the analyst models the decision problem in terms of assumed possible future outcomes, or scenarios, whose probabilities of occurrence can be estimated. A *decision under uncertainty*, by contrast, is a decision problem characterized by several unknown futures for which probabilities of occurrence cannot be estimated. In reality, the difference between risk and uncertainty is somewhat arbitrary.

The probability that a cost, revenue, useful life, or other factor value will occur, or that a particular equivalent-worth or rate-of-return value for a cash flow will occur, is usually considered to be the long-run relative frequency with which the event (value) occurs or the subjectively estimated likelihood that it will occur. Factors such as these, having probabilistic outcomes, are called *random variables*. For example, the cash-flow amounts for an alternative, such as CO_2 capture and sequestration at a power plant, often result from the sum, difference, product, or quotient of random variables. In such cases, the measures of profitability (e.g., equivalent-worth and rate-of-return values) of an alternative's cash flows will also be random variables.

The information about these random variables, which is particularly helpful in decision making, is their expected values and variances, especially for the economic measures of merit of the alternatives. These derived quantities for the random variables are used to make the uncertainty associated with each alternative more explicit, including any probability of loss. Thus, when uncertainty is considered, the variability in the economic measures of merit and the probability of loss associated with the alternatives are normally used in the decision-making process.

12.2 Sources of Uncertainty

It is useful to consider some of the factors that affect the uncertainty involved in the analysis of the future economic consequences of an engineering project. It would be almost impossible to list and discuss all of the potential factors. There are four major sources of uncertainty, however, that are nearly always present in engineering economy studies.

The first source that is always present is the *possible inaccuracy of the cash-flow estimates used in the study*. The accuracy of the cash-inflow estimates is difficult to determine. If they are based on past experience or have been determined by adequate market surveys, a fair degree of reliance may be placed on them. On the other hand, if they are based on limited information with a considerable element of hope thrown in, they probably contain a sizable element of uncertainty.

The second major source affecting uncertainty is the *type of business involved in relation to the future health of the economy*. Some types of business operations are less stable than others. For example, most mining enterprises are more risky than one engaged in manufactured homes. Whenever capital is to be invested in an engineering project, the nature of the business as well as expectations of future economic conditions (e.g., interest rates) should be considered in deciding what risk is present.

A third source affecting uncertainty is the *type of physical plant and equipment involved*. Some types of structures and equipment have rather definite lives and market values. A good engine lathe generally can be used for many purposes in nearly any fabrication shop. Quite different would be a special type of lathe that was built to do only one unusual job. Its value would be dependent almost entirely upon the demand for the special task that it can perform. Thus, the type of physical property involved affects the accuracy of the estimated cash-flow patterns.

The fourth important source of uncertainty that must always be considered is the *length of the study period* used in the analysis. A long study period naturally decreases the probability of all the factors turning out as estimated. Therefore, a long study period, all else being equal, generally increases the uncertainty of a capital investment.

12.3 The Distribution of Random Variables

Capital letters such as X, Y, and Z are usually used to represent random variables and lowercase letters (x, y, z) to denote the particular values that these variables take on in the sample space (i.e., in the set of all possible outcomes for each variable). When a random variable's sample space is discrete, its *probability mass function* is usually indicated by $p(x)$ and its *cumulative distribution function* by $P(x)$. When a random variable's sample space is continuous, its *probability density function* and its *cumulative distribution function* are usually indicated by $f(x)$ and $F(x)$, respectively.

12.3.1 Discrete Random Variables

A random variable X is said to be *discrete* if it can take on at most a countable (finite) number of values $(x_1, x_2, \ldots, x_L)$. The probability that a discrete random variable X takes on the value x_i is given by

$$\Pr\{X = x_i\} = p(x_i) \text{ for } i = 1, 2, \ldots, L \quad (i \text{ is a } sequential\ index \text{ of the}$$
$$\text{discrete values, } x_i, \text{ that the variable takes on}),$$

where $p(x_i) \geq 0$ and $\sum_i p(x_i) = 1$.

The probability of events about a discrete random variable can be computed from its probability mass function $p(x)$. For example, the probability of the event that the value of X is contained in the closed interval $[a, b]$ is given by (where the colon is read *such that*)

$$\Pr\{a \leq X \leq b\} = \sum_{i:a \leq X_i \leq b} p(x_i). \tag{12-1}$$

The probability that the value of X is less than or equal to $x = h$, the cumulative distribution function $P(x)$ for a discrete case, is given by

$$\Pr\{X \leq h\} = P(h) = \sum_{i:X_i \leq h} p(x_i). \tag{12-2}$$

In most practical applications, discrete random variables represent *countable* data such as the useful life of an asset in years, number of maintenance jobs per week, or number of employees as positive integers.

12.3.2 Continuous Random Variables

A random variable X is said to be continuous if there exists a nonnegative function $f(x)$ such that, for any set of real numbers $[c, d]$, where $c < d$, the probability of the event that the value of X is contained in the set is given by

$$\Pr\{c \leq X \leq d\} = \int_c^d f(x)\, dx \tag{12-3}$$

and

$$\int_{-\infty}^{\infty} f(x)\, dx = 1.$$

Thus, the probability of events about the continuous random variable X can be computed from its probability density function, and the probability that X assumes exactly any one of its values is zero. Also, the probability that the value of X is less than or equal to a value $x = k$, the cumulative distribution function $F(x)$ for a continuous case, is given by

$$\Pr\{X \leq k\} = F(k) = \int_{-\infty}^{k} f(x)\, dx. \tag{12-4}$$

Also, for a continuous case,

$$\Pr\{c \leq X \leq d\} = \int_{c}^{d} f(x)\, dx = F(d) - F(c). \tag{12-5}$$

In most practical applications, continuous random variables represent *measured* data such as time, cost, and revenue on a continuous scale. Depending upon the situation, the analyst decides to model random variables in engineering economic analysis as either discrete or continuous.

12.3.3 Mathematical Expectation and Selected Statistical Moments

The expected value of a single random variable X, $E(X)$, is a weighted average of the distributed values x that it takes on and is a measure of the central location of the distribution (central tendency of the random variable). The $E(X)$ is the first moment of the random variable about the origin and is called the *mean* (central moment) of the distribution. The expected value is

$$E(X) = \begin{cases} \displaystyle\sum_{i} x_i\, p(x_i) & \text{for } x \text{ discrete and } i = 1, 2, \ldots, L \\[2mm] \displaystyle\int_{-\infty}^{\infty} x f(x)\, dx & \text{for } x \text{ continuous.} \end{cases} \tag{12-6}$$

Although $E(X)$ provides a measure of central tendency, it does not measure how the distributed values x cluster around the mean. The *variance*, $V(X)$, which is nonnegative, of a single random variable X is a measure of the dispersion of the values it takes on around the mean. It is the expected value of the square of the difference between the values x and the mean, which is the second moment of the random variable about its mean:

$$E\{[X - E(X)]^2\} = V(X) = \begin{cases} \displaystyle\sum_{i} [x_i - E(X)]^2\, p(x_i) & \text{for } x \text{ discrete} \\[2mm] \displaystyle\int_{-\infty}^{\infty} [x - E(X)]^2 f(x)\, dx & \text{for } x \text{ continuous.} \end{cases} \tag{12-7}$$

From the binomial expansion of $[X - E(X)]^2$, it can be easily shown that $V(X) = E(X^2) - [E(X)]^2$. That is, $V(X)$ equals the second moment of the random variable around the origin, which is the expected value of X^2, minus the square of its mean. This is the form often used for calculating the variance of a random variable X:

$$V(X) = \begin{cases} \displaystyle\sum_{i} x_i^2\, p(x_i) - [E(X)]^2 & \text{for } x \text{ discrete} \\[2mm] \displaystyle\int_{-\infty}^{\infty} x_i^2 f(x)\, dx - [E(X)]^2 & \text{for } x \text{ continuous.} \end{cases} \tag{12-8}$$

The *standard deviation* of a random variable, SD(X), is the positive square root of the variance; that is, $SD(X) = [V(X)]^{1/2}$.

12.3.4   Multiplication of a Random Variable by a Constant

A common operation performed on a random variable is to multiply it by a constant, for example, the estimated maintenance labor expense for a time period, $Y = cX$, when the number of labor hours per period (X) is a random variable, and the cost per labor hour (c) is a constant. Another example is the present worth (PW) calculation for a project when the before-tax or after-tax net cash-flow amounts, F_k, are random variables, and then each F_k is multiplied by a constant $(P/F, i\%, k)$ to obtain the PW value.

When a random variable, X, is multiplied by a constant, c, the expected value, $E(cX)$, and the variance, $V(cX)$, are given by

$$E(cX) = c\,E(X) = \begin{cases} \displaystyle\sum_i cx_i\,p(x_i) & \text{for } x \text{ discrete} \\[2mm] \displaystyle\int_{-\infty}^{\infty} cx\,f(x)\,dx & \text{for } x \text{ continuous} \end{cases} \tag{12-9}$$

and

$$\begin{aligned} V(cX) &= E\left\{[cX - E(cX)]^2\right\} \\ &= E\left\{c^2X^2 - 2c^2X \cdot E(X) + c^2[E(X)]^2\right\} \\ &= c^2 E\{[X - E(X)]^2\} \\ &= c^2 V(X). \end{aligned} \tag{12-10}$$

12.3.5   Multiplication of Two Independent Random Variables

A cash-flow random variable, say Z, may result from the product of two other random variables, $Z = XY$. Sometimes, X and Y can be treated as statistically independent random variables. For example, consider the estimated annual expenses, $Z = XY$, for a repair part repetitively procured during the year on a competitive basis, when the unit price (X) and the number of units used per year (Y) are modeled as independent random variables.

When a random variable, Z, is a product of two independent random variables, X and Y, the expected value, $E(Z)$, and the variance, $V(Z)$, are

$$Z = XY$$
$$E(Z) = E(X)\,E(Y);$$
$$\begin{aligned} V(Z) &= E[XY - E(XY)]^2 \\ &= E\left\{X^2Y^2 - 2XY\,E(XY) + [E(XY)]^2\right\} \\ &= E(X^2)\,E(Y^2) - [E(X)\,E(Y)]^2. \end{aligned} \tag{12-11}$$

But the variance of any random variable, $V(\text{RV})$, is

$$V(\text{RV}) = E[(\text{RV})^2] - [E(\text{RV})]^2,$$
$$E[(\text{RV})^2] = V(\text{RV}) + [E(\text{RV})]^2.$$

Then, $V(Z) = \{V(X) + [E(X)]^2\}\{V(Y) + [E(Y)]^2\} - [E(X)]^2[E(Y)]^2$

or

$$V(Z) = V(X)[E(Y)]^2 + V(Y)[E(X)]^2 + V(X)\,V(Y). \qquad (12\text{-}12)$$

12.4 Evaluation of Projects with Discrete Random Variables

Expected value and variance concepts apply theoretically to long-run conditions in which it is assumed that the event is going to occur repeatedly. Application of these concepts, however, is often useful even when investments are not going to be made repeatedly over the long run. In this section, several examples are used to illustrate these concepts with selected economic factors modeled as discrete random variables.

EXAMPLE 12-1   Premixed-Concrete Plant Project (Chapter 11) Revisited

We now apply the expected value and variance concepts to the small premixed-concrete plant project discussed in Section 11.5. Suppose that the estimated probabilities of attaining various capacity utilizations are as follows:

Capacity (%)	Probability
50	0.10
65	0.30
75	0.50
90	0.10

It is desired to determine the expected value and variance of *annual revenue*. Subsequently, the expected value and variance of annual worth (AW) for the project can be computed. By evaluating both $E(\text{AW})$ and $V(\text{AW})$ for the concrete plant, indications of the venture's average profitability and its riskiness are obtained. The calculations are shown in Tables 12-1 and 12-2.

Solution

Expected value of annual revenue: $\sum(A \times B) = \$575{,}100.$

Variance of annual revenue: $\sum(A \times C) - (575{,}100)^2 = 6{,}360 \times 10^6\,(\$)^2.$

Expected value of AW: $\sum(A \times B) = \$41{,}028$

TABLE 12-1   Solution for Annual Revenue (Example 12-1)

i	Capacity (%)	(A) Probability $p(x_i)$	(B) Revenue[a] x_i	(A) × (B) Expected Revenue	$(C) = (B)^2$ x_i^2	(A) × (C)
1	50	0.10	$405,000	$40,500	1.64×10^{11}	0.164×10^{11}
2	65	0.30	526,500	157,950	2.77×10^{11}	0.831×10^{11}
3	75	0.50	607,500	303,750	3.69×10^{11}	1.845×10^{11}
4	90	0.10	729,000	72,900	5.31×10^{11}	0.531×10^{11}
				$575,100		$3.371 \times 10^{11} (\$)^2$

[a] From Table 11-4 with revenue for Capacity = 75% added.

TABLE 12-2   Solution for AW (Example 12-1)

i	Capacity (%)	(A) $p(x_i)$	(B) AW,[a] x_i	(A) × (B) Expected AW	$(C) = (B)^2$ $(AW)^2$	(A) × (C)
1	50	0.10	−$25,093	−$2,509	0.63×10^9	0.063×10^9
2	65	0.30	22,136	6,641	0.49×10^9	0.147×10^9
3	75	0.50	53,622	26,811	2.88×10^9	1.440×10^9
4	90	0.10	100,850	10,085	10.17×10^9	1.017×10^9
				$41,028		$2.667 \times 10^9 (\$)^2$

[a] From Table 11-4 with AW for Capacity = 75% added.

Variance of AW: $\sum (A \times C) - (41{,}028)^2 = 9{,}837 \times 10^5 (\$)^2$

Standard deviation of AW: $31,364

The standard deviation of AW, SD(AW), is less than the expected AW, E(AW), and only the 50% capacity utilization situation results in a negative AW. Consequently, with this additional information, the investors in this undertaking may well judge the venture to be an acceptable one.

There are projects, such as the flood-control situation in the next example, in which future losses due to natural or human-made risks can be decreased by increasing the amount of capital that is invested. Drainage channels or dams, built to control floodwaters, may be constructed in different sizes, costing different amounts. If they are correctly designed and used, the larger the size, the smaller will be the resulting damage loss when a flood occurs. As we might expect, the most economical size would provide satisfactory protection against most floods, although it could be anticipated that some overloading and damage might occur at infrequent periods.

EXAMPLE 12-2 **Channel Enlargement for Flash Flood Control**

A drainage channel in a community where flash floods are experienced has a capacity sufficient to carry 700 cubic feet per second. Engineering studies produced the following data regarding the probability that a given water flow in any one year will be exceeded and the cost of enlarging the channel:

Water Flow (ft^3/sec)	Probability of a Greater Flow Occurring in Any One Year	Capital Investment to Enlarge Channel to Carry This Flow
700	0.20	—
1,000	0.10	$20,000
1,300	0.05	30,000
1,600	0.02	44,000
1,900	0.01	60,000

Records indicate that the average property damage amounts to $20,000 when serious overflow occurs. It is believed that this would be the average damage whenever the storm flow is *greater* than the capacity of the channel. Reconstruction of the channel would be financed by 40-year bonds bearing 8% interest per year. It is thus computed that the capital recovery amount for debt repayment (principal of the bond plus interest) would be 8.39% of the capital investment, because $(A/P, 8\%, 40) = 0.0839$. It is desired to determine the most economical channel size (water-flow capacity).

Solution

The total expected equivalent uniform annual cost for the structure and property damage for all alternative channel sizes would be as shown in Table 12-3. These calculations show that the minimum expected annual cost would be achieved by enlarging the channel so that it would carry 1,300 cubic feet per second, with the expectation that a greater flood might occur in 1 year out of 20 on the average and cause property damage of $20,000.

TABLE 12-3 **Expected Equivalent Annual Cost (Example 12-2)**

Water Flow (ft^3/sec)	Capital Recovery Amount	Expected Annual Property Damage[a]	Total Expected Equivalent Uniform Annual Cost
700	None	$20,000(0.20) = $4,000	$4,000
1,000	$20,000(0.0839) = $1,678	20,000(0.10) = 2,000	3,678
1,300	30,000(0.0839) = 2,517	20,000(0.05) = 1,000	3,517
1,600	44,000(0.0839) = 3,692	20,000(0.02) = 400	4,092
1,900	60,000(0.0839) = 5,034	20,000(0.01) = 200	5,234

[a] These amounts are obtained by multiplying $20,000 by the probability of greater water flow occurring.

Note that when loss of life or limb might result, as in Example 12-2, there usually is considerable pressure to disregard pure economy and build such projects in recognition of the nonmonetary values associated with human safety.

The following example illustrates the same principles as in Example 12-2, except that it applies to safety alternatives involving electrical circuits.

EXAMPLE 12-3   **Investing in Circuit Protection to Reduce Probability of Failure**

Three alternatives are being evaluated for the protection of electrical circuits, with the following required investments and probabilities of failure:

Alternative	Capital Investment	Probability of Loss in Any Year
A	$90,000	0.40
B	100,000	0.10
C	160,000	0.01

If a loss does occur, it will cost $80,000 with a probability of 0.65 and $120,000 with a probability of 0.35. The probabilities of loss in any year are independent of the probabilities associated with the resultant cost of a loss if one does occur. Each alternative has a useful life of eight years and no estimated market value at that time. The minimum attractive rate of return (MARR) is 12% per year, and annual maintenance expenses are expected to be 10% of the capital investment. It is desired to determine which alternative is best based on expected total annual costs (Table 12-4).

Solution

The expected value of a loss, if it occurs, can be calculated as follows:

$$\$80,000(0.65) + \$120,000(0.35) = \$94,000.$$

Thus, Alternative B is the best based on total expected equivalent uniform annual cost, which is a long-run average cost. One might, however, rationally choose Alternative C to reduce significantly the chance of an $80,000 or $120,000 loss occurring in any year in return for a 24.3% increase in the total expected equivalent uniform annual cost.

TABLE 12-4   **Expected Equivalent Annual Cost (Example 12-3)**

Alternative	Capital Recovery Amount = Capital Investment × $(A/P, 12\%, 8)$	Annual Maintenance Expense = Capital Investment × (0.10)	Expected Annual Cost of Failure	Total Expected Equivalent Annual Cost
A	$90,000(0.2013) = $18,117	$9,000	$94,000(0.40) = $37,600	$64,717
B	100,000(0.2013) = 20,130	10,000	94,000(0.10) = 9,400	39,530
C	160,000(0.2013) = 32,208	16,000	94,000(0.01) = 940	49,148

In Examples 12-1 through 12-3, a revenue or cost factor was modeled as a discrete random variable, with project life assumed certain. A second type of situation involves the cash-flow estimates being certain, but the project life being modeled as a discrete random variable. This is illustrated in Example 12-4.

EXAMPLE 12-4	**Project Life as a Random Variable**

The heating, ventilating, and air-conditioning (HVAC) system in a commercial building has become unreliable and inefficient. Rental income is being hurt, and the annual expenses of the system continue to increase. Your engineering firm has been hired by the owners to (1) perform a technical analysis of the system, (2) develop a preliminary design for rebuilding the system, and (3) accomplish an engineering economic analysis to assist the owners in making a decision. The estimated capital-investment cost and annual savings in operating and maintenance expenses, based on the preliminary design, are shown in the following table. The estimated annual increase in rental revenue with a modern HVAC system has been developed by the owner's marketing staff and is also provided in the following table. These estimates are considered reliable because of the extensive information available. The useful life of the rebuilt system, however, is quite uncertain. The estimated probabilities of various useful lives are provided. Assume that MARR = 12% per year and the estimated market value of the rebuilt system at the end of its useful life is zero. Based on this information, what are $E(PW)$, $V(PW)$, and $SD(PW)$ of the project's cash flows? Also, what is the probability of $PW \geq 0$? What decision would you make regarding the project, and how would you justify your decision using the available information? Solve by hand and by spreadsheet.

Economic Factor	Estimate		Useful Life, Year (N)	p (N)	
Capital investment	$521,000		12	0.1	
Annual savings	48,600		13	0.2	
Increased annual revenue	31,000		14	0.3	
			15	0.2	$\sum = 1.00$
			16	0.1	
			17	0.05	
			18	0.05	

Solution by Hand

The PW of the project's cash flows, as a function of project life (N), is

$$PW(12\%)_N = -\$521{,}000 + \$79{,}600(P/A, 12\%, N).$$

The calculation of the value of $E(PW) = \$9{,}984$ and the value of $E[(PW)^2] = 577.527 \times 10^6 (\$)^2$ are shown in Table 12-5. Then, by using Equation (12-8), the

TABLE 12-5	Calculation of $E(PW)$ and $E[(PW)^2]$ (Example 12-4)				
(1) Useful Life (N)	(2) PW(N)	(3) p(N)	(4) = (2) × (3) E[PW(N)]	(5) = (2)² [PW(N)]²	(6) = (3) × (5) p(N)[PW(N)]²
12	−$27,926	0.1	−$2,793	779.86 × 10⁶	77.986 × 10⁶
13	−9,689	0.2	−1,938	93.88 × 10⁶	18.776 × 10⁶
14	6,605	0.3	1,982	43.63 × 10⁶	13.089 × 10⁶
15	21,148	0.2	4,230	447.24 × 10⁶	89.448 × 10⁶
16	34,130	0.1	3,413	1,164.86 × 10⁶	116.486 × 10⁶
17	45,720	0.05	2,286	2,090.32 × 10⁶	104.516 × 10⁶
18	56,076	0.05	2,804	3,144.52 × 10⁶	157.226 × 10⁶
		$E(PW) =$	$9,984	$E[(PW)^2] =$	577.527 × 10⁶($)²

variance of the PW is

$$V(\text{PW}) = E[(\text{PW})^2] - [E(\text{PW})]^2$$
$$= 577.527 \times 10^6 - (\$9{,}984)^2$$
$$= 477.847 \times 10^6 (\$)^2.$$

The SD(PW) is equal to the positive square root of the variance, $V(\text{PW})$:

$$\text{SD(PW)} = [V(\text{PW})]^{1/2} = (477.847 \times 10^6)^{1/2}$$
$$= \$21{,}859.$$

Based on the PW of the project as a function of N (column 2), and the probability of each PW(N) value occurring (column 3), the probability of the PW being ≥ 0 is

$$\Pr\{\text{PW} \geq 0\} = 1 - (0.1 + 0.2) = 0.7.$$

The results of the engineering economic analysis indicate that the project is a questionable business action. The $E(\text{PW})$ of the project is positive ($9,984) but small relative to the large capital investment. Also, even though the probability of the PW being greater than zero is somewhat favorable (0.7), the SD(PW) value is large [over two times the $E(\text{PW})$ value].

Spreadsheet Solution

Spreadsheets are well suited for the number crunching involved in computing expected values and variances associated with discrete probability functions. Figure 12-1 illustrates the use of a spreadsheet to perform the necessary calculations for this example. The basic format of Table 12-5 is used and the results are same, subject to rounding differences. Once the spreadsheet, however, is formulated, we can easily perform sensitivity analyses to support

$= E2$

$= PV(\$B\$1, A11, -(\$B\$3 + \$B\$4)) -\$B\$2$

	A	B	C	D	E	F
1	MARR =	12%		Useful Life	Probability	
2	Capital Investment =	$ 521,000		12	0.1	
3	Annual Savings =	$ 48,600		13	0.2	$= B11 * C11$
4	Increased Revenue =	$ 31,000		14	0.3	
5				15	0.2	$= C11 \wedge 2$
6				16	0.1	
7				17	0.05	
8				18	0.05	$= B11 * E11$
9						
10	Useful Life	prob (N)	PW	E(PW)	PW^2	p(N)[PW]^2
11	12	0.1	$ (27,928)	$ (2,793)	7.800E+08	7.800E+07
12	13	0.2	$ (9,686)	$ (1,937)	9.381E+07	1.876E+07
13	14	0.3	$ 6,602	$ 1,981	4.359E+07	1.308E+07
14	15	0.2	$ 21,145	$ 4,229	4.471E+08	8.942E+07
15	16	0.1	$ 34,129	$ 3,413	1.165E+09	1.165E+08
16	17	0.05	$ 45,723	$ 2,286	2.091E+09	1.045E+08
17	18	0.05	$ 56,074	$ 2,804	3.144E+09	1.572E+08
18	Totals =			$ 9,982		$ 577,477,384
19						
20	E(PW)	$ 9,982				
21	V(PW) =	$ 477,827,564				
22	SD(PW) =	$ 21,859				

$= D18$

$= SUM (D11:D17)$

$= F18 - D18 \wedge 2$

$= SQRT (B21)$

Figure 12-1 Spreadsheet Solution, Example 12-4

our recommendation. For example, a 5% decrease in annual savings results in a negative E(PW), supporting our conclusion that the profitability of the project is questionable.

12.4.1 Probability Trees

The discrete distribution of cash flows sometimes occurs in each time period. A *probability tree diagram* is useful in describing the prospective cash flows, and the probability of each value occurring, for this situation. Example 12-5 is a problem of this type.

EXAMPLE 12-5   **Project Analysis Using a Probability Tree**

The uncertain cash flows for a small improvement project are described by the probability tree diagram in Figure 12-2. (Note that the probabilities emanating from each node sum to unity.) The analysis period is two years, and MARR = 12% per year. Based on this information,

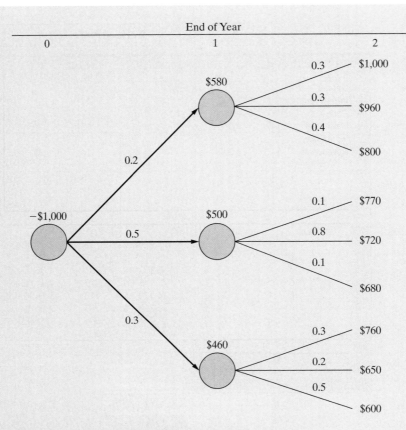

Figure 12-2 Probability Tree Diagram for Example 12-5

(a) What are the $E(PW)$, $V(PW)$, and SD(PW) of the project?
(b) What is the probability that PW $\leq$ 0?
(c) Which analysis result(s) favor approval of the project, and which ones appear unfavorable?

Solution

(a) The calculation of the values for $E(PW)$ and $E[(PW)^2]$ is shown in Table 12-6. In column 2, PW_j is the PW of branch j in the tree diagram. The probability of each branch occurring, $p(j)$, is shown in column 3. For example, proceeding from the right node for each cash flow in Figure 12-2 to the left node, we have $p(1) = (0.3)(0.2) = 0.06$ and $p(9) = (0.5)(0.3) = 0.15$. Hence,

$$E(PW) = \sum_j (PW_j)p(j) = \$39.56.$$

Then, $V(PW) = E[(PW)^2] - [E(PW)]^2$
$= 15,227 - (\$39.56)^2$
$= 13,662(\$)^2,$

| TABLE 12-6 | Calculation of $E(PW)$ and $E[(PW)^2]$ (Example 12-5) |

	(1) Net Cash Flow			(2)	(3)	(4) = (2) × (3)	(5) = (2)²	(6) = (3) × (5)
		EOY		PW_j	$p(j)$	$E(PW_j)$	$(PW_j)^2$	$E[(PW_j)^2]$
j	0	1	2					
1	−$1,000	$580	$1,000	$315	0.06	$18.90	99,225$²	5,953$²
2	−1,000	580	960	283	0.06	16.99	80,089	4,805
3	−1,000	580	800	156	0.08	12.45	24,336	1,947
4	−1,000	500	770	60	0.05	3.04	3,600	180
5	−1,000	500	720	20	0.40	8.17	400	160
6	−1,000	500	680	−11	0.05	−0.57	121	6
7	−1,000	460	760	17	0.09	1.49	289	26
8	−1,000	460	650	−71	0.06	−4.27	5,044	302
9	−1,000	460	600	−111	0.15	−16.64	12,321	1,848
						$E(PW) = \$39.56$		$E[(PW)^2] = 15,227\$^2$

and $\text{SD(PW)} = [V(PW)]^{1/2} = (13,662)^{1/2} = \$116.88.$

(b) Based on the entries in column 2, PW_j, and column 3, $p(j)$, we have

$$\Pr\{PW \leq 0\} = p(6) + p(8) + p(9)$$

$$= 0.05 + 0.06 + 0.15$$

$$= 0.26.$$

(c) The analysis results that favor approval of the project are $E(PW) = \$39.56$, which is greater than zero only by a small amount, and $\Pr\{PW > 0\} = 1 - 0.26 = 0.74$. The SD(PW) = \$116.92, however, is approximately three times the $E(PW)$. This indicates a relatively high variability in the measure of economic merit, PW of the project, and is usually an unfavorable indicator of project acceptability.

12.4.2 An Application Perspective

One of the major problems in computing expected values is the determination of the probabilities. In many situations, there is no precedent for the particular venture being considered. Therefore, probabilities seldom can be based on historical data and rigorous statistical procedures. In most cases, the analyst, or person making the decision, must make a judgment based on all available information in estimating the probabilities. This fact makes some people hesitate to use the expected-value concept, because they cannot see the value of applying such a technique to improve the evaluation of risk and uncertainty when so much apparent subjectivity is present.

Although this argument has merit, the fact is that engineering economy studies deal with future events and there must be an extensive amount of estimating. Furthermore, even if the probabilities could be based accurately on past history, there rarely is any assurance that the future will repeat the past. Thus, structured methods for assessing subjective probabilities are often used in practice.* Also, even if we must estimate the probabilities, the very process of doing so requires us to make explicit the uncertainty that is inherent in all estimates going into the analysis. Such structured thinking is likely to produce better results than little or no thinking about such matters.

12.5 Evaluation of Projects with Continuous Random Variables

In this section, we continue to compute the expected values and the variances of probabilistic factors, but we model the selected probabilistic factors as *continuous* random variables. In each example, simplifying assumptions are made about the distribution of the random variable and the statistical relationship among the values it takes on. When the situation is more complicated, such as a problem that involves probabilistic cash flows and probabilistic project lives, a second general procedure that utilizes Monte Carlo simulation is normally used. This is the subject of Section 12.6.

Two frequently used assumptions about uncertain cash-flow amounts are that they are distributed according to a normal distribution[†] and are statistically independent. Underlying these assumptions is a general characteristic of many cash flows, in that they result from a number of different and independent factors.

The advantage of using statistical independence as a simplifying assumption, when appropriate, is that no correlation between the cash-flow amounts (e.g., the net annual cash-flow amounts for an alternative) is being assumed. Consequently, if we have a linear combination of two or more independent cash-flow amounts, say $PW = c_0 F_0 + \cdots + c_N F_N$, where the c_k values are coefficients and the F_k values are periodic net cash flows, the expression for the $V(PW)$, based on Equation (12-10), reduces to

$$V(PW) = \sum_{k=0}^{N} c_k^2 V(F_k).$$

(12-13)

---

* For further information, see W. G. Sullivan and W. W. Claycombe, *Fundamentals of Forecasting* (Reston, VA: Reston Publishing Co., 1977), Chapter 6.

[†] This frequently encountered continuous probability function is discussed in any good statistics book, such as R. E. Walpole and R. H. Myers, *Probability and Statistics for Engineers and Scientists* (New York: Macmillan Publishing Co., 1989), 139–154.

And, based on Equation (12-9), we have

$$E(PW) = \sum_{k=0}^{N} c_k E(F_k).$$ (12-14)

EXAMPLE 12-6   Annual Net Cash Flow as a Continuous Random Variable

For the following annual cash-flow estimates, find the $E(PW)$, $V(PW)$, and $SD(PW)$ of the project. Assume that the annual net cash-flow amounts are normally distributed with the expected values and standard deviations as given and statistically independent and that the MARR $= 15\%$ per year.

End of Year, k	Expected Value of Net Cash Flow, F_k	SD of Net Cash Flow, F_k
0	−$7,000	0
1	3,500	$600
2	3,000	500
3	2,800	400

A graphical portrayal of these normally distributed cash flows is shown in Figure 12-3.

Solution

The expected PW, based on Equation (12-14), is calculated as follows, where $E(F_k)$ is the expected net cash flow in year k $(0 \le k \le N)$ and c_k is the single-payment

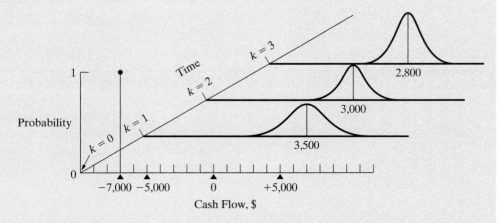

Figure 12-3   Probabilistic Cash Flows over Time (Example 12-6)

PW factor $(P/F, 15\%, k)$:

$$E(\text{PW}) = \sum_{k=0}^{3} (P/F, 15\%, k) E(F_k)$$

$$= -\$7{,}000 + \$3{,}500(P/F, 15\%, 1) + \$3{,}000(P/F, 15\%, 2)$$

$$+\$2{,}800(P/F, 15\%, 3)$$

$$= \$153.$$

To determine $V(\text{PW})$, we use the relationship in Equation (12-13). Thus,

$$V(\text{PW}) = \sum_{k=0}^{3} (P/F, 15\%, k)^2 V(F_k)$$

$$= 0^2 1^2 + 600^2 (P/F, 15\%, 1)^2 + 500^2 (P/F, 15\%, 2)^2$$

$$+ 400^2 (P/F, 15\%, 3)^2$$

$$= 484{,}324\$^2$$

and

$$\text{SD(PW)} = [V(\text{PW})]^{1/2} = \$696.$$

When we can assume that a random variable, say the PW of a project's cash flow, is normally distributed with a mean, $E(\text{PW})$, and a variance, $V(\text{PW})$, we can compute the probability of events about the random variable occurring. This assumption can be made, for example, when we have some knowledge of the shape of the distribution of the random variable and when it is appropriate to do so. Also, this assumption may be supportable when a random variable, such as a project's PW value, is a linear combination of other independent random variables (say, the cash-flow amounts, F_k), regardless of whether the form of the probability distribution(s) of these variables is known.[*]

EXAMPLE 12-7   **Probability of an Unfavorable Project Outcome**

Refer to Example 12-6. For this problem, what is the probability that the internal rate of return (IRR) of the cash-flow estimates is less than MARR, Pr{IRR < MARR}? Assume that the PW of the project is a normally distributed random variable, with its mean and variance equal to the values calculated in Example 12-6.

---

[*] The theoretical basis of this assumption is the central limit theorem of statistics. For a summary discussion of the supportability of this assumption under different conditions, see C. S. Park and G. P. Sharpe-Bette, *Advanced Engineering Economics* (New York: John Wiley & Sons, 1990), 420–421.

Solution

For a decreasing PW(*i*) function having a unique IRR, the probability that IRR is less than MARR is the same as the probability that PW is less than zero. Consequently, by using the standard normal distribution in Appendix E, we can determine the probability that PW is less than zero:*

$$Z = \frac{PW - E(PW)}{SD(PW)} = \frac{0 - 153}{696} = -0.22;$$

$$Pr\{PW \leq 0\} = Pr\{Z \leq -0.22\}.$$

From Appendix E, we find that $Pr\{Z \leq -0.22\} = 0.4129$.

| EXAMPLE 12-8 | **Linear Combination of Independent Random Variables** |

The estimated cash-flow data for a project are shown in the following table for the five-year study period being used. Each annual net cash-flow amount, F_k, is a linear combination of two statistically independent random variables, X_k and Y_k, where X_k is a revenue factor and Y_k is an expense factor. The X_k cash-flow amounts are statistically independent of each other, and the same is true of the Y_k amounts. Both X_k and Y_k are continuous random variables, but the form of their probability distributions is not known. MARR = 20% per year. Based on this information,

(a) What are the E(PW), V(PW), and SD(PW) values of the project's cash flows?

(b) What is the probability that PW is less than zero, that is, $Pr\{PW \leq 0\}$, and the project is economically attractive?

End of	Net Cash Flow	Expected Value		Standard Deviation (SD)	
Year, k	$F_k = a_k X_k - b_k Y_k$	X_k	Y_k	X_k	Y_k
0	$F_0 = X_0 + Y_0$	$0	-$100,000	$0	$10,000
1	$F_1 = X_1 + Y_1$	60,000	-20,000	4,500	2,000
2	$F_2 = X_2 + 2Y_2$	65,000	-15,000	8,000	1,200
3	$F_3 = 2X_3 + 3Y_3$	40,000	-9,000	3,000	1,000
4	$F_4 = X_4 + 2Y_4$	70,000	-20,000	4,000	2,000
5	$F_5 = 2X_5 + 2Y_5$	55,000	-18,000	4,000	2,300

---

* The standard normal distribution, $f(Z)$, of the variable $Z = (X - \mu)/\sigma$ has a mean of zero and a variance of one, when X is a normally distributed random variable with mean μ and standard deviation σ.

TABLE 12-7   Calculation of $E(F_k)$ and $V(F_k)$ (Example 12-8)

End of Year, k	F_k	$E(F_k) =$ $a_k E(X_k) + b_k E(Y_k)$			$V(F_k) =$ $a_k^2 V(X_k) + b_k^2 V(Y_k)$
0	F_0	\$0 −	\$100,000 =	−\$100,000	$0 + (1)^2(10,000)^2 = 100.0 \times 10^6 \$^2$
1	F_1	60,000 −	20,000 =	40,000	$(4,500)^2 + (1)^2 (2,000)^2 = 24.25 \times 10^6$
2	F_2	65,000 −	2(15,000) =	35,000	$(8,000)^2 + (2)^2 (1,200)^2 = 69.76 \times 10^6$
3	F_3	2(40,000) −	3(9,000) =	53,000	$(2)^2(3,000)^2 + (3)^2 (1,000)^2 = 45.0 \times 10^6$
4	F_4	70,000 −	2(20,000) =	30,000	$(4,000)^2 + (2)^2 (2,000)^2 = 32.0 \times 10^6$
5	F_5	2(55,000) −	2(18,000) =	74,000	$(2)^2(4,000)^2 + (2)^2 (2,300)^2 = 85.16 \times 10^6$

Solution

(a) The calculation of the $E(F_k)$ and $V(F_k)$ values of the project's annual net cash flows are shown in Table 12-7. The $E(PW)$ is calculated by using Equation (12-14) as follows:

$$E(PW) = \sum_{k=0}^{5} (P/F, 20\%, k)E(F_k)$$

$$= -\$100,000 + \$40,000(P/F, 20\%, 1)$$

$$+ \cdots + \$74,000(P/F, 20\%, 5)$$

$$= \$32,517.$$

Then, the $V(PW)$ is calculated using Equation (12-13) as follows:

$$V(PW) = \sum_{k=0}^{5} (P/F, 20\%, k)^2 V(F_k)$$

$$= 100.0 \times 10^6 + (24.25 \times 10^6)(P/F, 20\%, 1)^2$$

$$+ \cdots + (85.16 \times 10^6)(P/F, 20\%, 5)^2$$

$$= 186.75 \times 10^6 (\$)^2.$$

Finally, $SD(PW) = [V(PW)]^{1/2}$

$$= [186.75 \times 10^6]^{1/2}$$

$$= \$13,666.$$

(b) The PW of the project's net cash flow is a linear combination of the annual net cash-flow amounts, F_k, that are independent random variables. Each of these random variables, in turn, is a linear combination of the independent random variables X_k and Y_k. We can also observe in Table 12-7 that the $V(PW)$ calculation does not include any dominant $V(F_k)$ values. Therefore, we have a reasonable basis on which to assume that the PW of the project's net cash flow is approximately normally distributed, with $E(PW) = \$32,517$ and $SD(PW) = \$13,666$.

Based on this assumption, we have

$$Z = \frac{PW - E(PW)}{SD(PW)} = \frac{0 - \$32,517}{\$13,666} = -2.3794;$$

$$\Pr\{PW \le 0\} = \Pr\{Z \le -2.3794\}.$$

From Appendix E, we find that $\Pr\{Z \le -2.3794\} = 0.0087$. Therefore, the probability of loss on this project is negligible. Based on this result, the $E(PW) > 0$, and the $SD(PW) = 0.42[E(PW)]$; therefore, the project is economically attractive and has low risk of failing to add value to the firm.

12.6 Evaluation of Risk and Uncertainty by Monte Carlo Simulation

The modern development of computers and related software has resulted in the increased use of Monte Carlo simulation as an important tool for analysis of project uncertainties. For complicated problems, Monte Carlo simulation generates random outcomes for probabilistic factors so as to imitate the randomness inherent in the original problem. In this manner, a solution to a rather complex problem can be inferred from the behavior of these random outcomes.

To perform a simulation analysis, the first step is to construct an analytical model that represents the actual decision situation. This may be as simple as developing an equation for the PW of a proposed industrial robot in an assembly line or as complex as examining the economic effects of proposed environmental regulations on typical petroleum refinery operations. The second step is to develop a probability distribution from subjective or historical data for each uncertain factor in the model. Sample outcomes are randomly generated by using the probability distribution for each uncertain quantity and are then used to determine a *trial* outcome for the model. Repeating this sampling process a large number of times leads to a frequency distribution of trial outcomes for a desired measure of merit, such as PW or AW. The resulting frequency distribution can then be used to make probabilistic statements about the original problem.

This section has been adapted from W. G. Sullivan and R. Gordon Orr, "Monte Carlo Simulation Analyzes Alternatives in Uncertain Economy," *Industrial Engineering*, 14, no. 11 (November 1982): 42–49. Reprinted with permission from *Industrial Engineering* magazine. Copyright Institute of Industrial Engineers, Inc. 25 Technology Park/Atlanta, Norcross, Georgia, 30092–2988.

To illustrate the Monte Carlo simulation procedure, suppose that the probability distribution for the useful life of a piece of machinery has been estimated as shown in Table 12-8. The useful life can be simulated by assigning random numbers to each value such that they are proportional to the respective probabilities. (A random number is selected in a manner such that each number has an equal probability of occurrence.) Because two-digit probabilities are given in Table 12-8, random numbers can be assigned to each outcome, as shown in Table 12-9. Next, a single outcome is simulated by choosing a number at random from a table of random numbers.* For example, if any random number between and including 00 and 19 is selected, the useful life is three years. As a further example, the random number 74 corresponds to a life of seven years.

If the probability distribution that describes a random variable is *normal*, a slightly different approach is followed. Here the simulated outcome is based on the mean and standard deviation of the probability distribution and on a random normal deviate, which is a random number of standard deviations above or below the mean of a standard normal distribution. An abbreviated listing of typical random normal deviates is shown in Table 12-10. For normally distributed random variables, the simulated outcome is based on Equation (12-15):

$$\text{Outcome value} = \text{mean} + [\text{random normal deviate} \times \text{standard deviation}]. \qquad (12\text{-}15)$$

For example, suppose that an *annual* net cash flow is assumed to be normally distributed, with a mean of $50,000 and a standard deviation of $10,000, as shown in Figure 12-4.

TABLE 12-8	Probability Distribution for Useful Life	
Number of Years, N		$p(N)$
3	possible values	0.20
5		0.40
7		0.25
10		0.15

$\sum p(N) = 1.00$

TABLE 12-9	Assignment of Random Numbers
Number of Years, N	Random Numbers
3	00–19
5	20–59
7	60–84
10	85–99

---

* The last two digits of randomly chosen telephone numbers in a telephone directory are usually quite close to being random numbers.

TABLE 12-10	Random Normal Deviates (*RND*s)			
−1.565	0.690	−1.724	0.705	0.090
0.062	−0.072	0.778	−1.431	0.240
0.183	−1.012	−0.844	−0.227	−0.448
−0.506	2.105	0.983	0.008	0.295
1.613	−0.225	0.111	−0.642	−0.292

Figure 12-4 Normally Distributed Annual Cash Flow

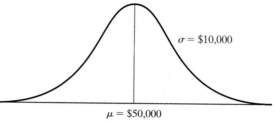

$\sigma = \$10,000$

$\mu = \$50,000$
Annual Net Cash Flow

TABLE 12-11	Example of the Use of *RND*s	
		Annual Net Cash Flow
Year	RND	[$50,000 + RND ($10,000)]
1	0.090	$50,900
2	0.240	52,400
3	−0.448	45,520
4	0.295	52,950
5	−0.292	47,080

Simulated cash flows for a period of five years are listed in Table 12-11. Notice that the average annual net cash flow is $248,850/5, which equals $49,770. This approximates the known mean of $50,000 with an error of 0.46%.

If the probability distribution that describes a random event is *uniform* and continuous, with a minimum value of A and a maximum value of B, another procedure should be followed to determine the simulated outcome. Here the outcome can be computed with the formula

$$\text{Simulation outcome} = A + \frac{RN}{RN_m}[B - A], \qquad (12\text{-}16)$$

where RN_m is the maximum possible random number (9 if one digit is used, 99 if two are used, etc.) and RN is the random number actually selected. This equation should be used when the minimum outcome, A, and the maximum outcome, B, are known.

For example, suppose that the market value in year N is assumed to be uniformly and continuously distributed between $8,000 and $12,000. A value of this random variable would be generated as follows with a random number of 74:

$$\text{Simulation outcome} = \$8,000 + \frac{74}{99}(\$12,000 - \$8,000) = \$10,990.$$

Proper use of these procedures, coupled with an accurate model, will result in an approximation of the actual outcome. But how many simulation trials are necessary for an *accurate* approximation of, for example, the average outcome? In general, the greater the number of trials, the more accurate the approximation of the mean and standard deviation will be. One method of determining whether a sufficient number of trials has been conducted is to keep a running average of results. At first, this average will vary considerably from trial to trial. The amount of change between successive averages should decrease as the number of simulation trials increases. Eventually, this running (cumulative) average should level off at an accurate approximation.

EXAMPLE 12-9   **Project Analysis Using Monte Carlo Simulation**

Monte Carlo simulation can also simplify the analysis of a more complex problem. The following estimates relate to an engineering project being considered by a large manufacturer of air-conditioning equipment. Subjective probability functions have been estimated for the four independent uncertain factors as follows:

(a) Capital investment: Normally distributed with a mean of $50,000 and a standard deviation of $1,000.

(b) Useful life: Uniformly and continuously distributed with a minimum life of 10 years and a maximum life of 14 years.

(c) Annual revenue:

$35,000 with a probability of 0.4

$40,000 with a probability of 0.5

$45,000 with a probability of 0.1

(d) Annual expense: Normally distributed, with a mean of $30,000 and a standard deviation of $2,000.

The management of this company wishes to determine whether the capital investment in the project will be a profitable one. The interest rate is 10% per year. To answer this question, the PW of the venture will be simulated.

Solution

To illustrate the Monte Carlo simulation procedure, five trial outcomes are computed manually in Table 12-12. The estimate of the average PW based on this very small sample is $19,010/5 = $3,802. For more accurate results, hundreds or even thousands of repetitions would be required.

TABLE 12-12 Monte Carlo Simulation of PW Involving Four Independent Factors (Example 12-9)

Trial Number	Random Normal Deviate (RND$_1$)	Capital Investment, I [$50,000 + RND$_1$($1,000)]	Three-Digit RNs	Project Life, N [$10 + \frac{RN}{999}(14-10)$]	Project Life, N (Nearest Integer)
1	−1.003	$48,997	807	13.23	13
2	−0.358	49,642	657	12.63	13
3	+1.294	51,294	488	11.95	12
4	−0.019	49,981	282	11.13	11
5	+0.147	50,147	504	12.02	12

	One-Digit RN	Annual Revenue, R $35,000 for 0–3 40,000 for 4–8 45,000 for 9	RND$_2$	Annual Expense, E [$30,000 + RND$_2$ ($2,000)]	PW $= -I +$ $(R-E)(P/A, 10\%, N)$
1	2	$35,000	−0.036	$29,928	−$12,969
2	0	35,000	+0.605	31,210	−22,720
3	4	40,000	+1.470	32,940	−3,189
4	9	45,000	+1.864	33,728	+23,232
5	8	40,000	−1.223	27,554	+34,656

Total +$19,010

The applications of Monte Carlo simulation for investigating risk and uncertainty are many and varied. Remember that the results, however, can be no more accurate than the model and the probability estimates used. In all cases, the procedure and rules are the same: careful study of the problem and development of the model; accurate assessment of the probabilities involved; true randomization of outcomes as required by the Monte Carlo simulation procedure; and calculation and analysis of the results. Furthermore, a sufficiently large number of Monte Carlo trials should always be used to reduce the estimation error to an acceptable level.

12.7 Performing Monte Carlo Simulation with a Computer

It is apparent from the preceding section that a Monte Carlo simulation of a complex project requiring several thousand trials can be accomplished only with the help of a computer. Indeed, there are numerous simulation programs that can be obtained from software companies and universities. MS Excel also has a simulation feature that can generate random data for seven different probability distributions. In this section, we will demonstrate that Monte Carlo simulation is not only feasible but also relatively easy to perform with spreadsheets.

At the heart of Monte Carlo simulation is the generation of random numbers. Most spreadsheet packages include a RAND( ) function that returns a random number between zero and one. Other advanced statistical functions, such as NORMSINV( ), will return the inverse of a cumulative distribution function (the standard normal distribution in this case). This function can be used to generate random normal deviates. The spreadsheet model shown in Figure 12-5 makes use of these functions in performing a Monte Carlo simulation for the project being evaluated in Example 12-9.

The probabilistic functions for the four independent uncertain factors are identified in the spreadsheet model. The input parameter names and distributions

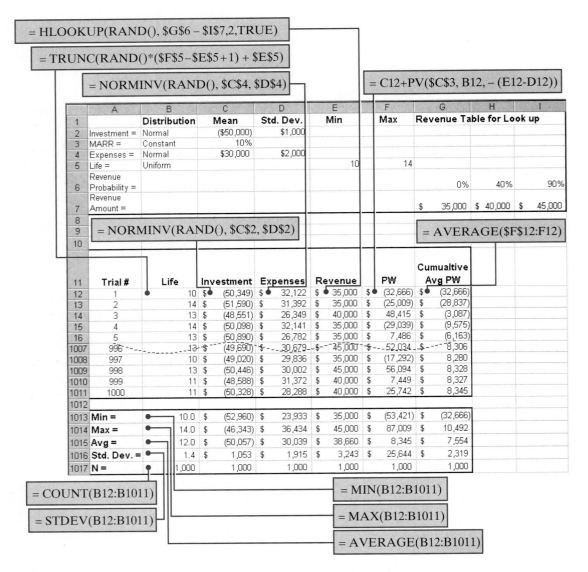

Figure 12-5   Monte Carlo Simulation Using a Spreadsheet, Example 12-9

are identified in cells A2:B7. The mean and standard deviation of the normally distributed factors (capital investment and annual expenses) are identified in the range C2:D4. The maximum and minimum values for the uniformly distributed project life are shown in E5:F5. A discrete probability function was compiled for annual revenues. The associated cumulative probability distribution is shown in G6:I7.

The formulas in cells B12:E12 return values for the uncertain factors, based on the underlying distribution and parameters. These values are used in cell F12 to calculate the PW. As these formulas are copied down the spreadsheet, each row becomes a new trial in the simulation. Each time the F9 (recalculate) key is pressed, the RAND( ) functions are reevaluated and a new simulation begins. The formulas in the range B1013:F1017 calculate basic statistics for the input parameters and the PW. By comparing these results with the input values, one can rapidly check for errors in the basic formulas.

A true simulation involves hundreds, or perhaps thousands, of trials. In this example, 1,000 trials were used. The first and last five trials are shown in Figure 12-5 (rows 17 through 1006 have been hidden for display purposes). The average PW is $8,345 (cell F1015), which is larger than the $3,802 obtained from Table 12-12. This underscores the importance of having a sufficient number of simulation trials to ensure reasonable accuracy in Monte Carlo analyses. The cumulative average PW calculated in column G is one simple way to determine whether enough trials have been run. A plot of these values will stabilize and show little change when the simulation has reached a steady state. Figure 12-6 shows the plot of the cumulative average PW for the 1,000 trials generated for this example.

The average PW is of interest, but the distribution of PW is often even more significant. A histogram of the PW values in column F displays the shape of the distribution, as shown in Figure 12-7. The dispersion of PW trial outcomes in this

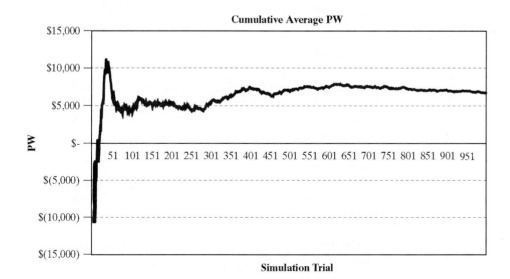

Figure 12-6 Plot of Cumulative Average PW to Determine whether Enough Simulation Trials Have Been Run

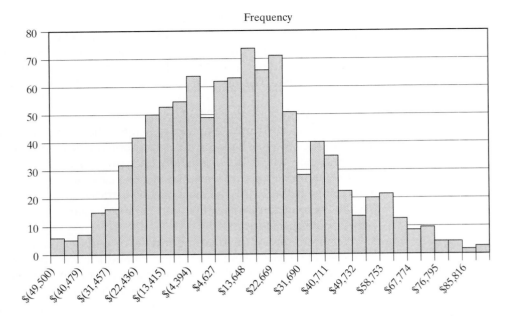

Figure 12-7   Histogram Showing Distribution of PW for 1,000 Simulation Trials

example is considerable. The standard deviation of simulated trial outcomes is one way to measure this dispersion. In addition, the proportion of values less than zero indicates the probability of a negative PW. Based on Figure 12-7, approximately 40% of all simulation outcomes have a PW less than zero. Consequently, this project may be too risky for the company to undertake, because the downside risk of failing to realize at least a 10% per year return on the capital investment is about 4 chances out of 10. Perhaps another project investment should be considered.

A typical application of simulation involves the analysis of several mutually exclusive alternatives. In such studies, how can one compare alternatives that have different expected values and standard deviations of, for instance, PW? One approach is to select the alternative that *minimizes* the probability of attaining a PW that is less than zero. Another popular response to this question utilizes a graph of expected value (a measure of the reward) plotted against standard deviation (an indicator of risk) for each alternative. An attempt is then made to assess subjectively the trade-offs that result from choosing one alternative over another in pairwise comparisons.

To illustrate the latter concept, suppose that three alternatives having varying degrees of uncertainty have been analyzed with Monte Carlo computer simulation and the results shown in Table 12-13 have been obtained. These results are plotted in Figure 12-8, where it is apparent that Alternative C is inferior to Alternatives A and B because of its lower E(PW) and larger standard deviation. Therefore, C offers a smaller PW that has a greater amount of risk associated with it! Unfortunately, the choice of B versus A is not as clear because the increased expected PW of B has to be balanced against the increased risk of B. This trade-off *may or may not* favor B, depending on management's attitude toward accepting the additional uncertainty

TABLE 12-13	Simulation Results for Three Mutually Exclusive Alternatives		
Alternative	E(PW)	SD(PW)	E(PW) ÷ SD(PW)
A	$37,382	$1,999	18.70
B	49,117	2,842	17.28
C	21,816	4,784	4.56

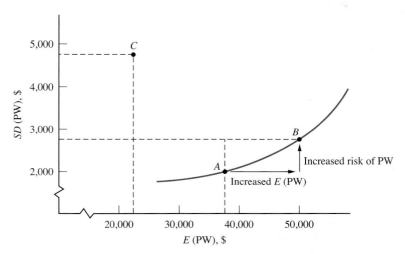

Figure 12-8 Graphical Summary of Computer Simulation Results

associated with a larger expected reward. The comparison also presumes that Alternative A is acceptable to the decision maker. One simple procedure for choosing between A and B is to rank alternatives based on the ratio of E(PW) to SD(PW). In this case, Alternative A would be chosen because it has the more favorable (larger) ratio.

12.8 Decision Trees

Decision trees, also called *decision flow networks* and *decision diagrams*, are powerful means of depicting and facilitating the analysis of important problems, especially those that involve sequential decisions and variable outcomes over time. Decision trees are used in practice because they make it possible to break down a large, complicated problem into a series of smaller simple problems, and they enable objective analysis and decision making that includes explicit consideration of the risk and effect of the future.

This section has been adapted (except Section 12.8.3) from John R. Canada and William G. Sullivan, *Economic and Multiattribute Evaluation of Advanced Manufacturing Systems* (Upper Saddle River, NJ: Prentice Hall, 1989), 341–343, 347. Reprinted by permission of the publisher.

The name *decision tree* is appropriate, because it shows branches for each possible alternative for a given decision and for each possible outcome (event) that can result from each alternative. Such networks reduce abstract thinking to a logical visual pattern of cause and effect. When costs and benefits are associated with each branch and probabilities are estimated for each possible outcome, analysis of the decision flow network can clarify choices and risks.

12.8.1 Deterministic Example

The most basic form of a decision tree occurs when each alternative can be assumed to result in a single outcome—that is, when certainty is assumed. The replacement problem in Figure 12-9 illustrates this. The problem illustrates that the decision on whether to replace the defender (old machine) with the new machine (challenger) is not just a one-time decision but rather one that recurs periodically. That is, if the decision is made to keep the old machine at decision point one, then later, at decision point two, a choice again has to be made. Similarly, if the old machine is chosen at decision point two, then a choice again has to be made at decision point three. For each alternative, the cash inflow and duration of the project are shown above the arrow, and the capital investment is shown below the arrow.

For this problem, the question initially seems to be which alternative to choose at decision point one. But an intelligent choice at decision point one should take into account the later alternatives and decisions that stem from it. Hence, the correct procedure is to start at the most distant decision point, determine the best alternative and quantitative result of that alternative, and then roll back to each preceding decision point, repeating the procedure until finally the choice at the initial or present decision point is determined. By this procedure, one can make a present decision that directly takes into account the alternatives and expected decisions of the future.

For simplicity in this example, timing of the monetary outcomes will first be neglected, which means that a dollar has the same value regardless of the year in which it occurs. Table 12-14 shows the necessary computations and decisions using

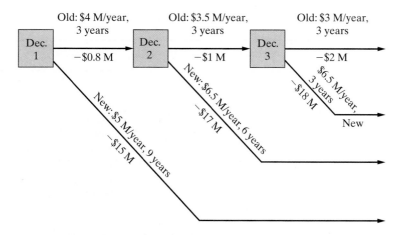

Figure 12-9 Deterministic Replacement Example

TABLE 12-14	Monetary Outcomes and Decisions at Each Point—Deterministic Replacement Example of Figure 12-9[a]			
Decision Point	Alternative	Monetary Outcome		Choice
3	Old	$3M(3) − $2M	= $7.0M	Old
	New	$6.5M(3) − $18M	= $1.5M	
2	Old	$7M + $3.5M(3) − $1M	= $16.5M	
	New	$6.5M(6) − $17M	= $22.0M	New
1	Old	$22.0M + $4M(3) − $0.8M	= $33.2M	Old
	New	$5M(9) − $15M	= $30.0M	

[a] Interest $= 0\%$ per year, that is, ignore timing of cash flows.

a nine-year study period. Note that the monetary outcome of the best alternative at decision point three ($7.0M for the *old*) becomes part of the outcome for the old alternative at decision point two. Similarly, the best alternative at decision point two ($22.0M for the *new*) becomes part of the outcome for the defender alternative at decision point one.

The computations in Table 12-14 show that the answer is to keep the old alternative now and plan to replace it with the new one at the end of three years (at decision point two). But this does not mean that the old machine should necessarily be kept for a full three years and a new machine bought without question at the end of that period. Conditions may change at any time, necessitating a fresh analysis—probably a decision tree analysis—based on estimates that are reasonable in light of conditions at that time.

12.8.1.1 Deterministic Example Considering Timing For decision tree analyses, which involve working from the most distant to the nearest decision point, the easiest way to take into account the timing of money is to use the PW approach and thus *discount all monetary outcomes to the decision points in question.* To demonstrate, Table 12-15 shows computations for the same replacement problem of Figure 12-9, using an interest rate of 25% per year.

Note from Table 12-15 that, when taking into account the effect of timing by calculating PWs at each decision point, the indicated choice is not only to keep the old alternative at decision point one but also to keep the old alternative at decision points two and three as well. This result is not surprising since the high interest rate tends to favor the alternatives with lower capital investments, and it also tends to place less weight on long-term returns (benefits).

12.8.2 General Principles of Diagramming

The proper diagramming of a decision problem is, in itself, generally very useful to the understanding of the problem, and it is essential to correct subsequent analysis.

The placement of decision points (nodes) and chance outcome nodes from the initial decision point to the base of any later decision point should give an accurate representation of the information that will and will not be available when the choice

TABLE 12-15	Decision at Each Point with Interest = 25% per Year for Deterministic Replacement Example of Figure 12-9			
Decision Point	Alternative	PW of Monetary Outcome	Choice	
3	Old	$3M($P/A$, 3) − $2M $3M(1.95) − $2M	= __$3.85M__	Old
	New	$6.5M($P/A$, 3) − $18M $6.5M(1.95) − $18M	= −$5.33M	
2	Old	$3.85M($P/F$, 3) + $3.5M($P/A$, 3) − $1M $3.85M(0.512) + $3.5M(1.95) − $1M	= __$7.79M__	Old
	New	$6.5M($P/A$, 6) − $17M $6.5M(2.95) − $17M	= $2.18M	
1	Old	$7.79M($P/F$, 3) + $4M($P/A$, 3) − $0.8M $7.79M(0.512) + $4M(1.95) − $0.8M	= __$10.99M__	Old
	New	$5.0M($P/A$, 9) − $15M $5.0M(3.46) − $15M	= $2.30M	

represented by the decision point in question actually has to be made. The decision tree diagram should show the following (normally, a square symbol is used to depict a decision node, and a circle symbol is used to depict a chance outcome node):

1. all initial or immediate alternatives among which the decision maker wishes to choose

2. all uncertain outcomes and future alternatives that the decision maker wishes to consider because they may directly affect the consequences of initial alternatives

3. all uncertain outcomes that the decision maker wishes to consider because they may provide information that can affect his or her future choices among alternatives and hence indirectly affect the consequences of initial alternatives

Note that the alternatives at any decision point and the outcomes at any chance outcome node must be

1. mutually exclusive, (i.e., no more than one can be chosen)

2. collectively exhaustive, (i.e., one event must be chosen or something must occur if the decision point or outcome node is reached)

12.8.3   Decision Trees with Random Outcomes

The deterministic replacement problem discussed in Section 12.8.1 introduced the concept of sequential decisions using assumed certainty for alternative outcomes. An engineering problem requiring sequential decisions, however, often includes random outcomes, and decision trees are very useful in structuring this type of situation. The decision tree diagram helps to make the problem explicit and assists in its analysis. This is illustrated in Examples 12-10 through 12-12.

| EXAMPLE 12-10 | **Probabilistic Decision Tree Analysis** |

The Ajax Corporation manufactures compressors for commercial air-conditioning systems. A new compressor design is being evaluated as a potential replacement for the most frequently used unit. The new design involves major changes that have the expected advantage of better operating efficiency. From the perspective of a typical user, the new compressor (as an assembled component in an air-conditioning system) would have an increased investment of $8,600 relative to the present unit and an annual expense saving dependent upon the extent to which the design goal is met in actual operations.

Estimates by the multidisciplinary design team of the new compressor achieving four levels (percentages) of the efficiency design goal and the probability and annual expense saving at each level are as follows:

Level (Percentage) of Design Goal Met (%)	Probability $p(L)$	Annual Expense Saving
90	0.25	$3,470
70	0.40	2,920
50	0.25	2,310
30	0.10	1,560

Based on a before-tax analysis (MARR = 18% per year, analysis period = 6 years, and market value = 0) and $E(\text{PW})$ as the decision criterion, is the new compressor design economically preferable to the current unit?

Solution

The single-stage decision tree diagram for the design alternatives is shown in Figure 12-10. The PWs associated with each of the efficiency design goal levels

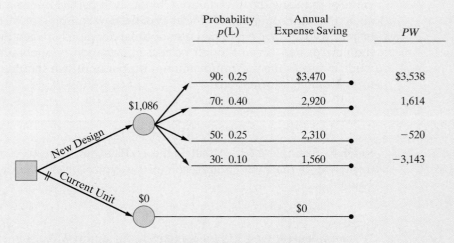

	Probability $p(L)$	Annual Expense Saving	PW
	90: 0.25	$3,470	$3,538
	70: 0.40	2,920	1,614
New Design — $1,086	50: 0.25	2,310	−520
	30: 0.10	1,560	−3,143
Current Unit — $0		$0	

Figure 12-10 Single-Stage Decision Tree (Example 12-10)

being met are as follows:

$$PW(18\%)_{90} = -\$8,600 + \$3,470(P/A, 18\%, 6) = \$3,538$$

$$PW(18\%)_{70} = -\$8,600 + \$2,920(P/A, 18\%, 6) = \$1,614$$

$$PW(18\%)_{50} = -\$8,600 + \$2,310(P/A, 18\%, 6) = -\$520$$

$$PW(18\%)_{30} = -\$8,600 + \$1,560(P/A, 18\%, 6) = -\$3,143$$

Based on these values, the $E(PW)$ of each installed unit of the new compressor is as follows:

$$E(PW) = 0.25(\$3,538) + 0.40(\$1,614)$$

$$+0.25(-\$520) + 0.10(-\$3,143)$$

$$= \$1,086.$$

The $E(PW)$ of the current unit is zero since the cash-flow estimates for the new design are incremental amounts relative to the present design. Therefore, the analysis indicates that the new design is economically preferable to the present design. (The parallel lines across the current unit path on the diagram indicate that it was not selected.)

12.8.3.1   The Expected Value of Perfect Information (EVPI)   The probability estimates of achieving the efficiency design goal levels, $p(L)$, developed by the design team in Example 12-10 reflect the uncertainty about the future operational performance of the new compressor. These probabilities are based on present information and are prior to obtaining any additional test data.

   If more test data were obtained to reduce the uncertainty, additional costs would be incurred. Therefore, these additional costs must be balanced against the value of reducing the uncertainty. Obviously, if perfect information were available about the future operational efficiency of the new compressor, all uncertainty would be removed and we could make an optimal decision between the current unit and the new design. Even though perfect information is unobtainable, its expected value indicates the maximum amount (upper limit) we should consider spending for additional information.

EXAMPLE 12-11   **Expected Value of Perfect Information**

Refer to Example 12-10. What is the EVPI regarding future operational performance of the new compressor to the typical user of an air-conditioning system?

Solution

We can calculate the EVPI by comparing the optimal decision based on perfect information with the original decision in Example 12-10 to select the new

TABLE 12-16   Expected Value of Perfect Information (Example 12-11)

Level (%) of Design Goal Met	Probability $p(L)$	Decision with Perfect Information		Prior Decision (New Design)
		Decision	Outcome	
90	0.25	New	$3,538	$3,538
70	0.40	New	1,614	1,614
50	0.25	Current	0	−520
30	0.10	Current	0	−3,143
		Expected value:	$1,530	$1,086

compressor. This comparison is shown in Table 12-16. Based on this comparison, the EVPI to the typical user of a unit is

$$\text{EVPI} = \$1,530 - \$1,086 = \$444.$$

12.8.3.2   The Use of Additional Information to Reduce Uncertainty The solution to Example 12-11 shows that there is some potential value to be obtained from additional test information about the operational performance of the new compressor. From the viewpoint of the typical user of a new unit versus the current unit, the maximum estimated value of additional information is $444.

The members of the management team of Ajax Corporation are strongly customer focused and want to exceed customers' expectations about the performance of their products. Thus, they asked the design team to estimate the value of data that could be obtained from an additional comprehensive test of prototypes of the new compressor. The information from the test will not be perfect, because the test cannot determine exactly the long-term operational performance of the new design under diverse customer applications. The imperfect test information, however, may reduce uncertainty and merit the additional cost of obtaining it.

We can evaluate the value of additional information before obtaining it only if we can estimate the reliability of the experiment being used. Therefore, the design team addressed the reliability of additional test information in predicting the future operational performance of the new compressor. The estimates developed by the design team, the calculated *revised probabilities* of achieving the new efficiency design goal levels, and the value of additional test information from a user's perspective are discussed in Example 12-12.

EXAMPLE 12-12	**Compressor Analysis Using Revised Probabilities**

Refer to Examples 12-10 and 12-11. The members of the design team are confident that data from the additional comprehensive test of prototype compressors will show whether future operational performance will be favorable (60% or more of the design goal met) or not favorable (60% of the design goal not met). Based on obtaining these results from the test, and by using current engineering data within Ajax Corporation, the design team developed the following *conditional probability* estimates:

Comprehensive Test Outcome	Conditional Probabilities of Test Outcome Given the Level (%) of the Design Goal Met			
	90	70	50	30
Favorable (F)	0.95	0.85	0.30	0.05
Not favorable (NF)	0.05	0.15	0.70	0.95
Sum	1.00	1.00	1.00	1.00

For example, given that the future operational performance of the new compressor is 90% of the design goal, the conditional probability of the comprehensive test results being favorable is estimated to be 0.95, and the conditional probability of the test results being not favorable is estimated to be 0.05. That is to say, $p(F \mid 90) = 0.95$, and $p(NF \mid 90) = 0.05$, where the | means given.

Based on these conditional probabilities (reliability estimates of test results) and the choice of either doing the test or not doing it,

(a) calculate the revised probabilities of the four efficiency design goal levels being met,

(b) estimate the value of performing the comprehensive test to the typical user of a new compressor unit.

Solution

(a) The two-stage decision tree diagram, which includes an initial decision on whether to do the test, is shown in Figure 12-11. In order to calculate the revised probabilities, we need to determine the joint probabilities of each level of the efficiency design goal being met and each test outcome occurring, as well as the marginal probability of each test outcome. These probabilities are shown in Table 12-17. As an example, the joint probabilities of the efficiency design goal being met at the 90% level and each test outcome occurring are calculated as follows:

$$p(90, \text{ F}) = p(\text{F} \mid 90) \cdot p(90) = (0.95)(0.25) = 0.2375;$$
$$p(90, \text{ NF}) = p(\text{NF} \mid 90) \cdot p(90) = (0.05)(0.25) = 0.0125.$$

The remaining six joint probabilities are determined in a similar manner. The sum of the joint probabilities over the four design goal levels gives the marginal probability of each test outcome occurring. That is to say, $p(F) =$ 0.6575 and $p(NF) = 0.3425$. Similarly, the sums of the joint probabilities over the test outcomes are the marginal probabilities of achieving the new efficiency design goal levels [and are the same as the prior probabilities, $p(L)$, in Example 12-10]. The revised probabilities of each level of the design goal being met, based on Table 12-17 [e.g., $p(90) = p(90, F)/p(F) =$ 0.2375/0.6575 = 0.3612], are also shown in Figure 12-11.

(b) The estimated value (to the typical user of a new compressor unit) of completing the additional comprehensive test can be determined by using the data shown in Figure 12-11. By starting on the right side of Figure 12-11 and working toward the left side, we can calculate the $E(PW)$ of the new design for both a favorable test outcome ($2,029) and an unfavorable test outcome (−$726). Based on these results, the choices at the two decision

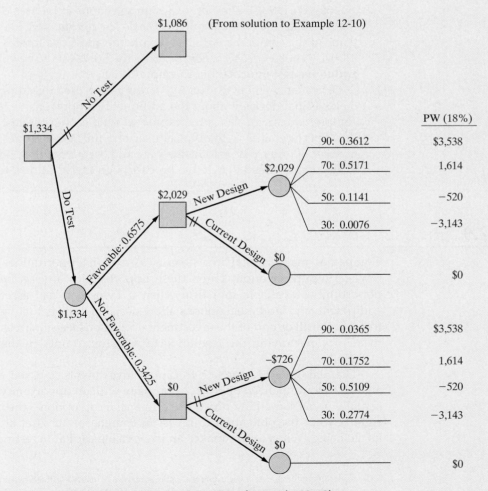

Figure 12-11   Two-Stage Decision Tree (Example 12-12)

TABLE 12-17   Joint and Marginal Probabilities (Example 12-12)

Level (%) of Design Goal Met	Joint Probabilities Favorable (F)	Not Favorable (NF)	Marginal Probabilities, $p(L)$
90	0.2375	0.0125	0.25
70	0.3400	0.0600	0.40
50	0.0750	0.1750	0.25
30	0.0050	0.0950	0.10
Marginal probabilities of test outcome:	0.6575	0.3425	1.00 (Sum)

nodes are the new design alternative for a favorable test outcome and the current design alternative for a not favorable test outcome, respectively.

Based on these alternative design selections at the two decision nodes, the $E(PW)$ at the chance node for the *do test* option is $1,334. The expected value of the comprehensive test, before the additional cost is considered, is $1,334 − $1,086 = $248, where $1,086 is the $E(PW)$ of the new design without additional test information (Example 12-10).

The management team at Ajax Corporation used this information to help make a final decision about the additional comprehensive test of the new compressor design. Since the estimated total cost of the test was less than the expected value to the typical user of a unit ($248) times the estimated number of units to be sold in one year, and because of their strong customer focus, management decided to proceed with the additional test.

12.9   Real Options Analysis

Companies make capital investments to exploit opportunities for shareholder (owner) wealth creation. Often these opportunities are *real options,* which are opportunities available to a firm when it invests in real assets such as plant, equipment, and land. Real options allow decision makers to invest capital now or to postpone all or part of the investment until later. Decision trees are a convenient means for portraying the resolution of risk/uncertainty in the analysis of real options.*

The real options approach to capital investments is based on an interesting analogy about financial options. A company with an opportunity to invest capital actually owns something much like a financial call option—the company has the right but not the obligation to invest in (purchase) an asset at a future time of its choosing. When a firm makes an irreversible capital investment that could be

---

* The reader interested in details regarding real options analysis is referred to these two articles: Luehrman, T. A., "Investment Opportunities as Real Options: Getting Started with the Numbers," *Harvard Business Review*, 76, no. 4 (July–August 1998a): 51–67. Luehrman, T. A., "Strategy as a Portfolio of Real Options," *Harvard Business Review*, 76, no.5 (September–October 1998b): 89–99.

postponed, it exercises its call option, which has value by virtue of the flexibility it gives the firm.

An example of a postponable investment is coal-fired generating capacity of an electric utility. Anticipated capacity needed for the next 10 years can be added in one large expansion project, or the capacity addition can be acquired in staggered stages, which permits the utility to better respond to future demand patterns and possibly different types of generating capacity, such as natural gas or nuclear power. If the utility company decides to go ahead with a single, large, irreversible expansion project, it eliminates the option of waiting for new information that might represent a more valuable phased approach to meeting customers' demands for electricity. The lost option's value is an *opportunity cost* that must be included in the overall evaluation of the investment. This is the essence of the real options approach to capital investment—to fairly value the option of waiting to invest in all or a part of the project and to include this value in the overall project profitability. Clearly, viewing capital investments as real options forces a greater emphasis to be placed on the value of information in risky situations facing a firm.

In the previous section, we have shown how decision trees are useful in determining the value of information of anticipated future outcomes. In this regard, decision trees enable the analyst to model real options that are hidden in classical present worth analysis. These options, not treated formally in classic single outcome analyses, are often understood by management and factored informally into their assessment of the analysis (and the assessment of the analyst!). This presence of hidden options is most easily understood with an example.

| EXAMPLE 12-13 | **Hytech Industries** |

Hytech is considering opening an entirely new electronic interface using radio frequency identification with a new coding system that deviates from some standards but offers advantages to bulk chemical manufacturers. The cost to develop the manufacturing capacity and build the market is estimated at $700 million, and the resulting net cash inflows after income taxes are estimated at $100 million per year (albeit with much potential variation depending upon acceptance of the product). The after-tax MARR is 15% per year, and the new product will have a life of 20 years.

Solution as Classical Single Outcome Analysis

The PW of this project is

$$\text{PW}(15\%) = -\$700 \text{ million} + \$100 \text{ million}(P/A, 15\%, 20) = -\$74.1 \text{ million,}$$

and the recommendation would be to avoid the investment.

Suppose, however, that the $100 million cash inflow per year is merely an average. If the device catches on, potential sales will exceed the annual capacity for the new facilities, and annual inflows will be limited to $200 million. In fact, if this happens, a second plant can be acquired one year later for an additional investment of $500 million, and it would be able to generate an additional $150 million cash inflow annually. On the other hand, if the new product fails to catch on, sales could effectively be zero, but the plant could be salvaged for $150 million.

Solution as a Decision Tree with Embedded Options

The decision tree appears as shown in Figure 12-12. If the end-of-path convention is used, then the tip values are as shown in Table 12-18. Note that path 3, abandon when first year inflows exceed $200 million, is not reasonable and is not evaluated. Similarly, path 4 and 7 would involve expansion when the market has not developed and are not evaluated.

As shown in the table, when first year net inflows are high ($200 million +), the better of paths 1 and 2 is to expand and realize a PW of $926 million. If response is tepid ($100 million inflow), then path 5 is better than path 6. The PW of path 5 is negative, but staying in business recoups some of the investment and is better than abandoning the project. Finally, if the market is negligible, path 9 is better than path 8, because the ability to salvage some of the investment mitigates the loss.

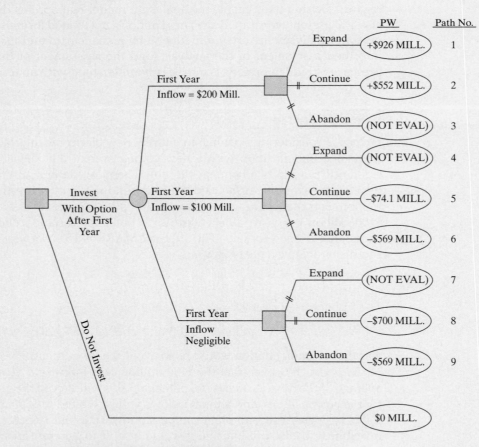

Figure 12-12 Hytech Industries Example

TABLE 12-18	Decision Tree Endpoint Evaluation			
Path	First Year Inflows	Option	Evaluation	Net PW
1	$200 million +	Expand	$-\$700 + \$200(P/A, 15\%, 20)$ $-\$500(P/F, 15\%, 1)$ $+\$150(P/A, 15\%, 19)(P/F, 15\%, 1)$	+$926 million
2	$200 million +	Continue	$-\$700 + \$200(P/A, 15\%, 20)$	+$552 million
3	$200 million +	Abandon	Not evaluated	
4	$100 million	Expand	Not evaluated	
5	$100 million	Continue	$-\$700 + \$100(P/A, 15\%, 20)$	-$74.1 million
6	$100 million	Abandon	$-\$700 + \$150(P/F, 15\%, 1)$	-$569 million
7	Negligible	Expand	Not evaluated	
8	Negligible	Continue	$-\$700$	-$700 million
9	Negligible	Abandon	$-\$700 + \$150(P/F, 15\%, 1)$	-$569 million

If each outcome of the first year cash flows were equally likely and management is risk neutral, then the expected monetary value is $94.3 (= [+$926 − $74.1 − $569]/3) million, an improvement of $168.4 million over the $74.1 million loss projected in the single outcome analysis. This improvement can be viewed as the value of the combined options to expand, if the market develops, or to salvage some investment if the project is a flop.

12.10 CASE STUDY—The Oils Well That Ends Well Company

Ms. Sameer Abdul is the vice president of operations for the Oils Well That Ends Well Company. She oversees operations at a plant that manufactures components for hydraulic systems. Ms. Abdul is concerned about the plant's present production capability. She has reduced the decision situation to three alternatives. The first alternative would result in significant changes in present operations, including increased automation. The second alternative involves fewer changes in present operations and would not include any new automation. The third alternative is to make no changes (do nothing).

As a member of the plant management team, you have been assigned the task of analyzing the alternatives and recommending a course of action. The incremental capital investment and incremental annual revenue for the first two alternatives, relative to present operations, are shown in the following table:

Alternative	Capital Investment	Future Sales	Annual Revenue
1	$300,000	Good	$142,000
		Average	119,000
		Poor	50,000
2	85,000	Good	66,000
		Average	46,000
		Poor	17,000

The annual revenue estimates are based on future sales of the components. The sales department estimates the probability of good, average, and poor future sales as 0.30, 0.60, and 0.10, respectively.

It is known that Ms. Abdul is a big fan of decision tree analysis. Based on a before-tax analysis (MARR = 20% per year, study period is five years, and zero market value for all alternatives) and the $E(PW)$ as a decision criterion, determine which alternative is preferred. What would be the EVPI in this case?

Solution

The single-stage decision tree for this study is shown in Figure 12-13. The PW associated with each of the sales conditions is calculated using the following equation:

$$PW(20\%) = -\text{Capital investment} + \text{Revenue }(P/A, 20\%, 5).$$

The expected values of PW for each alternative, based on these values, are as follows:

$$E(PW)_1 = 0.30(\$124,665) + 0.60(\$55,881) + 0.10(-\$150,470) = \$55,881$$

$$E(PW)_2 = 0.30(\$112,380) + 0.60(\$52,568) + 0.10(-\$34,160) = \$61,839$$

$$E(PW)_3 = 0$$

The expected value of making no changes (Alternative 3) is zero since the cash flow estimates for the first two alternatives are incremental amounts relative to present operations.

This analysis indicates that although the increased automation proposed in Alternative 1 is an improvement over present operations, the less radical changes

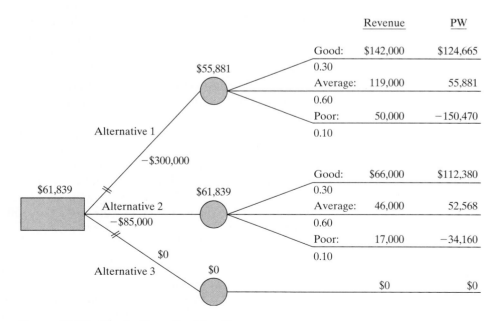

Figure 12-13 Single-Stage Decision Tree

TABLE 12-19   Expected Value of Perfect Information

Potential Sales	Probability	Decision with Perfect Information		PW for Alternative 2
		Alternative	PW	
Good	0.30	1	$124,665	$112,380
Average	0.60	1	55,881	52,568
Poor	0.10	3	0	−34,160
		Expected value:	$70,928	$61,839

proposed in Alternative 2 have more value to the company. Your recommendation to Ms. Abdul is to implement the second alternative.

As supplemental information, you decide to calculate the EVPI for this decision. To accomplish this, you calculate the difference between the PW of the best decision based on perfect information with the PW of Alternative 2. These calculations are shown in Table 12-19. Based on this comparison, the EVPI is $70,928 − $61,839 = $9,089. This amount represents an upper limit on what could be spent for additional information.

While Ms. Abdul was reviewing your analysis of the single-stage decision tree diagram, you realized that additional information about future sales of the hydraulic components should reduce the uncertainty involved. Therefore, the plant management team asked the sales department to survey their customers and improve the information about future sales conditions. The management team's estimates of the *conditional probabilities* of survey outcomes for each potential sales condition are shown in the following table:

	Conditional Probabilities of Survey Outcome Given the Future Sales Condition		
	Good	Average	Poor
Survey Outcome	(G)	(A)	(P)
Optimistic (O)	0.85	0.60	0.10
Not Favorable (NF)	0.15	0.40	0.90
Sum:	1.00	1.00	1.00

Based on this information, you develop the two-stage decision tree shown in Figure 12-14. This diagram includes an initial decision on whether to conduct the survey. To calculate the revised probabilities, you need to determine the joint probabilities of each potential sales condition and each survey prediction occurring, as well as the marginal probability of each survey prediction. As an example, the joint probabilities of the future sales condition being good and each survey prediction occurring are calculated as follows:

$$p(G,O) = p(O|G) \cdot p(G) = (0.85)(0.3) = 0.255$$
$$p(G,NF) = p(NF|G) \cdot p(G) = (0.15)(0.3) = 0.045.$$

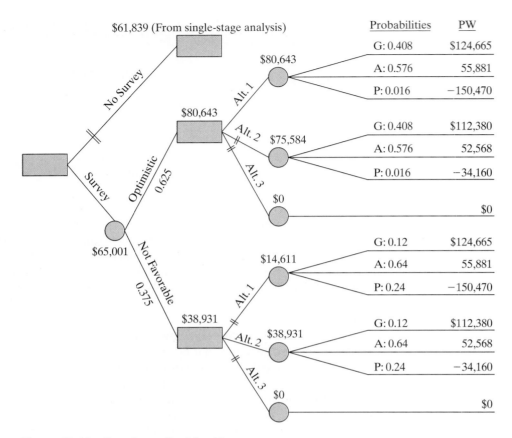

Figure 12-14   Two-Stage Decision Tree

The complete set of joint and marginal probabilities is shown in Table 12-20. Given the information in Table 12-20, the revised probabilities of each future sales condition occurring shown on the diagram were calculated as follows:

$$p(G|O) = p(G,O)/p(O) = 0.255/0.625 = 0.408$$

$$p(G|NF) = p(G,NF)/p(NF) = 0.045/0.375 = 0.120.$$

The remaining four revised probabilities were determined in a similar manner. The expected PW of taking the survey is as follows:

$$E(PW) = 0.625(\$80,643) + 0.375(\$38,931) = \$65,001.$$

Based on this result, and the $E(PW)$ of not taking the survey ($61,839 from the analysis of the single-stage decision tree), the value to the company of doing the survey (without consideration of the additional cost) is as follows:

$$E(\text{Value of survey}) = \$65,001 - \$61,839 = \$3,162.$$

The sales department is confident that the survey could be conducted for less than $2,500. Your final recommendation to Ms. Abdul is to conduct the

TABLE 12-20	Joint and Marginal Probabilities		
	Survey Prediction		Marginal
Potential Sales Condition	Optimistic (O)	Not Favorable (NF)	Probabilities of Potential Sales Condition
Good (G)	0.255	0.045	0.3
Average (A)	0.360	0.240	0.6
Poor (P)	0.010	0.090	0.1
Marginal Probability of Survey Prediction	0.625	0.375	1.0 (Sum)

customer survey before making a decision on which alternative to implement. If the results of the survey are optimistic, the automation alternative (Alternative 1) should be implemented. An unfavorable survey response would support the implementation of Alternative 2, which involves making minor changes to the present operations.

This case study illustrates how decision trees are useful in providing some analytical structure to ill-defined problems. In addition, in this case, the use of a two-stage decision tree provides additional insights for dealing with risk and uncertainty and highlights the value of obtaining survey data on projected sales before finalizing managerial decisions. ▬▬▬▬▬■

12.11 Summary

Engineering economy involves decision making among competing uses of scarce capital resources. The consequences of resultant decisions usually extend far into the future. In this chapter, we have presented various statistical and probability concepts that address the fact that the consequences (cash flows, project lives, etc.) of engineering alternatives can never be known with certainty, including Monte Carlo computer simulation techniques and decision tree analysis. Cash inflow and cash outflow factors, as well as project life, were modeled as discrete and continuous random variables. The resulting impact of uncertainty on the economic measures of merit for an alternative was analyzed. Included in the discussion were several considerations and limitations relative to the use of these methods in application.

Regrettably, there is no quick and easy answer to the question, How should risk best be considered in an engineering economic evaluation? Generally, simple procedures (e.g., breakeven analysis and sensitivity analysis, discussed in Chapter 11) allow some discrimination among alternatives to be made on the basis of the uncertainties present, and they are relatively inexpensive to apply. Additional discrimination among alternatives is possible with more complex procedures that utilize probabilistic concepts. These procedures, however, are more difficult to apply and require additional time and expense.

Problems

The number in parentheses () that follows each problem refers to the section from which the problem is taken.

12-1. A large mudslide caused by heavy rains will cost Sabino County $1,000,000 per occurrence in lost property tax revenues. In any given year, there is one chance in 100 that a major mudslide will occur.

A civil engineer has proposed constructing a culvert on a mountain where mudslides are likely. This culvert will reduce the likelihood of a mudslide to near zero. The investment cost would be $50,000, and annual maintenance expenses would be $2,000 in the first year, increasing by 5% per year thereafter. If the life of the culvert is expected to be 20 years and the cost of capital to Sabino County is 7% per year, should the culvert be built? (12.4)

12-2. A bridge is to be constructed now as part of a new road. An analysis has shown that traffic density on the new road will justify a two-lane bridge at the present time. Because of uncertainty regarding future use of the road, the time at which an extra two lanes will be required is currently being studied. The estimated probabilities of having to widen the bridge to four lanes at various times in the future are as follows:

Widen Bridge In	Probability
3 years	0.1
4 years	0.2
5 years	0.3
6 years	0.4

The present estimated cost of the two-lane bridge is $2,000,000. If constructed now, the four-lane bridge will cost $3,500,000. The future cost of widening a two-lane bridge will be an extra $2,000,000 plus $250,000 for every year that widening is delayed. If money can earn 12% per year, what would you recommend? (12.4)

12-3. In Problem 12-2, perform an analysis to determine how sensitive the choice of a four-lane bridge built now versus a four-lane bridge that is constructed in two stages is to the interest rate. Will an interest rate of 15% per year reverse the initial decision? At what interest rate would constructing the two-lane bridge now be preferred? (12.4)

12-4. It costs $250,000 to drill a natural gas well. Operating expenses will be 10% of the revenue from the sale of natural gas from this particular well. If found, natural gas from a highly productive well will amount to 260,000 cubic feet per day. The probability of locating such a productive well, however, is about 10%. (12.4)

a. If natural gas sells for $8 per thousand cubic feet, what is the $E(\text{PW})$ of profit to the owner/operator of this well? The life of the well is 10 years and MARR is 15% per year.

b. Repeat Part (a) when the life of the well is seven years.

c. Perform one-at-a-time sensitivity analyses for ±20% changes in daily well production and selling price of natural gas. Use a 10-year life for the well.

12-5. Annual profit (P) is the product of total annual sales (S) and profit per unit sold (X); that is, $P = X \cdot S$. It is desired to know the probability distribution of the random variable P when X and S have the following assumed probability mass functions (X and S are independent):

X (Unit Profit)		S (Annual Sales)	
Value	Probability	Value	Probability
0	0.05	5	0.4
1	0.45	6	0.2
2	0.30	8	0.4
3	0.20		

What are the mean, variance, and standard deviation of the probability distribution for annual profit, P? (12.4)

12-6. A very important levee spans a distance of 10 miles on the outskirts of a large metropolitan area. Hurricanes have hit this area twice in the past 20 years, so there is concern over the structural integrity of this particular levee. City engineers have proposed reinforcing and increasing the height of the levee by various designs to withstand the storm surge of numerous strength categories of hurricanes.

Study results regarding the probability that high flood water from a hurricane in any one year will exceed the increased height of the levee, and the cost of construction of the levee for each storm category, are summarized next.

Hurricane Category	Probability of a Storm Surge Exceeding the Levee Height	Capital Investment to Increase the Levee Height
5	0.005	$70,000,000
4	0.010	50,000,000
3	0.030	35,000,000
2	0.050	20,000,000
1	0.100	10,000,000

A panel of experts suggests that the average property damage will amount to $100,000,000 if a storm surge causes the levee to overflow. The capital investment to rebuild the levee for each hurricane category will be financed with 30-year municipal bonds earning 6% per year. These bonds will be retired as annuity payments each year. What is the most economical way to rebuild the levee to protect the city from flooding during a hurricane? What other factors might affect the decision in this situation? What if the average damage from a flood is $200,000,000? (12.4)

12-7. A diesel generator is needed to provide auxiliary power in the event that the primary source of power is interrupted. Various generator designs are available, and more expensive generators tend to have higher reliabilities should they be called on to produce power. Estimates of reliabilities, capital investment costs, operating and maintenance expenses, market value, and damages resulting from a complete power failure (i.e., the standby generator fails to operate) are given in Table P12-7 for three alternatives. If the life of each generator is 10 years and MARR = 10% per year, which generator should be chosen if you assume one main power failure per year? Does your choice change if you assume two main power failures per year? (Operating and maintenance expenses remain the same.) (12.4)

12-8. The owner of a ski resort is considering installing a new ski lift that will cost $900,000. Expenses for operating and maintaining the lift are estimated to be $1,500 per day when operating. The U.S. Weather Service estimates that there is a 60% probability of 80 days of skiing weather per year, a 30% probability of 100 days per year, and a 10% probability of 120 days per year. The operators of the resort estimate that during the first 80 days of adequate snow in a season, an average of 500 people will use the lift each day, at a fee of $10 each. If 20 additional days are available, the lift will be used by only 400 people per day during the extra period; and if 20 more days of skiing are available, only 300 people per day will use the lift during those days. The owners wish to recover any invested capital within five years and want at least a 25% per year rate of return before taxes. Based on a before-tax analysis, should the lift be installed? (12.4)

12-9. Refer to Problem 12-8. Assume the following changes: The study period is eight years; the ski lift will be depreciated by using the MACRS Alternative Depreciation System (ADS); the ADS recovery period is seven years; MARR = 15% per year (after-tax); and the effective income tax rate (t) is 40%. Based on this information, what is the $E(PW)$ and $SD(PW)$ of the ATCF? Interpret the analysis results and make a recommendation on installing the ski lift. (12.4)

12-10. An energy conservation project is being evaluated. Four levels of performance are considered feasible. The estimated probabilities of each performance level and the estimated before-tax cost savings in the first year are shown in the following table:

Performance Level (L)	$p(L)$	Cost Savings (1st yr; before taxes)
1	0.15	$22,500
2	0.25	35,000
3	0.35	44,200
4	0.25	59,800

TABLE P12-7 Three Generator Designs for Problem 12-7

Alternative	Capital Investment	O&M Expenses/Year	Reliability	Cost of Power Failure	Market Value
R	$200,000	$5,000	0.96	$400,000	$40,000
S	170,000	7,000	0.95	400,000	25,000
T	214,000	4,000	0.98	400,000	38,000

Assume the following:

- Initial capital investment: $100,000 [80% is depreciable property and the rest (20%) are costs that can be immediately expensed for tax purposes].
- The ADS under MACRS is being used. The ADS recovery period is four years.
- The before-tax cost savings are estimated to increase 6% per year after the first year.
- $MARR_{AT} = 12\%$ per year; the analysis period is five years; $MV_5 = 0$.
- The effective income tax rate is 40%.

Based on $E(PW)$ and after-tax analysis, should the project be implemented? (12.4)

12-11. The purchase of a new piece of electronic measuring equipment for use in a continuous metal-forming process is being considered. If this equipment were purchased, the capital cost would be $418,000, and the estimated savings are $148,000 per year. The useful life of the equipment in this application is uncertain. The estimated probabilities of different useful lives occurring are shown in the following table. Assume that MARR = 15% per year before taxes and the market value at the end of its useful life is equal to zero. Based on a before-tax analysis,

a. What are the $E(PW)$, $V(PW)$, and $SD(PW)$ associated with the purchase of the equipment?

b. What is the probability that the PW is less than zero? Make a recommendation and give your supporting logic based on the analysis results. (12.4)

Useful Life, Years (N)	p(N)
3	0.1
4	0.1
5	0.2
6	0.3
7	0.2
8	0.1

12-12. The tree diagram in Figure P12-12 describes the uncertain cash flows for an engineering project. The analysis period is two years, and MARR = 15% per year. Based on this information,

a. What are the $E(PW)$, $V(PW)$, and $SD(PW)$ of the project?

b. What is the probability that PW $\geq$ 0? (12.3)

12-13. A potential project has an initial capital investment of $100,000. Net annual revenues minus expenses are estimated to be $40,000 (A$) in the first year and to increase at the rate of 6.48% per year. The useful life of the primary equipment, however, is uncertain, as shown in the following table:

Useful Life, Years (N)	p(N)
1	0.03
2	0.10
3	0.30
4	0.30
5	0.17
6	0.10

Assume that i_m = MARR = 15% per year and f = 4% per year. Based on this information,

a. What are the $E(PW)$ and $SD(PW)$ for this project?

b. What is the $Pr\{PW \geq 0\}$?

c. What is the $E(AW)$ in R$?

Do you consider the project economically acceptable, questionable, or not acceptable, and why? (Chapter 8, 12.4)

12-14. A hospital administrator is faced with the problem of having a limited amount of funds available for capital projects. He has narrowed his choice down to two pieces of x-ray equipment, since the radiology department is his greatest producer of revenue. The first piece of equipment (project A) is a fairly standard piece of equipment that has gained wide acceptance and should provide a steady flow of income. The other piece of equipment (project B), although more risky, may provide a higher return. After deliberation with his radiologist and director of finance, the administrator has developed the following table:

Cash Flow per Year (thousands)			
Probability	Project A	Probability	Project B
0.6	$2,000	0.2	$4,000
0.3	1,800	0.5	1,200
0.1	1,000	0.3	900

Time Period

0	1	2

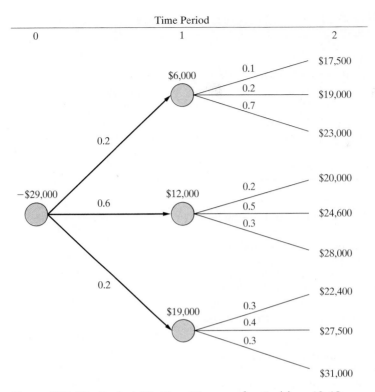

Figure P12-12   Probability Tree Diagram for Problem 12-12

Discovering that the budget director of the hospital is taking courses in engineering, the hospital administrator has asked him to analyze the two projects and make his recommendation. Prepare an analysis that will aid the budget director in making his recommendation. In this problem, do risk and reward travel in the same direction? (12.4)

12-15. In an industrial setting, process steam has been found to be normally distributed with an average value of 25,000 pounds per hour. There is an 80% probability that steam flow lies between 20,000 and 30,000 pounds per hour. What is the variance of the steam flow? (12.5)

12-16. Suppose that a random variable (e.g., market value for a piece of equipment) is normally distributed, with mean = $175 and variance = 25$^{2}. What is the probability that the actual market value is *at least* $171? (12.5)

12-17. The use of three estimates (defined here as H = high, L = low, and M = most likely) for random variables is a practical technique for modeling uncertainty in some engineering economy studies.

Assume that the mean and variance of the random variable, X_k, in this situation can be estimated by $E(X_k) = (1/6)(H + 4M + L)$ and $V(X_k) = [(H - L)/6]^2$. The estimated net cash-flow data for one alternative associated with a project are shown in Table P12-17.

The random variables, X_k, are assumed to be statistically independent, and the applicable MARR = 15% per year. Based on this information,

a. What are the mean and variance of the PW?

b. What is the probability that PW $\geq$ 0 (state any assumptions that you make)?

c. Is this the same as the probability that the IRR is acceptable? Explain. (12.5)

12-18. Two mutually exclusive investment alternatives are being considered, and one of them must be selected.

Alternative A requires an initial investment of $13,000 in equipment. Annual operating and maintenance costs are anticipated to be normally distributed, with a mean of $5,000 and a standard deviation of $500. The terminal salvage value at the end of the eight-year planning horizon is anticipated to be normally

TABLE P12-17   Estimates for Problem 12-17

End of Year, k	Net Cash Flow	Three-Point Estimates for X_k		
		L	M	H
0	$F_0 = X_0$	−$38,000	−$41,000	−$45,000
1	$F_1 = 2X_1$	−1,900	−2,200	−2,550
2	$F_2 = X_2$	9,800	10,600	11,400
3	$F_3 = 4X_3$	5,600	6,100	6,400
4	$F_4 = 5X_4$	4,600	4,800	5,100
5	$F_5 = X_5$	16,500	17,300	18,300

distributed, with a mean of $2,000 and a standard deviation of $800.

Alternative B requires end-of-year annual expenditures over the eight-year planning horizon, with the annual expenditure being normally distributed with a mean of $7,500 and a standard deviation of $750. Using a MARR of 15% per year, what is the probability that Alternative A is the most economic alternative (i.e., the least costly)? (12.5)

12-19. Two investment alternatives are being considered. The data below have been estimated by a panel of experts, and all cash flows are assumed to be independent. Life is *not* a variable.

	Alternative A		Alternative B	
End of Year	Expected Cash Flow	Std. Deviation of Cash Flow	Expected Cash Flow	Std. Deviation of Cash Flow
0	−$8,000	$0	−$12,000	$500
1	4,000	600	4,500	300
2	6,000	600	4,500	300
3	4,000	800	4,500	300
4	6,000	800	4,500	300

With MARR = 15% per year, determine the mean and standard deviation of the *incremental* PW [i.e., $\Delta(B - A)$]. (12.5)

12-20. The market value of a certain asset depends upon its useful life as follows:

Useful Life	Market Value
3 years	$18,000
4 years	$12,000
5 years	$6,000

a. The asset requires a capital investment of $120,000, and MARR is 15% per year. Use Monte Carlo simulation and generate four trial outcomes to find its expected equivalent AW if each useful life is equally likely to occur. (12.6)

b. Set up an equation to determine the variance of the asset's AW. (12.5)

12-21. A company that manufactures automobile parts is weighing the possibility of investing in an FMC (Flexible Manufacturing Cell). This is a separate investment and not a replacement of the existing facility. Management desires a good estimate of the distribution characteristics for the AW. There are three random variables; i.e., they have uncertainties in their values.

An economic analyst is hired to estimate the desired parameter, AW. He concludes that the scenario presented to him is suitable for Monte Carlo simulation (method of statistical trials) because there are uncertainties in three variables and the direct analytical approach is virtually impossible.

The company has done a preliminary economic study of the situation and provided the analyst with the following estimates:

Investment	Normally distributed with mean of $200,000 and standard deviation of $10,000
Life	Uniformly distributed with minimum of 5 years and maximum of 15 years
Market value	$20,000 (single outcome)
Annual net cash flow	$20,000 probability 0.3 $16,000 probability 0.5 $22,000 probability 0.2
Interest rate	10%

All the elements subject to variation vary independently. Use Monte Carlo simulation to generate 10 AW outcomes for the proposed FMC. What is the standard deviation of the outcomes? (12.6, 12.7)

12-22. Simulation results are available for two mutually exclusive alternatives. A large number of trials have been run with a computer, with the results shown in Figure P12-22.

Discuss the issues that may arise when attempting to decide between these two alternatives. (12.6, 12.7)

12-23. A company is considering an engineering improvement project with uncertain outcomes. The best present estimates, including prior probabilities of success, are as follows:

Success Category	Probability of Success	Net Annual Benefits
A	0.35	$200,000
B	0.35	100,000
C	0.30	20,000

The estimated net annual benefits are relative to current operations. Assume the project's initial investment is $280,000; taxes are not being considered; MARR (before taxes) is 15% per year; and a six-year analysis period applies to this type of project.

Due to the uncertain outcomes, the responsible manager has directed that a potential test experiment be evaluated prior to further consideration of the project. The estimated reliability of the test experiment is as follows:

	Conditional Probabilities of Test Outcome, Given the Success Category		
Test Outcome	A	B	C
Good (G)	0.90	0.25	0.05
Quite Poor (P)	0.10	0.75	0.95
Sum	1.0	1.0	1.0

Based on a decision tree analysis and the $E(PW)$ as the economic measure of interest, what is the estimated value of the additional information that would be provided by the test experiment in this case? (12.8)

12-24. An improved design of a computerized piece of continuous quality measuring equipment used to control the thickness of rolled sheet products is being developed. It is estimated to sell for $125,000 more than the current design. Based on present test data, however, the typical user has the following probabilities of achieving different performance results and cost savings (relative to the current unit) in the first year

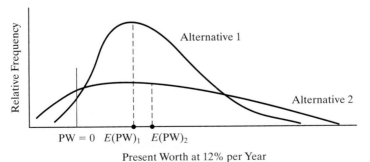

Figure P12-22   Simulation Results for Problem 12-22

of operation (assume these annual cost savings would escalate 5% per year thereafter; a five-year analysis period is appropriate for this situation; the before-tax market MARR is 18% per year; and the net market value at the end of five years is 0):

Performance Results	Probability	Cost Savings in First Year
Optimistic	0.30	$60,000
Most likely	0.55	40,000
Pessimistic	0.15	18,000

Based on the $E(PW)$, is the new design preferable to the current unit? (Show a single-stage decision tree diagram for this situation.) What is the EVPI? What does the EVPI tell you? (12.8)

12-25. If the interest rate is 8% per year, what decision would you make based on the decision tree diagram in Figure P12-25? (12.8)

12-26. *Extended Learning Exercise* The additional investment in a new computer system is a certain $300,000. It is likely to save an average of $100,000 per year compared to the old, outdated system. Because of uncertainty, this estimate is expected to be normally distributed, with a standard deviation of $7,000. The market value of the system at any time is its scrap value, which is $20,000 with a standard deviation of $3,000. MARR on such investments is 15% per year.

What is the smallest value of N (the life of the system) that can exist such that the probability of getting a 15% internal rate of return or greater is 0.90? Ignore the effects of income taxes. Also note that market value is independent of N. (*Hint*: Start with $N = 4$ years). (12.4)

12-27. *Extended Learning Exercise* A firm must decide between constructing a new facility or renting a comparable office space. There are two random outcomes for acquiring space, as shown in Figure P12-27. Each would accommodate the expected growth of this company over the next 10 years. The cost of rental space is expected to escalate over the 10 years for each rental outcome.

The option of constructing a new facility is also defined in Figure P12-27. An initial facility could be constructed with the costs shown. In five years, additional space will be required. At that time, there will be an option to build an office addition or rent space for the additional space requirements.

The probabilities for each alternative are shown. MARR for the situation is 10% per year. A PW analysis is to be conducted on the alternatives. Which course of action should be recommended? Note: At $\boxed{2}$, the PW(10%) of the upper branch is −$5,111.14 k and the PW(10%) for the lower branch is −$4,119.06 k.

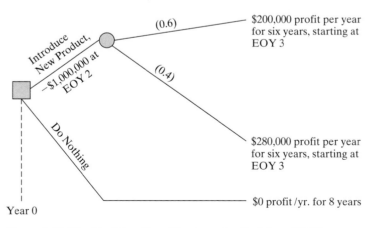

Figure P12-25 Decision Tree Diagram for Problem 12-25

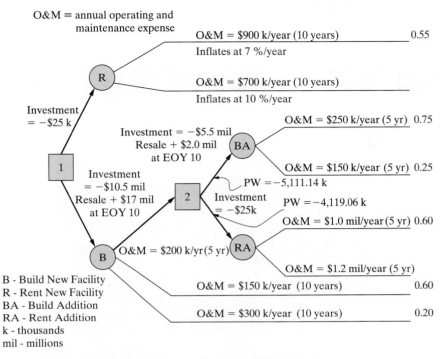

O&M ≡ annual operating and maintenance expense

O&M = $900 k/year (10 years)   0.55
Inflates at 7 %/year

O&M = $700 k/year (10 years)
Inflates at 10 %/year

R

Investment = −$25 k

Investment = −$5.5 mil
Resale + $2.0 mil
at EOY 10   BA

O&M = $250 k/year (5 yr) 0.75

O&M = $150 k/year (5 yr) 0.25

1

Investment
= −$10.5 mil
Resale + $17 mil
at EOY 10

Investment
= −$10.5 mil

2

PW = −5,111.14 k

Investment
= −$25k

PW = −4,119.06 k

O&M = $1.0 mil/year (5 yr) 0.60

O&M = $1.2 mil/year (5 yr)

B   O&M = $200 k/yr(5 yr) RA

O&M = $150 k/year (10 years)   0.60

O&M = $300 k/year (10 years)   0.20

B - Build New Facility
R - Rent New Facility
BA - Build Addition
RA - Rent Addition
k - thousands
mil - millions

Figure P12-27 Decision Tree Diagram for Problem 12-27

Spreadsheet Exercises

12-28. Refer to Example 12-4. After additional research, the range of probable useful lives has been narrowed down to the following three possibilities:

Useful Life, Years (N)	$p(N)$
14	0.3
15	0.4
16	0.3

How does this affect $E(PW)$ and $SD(PW)$? What are the benefits associated with the cost of obtaining more accurate estimates? (12.4)

12-29. Refer to Problem 12-21. Use a spreadsheet to extend the number of trials to 500. Compute $E(AW)$ and plot the cumulative average AW. Are 500 trials enough to reach steady state? (12.7)

CHAPTER 13

The Capital Budgeting Process*

Our aim in Chapter 13 is to give the student an understanding of the basic components of the capital budgeting process so that the important role of the engineer in this strategic function will be made clear.

Weighted Average Cost of Capital

C ompanies obtain the funds needed for their ventures from multiple sources. To evaluate potential projects, the cost of the different sources of capital must be accounted for in the interest rate used to measure profitability. Consider, for example, a large multinational company with an effective income tax rate of 49.24%. The percent of long-term debt in its capital structure is 49%, and its before-tax annual cost is 9.34%. In its capital structure, the firm also has 13% preferred stock paying 8.22% per year and 38% common equity valued at 16.5% per year. In this chapter, you will learn how to determine the weighted after-tax cost of capital for this company to be used in evaluating engineering projects.

---

* This chapter is adapted from J. R. Canada, W. G. Sullivan, D. J. Kulonda and J. A. White, *Capital Investment Analysis for Engineering and Management* (3rd ed.), (Upper Saddle River, NJ: Prentice Hall Inc., 2005). Reprinted by permission of the publisher.

> Money is the seed of money, and the first guinea is sometimes more difficult to acquire than the second million.
>
> —Jean Jacques Rousseau (1762)

13.1 Introduction

The decision by a company to implement an engineering project involves the expenditure of current capital funds to obtain future economic benefits or to meet safety, regulatory, or other operating requirements. This implementation decision is normally made, in a well-managed company, as part of a capital budgeting process. The capital financing and allocation functions are primary components of this process.

13.1.1 The Capital Financing and Allocation Functions

The capital financing and allocation functions are closely linked, as illustrated in Figure 13-1, and they are simultaneously managed as part of the capital budgeting process. The amounts of new funds needed from investors and lenders, as well as funds available from internal sources to support new capital projects, are determined in the capital financing function. Also, the *sources* of any new externally acquired funds—issuing additional stock, selling bonds, obtaining loans, and so on—are determined. These amounts, in total, as well as the ratio of debt to equity

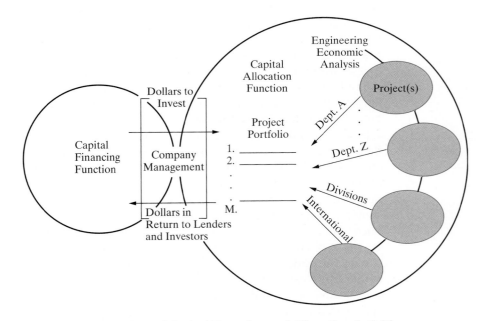

Figure 13-1 Overview of Capital Financing and Allocation Activities in a Typical Organization

capital, must be commensurate with the financial status of the firm and balanced with the current and future capital investment requirements.

The selection of the engineering projects for implementation occurs in the capital allocation function. The total capital investment in new projects is constrained by the amount decided upon for this purpose during the capital financing considerations. The capital allocation activities begin in the various organizations in the company—departments (say, engineering), operating divisions, research and development, and so on. During each capital budgeting cycle, these organizations plan, evaluate, and recommend projects for funding and implementation. Economic and other justification information is required with each project recommendation. Engineering economy studies are accomplished as part of this process to develop much of the information required.

As shown in Figure 13-1, the available capital is allocated among the projects selected on a company-wide basis in the capital allocation function. Management, through its integrated activities in both functions, is responsible for ensuring that a reasonable return (in dollars) is earned on these investments, so providers of debt and equity capital will be motivated to furnish more capital when the need arises. Thus, it should be apparent why the informed practice of engineering economy is an essential element in the foundation of an organization's competitive culture.

In sum, the capital financing and allocation functions are closely linked decision processes regarding *how much* and *where* financial resources will be obtained and expended on future engineering and other capital projects to achieve economic growth and to improve the competitiveness of a firm.

13.1.2   Sources of Capital Funds

Outside sources of capital include, for example, equity partners, stockholders, bondholders, banks, and venture capitalists, all of whom expect a return on their investment. If a company seeks to obtain new outside capital for investment in the business, it must attract new investors. It does so by paying a rate of return that is attractive in a competitive market in which many differing securities are offered. The variety of securities available and their special provisions are limited only by human creativity. A complete exploration is beyond the scope of this text, but an understanding of the cost of capital is a key concept. Hence, the theory of capital cost is developed for the most important equities and debts or liabilities that form the capitalization, or total capital base, for a company. Here, we focus on those financing concepts that guide the capital planning phase, as compared to the specific details needed in the execution phase (i.e., the actual acquisition of funds). Many refinements to the models we develop are appropriate when considering an action such as selling equity shares (stocks) or issuing new bonds. These kinds of actions and more complex models lie in the realm of finance and are not explicitly detailed here. The fundamental models presented in this chapter convey the understanding needed by engineers and nonfinancial managers. Those readers interested in the specifics of the capital acquisition process and more complex models should look to the references provided in Appendix F for details and procedures involved in the capital acquisition process.

Some perspective on the costs of various sources of capital can be gained by examining the returns provided to investors for various kinds of securities. To

TABLE 13-1	Rates of Return From 1926–2002 on Five Types of Securities	
Type of Security	Average Annual Return, R_m	Standard Deviation of Returns
Treasury bills	3.8%	4.4%
Long-term U.S. bonds	5.8	9.4
Corporate bonds	6.2	8.7
S&P 500 stocks	12.2	20.5
Small-firm stocks	16.9	33.2

Source: R. G. Ibbotson Associates, Chicago, IL, 2003.

provide some stable benchmarks, it is helpful to review historical values of rates of returns on various securities. A widely circulated reference is the yearbook, published by Ibbotson Associates,* which tracks the long-term performance of five different portfolios of securities.

Information from 1926 to 2002 is compiled in Table 13-1. Returns fluctuate significantly over time, but the long-term perspective used by Ibbotson provides some valuable insights. For example, it is obvious that the rates earned by investors increase as we move down the columns in Table 13-1 toward increasingly risky securities. The standard deviation of annual returns is our proxy measure of risk. Since the averages are the annual returns that investors expect, they are indicative of the rates that a company must offer in order to attract investors. During this same period, the average annual rate of inflation has been 3.1%. Investors, of course, are very aware of inflation, and the rates in the table are therefore nominal annual returns, already adjusted for the inflation expected by the marketplace.

13.2 Debt Capital

One source of capital is borrowed money. The money may be loaned by banks as a line of credit, or it may be obtained from the sale of bonds or debentures. The return to the buyer or bondholder is the interest on the bond and the eventual return of the principal. In order to market the bonds successfully, the return must be attractive to buyers. The attractiveness will depend upon the interest rate offered and the perceived riskiness of the offering. In the United States, two major investment companies, (1) Moody's and (2) Standard and Poor's Investment Services, publish ratings of various bond issues based upon their analyses of each company's financial health. The higher the rating, the lower the interest rate required to attract investors. The four highest rating categories are associated with investment grade bonds; the lowest, with *junk* bonds. The junk bonds have a significant risk of default and therefore offer higher interest rates to attract investors. Many investors pool that risk by developing investment portfolios that include such risky bonds. At the other end of the spectrum, treasury bills are backed by the U.S. government and are regarded as the safest investment, because they have never defaulted. As might

---

* R. G. Ibbotson Associates, *Stocks, Bonds, Bills and Inflation Yearbook* (Chicago, IL: R. G. Ibbotson Associates, 2003.)

be guessed, the interest rate paid by the government for T-bills is very low. The current treasury bill rate is often used as a proxy for the risk-free rate of return, R_F.

Regardless of the rate that companies pay for the money they borrow, the interest that they pay is tax deductible. This tax deductibility results in a tax savings that generally should be considered in engineering economic analysis of capital investments. There are basically two ways to accomplish this.

One way would be first to compute the interest during each year and deduct it from the project cash flows before computing income taxes and next to consider interest and principal reductions to the cash outflows in lieu of the lump-sum investment amount. This adds a computational chore, as the interest amount changes from period to period with level debt service or uniform annual amounts. More importantly, it is generally not easy to assign debt repayment to a specific project, as debt is determined at the corporate, not the project, level.

The second, and most commonly used, approach is to modify the cost of debt to account for its tax deductibility in the interest rate (MARR) used. To make it clear that the interest tax shield is being picked up in the MARR, rather than in the cash flows, we can modify the descriptors for the elements of after-tax cash flow. Specifically, in year k let

$$\text{ATCF}_k = \text{after-tax cash flow (excluding interest on debt);}$$

$$\text{NOPAT}_k = \text{the net operating profit after taxes; and}$$

$$\text{EBIT}_k = \text{the earnings before income taxes (R } - \text{ E).}$$

$$t = \text{the effective income tax rate.}$$

Then in year k, we have

$$\text{ATCF}_k = \text{NOPAT}_k + d_k = (1 - t)\text{EBIT}_k + t\,d_k. \tag{13-1}$$

When calculating the weighted average cost of capital (WACC), as in Section 13.4, the cost of debt capital (i_b) is offset by the tax deduction it provides, and the *after-tax* cost of a company's debt, as a percent of the amount borrowed, is $(1 - t)\,i_b$, where i_b is the cost of debt as an annual rate.

13.3   Equity Capital

Equity sources of funding include not only stockholder's capital but also earnings retained by the company for reinvestment in the business and the cash flow resulting from depreciation charges against income. Just as with bonds, the percentage cost of equity funds, e_a, can be determined by market forces whenever new equity is issued. This, however, is often an infrequent event. Furthermore, the cost of internal equity funds should be based upon opportunity costs associated with the best use of those funds within the firm. Here we must resort to some means of inferring that cost, on the basis of historical performance as a proxy for anticipated future costs. For example, one way might be to consider the trend in past values of return on equity (ROE). If these have been satisfactory, that might be an indication of the value of e_a. Another approach is to look at the return demanded by the shareholders; however, this fluctuates widely and creates an estimating challenge. Table 13-1 shows representative long-term values for e_a, but these are not especially useful as indicators at a given point in time. We need a way to obtain a fairly current value (yet to remove the impact of daily and other fluctuations). When this equity is issued as common stock, a theoretical construct, the capital asset pricing model (CAPM), offers a fruitful approach.

13.3.1   The Capital Asset Pricing Model (CAPM)

The CAPM appeared in the 1960s as a classic piece of economic research, developed independently by three eminent economists. Their work is based upon the classical portfolio theory developed by Harry Markowitz.* It forms the basis for much of contemporary thought about risk and return. The economic logic underpinning this work is elegant. Dedicated readers can consult the seminal works cited in the footnote or a good finance text. Our immediate needs are well served with an intuitive explanation of their results, as shown in Figure 13-2.

Markowitz showed that the best combinations of risk and return on investment can be achieved by investing in a portfolio that includes a mixture of stocks. Further, the most efficient portfolios lie on a curve that specifies the maximum return for a given level of risk. This concave-shaped curve is called the Markowitz efficient market portfolio and is shown in Figure 13-2. Return is measured as expected return, and risk is measured as the standard deviation of returns. An investor could then diversify his or her stock investments to achieve the maximum return, commensurate with the level of risk he or she is willing to tolerate. Or, conversely, the investor could minimize the risk associated with a required return by choosing a portfolio on the Markowitz curve in Figure 13-2.

Using some rather idealistic assumptions and sound economic reasoning, the CAPM asserts that the best combinations of return and risk must lie along a straight

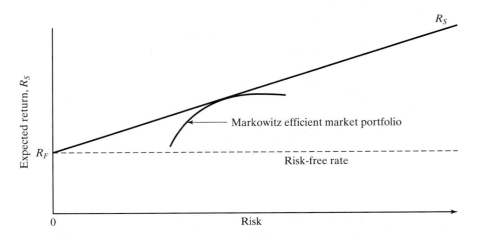

Figure 13-2   Markowitz Efficient Portfolio and the Security Market Line

---

* This work first appeared in the article, "Portfolio Selection," *Journal of Finance*, 7, no. 1 (March 1952): 77–91. The articles that developed the CAPM include W.F. Sharpe, "Capital Asset Prices: A Theory of Market Equilibrium under Conditions of Risk." *Journal of Finance*, 19, (September 1964): 425–442 and J. Lintner, "The Valuation of Risk Assets and the Selection of Risky Investments in Stock Portfolios and Capital Budgets," *Review of Economics and Statistics*, 47, no. 1 (February 1965): 13–37.

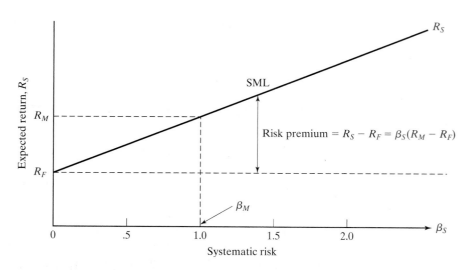

Figure 13-3   Risk and Return in the Capital Asset Pricing Model

line called the security market line, SML. This line is established by the risk-free rate R_F and the point of tangency with the Markowitz efficient market portfolio. That is, rather than choosing an equity portfolio based upon risk and return preference, investors can always achieve the best combination of risk and return by investing in the efficient market portfolio and then borrowing or lending at the risk-free rate to achieve their personal preference for combined risk and return values. Because of this ability, any security must lie on the SML. As shown in Figure 13-3, the SML can be used to develop a standardized estimate of the risk, and hence the return, associated with any security.

Using this model, risk is calibrated to the risk associated with a market portfolio (i.e., a portfolio consisting of all the stocks in the market). The return on the market portfolio, R_M, is keyed to a level of risk that we will call β_M and assign a value of 1.0. This is one point on the SML, which identifies the market return, R_M, and the level of market risk, β_M. The risk-free rate, R_F, and a beta value of zero indicate that the risk-free rate does not fluctuate with the market.

The CAPM asserts that the return R_S on any stock depends upon its risk relative to the market. That risk, β_S, is the stock's contribution to the riskiness of the market portfolio. CAPM suggests that the risk premium of any stock is proportional to its beta. That is,

$$R_S - R_F = \beta_S(R_M - R_F). \tag{13-2}$$

Equation (13-2) states that the quantity $(R_S - R_F)$ is the current *risk premium* associated with a stock S relative to a risk-free investment. The quantity $(R_M - R_F)$ is the *market premium* for the average stock market risk. Over many years, the market premium has steadily hovered at 8.4% above the risk-free rate, as suggested in Table 13-1 and confirmed in studies showing an annual range between 8% and 10%. In this chapter, the long-term average market premium of 8.4% is used.

β_s is the beta value of stock S. It is a widely published and periodically updated statistic for any stock. It is measured by regressing changes in the price of the stock against changes in the market index. The slope of that regression line is labeled as beta and is a widely used measure of a stock's riskiness. Many investment advisory services compute and publish estimates of beta.

13.3.2   Estimating the Cost of Equity

The bottom line of all this is that we can estimate the cost of equity, e_a, as the return on equity, R_s, required in the market by solving Equation (13-2).

EXAMPLE 13-1 **Microsoft's Cost of Equity**

Microsoft has a beta value of 1.68. They have no long-term debt. What is their cost of equity, based upon the CAPM?

Solution

Using CAPM, a long-term market premium $(R_M - R_F)$ of 8.4%, and an estimated risk-free rate (R_F) of 2% in 2002, we see that

$$e_a = R_S = R_F + \beta_S(R_M - R_F) = 0.02 + 1.68(0.084) = 0.161 = 16.1\%.$$

In 2002, Microsoft reported a ROE of 15.7%. This is consistent with the preceding estimate.

13.4   The Weighted Average Cost of Capital (WACC)

The WACC is the product of the fraction of total capital from each source and the cost of capital from that source, summed over all sources. To illustrate the concept, we will simply look at debt and equity sources and their respective costs. This usually provides an adequate figure for planning. For our purposes, Equation (13-3) is used to compute a firm's WACC.

Let

λ = the fraction of the total capital obtained from debt;

$(1 - \lambda)$ = the fraction of the total capital obtained from equity;

t = effective income tax rate as a decimal;

i_b = the cost of debt financing, as measured from appropriate bond rates;

e_a = the cost of equity financing, as measured from historical performance of the CAPM.

Then

$$\text{WACC} = \lambda(1 - t)i_b + (1 - \lambda)e_a. \tag{13-3}$$

Equation (13-3) says that the average cost of funds is a weighted average of the costs of each of its capital sources. Further, if we assess or value investment projects by discounting their returns at the WACC, then any project with a PW > 0 provides value in excess of the cost of the capital required to accomplish it.

EXAMPLE 13-2	**Weighted Average Cost of Capital**

Refer to the situation posed in the beginning of the chapter. Determine the weighted after-tax cost of capital (WACC) for the firm.

Solution

The weighted after-tax cost of long-term debt is $(0.49)(1 - 0.4924)(9.34\%) = 2.323\%$. The $(1 - t)$ term is necessary because 9.34% is the before-tax cost of long-term debt, and we desire an after-tax number. To determine the equity portion of the WACC for this firm, we must recognize that the cost of preferred stock and the cost of common equity are already after-tax values because they are determined after income taxes are paid. Their weighted after-tax values are $(0.13)(8.22\%) = 1.069\%$ and $(0.38)(16.5\%) = 6.270\%$, respectively. We calculate the WACC by adding these weighted components.

$$\text{WACC} = 2.323\% + 1.069\% + 6.270\% = 9.662\%$$

EXAMPLE 13-3	**More WACC Calculations**

What is the WACC for Microsoft? What is the WACC for Duke Energy?

Solution for Microsoft

Since Microsoft has no debt, it is 100% equity financed. Therefore,

$$\text{WACC} = e_a = R_S = 16.1\%, \text{as calculated in Example 13-1.}$$

Solution for Duke Energy

Since the Enron debacle, energy companies have been viewed cautiously by investors. Research sources show that Duke Energy's stock prices have been more stable than most, with a beta of 0.32, but their post-Enron bond rating of BBB has become much lower than the pre-Enron value. Their effective income tax rate is 0.35. As of 2003, Duke's balance sheet shows $20 billion in debt and roughly $15 billion in equity. But there are 900 million shares outstanding, and the price per share is approximately $20, resulting in a market value of the equity of $18 billion. Long-term bonds rated BBB currently earn 6% per year.

$$\lambda = \$20 \text{ billion} \div (\$20 \text{ billion} + \$18 \text{ billion}) = 0.526$$

$$i_b = 0.06$$

$$e_a = R_S = R_F + \beta_S(R_M - R_F) = 0.02 + 0.32(0.084) = 0.047$$

$$t = 0.35$$

$$\text{WACC} = 0.526(1 - 0.35)(0.06) + 0.474(0.047) = 0.043$$

Notice that the market value of the equity ($18 billion) rather than the book value ($15 billion) is used to compute λ, the debt–equity ratio. This reflects the relevance of current market information vis-a-vis the historical book value of equity. Market values of debt do not fluctuate as dramatically as equity values.

13.4.1   The Separation Principle

A related point is that, for every firm, there is an optimal mix of debt and equity financing that minimizes the WACC to that firm. The task of the company treasurer is to identify that mix and maintain it as a permanent part of the capital structure of the company. One difficulty is that it is not easy to determine or maintain the optimal mix. Because any new bond or stock issue incurs underwriting and marketing costs, there is a fixed charge of "going to the well." New capital is procured in chunks, rather than in small amounts, that would permit close adherence to the optimal ratio. It does not make business sense for the hurdle rate (MARR) for investments to be low when the cost of capital is low because this year's projects are to be funded with debt. A more consistent long-term view is obtained by evaluating investment projects at the WACC or the MARR, regardless of how they are being financed. This concept is referred to as the *separation principle*. This, in effect, separates the investment decision from the financing decision and requires projects to have a PW ≥ 0 at the WACC or MARR, or an IRR ≥ WACC or MARR, thus creating value for the investors.

13.4.2   WACC and Risk

The economic theory behind the CAPM was hardly needed to persuade experienced managers that some capital projects are riskier than others and that prudence is required for achieving a higher return (or very conservative estimates of the cash flows) for riskier projects. In reality, every project should be evaluated on the basis of the risk it adds to the company as well as its return. If the risks are roughly normal, and if there are no significant capital limitations, then WACC is an appropriate hurdle rate (i.e., the MARR). If the project is riskier than the current business, then an upward adjustment in the MARR might be appropriate. As intuitively appealing as this may be, the problem becomes one of operationalizing the rate adjustment.

An alternative is to use more formal techniques of risk assessment such as those discussed in Chapters 11 and 12. In those chapters we recommended that risk be attributed to mutually exclusive projects by quantifying the variability of the estimated cash flows and discounting at a single MARR value. In this chapter, however, we are frequently dealing with independent projects in capital budgeting studies, and risk is dealt with by discounting projects at multiple interest rates (MARRs) to account for different degrees of anticipated riskiness. This practice is widespread in industry and is used where appropriate in this chapter in the comparison of independent projects.

13.4.3   Relation of WACC to the MARR

Thus far, we have been comparing investment alternatives with the MARR used as the hurdle rate. The WACC merely establishes a floor on the MARR. In many circumstances, a higher MARR than the WACC may be chosen, because of a shortage of investment funds as well as differences in risk.

The theory developed thus far has implicitly assumed that the firm can obtain as much capital as it needs at the WACC and thus should go forward with any normal risk project that has a positive PW when evaluated at the WACC. In the real world, however, opportunities often occur in lumps rather than in a continuous stream. Management may wish to conserve capital in anticipation of a large future opportunity, or it may wish to encourage investment in a new division while discouraging investments in mature product lines. For that reason, the MARR may be deliberately raised above the WACC in some years or for some divisions. There may be instances in which unusually risk-free cash flows could be evaluated at a MARR lower than the WACC.

13.4.4 Opportunity Costs and Risk Categories in Determining MARRs

A commonly overlooked viewpoint on the determination of the MARR is the *opportunity cost* viewpoint; it comes as a direct result of the capital rationing necessary when there is a limitation of funds relative to prospective proposals to use the funds. This limitation may be either internally or externally imposed. Its parameter is often expressed as a fixed sum of capital; but when the prospective returns from investment proposals, together with the fixed sum of capital available to invest, are known, then the parameter can be expressed as a minimum acceptable rate of return or cut-off rate.

Ideally, the cost of capital by the opportunity cost principle can be determined by ranking prospective projects of similar risk according to a ladder of profitability and then establishing a cut-off point such that the capital is used on the better projects. The return earned by the last project after the cut-off point is the minimum attractive rate of return (MARR) by the opportunity cost principle.

To illustrate the preceding, Figure 13-4 ranks projects according to prospective internal rate of return (IRR) and the cumulative investment required. For purposes of illustration, the amount of capital shown available is $4 million. By connecting down (to the next whole project outside the $4 million) and across, one can read the minimum rate of return under the given conditions, which turns out to be 22%.

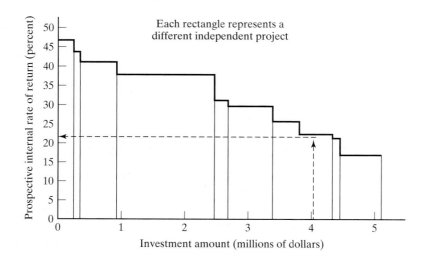

Figure 13-4 Schedule of Prospective Returns and Investment Amounts

It is not uncommon for firms to set two or more MARR levels, according to risk categories. For example, one major industrial firm defines risk categories for income-producing projects and normal MARR standards for each of these categories as follows:

1. *High Risk* (MARR = 40%)

   New products
   New business
   Acquisitions
   Joint ventures

2. *Moderate Risk* (MARR = 25%)

   Capacity increase to meet forecasted sales

3. *Low Risk* (MARR = 15%)

   Cost improvements
   Make versus buy
   Capacity increase to meet existing orders

To illustrate how the preceding set of MARR standards could be determined, the firm could rank prospective projects in each risk category according to prospective rates of return and investment amounts. After tentatively deciding how much investment capital should be allocated to each risk category, the firm could then determine the MARR for each risk category, as illustrated in Figure 13-4 for a single category. Of course, the firm might reasonably shift its initial allocation of funds according to the opportunities available in each risk category, thereby affecting the MARR for each category.

In principle, it would be desirable for a firm to invest additional capital as long as the return from that capital were greater than the cost of obtaining that capital. In such a case, the opportunity cost would equal the marginal cost (in interest and/or stockholder returns) of the best rejected investment. In practice, however, the amount of capital actually invested is more limited because of risk and conservative money policies; thus, the opportunity cost is higher than the marginal cost of the capital.

13.5   Project Selection

To the extent that project proposals can be justified through profitability measures, the most common basis of selection is to choose those proposals that offer the highest prospective profitability, subject to allowances for intangibles or nonmonetary considerations, risk considerations, and limitations on the availability of capital. If the minimum acceptable rate of return has been determined correctly, one can choose proposals according to the IRR, annual worth (AW) method, future worth (FW) method, or present worth (PW) method.

For certain types of project proposals, monetary justification is not feasible—or at least any monetary return is of minor importance compared with intangible or nonmonetary considerations. These types of projects should require careful judgment and analysis, including how they fit in with long-range strategy and plans.

The capital budgeting concepts discussed in this chapter are based on the presumption that the projects under consideration are *not* mutually exclusive. (That is, the adoption of one does not preclude the adoption of others, except with regard to the availability of funds.) Whenever projects are mutually exclusive, the alternative chosen should be based on justification through the incremental return on any incremental investment(s), as well as on proper consideration of nonmonetary factors.

13.5.1 Classifying Investment Proposals

For purposes of study of investment proposals, there should be some system or systems of classification into logical, meaningful categories. Investment proposals have so many facets of objective, form, and competitive design that no one classification plan is adequate for all purposes. Several possible classification plans are the following:

1. According to the kinds and amounts of scarce resources used, such as equity capital, borrowed capital, available plant space, the time required of key personnel, and so on

2. According to whether the investment is tactical or strategic: a *tactical investment* does not constitute a major departure from what the firm has been doing in the past and generally involves a relatively small amount of funds; *strategic investment* decisions, on the other hand, may result in a major departure from what a firm has done in the past and may involve large sums of money

3. According to the business activity involved, such as marketing, production, product line, warehousing, and so on

4. According to priority, such as absolutely essential, necessary, economically desirable, or general improvement

5. According to the type of benefits expected to be received, such as increased profitability, reduced risk, community relations, employee benefits, and so on

6. According to whether the investment involves facility replacement, facility expansion, or product improvement

7. According to the way benefits from the proposed project are affected by other proposed projects; this is generally a most important classification consideration, for there quite often exist interrelationships or dependencies among pairs or groups of investment projects

Of course, all of the preceding classification systems probably are not needed or desirable. As an example, one major corporation uses the following four major categories for higher management screening:

1. Expanded facilities
2. Research and development

3. Improved facilities—for process improvement, cost savings, or quality improvement

4. Necessity—for service facilities, emergency replacements, or for the removal or avoidance of a hazard or nuisance

13.5.2   Degrees of Dependency among Projects

Several main categories of dependency among projects are briefly defined in Table 13-2. Actually, the possible degrees of dependency among projects can be expressed as a continuum from *prerequisite* to *mutually exclusive*, with the degrees *complement*, *independent*, and *substitute* between these extremes, as shown in Figure 13-5.

In developing a project proposal to be submitted for review and approval, the sponsor should include whatever complementary projects seem desirable as part of a single package. Also, if a proposed project will be a partial substitute for any

TABLE 13-2	Degrees of Dependence Between Pairs of Projects	
"If the Results of the First Project Would _____ by Acceptance of the Second Project . . .	. . . then the Second Project is said to be _____ the First Project."	Example
be technically possible or would result in benefits only	a prerequisite of	Car stereo purchase feasible only with purchase of car
have increased benefits	a complement of	Additional hauling trucks more beneficial if automatic loader purchased
not be affected	independent of	A new engine lathe and a fence around the warehouse
have decreased benefits	a substitute for	A screw machine that would do part of the work of a new lathe
be impossible or would result in no benefits	mutually exclusive with	A brick building or a wooden building for a given need

Figure 13-5
Continuum of Degrees of Dependence between Pairs of Projects

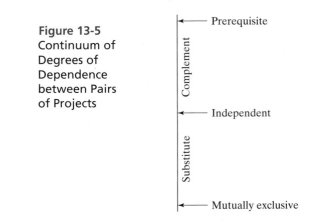

projects to which the firm is already committed or that are under consideration, this fact should be noted in the proposal.

For cases in which choices involved in planning a proposed project are considered sufficiently important that the final decision should be made by higher levels of management, the project proposal should be submitted in the form of a set of mutually exclusive alternatives. For example, if it is to be decided whether to move a plant to a new location and several alternative sites are possible, then separate proposals should be made for each site so as to facilitate the choice of which site, if any, should be chosen.

Whenever capital budgeting decisions involve several groups of mutually exclusive projects and independent projects to be considered within capital availability constraints, the mathematical programming models presented later in this chapter can be useful for selecting the optimal combination of projects.

13.5.3 Organization for Capital Planning and Budgeting

In most large organizations, project selections are accomplished by sequential review through various levels of the organization. The levels required for approval should depend on the nature and importance of the individual project, as well as on the particular organizational makeup of the firm. In general, a mix of central control and coordination, together with authority to make project commitments delegated to operating divisions, is considered desirable. Three typical basic plans for delegating investment decisions are as follows:

1. Whenever proposals are clearly good in terms of economic desirability according to operating division analysis, the division is given the power to commit, as long as appropriate controls can be maintained over the total amount invested by each division and as long as the division analyses are considered reliable.

2. Whenever projects represent the execution of policies already established by headquarters, such as routine replacements, the division is given the power to commit within the limits of appropriate controls.

3. Whenever a project requires a total commitment of more than a certain amount, this request is sent to higher levels within the organization. This is often coupled with a budget limitation regarding the maximum total investment that a division may undertake in a budget period.

To illustrate the concept of larger investments requiring higher administrative approval, the limitations for a small-sized firm might be as follows:

If the Total Investment Is ...		Then Approval Is Required Through
More Than	But Less Than	
$50	$5,000	Plant manager
5,000	50,000	Division vice president
50,000	125,000	President
125,000	—	Board of directors

13.5.4 Communication

The importance of effective communication of capital investment proposals is often overlooked. No matter how great are the merits of a proposed project, if those merits are not communicated to the decision maker(s) in understandable terms, with emphasis on the proper matters, that proposal may well be rejected in favor of less desirable, though better communicated, proposals. Klausner* provides good insight into this problem, with emphasis on the differing perspectives of engineers responsible for technical design and proposal preparation and the management decision makers responsible for monitoring the firm's capital resources. The proposal preparer should be as aware as possible of the decision maker's perspective and related informational needs. For example, in addition to basic information such as investment requirements, measures of merit, and other expected benefits, the decision maker may well want clearly presented answers to such questions as the following:

1. What bases and assumptions were used for estimates?
2. What level of confidence does the proposer have regarding these estimates?
3. How would the investment outcome be affected by variations in these estimated values?

   If project proposals are to be transmitted from one organizational unit to another for review and approval, there must be effective means for communication. The means can vary from standard forms to personal appearances. In communicating proposals to higher levels, it is desirable to use a format that is as standardized as possible to help assure uniformity and completeness of evaluation efforts. In general, the technical and marketing aspects of each proposal should be completely described in a format that is most appropriate to each individual case. The financial implications of all proposals, however, should be summarized in a standardized manner so that they may be uniformly evaluated.

13.6 Postmortem Review

The provision of a system for periodic postmortem reviews (postaudits) of the performance of consequential projects previously authorized is an important aspect of a capital budgeting system. That is, the earnings or costs actually realized on each such project undertaken should be compared with the corresponding quantities estimated at the time the project investment was committed.

   This kind of feedback review serves at least three main purposes, as follows:

1. It determines if planned objectives have been obtained.
2. It determines if corrective action is required.
3. It improves estimating and future planning.

---

* R. F. Klausner, "Communicating Investment Proposals to Corporate Decision Makers," *The Engineering Economist*, 17, no. 1 (Fall 1971): 45–55.

Postmortem reviews should tend to reduce biases in favor of what individual divisions or units preparing project proposals see as their own interests. When divisions of a firm have to compete with each other for available capital funds, there is a tendency for them to evaluate their proposals optimistically. Estimating responsibilities can be expected to be taken more seriously when the estimators know that the results of their estimates will be checked. This checking function, however, should not be overexercised, for there is a human tendency to become overly conservative in estimating when one fears severe accountability for unfavorable results.

It should be noted that a postmortem audit is inherently incomplete. That is, if only one of several alternative projects is selected, it can never be known exactly what would have happened if one of the other alternatives had been chosen. *What might have been if . . .* is at best conjecture, and all postmortem audits should be made with this reservation in proper perspective.

13.7 Budgeting of Capital Investments and Management Perspective

The approved capital budget is limited typically to a one- or two-year period or less, but this should be supplemented by a long-range capital plan, with provision for continual review and change as new developments occur. The long-range plan (or plans) can be for a duration of 2 to 20 years, depending on the nature of the business and the desire of the management to force preplanning.

Even when the technological and market factors in the business are so changeable that plans are no more than guesses to be continually revised, it is valuable to plan and budget as far ahead as possible. Planning should encourage the search for investment opportunities, provide a basis for adjusting other aspects of management of the firm as needed, and sharpen management's forecasting abilities. Long-range budget plans also provide a better basis for establishing minimum rate of return figures that properly take into account future investment opportunities.

An aspect of capital budgeting that is difficult and often important is deciding how much to invest now as opposed to later. If returns are expected to increase for future projects, it may be profitable to withhold funds from investment for some time. The loss of immediate return, of course, must be balanced against the anticipation of higher future returns.

In a similar vein, it may be advantageous to supplement funds available for present projects whenever returns for future projects are expected to become less than those for present projects. Funds for present investment can be supplemented by the reduction of liquid assets, the sale of other assets, and the use of borrowed funds.

The procedures and practices discussed in this book are intended to aid management in making sound investment decisions. Management's ability to sense the opportunities for growth and development and to time their investments to achieve optimum advantage is a primary ingredient of success for an organization.

13.8   Leasing Decisions

Leasing of assets is a business arrangement that makes assets available without incurring initial capital investment costs of purchase. By the term *lease*, we normally are referring to the financial type of lease; that is, a lease in which the firm has a legal obligation to continue making payments for a well-defined period of time for which it has use, but not ownership, of the asset(s) leased. Financial leases usually have fixed durations of at least one year. Many financial leases are very similar to debt and should be treated in essentially the same manner as debt. A significant proportion of assets in some firms is acquired by leasing, thus making the firm's capital available for other uses. Buildings, railroad cars, airplanes, production equipment, and business machines are examples of the wide array of facilities that may be leased.*

Lease specifications are generally detailed in a formal written contract. The contract may contain such specifications as the amount and timing of rental payments; cancellability and sublease provisions, if any; subsequent purchase provisions; and lease renewability provisions.

Although the financial lease provides an important alternative source of capital, it carries with it certain subtle, but important, disadvantages. Its impact is similar to that of added debt capital. The acquisition of financial leases or debt capital will reduce the firm's ability to attract further debt capital and will increase the variability (leverage) in prospective earnings on equity (owner) capital. Higher leverage results in more fixed charges for debt interest and repayment and thus will make good conditions even better and poor conditions even worse for the equity owners.

Figure 13-6 depicts the types of analyses that should be made for lease-related decisions and also shows what conclusion (final choice) should be made for various combinations of conditions (analysis outcomes). The *buy versus status quo* (do nothing) decision is to determine long-run economy or feasibility and is sometimes referred to as an *equipment* decision. In general practice, only if this is investigated and *buy* is found to be preferable should one be concerned with the *buy versus lease* question, which is considered a financing decision. We cannot, however, always separate the *equipment* and *financing* decisions. For example, if the *equipment* decision results in status quo being preferable and yet it is possible that the equipment might be favorable if leased, then we should compare *status quo* with *lease*. This, by definition, is a *mixed* decision involving both *equipment* and *financing* considerations.

The main point one should retain from the preceding is that one should not merely compare *buy* versus *lease* alternatives; one should also compare, if possible, against the *status quo* alternative to determine whether the asset is justified under any financing plan.

---

* So-called leases with an option to buy are treated as conditional sales contracts and are not considered to be true financial leases.

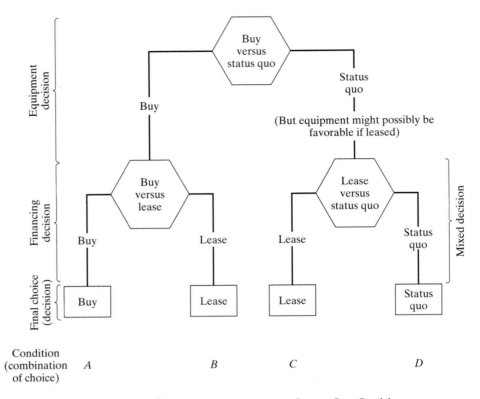

Figure 13-6 Separability of Buy versus Lease versus Status Quo Decisions, and the Choices that Should Result

A major factor in evaluating the economics of leasing versus buying is the tax deductions (reductions in taxable income) allowable. In the case of leasing, one is allowed to deduct the full cost of normal financial lease payments. In the case of buying (ownership), only depreciation charges and interest payments, if any, are deductible. Of course, other disbursements for operating the property are tax deductible under either financing plan.

EXAMPLE 13-4	After-Tax Analysis: Buy, Lease, or Status Quo

Suppose a firm is considering purchasing equipment for $200,000 that would last for five years and have zero market value. Alternatively, the same equipment could be leased for $52,000 at the *beginning* of each of those five years. The effective income tax rate for the firm is 55% and straight-line depreciation is used. The net before-tax cash benefits from the equipment are $56,000 at the end of each year for five years. If the after-tax MARR for the firm is 10%, use the PW method to show whether the firm should buy, lease, or maintain the status quo.

Solution

Alternative	Year	(A) Before-tax Cash Flow	(B) Depreciation	(C) = (A) − (B) Taxable Income	(D) = −0.55(C) Cash Flow For Income Taxes	(E) = (A) + (D) After-tax Cash Flow
Buy {	0	−$200,000				−$200,000
	1–5		−$40,000	−$40,000	+$22,000	+22,000
Lease {	0	−52,000		−52,000	+28,600	− 23,400
	1–4	−52,000		−52,000	+28,600	− 23,400
Status quo	{1–5	−56,000		−56,000	+30,800	− 25,200

Thus, the PWs (after taxes) for the three alternatives are as follows:

Buy: $-\$200,000 + \$22,000(P/A, 10\%, 5) = -\$116,600.$

Lease: $-\$23,400 - \$23,400(P/A, 10\%, 4) \ \ = -\$97,600.$

Status quo: $-\$25,200(P/A, 10\%, 5) \qquad\quad = -\$95,500.$

If the sequence of analysis follows that in Figure 13-6, we would first compare buy (PW = −$116,600) against status quo (PW = −$95,500) and find the status quo to be better. Then we would compare lease (PW = −$97,600) against status quo (PW = −$95,500) and find the status quo to be better. Thus, status quo is the better final choice. It should be noted that, had we merely compared buy versus lease, lease would have been the choice. But the final decision should not have been made without comparison against status quo.

Another way the problem could have been solved would have been to include the +$56,000 before-tax cash benefits with both the buy and the lease alternatives. Then the two alternatives being considered would have been buy rather than status quo (PW = −$116,600 + $95,500 = −$21,100) and lease rather than status quo (PW = −$97,600 + $95,500 = −$2,100). Of these alternatives, lease rather than status quo is the better, but lease is still not justified because of the negative PW, indicating that the costs of lease are greater than the benefits.

13.9 Capital Allocation

This section examines the capital expenditure decision-making process, also referred to as *capital allocation*. This process involves the planning, evaluation, and management of capital projects. In fact, much of this book has dealt with concepts and techniques required to make correct capital-expenditure decisions involving engineering projects. Now our task is to place them in the broader context of upper management's responsibility for proper planning, measurement, and control of the firm's overall portfolio of capital investments.

13.9.1 Allocating Capital among Independent Projects

Companies are constantly presented with independent opportunities in which they can invest capital across the organization. These opportunities usually represent a collection of the best projects for improving operations in all areas of the company (e.g., manufacturing, research and development). In most cases, the amount of available capital is limited, and additional capital can be obtained only at increasing incremental cost. Thus, companies have a problem of budgeting, or allocating, available capital to numerous possible uses.

One popular approach to capital allocation among independent projects uses the net PW criterion. If project risks are about equal, the procedure is to compute the PW for each investment opportunity and then to enumerate all feasible combinations of projects. *Such combinations will then be mutually exclusive.* Each combination of projects is mutually exclusive, because each is unique and the acceptance of one combination of investment projects precludes the acceptance of any of the other combinations. The PW of the combination is the sum of the PWs of the projects included in the combination. Now our capital allocation strategy is to select the mutually exclusive combination of projects that maximizes PW, subject to various constraints on the availability of capital. The next example provides a general overview of this procedure.

EXAMPLE 13-5 | **Mutually Exclusive Combinations of Projects**

Consider these four independent projects and determine the best allocation of capital among them if no more than $300,000 is available to invest:

Independent Project	Initial Capital Outlay	PW
A	$100,000	$25,000
B	125,000	30,000
C	150,000	35,000
D	75,000	40,000

Solution

Figure 13-7 displays a spreadsheet solution for this example. All possible combinations of the independent projects taken two, three, and four at a time are shown, along with their total initial capital outlay and total PW. After eliminating those combinations that violate the $300,000 funds constraint (labeled as not feasible in column D), the proper selection of projects would be ABD, and the maximum PW is $95,000. The process of enumerating combinations of projects having nearly identical risks is best accomplished with a computer when large numbers of projects are being evaluated.

$= \text{IF}(B11<=\$B\$7, C11, \text{"not feasible"})$

	A	B	C	D	E
1	Independent Project	Initial Capital Outlay	PW		
2	A	$ 100,000	$ 25,000		
3	B	$ 125,000	$ 30,000		
4	C	$ 150,000	$ 35,000		
5	D	$ 75,000	$ 40,000		
6					
7	Capital Constraint =	$ 300,000			
8					
9	Combination Enumeration Region				
10	Combination	Total Capital Required	Total PW	Feasible Combinations PW	Optimal Combination
11	AB	$ 225,000	$ 55,000	$ 55,000	
12	AC	$ 250,000	$ 60,000	$ 60,000	
13	AD	$ 175,000	$ 65,000	$ 65,000	
14	BC	$ 275,000	$ 65,000	$ 65,000	
15	BD	$ 200,000	$ 70,000	$ 70,000	
16	CD	$ 225,000	$ 75,000	$ 75,000	
17	ABC	$ 375,000	$ 90,000	not feasible	
18	ACD	$ 325,000	$ 100,000	not feasible	
19	BCD	$ 350,000	$ 105,000	not feasible	
20	ABD	$ 300,000	$ 95,000	$ 95,000	<<Optimal
21	ABCD	$ 450,000	$ 130,000	not feasible	

Left-side formula labels: $= B2 + B3$, $= B2 + B4$, $= B2 + B5$, $= B3 + B4$, $= B3 + B5$, $= B4 + B5$, $= B2 + B3 + B4$, $= B2 + B4 + B5$, $= B3 + B4 + B5$, $= B2 + B3 + B5$, $= B2 + B3 + B4 + B5$

$= \text{IF}(D11=\text{MAX}(\$D\$11:\$D\$21), \text{"<<Optimal"},\text{""})$

Figure 13-7 Spreadsheet Solution, Example 13-5

Methods for determining which possible projects should be allocated available funds seem to require the exercise of judgment in most realistic situations. Example 13-6 illustrates such a problem and possible methods of solution

EXAMPLE 13-6 **Realistic Considerations in Capital Allocation Problems**

Assume that a firm has five investment opportunities (projects) available, which require the indicated amounts of capital and which have economic lives and prospective after-tax IRRs as shown in Table 13-3. Further, assume that the five ventures are independent of each other, investment in one does not prevent investment in any other, and none is dependent upon the undertaking of another.

Now suppose that the company has unlimited funds available, or at least sufficient funds to finance all these projects, and that capital funds cost the company 6% per year after taxes. For these conditions, the company probably would decide to undertake all projects that offered a return of at least 6% per year, and thus projects A, B, C, and D would be acceptable. Such a conclusion, however, would assume that the risks associated with each project are reasonable

TABLE 13-3	Prospective Projects for a Firm[a]		
Project	Capital Investment	Life (Years)	Rate of Return (% per year)
A	$40,000	5	7
B	15,000	5	10
C	20,000	10	8
D	25,000	15	6
E	10,000	4	5

[a] Here we assume that the indicated rates of return for these projects can be repeated indefinitely by subsequent "replacements."

TABLE 13-4	Prospective Projects of Table 13-3 Ordered by IRR		
Project	Capital Investment	Life (Years)	Rate of Return (%)
B	$15,000	5	10
C	20,000	10	8
A	40,000	5	7
D	25,000	15	6

in light of the prospective IRR or are no greater than those encountered in the normal projects of the company.

Unfortunately, in most cases, the amount of capital is limited, either by absolute amount or by increasing cost. If the total of capital funds available is $60,000, the decision becomes more difficult. Here it would be helpful to list the projects in order of *decreasing* profitability, as we have done in Table 13-4 (omitting the undesirable project E). Here it is clear that a complication exists. We naturally would wish to undertake those ventures that have the greatest profit potential. If projects B and C are undertaken, however, there will not be sufficient capital for financing project A that offers the next greatest rate of return. Projects B, C, and D could be undertaken and would provide an approximate annual return of $4,600 (= $15,000 × 10% + $20,000 × 8% + $25,000 × 6%). If project A were undertaken, together with either B or C, the total annual return would not exceed $4,600.* A further complicating factor is the fact that project D involves a longer life than the others. It is thus apparent that we might *not* always decide to adopt the alternative that offers the greatest profit potential.

The problem of allocating limited capital becomes even more complex when the risks associated with the various available projects are not the same. Assume that the risks associated with project B are determined to be higher than the average risk associated with projects undertaken by the firm and that those

---

* This return amount is given, assuming that the leftover capital could earn no more than 6% per year.

TABLE 13-5   Prospective Projects of Table 13-4 Ordered According to Overall Desirability

Project	Capital Investment	Life (Years)	Rate of Return (%)	Risk Rating
C	$20,000	10	8	Lower
A	40,000	5	7	Average
B	15,000	5	10	Higher
D	25,000	15	6	Average

associated with project C are lower than average. The company thus might rank the projects according to their overall desirability, as in Table 13-5. Under these conditions, the company might decide to finance projects C and A, thus avoiding one project with a higher than average risk and another having the lowest prospective return and longest life of the group.

13.9.2   Linear Programming Formulations of Capital Allocation Problems

For large numbers of independent or interrelated investments, the "brute force" enumeration and evaluation of all combinations of projects, as illustrated in Example 13-5, is impractical. This section describes a mathematical procedure for efficiently determining the optimal *portfolio* of projects in industrial capital allocation problems (Figure 13-1). Only the formulations of these problems will be presented in this section; their solution is beyond the scope of this book.

Suppose that the goal of a firm is to maximize its net PW by adopting a capital budget that includes a large number of mutually exclusive combinations of projects. When the number of possible combinations becomes fairly large, manual methods for determining the optimal investment plan tend to become complicated and time consuming, and it is worthwhile to consider linear programming as a solution procedure. Linear programming is a mathematical procedure for maximizing (or minimizing) a linear objective function, subject to one or more linear constraint equations. Hopefully, the reader will obtain some feeling for how more involved problems might also be modeled.

Linear programming is a useful technique for solving certain types of multi-period *capital allocation problems* when a firm is not able to implement all projects that may increase its PW. For example, constraints often exist on how much investment capital can be committed during each fiscal year, and interdependencies among projects may affect the extent to which projects can be successfully carried out during the planning period.

The *objective function* of the capital allocation problem can be written as

$$\text{Maximize net PW} = \sum_{j=1}^{m} B_j^* X_j,$$

where $B_j^* = $ net PW of investment opportunity (project) j during the planning
           period being considered;

$X_j = $ fraction of project j that is implemented during the planning period
           (*Note:* In most problems of interest, X_j will be either zero or one;
           the X_j values are the decision variables);

$m = $ number of mutually exclusive combinations of projects under
           consideration.

In computing the net PW of each mutually exclusive combination of projects, a MARR must be specified.

   The following notation is used in writing the constraints for a linear programming model:

$c_{kj} = $ cash outlay (e.g., initial capital investment or annual operating budget)
           required for project j in time period k;

$C_k = $ maximum cash outlay that is permissible in time period k.

Typically, two types of constraints are present in capital budgeting problems:

1. *Limitations on cash outlays for period k of the planning horizon,*

$$\sum_{j=1}^{m} c_{kj} X_j \leq C_k.$$

2. *Interrelationships among the projects.* The following are examples:

   (a) If projects p, q, and r are mutually exclusive, then

   $$X_p + X_q + X_r \leq 1.$$

   (b) If project r can be undertaken only if project s has already been selected, then

   $$X_r \leq X_s \quad \text{or} \quad X_r - X_s \leq 0.$$

   (c) If projects u and v are mutually exclusive and project r is dependent (contingent) on the acceptance of u or v, then

   $$X_u + X_v \leq 1$$

   and

   $$X_r \leq X_u + X_v.$$

   To illustrate the formulation of linear programming models for capital allocation problems, Example 13-7 and Example 13-8 are presented.

EXAMPLE 13-7   **Linear Programming Formulation of the Capital Allocation Problem**

Five engineering projects are being considered for the upcoming capital budget period. The interrelationships among the projects and the estimated net cash flows of the projects are summarized in the following table:

Project	Cash Flow ($000s) for End of Year k					PW ($000s) at MARR = 10% per year
	0	1	2	3	4	
B1	−50	20	20	20	20	13.4
B2	−30	12	12	12	12	8.0
C1	−14	4	4	4	4	−1.3
C2	−15	5	5	5	5	0.9
D	−10	6	6	6	6	9.0

Projects B1 and B2 are mutually exclusive. Projects C1 and C2 are mutually exclusive *and* dependent on the acceptance of B2. Finally, project D is dependent on the acceptance of C1.

Using the PW method, and assuming that MARR = 10% per year, determine which combination (portfolio) of projects is best if the availability of capital is limited to $48,000.

Solution

The objective function and constraints for this problem are written as follows:
Maximize

$$\text{Net PW} = 13.4X_{B1} + 8.0X_{B2} - 1.3X_{C1} + 0.9X_{C2} + 9.0X_D,$$

subject to

$$50X_{B1} + 30X_{B2} + 14X_{C1} + 15X_{C2} + 10X_D \leq 48;$$

(constraint on investment funds)

$$X_{B1} + X_{B2} \leq 1;$$

(B1 and B2 are mutually exclusive)

$$X_{C1} + X_{C2} \leq X_{B2};$$

(C1 or C2 is contingent on B2)

$$X_D \leq X_{C1};$$

(D is contingent on C1)

$$X_j = 0 \text{ or } 1 \text{ for } j = \text{B1, B2, C1, C2, D.}$$

(No fractional projects are allowed.)

A problem such as this could be solved readily by using the simplex method of linear programming if the last constraint ($X_j = 0$ or 1) were not present. With that constraint included, the problem is classified as a linear *integer* programming problem. (Many computer programs are available for solving large linear integer programming problems.)

| EXAMPLE 13-8 | **Another Linear Programming Formulation** |

Consider a three-period capital allocation problem having the net cash flow estimates and PW values shown in Table 13-6. MARR is 12% per year, and the ceiling on investment funds available is $1,200,000. In addition, there is a constraint on operating funds for support of the combination of projects selected—$400,000 in year one. From these constraints on funds outlays and the interrelationships among projects indicated in Table 13-6, we shall formulate this situation in terms of a linear integer programming problem.

Solution

First, the net PW of each investment opportunity at 12% per year is calculated (Table 13-6). The objective function then becomes

$$\text{Maximize PW} = 135.3X_{A1} + 146.0X_{A2} + 119.3X_{A3} + 164.1X_{B1}$$
$$+ 151.9X_{B2} + 8.1X_{C1} - 13.1X_{C2} + 2.3X_{C3}.$$

TABLE 13-6 **Project Interrelationships and PW Values**

Project		Net Cash Flow ($000s), End of Year[a]				Net PW ($000s) at 12% per year[b]
		0	1	2	3	
A1		−225	150 (60)	150 (70)	150 (70)	+135.3
A2	mutually exclusive	−290	200 (180)	180 (80)	160 (80)	+146.0
A3		−370	210 (290)	200 (170)	200 (170)	+119.3
B1	independent	−600	100 (100)	400 (200)	500 (300)	+164.1
B2		−1,200	500 (250)	600 (400)	600 (400)	+151.9
C1	mutually exclusive and	−160	70 (80)	70 (50)	70 (50)	+8.1
C2	dependent on acceptance	−200	90 (65)	80 (65)	60 (65)	−13.1
C3	of A1 or A2	−225	90 (100)	95 (60)	100 (70)	+2.3

[a] Estimates in parentheses are annual operating expenses (which have already been subtracted in the determination of net cash flows).
[b] For example, net PW for A1 = −$225,000 + $150,000$(P/A, 12\%, 3)$ = +$135,300.

The budget constraints are the following:

Investment funds constraint:

$$225X_{A1} + 290X_{A2} + 370X_{A3} + 600X_{B1} + 1{,}200X_{B2}$$
$$+ 160X_{C1} + 200X_{C2} + 225X_{C3} \le 1{,}200$$

First year's operating cost constraint:

$$60X_{A1} + 180X_{A2} + 290X_{A3} + 100X_{B1} + 250X_{B2}$$
$$+ 80X_{C1} + 65X_{C2} + 100X_{C3} \le 400$$

Interrelationships among the investment opportunities give rise to these constraints on the problem:

$$
\begin{aligned}
X_{A1} + X_{A2} + X_{A3} &\le 1 && \text{A1, A2, A3 are mutually exclusive} \\
X_{B1} &\le 1 \\
X_{B2} &\le 1 && \text{B1, B2 are independent} \\
X_{C1} + X_{C2} + X_{C3} &\le X_{A1} + X_{A2} && \text{accounts for dependence of}
\end{aligned}
$$

C1, C2, C3 (which are mutually exclusive) on A1 or A2

Finally, if all decision variables are required to be either zero (not in the optimal solution) or one (included in the optimal solution), the last constraint on the problem would be written

$$X_j = 0, 1 \quad \text{for } j = \text{A1, A2, A3, B1, B2, C1, C2, C3.}$$

As can be seen in Example 13-7, a fairly simple problem such as this would require a large amount of time to solve by listing and evaluating all mutually exclusive combinations. Consequently, it is recommended that a suitable computer program be used to obtain solutions for all but the most simple capital allocation problems.

13.10 CASE STUDY—Financing an Automobile

The Holiday Car Company has offered to finance your new $24,000 automobile as follows:

> Purchase price = $24,000 (includes all taxes, etc.)
> Down payment = $4,000
> Amount to be financed = $20,000

You tell the credit manager that you think their 9% annual finance charge is too high and that you can raise $24,000 cash with which to purchase the automobile. Nonsense, she says, Why invest in a depreciating asset (the car) when you can finance it through us for 9% per year and then invest your $20,000 in a bond (a nondepreciating asset) that pays 6% per year?

She turns to her computer, enters the above information and comes back with the following proof that her claim is correct. (Note: BOY is an abbreviation for beginning of year. Years are used here to simplify the analysis—normally a monthly analysis is performed.)

	Your Loan Payments (at 9%)			Bond at 6%	
Year	Principal at BOY	Principal Repayment	Interest	Principal at BOY	Interest
1	$20,000	$4,000[a]	$1,800	$20,000	$1,200
2	16,000	4,000	1,440	21,200	1,272
3	12,000	4,000	1,080	22,472	1,348
4	8,000	4,000	720	23,820	1,429
5	4,000	4,000	360	25,249	1,515
		Total	$5,400	Total	$6,764[b]

[a] Her company's analysis assumes a uniform repayment each year of the loan principal.
[b] $20,000(1.06)^5 - $20,000 = $6,764$.

She states that the total interest expense on your car loan from her company is only $5,400, while the earned interest on a bond paying 6% per year accumulates to $6,764 at the end of five years. Therefore, you come out ahead by financing your automobile rather than paying cash for it!

Comment on the credit manager's analysis, and quantitatively show why you believe it to be correct or incorrect. Should the comparison be based solely on interest earned or paid? Should the $20,000 principal that is common to both plans be considered? What are the incremental cash flows between the two financing options? Your MARR is 6% per year.

Solution

If the *finance at 9%* alternative is chosen, the annual costs of *owning* the automobile (as opposed to operating it) are the sum of principal repayment and interest. A typical plan is:

EOY	Principal Repayment	9% Interest on Principal at BOY	Total
1	$4,000	$1,800	$5,800
2	4,000	1,440	5,440
3	4,000	1,080	5,080
4	4,000	720	4,720
5	4,000	360	4,360

Under the financing plan, the $24,000 cash that you have available can be viewed as *two* investments whose cash-flow diagrams are shown on page 577.

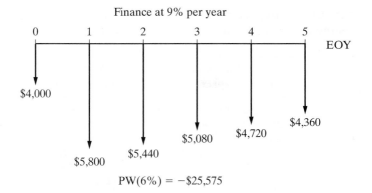

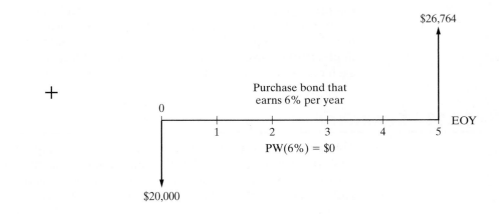

If the interest you can earn on your money is 6% per year, the total PW of both investments is −$25,575. But if you pay cash, the PW is −$24,000 (see below), so you come out ahead by purchasing the automobile for cash.

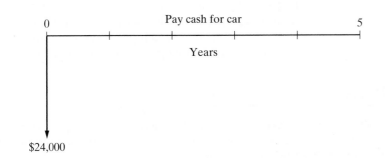

In this situation, one may wonder why the annual costs of owning the automobile are not explicitly dealt with in the pay cash alternative as they were in the finance alternative. The uniform loss in value (analogous to repayment of principal) and opportunity cost of interest forgone by paying cash rather than investing in a 6% bond can be seen to be equivalent to a PW of −$24,000 as follows:

EOY	Uniform Annual Loss in Value	6% Interest on BOY Principal	Total
1	$24,000/5 = −$4,800	−$1,440	−$6,240
2	−4,800	−1,152	−5,952
3	−4,800	−864	−5,664
4	−4,800	−576	−5,376
5	−4,800	−288	−5,088
	PW(6%) = −$24,000		

In conclusion, the credit manager's analysis is incorrect because it ignores the timing of the annual principal repayment and interest. When an incremental study is performed for the pay cash and 9% financing plan, the $20,000 common to both can be ignored. Remember that the automobile is going to lose value (i.e., depreciate) regardless of how it is purchased. Only the cash flows associated with each plan (pay cash versus finance) are relevant to the analysis. ■

13.11 Summary

This chapter has provided an overview of the capital financing and capital allocation functions, as well as the total capital-budgeting process. Our discussion of capital financing has dealt with where companies get money to continue to grow and prosper and the costs to obtain this capital. Also included was a discussion of the weighted average cost of capital. In this regard, differences between debt capital and owner's (equity) capital were made clear. Leasing, as a source of capital, was also described, and a lease-versus-purchase example was analyzed.

Our treatment of capital allocation among independent investment opportunities has been built on two important observations. First, the primary concern in capital expenditure activity is to ensure the survival of the company by implementing ideas to maximize future shareholder wealth, which is equivalent to maximization of shareholder PW. Second, engineering economic analysis plays a vital role in deciding which projects are recommended for funding approval and included in a company's overall capital investment portfolio.

Problems

13-1. In 2008, a firm has receipts of $8 million and expenses (excluding depreciation) of $4 million. Its depreciation for 2008 amounts to $2 million. If the effective income tax rate is 40%, what is this firm's net operating income after taxes (NOPAT)? (13.2)

13-2. The Caterpillar Company has a beta (a measure of common stock volatility) of 1.28. What is its estimated cost of equity capital based on the CAPM when the risk-free interest rate is 2.5%? (13.3)

13-3. Refer to the associated graph on page 579. Identify when the WACC approach to project acceptability agrees with the CAPM approach. When do recommendations of the two approaches differ? Explain why. (13.3)

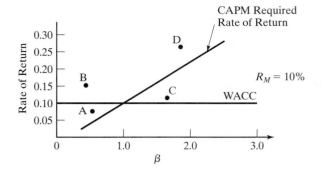

13-4. A firm is considering a capital investment. The risk premium is 0.04, and it is considered to be constant through time. Riskless investments may now be purchased to yield 0.06 (6%). If the project's beta (β) is 1.5, what is the expected return for this investment? (13.3)

13-5. The price to earnings (P/E) ratio is an after-tax metric reflecting growth potential of the common stock of a corporation. P is the selling price (per share) of the common stock, and E is the after-tax earnings per year of a share of stock. A high P/E ratio, for example, indicates that a firm is in a high-growth industry (such as biotechnology) and that annual earnings are not as important to investors as the growth rate of the price of common stock is. Because a corporation can be assumed to have an indefinitely long life, the P/E ratio can be likened to the ($P/A, i'\%, N$) factor when N approaches infinity. For a certain transportation company, the P/E ratio is 12. What is the implied IRR for this relatively stable company? (13.3)

13-6. Explain the various degrees of dependency among two or more projects. If one or more projects seem desirable and are complementary to a given project, should those complementary projects be included in a proposal package or kept separate for review by management? (13.5)

13-7. A four-year-old truck has a present net realizable value of $6,000 and is now expected to have a market value of $1,800 after its remaining three-year life. Its operating disbursements are expected to be $720 per year.

An equivalent truck can be leased for $0.40 per mile plus $30 a day for each day the truck is kept. The expected annual utilization is 3,000 miles and 30 days. If the before-tax MARR is 15%, find which alternative is better by comparing before-tax equivalent annual costs

a. using only the preceding information (13.8);

b. using further information that the annual cost of having to operate without a truck is $2,000. (13.8)

13-8. Work Problem 13-7 by comparing after-tax equivalent PWs if the effective income tax rate is 40%, the present book value is $5,000, and the depreciation charge is $1,000 per year if the firm continues to own the truck. Any gains or losses on disposal of the old truck affect taxes at the full 40% rate, and the after-tax MARR is 5%. (13.8)

13-9. A lathe costs $56,000 and is expected to result in net cash inflows of $20,000 at the end of each year for three years and then have a market value of $10,000 at the end of the third year. The equipment could be leased for $22,000 a year, with the first payment due immediately. (13.8)

a. If the organization does not pay income taxes and its MARR is 10%, show whether the organization should lease or purchase the equipment.

b. If the lathe is thought to be worth only, say, $18,000 per year to the organization, what is the better economic decision?

13-10. The Shakey Company can finance the purchase of a new building costing $2 million with a bond issue, for which it would pay $100,000 interest per year, and then repay the $2 million at the end of the life of the building. Instead of buying in this manner, the company can lease the building by paying $125,000 per year, the first payment being due one year from now. The building would be fully depreciated for tax purposes over an expected life of 20 years. The income tax rate is 40% for all expenses and capital gains or losses, and the firm's after-tax MARR is 5%. Use AW analysis based on equity (nonborrowed) capital to determine whether the firm should borrow and buy or lease if, at the end of 20 years, the building has the following market values for the owner: (a) nothing, (b) $500,000. Straight-line depreciation will be used but is allowable only if the company purchases the building. (13.8)

13-11. The Capitalpoor Company is considering purchasing a business machine for $100,000. An alternative is to rent it for $35,000 at the beginning of each year. The rental would include all repairs and service. If the machine is purchased, a comparable repair and service contract can be obtained for $1,000 per year.

The salesperson of the business machine firm has indicated that the expected useful service life of this machine is five years, with zero market value, but the company is not sure how long the machine will actually be needed. If the machine is rented, the company can cancel the lease at the end of any year. Assuming an income tax rate of 25%, a straight-line depreciation charge of $20,000 for each year the machine is kept, and an after-tax MARR of 10%, prepare an appropriate analysis to help the firm decide whether it is more desirable to purchase or rent. (13.8)

13-12. A firm is considering the development of several new products. The products under consideration are listed in the next table. Products in each group are mutually exclusive. At most, one product from each group will be selected. The firm has a MARR of 10% per year and a budget limitation on development costs of $2,100,000. The life of all products is assumed to be 10 years, with no salvage value. Formulate this capital allocation problem as a linear integer programming model. (13.9)

Group	Product	Development Cost	Annual Net Cash Income
A	A1	$500,000	$90,000
	A2	650,000	110,000
	A3	700,000	115,000
B	B1	600,000	105,000
	B2	675,000	112,000
C	C1	800,000	150,000
	C2	1,000,000	175,000

13-13. Four proposals are under consideration by your company. Proposals A and C are mutually exclusive; proposals B and D are mutually exclusive and cannot be implemented unless proposal A *or* C has been selected. No more than $140,000 can be spent at time zero. The before-tax MARR is 15% per year. The estimated cash flows are shown in the accompanying table. Form all mutually exclusive combinations in view of the specified contingencies, and formulate this problem as a linear integer programming model. (13.9)

End of year	Proposal			
	A	B	C	D
0	−$100,000	−$20,000	−$120,000	−$30,000
1	40,000	6,000	25,000	6,000
2	40,000	10,000	50,000	10,000
3	60,000	10,000	85,000	19,000

13-14. Three alternatives are being considered for an engineering project. Their cash-flow estimates are shown in the accompanying table. A and B are mutually exclusive, and C is an optional add-on feature to alternative A. Investment funds are limited to $5,000,000. Another constraint on this project is the engineering personnel needed to design and implement the solution. No more than 10,000 person-hours of engineering time can be committed to the project. Set up a linear integer programming formulation of this resource allocation problem. (13.9)

	Alternative		
	A	B	C
Initial investment (10^6)	4.0	4.5	1.0
Personnel requirement (hours)	7,000	9,000	3,000

	Alternative		
	A	B	C
After-tax annual savings, years one through four (10^6)	1.3	2.2	0.9
PW at 10% per year (10^6)	0.12	2.47	1.85

13-15. A deluxe home-entertainment system can be purchased for $3,000. The salesperson states, "There are no finance payments for 12 months. Then starting at the end of month 13, the monthly payment will be $125 for the next 24 months." What is the APR of this financing plan? (13.10)

13-16. Laura O'Kelly has agreed to pay $50 each month, starting one month from now, toward a $2,000 balance on a credit card. The credit card company charges 18% interest per year (APR), compounded monthly. (13.10)

a. How many months will it take Laura to pay off this $2,000 balance?

b. How much total interest will Laura have paid?

CHAPTER 14

Decision Making Considering Multiattributes

The aim of Chapter 14 is to present situations in which a decision maker must recognize and address multiple problem attributes.

The Aftermath of Hurricane Katrina

In 2005, the city of New Orleans and many areas in Louisiana, Mississippi, and Alabama were devastated by hurricane Katrina. The cost in lives and property damage was tremendous. Although there was much finger pointing and time spent determining whom to blame, the biggest questions were how to restore the area and what to do to prevent such devastation in the future. The answers are not simple. Consider the decision regarding what to do to prevent this from reoccurring. Of course, cost is a major element of the decision, but dollars are not the only consideration. Time is also important—how quickly can residents get some level of protection? What about environmental impact of the solution? What about the confidence of the residents who want to stay and rebuild? In this chapter, we consider how to make decisions when we take into account the many attributes present in this type of problem.

Ben Franklin developed a method for making difficult decisions. "My [Franklin's] way is to divide a sheet of paper by a line into two columns, writing over the one *PRO* and the other *CON.*" Then he would list all the arguments on each side and weigh how important each was. "Where I find two, one on each side, that seem equal, I strike them both out; if I find a reason *PRO* equal to some two reasons *CON*, I strike out the three." By repeating this bookkeeper's calculus for all *PROs* and *CONs*, it became clear to Franklin "where the balance lies."

—Walter Isaacson (2003)

14.1 Introduction

All chapters up to Chapter 14 dealt principally with the assessment of equivalent monetary worth of competing alternatives and proposals, but *few* decisions are based strictly on dollars and cents. In this chapter, our attention is directed at how diverse nonmonetary considerations (*attributes*) that arise from multiple objectives can be explicitly included in the evaluation of engineering and business ventures. By *nonmonetary*, we mean that a formal market mechanism does not exist in which value can be established for various aspects of a venture's performance such as aesthetic appeal, employee morale, and environmental enhancement.

Value is difficult to define because it is used in a variety of ways. In fact, in 350 BC, Aristotle conceived seven classes of value that are still recognized today: (1) economic, (2) moral, (3) aesthetic, (4) social, (5) political, (6) religious, and (7) judicial. Of these classes, only *economic value* can be measured in terms of (hopefully) objective monetary units such as dollars, yen, or euros. Economic value, however, is also established through an item's *use value* (properties that provide a unit of use, work, or service) and *esteem value* (properties that make something desirable). In highly simplified terms, we can say that use values cause a product to perform (e.g., a car serves as a reliable means of transportation), and esteem values cause it to sell (e.g., a convertible automobile has a look of sportiness). Use value and esteem value defy precise quantification in monetary terms, so we often resort to multiattribute techniques for evaluating the total value of complex designs and complicated systems.

14.2 Examples of Multiattribute Decisions

A common situation encountered by a new engineering graduate is selecting his or her first permanent professional job. Suppose that Mary Jones, a 22-year-old graduate engineer, is fortunate enough to have four acceptable job offers in writing. A choice among the four offers must be made within the next four weeks or they become void. She is not sure which offer to accept, but she decides to base her choice on these four important factors or attributes (not necessarily listed in their order of importance to her): (1) social climate of the town in which she will be working, (2) the opportunity for outdoor sports, (3) starting salary, and (4) potential for promotion and career advancement. Mary Jones next forms a table and fills it in with objective and subjective data relating to differences among the four offers.

The completed table (or matrix) is shown as Table 14-1. Notice that three attributes are rated subjectively on a scale ranging from poor to excellent.

It is not uncommon for monetary and nonmonetary data to be key ingredients in decision situations such as this rather elementary one. Take a minute or two to ponder which offer you would accept, given only the data in Table 14-1. Would starting salary dominate all other attributes, so that your choice would be the Apex Corporation in New York? Would you try to trade off poor social climate against excellent career advancement in Flagstaff to make the McGraw-Wesley offer your top choice?

Many decision problems in industry can be reduced to matrix form similar to that of the foregoing job selection example. To illustrate the wide applicability of such a tabular summary of data, consider a second example involving the choice of a computer-aided design (CAD) workstation by an architectural engineering firm. The data are summarized in Table 14-2. Three vendors and *do nothing* compose the list of feasible alternatives (choices) in this decision problem, and a total of seven attributes is judged sufficient for purposes of discriminating among the alternatives. Aside from the question of which workstation to select, other significant questions come to mind in multiattribute decision making: (1) How are the attributes chosen in the first place? (2) Who makes the subjective judgments regarding nonmonetary attributes such as quality or operating flexibility? (3) What response is required—a partitioning of alternatives or a rank-ordering of alternatives, for instance? Several simple, though workable and credible, models for selecting among alternatives such as those in Tables 14-1 and 14-2 are described in this chapter.

TABLE 14-1 Job Offer Selection Problem

Attributes	Alternatives (Offers and Locations)			
	Apex Corp., New York	Sycon, Inc., Los Angeles	Sigma Ltd., Macon, GA	McGraw-Wesley, Flagstaff, AZ
Social climate	Good	Good	Fair	Poor
Weather/outdoor sports	Poor	Excellent	Good	Very good
Starting salary (per annum)	$50,000	$45,000	$49,500	$46,500
Career advancement	Fair	Very good	Good	Excellent

TABLE 14-2 CAD Workstation Selection Problem

Attribute	Alternatives			
	Vendor A	Vendor B	Vendor C	Reference ("Do Nothing")
Cost of purchasing the system	$115,000	$338,950	$32,000	$0
Reduction in design time	60%	67%	50%	0
Flexibility	Excellent	Excellent	Good	Poor
Inventory control	Excellent	Excellent	Excellent	Poor
Quality	Excellent	Excellent	Good	Fair
Market share	Excellent	Excellent	Good	Fair
Machine utilization	Excellent	Excellent	Good	Poor

14.3 Choice of Attributes

The choice of attributes used to judge alternative designs, systems, products, processes, and so on is one of the most important tasks in multiattribute decision analysis. (The most important task, of course, is to identify the feasible alternatives from which to select.) The articulation of attributes for a particular decision can, in some cases, shed enough light on the problem to make the final choice obvious.

Consider again the data in Tables 14-1 and 14-2. Some general observations regarding the attributes used to discriminate among alternatives can immediately be made: (1) Each attribute distinguishes at least two alternatives—in no case should identical values for an attribute apply to all alternatives; (2) each attribute captures a unique dimension or facet of the decision problem (i.e., attributes are independent and nonredundant); (3) all attributes, in a collective sense, are assumed to be sufficient for the purpose of selecting the best alternative; and (4) differences in values assigned to each attribute are presumed to be meaningful in distinguishing among feasible alternatives.

Selecting a set of attributes is usually the result of group consensus and is clearly a subjective process. The final list of attributes, both monetary and nonmonetary, is heavily influenced by the decision problem at hand, as well as by an intuitive feel for which attributes will or will not pinpoint relevant differences among feasible alternatives. If too many attributes are chosen, the analysis will become unwieldy, unreliable, and difficult to manage. Too few attributes, on the other hand, will limit discrimination among alternatives. Again, judgment is required to decide what number is too few or too many. If some attributes in the final list lack specificity or cannot be quantified, it will be necessary to subdivide them into lower-level attributes that can be measured.

To illustrate these points, we might consider adding an attribute called cost of operating and maintaining the system in Table 14-2 to capture a vital dimension of the CAD system's life-cycle cost. The attribute *flexibility* should perhaps be subdivided into two other, more specific attributes such as *ability to communicate with computer-aided manufacturing equipment* (such as numerically controlled machine tools) and *ability to create and analyze solid geometry representations of engineering design concepts*. Finally, it would be constructive to aggregate two attributes in Table 14-2, namely *quality* and *market share*. Because there is no difference in the values assigned to these two attributes across the four alternatives, they could be combined into a single attribute, possibly named *achievement of greater market share through quality improvements*. But one should take great care in doing this. If it is possible that a new alternative might have differences in the values assigned to these two attributes, then they should be left as they are.

14.4 Selection of a Measurement Scale

Identifying feasible alternatives and appropriate attributes represents a large portion of the work associated with multiattribute decision analysis. The next task is to develop metrics, or measurement scales, that permit various states of each attribute to be represented. For example, in Table 14-1, "dollars" was an obvious choice for the metric of starting salary. A subjective assessment of career advancement was made on a metric having the gradations poor, fair, good, very good, and excellent. In many problems the metric is simply the scale upon which a physical measurement is made. For instance, anticipated noise pollution for various routings of an urban highway project might be a relevant attribute whose metric is decibels.

14.5 Dimensionality of the Problem

If you refer once again to Table 14-1, notice that there are two basic ways to process the information presented there. First, you could attempt to collapse each job offer into a single metric or dimension. For instance, all attributes could somehow be forced into their dollar equivalents, or they could be reduced to a *utility equivalent* ranging from, say, 0 to 100. Assigning a dollar value to good career advancement may not be too difficult, but how about placing a dollar value on a poor versus an excellent social climate? Similarly, translating all job offer data to a scale of worth or utility that ranges from 0 to 100 may not be plausible to most individuals. This first way of dealing with the data of Table 14-1 is called *single-dimension analysis*. (The dimension corresponds to the number of metrics used to represent the attributes that discriminate among alternatives.)

Collapsing all information into a single dimension is popular in practice because a complex problem can be made computationally tractable in this manner. Several useful models presented later are single-dimensioned. Such models are termed *compensatory* because changes in the values of a particular attribute can be offset by, or traded off against, opposing changes in another attribute.

The second basic way to process information in Table 14-1 is to retain the individuality of the attributes as the best alternative is being determined. That is, there is no attempt to collapse attributes into a common scale. This is referred to as *full-dimension analysis* of the multiattribute problem. For example, if r attributes have been chosen to characterize the alternatives under consideration, the predicted values for all r attributes are considered in the choice. If a metric is common to more than one attribute, as in Table 14-1, we have an intermediate-dimension problem that is analyzed with the same models as a full-dimension problem would be. Several of these models are illustrated in the next section, and they are often most helpful in eliminating inferior alternatives from the analysis. We refer to these models as *noncompensatory* because trade-offs among attributes are not permissible. Thus, comparisons of alternatives must be judged on an attribute-by-attribute basis.

14.6 Noncompensatory Models

In this section, we examine four noncompensatory models for making a choice when multiple attributes are present. They are (1) dominance, (2) satisficing, (3) disjunctive resolution, and (4) lexicography. In each model, an attempt is made to select the best alternative in view of the full dimensionality of the problem. Example 14-1 is presented after the description of these models and will be utilized to illustrate each.

14.6.1   Dominance

Dominance is a useful screening method for eliminating inferior alternatives from the analysis. When one alternative is better than another with respect to all attributes, there is no problem in deciding between them. In this case, the first alternative *dominates* the second one. By comparing each possible pair of alternatives to determine whether the attribute values for one are at least as good as those for the other, it may be possible to eliminate one or more candidates from further consideration or even to select the single alternative that is clearly superior to all the others. Usually it will not be possible to select the best alternative based on dominance.

14.6.2   Satisficing

Satisficing, sometimes referred to as the *method of feasible ranges*, requires the establishment of minimum or maximum acceptable values (the standard) for each attribute. Alternatives having one or more attribute values that fall outside the acceptable limits are excluded from further consideration.

The upper and lower bounds of these ranges establish two fictitious alternatives against which maximum and minimum performance expectations of feasible alternatives can be defined. By bounding the permissible values of attributes from two sides (or from one), information processing requirements are substantially reduced, making the evaluation problem more manageable.

Satisficing is more difficult to use than dominance because minimum acceptable attribute values must be determined. Furthermore, satisficing is usually employed to evaluate feasible alternatives in more detail and to reduce the number being considered rather than to make a final choice. The satisficing principle is frequently used in practice when *satisfactory* performance on each attribute, rather than *optimal* performance, is good enough for decision-making purposes.

14.6.3   Disjunctive Resolution

The disjunctive method is similar to satisficing in that it relies on comparing the attributes of each alternative to the standard. The difference is that the disjunctive method evaluates each alternative on the best value achieved for any attribute. If an alternative has *just one* attribute that meets or exceeds the standard, that alternative is kept. In satisficing, *all* attributes must meet or exceed the standard in order for an alternative to be kept in the feasible set.

14.6.4   Lexicography

This model is particularly suitable for decision situations in which a single attribute is judged to be more important than all other attributes. A final choice *might* be based solely on the most acceptable value for this attribute. Comparing alternatives with respect to one attribute reduces the decision problem to a single dimension (i.e., the measurement scale of the predominant attribute). The alternative having the highest value for the most important attribute is then chosen. When two or more alternatives have identical values for the most important attribute, however, the second most important attribute must be specified and used to break the deadlock.

If ties continue to occur, the analyst examines the next most important attribute until a single alternative is chosen or until all alternatives have been evaluated.

Lexicography requires that the importance of each attribute be specified to determine the order in which attributes are to be considered. If a selection is made by using one or a few of the attributes, lexicography does not take into account all the collected data. Lexicography does not require comparability across attributes, but it does process information in its original metric.

EXAMPLE 14-1	**Selecting a Dentist by Using Noncompensatory Models**

Mary Jones, the engineering graduate whose job offers were given in Table 14-1, has decided, based on comprehensive reasoning, to accept the Sigma position in Macon, Georgia. (Problem 14-9 will provide insight into why this choice was made.) Having moved to Macon, Mary now faces several other important multiattribute problems. Among them are (1) renting an apartment versus purchasing a small house, (2) what type of automobile or truck to purchase, and (3) whom to select for long-overdue dental work.

In this example, we consider the selection of a *dentist* as a means of illustrating noncompensatory (full-dimensioned) and compensatory (single-dimensioned) models for analyzing multiattribute decision problems.

After calling many dentists in the Yellow Pages, Mary finds that there are only four who are accepting new patients. They are Dr. Molar, Dr. Feelgood, Dr. Whoops, and Dr. Pepper. The alternatives are clear to Mary, and she decides that her objectives in selecting a dentist are to obtain high-quality dental care at a reasonable cost with minimum disruption to her schedule and little (or no) pain involved. In this regard, Mary adopts these attributes to assist in gathering data and making her final choice: (1) reputation of the dentist, (2) cost per hour of dental work, (3) available office hours each week, (4) travel distance, and (5) method of anesthesia. Notice that these attributes are more or less independent in that the value of one attribute cannot be predicted by knowing the value of any other attribute.

Mary collects data by interviewing the receptionist in each dental office, talking with local townspeople, calling the Georgia Dental Association, and so on. A summary of information gathered by Mary is presented in Table 14-3.

We are now asked to determine whether a dentist can be selected by using (a) dominance, (b) satisficing, (c) disjunctive resolution, and (d) lexicography.

Solution

(a) To check for dominance in Table 14-3, pairwise comparisons of each dentist's set of attributes must be inspected. There will be $4(3)/2 = 6$ pairwise comparisons necessary for the four dentists, and they are shown in Table 14-4. It is clear from Table 14-4 that Dr. Molar dominates Dr. Feelgood, so Dr. Feelgood will be dropped from further consideration. With dominance, it is not possible for Mary to select the best dentist.

TABLE 14-3 Summary Information for Choice of a Dentist

Attribute	Alternatives			
	Dr. Molar	Dr. Feelgood	Dr. Whoops	Dr. Pepper
Cost ($/hour)	$50	$80	$20	$40
Method of anesthesia[a]	Novocaine	Acupuncture	Hypnosis	Laughing Gas
Driving distance (mi)	15	20	5	30
Weekly office hours	40	25	40	40
Quality of work	Excellent	Fair	Poor	Good

☐ Best value ☐ Worst value

[a] Mary has decided that novocaine > laughing gas > acupuncture > hypnosis, where $a > b$ means that a is preferred to b.

TABLE 14-4 Check for Dominance among Alternatives

Attribute	Paired Comparison					
	Molar vs. Feelgood	Molar vs. Whoops	Molar vs. Pepper	Feelgood vs. Whoops	Feelgood vs. Pepper	Whoops vs. Pepper
Cost	Better	Worse	Worse	Worse	Worse	Better
Anesthesia	Better	Better	Better	Better	Worse	Worse
Distance	Better	Worse	Better	Worse	Better	Better
Office hours	Better	Equal	Equal	Worse	Worse	Equal
Quality	Better	Better	Better	Better	Worse	Worse
Dominance?	Yes	No	No	No	No	No

(b) To illustrate the satisficing model, acceptable limits (feasible ranges) must be established for each attribute. After considerable thought, Mary determines the feasible ranges given in Table 14-5.

   Comparison of attribute values for each dentist against the feasible range reveals that Dr. Whoops uses a less desirable type of anesthesia (hypnosis < acupuncture), and his quality rating is also not acceptable (poor < good). Thus, Dr. Whoops joins Dr. Feelgood on Mary's list of rejects. Notice that satisficing, by itself, did not produce the best alternative.

(c) By applying the feasible ranges in Table 14-5 to the disjunctive resolution model, all dentists would be acceptable because each has at least one attribute value that meets or exceeds the minimum expectation. For instance, Dr. Whoops scores acceptably on three out of five attributes, and Dr. Feelgood passes two out of five minimum expectations. Clearly, this model does not discriminate well among the four candidates.

(d) Many models, including lexicography, require that all attributes first be ranked in order of importance. Perhaps the easiest way to obtain a consistent

TABLE 14-5	Feasible Ranges for Satisficing		
Attribute	Minimum Acceptable Value	Maximum Acceptable Value	Unacceptable Alternative
Cost	—	$60	None (Dr. Feelgood already eliminated)
Anesthesia	Acupuncture	—	Dr. Whoops
Distance (miles)	—	30	None
Office hours	30	40	None (Dr. Feelgood already eliminated)
Quality	Good	Excellent	Dr. Whoops

TABLE 14-6   Ordinal Ranking of Dentists' Attributes

A. Results of Paired Comparisons

Cost > anesthesia	(Cost is more important than anesthesia)
Quality > cost	(Quality is more important than cost)
Cost > distance	(Cost is more important than distance)
Cost > office hours	(Cost is more important than office hours)
Anesthesia > distance	(Anesthesia is more important than distance)
Anesthesia > office hours	(Anesthesia is more important than office hours)
Quality > anesthesia	(Quality is more important than anesthesia)
Office hours > distance	(Office hours are more important than distance)
Quality > distance	(Quality is more important than distance)
Quality > office hours	(Quality is more important than office hours)

B. Attribute	Number of times on left of > (= Ordinal ranking)
Cost	3
Anesthesia	2
Distance	0
Office hours	1
Quality	4

ordinal ranking is to make paired comparisons between each possible attribute combination.* This is illustrated in Table 14-6. Each attribute can be ranked according to the number of times it appears on the left-hand side of the comparison when the preferred attribute is placed on the left as shown. In this case, the ranking is found to be quality > cost > anesthesia > office hours > distance.

Table 14-7 illustrates the application of lexicography to the ordinal ranking developed in Table 14-6. The final choice would be Dr. Molar because quality

---

* An ordinal ranking is simply an ordering of attributes from the most preferred to the least preferred.

TABLE 14-7	Application of Lexicography			
Attribute	Rank[a]	Alternative Rank[b]		
Cost	3	Whoops > Pepper	> Molar	> Feelgood
Anesthesia	2	Molar > Pepper	> Feelgood > Whoops	
Office hours	1	Molar = Whoops = Pepper	> Feelgood	
Distance	0	Whoops > Molar	> Feelgood > Pepper	
Quality	4	Molar > Pepper	> Feelgood > Whoops	

[a] Rank of 4 = most important, rank of 0 = least important.
[b] Selection is based on the highest-ranked attribute (Whoops and Feelgood included only to illustrate the full procedure).

is the top-ranked attribute, and Molar's quality rating is the best of all. If Dr. Pepper's work quality had also been rated as excellent, the choice would be made on the basis of cost. This would have resulted in the selection of Dr. Pepper. Therefore, lexicography does allow the best dentist to be chosen by Mary.

14.7 Compensatory Models

The basic principle behind all compensatory models, which involve a single dimension, is that the values for all attributes must be converted to a common measurement scale such as *dollars* or *utiles*,* from which it is possible to construct an overall dollar index or utility index for each alternative. The form of the function used to calculate the index can vary widely. For example, the converted attribute values may be added together, they may be weighted and then added, or they may be sequentially multiplied. Regardless of the functional form, the end result is that good performance in one attribute can *compensate* for poor performance in another. This allows trade-offs among attributes to be made during the process of selecting the best alternative. Because lexicography involves no trade-offs, it was classified as a full-dimensional model in Section 14.6.4.

In this section, we examine two compensatory models for evaluating multiattribute decision problems: nondimensional scaling and the additive weighting technique. Each model will be illustrated by using the data of Example 14-1. The interested reader can consult Canada et al. for a discussion of other methods.[†]

14.7.1 Nondimensional Scaling

A popular way to standardize attribute values is to convert them to nondimensional form. There are two important points to consider when doing this. First, the

---

* A utile is a dimensionless unit of worth.
[†] Canada, et al., *Capital Investment Decision Analysis for Engineering and Management*, 3rd ed. (Upper Saddle River, NJ: Prentice Hall, 2005).

nondimensional values should all have a common range, such as 0 to 1 or 0 to 100. Without this constraint, the dimensionless attributes will contain implicit weighting factors. Second, all of the dimensionless attributes should follow the same trend with respect to desirability; the most preferred values should be either all small or all large. This is necessary in order to have a believable overall scale for selecting the best alternative.

Nondimensional scaling can be illustrated with the data of Example 14-1. As shown in Table 14-8, the preceding constraints may require that different procedures be used to remove the dimension from each attribute. For example, a cost-related attribute is best when it is low, but office hours are best when they are high. The goal should be to devise a procedure that rates each attribute in terms of its fractional accomplishment of the best attainable value. Table 14-3, the original table of information for Example 14-1, is restated in dimensionless terms in Table 14-9. The procedure for converting the original data in Table 14-3 for a particular attribute to its dimensionless rating is

$$\text{Rating} = \frac{\text{Worst outcome} - \text{Outcome being made dimensionless}}{\text{Worst outcome} - \text{Best outcome}}. \quad (14\text{-}1)$$

Equation (14-1) applies when large numerical values, such as cost or driving distance, are considered to be *undesirable*. When large numerical values are considered to be *desirable* (anesthesia, office hours, quality), however, the relationship for converting original data to their dimensionless ratings is

$$\text{Rating} = \frac{\text{Outcome being made dimensionless} - \text{Worst outcome}}{\text{Best outcome} - \text{Worst outcome}}. \quad (14\text{-}2)$$

TABLE 14-8 Nondimensional Scaling for Example 14-1

Attribute	Value	Rating Procedure	Dimensionless Value
Cost	$20	$(80 - \text{cost})/60$	1.0
	40		0.67
	50		0.50
	80		0.0
Anesthesia	Hypnosis	$(\text{Relative rank}^a - 1)/3$	0.0
	Acupuncture		0.33
	Laughing gas		0.67
	Novocaine		1.0
Distance	5	$(30 - \text{distance})/25$	1.0
	15		0.60
	20		0.40
	30		0.0
Office hours	25	$(\text{Office hours} - 25)/15$	0.0
	40		1.0
Quality	Poor	$(\text{Relative rank}^a - 1)/3$	0.0
	Fair		0.33
	Good		0.67
	Excellent		1.0

[a] Scale of 1 to 4, is used, 4 being the best (from Table 14-3).

If all the attributes in Table 14-9 are of equal importance, a score for each dentist could be found by merely summing the nondimensional values in each column. The results would be Dr. Molar = 4.10, Dr. Feelgood = 1.06, Dr. Whoops = 3.00, and Dr. Pepper = 3.01. Presumably, Dr. Molar would be the best choice in this case.

Figure 14-1 displays a spreadsheet application of the nondimensional scaling procedure for Example 14-1. The attribute ratings for each alternative are entered into the table in the range B2:E6. Next, tables for the qualitative attributes (quality and method of anesthesia) are entered into conversion tables (A9:B12 and D9:E12). These tables must have the qualitative rating in alphabetical order.

The dimensionless rating for each alternative/attribute combination is determined by comparing the difference between the alternative under consideration with the lowest scoring alternative. This difference is then divided by the range between the best and worst scoring alternatives to arrive at the nondimensional value. The total score for each alternative is found by summing these values. The formulas in the highlighted cells in Figure 14-1 are given on the next page.

TABLE 14-9	Nondimensional Data for Example 14-1			
Attribute	Dr. Molar	Dr. Feelgood	Dr. Whoops	Dr. Pepper
Cost	0.50	0.0	1.0	0.67
Method of anesthesia	1.0	0.33	0.0	0.67
Driving distance	0.60	0.40	1.0	0.0
Weekly office hours	1.0	0.0	1.0	1.0
Quality of work	1.0	0.33	0.0	0.67

	A	B	C	D	E
1	**Attribute**	**Dr. Molar**	**Dr. Feelgood**	**Dr. Whoops**	**Dr. Pepper**
2	**Cost**	$ 50	$ 80	$ 20	$ 40
3	**Anesthesia**	Novocaine	Acupuncture	Hypnosis	Laughing Gas
4	**Distance**	15	20	5	30
5	**Office Hours**	40	25	40	40
6	**Quality**	Excellent	Fair	Poor	Good
7					
8		*Quality*			*Anesthesia*	
9	Excellent	4		Acupuncture	2
10	Fair	2		Hypnosis	1
11	Good	3		Laughing Gas	3
12	Poor	1		Novocaine	4
13					
14	**Attribute**	**Dr. Molar**	**Dr. Feelgood**	**Dr. Whoops**	**Dr. Pepper**
15	Cost	0.50	0.00	1.00	0.67
16	Anesthesia	1.00	0.33	0.00	0.67
17	Distance	0.60	0.40	1.00	0.00
18	Hours	1.00	0.00	1.00	1.00
19	Quality	1.00	0.33	0.00	0.67
20	Sum =	4.10	1.07	3.00	3.00
21		ʌ best choice			

Figure 14-1 Spreadsheet Solution, Nondimensional Scaling

Cell	Contents
B15	= ABS(B2-MAX($B$2:$E$2))/ABS(MAX($B$2:$E$2)-MIN($B$2:$E$2))
B16	= (VLOOKUP(B3,$D$9:$E$12,2)-MIN($E$9:$E$12))/(MAX($E$9:$E$12) −MIN($E$9:$E$12))
B17	= ABS(B4-MAX($B$4:$E$4))/ABS(MAX($B$4:$E$4)-MIN($B$4:$E$4))
B18	= ABS(B5-MIN($B$5:$E$5))/ABS(MAX($B$5:$E$5)-MIN($B$5:$E$5))
B19	= (VLOOKUP(B6,$A$9:$B$12,2)-MIN($B$9:$B$12))/(MAX($B$9:$B$12) −MIN($B$9:$B$12))
B20	= SUM(B15:B19)
B21	= IF(B20 = MAX($B$20:$E$20),"^ best choice","")

14.7.2 The Additive Weighting Technique

Additive weighting provides for the direct use of nondimensional attributes such as those in Table 14-9 and the results of ordinal ranking as illustrated in Table 14-6. The procedure involves developing weights for attributes (based on ordinal rankings) that can be multiplied by the appropriate nondimensional attribute values to produce a *partial contribution* to the overall score for a particular alternative. When the partial contributions of all attributes are summed, the resulting set of alternative scores can be used to compare alternatives directly. In the previous section, these partial contributions were assumed to be equal, but in this section they can be unequal based on how important they are believed to be.

Attribute weights should be determined in two steps following the completion of ordinal ranking. First, relative weights are assigned to each attribute according to its ordinal ranking. The simplest procedure is to use rankings of 1, 2, 3,..., based on position, with higher numbers signifying greater importance; but one might also include subjective considerations by using uneven spacing in some cases. For instance, in a case where there are four attributes, two of which are much more important than the others, the top two may be rated as 7 and 5 instead of 4 and 3. The second step is to normalize the relative ranking numbers by dividing each ranking number by the sum of all the rankings. Table 14-10 summarizes these steps for Example 14-1 and demonstrates how the overall score for each alternative is determined.

Additive weighting is probably the most popular single-dimensional method because it includes both the performance ratings and the importance weights of each attribute when evaluating alternatives. Furthermore, it produces recommendations that tend to agree with the intuitive feel of the decision maker concerning the best alternative. Perhaps its biggest advantage is that removing the dimension from data and weighting attributes are separated into two distinct steps. This reduces confusion and allows for precise definition of each of these contributions. From Table 14-10, it is apparent that the additive weighting score for Dr. Molar (0.84) makes him the top choice as Mary's dentist.

Figure 14-2 shows a spreadsheet application of the additive weighting technique for the dentist selection problem of Example 14-1. The relative ranking of attributes (Table 14-6) is entered in the range B3:B7. These values are normalized (put on a 0-1 scale) in column C. For each alternative, the normalized weight of the attribute is multiplied by the nondimensional attribute value (obtained from the

TABLE 14-10 The Additive Weighting Technique Applied to Example 14-1

| | Calculation of Weighting Factors | | Calculation of Scores for Each Alternative[b] | | | | | | | |
| | Step 1: | Step 2: | Dr. Molar | | Dr. Feelgood | | Dr. Whoops | | Dr. Pepper | |
Attribute	Relative Rank[a]	Normalized Weight (A)	(B)	(A) × (B)	(B)	(A) × (B)	(B)	(A) × (B)	(B)	(A) × (B)
Cost	4	4/15 = 0.27	0.50	0.14	0.00	0.00	1.00	0.27	0.67	0.18
Anesthesia	3	3/15 = 0.20	1.00	0.20	0.33	0.07	0.00	0.00	0.67	0.13
Distance	1	1/15 = 0.07	0.60	0.04	0.40	0.03	1.00	0.07	0.00	0.00
Office hours	2	2/15 = 0.13	1.00	0.13	0.00	0.00	1.00	0.13	1.00	0.13
Quality	5	5/15 = 0.33	1.00	0.33	0.33	0.11	0.00	0.00	0.67	0.22
	Sum = 15	Sum = 1.00	Sum = 0.84		Sum = 0.21		Sum = 0.47		Sum = 0.66	

[a] Based on Table 14-6, relative rank = ordinal ranking + 1. A rank of 5 is best.

[b] Data in column B are from Table 14-9.

594

	A	B	C	D	E	F	G	H	I	J	K
1	Attribute	Relative Rank	Normalized Weight	Dr. Molar		Dr. Feelgood		Dr. Whoops		Dr. Pepper	
2				Dimensionless Value	Score	Dimensionless Value	Score	Dimensionless Value	Score	Dimensionless Value	Score
3	Cost =	4	0.27	0.50	0.13	0.00	0.00	1.00	0.27	0.67	0.18
4	Anesthesia =	3	0.20	1.00	0.20	0.33	0.07	0.00	0.00	0.67	0.13
5	Distance =	1	0.07	0.60	0.04	0.40	0.03	1.00	0.07	0.00	0.00
6	Office Hours =	2	0.13	1.00	0.13	0.00	0.00	1.00	0.13	1.00	0.13
7	Quality =	5	0.33	1.00	0.33	0.33	0.11	0.00	0.00	0.67	0.22
8	TOTAL=	15	1.00		0.84		0.20		0.47		0.67
9					^ winner ^						

= B3 / $B$8

= D3 * $C3

= IF(E8 = MAX($E$8:$K$8), "^winner^" , "")

= SUM(B3:B7)

Figure 14-2   Spreadsheet Solution, Additive Weighting for Example 14-1

nondimensional scaling spreadsheet in Figure 14-1) to arrive at a weighted score for the attribute. These weighted scores are then summed to arrive at an overall score for each alternative.

| EXAMPLE 14-2 | **Decision Making in the Aftermath of Hurricane Katrina** |

Let us revisit hurricane Katrina. Decisions must be made about how to replace the flood protection and prevent the type of devastation that occurred. These decisions are based on myriad factors and certainly have many social and political consequences. There is no way to do justice to this type of problem in this text, but we will examine a fictitious example based on data available in the popular press (ABC News and *Time* magazine—www.time.com/time/specials/2007/article/0.28804, 1646611_1646683_1648904-2,00.html).

Plans must be put into place to protect life and property in and around the city of New Orleans, in the event of another hurricane event such as Katrina. Interestingly, the hurricane was not one of the most powerful and had been downgraded to a category 3 storm when she made landfall. This speaks of a more urgent need, knowing that the devastation could have been worse.

Assume that, after using the noncompensatory models, three primary alternatives remain available, as given in Table 14-11. Also shown in the table are a number of attributes to be considered in making the decision about which alternative to select. In reality, of course, there would likely be many more alternatives (and variations on alternatives) available after initial analysis, and many more attributes to consider.

We will use the additive weighting technique to recommend a course of action.

TABLE 14-11   Data for Example 14-2

	Standard Levee Repair	Massive Levee Construction	Restoration and Restructure through Nonstructural Methods
Cost (billion)	$1.3	$35	$20
Time until completion (years)	2	13	15
Environmental impact (low to high)	Moderate	High	Low
Resident confidence (1–10, 10 being highest confidence)	5	9	7
Category of hurricane alternative will withstand (1–5)	3	5	5
Economic development (low to high)	Moderate	High	Low

Solution

Section 14.7.2 presents the methods used in additive weighting. Table 14-12 contains ordinal rankings and the resulting weights of the various attributes (of course, these may vary from one decision maker to another).

The nondimensional scaling of the possible attribute values (using maximum and minimum values from the given alternatives) are presented in Table 14-13.

When the weights are combined with the performance for each alternative, we obtain the results shown in Table 14-14.

TABLE 14-12   Attribute Weight for Example 14-2

Attribute	Ordinal Ranking	Weight
Cost	6	0.29
Time until completion	3	0.14
Environmental impact	4	0.19
Resident confidence	2	0.09
Category of hurricane alternative will withstand	5	0.24
Economic development	1	0.05
Total	21	1.00

TABLE 14-13   Dimensionless Values for Example 14-2

Attribute	Value	Rating Procedure	Dimensionless Value
Cost	1.3	$(35 - \text{cost})/33.7$	1.00
	20		0.45
	35		0.00
Time until completion	2	$(15 - \text{time})/13$	1.00
	13		0.15
	15		0.00
Environmental impact	Low	$(3 - \text{Relative Rank})/2$	1.00
	Moderate		0.50
	High		0.00
Resident confidence	5	$(\text{confidence} - 5)/4$	0.00
	7		0.50
	9		1.00
Category of hurricane alternative will withstand	3	$(\text{category} - 3)/2$	0.00
	5		1.00
Economic development	Low	$(\text{Relative Rank} - 1)/2$	0.00
	Moderate		0.50
	High		1.00

TABLE 14-14   Weighted Scores for Example 14-2

Attributes	Weights	Standard Levee Repair		Massive Levee Construction		Restoration and Restructure through Nonstructural Methods	
		Perfor-mance	Weight Value	Perfor-mance	Weight Value	Perfor-mance	Weight Value
Cost	0.29	1.0	0.29	0.0	0.0	0.45	0.131
Time until completion	0.14	1.0	0.14	0.15	0.021	0.0	0.0
Environmental impact	0.19	0.5	0.095	0.0	0.0	1.0	0.19
Resident confidence	0.09	0.0	0.0	1.0	0.09	0.5	0.045
Category of hurricane alternative will withstand	0.24	0.0	0.0	1.0	0.24	1.0	0.24
Economic development	0.05	0.5	0.025	1.0	0.05	0.0	0.0
Total Score			0.55		0.401		0.606

Examining Table 14-14, and given the scaling and weights used, the decision would be restoration and restructure through nonstructural methods.

14.8 Summary

Several methods have been described for dealing with multiattribute decisions. Some key points are as follows:

1. When it is desired to maximize a single criterion of choice, such as PW, the evaluation of multiple alternatives is relatively straightforward.

2. For any decision, the objectives, available alternatives, and important attributes must be clearly defined at the beginning. Construction of a decision matrix such as that in Table 14-1 helps this process.

3. Decision making can become quite convoluted when multiple objectives and attributes must be included in an engineering economy study.

4. Multiattribute models can be classified as either multidimensional or singledimensional. Multidimensional techniques analyze the attributes in terms of their original metrics. Singledimensional techniques reduce attribute measurements to a common measurement scale.

5. Multidimensional, or noncompensatory, models are most useful for the initial screening of alternatives. In some instances, they can be used to make a

final selection, but this usually involves a high degree of subjectivity. Of the multidimensional methods discussed, dominance is probably the least selective, while satisficing is probably the most selective.

6. Single-dimensional, or compensatory, models are useful for making a final choice among alternatives. The additive weighting technique allows superb performance in some attributes to compensate for poor performance in others.

7. When dealing with multiattribute problems that have many attributes and alternatives to be considered, it is advisable to apply a combination of several models in sequence for the purpose of reducing the selection process to a manageable activity.

Problems

The number in parentheses () that follows each problem refers to the section from which the problem is taken.

14-1. As a recent college graduate you are ready for the same thing every recent graduate is ready for—a new car! You know this is a big decision, so you want to immediately put to use the tools you learned in engineering economy. Define five attributes you would use in the selection of a new car and rank them in order of importance. Assign approximate weights to the attributes by using a method discussed in this chapter. Be prepared to defend your attributes and their weights. (14.3, 14.7)

14-2. List two advantages and two disadvantages of noncompensatory models for dealing with multi-attribute decision problems. Do the same for compensatory models. (14.6, 14.7)

14-3. You have inherited a large sum of money from Aunt Bee. You think the purchase of a beach house would be a good investment. Besides a good investment, you could enjoy the house with your family for many years to come. You live about 150 miles from the ocean, and there are several communities that would be excellent candidates for your purchase (or for

building a new house). Identify five attributes, with at least three nonmonetary, that would be important in your decision of where to purchase or build. Specify appropriate values that you might assign to the nonmonetary attributes. Assign approximate weights to these attributes by using a method discussed in this chapter. (14.3, 14.7)

14-4. Discuss some of the difficulties of developing nonlinear functions for nondimensional scaling of qualitative (subjective) data. (14.7)

14-5. Given the matrix of outcomes in Table P14-5 for alternatives and attributes (with higher numbers being better), show what you can conclude, using the following methods: (14.6)

a. Satisficing

b. Dominance

c. Lexicography, with rank order of attributes D > C > B > A

14-6. With reference to the data provided in Table P14-6 on page 600, recommend the preferred alternative by using (a) dominance, (b) satisficing, (c) disjunctive resolution, and (d) lexicography. (14.6)

TABLE P14-5 Matrix of Outcomes for Problem 14-5

Attribute	Alternative 1	2	3	Ideal	(Minimum Acceptable)
A	60	75	90	100	70
B	7	8	8	10	6
C	Fair	Excellent	Poor	Excellent	Good
D	7	6	8	10	6

TABLE P14-6   Data for Problem 14-6

	Alternative				
Attribute	Vendor I	Vendor II	Vendor III	Retain Existing System	Worst Acceptable Value
A. Reduction in throughput time	75%	70%	84%	—	50%
B. Flexibility	Good	Excellent	Excellent	Poor	Good
C. Reliability	Excellent	Good	Good	—	Good
D. Quality	Good	Excellent	Excellent	Fair	Good
E. Cost of system (PW of life-cycle cost)	$270,000	$310,000	$280,000	$0	$350,000

Pairwise comparisons:	1. $A < B$	6. $B > D$
	2. $A = C$	7. $B < E$
	3. $A < D$	8. $C < D$
	4. $A < E$	9. $C < E$
	5. $B < C$	10. $D < E$

14-7. You are ready to finally plan your family summer vacation. After what seems like an eternity of long hours at work, you deserve a week of total relaxation. Now, where do you go? The mountains? The beach? Maybe somewhere with a lot of excitement, like Las Vegas? Table P14-7 contains the three alternatives and attribute values for each, some of which are not available in terms of money.

Use the noncompensatory methods to evaluate these alternatives. Can any be eliminated? Can any be selected? You will have some extra work to do, beyond Table P14-7, before you can complete the analysis. (14.6)

14-8. Three large industrial centrifuge designs are being considered for a new chemical plant.

TABLE P14-7   Data for Problem 14-7

	Alternative		
Attributes	Mountains	Beach	Vegas
Travel Cost	$1,300	$1,500	$1,000
Avg. lodging/night	$275	$250	$225
Avg. daily entertainment cost	Low	Moderate	Expensive
Relaxation level	Moderate	High	Low

TABLE P14-8   Data for Problem 14-8

		Design			
Attribute	Weight	A	B	C	Feasible Range
Initial cost	0.25	$140,000	$180,000	$100,000	$80,000–$180,000
Maintenance	0.10	Good	Excellent	Fair	Fair–excellent
Safety	0.15	Not known	Good	Excellent	Good–excellent
Reliability	0.20	98%	99%	94%	94–99%
Product quality	0.30	Good	Excellent	Good	Fair–excellent

TABLE P14-10 Data for Problem 14-10

Name	Total Years Experience	Years Experience	Project Management Skills	Years Previous Management Experience	General Attitude
Laff Alott	6	2	Yes	0	Excellent
I.B. Surley	5	4	Yes	1	Poor
Sven Busy	8	1	No	1	Good
Justin Wright	7	3	Yes	2	Excellent
Minimum acceptable performance	6	2	Yes	1	Good

a. By using the data in Table P14-8 on page 600, recommend a preferred design with each method that was discussed in this chapter for dealing with nonmonetary attributes.

b. How would you modify your analysis if two or more of the attributes were found to be dependent (e.g., maintenance and product quality)? (14.6, 14.7)

14-9. Mary Jones utilized the additive weighting technique to select a job with Sigma Ltd. in Macon, Georgia. The importance weights she placed on the four attributes in Table 14-1 were social climate = 1.00, starting salary = 0.50, career advancement = 0.33, and weather/sports = 0.25. Nondimensional values given to her ratings in Table 14-1 were excellent = 1.00, very good = 0.70, good = 0.40, fair = 0.25, and poor = 0.10.

a. Normalize Mary's importance weights.

b. Develop nondimensional values for the starting salary attribute.

c. Use the results of (a) and (b) in a decision matrix to see if Mary's choice was consistent with the results obtained from the additive weighting technique. (14.7)

14-10. Hiring an employee is always a multiattribute decision process. Most jobs possess a diversity of requirements, and most applicants have a diversity of skills to bring to the job. Table P14-10 matches a set of desired attributes with the capabilities of four job applicants for a job in a pharmaceutical company. Use the tools of this chapter to examine the candidates. (14.6)

Who would be chosen or eliminated using

a. dominance (consider more years of experience preferred to fewer)

b. satisficing

c. disjunctive resolution

d. lexicography with the following priorities: project management skills > general attitude > years manufacturing experience > previous management experience > total years experience

14-11. The town of Whoopup has decided that it needs a new public library. Although the city leaders have made this decision, the decision of location of the library is still an open question. Knowing that you have a background in economic decision making, they selected you to formulate their decision problem.

You gather data from the city leaders to determine the important attributes to consider in the decision; you then acquire information for both sites and present it in Table P14-11 (see page 602).

Use the tools introduced in this chapter to make a site recommendation. Prepare to defend your analysis. (14.6, 14.7)

14-12. You have volunteered to serve as a judge in a Midwestern contest to select Sunshine, the most wholesome pig in the world. Your assessments of the four finalists for each of the attributes used to distinguish among semifinalists are shown in Table P14-12.

a. Use dominance, feasible ranges, lexicography, and additive weighting to select your winner. Develop your own feasible ranges and weights for the attributes. (14.6, 14.7)

TABLE P14-11	Data for Problem 14-11	
	Sites	
Considerations	Site 1	Site 2
Monetary		
Land	$600,000	$950,000
Construction	$2,300,000	$2,200,000
Road construction for access	$1,500,000	$750,000
Nonmonetary		
Distance from population center	5 miles	2 miles
Maintenance	Medium	High
Homes displaced	3	5
Environmental impact	Medium	Low
Accessibility of property	Moderate	Moderate

b. If there were two other judges, discuss how the final selection of this year's Sunshine might be made. (14.7)

14-13. Consider the data from Problem 14-10. The human resources department requires that non-dimensional scaling be applied to make your decision. Rate the individual attributes using Equations (14-1) and (14-2) where appropriate. For the attributes Project Management Skills, use a score of 0 for No and 1 for Yes. For General Attitude, use a rating of 0 for Poor, 1 for Good, and 2 for Excellent. Who should be selected? (14.7)

14-14. The additive weighting model is a decision tool that aggregates information from different independent criteria to arrive at an overall score for each course of action being evaluated. The alternative with the highest score is preferred.

The general form of the model is

$$V_j = \sum_{i=1}^{n} w_i x_{ij},$$

where
V_j = the score of the jth alternative;
w_i = the weight assigned to the ith decision attribute ($1 \le i \le n$);
x_{ij} = the rating assigned to the ith attribute, which reflects the performance of alternative j relative to maximum attainment of the attribute.

Consider Table P14-14 (page 603) in view of these definitions and determine the value of each "?" shown. (14.7)

14-15. When traveling, one always has a choice of airlines. However, which should you choose? Many attributes are worthy of consideration, some of which are cost, airline miles, number of hops (intermediate stops between departure and final destination), and type of airplane (some people just do not like those small planes). Some people like more hops because each brings them more points in their frequent flyer program. Others want to minimize hops. On a particular trip, you have gathered data about the possible travel alternatives, and they are listed in Table P14-15.

Use four noncompensatory methods for dealing with multiple attributes (dominance, satisficing, disjunctive resolution, and lexicography) and determine whether a selection can be made, or alternatives eliminated, with each. You will need to develop additional data that reflect your preferences. (14.6)

14-16. For the data from Problem 14-15, apply the additive weighting technique, using weights you develop, to make a travel decision. (14.7)

TABLE P14-12	Assessment of Four Finalists for Problem 14-12			
	Contestant			
Attribute	I	II	III	IV
Facial quality	Cute but plump	Sad eyes, great snout	Big lips, small ears	A real killer!
Poise[a]	8	10	8	5
Body tone[a]	8	6	7	8
Weight (lb)	400	325	300	360
Coloring	Brown	Spotted, black and white	Gray	Brown and white
Disposition	Friendly	Tranquil	Easily excited	Sour

[a] Data scaled from 1 to 10, with 10 being the highest possible rating.

TABLE P14-14 Data for Problem 14-14

						Alternative j	
						(1) Keep Existing Machine Tool	(2) Purchase a New Machine Tool
i	w_i	Rank	Decision Factor				
1	1.0	1	Annual cost of ownership (capital recovery cost)	Rank		?	?
				x_{ij}		1.0	0.7
2	?	4	Flexibility in types of jobs scheduled	Rank		2	1
				x_{ij}		0.8	1.0
3	0.8	2	Ease of training and operation	Rank		1	2
				x_{ij}		?	0.5
4	0.7	?	Time savings per part produced	Rank		2	1
				x_{ij}		0.7	1.0
				V_j		2.69	2.30
				V_j (normalized)		1.00	?

TABLE P14-15 Data for Problem 14-15

	Alternatives			
Attribute	Fly-by-night	Puddle Jumper	Air 'R Us	Hopwe Makit
Air fare	$350	$280	$325	$300
Number of stops	0	2	1	2
Frequent flyer?	Yes	No	Yes	Yes
Airplane type	large	very small	small	small

14-17. "Paper or plastic?" asks a checkout clerk at your grocery store. According to *Time* magazine (August 13, 2007, page 50), paper bags cost 5.7 cents each as opposed to 2.2 cents per plastic bag. After leaving the store with several plastic bags, you decide to do some research on paper versus plastic bags. The *Time* article goes on to inform you that the United States uses 12 million barrels of oil to produce its plastic bags consumed annually. Furthermore, paper bags consumed in the United states equate to 14 million trees a year. What other factors might go into a multiattribute analysis of the pros and cons of paper versus plastic bags (e.g., littering, leak potential, biodegradability, transportation to landfills)? Use the additive weighting technique to assess the impact of imposing a 5 cent "finder's fee" for returning paper and plastic bags to an approved recycling center. (14.7)

Spreadsheet Exercises

14-18. Refer to Example 14-1. Mary Jones has been enlightened at a recent seminar and now rates hypnosis as her highest choice of anesthesia. The remaining anesthesia methods retain their relative order, just shifted down one. How does this impact the choice of a dentist via the nondimensional scaling technique? (14.7)

APPENDIX A

Using Excel to Solve Engineering Economy Problems

by James A. Alloway, Jr.
EMSQ Associates

This appendix presents techniques and specific functions needed to formulate spreadsheet models for solving engineering economy problems. The spreadsheet examples in the text use Microsoft Excel because of its widespread use. Other spreadsheet software, such as Lotus 1-2-3 and Corel Quattro Pro, can be used equally as well, with only minor changes to the formulas.

A.1 Spreadsheet Basics

The basic spreadsheet components and notation are shown in Figure A-1. The row immediately above the worksheet column headings is the formula bar. The *cell* is the basic building block of the spreadsheet and is identified by its column letter and row number. In Figure A-1, the text Sample Spreadsheet is in cell C2, the intersection of column C and row 2. The box on the left of the formula bar indicates the active cell, which in this case is C2. The contents of the active cell appear in the box to the right of f_x on the formula bar.

The other fundamental spreadsheet entity is the *range*, which can be a single cell, a portion of a row or column, or any uniform rectangular region. A range is identified by its first and last cells separated by a colon (e.g., H1:H7). A one-cell range has the same starting and ending cell (e.g., B1:B1), while a rectangular range uses the top left and bottom right cells (e.g., A1:G7). When prompted for a range by Excel, move the cell pointer to the first cell in the range and anchor the range by pressing the colon key. Then move to the last cell in the range and press the Enter key.

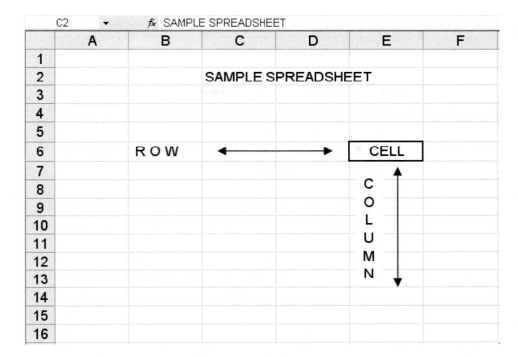

Figure A-1 Sample Excel Worksheet Showing Basic Terms and Relationships

A.2 Types of Input

A cell may contain *numbers* (e.g., −1.23, 123456789, 49E-15), *formulas* (e.g., = A23*5, = 7+4^2.3, = 5 + min(A1:A10)), or *labels* (e.g., Alternative Alpha). Cells can be edited by moving to the desired cell and pressing the F2 key. The keyboard arrow keys will now move the cursor within the expression. Use the Delete or Backspace key to delete characters. The Insert key toggles between insert and type-over mode.

It is helpful to document the input values and results with explanatory labels, which are left-justified by default. In Figure A-1, cell D2 appears to contain PREADSHEET, but in fact the cell is empty. This is because formulas that are longer than the column width spill over to adjacent blank cells.

Numbers are right-justified by default and are entered without commas or currency signs. These features are added separately by using formatting commands, which are described in Section A.4.

Formulas consist of algebraic expressions and functions. In Excel, the first character in the cell must be an equal (=) sign. Reference to other cells is done by either (a) clicking the cell with the mouse, (b) pointing to the cell by using the four keyboard arrow keys, (c) typing the cell address directly, or (d) entering the range name, if one has been assigned. Expressions are evaluated left to right and follow typical precedence rules. Exponentiation is denoted in Excel with the hat symbol (∧).

A.3 Financial Function Summary

The financial functions are based on the following assumptions, which agree with those presented in the text:

(a) The per period interest rate, i, remains constant.

(b) There is exactly one period between cash flows.

(c) The period length remains constant.

(d) The end of period cash-flow convention is utilized.

(e) The first cash flow in a range occurs at the end of the first period.

The last assumption needs to be emphasized, since most problems involve an investment at $k = 0$, which is the beginning of the first period. This only affects the NPV function (see page 615 for more details). Be sure to review the assumptions made in your particular spreadsheet package.

The most frequently used financial functions for engineering economy analysis are as follows:

$PV(i, N, A)$	returns the present worth of an annuity.
$FV(i, N, A)$	returns the future worth of an annuity.
$PMT(i, N, P)$	returns an annuity given a present worth.
$NPV(i, range)$	returns the present worth of any cash flow.
$IRR(guess, range)$	returns the internal rate of return for a range of cash flows.

There is no space preceding the left parenthesis in Excel functions. A, i, and N are single values and have the same definitions as in Chapter 4. *Range* is the cash-flow range, and *guess* is a decimal estimate of the internal rate of return (IRR), which is used internally by the IRR algorithm. The equivalent worth functions return monetary amounts, not the discount factor values. Other functions, including depreciation, are available. Check your particular spreadsheet program for their availability and usage.

A.4 Spreadsheet Appearance

Blank rows and columns are inserted into a worksheet with the commands Insert | Rows or Insert | Columns Rows and columns can be removed by highlighting their row identifier on the left and then using the Edit | Delete command. Be careful when deleting rows and columns: Remember, rows are horizontal, columns are vertical. Larger areas of the spreadsheet can be cleared by highlighting the range and using the Edit | Clear command. The dialog box allows one to delete cell contents and/or cell formatting.

The appearance of numbers, such as the number of significant digits and the presence of dollar and percent signs or commas, is changed with the Format | Cells command. Figure A-2 illustrates the results of several common formatting and alignment options.

Long text entries with nonblank cells to the right and large numbers may require wider columns to display their results. This can be done for each column individually by using the Format | Column | Width command or by moving the cursor to the edge of the column heading (the mouse pointer will change to a

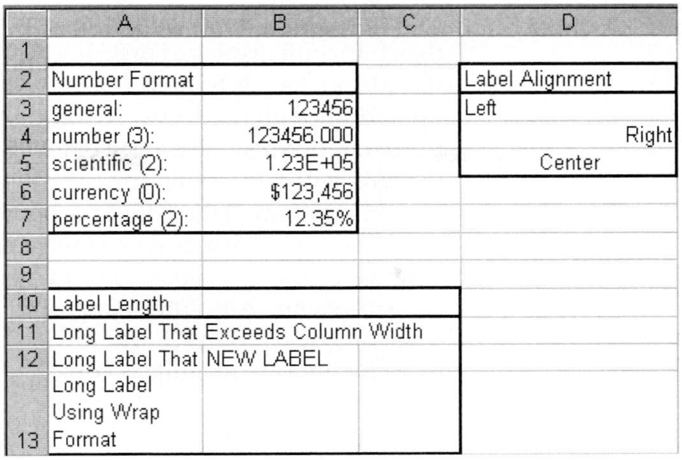

Figure A-2 Sample Screen Illustrating Typical Text and Numeric Formatting Options

	A	B	C	D
1				
2	Number Format			Label Alignment
3	general:	123456		Left
4	number (3):	123456.000		Right
5	scientific (2):	1.23E+05		Center
6	currency (0):	$123,456		
7	percentage (2):	12.35%		
8				
9				
10	Label Length			
11	Long Label That Exceeds Column Width			
12	Long Label That NEW LABEL			
13	Long Label Using Wrap Format			

double arrow) and dragging it in the appropriate direction. The alternative is to allow the text to wrap in the cell using Format | Cells | Alignment | Text and check the Wrap Text box. This will affect the row height.

A.4.1 Moving Cells

Cell contents are removed by highlighting the range and then using the Edit | Cut command. Cell contents are placed in a new location by highlighting the first cell in the new range and using the Edit | Paste command. Anything in the new range will be replaced with the new material.

The original location will be blank after the Cut command is issued. When a formula is cut and subsequently pasted to a new location, the cell references of the formula are automatically updated to reflect its new location. If a value is cut and pasted to a new location, all formulas that refer to it are automatically revised to reflect the new location.

A.4.2 Copying Cells

To copy the contents of a cell to a new location while leaving the original cell intact, highlight the cell or range and use the Edit | Copy command. Move the cursor to the new location and use the Edit | Paste command. Formulas in copied cells are modified, as described in section A.5.

A.5 Cell Referencing

The manner in which cells in formulas are referenced internally provides the power and flexibility of spreadsheet software. These features make it possible to construct complex models by using a minimal number of formulas and to get updated results as values change. Many problems encountered by novice spreadsheet users can be traced back to a lack of understanding of how formulas reference cells.

There are three methods to refer to the reference of a cell: relative, absolute, and mixed. The distinction between them becomes important only when copying formulas. If cell B5 contains the formula = A4 + 1 and cell A4 contains the value 7, then the result displayed in B5 is 8. Although the formula refers to cell A4, internally the spreadsheet interprets it as "Add 1 to the contents of the cell that is one row up and one column to the left of this cell." This is relative referencing, which is the default. When the formula in B5 is copied, the cell that receives the copy of the formula will display the sum of 1 plus the contents of the cell that is one row up and one column to the left of the new cell.

There are occasions when one always wants to refer to a specific cell, no matter where a formula is copied. The interest rate, i, is typically a value that is constant for a problem. Absolute referencing (a $ in front of both the column and row designation, e.g., $A$1) will always return the contents of that specific cell, no matter where the formula is copied.

Sometimes it is necessary to hold either the row or the column constant and allow the other portion of the reference to vary. This is achieved with mixed referencing. A $ is placed in front of the row or column that is to be kept constant. Thus, $A1 will make the column reference absolute and the row reference relative, while A$1 will make the row reference absolute and the column reference relative.

The F4 key can simplify the process of cell addressing. Each time the key is pressed, the highlighted cell reference cycles through absolute, mixed row, mixed column, and relative addressing.

Figure A-3 illustrates the results of the cut, copy, and paste commands for the three types of referencing. The range A5:C9 contains arbitrary values. Cells E5:E9 contain formulas to compute the respective row totals. The actual formulas are shown in column F. Results of cut, copy, and paste actions are shown in range E13:E17, and the resulting formulas are in column F.

	A	B	C	D	E	F
1						
2						
3						Original Formulas
4		Data Values			Result	Underlying Formula
5	1	1	1		3	=SUM(A5:C5)
6	2	2	2		6	=SUM(A6:C6)
7	3	3	3		9	=SUM($A$7:$C$7)
8	4	4	4		12	=SUM(A$8:C$8)
9	5	5	5		15	=SUM($A9:$C9)
10						
11						
12			Action		Result	Underlying Formula
13			Cut E5, Paste in E13		3	=SUM(A5:C5)
14			Copy E6 to E14		0	=SUM(A14:C14)
15			Copy E7 to E15		9	=SUM($A$7:$C$7)
16			Copy E8 to E16		12	=SUM(A$8:C$8)
17			Copy E9 to E17		0	=SUM($A17:$C17)

Figure A-3 Sample Screen Illustrating the Effects on Formula References of the Copy and Move Commands

A.6   Structured Spreadsheets

Using a structured approach for model formulation reduces the chance of overlooking information, aids in locating errors and making revisions, and makes it easier for others to interpret your analysis. Taking a few minutes to organize the data and plan the model prior to sitting down at the keyboard will reduce the total development time. There is no single best format, but the following tips work well in practice:

a. Treat everything as a variable, even if you do not expect it to change. Assign key parameters, such as the interest rate, life, and estimates of revenue and expenses, their own cells. Formulas should reference these cells, not the values themselves. This way, if these values change, it is only necessary to revise these few cells rather than try to locate all of the formulas that contain these values. If these cells are located in the upper left corner of the spreadsheet, you can return to them from anywhere in the spreadsheet by pressing the keyboard *Home* key.

b. Next, divide the problem into segments. Combining several calculations into a single formula saves space, but such an approach makes it more difficult to verify results and locate errors. Each segment should provide an intermediate result, which can be referenced in other segments and aids in isolating problems. This recommendation also applies for different types of cash flows. Rather than combine revenues and expenses for each period prior to entry on the spreadsheet, list each type of cash flow in a column and sum them to get a net figure for each period.

c. It is helpful to save each type of problem as a separate file, which is called a *template*. When a similar problem is encountered in the future, the appropriate template is retrieved, revised, and a new analysis is completed. If you are creating a new spreadsheet from a template, be sure to change the file name the first time you save it, or the template will be overwritten.

d. It is possible to create new functions, automate commands, and create new menus by using *macros*. These are beneficial if you will be using a template frequently, but macros are generally not productive in the classroom setting.

e. *Range names* are useful when entering long formulas and when using macros. They can be assigned to any valid range (including a single cell) with the command Insert | Name. Range names can be made absolute by prefixing the name with a dollar sign.

Three examples follow that illustrate useful spreadsheet commands and functions, in addition to basic model formulation techniques needed to solve engineering economy problems. Keep the following points in mind:

a. Remain familiar with the hand solutions so that the spreadsheet model can be verified.

b. Understand the course material first, have a plan for solving the problem, and then use the spreadsheet to help with the calculations.

c. Try the examples in this appendix to verify that they are working properly with your particular spreadsheet.

d. Blindly copying formulas and pressing keys without taking the time to look at what is going on doesn't aid learning.

e. Graphing your results can reveal relationships that may not be easily discernable when looking at columns of numbers.

f. Save your work frequently and keep backups of your files.

EXAMPLE A-1

This example uses breakeven analysis to illustrate many basic spreadsheet features. It includes examples of relative, mixed, and absolute cell addressing; a simple formula; the use of a spreadsheet function; and a graph. Spreadsheet commands used include formatting, label alignment, and the copy command. Because of the copy command, only five unique formulas are required.

Consider two options, Alpha and Beta, each with its own set-up (fixed) cost and incremental (variable) cost. We wish to determine the least expensive option for a given production quantity. The total cost (TC) for Q units produced by either method is

$$TC_m = FC_m + IC_m * Q,$$

where m is either Alpha or Beta, FC is the fixed cost, IC is the incremental cost, and Q is the number of units produced.

The spreadsheet for this problem is shown in Figure A-4. The data input area is at the top of the worksheet. Here, the specific fixed and variable costs are entered (cells B2 to C3), along with the name of each alternative under consideration (in cells B1 and C1). Cells B6 and B7 contain values that will be used later to bracket the breakeven quantity.

The total cost formula is entered in cell B12, using mixed addressing, and copied to the range B12:C22. The choice of cell in which to enter the initial formula is arbitrary, and any other cell (e.g., Z798) could have been used as well.

The Units Produced column is created with two formulas. Cell A12 uses relative addressing to refer to the starting value in cell B6. Cell A13 uses relative addressing to refer to the contents of the cell above it and absolute addressing to refer to the increment value to be added to it. While the start and increment values could be entered directly in the formulas, the current approach provides greater flexibility.

The general approach for solving a breakeven problem without using the solver function is to set the start value to 0 and the increment value to some large value, say 100. You may need to go higher or try other values, depending on the specifics of your particular problem. Once the breakeven point is bracketed, the start and increment values can be changed to get a more exact value. As the graph in Figure A-5 indicates, the breakeven point occurs between 25 and 30, so the start value can be changed to 25 and the increment reduced to 1 to locate the exact breakeven quantity.

An IF function is used in column D to make it easier to identify the breakeven point. It displays the name of the alternative with the lowest cost.

The graph is produced by highlighting the range A12:C22, then clicking on the *Chart Wizard* and selecting the XY(Scatter) option.

= B$1

= B$2 + B$3 * $A12

= IF(B12 > C12, $C$1, $B$1)

	A	B	C	D
1	Option:	**Alpha**	**Beta**	
2	Set-up cost =	$250	$800	
3	Per piece cost =	$75	$55	
4	- - - - - - - -	- - - - - - - -	- - - - - - - -	
5	Units produced			
6	Start =	0		
7	Increment =	5		
8				
9	**Units**	**Total Cost for N Units**		**Least Total**
10	**Produced (Q)**	**Alpha**	**Beta**	**Cost Option**
11				
12	0	$ 250	$ 800	Alpha
13	5	$ 625	$ 1,075	Alpha
14	10	$ 1,000	$ 1,350	Alpha
15	15	$ 1,375	$ 1,625	Alpha
16	20	$ 1,750	$ 1,900	Alpha
17	25	$ 2,125	$ 2,175	Alpha
18	30	$ 2,500	$ 2,450	Beta
19	35	$ 2,875	$ 2,725	Beta
20	40	$ 3,250	$ 3,000	Beta
21	45	$ 3,625	$ 3,275	Beta
22	50	$ 4,000	$ 3,550	Beta

= B6

= A12 + $B$7

Figure A-4   Spreadsheet and Unique Formulas for Example A-1

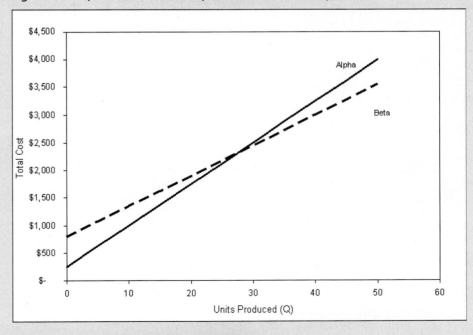

Figure A-5   Breakeven Graph for Example A-1

Since all values are treated as variables, a high degree of flexibility is inherent in the spreadsheet. Note that, as you change values, the spreadsheet is automatically revised, as is the graph. Even the labels in column D are updated if the names of the alternatives change!

EXAMPLE A-2

Cash flows can take many forms, such as single unique values, annuities, and gradients. Single values are entered individually into their respective cells. Annuities are created by entering the first period amount and then using the copy command to fill in the rest of the range. Both arithmetic and geometric gradients are created in a spreadsheet with simple formulas and the copy command, as shown in Figure A-6.

The formula in B8 copies the starting amount from which both gradients build. By using absolute addressing, we are able to apply the same starting amount to both columns. For an arithmetic gradient, cell B9 takes the amount in the cell above it and adds a fixed amount to it, as specified in B3. In a geometric gradient, cell C9 takes the cell above it and multiplies it by the fixed percentage specified in B4. Note the use of absolute addressing in both gradient formulas to keep the gradient quantity constant as it is copied down each column. Finally, B15 calculates the present worth of the gradient series. Note that absolute addressing is used to hold the interest rate and base (year 0) amount constant. To obtain a decreasing gradient, simply substitute a negative sign for the positive sign in the formulas in row 9.

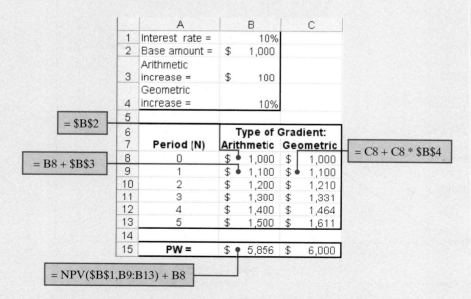

Figure A-6   Spreadsheet and Unique Formulas for Example A-2

EXAMPLE A-3

Once the cash flow values are entered into the spreadsheet, financial measures of merit can be determined (see Figure A-7). The equivalent worth measures are obtained with the following combination of functions:

$$PW: = NPV(i, \ range) + P_0$$

$$AW: = PMT(i, \ N, -PW)$$

$$FW: = FV(i, \ N, -AW)$$

$$IRR: = IRR(range, \ i)$$

Here, P_0 refers to the cash flow at time 0, PW is the result of the NPV function with P_0 added, AW is the result of the PMT function, and *range* is the cash-flow range excluding the amount P_0.

The NPV function is probably the most useful financial function. It works with any cash flow; it is not restricted to be an annuity, as with PMT and FV, and provides a unique result regardless of the number of sign changes. Once we have the PW, we can obtain any other equivalent worth measure.

In the past, many firms avoided the IRR method in project justification because it is difficult to compute by hand. With the spreadsheet, IRR is actually easier to compute than PW. The value *i* is an initial estimate of the IRR in decimal

	A	B	C
1	Interest Rate =	10%	
2	-------	-------	
3	Year	Net Cash Flow	
4	0	($10,000)	
5	1	$3,600	
6	2	$3,300	
7	3	$2,900	
8	4	$2,900	
9	5	$1,000	
10			
11	Present Value		$ 780
12	Annual Worth		$ 206
13	Future Worth		$ 1,257
14	Internal Rate of Return:		13.5%

= NPV (B1, B5:B9) + B4

= PMT (B1, A9, –C11)

= FV (B1, A9, –C12)

= IRR (B4:B9, B1)

Figure A-7   Spreadsheet and Unique Formulas for Example A-3

form. MARR is usually a good guess. For a unique answer, there must be only a single change of sign in the cash flow. The IRR function does not provide a warning if multiple rates of return exist.

A.7 Excel Formula Summary

ABS(value)
  Returns the absolute value of the argument.

AVERAGE(range)
  Returns the average of the values specified in *range*.

CEILING(number, significance)
  Rounds the specified number up to the significance specified. Significance is typically 1.

FV(i,N,A,{PW},type)
  Returns the future worth amount at the end of period N for an annuity A, beginning at the end of period 1. The interest rate is i per period. {PW} is optional.

HLOOKUP(class life, table range, k)
  This function returns a value from a specific location in a table. The location is specified by values in the header row and the number of rows below the header.

IF(condition, true, false)
  Condition is a logical expression, containing a logical operator ($<>$, $>$, $<$, $>=$, $<=$). If the result of the condition is true, the value or label in the *true* argument is returned. If the result is false, the value or label in the *false* argument is returned.

IRR(range, guess)
  Returns the IRR for cash flows in the specified range. Recall that more than one change of sign in the cash flow may generate multiple rates of return! The *guess* argument is a starting point used by the internal algorithm and is often specified as the MARR.

ISERR(cell)
  Computational errors, such as division by zero, display an error code in the cell containing the error and any cells that reference that cell. If the cell contains an error, this function returns a logical TRUE; otherwise, it returns a logical FALSE.

MAX(range)
  Returns the maximum value in the range. Text and logical results are ignored.

MIN(range)
> Returns the minimum value in the range. Text and logical results are ignored.

NORMINV(number, mean; std dev)
> Returns the inverse cumulative distribution for a normal distribution with the specified mean and standard deviation.

NPV(i,range)
> Computes present worth of any arbitrary cash flows in the specified range. Caution is required for handling the cash flow (P_0) at $T = 0$. Let

> > P_range = range containing the cash flows at the end of periods $1, 2, \ldots, N$

> > F_range = range containing the cash flows at the end of periods $0, 1, \ldots, N$

> Then the PW can be determined using either formula:

$$PW = P_0 + NPV(i, \ P\_range)$$

> or

$$PW = NPV(i, \ F\_range)^*(1 + i)$$

PMT($i,N,P,\{F\}$,type)
> Returns the annuity amount that is equivalent to P taken over N periods. $\{F\}$ is optional. Type = 1 denotes beginning of period cash flows, Type = 0 is end of period. End of period cash flows are assumed if *type* is blank.

PV($i,N,A,\{F\}$,type)
> Returns the present worth amount of an annuity A, beginning at the end of period 1 and continuing until the end of period N. The interest rate is i per period. $\{F\}$ is optional. Note that this function works only with an annuity, while NPV works with any cash-flow pattern.

RAND()
> Returns a continuous uniform random variable on the interval [0, 1). Note that the parentheses are required, but there is no argument.

ROUND(number, digits)
> Rounds the specified value to the number of digits.

STDEV(range)
> Calculates the sample standard deviation (divisor = $n - 1$) for the values specified in *range*.

SUM(range)
> Returns the sum of the values in the cells of the range specified.

VDB(B, SV_N, N, start, end, percent, no_switch)
> Returns d_k for the general declining balance depreciation method. The *percent* argument is 2 for double declining balance, 1.5 for 150% declining balance, and so on. If omitted, it is assumed to be 2 (200%). The *start* and *end* arguments permit use of the half-year (and half month) conventions. To permit switching to the straight line method, set the *switch* argument to FALSE. A value of TRUE for this argument disallows switching.

A.8 Selected Excel Menu Commands

Solver

The solver function is appropriate for problems where the analyst wants to optimize a dependent variable. The optimum could be a target value, a minimum, or a maximum. The analyst specifies a cell containing the relationship to optimize, the cells for the independent variables referenced in the optimization formula, and constraints on the independent variables. Access the solver functions with Tools | Solver . . .

The target cell contains a formula that relies on cells in the range specified in the *By Changing Cells* box. Constraints are entered individually in the large box at the bottom of the dialog box.

Because of its ability to evaluate multiple solutions, this is an excellent tool for nonlinear breakeven analysis.

NOTE: If *Solver* does not appear on the *Tools* menu, go to Tools | Add-Ins . . . and check the corresponding box. If the *Solver* add-in does not appear here, it must be installed from the original Excel disk.

Slider Control

To create a slider control (a good approach for breakeven analysis), select the Forms toolbar from the View | Toolbars menu.

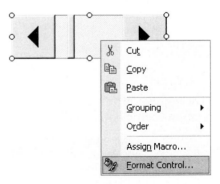

The Forms toolbar will appear undocked above the worksheet. Select the scrollbar icon from the Forms menu.

Use the crosshair curser to position, size, and align the slider icon. Right click on the slider icon to reveal the menu for Format Control.

Click on the Control tab, then specify the minimum and maximum values in the dialog box. The *cell link* identifies the cell that will display the value of the slider. This cell is referenced by a formula in your application.

The *incremental change* box sets how much the linked cell will change each time one of the scroll bar arrows is clicked. The *page change* box specifies the change to the linked cell each time the area between the scroll bar and the arrows is clicked.

Simple Linear Regression

Regression is a very useful estimation and prediction tool. To access it, select Tools | Data Analysis . . . | Regression.

The regression output from Excel includes the estimated regression parameters, correlation coefficient, and diagnostic statistics.

The range of cells containing the dependent variable is specified in the *Input Y Range*. The corresponding independent variable range goes in the *Input X Range* box. Recall that these values are paired, and corresponding *X* and *Y* values must appear in the same row.

Residuals are diagnostic values used to assess how well the regression assumptions are met by the data.

Histogram

Histograms provide a graphic summary showing the center, spread, and shape for large amounts of continuous numeric data. To access the histogram feature, use the Tools | Data Analysis | Histogram menu.

The *Input Range* contains the data for which you wish to create the histogram. The *Bin Range* can be left blank, and Excel will select the number of bins and the width of the bins to make the histogram look nice. You can exercise more control over the appearance of the histogram by specifying values for the Bin Range.

Check the box for *Chart Output* to view the histogram. Right click on any histogram bar and select *Format Data Series*, then specify a *Gap Width* of 0 on the *Options* tab to get a true histogram.

APPENDIX B

Abbreviations and Notation*

Chapter 2

C_F	total fixed cost
C_V	total variable cost
c_v	variable cost per unit
C_T	total cost
D	demand for a product or service in units
D^*	optimal demand or production volume that maximizes profit
D'	breakeven point
$\hat{D}$	demand or production volume that will produce maximum revenue

Chapter 3

$\bar{I}_n$	an unweighted or a weighted index number, dependent on the calculation
K	the number of input resource units needed to produce the first output unit
u	the output unit number
X	cost-capacity factor
Z_u	the number of input resource units needed to produce output unit number u

Chapter 4

A	equal and uniform end-of-period cash flows (or equivalent end-of-period values)
APR	annual percentage rate (nominal interest)
A_1	end of period 1 cash flow in a geometric sequence of cash flows
EOY	end of year

* Listed by the chapter in which they first appear.

F	a future equivalent sum of money
$\bar{f}$	a geometric change from one time period to the next in cash flows or equivalent values
G	an arithmetic (i.e., uniform) change from one period to the next in cash flows or equivalent values
$\underline{I}$	total interest earned or paid (simple interest)
i	effective interest rate per interest period
k	an index for time periods
P	principal amount of a loan; a present equivalent sum of money
M	number of compounding periods per year
N	number of interest periods
r	nominal interest rate per period (usually a year)
$\underline{r}$	a nominal interest rate that is continuously compounded

Chapter 5

AW($i\%$)	equivalent uniform annual worth, computed at $i\%$ interest, of one or more cash flows
CR($i\%$)	equivalent annual cost of capital recovery, computed at $i\%$ interest
CW($i\%$)	capitalized worth (a present equivalent), computed at $i\%$ interest, of cash flows occurring over an indefinite period of time
$\underline{E}$	equivalent annual expenses
ϵ	external reinvestment rate
ERR	external rate of return
EUAC($i\%$)	equivalent uniform annual cost, calculated at $i\%$ interest, of one or more cash flows
FW($i\%$)	future equivalent worth, calculated at $i\%$ interest, of one or more cash flows
I	initial investment for a project
IRR	internal rate of return, also designated i?%
MARR	minimum attractive rate of return
N	the length of the study period (usually years)
O&M	equivalent annual operating and maintenance expenses
PW($i\%$)	present equivalent worth, computed at i% interest, of one or more cash flows
$\underline{R}$	equivalent annual revenues (or savings)
S	salvage (market) value at the end of the study period
θ	payback period
θ'	discounted payback period
V_N	value (price) of a bond N periods prior to redemption
Z	face value of a bond

Chapter 6

$A \rightarrow B$	increment (or incremental net cash flow) between Alternative A and Alternative B (read: A to B)
$\Delta(B - A)$	incremental net cash flow (difference) calculated from the cash flow of Alternative B minus the cash flow of Alternative A (read: delta B minus A)

Chapter 7

ACRS	accelerated cost recovery system
ADS	alternative depreciation system
ATCF	after-tax cash flow
B	cost basis
BTCF	before-tax cash flow
BV	book value of an asset
d	depreciation deduction
d^*	cumulative depreciation over a specified period of time
E	annual expense
e_a	after-tax cost of equity capital
EVA	economic value added
GDS	general depreciation system
i_b	before-tax interest paid on borrowed capital
λ	fraction of capital that is borrowed from lenders
MACRS	modified accelerated cost recovery system
MV	market value of an asset; the price that a buyer will pay for a particular type of property
N	useful life of an asset (ADR life)
NOPAT	net operating profit after taxes
R	ratio of depreciation in a particular year to book value at the beginning of the same year
R	gross annual revenue
r_k	MACRS depreciation rate (a decimal)
SV_N	salvage value of an asset at the end of useful life
T	income taxes
t	effective income tax rate
WACC	tax adjusted weighted average cost of capital

Chapter 8

A\$	actual (current) dollars
b	base time period index
f_e	annual devaluation rate (rate of annual change in the exchange rate) between the currency of a foreign country and the U.S. dollar
f	general inflation rate
i_m	market (nominal) interest rate
i_{fm}	rate of return in terms of a combined (market) interest rate relative to the currency of a foreign country
i_r	real interest rate
i_{us}	rate of return in terms of a combined (market) interest rate relative to U.S. dollars
R\$	real (constant) dollars

Chapter 9

TC_k	total (marginal) cost for year k

Chapter 10

$B{-}C$	benefit–cost ratio
$\underline{B}$	equivalent uniform annual benefits of a proposed project

Chapter 11

EW	equivalent worth (annual, present, or future)

Chapter 12

$E(X)$	mean of a random variable
EVPI	expected value of perfect information
$f(x)$	probability density function of a continuous random variable
$F(x)$	cumulative distribution function of a continuous random variable
$p(x)$	probability mass function of a discrete random variable
$p(x_i)$	probability that a discrete random variable takes on the value x_i
$P(x)$	cumulative distribution function of a discrete random variable
$P_r\{\cdots\}$	probability of the described event occurring
$SD(X)$	standard deviation of a random variable
$V(X)$	variance of a random variable

Chapter 13

B^*	present worth of an investment opportunity in a specified budgeting period
β_M	market risk level
β_S	contribution of stock S to market risk
c	cash outlay for expenses in capital allocation problems
C_k	maximum cash outlay permissible in period k
CAPM	capital asset pricing model
EBIT	earnings before income taxes
m	number of mutually exclusive projects being considered
R_F	risk-free rate of return
R_M	market portfolio rate of return
R_S	stock S rate of return
SML	security market line
X_j	binary decision variable ($=0$ or 1) in capital allocation problems

Chapter 14

r^*	dimensionality of a multiattribute decision problem

APPENDIX C

Interest and Annuity Tables for Discrete Compounding

For various values of i from $1/4\%$ to 25%

i = effective interest rate per period (usually one year)
N = number of compounding periods

$$(F/P, i\%, N) = (1 + i)^N$$

$$(P/F, i\%, N) = \frac{1}{(1 + i)^N}$$

$$(F/A, i\%, N) = \frac{(1 + i)^N - 1}{i}$$

$$(P/A, i\%, N) = \frac{(1 + i)^N - 1}{i(1 + i)^N}$$

$$(A/F, i\%, N) = \frac{i}{(1 + i)^N - 1}$$

$$(A/P, i\%, N) = \frac{i(1 + i)^N}{(1 + i)^N - 1}$$

$$(P/G, i\%, N) = \frac{1}{i}\left[\frac{(1 + i)^N - 1}{i(1 + i)^N} - \frac{N}{(1 + i)^N}\right]$$

$$(A/G, i\%, N) = \frac{1}{i} - \frac{N}{(1 + i)^N - 1}$$

TABLE C-1 Discrete Compounding; $i = 1/4\%$

	Single Payment		Uniform Series				Uniform Gradient		
	Compound Amount Factor	Present Worth Factor	Compound Amount Factor	Present Worth Factor	Sinking Fund Factor	Capital Recovery Factor	Gradient Present Worth Factor	Gradient Uniform Series Factor	
N	To Find F Given P F/P	To Find P Given F P/F	To Find F Given A F/A	To Find P Given A P/A	To Find A Given F A/F	To Find A Given P A/P	To Find P Given G P/G	To Find A Given G A/G	N
1	1.0025	0.9975	1.0000	0.9975	1.0000	1.0025	0.000	0.0000	1
2	1.0050	0.9950	2.0025	1.9925	0.4994	0.5019	0.995	0.4994	2
3	1.0075	0.9925	3.0075	2.9851	0.3325	0.3350	2.980	0.9983	3
4	1.0100	0.9901	4.0150	3.9751	0.2491	0.2516	5.950	1.4969	4
5	1.0126	0.9876	5.0251	4.9627	0.1990	0.2015	9.901	1.9950	5
6	1.0151	0.9851	6.0376	5.9478	0.1656	0.1681	14.826	2.4927	6
7	1.0176	0.9827	7.0527	6.9305	0.1418	0.1443	20.722	2.9900	7
8	1.0202	0.9802	8.0704	7.9107	0.1239	0.1264	27.584	3.4869	8
9	1.0227	0.9778	9.0905	8.8885	0.1100	0.1125	35.406	3.9834	9
10	1.0253	0.9753	10.1133	9.8639	0.0989	0.1014	44.184	4.4794	10
11	1.0278	0.9729	11.1385	10.8368	0.0898	0.0923	53.913	4.9750	11
12	1.0304	0.9705	12.1664	11.8073	0.0822	0.0847	64.589	5.4702	12
13	1.0330	0.9681	13.1968	12.7753	0.0758	0.0783	76.205	5.9650	13
14	1.0356	0.9656	14.2298	13.7410	0.0703	0.0728	88.759	6.4594	14
15	1.0382	0.9632	15.2654	14.7042	0.0655	0.0680	102.244	6.9534	15
16	1.0408	0.9608	16.3035	15.6650	0.0613	0.0638	116.657	7.4469	16
17	1.0434	0.9584	17.3443	16.6235	0.0577	0.0602	131.992	7.9401	17
18	1.0460	0.9561	18.3876	17.5795	0.0544	0.0569	148.245	8.4328	18
19	1.0486	0.9537	19.4336	18.5332	0.0515	0.0540	165.411	8.9251	19
20	1.0512	0.9513	20.4822	19.4845	0.0488	0.0513	183.485	9.4170	20
21	1.0538	0.9489	21.5334	20.4334	0.0464	0.0489	202.463	9.9085	21
22	1.0565	0.9466	22.5872	21.3800	0.0443	0.0468	222.341	10.3995	22
23	1.0591	0.9442	23.6437	22.3241	0.0423	0.0448	243.113	10.8901	23
24	1.0618	0.9418	24.7028	23.2660	0.0405	0.0430	264.775	11.3804	24
25	1.0644	0.9395	25.7646	24.2055	0.0388	0.0413	287.323	11.8702	25
30	1.0778	0.9278	31.1133	28.8679	0.0321	0.0346	413.185	14.3130	30
36	1.0941	0.9140	37.6206	34.3865	0.0266	0.0291	592.499	17.2306	36
40	1.1050	0.9050	42.0132	38.0199	0.0238	0.0263	728.740	19.1673	40
48	1.1273	0.8871	50.9312	45.1787	0.0196	0.0221	1040.055	23.0209	48
60	1.1616	0.8609	64.6467	55.6524	0.0155	0.0180	1600.085	28.7514	60
72	1.1969	0.8355	78.7794	65.8169	0.0127	0.0152	2265.557	34.4221	72
84	1.2334	0.8108	93.3419	75.6813	0.0107	0.0132	3029.759	40.0331	84
100	1.2836	0.7790	113.4500	88.3825	0.0088	0.0113	4191.242	47.4216	100
∞				400.0000		0.0025			∞

TABLE C-2 Discrete Compounding: $i = 1/2\%$

| | Single Payment | | Uniform Series | | | | Uniform Gradient | | |
| | Compound Amount Factor | Present Worth Factor | Compound Amount Factor | Present Worth Factor | Sinking Fund Factor | Capital Recovery Factor | Gradient Present Worth Factor | Gradient Uniform Series Factor | |
N	To Find F Given P F/P	To Find P Given F P/F	To Find F Given A F/A	To Find P Given A P/A	To Find A Given F A/F	To Find A Given P A/P	To Find P Given G P/G	To Find A Given G A/G	N
1	1.0050	0.9950	1.0000	0.9950	1.0000	1.0050	0.000	0.0000	1
2	1.0100	0.9901	2.0050	1.9851	0.4988	0.5038	0.990	0.4988	2
3	1.0151	0.9851	3.0150	2.9702	0.3317	0.3367	2.960	0.9967	3
4	1.0202	0.9802	4.0301	3.9505	0.2481	0.2531	5.901	1.4938	4
5	1.0253	0.9754	5.0503	4.9259	0.1980	0.2030	9.803	1.9900	5
6	1.0304	0.9705	6.0755	5.8964	0.1646	0.1696	14.655	2.4855	6
7	1.0355	0.9657	7.1059	6.8621	0.1407	0.1457	20.449	2.9801	7
8	1.0407	0.9609	8.1414	7.8230	0.1228	0.1278	27.176	3.4738	8
9	1.0459	0.9561	9.1821	8.7791	0.1089	0.1139	34.824	3.9668	9
10	1.0511	0.9513	10.2280	9.7304	0.0978	0.1028	43.387	4.4589	10
11	1.0564	0.9466	11.2792	10.6770	0.0887	0.0937	52.853	4.9501	11
12	1.0617	0.9419	12.3356	11.6189	0.0811	0.0861	63.214	5.4406	12
13	1.0670	0.9372	13.3972	12.5562	0.0746	0.0796	74.460	5.9302	13
14	1.0723	0.9326	14.4642	13.4887	0.0691	0.0741	86.584	6.4190	14
15	1.0777	0.9279	15.5365	14.4166	0.0644	0.0694	99.574	6.9069	15
16	1.0831	0.9233	16.6142	15.3399	0.0602	0.0652	113.424	7.3940	16
17	1.0885	0.9187	17.6973	16.2586	0.0565	0.0615	128.123	7.8803	17
18	1.0939	0.9141	18.7858	17.1728	0.0532	0.0582	143.663	8.3658	18
19	1.0994	0.9096	19.8797	18.0824	0.0503	0.0553	160.036	8.8504	19
20	1.1049	0.9051	20.9791	18.9874	0.0477	0.0527	177.232	9.3342	20
21	1.1104	0.9006	22.0840	19.8880	0.0453	0.0503	195.243	9.8172	21
22	1.1160	0.8961	23.1944	20.7841	0.0431	0.0481	214.061	10.2993	22
23	1.1216	0.8916	24.3104	21.6757	0.0411	0.0461	233.677	10.7806	23
24	1.1272	0.8872	25.4320	22.5629	0.0393	0.0443	254.082	11.2611	24
25	1.1328	0.8828	26.5591	23.4456	0.0377	0.0427	275.269	11.7407	25
30	1.1614	0.8610	32.2800	27.7941	0.0310	0.0360	392.632	14.1265	30
36	1.1967	0.8356	39.3361	32.8710	0.0254	0.0304	557.560	16.9621	36
40	1.2208	0.8191	44.1588	36.1722	0.0226	0.0276	681.335	18.8359	40
48	1.2705	0.7871	54.0978	42.5803	0.0185	0.0235	959.919	22.5437	48
60	1.3489	0.7414	69.7700	51.7256	0.0143	0.0193	1448.646	28.0064	60
72	1.4320	0.6983	86.4089	60.3395	0.0116	0.0166	2012.348	33.3504	72
84	1.5204	0.6577	104.0739	68.4530	0.0096	0.0146	2640.664	38.5763	84
100	1.6467	0.6073	129.3337	78.5426	0.0077	0.0127	3562.793	45.3613	100
∞				200.0000		0.0050			∞

TABLE C-3 Discrete Compounding: i = 3/4%

| | Single Payment | | Uniform Series | | | | Uniform Gradient | | |
| | Compound Amount Factor | Present Worth Factor | Compound Amount Factor | Present Worth Factor | Sinking Fund Factor | Capital Recovery Factor | Gradient Present Worth Factor | Gradient Uniform Series Factor | |
N	To Find F Given P F/P	To Find P Given F P/F	To Find F Given A F/A	To Find P Given A P/A	To Find A Given F A/F	To Find A Given P A/P	To Find P Given G P/G	To Find A Given G A/G	N
1	1.0075	0.9926	1.0000	0.9926	1.0000	1.0075	0.000	0.0000	1
2	1.0151	0.9852	2.0075	1.9777	0.4981	0.5056	0.985	0.4981	2
3	1.0227	0.9778	3.0226	2.9556	0.3308	0.3383	2.941	0.9950	3
4	1.0303	0.9706	4.0452	3.9261	0.2472	0.2547	5.853	1.4907	4
5	1.0381	0.9633	5.0756	4.8894	0.1970	0.2045	9.706	1.9851	5
6	1.0459	0.9562	6.1136	5.8456	0.1636	0.1711	14.487	2.4782	6
7	1.0537	0.9490	7.1595	6.7946	0.1397	0.1472	20.181	2.9701	7
8	1.0616	0.9420	8.2132	7.7366	0.1218	0.1293	26.775	3.4608	8
9	1.0696	0.9350	9.2748	8.6716	0.1078	0.1153	34.254	3.9502	9
10	1.0776	0.9280	10.3443	9.5996	0.0967	0.1042	42.606	4.4384	10
11	1.0857	0.9211	11.4219	10.5207	0.0876	0.0951	51.817	4.9253	11
12	1.0938	0.9142	12.5076	11.4349	0.0800	0.0875	61.874	5.4110	12
13	1.1020	0.9074	13.6014	12.3423	0.0735	0.0810	72.763	5.8954	13
14	1.1103	0.9007	14.7034	13.2430	0.0680	0.0755	84.472	6.3786	14
15	1.1186	0.8940	15.8137	14.1370	0.0632	0.0707	96.988	6.8606	15
16	1.1270	0.8873	16.9323	15.0243	0.0591	0.0666	110.297	7.3413	16
17	1.1354	0.8807	18.0593	15.9050	0.0554	0.0629	124.389	7.8207	17
18	1.1440	0.8742	19.1947	16.7792	0.0521	0.0596	139.249	8.2989	18
19	1.1525	0.8676	20.3387	17.6468	0.0492	0.0567	154.867	8.7759	19
20	1.1612	0.8612	21.4912	18.5080	0.0465	0.0540	171.230	9.2516	20
21	1.1699	0.8548	22.6524	19.3628	0.0441	0.0516	188.325	9.7261	21
22	1.1787	0.8484	23.8223	20.2112	0.0420	0.0495	206.142	10.1994	22
23	1.1875	0.8421	25.0010	21.0533	0.0400	0.0475	224.668	10.6714	23
24	1.1964	0.8358	26.1885	21.8891	0.0382	0.0457	243.892	11.1422	24
25	1.2054	0.8296	27.3849	22.7188	0.0365	0.0440	263.803	11.6117	25
30	1.2513	0.7992	33.5029	26.7751	0.0298	0.0373	373.263	13.9407	30
36	1.3086	0.7641	41.1527	34.4468	0.0243	0.0318	524.992	16.6946	36
40	1.3483	0.7416	46.4464	34.4469	0.0215	0.0290	637.469	18.5058	40
48	1.4314	0.6986	57.5207	40.1848	0.0174	0.0249	886.840	22.0691	48
60	1.5657	0.6387	75.4241	48.1734	0.0133	0.0208	1313.519	27.2665	60
72	1.7126	0.5839	95.0070	55.4768	0.0105	0.0180	1791.246	32.2882	72
84	1.8732	0.5338	116.4269	62.1540	0.0086	0.0161	2308.128	37.1357	84
100	2.1111	0.4737	148.1445	70.1746	0.0068	0.0143	3040.745	43.3311	100
∞				133.3333		0.0075			∞

TABLE C-4 Discrete Compounding; $i = 1\%$

| | Single Payment | | Uniform Series | | | | Uniform Gradient | |
| | Compound Amount Factor | Present Worth Factor | Compound Amount Factor | Present Worth Factor | Sinking Fund Factor | Capital Recovery Factor | Gradient Present Worth Factor | Gradient Uniform Series Factor | |
N	To Find F Given P F/P	To Find P Given F P/F	To Find F Given A F/A	To Find P Given A P/A	To Find A Given F A/F	To Find A Given P A/P	To Find P Given G P/G	To Find A Given G A/G	N
1	1.0100	0.9901	1.0000	0.9901	1.0000	1.0100	0.000	0.0000	1
2	1.0201	0.9803	2.0100	1.9704	0.4975	0.5075	0.980	0.4975	2
3	1.0303	0.9706	3.0301	2.9410	0.3300	0.3400	2.922	0.9934	3
4	1.0406	0.9610	4.0604	3.9020	0.2463	0.2563	5.804	1.4876	4
5	1.0510	0.9515	5.1010	4.8534	0.1960	0.2060	9.610	1.9801	5
6	1.0615	0.9420	6.1520	5.7955	0.1625	0.1725	14.321	2.4710	6
7	1.0721	0.9327	7.2135	6.7282	0.1386	0.1486	19.917	2.9602	7
8	1.0829	0.9235	8.2857	7.6517	0.1207	0.1307	26.381	3.4478	8
9	1.0937	0.9143	9.3685	8.5660	0.1067	0.1167	33.696	3.9337	9
10	1.1046	0.9053	10.4622	9.4713	0.0956	0.1056	41.844	4.4179	10
11	1.1157	0.8963	11.5668	10.3676	0.0865	0.0965	50.807	4.9005	11
12	1.1268	0.8874	12.6825	11.2551	0.0788	0.0888	60.569	5.3815	12
13	1.1381	0.8787	13.8093	12.1337	0.0724	0.0824	71.113	5.8607	13
14	1.1495	0.8700	14.9474	13.0037	0.0669	0.0769	82.422	6.3384	14
15	1.1610	0.8613	16.0969	13.8651	0.0621	0.0721	94.481	6.8143	15
16	1.1726	0.8528	17.2579	14.7179	0.0579	0.0679	107.273	7.2886	16
17	1.1843	0.8444	18.4304	15.5623	0.0543	0.0643	120.783	7.7613	17
18	1.1961	0.8360	19.6147	16.3983	0.0510	0.0610	134.996	8.2323	18
19	1.2081	0.8277	20.8109	17.2260	0.0481	0.0581	149.895	8.7017	19
20	1.2202	0.8195	22.0190	18.0456	0.0454	0.0554	165.466	9.1694	20
21	1.2324	0.8114	23.2392	18.8570	0.0430	0.0530	181.695	9.6354	21
22	1.2447	0.8034	24.4716	19.6604	0.0409	0.0509	198.566	10.0998	22
23	1.2572	0.7954	25.7163	20.4558	0.0389	0.0489	216.066	10.5626	23
24	1.2697	0.7876	26.9734	21.2434	0.0371	0.0471	234.180	11.0237	24
25	1.2824	0.7798	28.2432	22.0232	0.0354	0.0454	252.895	11.4831	25
30	1.3478	0.7419	34.7849	25.8077	0.0287	0.0387	355.002	13.7557	30
36	1.4308	0.6989	43.0769	30.1075	0.0232	0.0332	494.621	16.4285	36
40	1.4889	0.6717	48.8863	32.8346	0.0205	0.0305	596.856	18.1776	40
48	1.6122	0.6203	61.2226	37.9740	0.0163	0.0263	820.146	21.5976	48
60	1.8167	0.5504	81.6697	44.9550	0.0122	0.0222	1192.806	26.5333	60
72	2.0471	0.4885	104.7099	51.1504	0.0096	0.0196	1597.867	31.2386	72
84	2.3067	0.4335	130.6723	56.6485	0.0077	0.0177	2023.315	35.7170	84
100	2.7048	0.3697	170.4814	63.0289	0.0059	0.0159	2605.776	41.3426	100
∞				100.0000		0.0100			∞

TABLE C-5 Discrete Compounding; $i = 2\%$

| | Single Payment | | Uniform Series | | | | Uniform Gradient | | |
| | Compound Amount Factor | Present Worth Factor | Compound Amount Factor | Present Worth Factor | Sinking Fund Factor | Capital Recovery Factor | Gradient Present Worth Factor | Gradient Uniform Series Factor | |
N	To Find F Given P F/P	To Find P Given F P/F	To Find F Given A F/A	To Find P Given A P/A	To Find A Given F A/F	To Find A Given P A/P	To Find P Given G P/G	To Find A Given G A/G	N
1	1.0200	0.9804	1.0000	0.9804	1.0000	1.0200	0.000	0.0000	1
2	1.0404	0.9612	2.0200	1.9416	0.4950	0.5150	0.961	0.4950	2
3	1.0612	0.9423	3.0604	2.8839	0.3268	0.3468	2.846	0.9868	3
4	1.0824	0.9238	4.1216	3.8077	0.2426	0.2626	5.617	1.4752	4
5	1.1041	0.9057	5.2040	4.7135	0.1922	0.2122	9.240	1.9604	5
6	1.1262	0.8880	6.3081	5.6014	0.1585	0.1785	13.680	2.4423	6
7	1.1487	0.8706	7.4343	6.4720	0.1345	0.1545	18.904	2.9208	7
8	1.1717	0.8535	8.5830	7.3255	0.1165	0.1365	24.878	3.3961	8
9	1.1951	0.8368	9.7546	8.1622	0.1025	0.1225	31.572	3.8681	9
10	1.2190	0.8203	10.9497	8.9826	0.0913	0.1113	38.955	4.3367	10
11	1.2434	0.8043	12.1687	9.7868	0.0822	0.1022	46.998	4.8021	11
12	1.2682	0.7885	13.4121	10.5753	0.0746	0.0946	55.671	5.2642	12
13	1.2936	0.7730	14.6803	11.3484	0.0681	0.0881	64.948	5.7231	13
14	1.3195	0.7579	15.9739	12.1062	0.0626	0.0826	74.800	6.1786	14
15	1.3459	0.7430	17.2934	12.8493	0.0578	0.0778	85.202	6.6309	15
16	1.3728	0.7284	18.6393	13.5777	0.0537	0.0737	96.129	7.0799	16
17	1.4002	0.7142	20.0121	14.2919	0.0500	0.0700	107.555	7.5256	17
18	1.4282	0.7002	21.4123	14.9920	0.0467	0.0667	119.458	7.9681	18
19	1.4568	0.6864	22.8406	15.6785	0.0438	0.0638	131.814	8.4073	19
20	1.4859	0.6730	24.2974	16.3514	0.0412	0.0612	144.600	8.8433	20
21	1.5157	0.6598	25.7833	17.0112	0.0388	0.0588	157.796	9.2760	21
22	1.5460	0.6468	27.2990	17.6580	0.0366	0.0566	171.380	9.7055	22
23	1.5769	0.6342	28.8450	18.2922	0.0347	0.0547	185.331	10.1317	23
24	1.6084	0.6217	30.4219	18.9139	0.0329	0.0529	199.631	10.5547	24
25	1.6406	0.6095	32.0303	19.5235	0.0312	0.0512	214.259	10.9745	25
30	1.8114	0.5521	40.5681	22.3965	0.0246	0.0446	291.716	13.0251	30
36	2.0399	0.4902	51.9944	25.4888	0.0192	0.0392	392.041	15.3809	36
40	2.2080	0.4529	60.4020	27.3555	0.0166	0.0366	461.993	16.8885	40
48	2.5871	0.3865	79.3535	30.6731	0.0126	0.0326	605.966	19.7556	48
60	3.2810	0.3048	114.0515	34.7609	0.0088	0.0288	823.698	23.6961	60
72	4.1611	0.2403	158.0570	37.9841	0.0063	0.0263	1034.056	27.2234	72
84	5.2773	0.1895	213.8666	40.5255	0.0047	0.0247	1230.419	30.3616	84
100	7.2446	0.1380	312.2323	43.0984	0.0032	0.0232	1464.753	33.9863	100
∞				50.0000		0.0200			∞

TABLE C-6 Discrete Compounding: $i = 3\%$

	Single Payment		Uniform Series					Uniform Gradient		
	Compound Amount Factor	Present Worth Factor	Compound Amount Factor	Present Worth Factor	Sinking Fund Factor	Capital Recovery Factor		Gradient Present Worth Factor	Gradient Uniform Series Factor	
N	To Find F Given P F/P	To Find P Given F P/F	To Find F Given A F/A	To Find P Given A P/A	To Find A Given F A/F	To Find A Given P A/P		To Find P Given G P/G	To Find A Given G A/G	N
1	1.0300	0.9709	1.0000	0.9709	1.0000	1.0300		0.000	0.0000	1
2	1.0609	0.9426	2.0300	1.9135	0.4926	0.5226		0.943	0.4926	2
3	1.0927	0.9151	3.0909	2.8286	0.3235	0.3535		2.773	0.9803	3
4	1.1255	0.8885	4.1836	3.7171	0.2390	0.2690		5.438	1.4631	4
5	1.1593	0.8626	5.3091	4.5797	0.1884	0.2184		8.889	1.9409	5
6	1.1941	0.8375	6.4684	5.4172	0.1546	0.1846		13.076	2.4138	6
7	1.2299	0.8131	7.6625	6.2303	0.1305	0.1605		17.955	2.8819	7
8	1.2668	0.7894	8.8923	7.0197	0.1125	0.1425		23.481	3.3450	8
9	1.3048	0.7664	10.1591	7.7861	0.0984	0.1284		29.612	3.8032	9
10	1.3439	0.7441	11.4639	8.5302	0.0872	0.1172		36.309	4.2565	10
11	1.3842	0.7224	12.8078	9.2526	0.0781	0.1081		43.533	4.7049	11
12	1.4258	0.7014	14.1920	9.9540	0.0705	0.1005		51.248	5.1485	12
13	1.4685	0.6810	15.6178	10.6350	0.0640	0.0940		59.420	5.5872	13
14	1.5126	0.6611	17.0863	11.2961	0.0585	0.0885		68.014	6.0210	14
15	1.5580	0.6419	18.5989	11.9379	0.0538	0.0838		77.000	6.4500	15
16	1.6047	0.6232	20.1569	12.5611	0.0496	0.0796		86.348	6.8742	16
17	1.6528	0.6050	21.7616	13.1661	0.0460	0.0760		96.028	7.2936	17
18	1.7024	0.5874	23.4144	13.7535	0.0427	0.0727		106.014	7.7081	18
19	1.7535	0.5703	25.1169	14.3238	0.0398	0.0698		116.279	8.1179	19
20	1.8061	0.5537	26.8704	14.8775	0.0372	0.0672		126.799	8.5229	20
21	1.8603	0.5375	28.6765	15.4150	0.0349	0.0649		137.550	8.9231	21
22	1.9161	0.5219	30.5368	15.9369	0.0327	0.0627		148.509	7.3186	22
23	1.9736	0.5067	32.4529	16.4436	0.0308	0.0608		159.657	9.7093	23
24	2.0328	0.4919	34.4265	16.9355	0.0290	0.0590		170.971	10.0954	24
25	2.0938	0.4776	36.4593	17.4131	0.0274	0.0574		182.434	10.4768	25
30	2.4273	0.4120	47.5754	19.6004	0.0210	0.0510		241.361	12.3141	30
35	2.8139	0.3554	60.4621	21.4872	0.0165	0.0465		301.627	14.0375	35
40	3.2620	0.3066	75.4012	23.1148	0.0133	0.0433		361.750	15.6502	40
45	3.7816	0.2644	92.7199	24.5187	0.0108	0.0408		420.633	17.1556	45
50	4.3839	0.2281	112.7969	25.7298	0.0089	0.0389		477.480	18.5575	50
60	5.8916	0.1697	163.0534	27.6756	0.0061	0.0361		583.053	21.0674	60
80	10.6409	0.0940	321.3630	30.2008	0.0031	0.0331		756.087	25.0353	80
100	19.2186	0.0520	607.2877	31.5989	0.0016	0.0316		879.854	27.8444	100
∞				33.3333		0.0300				∞

TABLE C-7 Discrete Compounding; $i = 4\%$

| | Single Payment | | Uniform Series | | | | Uniform Gradient | | |
| | Compound Amount Factor | Present Worth Factor | Compound Amount Factor | Present Worth Factor | Sinking Fund Factor | Capital Recovery Factor | Gradient Present Worth Factor | Gradient Uniform Series Factor | |
N	To Find F Given P F/P	To Find P Given F P/F	To Find F Given A F/A	To Find P Given A P/A	To Find A Given F A/F	To Find A Given P A/P	To Find P Given G P/G	To Find A Given G A/G	N
1	1.0400	0.9615	1.0000	0.9615	1.0000	1.0400	0.000	0.0000	1
2	1.0816	0.9246	2.0400	1.8861	0.4902	0.5302	0.925	0.4902	2
3	1.1249	0.8890	3.1216	2.7751	0.3203	0.3603	2.703	0.9739	3
4	1.1699	0.8548	4.2465	3.6299	0.2355	0.2755	5.267	1.4510	4
5	1.2167	0.8219	5.4163	4.4518	0.1846	0.2246	8.555	1.9216	5
6	1.2653	0.7903	6.6330	5.2421	0.1508	0.1908	12.506	2.3857	6
7	1.3159	0.7599	7.8983	6.0021	0.1266	0.1666	17.066	2.8433	7
8	1.3686	0.7307	9.2142	6.7327	0.1085	0.1485	22.181	3.2944	8
9	1.4233	0.7026	10.5828	7.4353	0.0945	0.1345	27.801	3.7391	9
10	1.4802	0.6756	12.0061	8.1109	0.0833	0.1233	33.881	4.1773	10
11	1.5395	0.6496	13.4864	8.7605	0.0741	0.1141	40.377	4.6090	11
12	1.6010	0.6246	15.0258	9.3851	0.0666	0.1066	47.248	5.0343	12
13	1.6651	0.6006	16.6268	9.9856	0.0601	0.1001	54.455	5.4533	13
14	1.7317	0.5775	18.2919	10.5631	0.0547	0.0947	61.962	5.8659	14
15	1.8009	0.5553	20.0236	11.1184	0.0499	0.0899	69.736	6.2721	15
16	1.8730	0.5339	21.8245	11.6523	0.0458	0.0858	77.744	6.6720	16
17	1.9479	0.5134	23.6975	12.1657	0.0422	0.0822	85.958	7.0656	17
18	2.0258	0.4936	25.6454	12.6593	0.0390	0.0790	94.350	7.4530	18
19	2.1068	0.4746	27.6712	13.1339	0.0361	0.0761	102.893	7.8342	19
20	2.1911	0.4564	29.7781	13.5903	0.0336	0.0736	111.565	8.2091	20
21	2.2788	0.4388	31.9692	14.0292	0.0313	0.0713	120.341	8.5779	21
22	2.3699	0.4220	34.2480	14.4511	0.0292	0.0692	129.202	8.9407	22
23	2.4647	0.4057	36.6179	14.8568	0.0273	0.0673	138.128	9.2973	23
24	2.5633	0.3901	39.0826	15.2470	0.0256	0.0656	147.101	9.6479	24
25	2.6658	0.3751	41.6459	15.6221	0.0240	0.0640	156.104	9.9925	25
30	3.2434	0.3083	56.0849	17.2920	0.0178	0.0578	201.062	11.6274	30
35	3.9461	0.2534	73.6522	18.6646	0.0136	0.0536	244.877	13.1198	35
40	4.8010	0.2083	95.0255	19.7928	0.0105	0.0505	286.530	14.4765	40
45	5.8412	0.1712	121.0294	20.7200	0.0083	0.0483	325.403	15.7047	45
50	7.1067	0.1407	152.6671	21.4822	0.0066	0.0466	361.164	16.8122	50
60	10.5196	0.0951	237.9907	22.6235	0.0042	0.0442	422.997	18.6972	60
80	23.0498	0.0434	551.2450	23.9154	0.0018	0.0418	511.116	21.3718	80
100	50.5049	0.0198	1237.6237	24.5050	0.0008	0.0408	563.125	22.9800	100
∞				25.0000		0.0400			∞

TABLE C-8 Discrete Compounding: $i = 5\%$

	Single Payment		Uniform Series				Uniform Gradient		
	Compound Amount Factor	Present Worth Factor	Compound Amount Factor	Present Worth Factor	Sinking Fund Factor	Capital Recovery Factor	Gradient Present Worth Factor	Gradient Uniform Series Factor	
N	To Find F Given P, F/P	To Find P Given F, P/F	To Find F Given A, F/A	To Find P Given A, P/A	To Find A Given F, A/F	To Find A Given P, A/P	To Find P Given G, P/G	To Find A Given G, A/G	N
1	1.0500	0.9524	1.0000	0.9524	1.0000	1.0500	0.000	0.0000	1
2	1.1025	0.9070	2.0500	1.8594	0.4878	0.5378	0.907	0.4878	2
3	1.1576	0.8638	3.1525	2.7232	0.3172	0.3672	2.635	0.9675	3
4	1.2155	0.8227	4.3101	3.5460	0.2320	0.2820	5.103	1.4391	4
5	1.2763	0.7835	5.5256	4.3295	0.1810	0.2310	8.237	1.9025	5
6	1.3401	0.7462	6.8019	5.0757	0.1470	0.1970	11.968	2.3579	6
7	1.4071	0.7107	8.1420	5.7864	0.1228	0.1728	16.232	2.8052	7
8	1.4775	0.6768	9.5491	6.4632	0.1047	0.1547	20.970	3.2445	8
9	1.5513	0.6446	11.0266	7.1078	0.0907	0.1407	26.127	3.6758	9
10	1.6289	0.6139	12.5779	7.7217	0.0795	0.1295	31.652	4.0991	10
11	1.7103	0.5847	14.2068	8.3064	0.0704	0.1204	37.499	4.5144	11
12	1.7959	0.5568	15.9171	8.8633	0.0628	0.1128	43.624	4.9219	12
13	1.8856	0.5303	17.7130	9.3936	0.0565	0.1065	49.988	5.3215	13
14	1.9799	0.5051	19.5986	9.8986	0.0510	0.1010	56.554	5.7133	14
15	2.0789	0.4810	21.5786	10.3797	0.0463	0.0963	63.288	6.0973	15
16	2.1829	0.4581	23.6575	10.8378	0.0423	0.0923	70.160	6.4736	16
17	2.2920	0.4363	25.8404	11.2741	0.0387	0.0887	77.141	6.8423	17
18	2.4066	0.4155	28.1324	11.6896	0.0355	0.0855	84.204	7.2034	18
19	2.5270	0.3957	30.5390	12.0853	0.0327	0.0827	91.328	7.5569	19
20	2.6533	0.3769	33.0660	12.4622	0.0302	0.0802	98.488	7.9030	20
21	2.7860	0.3589	35.7193	12.8212	0.0280	0.0780	105.667	8.2416	21
22	2.9253	0.3418	38.5052	13.1630	0.0260	0.0760	112.846	8.5730	22
23	3.0715	0.3256	41.4305	13.4886	0.0241	0.0741	120.009	8.8971	23
24	3.2251	0.3101	44.5020	13.7986	0.0225	0.0725	127.140	9.2140	24
25	3.3864	0.2953	47.7271	14.0939	0.0210	0.0710	134.228	9.5238	25
30	4.3219	0.2314	66.4388	15.3725	0.0151	0.0651	168.623	10.9691	30
35	5.5160	0.1813	90.3203	16.3742	0.0111	0.0611	200.581	12.2498	35
40	7.0400	0.1420	120.7998	17.1591	0.0083	0.0583	229.545	13.3775	40
45	8.9850	0.1113	159.7002	17.7741	0.0063	0.0563	255.315	14.3644	45
50	11.4674	0.0872	209.3480	18.2559	0.0048	0.0548	277.915	15.2233	50
60	18.6792	0.0535	353.5837	18.9293	0.0028	0.0528	314.343	16.6062	60
80	49.5614	0.0202	971.2288	19.5965	0.0010	0.0510	359.646	18.3526	80
100	131.5013	0.0076	2610.0252	19.8479	0.0004	0.0504	381.749	19.2237	100
∞				20.0000		0.0500			∞

TABLE C-9 Discrete Compounding; $i = 6\%$

	Single Payment		Uniform Series				Uniform Gradient		
	Compound Amount Factor	Present Worth Factor	Compound Amount Factor	Present Worth Factor	Sinking Fund Factor	Capital Recovery Factor	Gradient Present Worth Factor	Gradient Uniform Series Factor	
N	To Find F Given P F/P	To Find P Given F P/F	To Find F Given A F/A	To Find P Given A P/A	To Find A Given F A/F	To Find A Given P A/P	To Find P Given G P/G	To Find A Given G A/G	N
1	1.0600	0.9434	1.0000	0.9434	1.0000	1.0600	0.000	0.0000	1
2	1.1236	0.8900	2.0600	1.8334	0.4854	0.5454	0.890	0.4854	2
3	1.1910	0.8396	3.1836	2.6730	0.3141	0.3741	2.569	0.9612	3
4	1.2625	0.7921	4.3746	3.4651	0.2286	0.2886	4.946	1.4272	4
5	1.3382	0.7473	5.6371	4.2124	0.1774	0.2374	7.935	1.8836	5
6	1.4185	0.7050	6.9753	4.9173	0.1434	0.2034	11.459	2.3304	6
7	1.5036	0.6651	8.3938	5.5824	0.1191	0.1791	15.450	2.7676	7
8	1.5938	0.6274	9.8975	6.2098	0.1010	0.1610	19.842	3.1952	8
9	1.6895	0.5919	11.4913	6.8017	0.0870	0.1470	24.577	3.6133	9
10	1.7908	0.5584	13.1808	7.3601	0.0759	0.1359	29.602	4.0220	10
11	1.8983	0.5268	14.9716	7.8869	0.0668	0.1268	34.870	4.4213	11
12	2.0122	0.4970	16.8699	8.3838	0.0593	0.1193	40.337	4.8113	12
13	2.1329	0.4688	18.8821	8.8527	0.0530	0.1130	45.963	5.1920	13
14	2.2609	0.4423	21.0151	9.2950	0.0476	0.1076	51.713	5.5635	14
15	2.3966	0.4173	23.2760	9.7122	0.0430	0.1030	57.555	5.9260	15
16	2.5404	0.3936	25.6725	10.1059	0.0390	0.0990	63.459	6.2794	16
17	2.6928	0.3714	28.2129	10.4773	0.0354	0.0954	69.401	6.6240	17
18	2.8543	0.3503	30.9057	10.8276	0.0324	0.0924	75.357	6.9597	18
19	3.0256	0.3305	33.7600	11.1581	0.0296	0.0896	81.306	7.2867	19
20	3.2071	0.3118	36.7856	11.4699	0.0272	0.0872	87.230	7.6051	20
21	3.3996	0.2942	39.9927	11.7641	0.0250	0.0850	93.114	7.9151	21
22	3.6035	0.2775	43.3923	12.0416	0.0230	0.0830	98.941	8.2166	22
23	3.8197	0.2618	46.9958	12.3034	0.0213	0.0813	104.701	8.5099	23
24	4.0489	0.2470	50.8156	12.5504	0.0197	0.0797	110.381	8.7951	24
25	4.2919	0.2330	54.8645	12.7834	0.0182	0.0782	115.973	9.0722	25
30	5.7435	0.1741	79.0582	13.7648	0.0126	0.0726	142.359	10.3422	30
35	7.6861	0.1301	111.4348	14.4982	0.0090	0.0690	165.743	11.4319	35
40	10.2857	0.0972	154.7620	15.0463	0.0065	0.0665	185.957	12.3590	40
45	13.7646	0.0727	212.7435	15.4558	0.0047	0.0647	203.110	13.1413	45
50	18.4202	0.0543	290.3359	15.7619	0.0034	0.0634	217.457	13.7964	50
60	32.9877	0.0303	533.1282	16.1614	0.0019	0.0619	239.043	14.7909	60
80	105.7960	0.0095	1746.5999	16.5091	0.0006	0.0606	262.549	15.9033	80
100	339.3021	0.0029	5638.3681	16.6175	0.0002	0.0602	272.047	16.3711	100
∞				16.6667		0.0600			∞

TABLE C-10 Discrete Compounding; $i = 7\%$

	Single Payment		Uniform Series				Uniform Gradient		
	Compound Amount Factor	Present Worth Factor	Compound Amount Factor	Present Worth Factor	Sinking Fund Factor	Capital Recovery Factor	Gradient Present Worth Factor	Gradient Uniform Series Factor	
N	To Find F Given P F/P	To Find P Given F P/F	To Find F Given A F/A	To Find P Given A P/A	To Find A Given F A/F	To Find A Given P A/P	To Find P Given G P/G	To Find A Given G A/G	N
1	1.0700	0.9346	1.0000	0.9346	1.0000	1.0700	0.000	0.0000	1
2	1.1449	0.8734	2.0700	1.8080	0.4831	0.5531	0.873	0.4831	2
3	1.2250	0.8163	3.2149	2.6243	0.3111	0.3811	2.506	0.9549	3
4	1.3108	0.7629	4.4399	3.3872	0.2252	0.2952	4.795	1.4155	4
5	1.4026	0.7130	5.7507	4.1002	0.1739	0.2439	7.647	1.8650	5
6	1.5007	0.6663	7.1533	4.7665	0.1398	0.2098	10.978	2.3032	6
7	1.6058	0.6227	8.6540	5.3893	0.1156	0.1856	14.715	2.7304	7
8	1.7182	0.5820	10.2598	5.9713	0.0975	0.1675	18.789	3.1465	8
9	1.8385	0.5439	11.9780	6.5152	0.0835	0.1535	23.140	3.5517	9
10	1.9672	0.5083	13.8164	7.0236	0.0724	0.1424	27.716	3.9461	10
11	2.1049	0.4751	15.7836	7.4987	0.0634	0.1334	32.467	4.3296	11
12	2.2522	0.4440	17.8885	7.9427	0.0559	0.1259	37.351	4.7025	12
13	2.4098	0.4150	20.1406	8.3577	0.0497	0.1197	42.330	5.0648	13
14	2.5785	0.3878	22.5505	8.7455	0.0443	0.1143	47.372	5.4167	14
15	2.7590	0.3624	25.1290	9.1079	0.0398	0.1098	52.446	5.7583	15
16	2.9522	0.3387	27.8881	9.4466	0.0359	0.1059	57.527	6.0897	16
17	3.1588	0.3166	30.8402	9.7632	0.0324	0.1024	62.592	6.4110	17
18	3.3799	0.2959	33.9990	10.0591	0.0294	0.0994	67.622	6.7225	18
19	3.6165	0.2765	37.3790	10.3356	0.0268	0.0968	72.599	7.0242	19
20	3.8697	0.2584	40.9955	10.5940	0.0244	0.0944	77.509	7.3163	20
21	4.1406	0.2415	44.8652	10.8355	0.0223	0.0923	82.339	7.5990	21
22	4.4304	0.2257	49.0057	11.0612	0.0204	0.0904	87.079	7.8725	22
23	4.7405	0.2109	53.4361	11.2722	0.0187	0.0887	91.720	8.1369	23
24	5.0724	0.1971	58.1767	11.4693	0.0172	0.0872	96.255	8.3923	24
25	5.4274	0.1842	63.2490	11.6536	0.0158	0.0858	100.677	8.6391	25
30	7.6123	0.1314	94.4608	12.4090	0.0106	0.0806	120.972	9.7487	30
35	10.6766	0.0937	138.2369	12.9477	0.0072	0.0772	138.135	10.6687	35
40	14.9745	0.0668	199.6351	13.3317	0.0050	0.0750	152.293	11.4233	40
45	21.0023	0.0476	285.7495	13.6055	0.0035	0.0735	163.756	12.0360	45
50	29.4570	0.0339	406.5289	13.8007	0.0025	0.0725	172.905	12.5287	50
60	57.9464	0.0173	813.5204	14.0392	0.0012	0.0712	185.768	13.2321	60
80	224.2344	0.0045	3189.0627	14.2220	0.0003	0.0703	198.075	13.9273	80
100	867.7163	0.0012	12381.6618	14.2693	0.0001	0.0701	202.200	14.1703	100
∞				14.2857		0.0700			∞

TABLE C-11 Discrete Compounding; $i = 8\%$

| | Single Payment | | Uniform Series | | | | Uniform Gradient | | |
| | Compound Amount Factor | Present Worth Factor | Compound Amount Factor | Present Worth Factor | Sinking Fund Factor | Capital Recovery Factor | Gradient Present Worth Factor | Gradient Uniform Series Factor | |
N	To Find F Given P, F/P	To Find P Given F, P/F	To Find F Given A, F/A	To Find P Given A, P/A	To Find A Given F, A/F	To Find A Given P, A/P	To Find P Given G, P/G	To Find A Given G, A/G	N
1	1.0800	0.9259	1.0000	0.9259	1.0000	1.0800	0.000	0.0000	1
2	1.1664	0.8573	2.0800	1.7833	0.4808	0.5608	0.857	0.4808	2
3	1.2597	0.7938	3.2464	2.5771	0.3080	0.3880	2.445	0.9487	3
4	1.3605	0.7350	4.5061	3.3121	0.2219	0.3019	4.650	1.4040	4
5	1.4693	0.6806	5.8666	3.9927	0.1705	0.2505	7.372	1.8465	5
6	1.5869	0.6302	7.3359	4.6229	0.1363	0.2163	10.523	2.2763	6
7	1.7138	0.5835	8.9228	5.2064	0.1121	0.1921	14.024	2.6937	7
8	1.8509	0.5403	10.6366	5.7466	0.0940	0.1740	17.806	3.0985	8
9	1.9990	0.5002	12.4876	6.2469	0.0801	0.1601	21.808	3.4910	9
10	2.1589	0.4632	14.4866	6.7101	0.0690	0.1490	25.977	3.8713	10
11	2.3316	0.4289	16.6455	7.1390	0.0601	0.1401	30.266	4.2395	11
12	2.5182	0.3971	18.9771	7.5361	0.0527	0.1327	34.634	4.5957	12
13	2.7196	0.3677	21.4953	7.9038	0.0465	0.1265	39.046	4.9402	13
14	2.9372	0.3405	24.2149	8.2442	0.0413	0.1213	43.472	5.2731	14
15	3.1722	0.3152	27.1521	8.5595	0.0368	0.1168	47.886	5.5945	15
16	3.4259	0.2919	30.3243	8.8514	0.0330	0.1130	52.264	5.9046	16
17	3.7000	0.2703	33.7502	9.1216	0.0296	0.1096	56.588	6.2037	17
18	3.9960	0.2502	37.4502	9.3719	0.0267	0.1067	60.843	6.4920	18
19	4.3157	0.2317	41.4463	9.6036	0.0241	0.1041	65.013	6.7697	19
20	4.6610	0.2145	45.7620	9.8181	0.0219	0.1019	69.090	7.0369	20
21	5.0338	0.1987	50.4229	10.0168	0.0198	0.0998	73.063	7.2940	21
22	5.4365	0.1839	55.4568	10.2007	0.0180	0.0980	76.926	7.5412	22
23	5.8715	0.1703	60.8933	10.3711	0.0164	0.0964	80.673	7.7786	23
24	6.3412	0.1577	66.7648	10.5288	0.0150	0.0950	84.300	8.0066	24
25	6.8485	0.1460	73.1059	10.6748	0.0137	0.0937	87.804	8.2254	25
30	10.0627	0.0994	113.2832	11.2578	0.0088	0.0888	103.456	9.1897	30
35	14.7853	0.0676	172.3168	11.6546	0.0058	0.0858	116.092	9.9611	35
40	21.7245	0.0460	259.0565	11.9246	0.0039	0.0839	126.042	10.5699	40
45	31.9204	0.0313	386.5056	12.1084	0.0026	0.0826	133.733	11.0447	45
50	46.9016	0.0213	573.7702	12.2335	0.0017	0.0817	139.593	11.4107	50
60	101.2571	0.0099	1253.2133	12.3766	0.0008	0.0808	147.300	11.9015	60
80	471.9548	0.0021	5886.9354	12.4735	0.0002	0.0802	153.800	12.3301	80
100	2199.7613	0.0005	27484.5157	12.4943	a	0.0800	155.611	12.4545	100
∞				12.5000		0.0800			∞

^{a} Less than 0.0001.

TABLE C-12 Discrete Compounding; $i = 9\%$

	Single Payment		Uniform Series				Uniform Gradient		
	Compound Amount Factor	Present Worth Factor	Compound Amount Factor	Present Worth Factor	Sinking Fund Factor	Capital Recovery Factor	Gradient Present Worth Factor	Gradient Uniform Series Factor	
N	To Find F Given P F/P	To Find P Given F P/F	To Find F Given A F/A	To Find P Given A P/A	To Find A Given F A/F	To Find A Given P A/P	To Find P Given G P/G	To Find A Given G A/G	N
1	1.0900	0.9174	1.0000	0.9174	1.0000	1.0900	0.000	0.0000	1
2	1.1881	0.8417	2.0900	1.7591	0.4785	0.5685	0.842	0.4785	2
3	1.2950	0.7722	3.2781	2.5313	0.3051	0.3951	2.386	0.9426	3
4	1.4116	0.7084	4.5731	3.2397	0.2187	0.3087	4.511	1.3925	4
5	1.5386	0.6499	5.9847	3.8897	0.1671	0.2571	7.111	1.8282	5
6	1.6771	0.5963	7.5233	4.4859	0.1329	0.2229	10.092	2.2498	6
7	1.8280	0.5470	9.2004	5.0330	0.1087	0.1987	13.375	2.6574	7
8	1.9926	0.5019	11.0285	5.5348	0.0907	0.1807	16.888	3.0512	8
9	2.1719	0.4604	13.0210	5.9952	0.0768	0.1668	20.571	3.4312	9
10	2.3674	0.4224	15.1929	6.4177	0.0658	0.1558	24.373	3.7978	10
11	2.5804	0.3875	17.5603	6.8052	0.0569	0.1469	28.248	4.1510	11
12	2.8127	0.3555	20.1407	7.1607	0.0497	0.1397	32.159	4.4910	12
13	3.0658	0.3262	22.9534	7.4869	0.0436	0.1336	36.073	4.8182	13
14	3.3417	0.2992	26.0192	7.7862	0.0384	0.1284	39.963	5.1326	14
15	3.6425	0.2745	29.3609	8.0607	0.0341	0.1241	43.807	5.4346	15
16	3.9703	0.2519	33.0034	8.3126	0.0303	0.1203	47.585	5.7245	16
17	4.3276	0.2311	36.9737	8.5436	0.0270	0.1170	51.282	6.0024	17
18	4.7171	0.2120	41.3013	8.7556	0.0242	0.1142	54.886	6.2687	18
19	5.1417	0.1945	46.0185	8.9501	0.0217	0.1117	58.387	6.5236	19
20	5.6044	0.1784	51.1601	9.1285	0.0195	0.1095	61.777	6.7674	20
21	6.1088	0.1637	56.7645	9.2922	0.0176	0.1076	65.051	7.0006	21
22	6.6586	0.1502	62.8733	9.4424	0.0159	0.1059	68.205	7.2232	22
23	7.2579	0.1378	69.5319	9.5802	0.0144	0.1044	71.236	7.4357	23
24	7.9111	0.1264	76.7898	9.7066	0.0130	0.1030	74.143	7.6384	24
25	8.6231	0.1160	84.7009	9.8226	0.0118	0.1018	76.927	7.8316	25
30	13.2677	0.0754	136.3075	10.2737	0.0073	0.0973	89.028	8.6657	30
35	20.4140	0.0490	215.7108	10.5668	0.0046	0.0946	98.359	9.3083	35
40	31.4094	0.0318	337.8824	10.7574	0.0030	0.0930	105.376	9.7957	40
45	48.3273	0.0207	525.8587	10.8812	0.0019	0.0919	110.556	10.1603	45
50	74.3575	0.0134	815.0836	10.9617	0.0012	0.0912	114.325	10.4295	50
60	176.0313	0.0057	1944.7921	11.0480	0.0005	0.0905	118.968	10.7683	60
80	986.5517	0.0010	10950.5741	11.0998	0.0001	0.0901	122.431	11.0299	80
100	5529.0408	0.0002	61422.6755	11.1091	a	0.0900	123.234	11.0930	100
∞				11.1111		0.0900			∞

^{a}Less than 0.0001.

TABLE C-13 Discrete Compounding; $i = 10\%$

	Single Payment		Uniform Series				Uniform Gradient		
	Compound Amount Factor	Present Worth Factor	Compound Amount Factor	Present Worth Factor	Sinking Fund Factor	Capital Recovery Factor	Gradient Present Worth Factor	Gradient Uniform Series Factor	
N	To Find F Given P F/P	To Find P Given F P/F	To Find F Given A F/A	To Find P Given A P/A	To Find A Given F A/F	To Find A Given P A/P	To Find P Given G P/G	To Find A Given G A/G	N
1	1.1000	0.9091	1.0000	0.9091	1.0000	1.1000	0.000	0.0000	1
2	1.2100	0.8264	2.1000	1.7355	0.4762	0.5762	0.826	0.4762	2
3	1.3310	0.7513	3.3100	2.4869	0.3021	0.4021	2.329	0.9366	3
4	1.4641	0.6830	4.6410	3.1699	0.2155	0.3155	4.378	1.3812	4
5	1.6105	0.6209	6.1051	3.7908	0.1638	0.2638	6.862	1.8101	5
6	1.7716	0.5645	7.7156	4.3553	0.1296	0.2296	9.684	2.2236	6
7	1.9487	0.5132	9.4872	4.8684	0.1054	0.2054	12.763	2.6216	7
8	2.1436	0.4665	11.4359	5.3349	0.0874	0.1874	16.029	3.0045	8
9	2.3579	0.4241	13.5795	5.7590	0.0736	0.1736	19.422	3.3724	9
10	2.5937	0.3855	15.9374	6.1446	0.0627	0.1627	22.891	3.7255	10
11	2.8531	0.3505	18.5312	6.4951	0.0540	0.1540	26.396	4.0641	11
12	3.1384	0.3186	21.3843	6.8137	0.0468	0.1468	29.901	4.3884	12
13	3.4523	0.2897	24.5227	7.1034	0.0408	0.1408	33.377	4.6988	13
14	3.7975	0.2633	27.9750	7.3667	0.0357	0.1357	36.801	4.9955	14
15	4.1772	0.2394	31.7725	7.6061	0.0315	0.1315	40.152	5.2789	15
16	4.5950	0.2176	35.9497	7.8237	0.0278	0.1278	43.416	5.5493	16
17	5.0545	0.1978	40.5447	8.0216	0.0247	0.1247	46.582	5.8071	17
18	5.5599	0.1799	45.5992	8.2014	0.0219	0.1219	49.640	6.0526	18
19	6.1159	0.1635	51.1591	8.3649	0.0195	0.1195	52.583	6.2861	19
20	6.7275	0.1486	57.2750	8.5136	0.0175	0.1175	55.407	6.5081	20
21	7.4002	0.1351	64.0025	8.6487	0.0156	0.1156	58.110	6.7189	21
22	8.1403	0.1228	71.4027	8.7715	0.0140	0.1140	60.689	6.9189	22
23	8.9543	0.1117	79.5430	8.8832	0.0126	0.1126	63.146	7.1085	23
24	9.8497	0.1015	88.4973	8.9847	0.0113	0.1113	65.481	7.2881	24
25	10.8347	0.0923	98.3471	9.0770	0.0102	0.1102	67.696	7.4580	25
30	17.4494	0.0573	164.4940	9.4269	0.0061	0.1061	77.077	8.1762	30
35	28.1024	0.0356	271.0244	9.6442	0.0037	0.1037	83.987	8.7086	35
40	45.2593	0.0221	442.5926	9.7791	0.0023	0.1023	88.953	9.0962	40
45	72.8905	0.0137	718.9048	9.8628	0.0014	0.1014	92.454	9.3740	45
50	117.3909	0.0085	1163.9085	9.9148	0.0009	0.1009	94.889	9.5704	50
60	304.4816	0.0033	3034.8164	9.9672	0.0003	0.1003	97.701	9.8023	60
80	2048.4002	0.0005	20474.0021	9.9951	a	0.1000	99.561	9.9609	80
100	13780.6123	0.0001	137796.1234	9.9993	a	0.1000	99.920	9.9927	100
∞				10.0000		0.1000			∞

a Less than 0.0001.

TABLE C-14 Discrete Compounding: $i = 12\%$

	Single Payment		Uniform Series					Uniform Gradient		
	Compound Amount Factor	Present Worth Factor	Compound Amount Factor	Present Worth Factor	Sinking Fund Factor	Capital Recovery Factor		Gradient Present Worth Factor	Gradient Uniform Series Factor	
N	To Find F Given P F/P	To Find P Given F P/F	To Find F Given A F/A	To Find P Given A P/A	To Find A Given F A/F	To Find A Given P A/P		To Find P Given G P/G	To Find A Given G A/G	N
1	1.1200	0.8929	1.0000	0.8929	1.0000	1.1200		0.000	0.0000	1
2	1.2544	0.7972	2.1200	1.6901	0.4717	0.5917		0.797	0.4717	2
3	1.4049	0.7118	3.3744	2.4018	0.2963	0.4163		2.221	0.9246	3
4	1.5735	0.6355	4.7793	3.0373	0.2092	0.3292		4.127	1.3589	4
5	1.7623	0.5674	6.3528	3.6048	0.1574	0.2774		6.397	1.7746	5
6	1.9738	0.5066	8.1152	4.1114	0.1232	0.2432		8.930	2.1720	6
7	2.2107	0.4523	10.0890	4.5638	0.0991	0.2191		11.644	2.5515	7
8	2.4760	0.4039	12.2997	4.9676	0.0813	0.2013		14.471	2.9131	8
9	2.7731	0.3606	14.7757	5.3282	0.0677	0.1877		17.356	3.2574	9
10	3.1058	0.3220	17.5487	5.6502	0.0570	0.1770		20.254	3.5847	10
11	3.4785	0.2875	20.6546	5.9377	0.0484	0.1684		23.129	3.8953	11
12	3.8960	0.2567	24.1331	6.1944	0.0414	0.1614		25.952	4.1897	12
13	4.3635	0.2292	28.0291	6.4235	0.0357	0.1557		28.702	4.4683	13
14	4.8871	0.2046	32.3926	6.6282	0.0309	0.1509		31.362	4.7317	14
15	5.4736	0.1827	37.2797	6.8109	0.0268	0.1468		33.920	4.9803	15
16	6.1304	0.1631	42.7533	6.9740	0.0234	0.1434		36.367	5.2147	16
17	6.8660	0.1456	48.8837	7.1196	0.0205	0.1405		38.697	5.4353	17
18	7.6900	0.1300	55.7497	7.2497	0.0179	0.1379		40.908	5.6427	18
19	8.6128	0.1161	63.4397	7.3658	0.0158	0.1358		42.998	5.8375	19
20	9.6463	0.1037	72.0524	7.4694	0.0139	0.1339		44.968	6.0202	20
21	10.8038	0.0926	81.6987	7.5620	0.0122	0.1322		46.819	6.1913	21
22	12.1003	0.0826	92.5026	7.6446	0.0108	0.1308		48.554	6.3514	22
23	13.5523	0.0738	104.6029	7.7184	0.0096	0.1296		50.178	6.5010	23
24	15.1786	0.0659	118.1552	7.7843	0.0085	0.1285		51.693	6.6406	24
25	17.0001	0.0588	133.3339	7.8431	0.0075	0.1275		53.105	6.7708	25
30	29.9599	0.0334	241.3327	8.0552	0.0041	0.1241		58.782	7.2974	30
35	52.7996	0.0189	431.6635	8.1755	0.0023	0.1223		62.605	7.6577	35
40	93.0510	0.0107	767.0914	8.2438	0.0013	0.1213		65.116	7.8988	40
45	163.9876	0.0061	1358.2300	8.2825	0.0007	0.1207		66.734	8.0572	45
50	289.0022	0.0035	2400.0182	8.3045	0.0004	0.1204		67.762	8.1597	50
60	897.5969	0.0011	7471.6411	8.3240	0.0001	0.1201		68.810	8.2664	60
80	8658.4831	0.0001	72145.6925	8.3324	*a*	0.1200		69.359	8.3241	80
100	83522.2657	*a*	696010.5477	8.3332	*a*	0.1200		69.434	8.3321	100
∞				8.3333		0.1200				∞

a Less than 0.0001.

TABLE C-15 Discrete Compounding; i = 15%

	Single Payment		Uniform Series				Uniform Gradient		
	Compound Amount Factor	Present Worth Factor	Compound Amount Factor	Present Worth Factor	Sinking Fund Factor	Capital Recovery Factor	Gradient Present Worth Factor	Gradient Uniform Series Factor	
N	To Find F Given P, F/P	To Find P Given F, P/F	To Find F Given A, F/A	To Find P Given A, P/A	To Find A Given F, A/F	To Find A Given P, A/P	To Find P Given G, P/G	To Find A Given G, A/G	N
1	1.1500	0.8696	1.0000	0.8696	1.0000	1.1500	0.000	0.0000	1
2	1.3225	0.7561	2.1500	1.6257	0.4651	0.6151	0.756	0.4651	2
3	1.5209	0.6575	3.4725	2.2832	0.2880	0.4380	2.071	0.9071	3
4	1.7490	0.5718	4.9934	2.8550	0.2003	0.3503	3.786	1.3263	4
5	2.0114	0.4972	6.7424	3.3522	0.1483	0.2983	5.775	1.7228	5
6	2.3131	0.4323	8.7537	3.7845	0.1142	0.2642	7.937	2.0972	6
7	2.6600	0.3759	11.0668	4.1604	0.0904	0.2404	10.192	2.4498	7
8	3.0590	0.3269	13.7268	4.4873	0.0729	0.2229	12.481	2.7813	8
9	3.5179	0.2843	16.7858	4.7716	0.0596	0.2096	14.755	3.0922	9
10	4.0456	0.2472	20.3037	5.0188	0.0493	0.1993	16.980	3.3832	10
11	4.6524	0.2149	24.3493	5.2337	0.0411	0.1911	19.129	3.6549	11
12	5.3503	0.1869	29.0017	5.4206	0.0345	0.1845	21.185	3.9082	12
13	6.1528	0.1625	34.3519	5.5831	0.0291	0.1791	23.135	4.1438	13
14	7.0757	0.1413	40.5047	5.7245	0.0247	0.1747	24.973	4.3624	14
15	8.1371	0.1229	47.5804	5.8474	0.0210	0.1710	26.693	4.5650	15
16	9.3576	0.1069	55.7175	5.9542	0.0179	0.1679	28.296	4.7522	16
17	10.7613	0.0929	65.0751	6.0472	0.0154	0.1654	29.783	4.9251	17
18	12.3755	0.0808	75.8364	6.1280	0.0132	0.1632	31.157	5.0843	18
19	14.2318	0.0703	88.2118	6.1982	0.0113	0.1613	32.421	5.2307	19
20	16.3665	0.0611	102.4436	6.2593	0.0098	0.1598	33.582	5.3651	20
21	18.8215	0.0531	118.8101	6.3125	0.0084	0.1584	34.645	5.4883	21
22	21.6447	0.0462	137.6316	6.3587	0.0073	0.1573	35.615	5.6010	22
23	24.8915	0.0402	159.2764	6.3988	0.0063	0.1563	36.499	5.7040	23
24	28.6252	0.0349	184.1678	6.4338	0.0054	0.1554	37.302	5.7979	24
25	32.9190	0.0304	212.7930	6.4641	0.0047	0.1547	38.031	5.8834	25
30	66.2118	0.0151	434.7451	6.5660	0.0023	0.1523	40.753	6.2066	30
35	133.1755	0.0075	881.1702	6.6166	0.0011	0.1511	42.359	6.4019	35
40	267.8635	0.0037	1779.0903	6.6418	0.0006	0.1506	43.283	6.5168	40
45	538.7693	0.0019	3585.1285	6.6543	0.0003	0.1503	43.805	6.5830	45
50	1083.6574	0.0009	7217.7163	6.6605	0.0001	0.1501	44.096	6.6205	50
60	4383.9987	0.0002	29219.9916	6.6651	a	0.1500	44.343	6.6530	60
80	71750.8794	a	478332.5293	6.6666	a	0.1500	44.436	6.6656	80
100	1174313.4507	a	7828749.6713	6.6667	a	0.1500	44.444	6.6666	100
∞				6.6667		0.1500			∞

a Less than 0.0001.

TABLE C-16 Discrete Compounding; $i = 18\%$

	Single Payment		Uniform Series					Uniform Gradient	
	Compound Amount Factor	Present Worth Factor	Compound Amount Factor	Present Worth Factor	Sinking Fund Factor	Capital Recovery Factor	Gradient Present Worth Factor	Gradient Uniform Series Factor	
N	To Find F Given P F/P	To Find P Given F P/F	To Find F Given A F/A	To Find P Given A P/A	To Find A Given F A/F	To Find A Given P A/P	To Find P Given G P/G	To Find A Given G A/G	N
1	1.1800	0.8475	1.0000	0.8475	1.0000	1.1800	0.000	0.0000	1
2	1.3924	0.7182	2.1800	1.5656	0.4587	0.6387	0.718	0.4587	2
3	1.6430	0.6086	3.5724	2.1743	0.2799	0.4599	1.935	0.8902	3
4	1.9388	0.5158	5.2154	2.6901	0.1917	0.3717	3.483	1.2947	4
5	2.2878	0.4371	7.1542	3.1272	0.1398	0.3198	5.231	1.6728	5
6	2.6996	0.3704	9.4420	3.4976	0.1059	0.2859	7.083	2.0252	6
7	3.1855	0.3139	12.1415	3.8115	0.0824	0.2624	8.967	2.3526	7
8	3.7589	0.2660	15.3270	4.0776	0.0652	0.2452	10.829	2.6558	8
9	4.4355	0.2255	19.0859	4.3030	0.0524	0.2324	12.633	2.9358	9
10	5.2338	0.1911	23.5213	4.4941	0.0425	0.2225	14.353	3.1936	10
11	6.1759	0.1619	28.7551	4.6560	0.0348	0.2148	15.972	3.4303	11
12	7.2876	0.1372	34.9311	4.7932	0.0286	0.2086	17.481	3.6470	12
13	8.5994	0.1163	42.2187	4.9095	0.0237	0.2037	18.877	3.8449	13
14	10.1472	0.0985	50.8180	5.0081	0.0197	0.1997	20.158	4.0250	14
15	11.9737	0.0835	60.9653	5.0916	0.0164	0.1964	21.327	4.1887	15
16	14.1290	0.0708	72.9390	5.1624	0.0137	0.1937	22.389	4.3369	16
17	16.6722	0.0600	87.0680	5.2223	0.0115	0.1915	23.348	4.4708	17
18	19.6733	0.0508	103.7403	5.2732	0.0096	0.1896	24.212	4.5916	18
19	23.2144	0.0431	123.4135	5.3162	0.0081	0.1881	24.988	4.7003	19
20	27.3930	0.0365	146.6280	5.3527	0.0068	0.1868	25.681	4.7978	20
21	32.3238	0.0309	174.0210	5.3837	0.0057	0.1857	26.300	4.8851	21
22	38.1421	0.0262	206.3448	5.4099	0.0048	0.1848	26.851	4.9632	22
23	45.0076	0.0222	244.4868	5.4321	0.0041	0.1841	27.339	5.0329	23
24	53.1090	0.0188	289.4945	5.4509	0.0035	0.1835	27.773	5.0950	24
25	62.6686	0.0160	342.6035	5.4669	0.0029	0.1829	28.156	5.1502	25
30	143.3706	0.0070	790.9480	5.5168	0.0013	0.1813	29.486	5.3448	30
35	327.9973	0.0030	1816.6516	5.5386	0.0006	0.1806	30.177	5.4485	35
40	750.3783	0.0013	4163.2130	5.5482	0.0002	0.1802	30.527	5.5022	40
45	1716.6839	0.0006	9531.5771	5.5523	0.0001	0.1801	30.701	5.5293	45
50	3927.3569	0.0003	21813.0937	5.5541	a	0.1800	30.786	5.5428	50
60	20555.1400	a	114189.6665	5.5553	a	0.1800	30.847	5.5526	60
80	563067.6604	a	3128148.1133	5.5555	a	0.1800	30.863	5.5554	80
∞				5.5556		0.1800			∞

TABLE C-17 Discrete Compounding; $i = 20\%$

	Single Payment		Uniform Series				Uniform Gradient		
	Compound Amount Factor	Present Worth Factor	Compound Amount Factor	Present Worth Factor	Sinking Fund Factor	Capital Recovery Factor	Gradient Present Worth Factor	Gradient Uniform Series Factor	
N	To Find F Given P F/P	To Find P Given F P/F	To Find F Given A F/A	To Find P Given A P/A	To Find A Given F A/F	To Find A Given P A/P	To Find P Given G P/G	To Find A Given G A/G	N
1	1.2000	0.8333	1.0000	0.8333	1.0000	1.2000	0.000	0.0000	1
2	1.4400	0.6944	2.2000	1.5278	0.4545	0.6545	0.694	0.4545	2
3	1.7280	0.5787	3.6400	2.1065	0.2747	0.4747	1.852	0.8791	3
4	2.0736	0.4823	5.3680	2.5887	0.1863	0.3863	3.299	1.2742	4
5	2.4883	0.4019	7.4416	2.9906	0.1344	0.3344	4.906	1.6405	5
6	2.9860	0.3349	9.9299	3.3255	0.1007	0.3007	6.581	1.9788	6
7	3.5832	0.2791	12.9159	3.6046	0.0774	0.2774	8.255	2.2902	7
8	4.2998	0.2326	16.4991	3.8372	0.0606	0.2606	9.883	2.5756	8
9	5.1598	0.1938	20.7989	4.0310	0.0481	0.2481	11.434	2.8364	9
10	6.1917	0.1615	25.9587	4.1925	0.0385	0.2385	12.887	3.0739	10
11	7.4301	0.1346	32.1504	4.3271	0.0311	0.2311	14.233	3.2893	11
12	8.9161	0.1122	39.5805	4.4392	0.0253	0.2253	15.467	3.4841	12
13	10.6993	0.0935	48.4966	4.5327	0.0206	0.2206	16.588	3.6597	13
14	12.8392	0.0779	59.1959	4.6106	0.0169	0.2169	17.601	3.8175	14
15	15.4070	0.0649	72.0351	4.6755	0.0139	0.2139	18.510	3.9588	15
16	18.4884	0.0541	87.4421	4.7296	0.0114	0.2114	19.321	4.0851	16
17	22.1861	0.0451	105.9306	4.7746	0.0094	0.2094	20.042	4.1976	17
18	26.6233	0.0376	128.1167	4.8122	0.0078	0.2078	20.681	4.2975	18
19	31.9480	0.0313	154.7400	4.8435	0.0065	0.2065	21.244	4.3861	19
20	38.3376	0.0261	186.6880	4.8696	0.0054	0.2054	21.740	4.4643	20
21	46.0051	0.0217	225.0256	4.8913	0.0044	0.2044	22.174	4.5334	21
22	55.2061	0.0181	271.0307	4.9094	0.0037	0.2037	22.555	4.5941	22
23	66.2474	0.0151	326.2369	4.9245	0.0031	0.2031	22.887	4.6475	23
24	79.4968	0.0126	392.4842	4.9371	0.0025	0.2025	23.176	4.6943	24
25	95.3962	0.0105	471.9811	4.9476	0.0021	0.2021	23.428	4.7352	25
30	237.3763	0.0042	1181.8816	4.9789	0.0008	0.2008	24.263	4.8731	30
35	590.6682	0.0017	2948.3411	4.9915	0.0003	0.2003	24.661	4.9406	35
40	1469.7716	0.0007	7343.8578	4.9966	0.0001	0.2001	24.847	4.9728	40
45	3657.2620	0.0003	18281.3099	4.9986	0.0001	0.2001	24.932	4.9877	45
50	9100.4382	0.0001	45497.1908	4.9995	a	0.2000	24.970	4.9945	50
60	56347.5144	a	281732.5718	4.9999	a	0.2000	24.994	4.9989	60
80	216228.4620	a	10801137.3101	5.0000	a	0.2000	25.000	5.0000	80
∞				5.0000		0.2000			∞

a Less than 0.0001.

TABLE C-18 Discrete Compounding; $i = 25\%$

N	Single Payment		Uniform Series				Uniform Gradient		N
	Compound Amount Factor	Present Worth Factor	Compound Amount Factor	Present Worth Factor	Sinking Fund Factor	Capital Recovery Factor	Gradient Present Worth Factor	Gradient Uniform Series Factor	
	To Find F Given P F/P	To Find P Given F P/F	To Find F Given A F/A	To Find P Given A P/A	To Find A Given F A/F	To Find A Given P A/P	To Find P Given G P/G	To Find A Given G A/G	
1	1.2500	0.8000	1.0000	0.8000	1.0000	1.2500	0.000	0.0000	1
2	1.5625	0.6400	2.2500	1.4400	0.4444	0.6944	0.640	0.4444	2
3	1.9531	0.5120	3.8125	1.9520	0.2623	0.5123	1.664	0.8525	3
4	2.4414	0.4096	5.7656	2.3616	0.1734	0.4234	2.893	1.2249	4
5	3.0518	0.3277	8.2070	2.6893	0.1218	0.3718	4.204	1.5631	5
6	3.8147	0.2621	11.2588	2.9514	0.0888	0.3388	5.514	1.8683	6
7	4.7684	0.2097	15.0735	3.1611	0.0663	0.3163	6.773	2.1424	7
8	5.9605	0.1678	19.8419	3.3289	0.0504	0.3004	7.947	2.3872	8
9	7.4506	0.1342	25.8023	3.4631	0.0388	0.2888	9.021	2.6048	9
10	9.3132	0.1074	33.2529	3.5705	0.0301	0.2801	9.987	2.7971	10
11	11.6415	0.0859	42.5661	3.6564	0.0235	0.2735	10.846	2.9663	11
12	14.5519	0.0687	54.2077	3.7251	0.0184	0.2684	11.602	3.1145	12
13	18.1899	0.0550	68.7596	3.7801	0.0145	0.2645	12.262	3.2437	13
14	22.7374	0.0440	86.9495	3.8241	0.0115	0.2615	12.833	3.3559	14
15	28.4217	0.0352	109.6868	3.8593	0.0091	0.2591	13.326	3.4530	15
16	35.5271	0.0281	138.1085	3.8874	0.0072	0.2572	13.748	3.5366	16
17	44.4089	0.0225	173.6357	3.9099	0.0058	0.2558	14.109	3.6084	17
18	55.5112	0.0180	218.0446	3.9279	0.0046	0.2546	14.415	3.6698	18
19	69.3889	0.0144	273.5558	3.9424	0.0037	0.2537	14.674	3.7222	19
20	86.7362	0.0115	342.9447	3.9539	0.0029	0.2529	14.893	3.7667	20
21	108.4202	0.0092	429.6809	3.9631	0.0023	0.2523	15.078	3.8045	21
22	135.5253	0.0074	538.1011	3.9705	0.0019	0.2519	15.233	3.8365	22
23	169.4066	0.0059	673.6264	3.9764	0.0015	0.2515	15.363	3.8634	23
24	211.7582	0.0047	843.0329	3.9811	0.0012	0.2512	15.471	3.8861	24
25	264.6978	0.0038	1054.7912	3.9849	0.0009	0.2509	15.562	3.9052	25
30	807.7936	0.0012	3227.1743	3.9950	0.0003	0.2503	15.832	3.9628	30
35	2465.1903	0.0004	9856.7613	3.9984	0.0001	0.2501	15.937	3.9858	35
40	7523.1638	0.0001	30088.6554	3.9995	*a*	0.2500	15.977	3.9947	40
45	22958.8740	*a*	91831.4962	3.9998	*a*	0.2500	15.992	3.9980	45
50	70064.9232	*a*	280255.6929	3.9999	*a*	0.2500	15.997	3.9993	50
60	652530.4468	*a*	2610117.7872	4.0000	*a*	0.2500	16.000	3.9999	60
∞				4.0000		0.2500			∞

a Less than 0.0001.

APPENDIX D

Interest and Annuity Tables for Continuous Compounding

For various values of r from 8% to 20%,

r = nominal interest rate per period, compounded continuously;
N = number of compounding periods;

$(F/P, r\%, N) = e^{rN};$

$(P/F, r\%, N) = e^{-rN} = \dfrac{1}{e^{rN}};$

$(F/A, r\%, N) = \dfrac{e^{rN} - 1}{e^{r} - 1};$

$(P/A, r\%, N) = \dfrac{e^{rN} - 1}{e^{rN}(e^{r} - 1)}.$

TABLE D-1 Continuous Compounding; $r = 8\%$

Discrete Flows

	Single Payment		Uniform Series		
	Compound Amount Factor	Present Worth Factor	Compound Amount Factor	Present Worth Factor	
N	To Find F Given P F/P	To Find P Given F P/F	To Find F Given A F/A	To Find P Given A P/A	N
1	1.0833	0.9231	1.0000	0.9231	1
2	1.1735	0.8521	2.0833	1.7753	2
3	1.2712	0.7866	3.2568	2.5619	3
4	1.3771	0.7261	4.5280	3.2880	4
5	1.4918	0.6703	5.9052	3.9584	5
6	1.6161	0.6188	7.3970	4.5771	6
7	1.7507	0.5712	9.0131	5.1483	7
8	1.8965	0.5273	10.7637	5.6756	8
9	2.0544	0.4868	12.6602	6.1624	9
10	2.2255	0.4493	14.7147	6.6117	10
11	2.4109	0.4148	16.9402	7.0265	11
12	2.6117	0.3829	19.3511	7.4094	12
13	2.8292	0.3535	21.9628	7.7629	13
14	3.0649	0.3263	24.7920	8.0891	14
15	3.3201	0.3012	27.8569	8.3903	15
16	3.5966	0.2780	31.1770	8.6684	16
17	3.8962	0.2567	34.7736	8.9250	17
18	4.2207	0.2369	38.6698	9.1620	18
19	4.5722	0.2187	42.8905	9.3807	19
20	4.9530	0.2019	47.4627	9.5826	20
21	5.3656	0.1864	52.4158	9.7689	21
22	5.8124	0.1720	57.7813	9.9410	22
23	6.2965	0.1588	63.5938	10.0998	23
24	6.8120	0.1466	69.8903	10.2464	24
25	7.3891	0.1353	76.7113	10.3817	25
26	8.0045	0.1249	84.1003	10.5067	26
27	8.6711	0.1153	92.1048	10.6220	27
28	9.3933	0.1065	100.776	10.7285	28
29	10.1757	0.0983	110.169	10.8267	29
30	11.0232	0.0907	120.345	10.9174	30
35	16.4446	0.0608	185.439	11.2765	35
40	24.5325	0.0408	282.547	11.5172	40
45	36.5982	0.0273	427.416	11.6786	45
50	54.5982	0.0183	643.535	11.7868	50
55	81.4509	0.0123	965.947	11.8593	55
60	121.510	0.0082	1446.93	11.9079	60
65	181.272	0.0055	2164.47	11.9404	65
70	270.426	0.0037	3234.91	11.9623	70
75	403.429	0.0025	4831.83	11.9769	75
80	601.845	0.0017	7214.15	11.9867	80
85	897.847	0.0011	10768.1	11.9933	85
90	1339.43	0.0007	16070.1	11.9977	90
95	1998.20	0.0005	23979.7	12.0007	95
100	2980.96	0.0003	35779.3	12.0026	100

TABLE D-2   Continuous Compounding; _r_ = 10%

Discrete Flows

| | Single Payment | | Uniform Series | | |
| | Compound Amount Factor | Present Worth Factor | Compound Amount Factor | Present Worth Factor | |
N	To Find F Given P F/P	To Find P Given F P/F	To Find F Given A F/A	To Find P Given A P/A	N
1	1.1052	0.9048	1.0000	0.9048	1
2	1.2214	0.8187	2.1052	1.7236	2
3	1.3499	0.7408	3.3266	2.4644	3
4	1.4918	0.6703	4.6764	3.1347	4
5	1.6487	0.6065	6.1683	3.7412	5
6	1.8221	0.5488	7.8170	4.2900	6
7	2.0138	0.4966	9.6391	4.7866	7
8	2.2255	0.4493	11.6528	5.2360	8
9	2.4596	0.4066	13.8784	5.6425	9
10	2.7183	0.3679	16.3380	6.0104	10
11	3.0042	0.3329	19.0563	6.3433	11
12	3.3201	0.3012	22.0604	6.6445	12
13	3.6693	0.2725	25.3806	6.9170	13
14	4.0552	0.2466	29.0499	7.1636	14
15	4.4817	0.2231	33.1051	7.3867	15
16	4.9530	0.2019	37.5867	7.5886	16
17	5.4739	0.1827	42.5398	7.7713	17
18	6.0496	0.1653	48.0137	7.9366	18
19	6.6859	0.1496	54.0634	8.0862	19
20	7.3891	0.1353	60.7493	8.2215	20
21	8.1662	0.1225	68.1383	8.3440	21
22	9.0250	0.1108	76.3045	8.4548	22
23	9.9742	0.1003	85.3295	8.5550	23
24	11.0232	0.0907	95.3037	8.6458	24
25	12.1825	0.0821	106.327	8.7278	25
26	13.4637	0.0743	118.509	8.8021	26
27	14.8797	0.0672	131.973	8.8693	27
28	16.4446	0.0608	146.853	8.9301	28
29	18.1741	0.0550	163.298	8.9852	29
30	20.0855	0.0498	181.472	9.0349	30
35	33.1155	0.0302	305.364	9.2212	35
40	54.5981	0.0183	509.629	9.3342	40
45	90.0171	0.0111	846.404	9.4027	45
50	148.413	0.0067	1401.65	9.4443	50
55	244.692	0.0041	2317.10	9.4695	55
60	403.429	0.0025	3826.43	9.4848	60
65	665.142	0.0015	6314.88	9.4940	65
70	1096.63	0.0009	10417.6	9.4997	70
75	1808.04	0.0006	17182.0	9.5031	75
80	2980.96	0.0003	28334.4	9.5051	80
85	4914.77	0.0002	46721.7	9.5064	85
90	8103.08	0.0001	77037.3	9.5072	90
95	13359.7	a	127019.0	9.5076	95
100	22026.5	a	209425.0	9.5079	100

[a]Less than 0.0001.

TABLE D-3   Continuous Compounding; $r = 20\%$

Discrete Flows

	Single Payment		Uniform Series		
	Compound Amount Factor	Present Worth Factor	Compound Amount Factor	Present Worth Factor	
N	To Find F Given P F/P	To Find P Given F P/F	To Find F Given A F/A	To Find P Given A P/A	N
1	1.2214	0.8187	1.0000	0.8187	1
2	1.4918	0.6703	2.2214	1.4891	2
3	1.8221	0.5488	3.7132	2.0379	3
4	2.2255	0.4493	5.5353	2.4872	4
5	2.7183	0.3679	7.7609	2.8551	5
6	3.3201	0.3012	10.4792	3.1563	6
7	4.0552	0.2466	13.7993	3.4029	7
8	4.9530	0.2019	17.8545	3.6048	8
9	6.0496	0.1653	22.8075	3.7701	9
10	7.3891	0.1353	28.8572	3.9054	10
11	9.0250	0.1108	36.2462	4.0162	11
12	11.0232	0.0907	45.2712	4.1069	12
13	13.4637	0.0743	56.2944	4.1812	13
14	16.4446	0.0608	69.7581	4.2420	14
15	20.0855	0.0498	86.2028	4.2918	15
16	24.5325	0.0408	106.288	4.3325	16
17	29.9641	0.0334	130.821	4.3659	17
18	36.5982	0.0273	160.785	4.3932	18
19	44.7012	0.0224	197.383	4.4156	19
20	54.5981	0.0183	242.084	4.4339	20
21	66.6863	0.0150	296.682	4.4489	21
22	81.4509	0.0123	363.369	4.4612	22
23	99.4843	0.0101	444.820	4.4713	23
24	121.510	0.0082	544.304	4.4795	24
25	148.413	0.0067	665.814	4.4862	25
26	181.272	0.0055	814.227	4.4917	26
27	221.406	0.0045	995.500	4.4963	27
28	270.426	0.0037	1216.91	4.5000	28
29	330.299	0.0030	1487.33	4.5030	29
30	403.429	0.0025	1817.63	4.5055	30
35	1096.63	0.0009	4948.60	4.5125	35
40	2980.96	0.0003	13459.4	4.5151	40
45	8103.08	0.0001	36594.3	4.5161	45
50	22026.5	[a]	99481.4	4.5165	50
55	59874.1	[a]	270426.0	4.5166	55
60	162755.0	[a]	735103.0	4.5166	60

[a]Less than 0.0001.

APPENDIX E

Standard Normal Distribution

The standard normal distribution is a normal (Gaussian) distribution with a mean of 0 and a variance of 1. It is a continuous distribution with a range of $-\infty$ to $+\infty$. The tabled values denote the probability of observing a value from minus infinity to the Z value indicated by the left column and top row. The Z value is determined by applying the following formula to the observed data:

$$Z = \frac{(X - \mu)}{\sigma}$$

Interested readers are referred to any introductory statistics text book for an in-depth discussion of the use of the standard normal distribution function.

| TABLE E-1 | Areas under the Normal Curve | | | | | | | | |
z	0.00	0.01	0.02	0.03	0.04	0.05	0.06	0.07	0.08	0.09
−3.4	0.0003	0.0003	0.0003	0.0003	0.0003	0.0003	0.0003	0.0003	0.0003	0.0002
−3.3	0.0005	0.0005	0.0005	0.0004	0.0004	0.0004	0.0004	0.0004	0.0004	0.0003
−3.2	0.0007	0.0007	0.0006	0.0006	0.0006	0.0006	0.0006	0.0005	0.0005	0.0005
−3.1	0.0010	0.0009	0.0009	0.0009	0.0008	0.0008	0.0008	0.0007	0.0007	0.0007
−3.0	0.0013	0.0013	0.0013	0.0012	0.0012	0.0011	0.0011	0.0011	0.0010	0.0010
−2.9	0.0019	0.0018	0.0017	0.0017	0.0016	0.0016	0.0015	0.0015	0.0014	0.0014
−2.8	0.0026	0.0025	0.0024	0.0023	0.0023	0.0022	0.0021	0.0021	0.0020	0.0019
−2.7	0.0035	0.0034	0.0033	0.0032	0.0031	0.0030	0.0029	0.0028	0.0027	0.0026
−2.6	0.0047	0.0045	0.0044	0.0043	0.0041	0.0040	0.0039	0.0038	0.0037	0.0036
−2.5	0.0062	0.0060	0.0059	0.0057	0.0055	0.0054	0.0052	0.0051	0.0049	0.0048
−2.4	0.0082	0.0080	0.0078	0.0075	0.0073	0.0071	0.0069	0.0068	0.0066	0.0064
−2.3	0.0107	0.0104	0.0103	0.0099	0.0096	0.0094	0.0091	0.0089	0.0087	0.0084
−2.2	0.0139	0.0136	0.0132	0.0129	0.0125	0.0122	0.0119	0.0116	0.0113	0.0110
−2.1	0.0179	0.0174	0.0170	0.0166	0.0162	0.0158	0.0154	0.0150	0.0146	0.0143
−2.0	0.0228	0.0222	0.0217	0.0212	0.0207	0.0202	0.0197	0.0192	0.0118	0.0183
−1.9	0.0287	0.0281	0.0274	0.0268	0.0262	0.0256	0.0250	0.0244	0.0239	0.0233
−1.8	0.0359	0.0352	0.0344	0.0336	0.0329	0.0322	0.0314	0.0307	0.0301	0.0294
−1.7	0.0446	0.0436	0.0427	0.0418	0.0409	0.0401	0.0392	0.0384	0.0375	0.0367
−1.6	0.0548	0.0537	0.0526	0.0516	0.0505	0.0495	0.0485	0.0475	0.0465	0.0455
−1.5	0.0668	0.0655	0.0643	0.0630	0.0618	0.0606	0.0594	0.0582	0.0571	0.0559
−1.4	0.0808	0.0793	0.0778	0.0764	0.0749	0.0735	0.0722	0.0708	0.0694	0.0681
−1.3	0.0968	0.0951	0.0934	0.0918	0.0901	0.0885	0.0869	0.0853	0.0838	0.0823
−1.2	0.1151	0.1131	0.1112	0.1093	0.1075	0.1056	0.1038	0.1020	0.1003	0.0985
−1.1	0.1357	0.1335	0.1314	0.1292	0.1271	0.1251	0.1230	0.1210	0.1190	0.1170
−1.0	0.1587	0.1562	0.1539	0.1515	0.1492	0.1469	0.1446	0.1423	0.1401	0.1379
−0.9	0.1841	0.1841	0.1788	0.1762	0.1736	0.1711	0.1685	0.1660	0.1635	0.1611
−0.8	0.2119	0.2090	0.2061	0.2033	0.2005	0.1977	0.1949	0.1922	0.1894	0.1867
−0.7	0.2420	0.2389	0.2358	0.2327	0.2296	0.2266	0.2236	0.2206	0.2177	0.2148
−0.6	0.2743	0.2709	0.2676	0.2643	0.2611	0.2578	0.2546	0.2514	0.2483	0.2451
−0.5	0.3085	0.3050	0.3015	0.2981	0.2946	0.2912	0.2877	0.2843	0.2810	0.2776
−0.4	0.3446	0.3409	0.3372	0.3336	0.3300	0.3264	0.3228	0.3192	0.3156	0.3121
−0.3	0.3821	0.3783	0.3745	0.3707	0.3669	0.3632	0.3594	0.3557	0.3520	0.3483
−0.2	0.4207	0.4168	0.4129	0.4090	0.4052	0.4013	0.3974	0.3936	0.3897	0.3859
−0.1	0.4602	0.4562	0.4522	0.4483	0.4443	0.4404	0.4364	0.4325	0.4286	0.4247
−0.0	0.5000	0.4960	0.4920	0.4880	0.4840	0.4801	0.4761	0.4721	0.4681	0.4641

Source: From R. E. Walpole and R. H. Myers, *Probability and Statistics for Engineers and Scientists*, 2nd ed. (New York: Macmillan, 1978), 513.

TABLE E-1	(Continued)								
z	0.00	0.01	0.02	0.03	0.04	0.05	0.06	0.07	0.08	0.09
0.0	0.5000	0.5040	0.5080	0.5120	0.5160	0.5199	0.5239	0.5279	0.5319	0.5359
0.1	0.5398	0.5438	0.5478	0.5517	0.5557	0.5596	0.5636	0.5675	0.5714	0.5753
0.2	0.5793	0.5832	0.5871	0.5910	0.5948	0.5987	0.6026	0.6064	0.6103	0.6141
0.3	0.6179	0.6217	0.6255	0.6293	0.6331	0.6368	0.6406	0.6443	0.6480	0.6517
0.4	0.6554	0.6591	0.6628	0.6664	0.6700	0.6736	0.6772	0.6808	0.6844	0.6879
0.5	0.6915	0.6950	0.6985	0.7019	0.7054	0.7088	0.7123	0.7157	0.7190	0.7224
0.6	0.7257	0.7291	0.7324	0.7357	0.7389	0.7422	0.7454	0.7486	0.7517	0.7549
0.7	0.7580	0.7611	0.7642	0.7673	0.7704	0.7734	0.7764	0.7794	0.7823	0.7852
0.8	0.7881	0.7910	0.7939	0.7967	0.7995	0.8023	0.8051	0.8078	0.8106	0.8133
0.9	0.8159	0.8186	0.8212	0.8238	0.8264	0.8289	0.8315	0.8340	0.8365	0.8389
1.0	0.8413	0.8438	0.8461	0.8485	0.8508	0.8531	0.8554	0.8577	0.8599	0.8621
1.1	0.8643	0.8665	0.8686	0.8708	0.8729	0.8749	0.8770	0.8790	0.8810	0.8830
1.2	0.8849	0.8869	0.8888	0.8907	0.8925	0.8944	0.8962	0.8980	0.8997	0.9015
1.3	0.9032	0.9049	0.9066	0.9082	0.9099	0.9115	0.9131	0.9147	0.9162	0.9177
1.4	0.9192	0.9207	0.9222	0.9236	0.9251	0.9265	0.9278	0.9292	0.9306	0.9319
1.5	0.9332	0.9345	0.9357	0.9370	0.9382	0.9394	0.9406	0.9418	0.9429	0.9441
1.6	0.9452	0.9463	0.9474	0.9484	0.9495	0.9505	0.9515	0.9525	0.9535	0.9545
1.7	0.9554	0.9564	0.9573	0.9582	0.9591	0.9599	0.9608	0.9616	0.9625	0.9633
1.8	0.9641	0.9649	0.9656	0.9664	0.9671	0.9678	0.9686	0.9693	0.9699	0.9706
1.9	0.9713	0.9719	0.9726	0.9732	0.9738	0.9744	0.9750	0.9756	0.9761	0.9767
2.0	0.9772	0.9778	0.9783	0.9788	0.9793	0.9798	0.9803	0.9808	0.9812	0.9817
2.1	0.9821	0.9826	0.9830	0.9834	0.9838	0.9842	0.9846	0.9850	0.9854	0.9857
2.2	0.9861	0.9864	0.9868	0.9871	0.9875	0.9878	0.9881	0.9884	0.9887	0.9890
2.3	0.9893	0.9896	0.9898	0.9901	0.9904	0.9906	0.9909	0.9911	0.9913	0.9916
2.4	0.9918	0.9920	0.9922	0.9925	0.9927	0.9929	0.9931	0.9932	0.9934	0.9936
2.5	0.9938	0.9940	0.9941	0.9943	0.9945	0.9946	0.9948	0.9949	0.9951	0.9952
2.6	0.9953	0.9955	0.9956	0.9957	0.9959	0.9960	0.9961	0.9962	0.9963	0.9964
2.7	0.9965	0.9966	0.9967	0.9968	0.9969	0.9970	0.9971	0.9972	0.9973	0.9974
2.8	0.9974	0.9975	0.9976	0.9977	0.9977	0.9978	0.9979	0.9979	0.9980	0.9981
2.9	0.9981	0.9982	0.9982	0.9983	0.9984	0.9984	0.9985	0.9985	0.9986	0.9986
3.0	0.9987	0.9987	0.9987	0.9988	0.9988	0.9989	0.9989	0.9989	0.9990	0.9990
3.1	0.9990	0.9991	0.9991	0.9991	0.9992	0.9992	0.9992	0.9992	0.9993	0.9993
3.2	0.9993	0.9993	0.9994	0.9994	0.9994	0.9994	0.9994	0.9995	0.9995	0.9995
3.3	0.9995	0.9995	0.9995	0.9996	0.9996	0.9996	0.9996	0.9996	0.9996	0.9997
3.4	0.9997	0.9997	0.9997	0.9997	0.9997	0.9997	0.9997	0.9997	0.9997	0.9998

APPENDIX F

Selected References

Au, T., and T.P. Au. *Engineering Economics for Capital Investment Analysis*, 2nd ed. (Boston: Allyn and Bacon, 1992).

Barish, N.N., and S. Kaplan. *Economic Analysis for Engineering and Managerial Decision Making* (New York: McGraw-Hill, 1978).

Bierman, H., Jr., and S. Smidt. *The Capital Budgeting Decision*, 8th ed. (New York: Macmillan, 1993).

Blank, L.T., and A.J. Tarquin. *Engineering Economy*, 6th ed. (New York: McGraw-Hill, 2005).

Bodie, Z., and R.C. Merton, *Finance* (Upper Saddle River, NJ: Prentice Hall, 2000).

Bowman, M.S. *Applied Economic Analysis for Technologists, Engineers, and Managers*, 2nd ed. (Upper Saddle River, NJ: Prentice Hall, 2003).

Brimson, J.A. *Activity Accounting: An Activity-Based Approach* (New York: John Wiley & Sons, 1991).

Bussey, L.E., and T.G. Eschenbach. *The Economic Analysis of Industrial Projects*, 2nd ed. (Upper Saddle River, NJ: Prentice Hall, 1992).

Campen, J.T. *Benefit, Cost, and Beyond* (Cambridge, MA: Ballinger Publishing Company, 1986).

Canada, J.R., and W.G. Sullivan. *Economic and Multiattribute Analysis of Advanced Manufacturing Systems* (Upper Saddle River, NJ: Prentice Hall, 1989).

Canada, J.R., W.G. Sullivan, D.J. Kulonda and J.A. White. *Capital Investment Decision Analysis for Engineering and Management*, 3rd ed. (Upper Saddle River, NJ: Prentice Hall, 2005).

Clark, J.J., T.J. Hindelang, and R.E. Pritchard. *Capital Budgeting: Planning and Control of Capital Expenditures* (Upper Saddle River, NJ: Prentice Hall, 1979).

Collier, C.A., and C.R. Gragada. *Engineering Cost Analysis*, 3rd ed. (New York: Harper & Row, 1998).

Engineering Economist, The. A quarterly journal jointly published by the Engineering Economy Division of the American Society for Engineering Education and the Institute of Industrial Engineers. Published by IIE, Norcross, GA.

Engineering News-Record. Published monthly by McGraw-Hill, New York.

ESCHENBACH, T. *Engineering Economy: Applying Theory to Practice,* 2nd ed. (New York: Oxford University Press, 2003).

FABRYCKY, W.J., G.J. THUESEN, and D. VERMA. *Economic Decision Analysis,* 3rd ed. (Upper Saddle River, NJ: Prentice Hall, 1998).

FLEISCHER, G.A. *Introduction to Engineering Economy* (Boston, MA: PWS Publishing Company, 1994).

GOICOECHEA, A., D.R. HANSEN, and L. DUCKSTEIN. *Multiobjective Decision Analysis with Engineering and Business Applications* (New York: John Wiley & Sons, 1982).

GRANT, E.L., W.G. IRESON, and R.S. LEAVENWORTH. *Principles of Engineering Economy,* 8th ed. (New York: John Wiley & Sons, 1990).

HARTMAN, J.C. *Engineering Economy and the Decision Making Process* (Upper Saddle River, NJ: Pearson Prentice Hall, 2007).

HORNGREN, C.T., G.L. SUNDEM, and W.O. STRATTON, *Introduction to Management Accounting,* 11th ed. (Upper Saddle River, NJ: Prentice Hall, 1999).

HULL, J.C. *The Evaluation of Risk in Business Investment* (New York: Pergamon Press, 1980).

Industrial Engineering. A monthly magazine published by the Institute of Industrial Engineers, Norcross, GA.

Internal Revenue Service Publication 534. *Depreciation.* U.S. Government Printing Office, revised periodically.

JELEN, F.C., and J.H. BLACK. *Cost and Optimization Engineering,* 3rd ed. (New York: McGraw-Hill, 1991).

JONES, B.W. *Inflation in Engineering Economic Analysis* (New York: John Wiley & Sons, 1982).

KAHL, A.L., and W.F. RENTZ. *Spreadsheet Applications in Engineering Economics* (St. Paul, MN West Publishing Company, 1992).

KAPLAN, R.S., and R. COOPER. *The Design of Cost Management Systems* (Upper Saddle River, NJ: Prentice Hall, 1999).

KEENEY, R.L., and H. RAIFFA. *Decisions with Multiple Objectives: Preferences and Value Trade-offs* (New York: John Wiley & Sons, 1976).

LASSER, J.K. *Your Income Tax* (New York: Simon & Schuster, 2007).

MALLIK, A.K. *Engineering Economy with Computer Applications* (Mahomet, IL: Engineering Technology, 1979).

MATTHEWS, L.M. *Estimating Manufacturing Costs: A Practical Guide for Managers and Estimators* (New York: McGraw-Hill, 1983).

MICHAELS, J.V., and W.P. WOOD. *Design to Cost* (New York: John Wiley & Sons, 1989).

MERRETT, A.J., and A. SYKES. *The Finance and Analysis of Capital Projects* (New York: John Wiley & Sons, 1963).

MISHAN, E.J. *Cost-Benefit Analysis* (New York: Praeger Publishers, 1976).

MORRIS, W.T. *Engineering Economic Analysis* (Reston, VA: Reston Publishing, 1976).

Newnan, D.G., T.G. Eschenbach, and J.P. Lavelle, *Engineering Economic Analysis*, 9th ed. (New York: Oxford University Press, 2004).

Ostwald, P.F. *Engineering Cost Estimating*, 3rd ed. (Upper Saddle River, NJ: Prentice Hall, 1992).

Park, C.S. *Contemporary Engineering Economics* (Upper Saddle River, NJ: Prentice Hall, 2002).

Park, C.S., and G.P. Sharp-Bette. *Advanced Engineering Economics* (New York: John Wiley & Sons, 1990).

Park, W.R., and D.E. Jackson. *Cost Engineering Analysis: A Guide to Economic Evaluation of Engineering Projects*, 2nd ed. (New York: John Wiley & Sons, 1984).

Porter, M.E. *Competitive Strategy: Techniques for Analyzing Industries and Competitors* (New York: The Free Press, 1980).

Riggs, H.E. *Financial and Cost Analysis for Engineering and Technology Management* (New York: John Wiley & Sons, 1994).

Riggs, J.L., D.D. Bedworth, and S.V. Randhawa. *Engineering Economics*, 4th ed. (New York: McGraw-Hill, 1996).

Sassone, Peter G., and William A. Schaffer. *Cost-Benefit Analysis: A Handbook* (New York: Academic Press, 1978).

Smith, G.W. *Engineering Economy: The Analysis of Capital Expenditures*, 4th ed. (Ames, IO: Iowa State University Press, 1987).

Steiner, H.M. *Engineering Economic Principles*, 2nd ed. (New York: McGraw-Hill, 1996).

Stermole, F.J., and J.M. Stermole. *Economic Evaluation and Investment Decision Methods*, 6th ed. (Golden, CO: Investment Evaluations Corp., 1987).

Stewart, R.D. *Cost Estimating* (New York: John Wiley & Sons, 1982).

Stewart, R.D., R.M. Wyskida, and J.D. Johannes, eds. *Cost Estimators? Reference Manual*, 2nd ed. (New York: John Wiley & Sons, 1995).

Taylor, G.A. *Managerial and Engineering Economy*, 3rd ed. (New York: Van Nostrand Reinhold, 1980).

Thuesen, G.J., and W.J. Fabrycky. *Engineering Economy*, 9th ed. (Upper Saddle River, NJ: Prentice Hall, 2001).

VanHorne, J.C. *Financial Management and Policy*, 8th ed. (Upper Saddle River, NJ: Prentice Hall, 1989).

Weingartner, H.M. *Mathematical Programming and the Analysis of Capital Budgeting Problems* (Englewood Cliffs, NJ: Prentice-Hall, 1975).

White, J.A., K.E. Case, D.B. Pratt, and M.H. Agee. *Principles of Engineering Economic Analysis*, 4th ed. (New York: John Wiley & Sons, 1998).

Woods, D.R. *Financial Decision Making in the Process Industry* (Englewood Cliffs, NJ: Prentice-Hall, 1975).

Zeleny, M. *Multiple Criteria Decision Making* (New York: McGraw-Hill, 1982).

APPENDIX G

Answers to Selected Problems

Chapter 2

2-12 a. $D^* = 2{,}425$ circuit boards per month
 b. Maximum profit $= \$75{,}613$ per month
 c. $D'_1 = 481$ boards per month
 d. $D'_2 = 4{,}369$ boards per month

2-14 a. Maximum profit $= \$197{,}461$ per month
 b. $D'_1 = 2{,}698$ units per month
 $D'_2 = 6{,}672$ units per month

2-16 $X = 15.64$ megawatts

2-18 $D' = 2{,}404$ pumps per month; 22.75% reduction

2-20 a. $D' = 40{,}000$ units per year
 b. Profit (90%) $= \$2{,}500{,}000$ per year
 Profit (100%) $= \$2{,}800{,}000$ per year

2-23 Cost of E85 $= \$2.268$ per gallon

2-25 Harvest in 2.5 weeks; Revenue $= \$6{,}125$

2-26 Minimum cost velocity $= 10.25$ miles per hour

2-30 Total extra amount $= \$5{,}422$

2-33 Choose brass–copper alloy; $\Delta = \$28.25$ per unit

2-36 Process 1; Profit $= \$2{,}640$ per day

2-39 Speed C; cost per piece $= \$0.0195$

2-42 Method 1; Profit $= \$10{,}974{,}000$

2-45 a. 486 pounds of coal
 b. 889 pounds of CO_2

2-54 (a)
2-57 (e)
2-60 (d)

Chapter 3

3-4 a. \$387 per ton
 b. \$34.84 billion

3-5 \$12,960

3-8 $C_{2009} = \$262{,}780$

3-11 \$354,879

3-14 Total cost $= \$2{,}575{,}000$

3-16 Cost in 10 years $= \$2{,}609$

3-18 31.4% reduction

3-19 $n = -0.152$

3-22 a. $y = 31.813 + 0.279x$
 b. $R = 0.99$
 c. $y = \$101.56$

3-26 4,497 units

3-29 Total cost $= \$2{,}239{,}046$

3-38 (d)
3-40 (c)
3-41 (a)

Chapter 4

4-1 $I = \$4{,}250$

4-6 Total interest $= \$510$

4-8 $F = \$13{,}382$

4-10 Difference $= \$1{,}293$

4-13 $P = \$7{,}835$

4-17 $N = 10$ years

4-19 Gain $= \$511,000$

4-21 **a.** $i = 5.7\%$ per year
       b. $i = 10.1\%$ per year

4-25 $F = \$21,579$

4-28 $F = \$100,913$

4-32 $P = \$8,865$

4-34 $P = \$73,748$

4-37 $P = \$501,880$

4-39 $A = \$720$

4-42 $A = \$3,503,000$

4-45 $A = \$132,100$

4-47 $i = 1.89\%$ per year

4-51 $N = 55$ months

4-53 $F = \$16,535$

4-56 $A = \$60.20$

4-58 $F_4 = \$124,966$

4-62 $Z = \$3,848$

4-64 $A_1 = \$1,994; A_2 = \$1,543.50$

4-67 $A = \$27,358$

4-70 $P_0 = -\$165,104$

4-72 $N = 11$ years

4-74 $A = \$1,204$

4-76 $Z = \$608$

4-79 $K = \$1034.25$

4-81 $A = \$437.14$

4-83 $N = 8$ years

4-87 $P = \$4,685$

4-90 $F = \$644,128$

4-92 $P = \$8,012$

4-95 $F = \$2,850$

4-99 **a.** $i = 10.25\%$ per year
       b. $i = 10.38\%$ per year
       c. $i = 10.51\%$ per year

4-101 $A = \$250$

4-104 $A = \$260,000$ per quarter

4-107 **a.** $A = \$332$
        b. APR $= 13.38\%$ compounded monthly

4-110 $i = 11.55\%$ per year

4-113 $i = 12\%$ per year

4-116 $Z = \$1,422$

4-119 $A = \$1,321$

4-122 **a.** $F = \$362,944$
        b. $A = \$60,386$

4-125 F (today's dollars) $= \$64,884$

4-128 Three points

4-136 (e)

4-138 (d)

4-139 (c)

4-142 (c)

4-144 (c)

4-145 (a)

4-148 (c)

Chapter 5

5-2 **a.** PW $= \$82,083$
      b. Acceptable, PW > 0

5-4 Not acceptable; PW $= -\$13,424$

5-7 Not acceptable; PW $= -\$21,043$

5-10 $V_N = \$1,245.74$

5-13 $i = 11.77\%$

5-16 **a.** CW $= \$34,591$
       b. $N = 80$ years

5-19 FW $= \$31,546$

5-21 FW $= \$5,856$

5-25 AW $= -\$1,720$; reject

5-28 **a.** EUAC $= \$35,570$
       b. 63.3%

5-30 $A = \$4,490$ per year

5-33 $N = 50$ months

5-35 IRR $= 14\%$

5-38 APR $= 38.4\%; i_{\text{eff}} = 45.9\%$

5-40 $i = 8.6\%$ per year

5-44 APR $= 22.8\%$ compounded monthly

5-46 **a.** IRR $= 10\%$
       b. FW $= -\$27,000$
       c. ERR $= 10.74\%$

5-49 $\theta = 6$ years

5-52 **a.** Maximum investment $= \$2,100$
       b. AW $= \$495$

5-55 **a.** FW $= \$84,028$
       b. IRR $= 38.4\%$
       c. $\theta' = 4$ years

5-58 **a.** MV $= \$285$
       b. IRR $= 8.15\%$

5-61 **a.** Loss $= \$315,000$ per year
       b. Profit $= \$225,000$ per year

5-72 (d)

5-74 (a)

5-75 (a)

5-77 (c)
5-80 (c)
5-81 (d)

Chapter 6

6-3 2 inches of insulation
6-5 Design C; $FW_C = \$273,758$
6-7 Alternative B; $PW_B = \$9,180$
6-9 Alternative B; $FW_B = \$184,123$
6-12 Design A; $PW_A = -\$557,273$ per mile
6-14 **a.** $AW_1 = \$371.30$
    b. $AW_2 = \$405.20$
    c. $AW_\Delta = \$34$
    d. Alternative 2
6-18 **a.** $C > A > B$
    b. $i' = 26.19\%$
6-21 **a.** $\theta = 2.5$ or 3 years
    b. $\theta' = 4$ years
    c. $IRR = 30\%$
6-24 $IRR_\Delta = 14.3\% > MARR$; Deere
6-28 **a.** X; $AW_x = -\$3,177$
    b. Y; $PW_y = -\$40,886$
6-30 Skyline; Capitalized cost $= \$749,864$
6-33 A; $CW_A = -\$66,590$
6-36 C; $AW_C = \$1,254$
6-38 2 cm; $AW_{2\,cm} = -\$2,006$
6-41 **a.** Alternative A; $AW_A = -\$12,054$
    b. Alternative A; $AW_B = -\$14,691$
6-44 **a.** Three Lathes L1
    b. $CR_{L2} = \$10,740$
    c. $\$19,000$
    d. Lathe L1
6-46 Auxiliary equipment $= \$19,925$
6-48 E2; $PW_{E2} = -\$273,100$
6-51 S1; $CW_{S1} = -\$150,927$
6-54 Select A
6-56 Gravity-fed; $IRR_\Delta = 12.8\%$
6-59 Alternative 1; $FW_{Alt.\,1} = \$3,662$
6-61 **b.** $PW = \$261$
6-64 CD
6-67 Design 4; $PW_4 = \$23,433$
6-75 (c)
6-78 (a)
6-80 (c)
6-83 (a)
6-85 (b)

Chapter 7

7-6 **a.** $d_2 = \$11,000$
    b. $BV_1 = \$109,000$
    c. $BV_{10} = \$10,000$
7-7 **a.** $BV_3 = \$100,000$
    b. $BV_4 = \$14,444$
7-9 **a.** $\$7,143$
    b. $\$11,200$
    c. $\$5,000$
7-11 **a.** $d_3^* = \$180,550.50$
    b. $d_4 = \$14,449.50$
    c. $BV_2 = \$43,329$
7-14 **a.** $d_3 = \$21,984$
    b. $BV_4 = \$19,786$
    c. $d_4^* = \$70,015$
7-17 $d_3 = \$18,000$; $BV_2 = \$40,000$
7-20 $t = 37.96\%$; $t = 41.92\%$
7-22 **a.** $\$62,000$
    b. $\$32,000$
7-24 **a.** $\$3.4$ million each year
    b. $PW = -\$293,000$; $IRR = 13.54\%$
7-27 $PW = \$393,690$; make investment
7-29 $EUAC = \$83,989$
7-32 4,428 units per year
7-35 $\$864,135$ per year
7-38 Maximum leasing cost $= \$8,733$ per year
7-41 Machine B; $AW_B = \$6,678$
7-43 **a.** Method II; $PW_{II} = -\$54,211$
    b. Method II; $PW_{II} = -\$51,870$
7-46 Accept Quotation II;
    $PW_{II} = -\$136,848$
7-49 $AW = \$3,468$ for both ATCF and EVA
7-51 After taxes, $F = \$214,614$
7-53 $i_{eff} = 8.99\%$ per year
7-68 (b)
7-70 (c)
7-71 (a)
7-74 (d)
7-75 (c)
7-77 (b)
7-79 (e)
7-80 (e)
7-82 (c)

Chapter 8

8-2 a. $6,082
   b. $8,111
8-5 Situation b; $FW_a = \$3,673$
8-6 Alternative B; $PW_B = -\$369,080$
8-8 Alternative I; $PW_I(\text{costs}) = \$10,000$
8-11 2.77% per year
8-14 a. Investment $= \$24,230,790$
   b. PW $= -\$28,584,102$
   c. $AW_{R\$} = -\$3,524,420$
8-17 1.85% per year
8-20 a. $157,263
   b. $1,214 per month
8-22 a. $187,978
   b. $337,629
8-25 $A = \$3.67$ billion per year
8-27 $N = 5$ years
8-30 $39,836
8-33 a. 6.08 units of X
   b. 6.91 units of X
8-35 a. 36.08% per year
   b. 18.44% per year
8-37 a. No; PW $= -\$19,635$
   b. 28% per year
   c. 14.29% per year
8-40 b. EUAC $= \$38,595$
8-42 a. $-\$1,859$
   b. $-\$1,309$
8-44 a. Total investment $= \$356,557$
8-55 (d)
8-57 (b)
8-59 (a)
8-61 (b)

Chapter 9

9-1 Keep defender; PW $= -\$30,925$
9-5 $N^* = 5$ years
9-9 Replace immediately
9-11 Reinforce existing bridge
9-15 Abandon after three years
9-17 $13,938
9-21 Sell station; EUAC $= \$5,000$
9-23 ΔIRR $= 4.5\%$; Lease challenger
9-25 Alternative B; CW $= -\$4,239$
9-32 (a)

9-33 (c)
9-35 (d)

Chapter 10

10-1 Attractive; B–C $= 1.29$
10-3 a. Acceptable; B–C $= 1.16$
   b. Unacceptable; B–C $= 0.79$
10-6 a. Construct tunnel; B–C $= 1.16$
   b. $706,053
10-8 a. CW $= \$3,639,750$
   b. B–C $= 1.14$
   c. Keep initial design
10-11 Alternative C
10-13 Alternative B; $\Delta B/\Delta C = 1.34$
10-17 Alternative A
10-21 Recommend building the dam (II)
10-24 Construct the levee
10-25 Select Design B
10-27 (b)
10-29 (b)

Chapter 11

11-1 $X = 10,966$ units per year
11-4 Breakeven time $= 7$ years
11-7 Select B for $0 \leq X \leq \$6,472$
11-9 $X = \$933,953$ in revenues per year
11-11 a. 20.3% reduction
11-16 $X = \$3,318$
11-18 $X = \$362,500,000$ Btu per year
11-30 (b)
11-31 (a)
11-34 (b)
11-38 True
11-39 False

Chapter 12

12-2 Build four-lane bridge now;
   PW $= -\$3,500,000$
12-5 $E(P) = 10.56$; $V(P) = 36.146$;
   $SD(P) = 6.012$
12-7 Select T for one or two main failures
   per year
12-10 $E(PW_{AT}) = \$33,403$; implement
   project

12-13 **a.** $E(PW) = \$16,971$;
$SD(PW) = \$33,131$
    b. $Pr(PW > 0) = 0.57$
    c. $E(AW)_{R\$} = \$1,866$

12-16 $Pr(X \geq 171) = 0.7881$

12-19 $E(\Delta PW) = -\$5,228$;
$SD(\Delta PW) = \$1,184$

12-25 Select new product; $PW = \$62,165$

12-27 Choose to build

Chapter 13

13-3 Recommendations differ for projects A and C

13-4 $R_S = 10\%$

13-7 **a.** Lease truck
    b. Operate without a truck

13-10 Case (a)– lease; Case (b)– buy

13-11 Lease if life $\leq$ 3 years

13-12 Objective function value = $219.89

13-15 APR = 0%

Chapter 14

14-5 **a.** Alternative 2
    b. None dominate
    c. Alternative 3

14-8 **a.** Dominance – nothing eliminated
Satisficing – Alternative A eliminated
Lexicography – nothing eliminated
Additive weighting – Alternative B selected

14-9 Select Apex

INDEX

Excel Formula Summary

ABS(value)
    Returns the absolute value of the argument.

AVERAGE(range)
    Returns the average of the values specified in *range*.

CEILING(number, significance)
    Rounds the specified number up to the significance specified. Significance is typically 1.

FV(*i*,*N*,*A*,{PW},type)
    Returns the future worth amount at the end of period N for an annuity A, beginning at the end of period 1. The interest rate is i per period. {PW} is optional.

HLOOKUP(class life, table range, *k*)
    This function returns a value from a specific location in a table. The location is specified by values in the header row and the number of rows below the header.

IF(condition, true, false)
    Condition is a logical expression, containing a logical operator ($<>$, $>$, $<$, $>=$, $<=$). If the result of the condition is true, the value or label in the *true* argument is returned. If the result is false, the value or label in the *false* argument is returned.

IRR(range, guess)
    Returns the IRR for cash flows in the specified range. Recall that more than one change of sign in the cash flow may generate multiple rates of return! The *guess* argument is a starting point used by the internal algorithm and is often specified as the MARR.

ISERR(cell)
    Computational errors, such as division by zero, display an error code in the cell containing the error and any cells that reference that cell. If the cell contains an error, this function returns a logical TRUE; otherwise, it returns a logical FALSE.

MAX(range)
    Returns the maximum value in the range. Text and logical results are ignored.

MIN(range)
    Returns the minimum value in the range. Text and logical results are ignored.

NORMINV(number, mean; std dev)
    Returns the inverse cumulative distribution for a normal distribution with the specified mean and standard deviation.